Emerging Trends in Energy Storage Systems and Industrial Applications

Emerging Trends in Energy Storage Systems and Industrial Applications

Edited by

Prabhansu

Department of Mechanical Engineering, Sardar Vallabhbhai National Institute of Technology Surat, Gujarat, India

Nayan Kumar

Department of Electrical Engineering, Muzaffarpur Institute of Technology, Muzaffarpur, Bihar, India

ELSEVIER

ACADEMIC PRESS

An imprint of Elsevier

Academic Press is an imprint of Elsevier
125 London Wall, London EC2Y 5AS, United Kingdom
525 B Street, Suite 1650, San Diego, CA 92101, United States
50 Hampshire Street, 5th Floor, Cambridge, MA 02139, United States
The Boulevard, Langford Lane, Kidlington, Oxford OX5 1GB, United Kingdom

ISBN: 978-0-323-90521-3

For Information on all Academic Press publications
visit our website at https://www.elsevier.com/books-and-journals

Publisher: Charlotte Cockle
Acquisitions Editor: Edward Payne
Editorial Project Manager: Zsereena Rose Mampusti
Production Project Manager: Nirmala Arumugam
Cover Designer: Matthew Limbert

Typeset by MPS Limited, Chennai, India

Contents

List of contributors

Emmanuel Ackom UNEP DTU Partnership, Technical University of Denmark, Copenhagen, Denmark

Mohsen Alizadeh Bidgoli Department of Electrical Engineering, Yadegar-e-Imam Khomeini (RAH) Share Rey Branch, Islamic Azad University, Tehran, Iran

Ruhul Amin Electrification and Energy Infrastructures Division, Oak Ridge National Laboratory, Oak Ridge, TN, United States

Syed Muhammad Amrr Department of Electrical Engineering, Indian Institute of Technology Delhi, New Delhi, India

M.S. Jamil Asghar Department of Electrical Engineering, Aligarh Muslim University, Aligarh, Uttar Pradesh, India

Mohammad Ashiq CSIR—Advanced Materials and Processes Research Institute (AMPRI), Bhopal, Madhya Pradesh, India; Academy of Scientific and Innovative Research (AcSIR), Ghaziabad, Uttar Pradesh, India

Y.A.O. Assagra Department of Electrical Engineering (SEL), University of São Paulo (USP), São Carlos, Brazil

Ammar Atif Electrical Engineering Department, King Fahd University of Petroleum and Minerals (KFUPM), Dhahran, Saudi Arabia

Ayhan Atiz Department of Physics, Faculty of Sciences and Letters, Cukurova University, Adana, Turkey

Muhammad Aziz Institute of Industrial Science, The University of Tokyo, Tokyo, Japan

Ilias Belharouak Electrification and Energy Infrastructures Division, Oak Ridge National Laboratory, Oak Ridge, TN, United States; Bredesen Center for Interdisciplinary Research and Graduate Education, University of Tennessee, Knoxville, TN, United States

Madhushri Bhar Department of Chemistry, Indian Institute of Technology Hyderabad, Sangareddy, Telangana, India

A. Bharadwaj Department of Electrical Engineering, Indian Institute of Technology Kharagpur, West Bengal, India

Udita Bhattacharjee Department of Chemistry, Indian Institute of Technology Hyderabad, Sangareddy, Telangana, India

Bentang Arief Budiman National Center for Sustainable Transportation Technology, Bandung, Indonesia; Faculty of Mechanical and Aerospace Engineering, Institut Teknologi Bandung, Bandung, Indonesia

J.P. Carmo Department of Electrical Engineering (SEL), University of São Paulo (USP), São Carlos, Brazil

Viplov Chauhan CSIR—Advanced Materials and Processes Research Institute (AMPRI), Bhopal, Madhya Pradesh, India

Bowen Chen School of Electrical and Information Engineering, Changsha University of Science and Technology, International Joint Laboratory of Ministry of Education for Operation and Planning of Energy Internet Based on Distributed Photovoltaic-Storage Energy, Hunan Provincial Engineering Research Center for Electric Transportation and Smart Distribution Network, Changsha, P.R. China

Mohammadhossein Deihimi Department of Electrical Engineering, University of Kurdistan, Sanandaj, Kurdistan, Iran

Marm Dixit Electrification and Energy Infrastructures Division, Oak Ridge National Laboratory, Oak Ridge, TN, United States

Z.Y. Dong School of Electrical & Electronic Engineering, Nanyang Technological University, Singapore

Mamdouh El Haj Assad Sustainable and Renewable Energy Engineering Department, University of Sharjah, Sharjah, United Arab Emirates

Mustafa Erden Department of Physics, Faculty of Sciences and Letters, Cukurova University, Adana, Turkey

Rachid Essehli Electrification and Energy Infrastructures Division, Oak Ridge National Laboratory, Oak Ridge, TN, United States

Mehrdad Gholami Department of Electrical Engineering, University of Kurdistan, Sanandaj, Kurdistan, Iran

Shuvajit Ghosh Department of Chemistry, Indian Institute of Technology Hyderabad, Sangareddy, Telangana, India

L.M. Gonçalves Department of Industrial Electronics (DEI), University of Minho (UMinho), Guimarães, Portugal

Manoj Goswami CSIR—Advanced Materials and Processes Research Institute (AMPRI), Bhopal, Madhya Pradesh, India; Academy of Scientific and Innovative Research (AcSIR), Ghaziabad, Uttar Pradesh, India

R.H. Gounella Department of Electrical Engineering (SEL), University of São Paulo (USP), São Carlos, Brazil

Amin Hajizadeh AAU ENERGY, Aalborg University, Esbjerg, Denmark

Siamak Hoseinzadeh Department of Planning, Design, Technology of Architecture, Sapienza University of Rome, Rome, Italy

Metha Islameka National Center for Sustainable Transportation Technology, Bandung, Indonesia; Department of Design Engineering, Politeknik Manufaktur Bandung, Bandung, Indonesia

Trapti Jain Discipline of Electrical Engineering, IIT Indore, Indore, Madhya Pradesh, India

Satyaranjan Jena School of Electrical Engineering, KIIT, Deemed to be University, Bhubaneswar, Odisha, India

Firman Bagja Juangsa Faculty of Mechanical and Aerospace Engineering, Institut Teknologi Bandung, Bandung, Indonesia

Hatice Karakilcik Department of Geology Engineering, Faculty of Engineering, Cukurova University, Adana, Turkey

Mehmet Karakilcik Department of Physics, Faculty of Sciences and Letters, Cukurova University, Adana, Turkey

Muhammad Khalid Center for Renewable Energy and Power Systems, KFUPM, Dhahran, Saudi Arabia; SDAIA-KFUPM Joint Research Center for Artificial Intelligence, Dhahran, Saudi Arabia; Electrical Engineering Department, King Fahd University of Petroleum and Minerals (KFUPM), Dhahran, Saudi Arabia

Khalid Abdullah Khan Electrical Engineering Department, King Fahd University of Petroleum and Minerals (KFUPM), Dhahran, Saudi Arabia

Nayan Kumar Department of Electrical Engineering, Muzaffarpur Institute of Technology, Muzaffarpur, Bihar, India

Satendra Kumar CSIR—Advanced Materials and Processes Research Institute (AMPRI), Bhopal, Madhya Pradesh, India; Academy of Scientific and Innovative Research (AcSIR), Ghaziabad, Uttar Pradesh, India

Surender Kumar CSIR—Advanced Materials and Processes Research Institute (AMPRI), Bhopal, Madhya Pradesh, India; Academy of Scientific and Innovative Research (AcSIR), Ghaziabad, Uttar Pradesh, India

Chun Sing Lai Brunel Interdisciplinary Power Systems Research Center, Department of Electronic and Electrical Engineering, Brunel University London, London, United Kingdom

Giorgio Locatelli School of Management, Politecnico di Milano, Italy

S. Maiti Department of Electrical Engineering, Indian Institute of Technology Kharagpur, West Bengal, India

Surendra Kumar Martha Department of Chemistry, Indian Institute of Technology Hyderabad, Sangareddy, Telangana, India

Nitin Muralidharan Electrification and Energy Infrastructures Division, Oak Ridge National Laboratory, Oak Ridge, TN, United States

Venkataramana Naik N Department of Electrical Engineering, NIT Rourkela, Rourkela, Odisha, India

Ankireddy Narendra Department of Electrical Engineering, NIT Rourkela, Rourkela, Odisha, India

Anup Kumar Panda Department of Electrical Engineering, NIT Rourkela, Rourkela, Odisha, India

Anand Parejiya Electrification and Energy Infrastructures Division, Oak Ridge National Laboratory, Oak Ridge, TN, United States; Bredesen Center for Interdisciplinary Research and Graduate Education, University of Tennessee, Knoxville, TN, United States

Prabhansu Department of Mechanical Engineering, Sardar Vallabhbhai National Institute of Technology Surat, Surat, Gujarat, India

E.S.N. Raju P School of Energy Science & Engineering, IIT Guwahati, Guwahati, Assam, India

Navid Rezaei Department of Electrical Engineering, University of Kurdistan, Sanandaj, Kurdistan, Iran

Madan Mohan Sahu Department of Electrical Engineering, National Institute of Technology Rourkela, Rourkela, Odisha, India

Pradeep Kumar Sahu School of Electrical Engineering, KIIT, Deemed to be University, Bhubaneswar, Odisha, India

Abdelaziz Salah Saidi Department of Electrical Engineering, King Khalid University, Abha City, Saudi Arabia; University of Tunis El Manar, National School of Engineers of Tunis, Laboratory of Electrical Systems, Tunis, Tunisia

Saygin San Department of Physics, Faculty of Sciences and Letters, Cukurova University, Adana, Turkey

N. Sathish CSIR—Advanced Materials and Processes Research Institute (AMPRI), Bhopal, Madhya Pradesh, India; Academy of Scientific and Innovative Research (AcSIR), Ghaziabad, Uttar Pradesh, India

Mohammad Hassan Shahverdian Lab of Optimization of Thermal Systems' Installations, Faculty of Mechanical Engineering-Energy Division, K.N. Toosi University of Technology, Tehran, Iran

Hafsa Siddiqui CSIR—Advanced Materials and Processes Research Institute (AMPRI), Bhopal, Madhya Pradesh, India

Netrapal Singh CSIR—Advanced Materials and Processes Research Institute (AMPRI), Bhopal, Madhya Pradesh, India; Academy of Scientific and Innovative Research (AcSIR), Ghaziabad, Uttar Pradesh, India

Rahul Singh Department of Electrical Engineering, Indian Institute of Technology Jodhpur, Jodhpur, Rajasthan, India

Ali Sohani Lab of Optimization of Thermal Systems' Installations, Faculty of Mechanical Engineering-Energy Division, K.N. Toosi University of Technology, Tehran, Iran

Hadi Tarimoradi Department of Electrical Engineering, University of Kurdistan, Sanandaj, Kurdistan, Iran

Sheng Xiang School of Electrical and Information Engineering, Changsha University of Science and Technology, International Joint Laboratory of Ministry of Education for Operation and Planning of Energy Internet Based on Distributed Photovoltaic-Storage Energy, Hunan Provincial Engineering Research Center for Electric Transportation and Smart Distribution Network, Changsha, P.R. China

Y. Xu School of Electrical and Information Engineering, Changsha University of Science and Technology, International Joint Laboratory of Ministry of Education for Operation and Planning of Energy Internet Based on Distributed Photovoltaic-Storage Energy, Hunan Provincial Engineering Research Center for Electric Transportation and Smart Distribution Network, Changsha, P. R. China

Yan Xu Nanyang Technol Univ, Sch Elect & Elect Engn, Singapore, Singapore

Hongming Yang School of Electrical and Information Engineering, Changsha University of Science and Technology, International Joint Laboratory of Ministry of Education for Operation and Planning of Energy Internet Based on Distributed Photovoltaic-Storage Energy, Hunan Provincial Engineering Research Center for Electric Transportation and Smart Distribution Network, Changsha, P. R. China

Bangzhe Yin School of Electrical and Information Engineering, Changsha University of Science and Technology, International Joint Laboratory of Ministry of Education for Operation and Planning of Energy Internet Based on Distributed Photovoltaic-Storage Energy, Hunan Provincial Engineering Research Center for Electric Transportation and Smart Distribution Network, Changsha, P. R. China

Shijie Zhang School of Electrical and Information Engineering, Changsha University of Science and Technology, International Joint Laboratory of Ministry of Education for Operation and Planning of Energy Internet Based on Distributed Photovoltaic-Storage Energy, Hunan Provincial Engineering Research Center for Electric Transportation and Smart Distribution Network, Changsha, P. R. China

Yongxi Zhang School of Electrical and Information Engineering, Changsha University of Science and Technology, International Joint Laboratory of Ministry of Education for Operation and Planning of Energy Internet Based on Distributed Photovoltaic-Storage Energy, Hunan Provincial Engineering Research Center for Electric Transportation and Smart Distribution Network, Changsha, P. R. China

Preface

It is our pleasure to present this book on up-to-date energy storage technologies and advancements in their use in the renewable energy sector.

The purpose of this book is to help interested and budding researchers understand the importance of Energy Storage systems. At the same time, the intention is to disseminate energy storage technologies to the wider community who are interested to consider them for some specific application. In Chapter 1, the overview of the application of artificial intelligence and machine learning in Energy Storage systems is addressed. This chapter discusses current state-of-the-art and technology trends use of artificial intelligence (AI) and machine learning (ML) and provides readers with a comprehensive and insightful review. Chapter 2 presents a detailed description of the design of power electronic devices in the domain of battery energy storage. It is shown how Power Electronic Converters make the battery to be charge faster and control the amount of energy flow more effectively. In Chapter 3, an investigation is made on the ratio of cushion gas to working gas in the storage of hydrogen gas at different depths in underground salt caverns. Energy storage capacitor banks are widely used in pulsed power for high-current applications, including exploding wire phenomena, sockless compression, and the generation, heating, and confinement of high-temperature, high-density plasmas, and their many uses are briefly highlighted in Chapter 4.

Chapter 5 describes energy conversion and balance performance in a battery electric vehicle. The mechanical and electrical power flows in the vehicle along with an explanation of how the system and configuration can be selected from the use of regenerative braking and range extender. Electrochemical energy storage systems are crucial components for the realization of a carbon-neutral/carbon-negative energy sector globally and are highlighted in Chapter 6. Chapter 7 discusses the overview of the various types of electrochemical energy storage systems explored so far, emphasizing the basic operating principle, history of the development of electrochemical energy storage (EES) devices from the research, as well as commercial success point of view. The thermodynamic, energy conversion, polarizability, and electrochemical aspects of the devices are touched upon in detail. Chapter 8 explores the hybrid and future systems based on electrochemical energy storage. Thermal energy storage systems are elaborated on in Chapter 9. In it, three different kinds of thermal energy storage systems, namely sensible, latent, and chemical reactions are introduced. Many widely-used thermal energy storage technologies like water tanks, underground, packed (pebble) beds, Trombe walls, solar façade, ice storage, molten salt, and solar thermochemical thermal energy storage systems are described.

In Chapter 10, Li-ion and Na-ion capacitors are broadly discussed for Hybrid energy storage devices. This provides an in-depth understanding state-of-the-art of these technologies in terms of materials, processing, working principle, and architecture. Chapter 11 presents an overview of energy harvesting systems for powering stand-alone microsystems (MSTs). This chapter complements the topic of energy harvesting, through the presentation of electronic conversion/management systems and energy storage technologies.

Chapter 12 focuses on large-scale energy storage techno-economic appraisal with special emphasis on how smart grids can benefit from large-scale energy storage technologies. The financial model and research agenda for techno-economic appraisal of energy storage systems are also presented and discussed. The utilization of a battery energy storage system (BESS) to enhance the dynamic performance of islanded AC microgrids (IACMG) against large load disturbances is presented in Chapter 13. To explore this, a complete dynamic model of BESS has been developed and incorporated into the studied IACMG system. Chapter 14 introduces the integration of battery energy storage systems (BESS) into the Microgrid to improve the grid's economic efficiency and sustainability. Also, along with two different time frames, this chapter presents methodologies respectively for long-term BESS expansion planning in microgrids considering battery degradation effects and short-term operational planning of BESS to cope with high-level renewable power generation. Chapter 15 intends to present modeling, controller, and simulation of doubly fed induction machine (DFIM)-based pumped storage power plant (PSPP) under generating operation mode. Chapter 16 discusses the use of energy storage systems (ESSs) in smart distribution networks to increase the efficiency and effectiveness of smart grids. ESSs have different characteristics in terms of power and energy density, response time, and techno-economic features. Chapter 17 applies the energy storage technology to large-scale grid-connected PV generation, and designs energy storage configuration. Furthermore, solar cell–supercapacitor devices (SCSD) are introduced as a series array to solve the problem that the solar cell cannot work on the maximum power point (MPP) under partial shading conditions.

A novel fuzzy-based control technique to smooth the transient solar output power generation by designing a varying low pass filter is presented in Chapter 18. The proposed controller improves on the conventional low pass filter technique to appropriately obviate system power variability as well as reduce overworking of the battery. Chapter 19 throws light on Multilevel Converter-based STATCOM with Hybrid Storage System.

Chapter 20 proposes the development of a decoupled controller framework using k-type compensators for battery energy storage systems and a nonlinear PI controller (NPIC) for supercapacitor energy storage systems (SCSS). Further, the proposed controller framework was tested and validated on a solar PV-powered microgrid system.

Chapter 21 proposes a supervisory control strategy with hardware realization for the grid-connected solar PV-based microgrid system. The objectives were to increase the utilization share of solar power, provide an uninterrupted power supply, reduce the stress on the utility grid, and ensure backup power. The proposed system presents a satisfactory performance in terms of effective solar energy usage and storage.

Lastly, Chapter 22 takes the reader towards sustainability through the promotion of electric vehicles and explains its pros and cons.

Artificial intelligence and machine learning applications in energy storage system: technology overview and perspectives

1

Nayan Kumar[1] *and Prabhansu*[2]
[1]Department of Electrical Engineering, Muzaffarpur Institute of Technology, Muzaffarpur, Bihar, India, [2]Department of Mechanical Engineering, Sardar Vallabhbhai National Institute of Technology Surat, Surat, Gujarat, India

Chapter Outline

1.1 Introduction

Today, smart energy storage systems (SESSs) are gaining popularity as a result of increased energy demand in industries and residential areas. The energy storage system converts electrical energy into a sustainable form and converts stored energy into electricity during energy demand. Energy conservation is an effective solution

Emerging Trends in Energy Storage Systems and Industrial Applications. DOI: https://doi.org/10.1016/B978-0-323-90521-3.00014-4

for using renewable energy as well as an attractive option for using clean energy sources. SESSs have been used successfully in many applications, including load rating, system stability, power stability, frequency control, transmission power enhancement, power quality improvement, automatic production control, off-grid service, building power storage systems, and uninterrupted electrical equipment [1−3]. In addition, they can also store additional energy generated by renewable energy sources, such as wind, solar, etc. It will not be the only energy provider without a related energy storage facility. Even if they are connected to a grid, a storage device is required for smooth output.

Over the past decade, there has been renewed attention to electrification in the transport sector; electric vehicles dominated transportation for a short time in the early 21st century. In the past few months, the novel coronavirus (COVID-19) has caused an unprecedented economic and social crisis. The epidemic has profoundly affected all aspects of life, including the energy sector. Energy conservation was reduced by the beginning of the 2019 coronavirus disaster (COVID-19): for the first time in almost a decade, the annual introduction of energy storage technologies declined year by year in 2019 markets, and uncertainty about battery safety contributed to growth, with grid-scale reduction by 20% [4]. Many researchers have focused on the different schemes of SESSs because they can store energy at off-peak hours and supply energy at peak hours. However, existing ESSs technologies face challenges in energy conservation due to several issues, such as charging/discharging, safety, reliability, size, cost, life cycle, and overall management. Storage also plays a key role in ensuring the dynamic performance of energy systems: India will become the largest market for scale battery storage by 2040 as its investment in solar PV grows rapidly [4].

This chapter describes a system that does not have the ability to conserve intelligent energy and can use that energy stored in a future energy supply called an intelligent energy storage system. In order to improve energy conservation, it is important to differentiate between different energy storage systems, as shown in Fig. 1.1. It also discusses various types of energy storage systems, different energy management, and engineering aspects. Furthermore, it also incorporates huge applications of energy storage systems. The environmental impacts due to the distribution of ESSs should be discussed in detail. Finally, the chapter is more faced on the present technology like artificial intelligence (AI) and machine learning (ML) overview and perspectives.

1.2 Classification of energy storage systems

Energy storage systems offer a wide range of technological approaches to managing power supplies to create a more resilient energy infrastructure and bring cost savings to utilities. Energy storage systems are classified into mechanical, electrochemical, chemical, electrical, and thermal, as shown in Fig. 1.1.

1.2.1 Mechanical energy storage systems

Electricity is probably a type of energy that people often use for various purposes. The advantage of mechanical power is that it is often placed away efficiently and

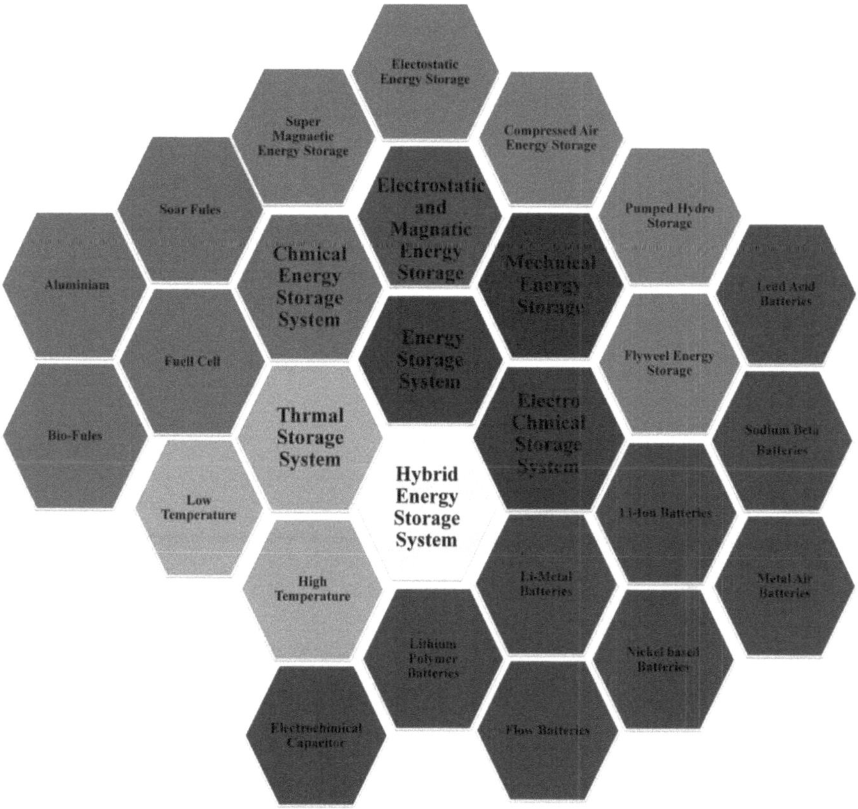

Figure 1.1 Classification of energy storage systems [5–7].

for a long time. It is completely flexible as it is typically converted successfully into and out of other energy structures [7]. Machine power may indicate potential power or vehicle potential. Pressurized gas and blocked spring are two potentially powerful types of energy, while active energy can be removed within body movements.

1.2.2 Pumped hydro storage systems

Pumped hydro storage systems (PHSSs) are an innovative and powerful innovation with a high capacity of 76%–85%, low cost per unit of energy, long duration, unlimited life cycle, and 50 years or more [8,9]. They are mainly used in scheduled applications, for example, night loading and high shaving [10]. PHS is a transformation of a standard hydroelectric power plant, as depicted in Fig. 1.2 [10]. In PHS, there are two storage areas, one on top of the other, accessible near the turbine / siphoning corridor [10]. The energy created by a given amount of water depends on the head, the distance between turbines, and above. If the head is large, the energy produced by the water volume will be greater. Table 1.1 represents the PHP development and future aspects in India.

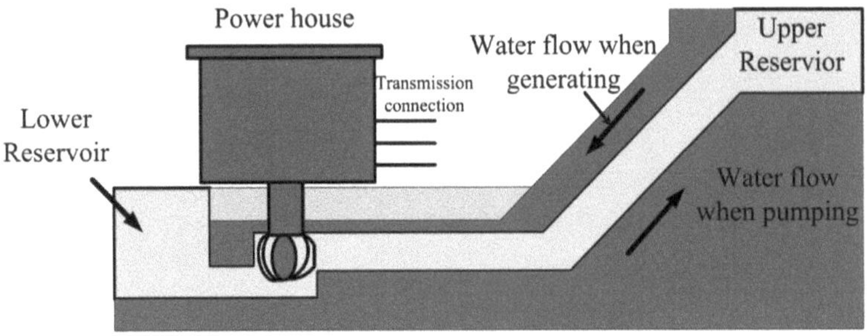

Figure 1.2 Pumped hydro storage system.

Table 1.1 Pumped hydro storage systems in India [11].

S. no.	Project name	Indian state	Capacity (MW)
1.	Bhira	Maharashtra	150
2.	Ghatgar	Maharashtra	250
3.	Kadamparai	Tamil Nadu	400
4.	Kadana St.I & II	Gujarat	240
5.	Nagarjuna Sagar	Telangana	705
6.	Purlia	West Bengal	900
7.	Poithan (Jayakwadi Dam)	Maharashtra	12
8.	Srisailam	Andhra Pradesh	900
9.	Sardar Sarovar	Gujarat	1200
10.	Tehri	Uttarakhand	1000

Advantage

1. Need economic power to start the pumps
2. PHSSs are reliable system
3. Its storage capacity is very high
4. Long life cycle and high efficiency is around 87%

Disadvantage

1. The electric power and energy density are low, respectively
2. Costly power generation

Industrial application

1. AS backup power

1.2.3 Compressed air energy storage systems

The compressed air energy storage systems (CAESSs) industry has two main components for the carrying capacity, the blower/expander. A cargo vessel is utilized to store air at its peak without severe misfortune. The air is compacted using a fan

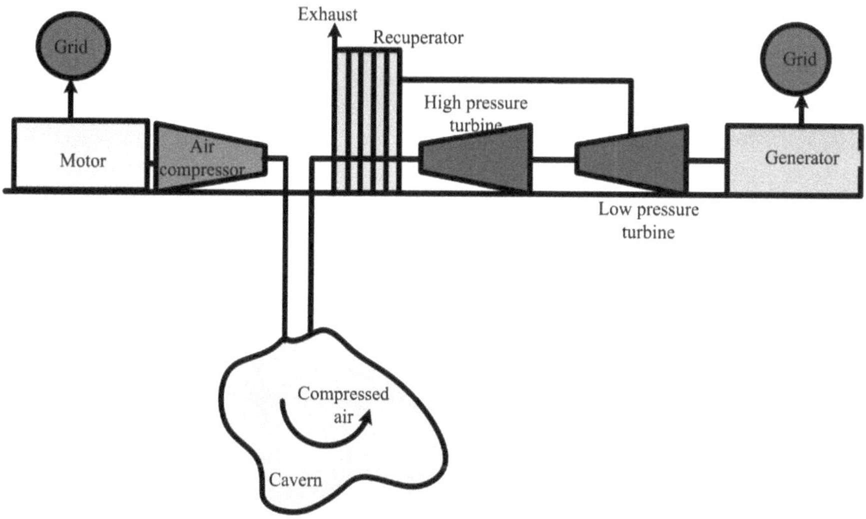

Figure 1.3 Block diagram of compressed air energy storage systems [7].

and, if necessary, expanded by an expander. The CAESSs power station is similar to PHS when considering its function. The accumulated energy during the idle hour, that is when the application and energy level are low, is used to pack air when it is too thick and store it in a volume chamber. This removed energy is obtained during the running hour by expanding the air filled with a slate. In this frame type, compensation for a revolving gas turbine is used. In some complex CAES structures, two blowers are used when the effect of one blower is cared for in the next to be used in production. The CAESSs plant is shown in Fig. 1.3.

Advantage

1. High energy storage capacity
2. Low cost and long life cycle
3. Regenerations are possible
4. System overall efficiency is high at approximately 80%

Disadvantage

1. Environmental issues and location limitations
2. Variable system efficiency

Industrial application

1. For load balancing and during power outages

1.2.4 Liquid air energy storage systems

Recently, liquid air energy storage systems (LAESSs), similar to the new CAESSs, have received much consideration [10]. In this type of volume, the liquid rather than air is saturated; this is more important than the CAESSs framework for local applications. The use of air-soluble liquids creates a dense energy reserve in the frame [12]. However, it requires the establishment of cryogenic liquefaction, which makes this framework more confusing than CAESSs [10]. During the charging phase, the air is compressed due to the solid state of the fuel and then placed in a low-temperature energy tank. In the release phase, condensed air is converted into high-pressure air and is allowed to grow with a high-temperature amplifier to drive engines. The LAESSs system, integrated with wind turbines, tackles wind turbines and other sustainable energy sources. The modified LAESSs, in view of the gas turbine industry, were proposed by Chino et al. also achieved a 73% energy efficiency product [12].

Supercritical compressed air energy storage systems (SC-CAESSs) framework is more robust, efficient, and less weather-sensitive than other previously mentioned CAESSs components [12]. It has integrated benefits for both AA-CAESSs and LAESSs frameworks. In this type of frame, the air is filled with a more sensitive atmosphere with a higher concentration than 37.9 bars and a significantly higher temperature than 132K. the wind. The hotness exchanger collects heat from very high temperatures, which are used to generate energy [13]. There are two ways to blend air-using a melting valve to squeeze or using a melted expander. The choke immersion valve uses the cooling effect of air to obtain liquid air through the choking system. Temperature decreases due to congestion, which is an irreversible cycle. Therefore, this interaction consumes much energy and causes cavitation [12]. Then again, by replacing the compression valve with a water expander, compressing the stress effect can be achieved [13].

Advantage

1. Excellent power density and energy compare to battery energy storage systeMs
2. Faster and short response
3. High efficiency

Disadvantage

1. Independent supercapacitor cells

Industrial applications

1. Regulation of voltage and frequency.

1.2.5 Flywheel energy storage systems

A flywheel energy storage system (FESS) has a high power range from KW to GW range [14–19]. The several features of FESS make for a wide application in force framework, military framework, satellites, and transportation [16]. Fig. 1.4 depicts the structure and components of the FES energy storage system. Table 1.2 addresses the comparative analysis of the low and high-speed FES.

Advantage

1. Required low maintenance and long life cycle

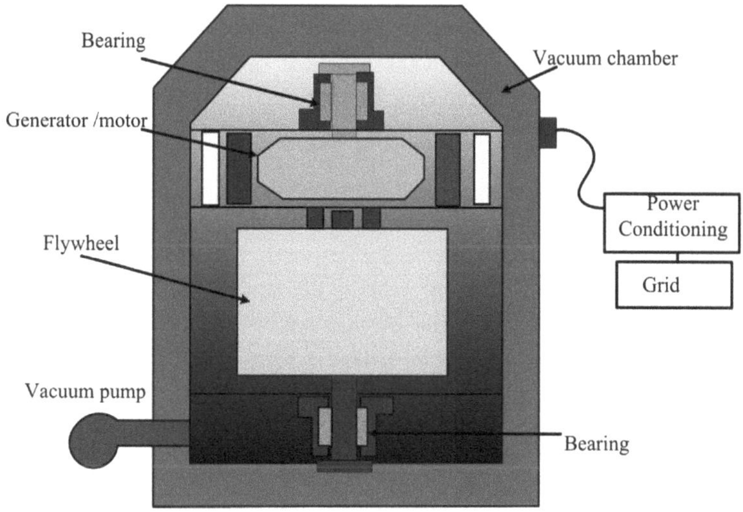

Figure 1.4 Systematic diagram of flywheel energy storage system [7].

Table 1.2 Comparison of low and high-speed flywheel energy storage system [7].

S. no.	Parameters	Low-speed flywheel energy storage system (FESS)	High-speed FESS
1.	Material	Steel	Composite material
2.	Electrical machine (EM)	Asynchronous, PMSM and reluctance machines	PMSM and reluctance machines
3.	Flywheel interface with EM	Partial integration	Full or partial integration
4.	Environment	Partial vacuum	Complete vacuum
5.	Enclosure weight	Dual flywheel weight	Single flywheel weight
6.	Main applications	Power quality improvement	Aerospace and Traction
7.	Cost	Economical	Costly

2. Ecofriendly
3. Excellent power response
4. High efficiency up to 90% and high capability of energy storage

Disadvantage

1. High mechanical stress and more cost

Industrial applications

1. Renewables, FACTS devices, and transportation

1.2.6 Thermal energy storage systems

Heat energy can be removed, such as inactive heat, rational heat, and thermochemical reversible reactions. Thermal energy storage systems (TESSs) have a long-term need for energy redistribution and energy production in a short- or long-term drag [20–22]. In TESSs, energy is stored by cooling or heating the medium, which can be used to cool or burn various substances, or in any case, to produce energy [23]. TES tracks down applications in businesses or buildings and can create general productivity, consistent quality, and financial issues. This framework is fully eco-accommodating. The most commonly used TES are related to photovoltaic (PV)-based solar panels, in which the solar energy is effectively removed. Therefore, in addition to bringing energy to photography, PV-based solar panels can ensure the collection of the most remarkable amount of energy from the sun, which can be used in the evening or now and often in extreme weather conditions. Solar-based concentrators also use TES to store heat concentrated during the day. TES frames can successfully prepare for the future with solar-powered power for 24 hours. Besides, they see widespread use in load movement and generating thermal energy. TES has unfortunate low emissions (0.05%–1%), high density (80–500 Wh/L), clear energy (80–250 Wh/kg), and low-cost costs (3%–60%), although more has a low cycle efficiency of 30%–60% [20].

Advantage

1. Greenhouse gas emissions decrease
2. Excellent in energy storage density
3. Low capital cost

Disadvantage

1. The systems capable of limited energy storage

Very low efficiency 30%–60%

Industrial applications

1. Renewable energy power supply
2. Energy management

1.2.7 Electrostatic and magnetic energy storage systems

Energy can be put away as electric and attractive fields separately through supercapacitors and superconducting magnets. They are capacity gadgets with high power and medium energy thickness and are helpful in satisfying pinnacle power needs, yielding power smoothing, and recuperation of energy in mass travel gadgets [7,20]. This segment can be portrayed in two primary subsections-electrostatic and attractive energy stockpiling. Electrostatic energy stockpiling is fundamental for capacitors and supercapacitors. Supercapacitors are additionally prevalently known as electric double-layer capacitors (EDLC).

1. Super capacitor energy storage (SCES)

Supercapacitor energy storage is a very popular name for ultra-capacitor energy storage or electric double-layer capacitor as depicted in Fig. 1.5. They store energy in the form of an electrostatic field produced due to the constant dc voltage used between two electrodes separated by a thin layer of the insulator or a dielectric medium [5,7,24,25].

Advantage

1. High power density and fast response
2. Adjustable energy storage capacity

Disadvantage

1. Supercapacitor cells are independent (shown in Fig. 1.5 [24])

Industrial applications

1. Regulation of voltage and frequency.
2. Super magnetic energy storage system (SMESS)

As shown in Fig. 1.6, SMESS stores electrical energy in the form of a superconducting coil magnet created for direct energy flow. The ohmic loss in SMES once charged is almost zero due to the negligible resistance of the superconducting coil

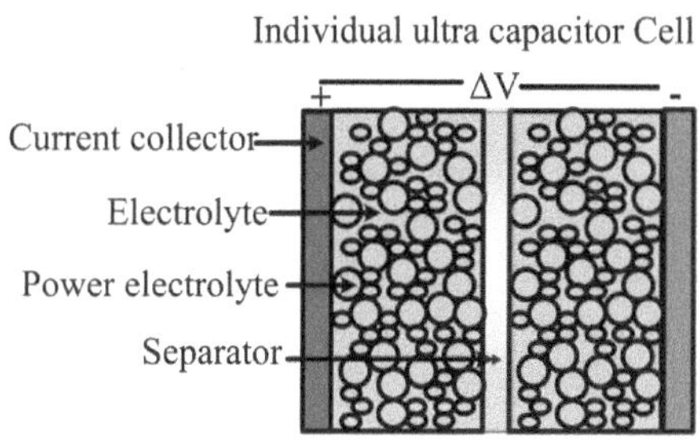

Figure 1.5 Schematic diagram for a doubled layered supercapacitor [24].

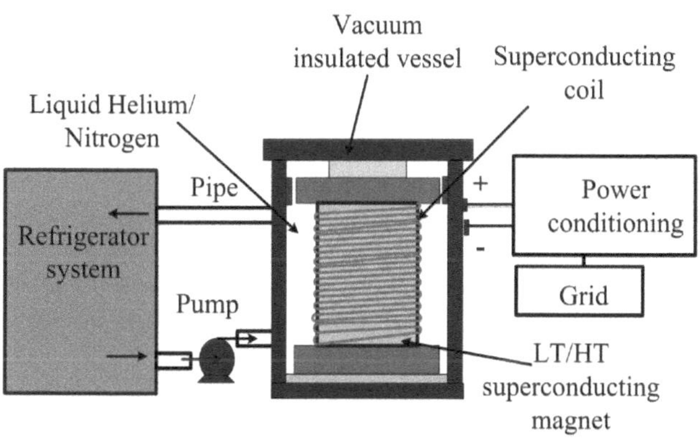

Figure 1.6 Structure of a super magnetic energy storage system and its components [20].

[20,26,27]. A cryogenic coil is used in SMES to cool it under critical superconducting temperatures. While in output mode, SMES can deliver very high power (several megawatts) to the electrical network within a few milliseconds.

Advantages

1. Ecofriendly, high reliable
2. Faster and short response time
3. High power with excellent stability
4. Better energy efficiency (90%)

Disadvantages

1. Costly required high cooling and temperature-sensitive

Industrial applications

1. Improvements for transient stability

1.2.8 Chemical energy storage systems

Nowadays chemical energy storage systems (CESSs) based energy storage systems are very popular due to long-time energy storage capability. They are classified into two categories well-known as hydrogen and biofuels [7,20].

1. Hydrogen storage system

The hydrogen storage system is one of the emerging storage technics to store the energy for the long-term with a flexible energy carrier. Hydrogen can be pressed, hydrated, or stored in a solid or liquid state for use in fuel cells, and industrial and transport applications. A major challenge related to hydrogen is its production and storage. In [20] addressed the electrical power generation of the hydrogen storage system. In this process, fuel plays a vital role. For a reliable and highly efficient system, the choice of the fuel cell is very important. Table 1.3 addressed several types of fuel cells.

Advantages

1. Low pressure and compatibility
2. Higher absorption rates

1. Biofuels

   Natural energy commonly referred to as biofuels are nonpolluting, ecofriendly, and reliable to fossil fuels [28,29]. Sources of biofuels are plants in particular, as they take the first part of the energy harvest from daylight as energy is generated inside their bodies during photosynthesis. This captured energy can be converted into suitable structures and used for multifaceted energy needs. This is a form of common energy conservation, in which plants are the main heroes. Biofuels provide 3% of global transportation.

2. Solar fuels

   Lately, sunlight based energizes stand out, where the sun-oriented energy is used for delivering clean alternate powers [30]. Here, high temperatures are acquired by utilizing explanatory mirrors to think the weakened daylight over a little region and removing the radiation energy from it by appropriate collectors [31]. This high temperature is used in

Table 1.3 Chemical reactions of different types of fuel cells [20].

Fuel cell	Fuel	Oxidant	Cell voltage	Chemical reactions	Industrial applications
Alkaline fuel cell	Pure H_2	O_2 in air	1	$2H_2 + 4OH^- \rightarrow 4H_2O + 4e^-$ $O_2 + 2H_2O + 4e^- \rightarrow 4OH^-$	Space technology
Solid oxide fuel cell	H_2, CO, CH_4 and other	O_2 in air	0.8–1.0	$O^{2-}(S) + H_2(g) \rightarrow H_2O(g) + 2e^-$ $0.5O_2(g) + 2e^- \rightarrow O^{2-}(s)$	Distributed generation
Phosphoric acid fuel cell	Pure H_2	O_2 in air	1.1	$2H_2 \rightarrow 4H^+ + 4e^-$ $O_2 + 4H^+ + 4e^- \rightarrow H_2O$	Distributed generation
Molten carbonate fuel cell	H_2, CO, CH_4 and other	O_2 in air	0.7–1.0	$H_2O + CO_3^{2-} \rightarrow H_2O + CO_2 + 2e^-$ $2H_2 + 4OH^- \rightarrow 4H_2O + 4e^-$	Distributed generation
Direct methanol fuel cell	CH_3OH	O_2 in air	0.2–0.4	$CH_3OH + H_2O \rightarrow CO_2 + 6H^+ + 6e^-$ $3/2O_2 + 6e^- + 6H^+ \rightarrow 3H_2O$	Transportation
Proton exchange Membrane fuel cell	Pure H_2	O_2 in air	1.1	$H_2(g) \rightarrow 2H^+ + 2e^-$ $0.5O_2(g) + 2H^+ + 2e^- \rightarrow H_2O$	Power backup

endothermic substance responses to create clean elective powers, which can undoubtedly be put away and shipped [32]. The discontinuous idea of sunlight-based energy negatively affects the sun oriented to electrical energy transformation as it compromises the security and unwavering quality of the network [30]. In any case, for the instance of sunlight-based fuel innovation, the discontinuous idea of the sustainable power sources no affects the sun-based fuel creation. The significant benefit of this framework is that it has capacities as a virtual stockpiling framework with near 100% virtual productivity and replaces the parts, which cause more energy losses, along these lines working on the general effectiveness of the framework [33]. Besides, it can scale back the hotness motor so it can run at a higher limit. These characteristics add to keeping the sunlight-based fuel framework more monetary.

The most essential fuel-hydrogen has been drawn into the consideration due to having a spotless burning, where just water is created as by item. This hydrogen can be created using sun-powered energy. There exist five particular ways of delivering sunlight-based hydrogen. Hydrogen can be created from water, other petroleum products, or a blend of these two. Water is utilized in the sun-oriented thermolysis and sun-based thermo-compound cycles for creating hydrogen, though a nonrenewable energy source is utilized in the sun-powered breaking procedure for hydrogen creation. For sun-oriented changing and sun-powered gasification strategy, a blend of water and petroleum products is required for creating hydrogen fuel. The delivered sun-oriented hydrogen can be promptly utilized as the fuel for the power device to change its substance energy over to electrical energy.

Although hydrogen has many promising properties to be utilized as sunlight-based powers, it has a low volumetric energy thickness. Besides, the capacity and dispersing component of hydrogen is not quite as simple as the hydrocarbon powers [30]. From numerous perspectives, hydrocarbons dominate all suitable fuel structures, which is the explanation for their far and wide use. Nonetheless, hydrocarbons experience the ill effects of one significant impediment, which is the outflow of CO_2 upon burning, making them unsafe for the climate. This made established researchers decide on carbon-nonpartisan hydrocarbons, which are acknowledged as sunlight-based hydrocarbons. Sun-based hydrocarbons are fabricated in a sun-oriented treatment facility, in which the climatic CO_2 and H_2O are separated utilizing concentrated sunlight-based energy to deliver hydrocarbons. This cycle produces hurtful hydrocarbons as well, however, the ignition of these hydrocarbons delivers a similar measure of CO_2 as was initially utilized for assembling the hydrocarbon, making the net CO_2 outflows zero. Besides, hydrocarbons are a lot simpler to store, and transport, and can be used inside the current energy supply foundation. ETH Zurich in Switzerland has effectively created syngas from their sun-powered processing plant. Syngas, a combination of hydrogen and carbon monoxide, can be handled into fluid hydrocarbon powers that can be utilized for motivations behind transportation. Different powers, like methane, methanol, and so on, can likewise be delivered from sun-oriented processing plants [34].

To for all intents and purposes execute sunlight-based processing plants at large scales, it is important to reduce the catch expenses of CO_2 and productively use sun-oriented energy to drive the compound responses. Utilizing sun-powered hydrocarbons can brace the infiltration of RESs into the transportation area. Battery-powered electric vehicles (EV), notwithstanding being perfect types of transportation, establish under 3% of all vehicles sold in the United States [30]. Along these lines, the use of sun-powered hydrocarbons can be promising in making a cleaner, if not the cleanest, transportation area.

3. Graphite solar storage

Graphite is a stable form of carbon under normal conditions and is considered the highest grade of coal, also known as met-anthracite, with the potential to provide clean energy in the future. The idea is to collect and store solar thermal energy in graphite blocks that can be used to generate electricity through the steam Rankine cycle or other heat engine cycles [5,34–36]. The site has a large graphite thermal cell to store solar heat and a heat exchanger controller to capture and control thermal energy from the graphite.

Advantages

1. High economically
2. Strong and stable at high temperatures
3. High electrical and thermal conductivity

Applications

Electric vehicle (EV)

1. Paper batteries

Paper batteries are ultrathin, flexible, and lightweight devices that can be used as batteries and capacitors. They are composed of a composite of composite nanopaper (90% cellulose) and carbon nanotubes [37–39]. Paper batteries have the potential to provide both short-term and long-term continuous power generation and power bursts (if needed) to the next generation of electronic, medical devices and integrated vehicles.

Advantages

1. High economic
2. Easy fabrication
3. Ecofriendly

Disadvantages

1. Low performance (caused by soaking of electrons by oxygen before they reach the electrode)

1.2.9 Battery energy storage systems

Battery energy storage technology gain popularity in energy storage devices because of the ease to store and converter electrical energy [40–42]. Table 1.4 present the different electrical attributes of BESS.

Advantages

1. The BESS is well known for its good power density and high energy capacity.
2. Its electrical connection is very simple in series-parallel
3. Excellent storage and power capability
4. Ease of use with high efficiency (70%–90%)

Disadvantages

1. The life cycle decides the battery charging and discharging

Table 1.4 Different types of battery energy storage systems' electrical features [42].

Property	Lead acid	Li-ion	NaS	NiCd
Lifetime (years.)	5–15	5–15	10–15	10–15
Cycling times	500–1000	1000–10,000	2500	2000–2500
Discharge efficiency	85	85	85	85
Cycle efficiency	70–80	90–97	75–90	60–70
Self- discharge	0.1–0.3	0.1–0.3	~0	0.2–0.6
Response time	ms	ms	ms	ms
Energy density (Wh/L)	50–80	200–500	150–250	60–150
Power density (W/L)	10–100	1500–10,000	140–180	80–600

2. Costly and high maintenance

Industrial application

1. Renewable energy
2. Electric vehicle (EV)

1.3 Hybrid energy storage system

Storage systems are being used for a very long time however it so happens frequently that one ESS can't do the trick every one of the requirements of a specific client. Thus, it becomes important to consolidate different energy storage schemes to take care of all necessities. Such a framework, which utilizes at least two ESS consolidated together, is known as a half and half ESS (HESS). ESSs can work in both two modes-high-power modes or runner mode and high-energy mode or long-distance race mode [43].

High-power ESSs answer quick and incorporate flywheel ES, SMES, SCES, BESS, and so on, though high-energy ESSs answer gradually and incorporate PHS, CAES, power devices, BESS, and so forth Batteries, for example, Li-particle batteries are reasonable for both high-power and high-energy applications [44]. For the most part, HESSs have planned with the end goal that the two modes are accessible for a solitary application. This altogether diminishes the establishment and support costs because similar power electronic gadgets and matrix associations are adequate for the part ESSs. HESS might bring further developed exhibitions, expanded productivity, longer life expectancy, decreased expenses, and more reasonable plans and estimates [45]. Energy the executives are of essential thought if there should be an occurrence of HESS [46]. It decides the dissemination of force among the different ESSs utilized in the HESS and directions them to satisfy one shared objective. It additionally characterizes the size of the ESSs, their life cycle, effectiveness, and a few different attributes. In this way, the decision of the right energy the board situation calculation is critical to advanced usage of the HESS. In view of how the different ESSs are associated, there are various geographies of HESSs, fundamentally named series or equal; the equal geography is for the most part utilized. In inequality, the ESSs can be either straightforwardly associated (detached arrangement), or through power converters in the middle (dynamic setup). A typical transport, normally DC due for its inborn potential benefits over AC transport, associates the entire framework. HESSs can be utilized in a few applications, for example, in crossover electric vehicles (HEV), power device fueled electric vehicles, inexhaustible independent energy supply frameworks utilizing battery/hydrogen mix, matrix tied HESS in homegrown or territorial use, enormous scope wind and sunlight based ranches, and so on [44].

1.4 Artificial intelligence-based energy storage systems

Artificial intelligence (AI) techniques gain high attention in the energy storage industry. Smart energy storage technology demands high performance, life cycle long, reliability, and smarter energy management. AI can dramatically accelerate calculations, improve prediction accuracy, optimize information, and enhanced system performance. Many studies addressed the several AI techniques such as human brain-inspired based fuzzy logic (FL), artificial neural network (ANN), animal-inspired based particle swarm optimization (PSO) and teaching-learning, bee colony algorithm or natural inspired based genetic algorithm (GA), and simulated annealing [47].

Some popular AI case studies are addressed in Table 1.5 [48–55].

Table 1.5 Artificial intelligence techniques -based energy storage system.

AI techniques	Energy storage type	Factors	Findings
PSO [48]	Hydrogen, and battery	Costs, lifetime, and efficiency	• Minimize the operating costs (29.36%) • Maximize the system efficiency (27.21%) • Improve the devices life cycle (43.43%)
multi-objective PSO [49]	Thermal energy and battery system	Outcome-based on power	• Increase the performance and stability of the transmission system
PSO [50]	Electrochemical energy storage system	Power output	• Voltage fluctuation following the connection of the electrochemical energy storage
Hybrid PSO and Teaching Learning [51]	Battery	Emission, time, and cost	• PSO-TLBO algorithm successful success in addressing the suggested shipping problems.

(*Continued*)

Table 1.5 (Continued)

AI techniques	Energy storage type	Factors	Findings
Bee colony algorithm, PSO [52]	Fuel cell	Power loss, voltage stability index, and emissions	Sufficient consistency and greater variety relative to the updated PSO process
Agglomerative hierarchical clustering Algorithm [53]	Battery	Time	• Time series data of (5–15 days) sufficiently to improve the system performance by 5%.
Nondominated sorting GA [54]	Pumped storage, battery	Cost and emissions	• System economic and environmental Performance improved.
Harmony search optimization [55]	Battery	Time-of-use pricing	• Optimal ESS scheduling at demand-free time-of-use rates. • Economical
Hybrid GA and ANN [55]	Battery	Cost, power output	• Average capital cost minimized • System stability enhanced

AI, artificial intelligence; *ANN*, artificial neural network; *PSO*, particle swarm optimization; *GA*, genetic algorithm.

1.5 Energy storage system control strategy

Energy storage systems have two-way power regulations such as absorb power and release power. AI-based control techniques are mainly used to enhance the system's power generation capacity, performance, power stability, and reliability. Table 1.6 addressed several important pieces of research in terms of AI-based control strategies, and outcomes.

1.6 Machine learning-based energy storage system

Machine learning (ML) has been popular and widely used in the energy storage industry. Many researchers reported different applications such as batteries, capacitors/supercapacitors, and fuel cells. Integrating human inelegancy into machine learning can significantly enhance the robustness and reliability, and performance of the systems. Several ML algorithms are very well known such as

1. Unsupervised learning
2. Supervised learning

Table 1.6 AI-based energy storage system control and outcomes.

Schemes	Outcomes
Hybrid differential evolution optimization and ANN [56]	• The system consists of a synchronous generator (SG), PV, and battery-based energy storage system • The proposed control scheme effectively optimizes the phase of the two controllers.
Polyline fuzzy neural network (PFNN) [57]	The PFNN control scheme suggests adjusting frequency frequencies as the ES device input and state as output.
PSO [58]	• The proposed control schemes effectively controlled a 100-megawatt energy storage system.
Fuzzy logic controller (FLC) [59]	• FLCs reduce the operating hours of fuel cells and electrical gasses and SOC variability in the battery stack.
ANN and dynamic programming (DP) [60]	• Power flow is balanced • Minimizing power loss • Increasing battery life cycle • Minimizing battery current variations.
ANN [61]	• The ANN-based fast response capability is represented

3. Deep learning
4. Reinforcement learning
5. Data pre-processing

Table 1.7 present the ML-based energy storage system enhanced performance, and reliability, improve life cycle, and fast and robust computations (Table 1.8).

1.7 Energy storage policies and standards

The ESS, which is characterized as a fundamental network resource, has acquired plentiful consideration on account of its multipurpose usefulness. The ESSs can diminish discontinuities brought about by the incorporation of RESs into the framework, hence expanding the strength and dependability of the network. Moreover, as talked about prior, the ESS additionally can offer different types of assistance, for example, top shaving, recurrence guideline, dark beginning, receptive power remuneration, and so forth The overall energy stockpiling sending has expanded to 9 GW/17 GWh in 2018, which is practically twofold of that of 2017 [78]. In any case, this records for just 0.0000728% of the complete energy utilization of 2018. Accordingly, to work with and cultivate the inescapable sending of energy stockpiling, various kinds of government and state guidelines are required, which will guarantee charge impetuses and refunds for the ESS. Besides, strategy and administrative specialists should guarantee that the ESS gets sufficient to pay for various sorts of

Table 1.7 Machine learning use in an energy storage system.

Algorithm	Drawbacks	Advantages	Power transmission system
Reinforcement learning and Markov decision system [62,63]	Data requirements	Improved battery	Hybrid electric vehicles and plug-in hybrid electric vehicles
Q-learning [64]	Simplification of complex models	Adaptive to riding conditions	Electric bike
Q-learning [65]	Sporadic local optimization	Ability to run online	Plug-in hybrid electric vehicle
Q-learning [66]	Computational load	Multiple control	Hybrid electric vehicle
Temporal-difference learning [67,68]	Dependence on driving data	High accuracy	Hybrid electric vehicle
Deep reinforcement learning and dynamic neural programming [69]	Design complexity	Robust against	Hydraulic hybrid vehicle
Dynamic learning [70]	Complex mathematics	Real-time control	Hybrid truck
Deep learning [70]	Needs specific training	Data-driven model	Plug-in hybrid electric vehicle
Online reinforcement learning [63]	Sensitivity to driving cycle	Fast computation	Electric vehicle

Table 1.8 ML research contribution to the energy storage system.

Reference	Results	Application
[71]	The battery management system state of charge (SOC) and state of health (SOH) are plays a vital role in battery performance enhancement and safety and lifetime.	Battery state estimation
[72]	Forecast the SOH	
[73]	Load-classifying neural network (NN) model is also developed to estimate SOC	
[74]	Radial basis function neural network (RBFNN) model is applied to estimate the capacity and internal resistance of Li-ion batteries	
[75]	Predict the battery cycle life before capacity degradation based on the early cycle discharge data	Battery lifetime estimation and prediction

(*Continued*)

Table 1.8 (Continued)

Reference	Results	Application
[76]	Classify the unbalance and damage of Ni-MH battery cells	Battery fault diagnosis
[77]	Model the cyclic voltammetry for supercapacitors with $Ni(OH)_2$ electrodes	capacitors/ supercapacitors

administrations they serve. Furthermore, the approach creators ought to security regulations for making appropriate business sectors, which will energize the financial backers in putting resources into ESSs.

1.8 Barriers and potential solutions

Although the ESSs have different applications and advantages, there exist various boundaries, which are obstructing the inescapable organization of energy stockpiling on the lookout. The creators of [79] classify these hindrances into five sorts administrative obstructions, mechanical boundaries, monetary hindrances, cross-cutting obstructions, and utility and engineer plan of action boundaries. Every hindrance presents a few issues, which ought to be managed for the far-reaching implantation of the ESS. Making markets for the ESS, giving sufficient pay system, moderating various kinds of procedural deferrals, teaching the financial backers about the various administrations ESS can give, and so forth, are required for the inescapable execution of ESS. Different government and state associations have recognized these previously mentioned obstructions and have taken a drive in the structure acts, bills, and requests to defeat them.

1.9 Environmental impacts of energy storage systems

Energy storage systems (ESS) are without a doubt bright green since they support the utilization of sustainable wellsprings of energy, along these lines diminishing the emanation of greenhouse gases (GHG) and safeguarding the restricted stores of petroleum products. Ecological contamination is a tremendous concern for mainstream researchers yet in addition among the overall cognizant residents all over the planet. The energy requests of individuals are dramatically expanding, which brought about the worldwide CO_2 emanations ascending by 1.7% to a record high of 33.1 Gigaton. This was the most elevated development rate beginning around 2013, and practically 70% higher than the normal development rate beginning around 2010. Around 66% of the emanations have emerged from the power area alone. The broad utilization of coal in the power area surpassed 10 Gt in CO_2 emanations in Asia. The main nations on the planet with the most elevated energy requests, that is the USA, India, and China, are liable for 85% of the general expansion in discharges. Then again, nations like France, Germany, the United Kingdom, Mexico, and Japan have decreased their outflows.

As individuals' way of life is improving, with expanded present-day conveniences and a lift in the economy, the energy requests are on an emotional ascent. Besides, the changing climatic circumstances are liable for expanding the energy interest for warming and cooling applications. Subsequently, the worldwide CO_2 outflows are challenging to control. Notwithstanding, the outflows were consistent somewhere in the range of 2014 and 2016 because of sending of low-carbon innovations and enhancements in energy efficiencies. Be that as it may, in 2017 and 2018, the outflows rose by around 0.5% for each 1% expansion in a worldwide economy, a figure a lot higher than the 0.3% normal ascent beginning around 2010. This speed increase was because of the way that the energy creations couldn't get the job done the higher energy requests, and the low-carbon advances didn't develop apace the rising energy requests.

ES isn't just a savvy technique for using sustainable wellsprings of energy at their fullest, yet additionally helps the climate in numerous ways. Since ESSs assist with battling the climatic caprices and irregularities of sustainable sources, they can support the coordination of more renewables in the power matrix. Further, ESSs can guarantee that the age framework works at ideal levels and cut down on the use of less productive creating units, which run distinctly to satisfy top needs. Also, ESSs increment the limit of a powerful framework, wiping out the requirements for extra creating units, or transmission and dispersion frameworks [80].

There can be a few potential negative parts of ES, yet it relies upon the kind, materials utilized, and productivity of the ESSs. For example, a few batteries, like lithium and lead, are regularly unsafe according to a natural point of view and could cause well-being perils on the off chance that not discarded as expected. These harmful materials should be reused successfully to guarantee that ESSs stay a protected and solid choice for long-haul pertinence. Furthermore, the capacity interaction includes energy transformation, which can cause a loss of energy. Some energy can likewise be lost during capacity itself [80]. The carbon impression of the ESS ought to be kept as low as could be expected. There are various advantages of ESS concerning the ecological angle, making the reception of ESSs an incredible decision for meeting the sustainable development goals. ESSs are fit for upgrading the general proficiency of force networks, and hence, they can increase the shift towards the utilization of sustainable wellsprings of energy in the power framework. ESS is liberated from any kind of emanation that could corrupt the air or compromise the climate. Sun-based PV and ESS are an extraordinary couple, together with clearing the way towards a without GHG, clean, and beneficial power age framework, qualified for government venture tax breaks [81].

1.10 Conclusions

This book chapter is a critical review of available technology in the field of energy storage. Energy storage is a key to making modern-day renewables more reliable.

The COVID-19 pandemic era resulted in social and economic crises. Looking into the exponential growth that India has taken in the manufacturing of PV modules, it is competing with China for low-cost modules. Based on the specific requirements, numerous energy storage options are available now, in particular for grid integration. Nowadays, the focus is not only on the large reservoirs of energy, rather, but mini and micro-level projects are also upcoming to take the lead. Mechanical, Chemical, Electrical, and Electro-magnetic based energy storage systems are the backbone. In recent years, because of the need to shift to some alternative to internal combustion engines, battery storage, and hydrogen storage are of prime importance. Fuel cells and Electric vehicles are the future of transportation. Hydrogen fuel cell vehicles have the added advantage of generating only water as the by-product with zero emission, but the challenge lies in hydrogen storage. Researchers are continuously updating the technologies for the utilization of hydrogen in a safer manner. Biofuels are also being appreciated nowadays as an alternative to the existing petroleum products, but again, a lot of challenges are there for their full capacity utilization. Solar-based plants are also upcoming which are not only helping in generating biofuels but also provides zero or negligible running cost. As the demand for reliable, high-performing storage technology is the need of the hour, many researchers are using AI techniques like FL, ANN to provide a better solution and in a quick time. Also with AI, Machine Learning is gradually becoming popular in the energy storage industry. The reliability and robustness of machine learning can take the energy storage technology to a greater height. Of course, some technological barriers depend on government policies and market ups and downs. It is certain that in the years to come, energy storage will do wonders and will be a part of the life and culture of mankind.

References

[1] S. Vazquez, S.M. Lukic, E. Galvan, L.G. Franquelo, J.M. Carrasco, Energy storage systems for transport and grid applications, IEEE Trans. Ind. Electron. 57 (12) (2010) 3881–3895. Available from: https://doi.org/10.1109/TIE.2010.2076414.

[2] S.M. Lukic, J. Cao, R.C. Bansal, F. Rodriguez, A. Emadi, Energy storage systems for automotive applications, IEEE Trans. Ind. Electron. 55 (6) (2008) 2258–2267. Available from: https://doi.org/10.1109/TIE.2008.918390.

[3] L. Chang, W. Zhang, S. Xu, K. Spence, Review on distributed energy storage systems for utility applications, CPSS Trans. Power Electron. Appl. 2 (4) (2017) 267–276. Available from: https://doi.org/10.24295/CPSSTPEA.2017.00025.

[4] Battery Energy storage, IEA, available at: https://www.iea.org/reports/the-covid-19-crisis-and-clean-energy-progress/energy-integration.

[5] A.G. Olabi, C. Onumaegbu, Tabbi Wilberforce, Mohamad Ramadan, Mohammad Ali Abdelkareem, Abdul Hai Al-Alami, Critical review of energy storage systems, Energy 214 (2021) 1–22. 1 January.

[6] M. S. Whittingham, History, evolution, and future status of energy storage, in: Proceedings of the IEEE, vol. 100, no. Special Centennial Issue, 13 May 2012, pp. 1518–1534. doi: 10.1109/JPROC.2012.2190170.

[7] F. Nadeem, S.M.S. Hussain, P.K. Tiwari, A.K. Goswami, T.S. Ustun, Comparative review of energy storage systems, their roles, and impacts on future power systems, IEEE Access 7 (2019) 4555–4585.
[8] H. Chen, T.N. Cong, W. Yang, C. Tan, Y. Li, Y. Ding, Progress in electrical energy storage system: a critical review, Prog. Nat. Sci. 19 (2009) 291–312.
[9] U.S. Department of Energy Office of Electricity Delivery and Energy Reliability. Energy storage program planning document. Available from: https://www.energy.gov/sites/prod/, 2011.
[10] P. Breeze, Power system energy storage technologies, Power Generation Technologies, Elsevier: Amsterdam, The Netherlands, 2019, pp. 219–249. ISBN 978-0-08-102631-1.
[11] Pumped hydro storage, IEEFA, available at: https://ieefa.org/wp-content/uploads/2019/03/IEEFA-India_Pumped-Hydro-Storage_Mar-2019.pdf.
[12] J. Wang, K. Lu, L. Ma, J. Wang, M. Dooner, S. Miao, et al., Overview of compressed air energy storage and technology development, Energies 10 (2017) 991.
[13] S. Kalaiselvam, R. Parameshwaran, Energy storage, Thermal Energy Storage Technologies for Sustainability, Elsevier: Amsterdam, The Netherlands, 2014, pp. 21–56. ISBN 978-0-12-417291-3.
[14] Vafakhah, B.; Masiala, M.; Salmon, J.; Knight, A., Emulation of flywheel energy storage systems with a PMDC machine, in: Proceedings of the Eighteenth International Conference on Electrical Machines, Vilamoura, Portugal, 6–9 September 2008, pp. 1–6.
[15] H. Liu, J. Jiang, Flywheel energy storage—An upswing technology for energy sustainability, Energy Build. 39 (2007) 599–604.
[16] R. Hebner, J. Beno, A. Walls, Flywheel batteries come around again, IEEE Spectr. 39 (2002) 46–51.
[17] M. Amiryar, K. Pullen, A review of flywheel energy storage system technologies and their applications, Appl. Sci. 7 (2017) 286.
[18] R. Sebastián, Peña Alzola, R. Flywheel energy storage systems: review and simulation for an isolated wind power system, Renew. Sustain. Energy Rev. 16 (2012) 6803–6813.
[19] S. Choudhury, Flywheel energy storage systems: a critical review on technologies, applications, and future prospects, Int. Trans. Electr. Energ. Syst. 31 (9) (2021) e13024.
[20] Xing Luo, Jihong Wang, Mark Dooner, Jonathan Clarke, Overview of current development in electrical energy storage technologies and the application potential in power system operation, Appl. Energy 137 (2015) 511–536. vol.
[21] L.F. Cabeza, I. Martorell, L. Miro, A.I. Fernandez, C. Barreneche, Introduction to Thermal Energy Storage (TES) Systems, Woodhead Publishing Limited, 2015.
[22] F. Wei, et al., A novel thermal energy storage system in smart building based on phase change material, IEEE Trans. Smart Grid 10 (3) (2019) 2846–2857. Available from: https://doi.org/10.1109/TSG.2018.2812160.
[23] I. Sarbu, C. Sebarchievici, A comprehensive review of thermal energy storage, Sustainability 10 (2018) 191.
[24] G. Wang, L. Zhang, J. Zhang, A review of electrode materials for electrochemical supercapacitors, Chem. Soc. Rev. 41 (2) (2012) 797e828.
[25] Wagner L. Overview of Energy Storage Methods, 2007, 0e15.
[26] Guney MS, Tepe Y. Classification assess. Energy Storage Syst. 2017;75:1187e97.
[27] B. Zakeri, S. Syri, Electrical energy storage systems: a comparative life cycle cost analysis, Renew. Sustain. Energy Rev. 42 (2015) 569e96.
[28] A. Demirbas, Biofuels sources, biofuel policy, biofuel economy and global biofuel projections, Energy Convers. Manag. 49 (2008) 2106–2116.

[29] Global Bioenergy Statistics 2019; World Bioenergy Association: Stockholm, Sweden. Available from: https://worldbioenergy.org/uploads/191129%20WBA%20GBS%202019_LQ.pdf, 2019.
[30] H.L. Tuller, Solar to fuels conversion technologies: a perspective, Mater. Renew. Sustain. Energy 6 (2017) 3.
[31] A. Steinfeld, P. Kuhn, A. Reller, R. Palumbo, J. Murray, Y. Tamaura, Solar-processed metals as clean energy carriers and water-splitters, Int. J. Hydrog. Energy 23 (1998) 767–774.
[32] A. Steinfeld, Solar thermochemical production of hydrogen—a review, Sol. Energy 78 (2005) 603–615.
[33] E. OguzKoroglu, H. CivelekYoruklu, A. Demir, B. Ozkaya, Scale-up and commercialization issues of the MFCs, Microbial Electrochemical Technology, Elsevier: Amsterdam, The Netherlands, 2019, pp. 565–583.
[34] Graphite Solar Power Pty Ltd. Graphite Thermal Battery [Online]. Available from: http://gsplicensing.com.au/graphitethermal-battery. (accessed 18.10.18).
[35] D. Drake. Is graphite the next big thing in the world of alternative energy Huffingtonpost. [Online]. Available from: https://www.huffingtonpost.com/david-drake/is-graphite-the-nextbig-_b_10588420.html; 2017 (accessed 18.10.18).
[36] A. Theron. ESI-Africa. Graphite Power Generator (GPG) Technology [Online]. Available from: https://www.esiafrica.com/canadian-firm-turns-coal-to-graphite; 2018 (accessed 19.10.18).
[37] D. Wagman, How paper batteries charged by bacteria could power the internet of things, in: IEEE Spectr., Aug. [Online]. Available from: https://spectrum.ieee.org/energywise/energy/renewables/paper-batterythat-could-power-the-internet-of-things; 2018.
[38] T. Agarwal. Construction of a paper battery. EDGEFX. Accessed: Oct. 19, 2018. [Online]. Available from: https://www.edgefx.in/paper-batteryworking-and-construction.
[39] D. S. Ebnesajjad. Paper battery: a thin claim of the storage technology. Chemicals & Materials/Elsevier. [Online]. Available from: https://chemical-materials.elsevier.com/newmaterials-applications/paper-battery-thin-claim-storage-technology; 2017 (accessed 19.10.18).
[40] P. K. Pathak and A. R. Gupta, Battery energy storage system, in: Proceedings of the Fourth International Conference on Computational Intelligence & Communication Technology (CICT), 2018, pp. 1–9, doi: 10.1109/CIACT.2018.8480377.
[41] N. Kawakami and Y. Iijima, Overview of battery energy storage systems for stabilization of renewable energy in Japan, in: Proceedings of the International Conference on Renewable Energy Research and Applications (ICRERA), 2012, pp. 1–5, doi: 10.1109/ICRERA.2012.6477391.
[42] Agwu Daberechi D., Opara F. K., Chukwuchekwa N, Dike. D. O., and Uzoechi L, Review of comparative battery energy storage systems (Bess) for energy storage applications in tropical enviroments, in: Proceedings of the IEEE Third International Conference on Electro-Technology for National Development (NIGERCON), 2017, pp. 1000–1005.
[43] Hybrid energy storage: are combined solutions gaining ground? Energy Storage Forum. Available from: https://energystorageforum.com/news/energy-storage/hybrid-energy-storage-combined-solutions-gaining-ground, 2018.
[44] T. Bocklisch, Hybrid energy storage systems for renewable energy applications, Energy Procedia 73 (2015) 103–111.
[45] Serpi, A.; Porru, M.; Damiano, A., A Novel Highly Integrated Hybrid Energy Storage System for Electric Propulsion and Smart Grid Applications, Advancements in Energy

Storage Technologies, Xiangping Chen and Wenping Cao; IntechOpen: London, UK, 2 May 2018.

[46] Etxeberria, A.; Vechiu, I.; Camblong, H.; Vinassa, J.M.; Camblong, H., Hybrid energy storage systems for renewable energy sources integration in microgrids: a review, in: Proceedings of the Conference Proceedings IPEC, Singapore, 27–29 October 2010, pp. 532–537.

[47] Ahmed N. Abdalla, Muhammad Shahzad Nazir, Hai Tao, Suqun Cao, Rendong Ji, Mingxin Jiang, et al., Integration of energy storage system and renewable energy sources based on artificial intelligence: an overview, J. Energy Storage 40 (2021) 1–13. Vol.

[48] P. García-Trivino, et al., Optimized operation combining costs, efficiency and lifetime of a hybrid renewable energy system with energy storage by battery and hydrogen in grid-connected applications, Int. J. Hydrog. Energy 41 (48) (2016) 23132–23144.

[49] S. Guo, et al., The multi-objective capacity optimization of wind-photovoltaicthermal energy storage hybrid power system with electric heater, Sol. Energy 195 (2020) 138–149.

[50] Y. Hao, et al., Active reactive power control strategy based on electrochemical energy storage power station, in: IEEE Third Conference on Energy Internet and Energy System Integration (EI2), IEEE, 2019.

[51] T. Cheng, et al., Adaptive robust method for dynamic economic emission dispatch incorporating renewable energy and energy storage, Complexity 2018 (2018).

[52] H. Nasiraghdam, S. Jadid, Optimal hybrid PV/WT/FC sizing and distribution system reconfiguration using multi-objective artificial bee colony (MOABC) algorithm, Sol. Energy 86 (10) (2012) 3057–3071.

[53] W.W. Tso, et al., A hierarchical clustering decomposition algorithm for optimizing renewable power systems with storage, Appl. Energy 270 (2020) 115190.

[54] A.M. Abdelshafy, et al., Optimized energy management strategy for grid connected double storage (pumped storage-battery) system powered by renewable energy resources, Energy 192 (2020) 116615.

[55] Z.W. Geem, Y. Yoon, Harmony search optimization of renewable energy charging with energy storage system, Int. J. Electr. Power Energy Syst. 86 (2017) 120–126.

[56] S. Zhou, et al., The combinatorial optimization by genetic algorithm and neural network for energy storage system in solar energy electric vehicle, in: Proceedings of the Seventh World Congress on Intelligent Control and Automation, IEEE, 2008.

[57] Y. Wu, et al., Study on energy storage system participating in frequency regulation of wind farm based on polyline fuzzy neural network, in: Proceedings of the Fifth IEEE International Conference on Cloud Computing and Intelligence Systems (CCIS), IEEE, 2018.

[58] R. Zahedi, M.M. Ardehali, Power management for storage mechanisms including battery, supercapacitor, and hydrogen of autonomous hybrid green power system utilizing multiple optimally-designed fuzzy logic controllers, Energy 204 (2020) 117935.

[59] M. Athari, M. Ardehali, Operational performance of energy storage as function of electricity prices for on-grid hybrid renewable energy system by optimized fuzzy logic controller, Renew. Energy 85 (2016) 890–902.

[60] J. Shen, A. Khaligh, A supervisory energy management control strategy in a battery/ultracapacitor hybrid energy storage system, IEEE Trans. Transp. Electr. 1 (3) (2015) 223–231.

[61] S.F. Osmanaj, R.A. Selimaj, Hybrid renewable energy systems: case study-based analysis considering varying seasonal conditions, in: Proceedings of the Eighteenth International Research/ Expert Conference, 2014.

[62] H. Yu, M. Kuang, R. McGee, Trip-oriented energy management control strategy for plug-in hybrid electric ve- hicles, in Proceedings of the Fiftieth IEEE Conference on Decision and Control and European Control Conference, Orlando, FL, USA, January 2013, pp. 1323–1336.

[63] X. Lin, Z. Wang, J. Wu, Energy management strategy based on velocity prediction using back propagation neural network for a plug-in fuel cell electric vehicle, Int. J. Energy Res. 45 (2) (2021) 2629–2643.

[64] R.C. Hsu, C.T. Liu, D.Y. Chan, A reinforcement- learning-based assisted power management with QoR pro- visioning for human–electric hybrid bicycle, IEEE Trans. Ind. Electron. 59 (8) (2011) 3350–3359.

[65] T. Liu, Y. Zou, D. Liu, F. Sun, Reinforcement learning of adaptive energy management with transition probability for a hybrid electric tracked vehicle, IEEE Trans. Indust. Electron. 62 (12) (2015) 7837–7846.

[66] J. Hu, D. Liu, C. Du, F. Yan, C. Lv, Intelligent energy management strategy of hybrid energy storage system for electric vehicle based on driving pattern recognition, Energy 198 (2020) 117298. vol.

[67] X. Qi, G. Wu, K. Boriboonsomsin, M.J. Barth, J. Gonder, Data-driven reinforcement learning-based real-time energy management system for plug-in hybrid electric vehicles, Trans. Res. Record: J. Trans. Res. Board. 2572 (1) (2016) 1–8.

[68] C. Liu, Y.L. Murphey, Power management for plug-in hybrid electric vehicles using reinforcement learning with trip informationJune Proceedings of the 2014 IEEE Transportation Electrification Conference and Expo (ITEC), IEEE, Dearborn, MI, USA, 2014pp. 1–6.

[69] A.A. Kamel, H. Rezk, M.A. Abdelkareem, Enhancing the operation of fuel cell-photovoltaic-battery-supercapacitor renewable system through a hybrid energy management strategy, Int. J. Hydrog. Energy 46 (8) (2021) 6061–6075.

[70] P. Zhao, Y. Wang, N. Chang, Q. Zhu, X. Lin, A deep reinforcement learning framework for optimizing fuel economy of hybrid electric vehiclesJanuary Proceedings of the 2018 23rd Asia and South Pacific design automation conference (ASP-DAC), IEEE, Jeju, South Korea, 2018pp. 196–202.

[71] Murnane, M., Ghazel, A., A closer look at state of charge (SOC) and state of health (SOH) estimation techniques for batteries. Analog devices, 2017. Available from: https://www.analog.com/media/en/technical-documentation/technical-articles/acloser-look-at-state-of-charge-and-state-health-estimation-techniques.pdf (accessed 03.07.17).

[72] R.R. Richardson, M.A. Osborne, D.A. Howey, Gaussian process regression for forecasting battery state of health, J. Power Sources. 2017 (2017). Available from: https://doi.org/10.1016/j.jpowsour.

[73] S. Tong, J.H. Lacap, J.W. Park, Battery state of charge estimation using a load classifying neural network, J. Energy Storage 7 (2016) 236–243.

[74] P. Zhou, Z. He, T. Han, X. Li, X. Lai, L. Yan, et al., A rapid classification method of the retired LiCoxNiyMn1-x-yO2 batteries for electric vehicles, Energy Rep. 6 (2020) 672–683.

[75] K.A. Severson, P.M. Attia, N. Jin, N. Perkins, B. Jiang, Z. Yang, et al., Data- driven prediction of battery cycle life before capacity degradation, Nat. Energy (2019).

[76] J.P. Ortiz, J.D. Valladolid, C.L. Garcia, G. Novillo, F. Berrezueta, Analysis of machine learning techniques for the intelligent diagnosis of Ni-MH battery cells. 2018 IEEE Int. Autumn Meet, Power, Electron. Comput. ROPEC 2018 (2019).

[77] P.E. Lokhande, Cyclic voltammetry behavior modeling of fabricated nanostructured Ni (OH) 2 electrode using artificial neural network for supercapacitor application, Proc. Inst. Mech. Eng. C. 234 (2020) 2563. –256.

[78] Power & Renewable Reports, GreenTechmedia. Available from: https://www.greentechmedia.com/articles/read/global-energy-storage-to-hit-158-gigawatt-hours-by-2024-with-u-s-and-china/.

[79] Patel, S. IEA World Energy Outlook: Solar Capacity Surges Past Coal and Gas by 2040. *POWER Magazine*, 14 November 2019. Available from: https://www.powermag.com/iea-world-energy-outlook-solar-capacity-surges-past-coal-and-gas-by-2040/.

[80] US EPA. Electricity Storage. Available from: https://www.epa.gov/energy/electricity-storage.

[81] Five Benefits of Energy Storage: The Holy Grail of Energy. Available from: https://www.facilitiesnet.com/energyefficiency/article/Five-Benefits-of-Energy-Storage-The-Holy-Grail-of-Energy-16907.

Design of power electronic devices in the domain of energy storage

2

Ankireddy Narendra, Venkataramana Naik N and Anup Kumar Panda
Department of Electrical Engineering, NIT Rourkela, Rourkela, Odisha, India

Chapter Outline

Nomenclature

CAES	compressed air energy storage
CSI	current source inverter
CSR	current source rectifier
DFIG	doubly-fed induction generator
ES	energy storage

Emerging Trends in Energy Storage Systems and Industrial Applications. DOI: https://doi.org/10.1016/B978-0-323-90521-3.00008-9

FB-DCBP full-bridge DC bypass
FB-ZVR full-bridge zero voltage rectifier
FSWPG fixed speed wind power generation
GES green energy source
HERIC highly efficient realizable inverter configuration
HVDC high voltage direct current
IG induction generator
NPC neutral point clamping
PEC power electronic converter
PED power electronic device
PWM pulse width modulation
SCMES superconducting magnetic energy storage
SEM scanning electronic microscope
SEPIC single-ended primary inductor converter
SG synchronous generator
SPV solar photovoltaic
STEM scanning tunneling electron microscope
SThM scanning thermal microscope
TEM transmission electron microscope
TRE time-resolved electroluminescence
VSI voltage source inverter
VSR voltage source rectifier
VSWPG variable speed wind power generation
ZCS zero current switching
ZVS zero voltage switching

Index

flywheel energy storage
FSWPG
fuel cells
green energy source
GTO
HERIC
IGBT
isolated dc-dc converter
MCT
molten carbonate fuel cell
nonisolated dc-dc converter
NPC
phosphoric acid fuel cell
power BJT
power diode
power electronic devices
power MOSFET
proton exchange membrane fuel cell
pumped storage hydro
push-pull converter
PWM
rechargeable battery
resonant converter
SEPIC converter
single active bridge
solar photovoltaic
solar pond
solid oxide fuel cell
spring energy storage
stream accumulator
supercapacitor
superconducting magnetic energy storage
super lift converter
thyristor
TRAIC
VSI
VSR
ZCS
ZVS

Highlights of the chapter

This chapter focuses on various classifications of energy storage systems and the crucial role of power electronic devices in the energy storage system. Further, this chapter emphasizes classifications, specifications, and tests to be performed on power electronic devices. Moreover, the converter circuits are formed with the help of power semiconductor devices only. So, the converter circuits that are associated with renewable power generation such as wind, solar, and tidal are studied. Finally, the power conditioning converters used in energy storage are also elaborated.

2.1 Introduction

The world development index considers the consumption of electrical power as one of the parameters. Moreover, the utilization of electrical power increases due to industrialization and urbanization. Hence, the demand for electrical power keeps on increasing. So, to meet the demand pressure on the generation of conventional energy sources is increasing. However, conventional sources are causing global warming, climate change, and pollution. So, power generation is working towards renewable/green energy sources (GES) such as solar, wind, tidal, etc. However, the major problem with these green energy sources is fragile.

The green sources develop power only during certain conditions. For example, the solar photovoltaic (SPV) system generates power during sunny times; in the wind system, the power generation is based on the availability of wind, and in tidal also power generation is based on tidal waves. So, most green energy sources generate power at irregular periods at a variable amount. Moreover, the demand may not meet these circumstances. To meet the load demand and utilization of maximum power from GES, a storage system is required. However, the storage system keeps the power generated from GES and it can be utilized as per the load demand.

So, the solution for problems associated with GES can be overcome using energy storage (ES). Now, the issue is how to deal with energy sources and ES. What are various ES systems and the power management that takes place between GES and ES? The power management and conversions of energy are taken care of using power electronic devices (PED) [1]. This chapter discloses various converter circuits for energy storage systems are analyzed their performance and operation under various operating conditions of the storage device. Most generally, the energy is a transfer between DC-DC or DC-AC or AC-AC or AC-DC [2]; all these can be achieved with the help of PEC. This chapter elaborates on what various converter circuits are there and how their classification takes place. However, the crucial part in the selection of the PEC for various applications, after reading this chapter, the reader can become aware of the selection of the converter based on the available storage.

2.2 Classification and importance of power electronic devices in energy storage

The power generation from renewable energy sources requires energy storage to maximize proper utilization. Here, the various energy storage [3,4] is disclosed as follows: (Fig. 2.1)

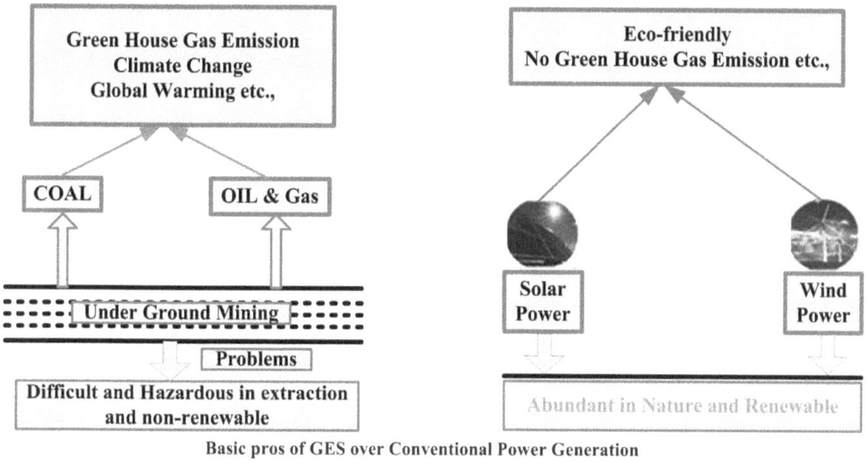

Figure 2.1 Basic pros of green energy sources over Conventional Power Generation.

2.2.1 Classification of energy storage

The electrical energy has stored in any form that can be converted back to electrical form. The ES system can be classified based on the fast recovery, long-lasting energy, size, etc. however, the ES system can also be classified as per the form of the energy stored such as mechanical, thermal, electromagnetic, electro-chemical, and chemical [5,6] as shown in Fig. 2.2.

1. Electro-thermal energy storage system

   Electrical energy is generated according to load demand. However, sometimes the demand varies from low to high and vice versa. If the load demand is high then peak load generation has to turn on immediately along with baseload generation. Moreover, when the load demand is low then the baseload power generation has to decrease. The cost of the peak load power generation is higher and also shutting down baseload power generation immediately is difficult. To overcome these problems the energy is stored from baseload plants when the demand is low and used back during high demand. Here, some of the important electro-thermal energy storage are explained as follows:

   a. Cryogenic energy storage

      Cryogenic energy storage utilizes low-temperature liquids such as liquid nitrogen or liquid air to store electrical energy. The surplus electric power is used to transform the nitrogen/air into its liquid form and is stored in a vacuum [7]. During the peak demand the liquid gases to a heat exchanger (which acts as a boiler) at high pressure to generate the electrical energy.

   b. Solar pond

      A solar pond is a technology that utilizes the heat energy from the sun. The solar pond consists of salt water in a large area, the saltwater is heated and the density becomes low. The hot water reaches the top which can be used to drive the turbine for electrical energy generation [8,9].

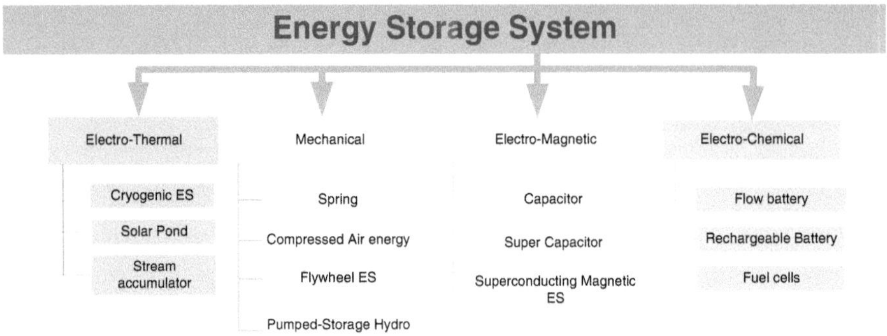

Figure 2.2 Classification of energy storage system.

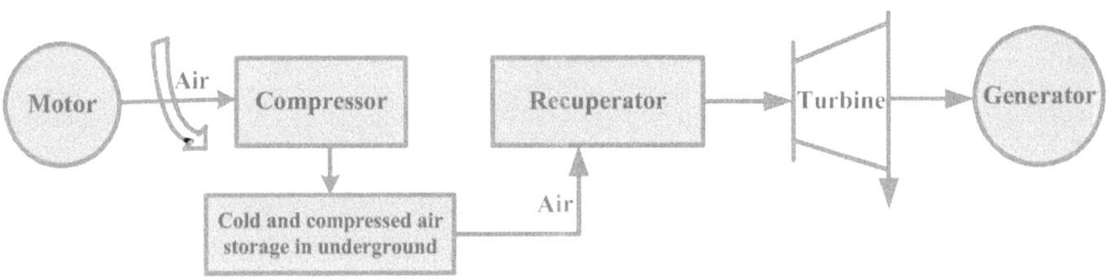

Figure 2.3 Compressed air energy storage.

c. Stream accumulator

A stream accumulator is also known as a solar tower. The tower is a steel container to store water and the water is heated using solar. During the availability of sun solar tower stores the heated water at high pressure and temperature. Later, it can be used for electrical power generation for the peak demand of load [10].

d. Mechanical energy storage system

i. Mechanical springs

Spring stores mechanical energy in terms of potential energy. When surplus power is available, it can be used to move the springs in a clockwise direction to store the energy. However, during the peak demand for power electricity is generated from spring using dynamo [11].

ii. Compressed air energy storage

Compressed air energy storage (CAES) is one of the large energy storage systems (above 100 MW units) [12]. The complete block diagram of CAES [13] is shown in Fig. 2.3.

During the off-peak period, the available electrical power is used to compress the air. Moreover, the cold and compressed air is stored underground. However, while peak-load demands the cold compressed air is transferred to recuperate the compressed air and flown in a turbine to drive the generator. Hence, required electrical power is generated for peak load demand.

iii. Flywheel energy storage

Flywheel energy storage system is one of the high energy storage systems. It stores the electrical energy in terms of kinetic (rotational) energy [14]. The amount of energy stored depends upon the size and material of the rotor. Moreover, the amount of energy stored is given by

$$E = \frac{1}{2} J\omega^2$$

where, E, J, and ω are kinetic energy stored, the moment of inertia, and angular velocity of the rotor respectively.

The flywheel energy storage is also known as a mechanical battery. During the off-peak periods, the electrical energy is used to drive the motor and the kinetic energy is stored. However, during the peak-load demand, the kinetic energy is transformed into electrical energy using a generator and fed to the grid using a bi-directional converter.

iv. Pumped storage hydro

A pumped storage hydro system is also one of the well-practiced energy storage techniques [15]. The pumped storage hydro system consists of two simple reservoirs as shown in Fig. 2.4 [16].

During the low demand for electric power, water is pumped from the lower reservoir to the upper reservoir using a motor. Here, the electrical power is stored as potential energy and it can be used when the demand is high. Moreover, during peak load demand the potential energy is converted to electrical energy with the help of turbines and generators.

However, the pumped storage hydro system is classified as an open-loop or closed-loop hydro system. In an open-loop hydro system, the water from the lower reservoir is released to down steam whereas in closed-loop same water is circulated back to the upper reservoir to continue the cycle.

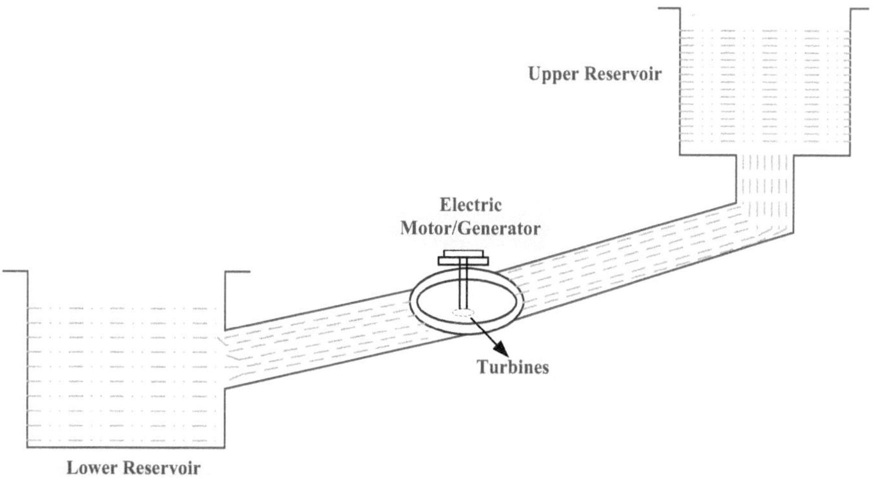

Figure 2.4 Pumped storage hydro system.

v. Electro-magnetic energy storage system

Magnetic energy is one of the high-capacity energy storage. The following storages are explored as given below:

1. *Capacitor*

The capacitor stores the energy in terms of charge. Whenever there is the availability of electrical energy the capacitors are getting charged and discharged during the nonavailability of power like a battery [17]. However, the capacitor discharges faster than the battery. The amount of energy stored in the capacitor depends upon the voltage across the plates and capacitance as given in the following equation:

$$W = \frac{1}{2}CV^2$$

where W is the energy stored, C is the capacitance and V is the voltage applied across it.

The capacitor-based energy stored is used in UPS, camera flashes and audio equipment, etc.

2. Super capacitors

Supercapacitors are double-layer capacitors to store high energy. The supercapacitors have a capacitance of up to 2 kF. These capacitors are used for regenerative braking of motors, computers for backup during power loss, industrial automation, etc. Supercapacitors have the ability to store large amounts of energy as well as discharge quickly [17].

3. Superconducting magnetic energy storage

Superconducting magnetic energy storage (SCMES) stores a larger amount of energy and releases it rapidly [18]. The basic components of the SCMES are shown in Fig. 2.5 [19].

The basic principle of SCMES is with the help of a superconducting coil the nitrogen/helium gas is liquefied. However, the liquefied gas maintained the same temperature using a cooling system. The DC power is supplied to the superconducting coil. For SCMES operation either helium or nitrogen is used due to their low boiling points (4.2K and 130K respectively). Even though helium has a low boiling point it is more costly than nitrogen. If AC power is required then it can achieve using a power semiconductor circuit.

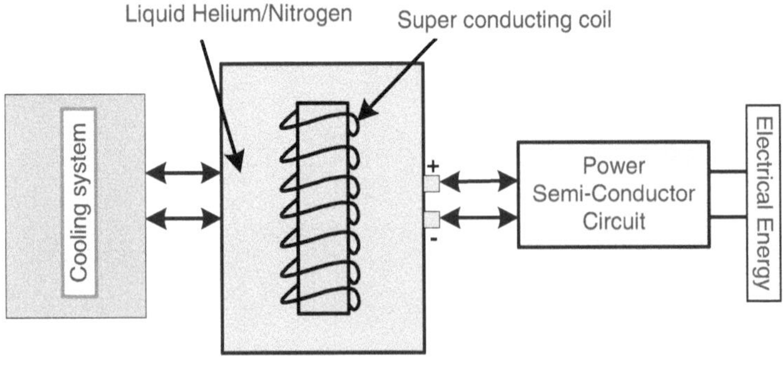

Figure 2.5 Super conducting magnetic energy storage system.

However, superconducting is classified as low-temperature and high-temperature superconductors. The low-temperature superconductors are used for magnetic resonance imaging, nuclear magnetic resonance, particle accelerators, etc., however, the high-temperature superconductors are used for electrical equipment such as motors, generators, transformers, transmission cables, etc.

The superconductor magnetic energy storage attains greater importance due to the implementation of renewable energy sources. The renewable energy sources increase the instability of the grid due to their irregular amount of power generation and hence, the stability of the grid will effect. To overcome this, the SCMES stores the power during the surplus period and reverts during the deficit in power such that the grid stability is maintained.

vi. Electro-chemical energy storage system

The electrochemical energy storage system utilizes surplus electrical energy to perform chemical reactions to store the electrical energy. Moreover, the chemical energy transformed back to electrical energy during a surge in electrical power. Electrochemical energy is classified as a flow battery, rechargeable battery, and fuel cell. However, the charging of the battery has to take at a specified voltage and current range. Moreover, this can be achieved with the help of power semiconductor device-based circuits [20].

1. Flow battery

The batteries are primarily classified into two types primary battery and secondary battery. The primary battery is one-time usable. However, the secondary battery can be reused by recharging it. Coming to flow battery is also a secondary battery, it can be rechargeable. In the flow battery system, the energy-stored electrodes are dissolved in liquids and stored in tanks [21].

Their various flow batteries are available in which the most used are zinc-bromine hybrid, all vanadium, and all slurry iron flow batteries.

2. Rechargeable battery

The rechargeable battery is also a secondary battery. However, a rechargeable battery is one of the widely used energy storage systems for renewable energy applications. The rechargeable battery consists of a number of series cells to store the energy. The electrical energy is stored during charging and delivered the energy during discharging. There are various rechargeable batteries such as lithium, lead-acid, sodium Sulfur and nickel-metal hydrate, etc.

3. Fuel cells

Fuel cells are voltaic cells, which not only store the energy but also deliver the energy at a time. The fuel cells give electrical energy with the help of hydrogen and oxygen through a pair of redox reactions [22]. The fuel cells classification is as shown in Fig. 2.6 [23].

2.2.2 Necessity of energy storage

The power grid does not have any storage capacity. So, whenever there is an increase in power demand, then the conventional power generation has to increase to balance the load demand. However, it is not a viable solution all the time due to the requirement of burning fossil fuels. In recent times, power generation is moved to green energy generation. Solar and wind power generations are contributing most of the part in it.

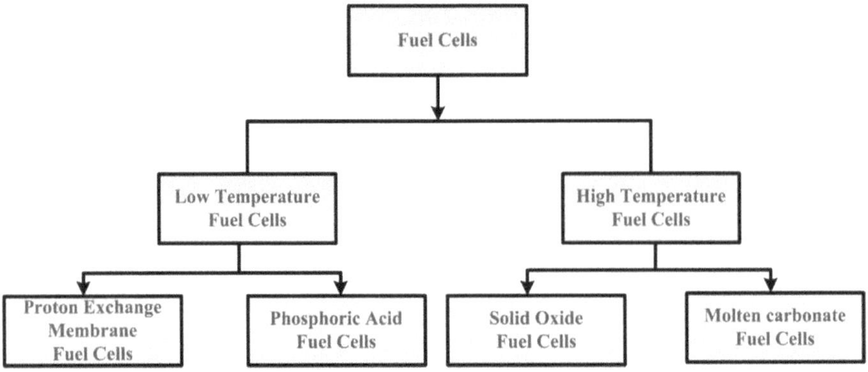

Figure 2.6 Classification of fuel cells.

However, these solar and wind power generations are suffering from some problems. As solar power generation is based on insolation and wind power generation is on the speed of the wind. These problems affect the frequency of the power grid. For example, due to environmental effects, any of these sources may give less power to the grid. In that condition, the grid becomes unstable to maintain the power balance.

In all the above aspects energy storage comes into the picture to play a key role. The energy storage accepts the energy during low demand periods and returns it during peak demand. Moreover, the energy storage systems keep the grid stable and balance load demand. The energy storage also acts as a backup power supply during grid failure to restore the grid [24].

2.2.3 Role of power electronic device in energy storage

The energy storage systems are playing a vital role in improving the reliability of the grid, saving money, reduction of environmental impact, and integrating various sources. However, the various types of energy storage systems are as shown in Fig. 2.2. Moreover, the basic idea behind the energy storage systems is to store the electrical energy during off-peak periods and utilize it back during peak-period.

However, there must be a conversion of electrical energy into another form (chemical, thermal, mechanical and magnetic, etc.) and vice versa. In most cases, energy conversion is done with the help of power semiconductor device-based circuits.

The power electronic device-based circuits are used to interface the energy systems with renewable energy sources to control and improve the power transfer between them. Most renewable energy sources are volatile in power generations hence these PED circuits control the changeover of the sources to maintain grid stability. For example, PV panels give only DC voltage; for grid connection, DC power is transformed into AC using PED circuits. Moreover, in the case of wind power generation, the frequency of the output varies according to the speed of the turbine so using the PED circuit the frequency is maintained at the desired value. So, the PED circuits are playing a crucial role in power conversion and energy storage as well.

2.3 Power electronic devices

2.3.1 History of semiconductor device development

In long-back, the power conversion is based on electromechanically or electromagnetic equipment. Since the invention of power electronic devices, they play a vital role in power conversion. The PED combines transistors and micro-electronics with large power handling capability [25]. However, the PEDs are robust and faster than electromechanical or magnetic devices for electric energy conversion, control, and delivery. The invention of semiconductor devices is shown in Fig. 2.7.

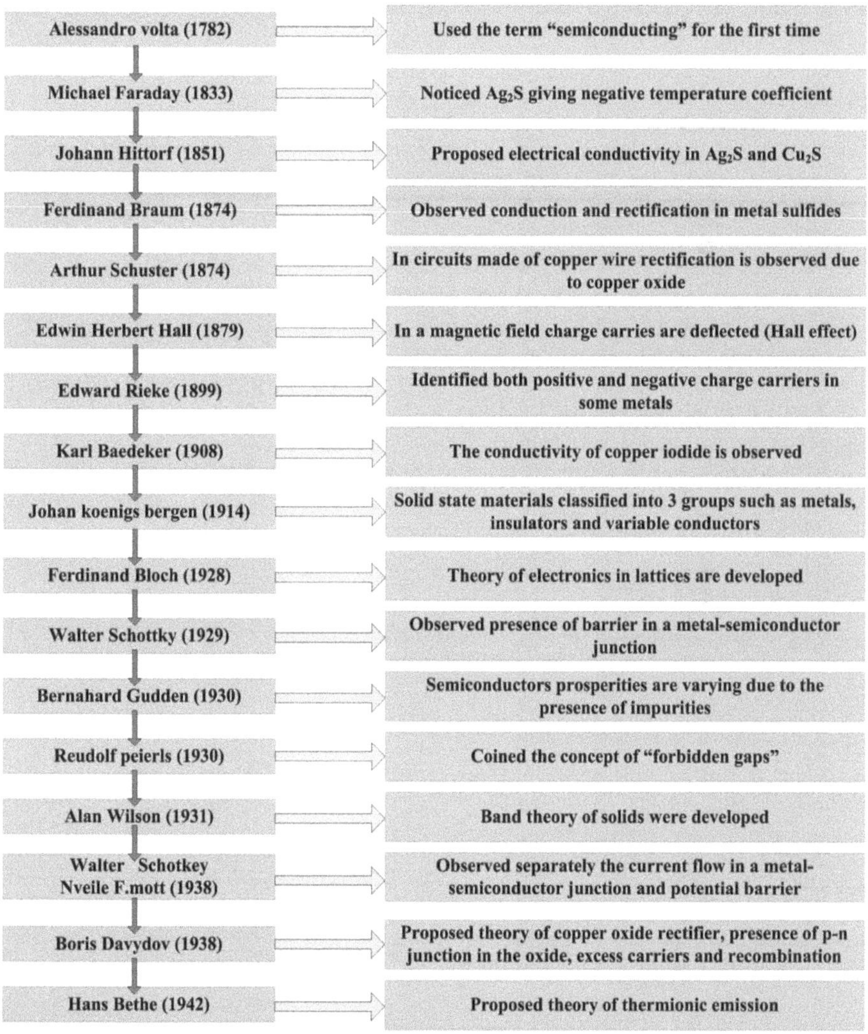

Figure 2.7 Historical development of semiconductor devices.

2.3.2 Various power electronic devices

The operation of a power system always requires control and delivery. In previous days the control and delivery are done with the help of transformers and transmission lines. However, the control of power using mechanical switches is slow and unable to complete power control [26]. Moreover, the frequency control cannot achieve using traditional equipment. In PV and wind power generation frequency control is the key factor. Hence, with the invention of power electronic devices, the control aspect of power becomes easier. The historical development in the PEDs is also taken place as shown in Fig. 2.8.

The power electronic devices have high power capacity, robust, faster, and flexible. The major part of the power electronic devices is a power semiconductor device. The basic operation of power semiconductor devices is switching on/off to maintain control over the power at minimum losses. Since the inception of power semiconductor devices, their use has varied based on the application. Moreover, the various power electronic devices are depicted in Table 2.1 [27–31].

1. *Power diode:* Power diode is similar to the normal diode. It conducts in only one direction, which allows the current in one direction. However, the power diode can handle high currents and voltages. Moreover, in the construction also there is a slight difference from the diode as it contains a drift region between p-n junctions.

   The power diodes are classified as Schottky, fast recovery, and general-purpose diodes. Moreover, the power diodes are used in rectifiers to convert AC power into DC power.
2. *Thyristor*: Thyristor is also known as silicon controlled rectifier (SCR). SCR is a three-terminal device and has the property of controlled turn-on. Moreover, the thyristor will turn on by giving a sufficient gate signal and once it reaches above the latching current, it is on. For the turning off, there must be an application of a reverse voltage across the junction and the current must be below the holding current.

   The property of controlled turn-on SCRs is used in controlled rectifiers in digital circuits. These rectifiers are also known as "phase-controlled rectifiers." The thyristors are used in television, motion pictures, and theaters long back.

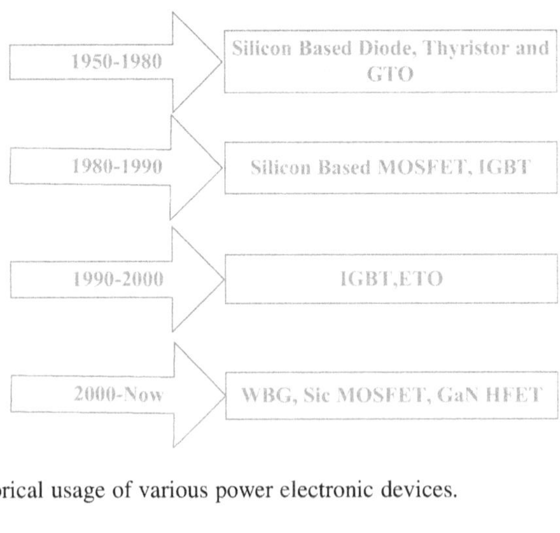

Figure 2.8 Historical usage of various power electronic devices.

Table 2.1 Power electronic devices and their symbols.

Power electronic devices	**Symbol**
Power diode	
Thyristor	
Power BJT	
Power MOSFET	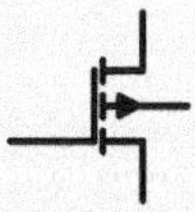
GTO	
IGBT	
TRIAC	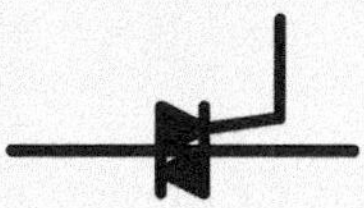

3. *Power BJT*: The bipolar junction transistor (BJT) is the first switch capable of controlled turn-on and turn-off. However, the power BJT is also a three-terminal device and it replaces forced turn-off SCRs in some applications. The power BJT will handle higher power applications as compared to BJT. Moreover, the power BJTs are used in SMPS, power amplifiers, relays and drivers, power control circuits, etc., but the MOSFETs and IGBTs are replacing the power BJT due to their faster switching and more efficiency.
4. *Power MOSFET*: The power (metal oxide semiconductor field-effect transistor) MOSFET has the advantages of high efficiency, and faster switching at low voltages. It will handle large power circuits and here the turn-on of the power MOSFET is based upon the voltage supplied to the field. The MOSFET is classified as n-channel enhancement and p-channel enhancement. The power MOSFETs are widely used for switching and amplifying applications.
5. *GTO*: The major drawback of SCR is turning it off is difficult. However, the gate turn-off thyristors (GTO) are attaining the ability to turn off with the negative signal to the gate terminal. The major applications of GTOs are in DC circuit breakers, DC choppers, AC drives, AC is stabilizing power supplies, etc.
6. *IGBT*: The insulated gate bipolar transistor (IGBT) is a semiconductor device having both the properties of BJT and MOSFET together. So IGBT will attain high efficiency and fast

switching capabilities. The IGBT semiconductors are widely used in inverters, converters, regulated power supplies, etc.

7. *MCT*: The MOS-controlled thyristor (MCT) is a semiconductor device used for high-power applications. However, it looks similar to GTO but MCT is a voltage-controlled device. Due to the insulated gate terminal to the MCT switching is easy. Moreover, it is faster than GTO.
8. *TRIAC*: Triode for alternating current (TRIAC) is a three-terminal device. It is similar to back-to-back connected SCRs conducted in both directions based on the gate signal. These are widely used in voltage controllers, universal motor control and dimming lamps, etc.

The semiconductor devices discussed above are all silicon-based. However, the silicon-based semiconductor material has limitations in conduction, blocking voltage, temperature, and switching characteristics. However, with the development of material science wide bandgap (WBG) based semiconductors come into usage. The WBG-based semiconductor has more than 1 eV energy bandgap and hence it has a high electric field, faster switching, and high junction temperature. Nowadays the WBG has become mostly used. Moreover, WBG-based semiconductors are silicon carbide (SiC) and gallium nitride (GaN) devices, these are used in high voltage and low voltage applications respectively.

SiC is greatly used in renewable energy systems and grid-connected systems [32]. However, GaN semiconductors have high voltage and high frequency but fabrication is difficult [33]. The GaN semiconductors are used for limited applications such as low-power PV system-based inverters [34].

Except for the diode, all other power semiconductor switches are three terminals and the diode is a two-terminal device. Moreover, thyristor, BJT, and IGBT are majority carrier devices and MOSFET is a minority charge carrier semiconductor switch.

2.3.3 Specifications of power electronic device

One of the selection criteria for power semiconductor devices is based on the ratings. The choice is based on how much voltage, current, and switching frequency can achieve using them. Here, Table 2.2 provide the information regarding all the power semiconductor device specification.

Table 2.2 Specifications of power electronic devices.

Semiconductor device	Voltage	Current	Switching frequency	Power capability
Power diode	Up to 10 kV	Up to 10 kA	variable	Medium
SCR	8 kV	2.5 kA	50 or 60 Hz	High
Power BJT	1.4 kV	0.4 kA	10–10 kHz	Medium
Power MOSFET	4.5 kV	1.6 kA	25–500 kHz	Low
MCT	1.5 kV	0.1 kA	Up to 500 Hz	Medium
GTO	10 kV	6.5 kA	Up to 500 Hz	High
IGBT	6.5 kV	2.4 kA	10–20 kHz	Medium

Table 2.3 Parameters associated with power electronic device along with tests to determine.

S. no	Parameters	Tests
1.	Semiconductor conductivity type (n- or p-type)	a. Hot Probe [35] b. Hall coefficient method
2.	Charge carrier concentration	a. Hall coefficient method b. Conductivity relations c. Doping concentrations
3.	The resistivity of a semiconductor	a. Conductivity relations b. Four-point probe method [36] c. Van der-Pauw method [37]
4.	The doping profile of a semiconductor	a. C-V method [38] b. Scanning electron microscope c. Transmission electron microscope d. Scanning tunneling electron microscope e. Raman spectroscopy f. Secondary ion mass spectroscopy g. X-ray photoelectron spectroscopy
5.	Drift mobility of charge carriers in a semiconductor	a. Hall effect method b. Time-resolved electroluminescence c. Time of flight method d. Haynes-Shockley experiment [39,40]
6.	Measurement of the minority carriers' lifetime	a. Open circuit voltage decay method b. Photoconductive decay method
7.	Diffusion coefficient of charge carriers in a semiconductor	a. Haynes-Shockley experiment
8.	The energy gap of the semiconductor	a. From electrical conductivity b. Spectroscopic method
9.	Thermal conductivity of a semiconductor	a. Scanning thermal microscope b. Time-domain thermos-reflectance c. The 3-omega method
10.	Ionization coefficient impact on semiconductor	a. Modeling based approach

From Table 2.2, for high voltage range conversion, are there then GTO gives better performance. However, for high current operation power diode is used. Moreover, the MOSFET gives a higher switching frequency than other power electronic devices.

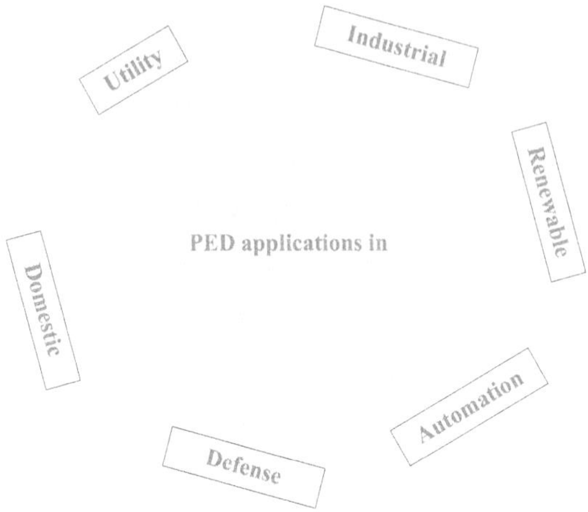

Figure 2.9 Power electronic device application in various fields.

2.3.4 Parameters associated with power electronic device

Power electronic devices which are used in converter circuits have gone through certain tests for determining the parameters. The parameters associated with PED are measured using their electrical, physical, and optical characteristics. However, the major parameters are conductivity type, resistivity, mobility, carrier concentration, semiconductor energy gap, impact ionization coefficient, etc. There are several methods to determine the parameters associated with PED which are given in Table 2.3.

2.3.5 Various applications of power electronic device

Power electronic devices have wider applications [41]. The demand for utilization of PED is increasing in domestic, industrial, deference, automation, and renewable energy systems, as shown in Fig. 2.9.

The various real-time applications of power electronic devices are voltage stabilizers, uninterrupted power supply, air condition, satellites, flexible AC transmission systems, smart grids, HVDC transmission, solar power generation, wind power generation, motor drive applications, pumps, hybrid vehicles, and electric vehicles, etc.

2.4 Power electronic converter circuits

The heart of the power electronic converter circuit is power semiconductor devices, here the basic converter [42] topologies and converters that are used in specific (wind, solar, fuel cells, and tidal power generations) applications are disclosed [43,44]:

2.4.1 Basic converter topology

The basic converter topologies are classified based on the conversion process such as DC–DC converters, AC–DC converters, DC–AC Converters, and AC–AC converters [45]. These basic converters are further modified and improved based on the application.

1. DC–DC converter
   The DC–DC converter is also called a chopper. The DC–DC converters change the input DC voltage to a variable DC voltage of the desired level. However, there are many DC–DC converters such as buck, boost, *C*'uk, SEPIC, Luo, super lift Luo converters, and many more. Each converter configuration has power semiconductor devices; based upon the switching of the PED converter provides the desired operation.
2. AC–DC converter
   The AC–DC converter is also known as a rectifier. The rectifier converts available AC power into DC power. However, the rectifier converts AC power to controlled or uncontrolled DC power. Moreover, the uncontrolled operation is performed using power diodes and the controlled one using SCR, MOSFET, and GTO-based circuits. The rectifier circuits can convert either three-phase or single-phase AC power into DC power.
3. DC–AC converter
   The DC–AC converter is also known as an inverter. The inverter is used to transform the DC power into AC power. However, either single-phase or three-phase AC power is achieved based on the number of semiconductors and configuration. Moreover, uncontrolled or controlled AC power is also obtained. The control of AC voltage and frequency of an inverter is performed using switching (pulse width modulation) techniques. The inverter is also modified to achieve better quality AC power using multilevel inverter configurations.
4. AC–AC converter
   The AC–AC converters are also known as voltage regulators or frequency changers. The AC–AC conversion is achieved only by controlled switches.

2.4.2 Converters in wind energy generations

Wind power generation is one of the alternatives or renewable electric power generation. Generally, wind power generation is classified as fixed speed wind power generation (FSWPG) and variable speed wind power generation (VSWPG) as shown in Fig. 2.10. However, for the FSWPG, the converter [46] is used for smooth starting to reduce the inrush currents and torque oscillation at starting. Coming to VSWPG, the converter circuits are used for speed and torque control in standalone mode and converters to balance active and reactive power for a grid-connected mode of operation [47,48].

1. Fixed speed wind power generation converters
   The fixed speed wind power generator uses a synchronous generator (SG), starter (AC voltage controller), and static VAR compensator connected to the grid via transformer as shown in Fig. 2.11.

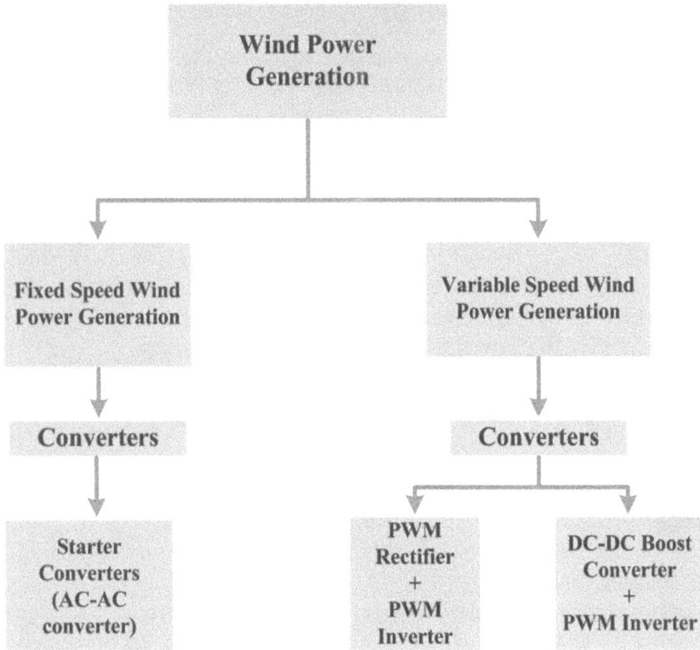

Figure 2.10 Converters used in wind power generation.

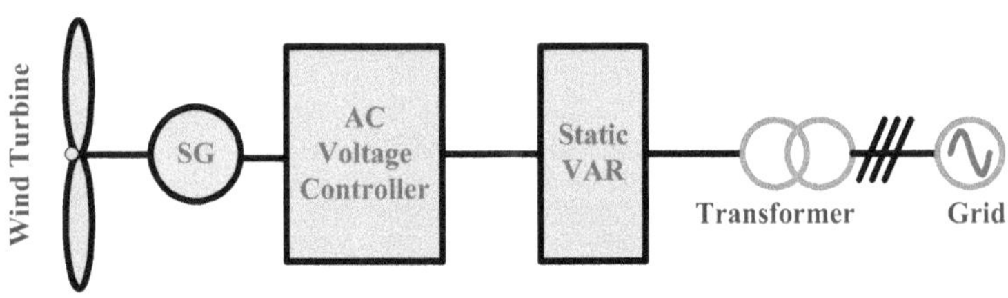

Figure 2.11 Fixed speed wind power generator connected to the grid.

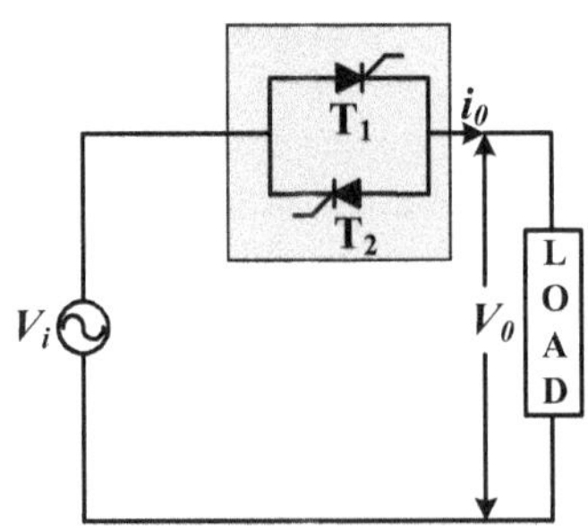

Figure 2.12 Single-phase AC voltage controller.

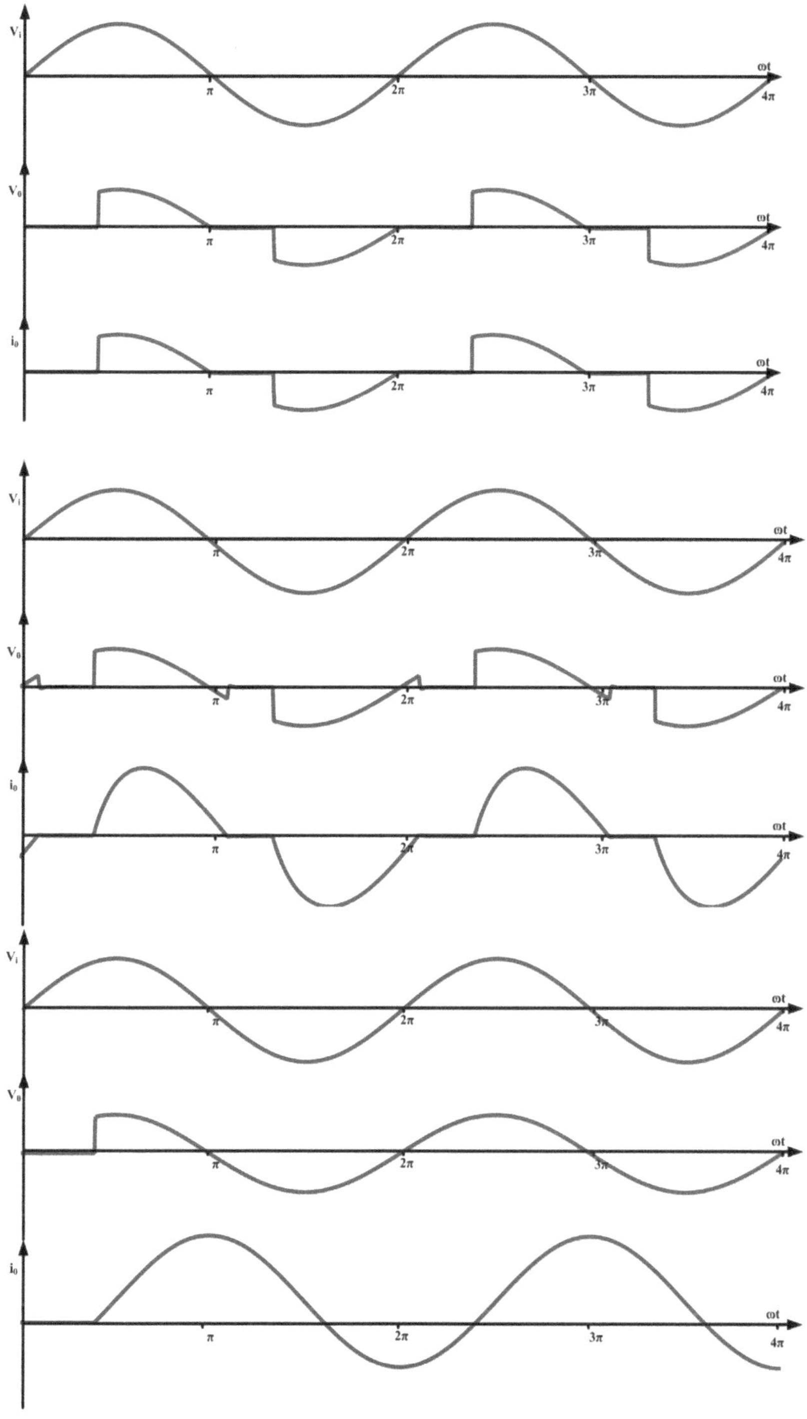

Figure 2.13 Output waveforms for single-phase AC voltage controller.

The major converters used in fixed-speed wind generation are AC voltage controllers. However, the AC voltage controller is either a single-phase or three-phase. Moreover, the AC voltage controller uses SCRs.

a. Single-phase AC Voltage controller

In a single-phase AC voltage controller, two SCRs are connected back to back as shown in Fig. 2.12. Here, the controlled SCRs give the voltage as per the gating signal.

Here, the input voltage (V_i) is controlled using thyristors T_1 and T_2 using a gating signal, and the corresponding output voltage (V_O) and current (i_O) are obtained as shown in Fig. 2.13.

b. Three-phase AC voltage controller

In a three-phase ac voltage controller, the output voltage is controlled using six thyristor switches as shown in Fig. 2.14.

Here, the star-connected load is considered for analysis, and the corresponding output waveforms are shown in Fig. 2.15.

2. Variable speed wind power generation converters

The variable speed wind power generations have two types of configurations are as shown in Fig. 2.16A and B [49,50]. The doubly-fed induction generator (DFIG) is fed via PWM-based converters. However, the PWM rectifier gives controlled DC power and the PWM inverter gives controlled AC power as shown in Fig. 2.16A. The output from the PED-based harmonics produced by the inverter. Moreover, the harmonics are filtered and fed to the grid using a transformer.

Instead of using two PWM-based converters in wind power generation, the output from the induction generator (IG) is fed to a diode rectifier then the DC–DC boost converter (used to track maximum power from the wind turbine) is fed to the PWM inverter and filter the harmonics before supplying power to the grid via a transformer.

The following section explains more details of the various converter and their operations that are used in variable speed wind power generation.

1. PWM rectifier

The rectifier is a circuit that converts the AC power into DC power. In wind power generation, the generated AC power has variable nature due to its variable speed. Hence,

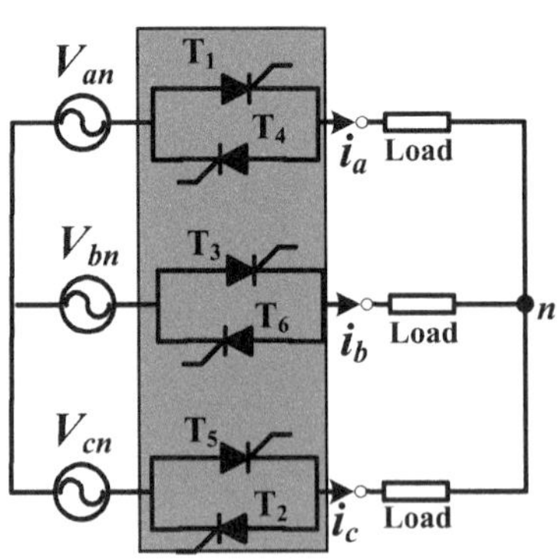

Figure 2.14 Three-phase AC voltage controller.

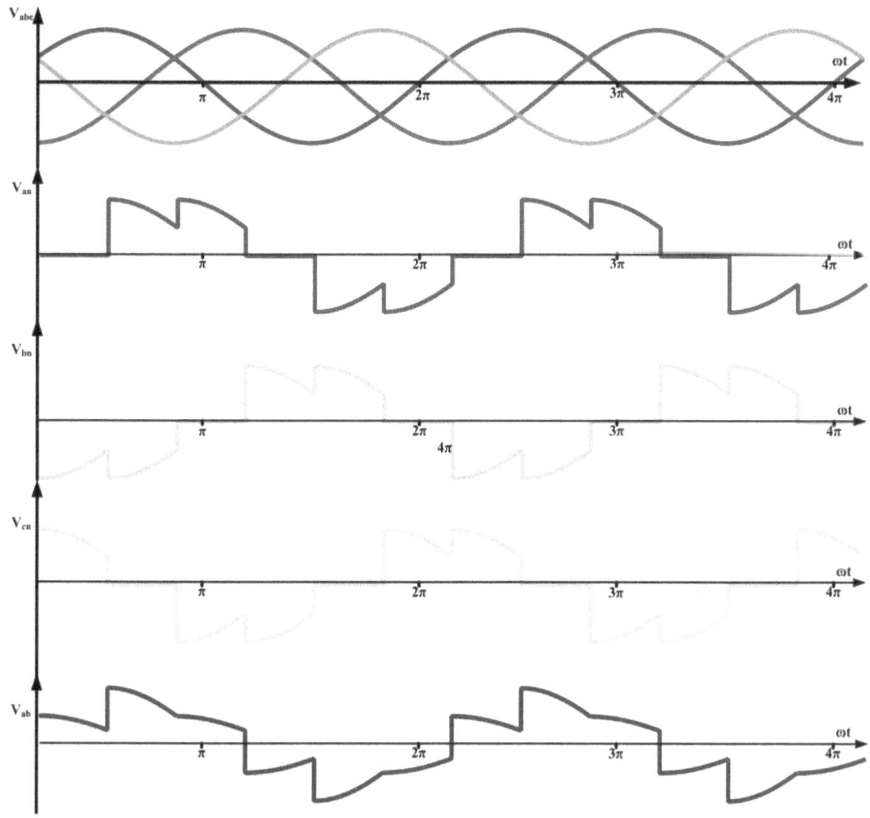

Figure 2.15 Output waveforms for single-phase AC voltage controller.

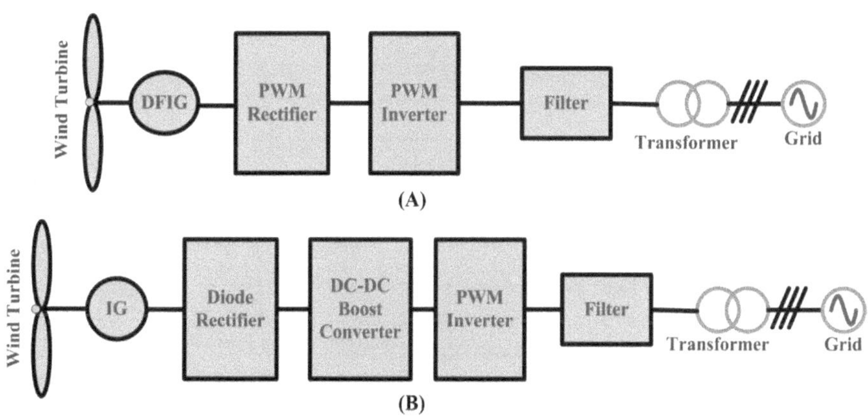

Figure 2.16 (A) Variable speed wind generator with pulse width modulation (PWM) converters (B) Variable speed wind generator with a DC–DC boost converter and PWM inverter.

the available AC power is transformed to controlled DC power using pulse width modulation (PWM) based rectifiers.

The PWM rectifiers are classified as multi-pulse voltage source rectifier (VSR) and current source rectifier (CSR) respectively and also further classified as shown in Fig. 2.17.

a. 6-pulsed based VSR

The six pulse-based controlled voltage source rectifier is formed using 6 SCRs as shown in Fig. 2.18. The input 3-phase AC voltage is converted into controlled DC using the gate signals.

b. 12- pulsed based VSR

The 12-pulse-based voltage source rectifier is achieved using a phase-shifting transformer and two 6-pulse VSR as shown in Fig. 2.19.

Here, the input supply voltage connected to star connected transformer divides the voltage into two halves to star and delta-connected transformers. The inductances are showing source inductances in an ideal case those values are zero. However, each transformer input is fed to 6-pulse rectifiers and the corresponding voltages are supplied load in a series combination.

c. 18 or 24 pulse VSR

As seen from 12-pulse VSR, it is formed using a phase-shifting transformer and a combination of six-pulse rectifier circuits. Similarly, 18 or 24 pulses are achieved using

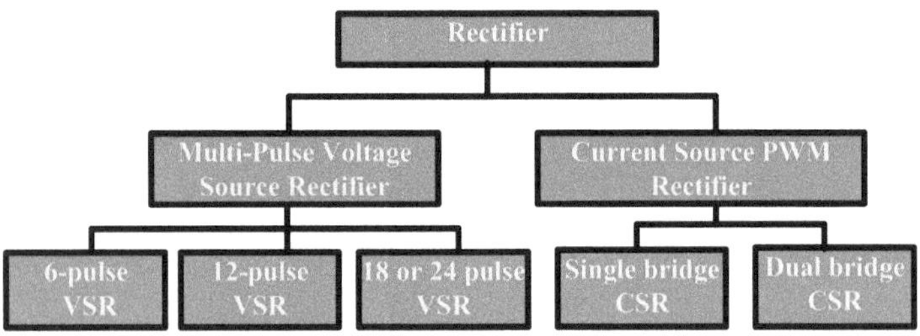

Figure 2.17 Pulse width modulation rectifiers classification.

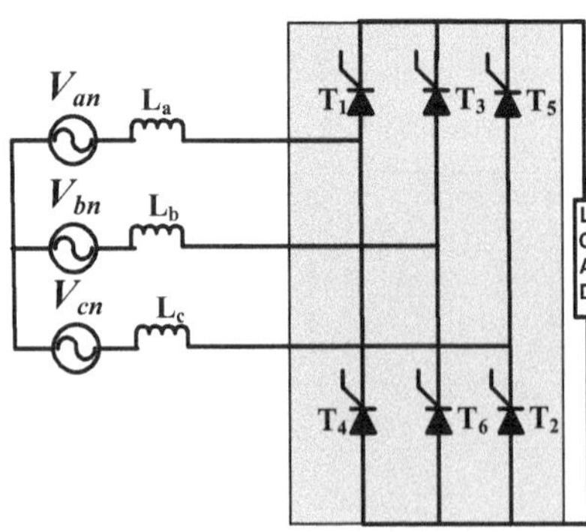

Figure 2.18 Six pulse controlled rectifier.

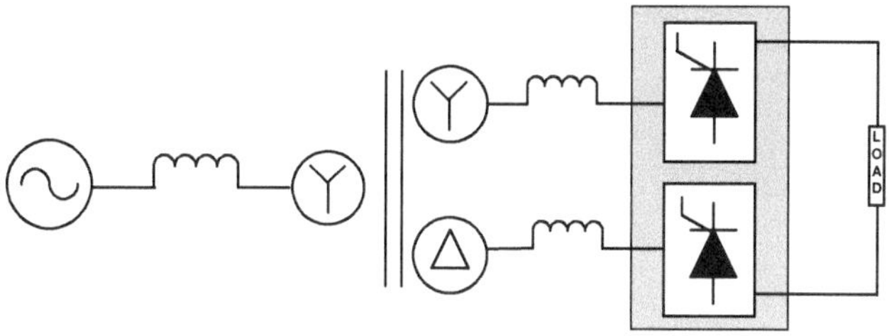

Figure 2.19 Twelve pulse controlled rectifier.

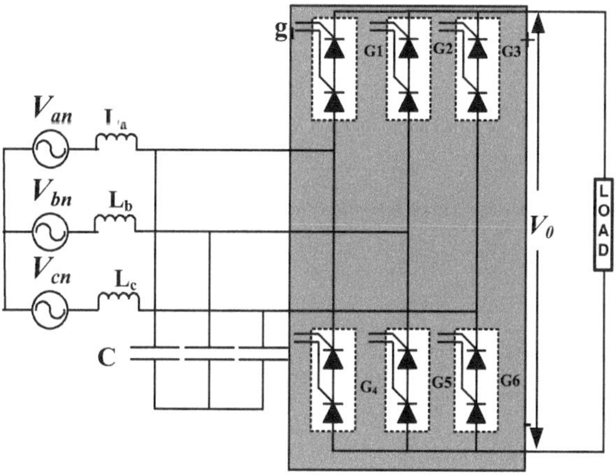

Figure 2.20 Single bridge current source rectifier.

transformers and six-pulse rectifiers. However, increasing the pulse number gives quality DC power, increasing the cost and maintenance.

d. Single-bridge CSR

The current source rectifiers are used for input current source fed medium voltage motor drives.

However, the CSR is performed using GCT (gate-commutated thyristors). Moreover, the CSR requires a capacitor bank at the input to filter the harmonics as well as assist in the commutation of GCT. The single bridge CSR is as shown in Fig. 2.20.

The capacitor value depends upon the switching frequency, input power factor, and line current total harmonic distortion. However, the switching of the current source rectifier is performed using pulse width modulation (PWM) techniques. Generally, the CSR uses the selective harmonic elimination (SHE) method or space vector modulation (SVM) to reduce the overall THD of the system.

e. Dual-bridge CSR

The dual bridge current source rectifier is formed using two single bridges CSR is connected using a phase-shifting transformer as shown in Fig. 2.21.

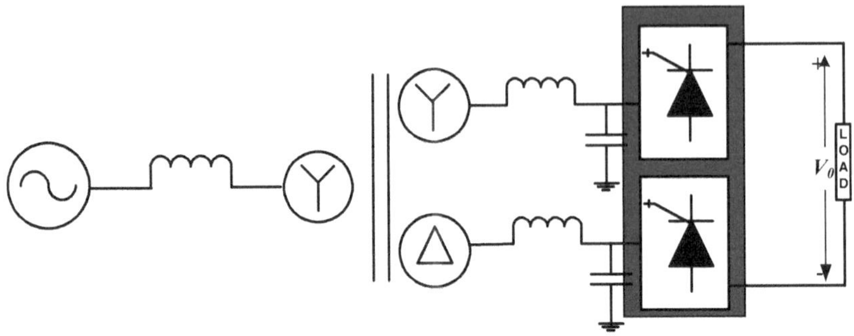

Figure 2.21 Dual bridge current source rectifier.

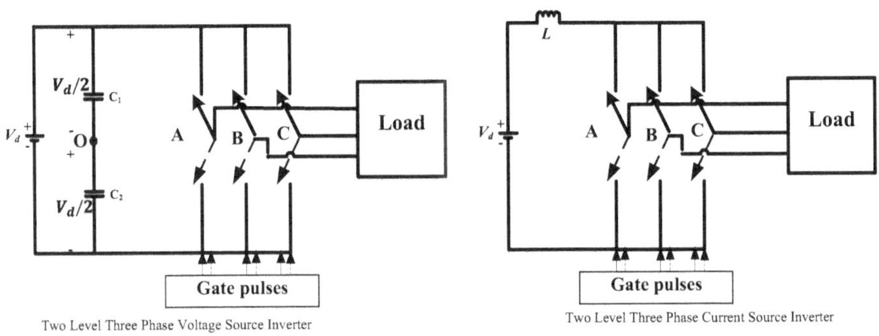

Figure 2.22 2-level 3-phase inverter.

The input supply is connected to a transformer and which divides the voltage by half of the input. The transformed supply is fed to the current source rectifier after filtering current harmonics.

i. *PWM inverter*

The grid-tied wind energy systems first convert the variable AC voltage into a controlled DC voltage. However, for the grid-connected mode, the voltage and frequency of the AC supply have to synchronize.

Moreover, the synchronization (required voltage and frequency) has been obtained with the help of a pulse width modulation (PWM) based inverter. The PWM inverters are classified based on the input supply as current source inverter (CSI) and voltage source inverter (VSI). The basic 2-level inverter is formed using 6 SCRs or GCTs as shown in Fig. 2.22.

ii. *Diode rectifier*

The variable speed grid-tied wind systems use a PWM rectifier and PWM inverter. However, the converter costs are high. To reduce the cost and complexity of the converter circuit, a diode rectifier along with a boost converter is implemented. The basic three-phase diode rectifier is shown in Fig. 2.23.

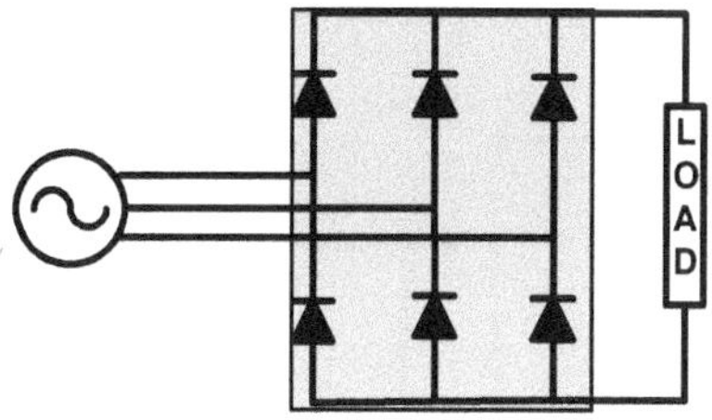

Figure 2.23 Three-phase diode bridge rectifier.

iii. *Boost converter*

The boost converter is used to achieve the required DC voltage after rectifier to an inverter for a low-speed wind turbine. In a variable speed wind generator, the major control is DC voltage regulation and active and reactive power management. However, the DC–DC converter is used for DC voltage regulation and the active power balance. Moreover, the inverter is taken care of reactive power management.

2.4.3 Converters in solar power generation

A photovoltaic system (PVS) converts solar energy into a direct electric current. However, the generated electrical power is either used for stand-alone or grid-connected mode. In PVS grid-connected/ stand-alone AC output mode, the system is further classified as a single-stage or two-stage based on the converters used, as shown in Fig. 2.24 [44].

The major converters used in PV power generation are DC–DC converters and inverters. However, the DC–DC converters are optional as they boost the DC voltage and track the maximum power point from the PV system [51].

1. *DC–DC converters*

   The DC–DC converters that are used in PV applications are classified as nonisolated and isolated converters as shown in Fig. 2.25. However, the nonisolated converters such as buck, boost, buck-boost, SEPIC, Cuk, and super lift converters. Moreover, isolated converters are fly back, resonant, push-pull, single active bridge and dual active bridge, etc. [52].
2. *DC–AC converters*

   The grid-connected PV inverters are more complicated than inverters for motor drive applications. The PV inverters take care of grid synchronization, reactive power management, anti-islanding, and maximum power point tracking. However, the PV inverters are categorized based on power plant configuration along with power range as shown in Fig. 2.26.

   Earlier, only SCR-based inverters are used. Later, due to the development of IGBT and MOSFETs are widely used. Except in module-based inverters, MOSFETs are dominating still now. However, many new PV inverters are coming into the market day by day. As the lifetime of PV panels is around 20 years, the converter circuits are designed to increase the efficiency and extended lifetime.

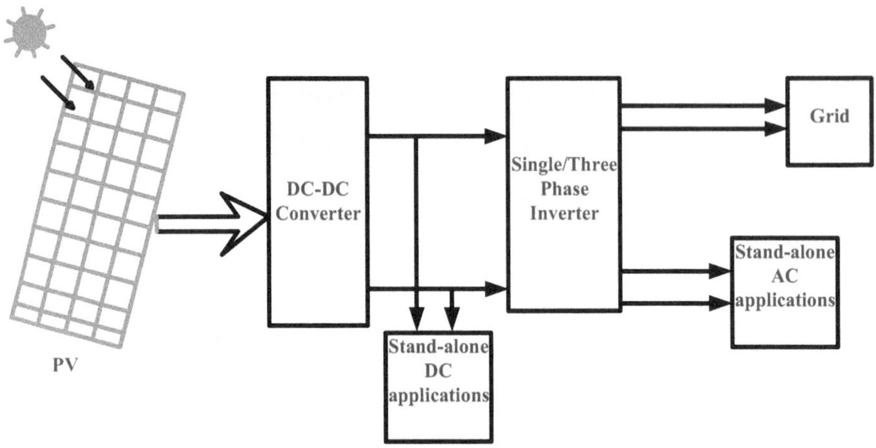

Figure 2.24 Block diagram of PV power generation.

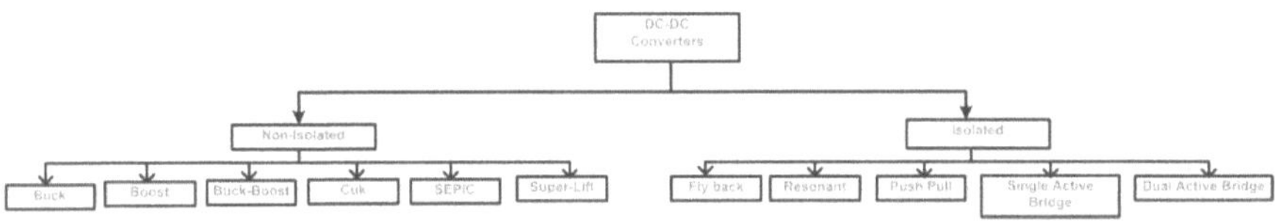

Figure 2.25 Classification of DC–DC converter.

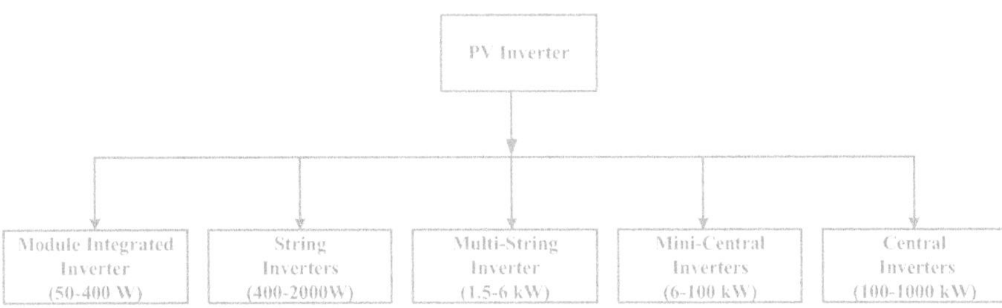

Figure 2.26 PV inverter configurations along with power range.

To improve the inverter efficiency, a high-frequency transformer is used in front of a DC–DC boost converter or a low-frequency transformer after the inverter. This will increase the overall efficiency by 1%–2%, however, new technology derived towards transformerless inverter configuration. Moreover, the new technology focuses on maximum efficiency under partial loading conditions also. The new converters are derived using conventional H-bridge and neural point clamping (NPC) inverters. The basic full-bridge inverter is classified as bipolar modulation, unipolar modulation, and hybrid modulation each has its own merits and demerits. Some of the new inverters are exposed given below:

Table. 2.4 PV inverters along with the description.

S. no	Inverter configuration	Circuit diagram	Circuit description	Merits	Demerits
1.	*H5 inverter* [53]	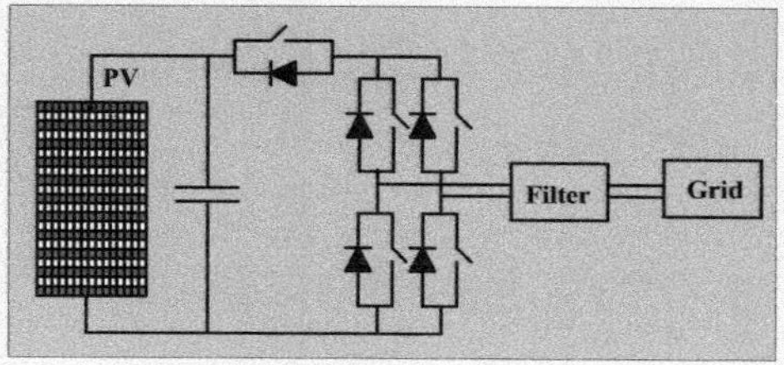	An extra switch is there as compared to H-bridge	(i). High efficiency up to 98% (ii). Low EMI	*One extra switch *Extra switches cause more switching losses
2.	*HERIC* [54] *(highly efficient reliable inverter configuration)*	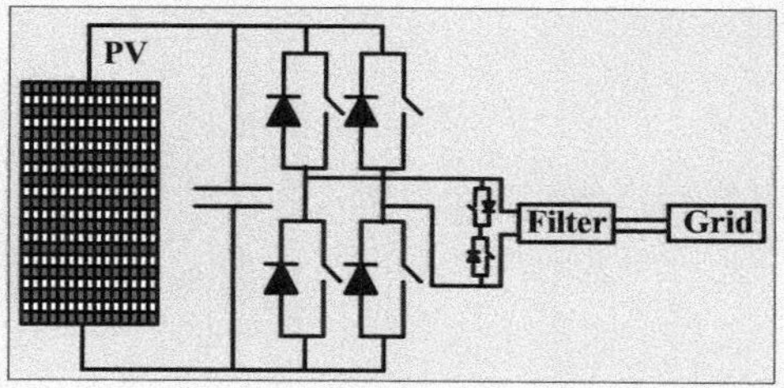	By adding bypass leg in AC side to H-bridge	(i). High efficiency of 97% (ii). Low leakage current	*Two extra switches are used *More switching losses
3.	*REFU inverter*	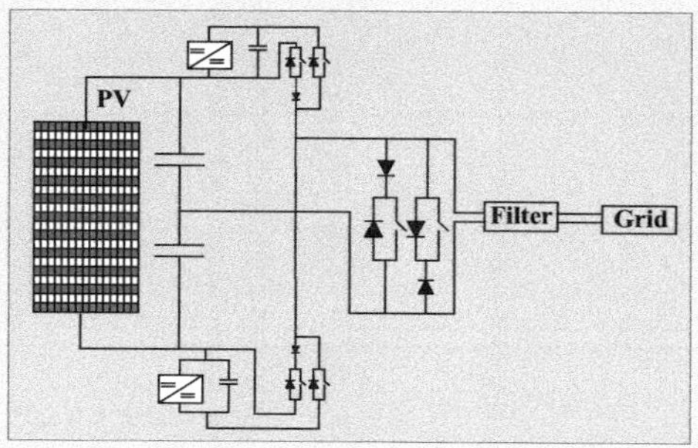	It is obtained from H-bridge with a half bypass DC–DC converter	(i). High efficiency (98%) and no exchange of reactive power (ii). Very low leakage current	*Need double DC source *It works at a low switching frequency

(*Continued*)

Table. 2.4 (Continued)

S. no	Inverter configuration	Circuit diagram	Circuit description	Merits	Demerits
4.	*Full-bridge inverter with DC bypass (FB-DCBP)*	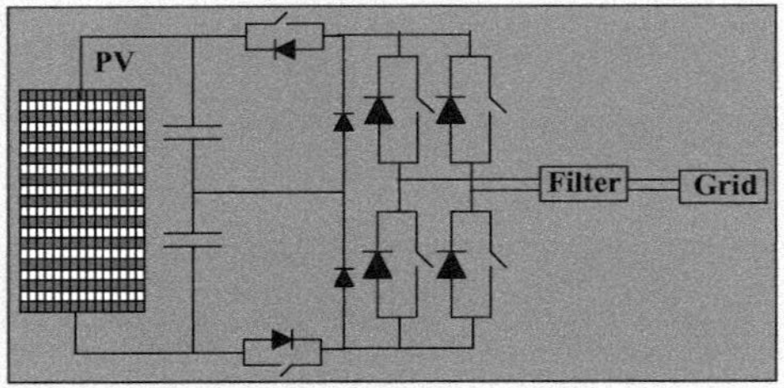	It is a modification of the H-bridge using a bypass DC side with the two extra switches	(i). Rating of bypass switches is half of DC voltage (ii). Very low EMI	*Two switches and diode each *More conduction losses
5.	*Full-bridge zero voltage rectifier (FB-ZVR)*	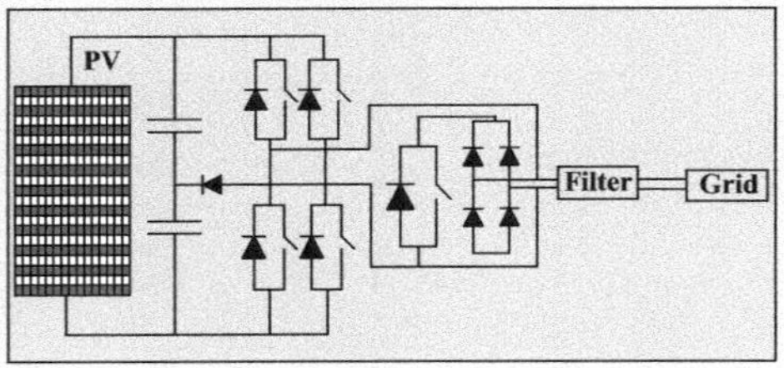	H-bridge modified by adding grid short-circuiting obtained using 5th switch through diode bridge	(i). No reactive power transfer during zero and low switching	*One extra switch and four diodes *During dead time increased losses across the filter
6.	*Neutral point clamped (NPC) half-bridge inverter*	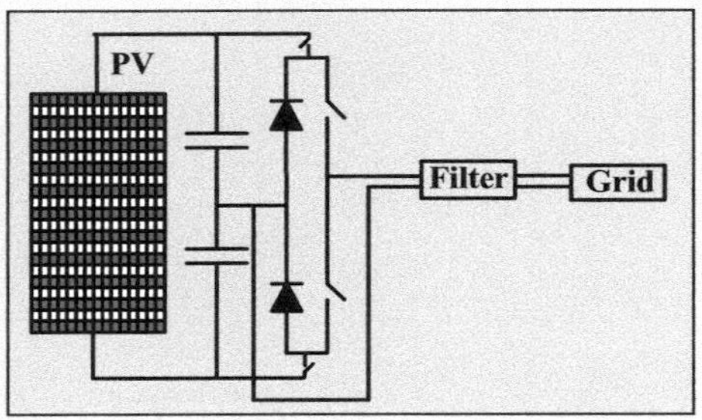	It is an NPC-based inverter. Moreover, inversion is obtained using clamping	(i). Voltage rating of outer switches reduced by ¼ provides fewer losses (ii). Low EMI	*Requires two extra diodes *Switching losses are unbalanced

7.	*Conergy NPC inverter*	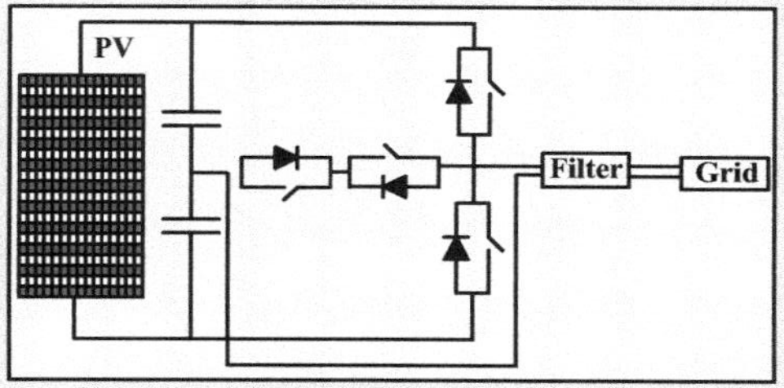	A modified NPC contains output clamped using back-to-back connected switches	(i). Balanced switching losses (ii). Very low leakage current	*Double voltage rating switches are used *Chances of increase in leakage current is more

The single-phase inverters are explained in Table 2.4. However, similarly, the three-phase inverters are formed easily. Moreover, the H-bridge and NPC are the basic inverter topologies for grid-connected PV systems.

2.4.4 Converters used in fuel cells

A fuel cell is an electro-chemical device, which converts the chemical energy into electrical power continuously using the isothermal process as shown in Fig. 2.27 [55].

The hydrogen fuel cell takes hydrogen as fuel and oxygen from the air. When the hydrogen and oxygen from water the electrons from hydrogen travel to oxygen with thermal emission. The fuel cells are classified as low-temperature and high-temperature fuel cells.

Low-temperature fuel cells are operating below or around 200°C only due to the rapid degradation of the fuel membrane at high temperatures. However, the low-temperature fuel cells are proton exchange membrane fuel cells (PEMFC) and phosphoric acid fuel cells (PAFC). Moreover, in the case of low-temperature fuel cells, the hydrogen has to separate itself initially, and the catalyst used in fuel cells is platinum.

High-temperature fuel cells are operating to 600°C–1000°C; here, in high-temperature fuel cells, the input may be methane gas itself, which means it is not necessary to take pure gas as input. Moreover, the high-temperature fuel cells are solid oxide fuel cells (SOFC) and molten carbonate fuel cells (MCFC).

The most commonly used fuel in distributed generation is proton exchange membrane fuel cells. However, the power generated from fuel cells is supplied to the grid using power semiconductor device-based converters. All possible configurations of the fuel-tied grid system along with various converters [56] are shown in Fig. 2.28.

The low voltage DC power generated from the fuel cells is transformed into usable DC or AC power. The fuel cell output is (1) directly transformed to AC using a high-frequency DC–AC inverter, or (2) using a DC–DC converter and high-frequency DC–AC inverter to get AC power, or (3) DC–AC converter followed by an AC–AC converter to feed into the grid.

1. DC–DC power converters

   The general voltage generated with the help of fuel cells is around 20–50 V. To connect it to a DC–AC converter, the DC voltage has to update as per the requirements.

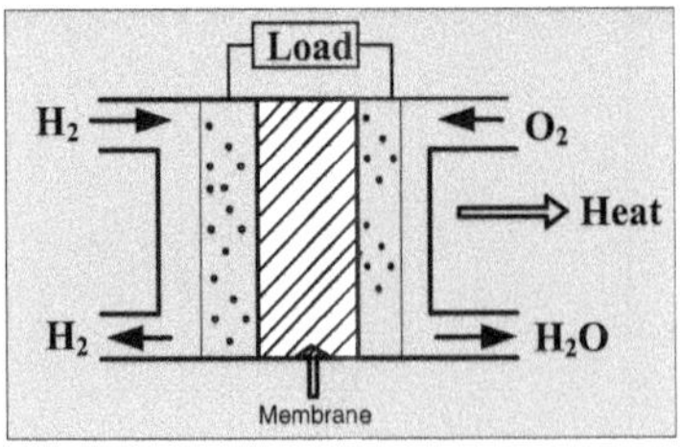

Figure 2.27 Fuel cell.

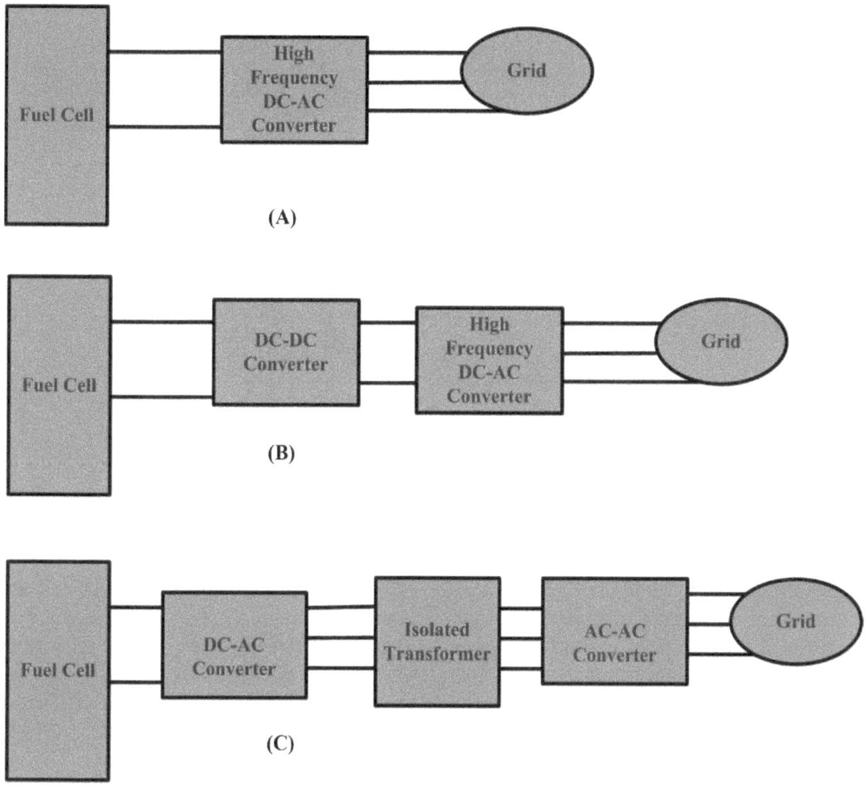

Figure 2.28 Grid-connected Fuel Cells along with the Converters.

However, this conversion is done using DC–DC converters, and there are various DC–DC converters available. The basic classification of DC–DC converters are linear, hard switching, and soft switching-based converters, and the further classification is shown in Fig. 2.29.

The resonant switching converters are more advantageous than linear due to their high efficiency, operating high voltage, and smooth operation. The resonant or soft switching converters use a ferrite core transformer to operate high voltages. However, the control circuit is complicated and noise in the supply due to switching is more in soft switching than linear mode. Moreover, the hard switching converters are classified as nonisolated and isolated ones. The nonisolated converters are a buck, boost, buck-boost, and Cuk converters. However, in some applications, isolation is required between the converter circuit and the power circuit. In that case, isolated converters such as push-pull, half-bridge, and full-bridge converters are used.

Selection of the DC–DC converter is based on the converter's efficiency, conversion range of voltage, current ripple, and THD, operating under all the conditions, and storing possibilities. Coming to the major task of the DC–DC converter in a fuel cell is to generate sufficient DC voltage to the inverter and also provide the isolation between the DC–DC converter and the inverter.

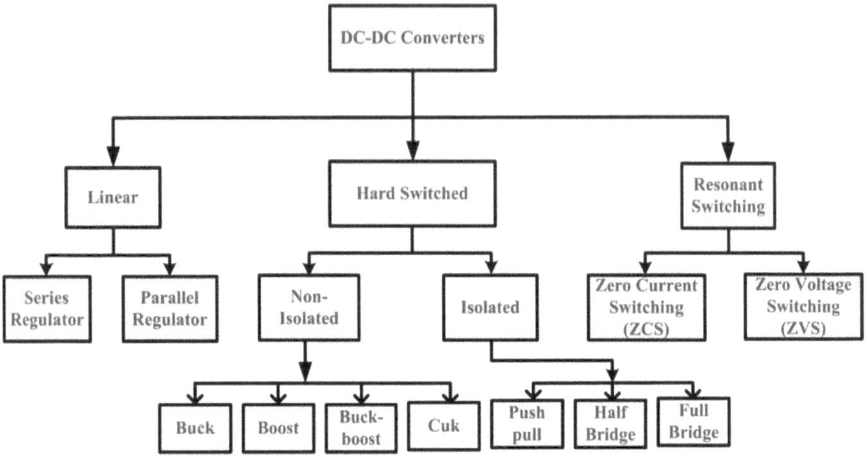

Figure 2.29 DC–DC Converters.

2. PWM inverters

Grid-connected fuel cells require a DC–AC converter. However, grid-connected fuel cells are either single-stage or multi-stage. In the single stage, the output of the fuel cells is transformed to AC using a high-frequency DC–AC converter. The inverter also takes care of all the control aspects (grid synchronization, power management). However, in a multi-stage configuration, an extra DC–DC converter or AC–AC converter is used to balance the voltage or frequency.

2.4.5 Power electronic converters in tidal power generation

Global green energy generation is increasing day by day. Tidal energy is one of the best alternatives to electrical power generation. However, constant power is generated from tidal instead of variable in the case of solar. Moreover, the tidal power generation is done during both the movements (forward or reverse direction) of water waves at the sea shore [57,58].

The basic block diagram of the grid-connected tidal power generation system is shown in Fig. 2.30. Here, the power generated by a permanent magnet synchronous generator (PMSG) is controlled by a series of power electronics converters. Firstly, the hydrodynamic energy is converted into electrical energy with the help of PMSG and the variable AC is fed to a diode rectifier. Once the DC power is supplied to the DC–DC boost converter the required DC voltage is obtained. Finally, with the help of a controlled inverter the desired AC voltage and frequency are achieved and fed to the grid through a filter and transformer.

1. AC–DC converter

The power generated from PMSG is fed to a diode rectifier to obtain the DC voltage as shown in Fig. 2.31. However, the tidal energy is variable hence due to the usage of the diode bridge rectifier, the DC link voltage becomes a variable one.

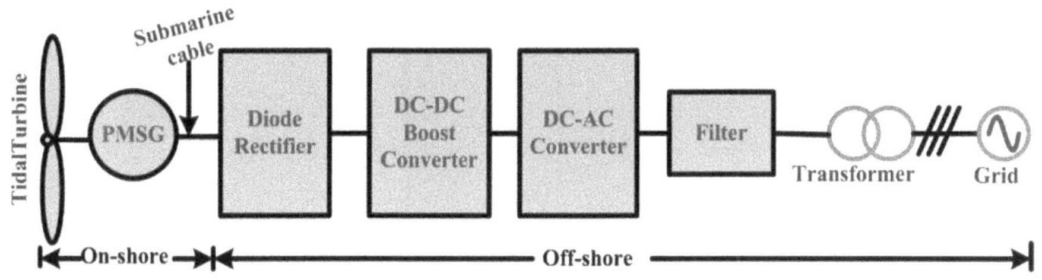

Figure 2.30 Basic block diagram of grid connected tidal power generation.

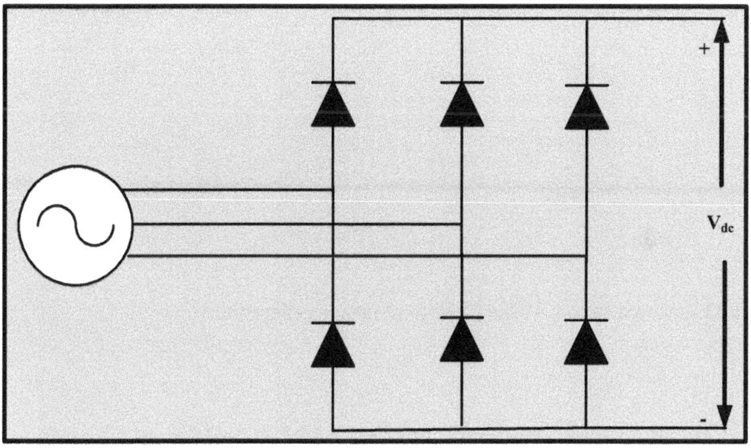

Figure 2.31 Diode bridge rectifier connected to permanent magnet synchronous generator.

The tidal speed is a variable one; hence the AC power generated is variable. If a controlled rectifier is used then the constant DC link voltage is achieved.

2. DC–DC boost converter

   The variable DC voltage obtained from the diode bridge rectifier has to transform into constant DC bus voltage before it is fed to an inverter. However, the constant DC bus voltage is achieved using a DC–DC converter as shown in Fig. 2.32. The DC–DC converter transforms the variable speed of tidal into a constant DC voltage and also matches the load characteristics of PMSG to obtain the maximum power.

3. DC–AC inverter

   The controlled DC bus voltage from the DC–DC boost converter is fed to the grid via an inverter as shown in Fig. 2.33. Here, the inverter takes care of active and reactive power independently and instantaneously.

   The grid-connected inverter has a ride-through capability and provides quality power towards the grid side. The tidal power generation requires droop control voltage/reactive power control. However, to increase the quality of power fed to the grid, three-level neutral point clamped-based inverters are implemented along with a third harmonic injection pulse width modulation technique.

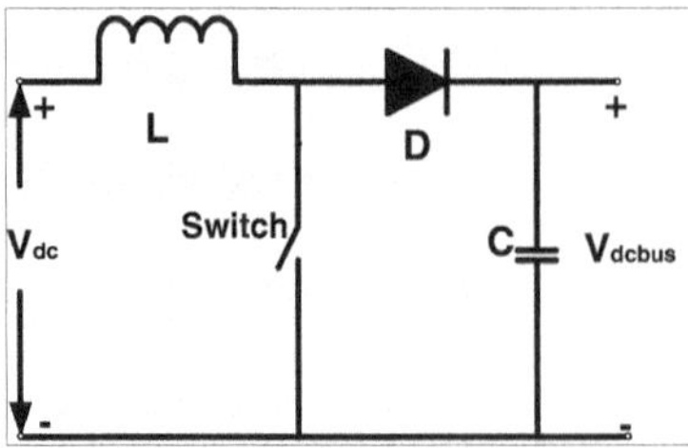

Figure 2.32 DC–DC boost converter.

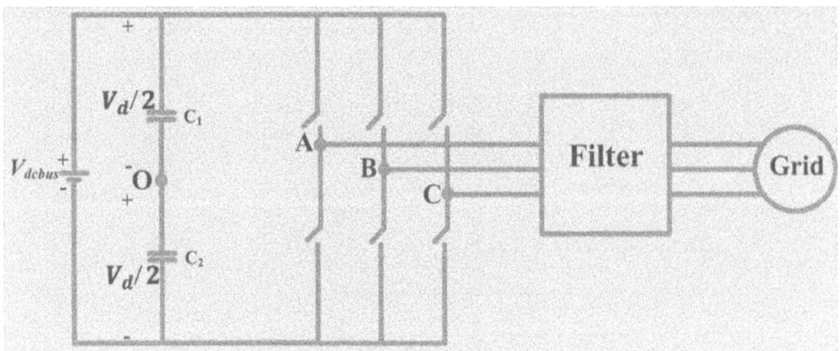

Figure 2.33 Inverter circuit for grid-tied tidal power generation.

2.5 Power conditioning system for energy storage

2.5.1 Battery management system using power electronic device

The power generation from renewable energy sources is not stable as it varies from time to time. To achieve continuous operation of the systems, energy storage is required. Battery storage is one of the best energy storage for renewable applications. However, the energy generated from renewable sources is harnessed using a power electronic device-based converter as shown in Fig. 2.34.

Here the major power electronic converters are bi-directional DC–DC converters. Here, the power from the renewable energy sources is utilized to store in a battery during excess availability and fed back to load during off conditions of renewable energy sources.

DC–DC bidirectional converter:

The conventional DC–DC converters transfer power in only one direction, whereas these bidirectional DC–DC converters transfer the power in both directions [59]. However, the major classification of bidirectional converters is nonisolated and isolated converters. Moreover, further, subclassification is shown in Fig. 2.35 [60].

The nonisolated DC–DC converter does not use any transformer for isolation; hence its size is compact and easy to control. However, this cannot be used for

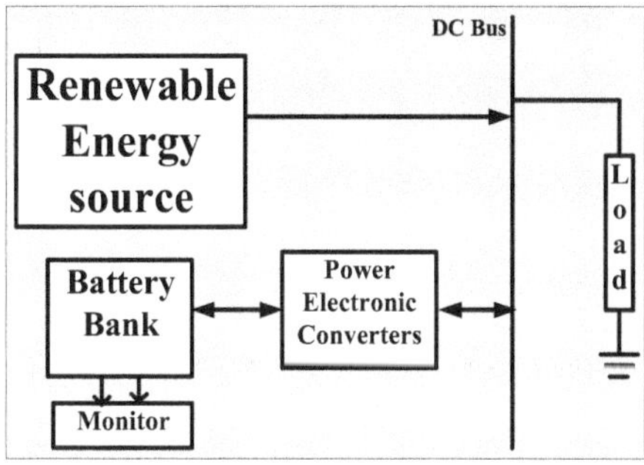

Figure 2.34 Battery management system.

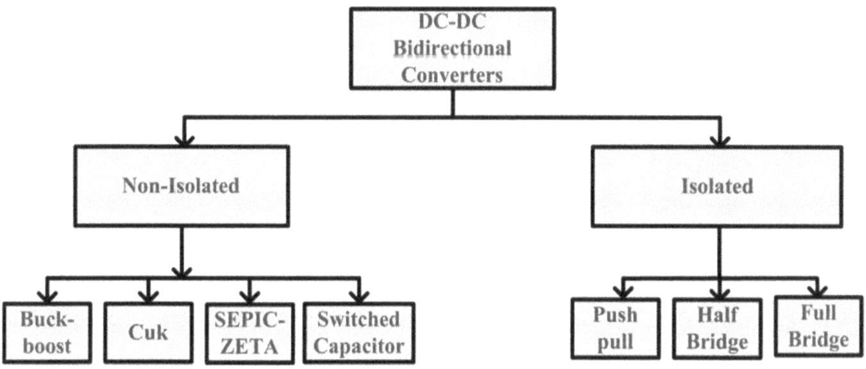

Figure 2.35 Classification of DC−DC bidirectional converters.

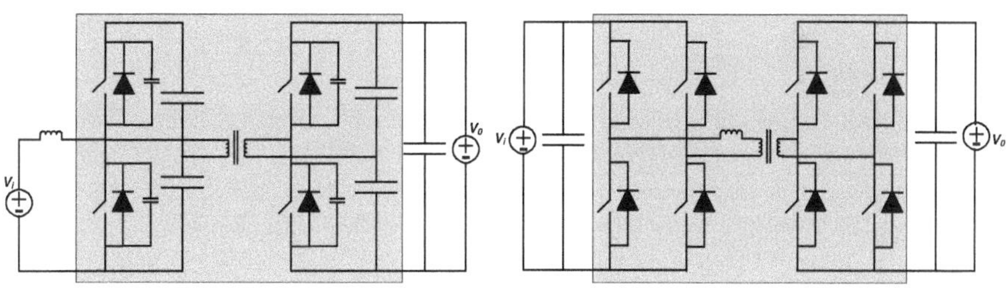

Figure 2.36 Half-bridge and full-bridge bi-directional DC−DC converter.

high-power applications due to operational safety. Some of the nonisolated converters are buck-boost, Cuk, SEPIC-ZETA and switched capacitors, etc.

The isolated DC−DC converters are using high-frequency transformers to provide the isolation and hence these are widely used for high-power applications. However, the basic operation of an isolated DC−DC converter is similar to the power transfer in a transmission system. The direction of the inductor current changes as per the power transfer of buck-boost. The half-bridge and full-bridge isolated converters are shown in Fig. 2.36. The control of switching can be performed either hard or soft switching. However, mostly soft switching (ZVS, ZCS) gives better performance rather than hard switching.

2.6 Conclusions

The continuous and surplus power can be utilized only with the help of energy storage systems. However, the significance of energy storage systems is enhanced due to new alternative power generations of solar, wind, fuel cells, and tidal. But, all these new alternative energy sources can generate power in discontinuous nature. Hence, the power generated from alternative energy sources is either used during their availability or else by storing energy. The storing of energy is performed using energy transformation from one form to another and later utilized using the reversible procedure.

Power semiconductor devices are used in energy storage systems with the help of power electronic converters. The entire chapter presents various classifications of energy storage systems and the importance of power semiconductor devices in energy storage. Also, the power semiconductor devices such as Diode, BJT, MOSFET, IGBT, GTO, and GCT are used in an energy storage system and their specifications, parameters, and applications. However, the control and conversion of electrical power are performed using power semiconductor devices-based converter circuits. Moreover, the chapter focused on the various power conversions that take place in the wind, solar, and tidal using power electronic converters and is also explained in detail. Finally, it provides the power semiconductor converters that are used in battery energy management systems as well.

References

[1] M.G. Molina, Energy storage and power electronics technologies: a strong combination to empower the transformation to the smart grid, Proc. IEEE 105 (11) (2017) 2191–2219. Available from: https://doi.org/10.1109/JPROC.2017.2702627.

[2] A. Narendra, N. Venkataramana Naik, A.K. Panda, N. Tiwary, A comprehensive review of PV driven electrical motors, Sol. Energy 195 (2020) 278–303. Available from: https://doi.org/10.1016/j.solener.2019.09.078.

[3] M.R. Sheibani, G.R. Yousefi, M.A. Latify, S.H. Dolatabadi, Energy storage system expansion planning in power systems: a review, IET Renew. Power Gener. 12 (11) (2018) 1203–1221. Available from: https://doi.org/10.1049/iet-rpg.2018.0089.

[4] A.K. Rohit, K.P. Devi, S. Rangnekar, An overview of energy storage and its importance in Indian renewable energy sector: part I – technologies and comparison, J. Energy Storage 13 (2017) 10–23. Available from: https://doi.org/10.1016/j.est.2017.06.005. vol.

[5] M. Jarnut, S. Wermiński, B. Waśkowicz, Comparative analysis of selected energy storage technologies for prosumer-owned microgrids, Renew. Sustain. Energy Rev. 74 (2017) 925–937. Available from: https://doi.org/10.1016/j.rser.2017.02.084. vol.

[6] E. Bullich-Massagué, et al., A review of energy storage technologies for large scale photovoltaic power plants, Appl. Energy 274 (2020) 115213. Available from: https://doi.org/10.1016/j.apenergy.2020.115213.
[7] Y. Ding, Y. Li, D. Li, J. Radcliffe, Y. Huang, Cryogenic energy storage, Handbook of Clean Energy Systems, John Wiley & Sons, Ltd, Chichester, UK, 2015, pp. 1–15.
[8] C. Valderrama, J. Luis Cortina, A. Akbarzadeh, Solar ponds, Storing Energy: With Special Reference to Renewable Energy Sources, Elsevier Inc, 2016, pp. 273–289.
[9] S.A. Kalogirou, Solar thermal power systems, Sol. Energy Eng. (2014) 541–581. Available from: https://doi.org/10.1016/B978-0-12-397270-5.00010-8.
[10] F. Bai, C. Xu, Performance analysis of a two-stage thermal energy storage system using concrete and steam accumulator, Appl. Therm. Eng. 31 (14–15) (2011) 2764–2771. Available from: https://doi.org/10.1016/j.applthermaleng.2011.04.049.
[11] F. Rossi, B. Castellani, A. Nicolini, Benefits and challenges of mechanical spring systems for energy storage applications, Energy Procedia 82 (2015) 805–810. Available from: https://doi.org/10.1016/j.egypro.2015.11.816. Dec.
[12] X. Luo, J. Wang, M. Dooner, J. Clarke, C. Krupke, Overview of current development in compressed air energy storage technology, Energy Procedia 62 (2014) 603–611. Available from: https://doi.org/10.1016/j.egypro.2014.12.423. vol.
[13] P. Breeze, Compressed air energy storage, Power System Energy Storage Technologies, Elsevier, 2018, pp. 23–31.
[14] K.A. Kavadias, Integration of stand-alone and hybrid wind energy systems into buildings, Stand-Alone and Hybrid Wind Energy Systems, Elsevier, 2010, pp. 475–505.
[15] Amos Kabo-Bah, Felix Diawuo, Eric Antwi, Pumped Hydro Energy Storage for Hybrid Systems, Elsevier, 2022.
[16] Z. Glasnovic, K. Margeta, Revitalization of hydro energy: a new approach for storing solar energy, Sol. Energy Storage (2015) 189–206. Available from: https://doi.org/10.1016/B978-0-12-409540-3.00008-6.
[17] W. Gu, L. Wei, G. Yushin, Capacitive Energy Storage, 2017.
[18] C. Shaw, Superconducting Magnetic Energy Storage System—Commercialisation and Marketing Challenges Final Report, no. 1, 2016, pp. 1–12.
[19] V.M. Fthenakis, T. Nikolakakis, Storage options for photovoltaics, Compr. Renew. Energy 1 (2012) 199–212. Available from: https://doi.org/10.1016/B978-0-08-087872-0.00106-2. vol.
[20] A. Ter-Gazarian, Electrochemical energy storage, Energy Storage for Power Systems, Institution of Engineering and Technology, 2011, pp. 131–147.
[21] G. Tomazic, M. Skyllas-Kazacos, Redox flow batteries, Electrochemical Energy Storage for Renewable Sources and Grid Balancing, Elsevier Inc, 2015, pp. 309–336.
[22] T.A. Semelsberger, Fuels - hydrogen storage | chemical carriers, Encycl. Electrochem. Power Sources (2009) 504–518. Available from: https://doi.org/10.1016/B978-044452745-5.00331-2. Jan.
[23] Y. Wang, K.S. Chen, J. Mishler, S.C. Cho, X.C. Adroher, A review of polymer electrolyte membrane fuel cells: technology, applications, and needs on fundamental research, Appl. Energy 88 (4) (2011) 981–1007. Available from: https://doi.org/10.1016/J.APENERGY.2010.09.030.
[24] A.A. Khodadoost Arani, G.B. Gharehpetian, M. Abedi, Review on energy storage systems control methods in microgrids, Int. J. Electr. Power Energy Syst. 107 (2019) 745–757. Available from: https://doi.org/10.1016/j.ijepes.2018.12.040. no. December 2018.
[25] B.K. Bose, Power electronics-a technology review, Proc. IEEE 80 (8) (1992) 1303–1334. Available from: https://doi.org/10.1109/5.158603.
[26] M. H. Rashid, Power Electronics Handbook, 2011.

[27] B.K. Bose, Power electronics and motor drives recent progress and perspective, IEEE Trans. Ind. Electron. 56 (2) (2009) 581–588. Available from: https://doi.org/10.1109/TIE.2008.2002726.
[28] K. Satoh, M. Yamamoto, The present state of the art in high-power semiconductor dvices, Proc. IEEE 89 (6) (2001) 813–820. Available from: https://doi.org/10.1109/5.931470.
[29] B. Jayant Baliga, The future of power semiconductor device technology, Proc. IEEE 89 (6) (2001) 822–832. Available from: https://doi.org/10.1109/5.931471.
[30] Z.J. Shen, I. Omura, Power semiconductor devices for hybrid, electric, and fuel cell vehicles, Proc. IEEE 95 (4) (2007) 778–789. Available from: https://doi.org/10.1109/JPROC.2006.890118.
[31] K. Shenai, R.S. Scott, B.J. Baliga, Optimum semiconductors for high-power electronics, IEEE Trans. Electron. Dev. 36 (9) (1989) 1811–1823. Available from: https://doi.org/10.1109/16.34247.
[32] J.A. Cooper, M.R. Melloch, R. Singh, A. Agarwal, J.W. Palmour, Status and prospects for SiC power MOSFETs, IEEE Trans. Electron. Dev. 49 (4) (2002) 658–664. Available from: https://doi.org/10.1109/16.992876. Apr.
[33] K. Ahi, Review of GaN-based devices for terahertz operation, Opt. Eng. 56 (09) (2017) 1. Available from: https://doi.org/10.1117/1.OE.56.9.090901. Sep.
[34] L. Garcia-Rodriguez, V. Jones, J.C. Balda, E. Lindstrom, A. Oliva, J. Gonzalez-Llorente, Design of a GaN-based microinverter for photovoltaic systems," in: Proceedings of the IEEE Fifth International Symposium on Power Electronics for Distributed Generation Systems PEDG 2014, 2014, Available from: https://doi.org/10.1109/PEDG.2014.6878639.
[35] G. Golan, A. Axelevitch, B. Gorenstein, V. Manevych, Hot-Probe method for evaluation of impurities concentration in semiconductors, Microelectron. J. 37 (9) (2006) 910–915. Available from: https://doi.org/10.1016/J.MEJO.2006.01.014. Sep.
[36] T.G. Santos, Characterization of FSP by electrical conductivity, Surf. Modif. Solid. State Process. (2014) 153–176. Available from: https://doi.org/10.1533/9780857094698.153. Jan.
[37] A.A. Ramadan, R.D. Gould, A. Ashour, On the Van der Pauw method of resistivity measurements, Thin Solid. Films 239 (2) (1994) 272–275. Available from: https://doi.org/10.1016/0040-6090(94)90863-X.
[38] S.T. Lin, J. Reuter, The complete doping profile using MOS CV technique, Solid State Electron. 26 (4) (1983) 343–351. Available from: https://doi.org/10.1016/0038-1101(83)90134-X.
[39] J.E. Parrott, The theory of majority-carrier motion in the Haynes-Shockley experiment, Solid State Electron. 28 (11) (1985) 1065–1075. Available from: https://doi.org/10.1016/0038-1101(85)90186-8.
[40] B. Krüger, T. Friese, F. El-Ratel, W. Sharabati, H.G. Wagemann, Minority carrier behavior in silicon—measurement results derived from the Shockley-Haynes experiment, Solid. State. Electron. 39 (6) (1996) 897–900. Available from: https://doi.org/10.1016/0038-1101(95)00407-6.
[41] O. Ellabban, H. Abu-Rub, F. Blaabjerg, Renewable energy resources: current status, future prospects and their enabling technology, Renew. Sustain. Energy Rev. 39 (2014) 748–764. Available from: https://doi.org/10.1016/J.RSER.2014.07.113. Nov.
[42] K. Bose Bimal, Power Electronics in Renewable Energy Systems and Smart Grid: Technology and Applications, Wiley-IEEE, 2019.
[43] R.W. De Doncker, C. Meyer, R.U. Lenke, F. Mura, Power electronics for future utility applications, Proc. Int. Conf. Power Electron. Drive Syst. (2007) 3–10. Available from: https://doi.org/10.1109/PEDS.2007.4487666.

[44] M. Vilathgamuwa, D. Nayanasiri, S. Gamini, Power electronics for photovoltaic power systems, Synth. Lect. Power Electron. 8 (2015) 1–131. Available from: https://doi.org/10.2200/S00638ED1V01Y201504PEL008. vol.
[45] M. Narimani, B. Wu, N.R. Zargari, A novel five-level voltage source inverter with sinusoidal pulse width modulator for medium-voltage applications, IEEE Trans. Power Electron. 31 (3) (2016) 1959–1967. Available from: https://doi.org/10.1109/TPEL.2015.2440656.
[46] F. Blaabjerg, Z. Chen, R. Teodorescu, F. Iov, Power electronics in wind turbine systems, Conf. Proc. - IPEMC 2006 CES/IEEE 5th Int. Power Electron. Motion Control. Conf. 1 (2007) 46–56. Available from: https://doi.org/10.1109/IPEMC.2006.297043. vol.
[47] V. Yaramasu, B. Wu, Model Predictive Control of Wind Energy Conversion Systems, 2016.
[48] Bin Wu, Yongqiang Lang, Navid Zargari, Samir Kouro, Power Conversion and Control of Wind Energy Systems, Wiley-IEEE Press, 2011.
[49] F. Blaabjerg, Z. Chen, Power electronics for modern wind turbines, Synth. Lect. Power Electron. 1 (2005) 1–68. Available from: https://doi.org/10.2200/S00014ED1V01Y200602PEL001. vol.
[50] V. Yaramasu, B. Wu, Power electronics for high-power wind energy conversion systems, Encycl. Sustain. Technol. (2017) 37–49. Available from: https://doi.org/10.1016/B978-0-12-409548-9.10094-6.
[51] N. Femia, G. Petrone, G. Spagnuolo, M. Vitelli, Power Electronics and Control Techniques for Maximum Energy Harvesting in Photovoltaic Systems. 2017.
[52] V. G. R. Kummara et al., A comprehensive review of DC–DC converter topologies and modulation strategies with recent advances in solar photovoltaic systems, 9, 1, 2020.
[53] H. Li, Y. Zeng, B. Zhang, T.Q. Zheng, R. Hao, Z. Yang, An improved H5 topology with low common-mode current for transformerless PV grid-connected inverter, IEEE Trans. Power Electron. 34 (2) (2019) 1254–1265. Available from: https://doi.org/10.1109/TPEL.2018.2833144.
[54] S. Jain, V. Sonti, A highly efficient and reliable inverter configuration based cascaded multilevel inverter for PV systems, IEEE Trans. Ind. Electron. 64 (4) (2017) 2865–2875. Available from: https://doi.org/10.1109/TIE.2016.2633537.
[55] M.S. Ali, S.K. Kamarudin, M.S. Masdar, A. Mohamed, An overview of power electronics applications in fuel cell systems: DC and AC converters, Sci. World J. 2014 (2014). Available from: https://doi.org/10.1155/2014/103709. vol.
[56] M.H. Nehrir, C. Wang, Modeling and Control of Fuel Cells: Distributed Generation Applications, Wiley-IEEE Press, 2009.
[57] G. Mademlis, Y. Liu, P. Chen, E. Singhroy, Design of maximum power point tracking for dynamic power response of tidal undersea kite systems, IEEE Trans. Ind. Appl. 56 (2) (2020) 2048–2060. Available from: https://doi.org/10.1109/TIA.2020.2966189.
[58] M.J. Jahromi, A.I. Maswood, K.J. Tseng, Design and evaluation of a new converter control strategy for near-shore tidal turbines, IEEE Trans. Ind. Electron. 60 (12) (2013) 5648–5659. Available from: https://doi.org/10.1109/TIE.2012.2230601.
[59] D. Ravi, B.M. Reddy, S.L. Shimi, P. Samuel, Bidirectional dc to dc converters: an overview of various topologies, switching schemes and control techniques, Int. J. Eng. Technol. 7 (4) (2018) 360–365. Available from: https://doi.org/10.14419/ijet.v7i4.5.20107.
[60] A.K. Panda, N. Tiwary, N.V. Naik, Overview of Control Strategies and Design of Isolated Bidirectional Dual Active Bridge Converter for Renewable Energy Systems, 2022, pp. 253–273. Available from: https://doi.org/10.1007/978-981-16-4388-0_13.

Investigation of cushion gas/working gas ratios of underground salt caverns for hydrogen storage

3

Saygin San[1], *Hatice Karakilcik*[2], *Mehmet Karakilcik*[1], *Mustafa Erden*[1] *and Ayhan Atiz*[1]
[1]Department of Physics, Faculty of Sciences and Letters, Cukurova University, Adana, Turkey, [2]Department of Geology Engineering, Faculty of Engineering, Cukurova University, Adana, Turkey

Chapter Outline

Highlights

1. The importance of hydrogen storage in salt caverns
2. Determination of cushion gas/hydrogen ratios
3. Positions of the salt caverns for hydrogen storage
4. Find out of energy values (GWh) of cushion gas and hydrogen

Nomenclature

UHSs	underground hydrogen storage areas
LCCS	last cemented casing shoe point
P	pressures (Pa)
h	depth (m)
T	temperature (°C)

Emerging Trends in Energy Storage Systems and Industrial Applications. DOI: https://doi.org/10.1016/B978-0-323-90521-3.00003-X

V volume (Nm^3)
m mass (kg)

Greek symbols

ρ density ($kg\ m^{-3}$)

Subscripts

c cavern center
L last cemented casing shoe
max maximum
min minimum
N normal condition (standard condition; 1 atm, 0°C).
z compressibility factor of the cavern
i, j parameter

Index

cushion gas
working gas
underground
hydrogen storage
salt caverns
pressure
salt dome
collapse
metal hybrid
geological structures
methanol
ammonia
dibenzyltolue
solution mining
natural gas storage
anhydrite
UHS
rock salt
cemented
casing
shoe point
LCCS
Tuz Gölü
geothermal gradient
temperature
compressibility
normal condition

standard condition
geological

3.1 Introduction

Hydrogen gas can be stored as a liquid or as a solid combined with a metal hybrid. The optimal storage method depends on safety considerations, environmental issues, economic criteria, and the end-use of hydrogen. Hydrogen is produced in large quantities with renewable energy sources and stored underground is both economical and safe.

If hydrogen can be economically stored in large quantities, it could become an important energy player in the future. Although the gaseous hydrogen is stored in salt caverns, this method cannot be applied everywhere due to the different geological structures. All options related with the large-scale storage of hydrogen must be reviewed and compared by using the thermodynamic and engineering approach. Considering the cost of safe storage at high density, storage applications such as liquid hydrogen, methanol, ammonia, and dibenzyl toluene are advantageous. However, it should be noted that the need for electricity also increases in these high-density storage technologies. To overcome this difficulty, the production and storage of hydrogen may be a suitable way when electricity prices are low [1]. Today, electricity is produced in many nuclear power plants. It is possible to produce hydrogen with the excess electricity produced and store it. Thus, electricity can be produced again from stored hydrogen when the electricity demand is unexpectedly high [2]. However, the difficulties associated with storing the hydrogen produced in large quantities must be eliminated. Hydrogen storage in salt domes may be a viable option for this. Salt domes are underground geological formations. These formations are very suitable for the formation of underground salt caverns by the solution mining method. In these structures, hydrogen gas is stored and reinjected under pressure. This causes a sudden increase and decrease in cavern pressure. This can lead to negativities such as collapses and cracks in the salt structure in the cavern. Determination of cushion gas and working gas rates is very important for hydrogen storage [3]. Salt caverns are normally operated by compression and decompression between the minimum and maximum pressure. Thus, the gas can be separated into working gas, which can be drawn from the cavern during normal operation, and cushion gas as a buffer gas, which must remain in the cave to ensure its stability. Roughly estimated, about one-third of the gas inventory is required as a buffer gas, depending on the geology, and some factors such as depth. While injecting and extracting the gas, the operating pressure of the tank changes between certain pressure limits. Besides the maximum and minimum cavity pressure, salt well operations are mainly limited by the maximum pressure change rate per unit time as well as the maximum flow rates in the well to ensure stability. Also, recovery times are required at higher storage pressures after very low pressures [4]. Therefore, determining the optimal cushion gas/working gas ratio has a key role in preserving the cavern structure. To store hydrogen gas in the underground salt cavern more efficiently, it is recommended that the gas and cushion rates are strongly considered. There are more than 1500 salt caverns in the world. Although these caverns are used for natural gas storage, they can be used to store hydrogen gas [5]. Although underground hydrogen storage is a promising technology with which large quantities of H_2 can be stored safely and economically, it is a relatively new application. For this, all parameters related to the system should be well known. For example, H_2 density increases from 0.5 to 12 kg cm^{-3} when pressure increases from 0.6 to 16 MPa, while H_2 density decreases slightly with increasing temperature [6]. The numerical simulations of seasonal hydrogen storage are examined by injecting pure hydrogen into the gas, oil, and water zones in the Norne hydrocarbon field in Norway. The case studies show that the pre-injection of formation gas as a cushion gas efficiently increases hydrogen recovery. For example, the gas zone is the most advantageous system for pure hydrogen storage with final recovery factor of 87%. Hydrogen storage is not recommended in the water zone, due to lower final recovery factor of 49% [7]. A cushion gas is needed to keep the reservoir pressure at a sufficient level while the hydrogen is drawn from the underground storage. Gaseous forms of carbon dioxide, nitrogen, and methane can be used as a cushion gas. Various scenarios with respect to the properties of the cushion gas are simulated and evaluated in terms of the hydrogen recovery factor. The results show that the maximum hydrogen recovery can be realized as 89.7% when methane is injected as a cushion gas [8]. Hydrogen obtained with electricity produced from various energy sources can be stored in salt caverns. Therefore, increasing the storage capacity of these caverns will provide more hydrogen storage. It is observed that the amount of gas stored increased with the increase in the number of layers in the salt caverns [9]. However, the low cost of the technology used for storage is extremely important in terms of usability. Since hydrogen production and salt caverns have relatively low investment costs, excess electricity generated can be converted into hydrogen and stored for later use in salt caverns [10].

In this study, some mathematical equations are obtained to determine the cushion gas/working gas ratios for six salt caverns at different depths as given in Fig. 3.1. Using these equations, it is seen that the most suitable cushion gas/working gas ratios can be calculated. Cushion gas is needed for the long-lasting use of underground hydrogen storage systems. Thus, the cushion gas/working gas ratios are tried to be determined for salt caverns constructed at different depths within the same salt formation. As a result, the determination of the optimal cushion gas/working gas ratio is investigated not only for the protection of the caverns but also for the effect on the energy efficiency of hydrogen gas storage and retrieval in salt caverns. Since underground hydrogen storage systems are relatively new applications, studies on the development of these systems are important. It is thought that this study will contribute to the studies on underground hydrogen storage.

3.2 Materials and methods

Fig. 3.1 shows the locations of the six salt caverns at the various depths that are used as underground hydrogen storage areas (UHSs) in the salt dome. It can be

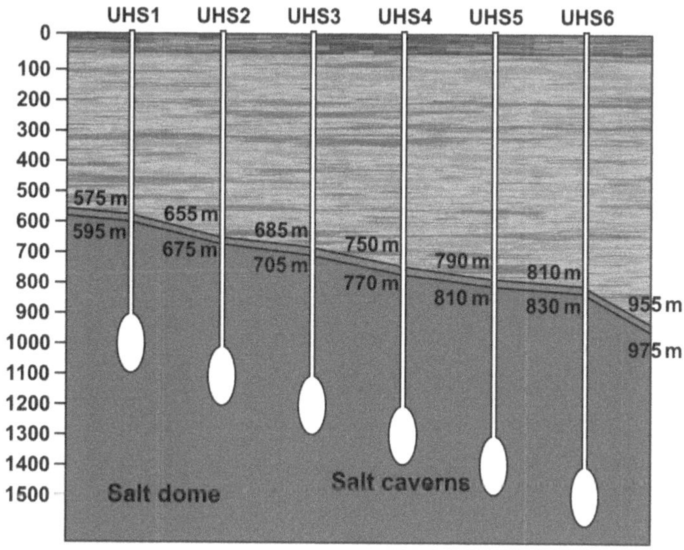

Figure 3.1 Underground salt caverns at various depths.

seen that the salt formation takes place at depths varying between 600 and 975 m. On top of the rock salt dome, there is an also impermeable form of anhydrite with a thickness of nearly 20 m. Depths of the UHS1 to UHS6 are selected to be between 1000 to 1500 m at an interval of 100 m. The area and underground rock salt measurements have been chosen as a suitable region for the formation of a salt cavern for hydrogen gas storage. Thus, the parameters affecting the hydrogen storage potential of the salt caverns constructed at various depths in the salt dome are determined.

The maximum and minimum pressures of the caverns formed in this region are determined according to the equations given below [11].

$$P_{max} = 0.8 \times P_L \tag{3.1}$$

where P_L is the pressure of the last cemented casing shoe point (LCCS) of the cavern and it is given with Eq. 3.1. The depth of the LCCS, as in Tuz Gölü caverns, was accepted as 30 m above the top of the cavern. The maximum gas pressure inside the cavern can not be greater than 80% of the geological pressure value at the LCCS point [12].

$$P_L = 0.022 \times h_L \tag{3.2}$$

where h_L is the depth of the LCCS, it is given with Eq. (3.1). The minimum pressure of the cavern is given in Eq. (3.3) [11].

$$P_{min} = 0.3 \times P_{max} \tag{3.3}$$

The geothermal gradient temperature of the region is given in Eq. (3.4) [12]:

$$T_C = 12 + 0.03h_C \tag{3.4}$$

where h_C indicates the cavern center depth and T_C indicates the cavern center temperature.

The compressibility factor of the cavern is given as [13]:

$$z(P,T) = 1 + \sum_{i=2}^{6}\sum_{j=1}^{2} V_{ij}(P/1\text{MPa})^{i-1}(T/100\text{K})^{n_{ij}} \tag{3.5}$$

where i, j, V_{ij} ve n_{ij} parameters are given in Table 3.1.

The density of hydrogen gas stored in the cavern was calculated by Eq. (3.7).

$$\text{PM}_a = z\rho\text{RT} \tag{3.7}$$

where P (atm) is the pressure value of the gas, M_a (g mol^{-1}) molecular weight value, T (Kelvin) the temperature value of the gas, z is the compression factor value expressing the deviation from ideality under the gas conditions, R is the universal gas constant value (0.082 Latm)kmol^{-1}. The ρ value to be obtained using the Eq. (3.7) gives the density of the gas and opens the way to calculate the mass and energy value of the gas for the determined volume.

The mass of the hydrogen gas stored in the cavern is determined by Eq. (3.8).

$$m_{H_2} = \rho_{H_2}V \tag{3.8}$$

where V is the volume of the cavern. The volume of the hydrogen gas storage is found in Eq. (3.9).

$$V_{Nm^3} = \frac{m_{H_2}}{\rho_N} \tag{3.9}$$

Table 3.1 Some constants to calculate the compressibility factors for hydrogen [13].

i	j	V_{ij}	n_{ij}
2	1	0.036719	− 1.23
2	2	− 0.039839	− 2.22
3	1	− 0.0014722	− 2.68
3	2	0.0024083	− 3.1
4	1	0.65994×10^{-5}	− 2.7
4	2	-0.15469×10^{-4}	− 4.3
5	1	-0.13383×10^{-6}	− 3.3
6	1	0.15608×10^{-8}	− 4.1

Table 3.2 Maximum and minimum compressibility factors under different temperatures and pressure in the caverns at different depths.

Depth *h* (m)	***P* (MPa)**		***T* (°C)**		***z***	
	Min.	**Max.**	**Min.**	**Max.**	**Min.**	**Max.**
1000	5.122	17.072	41.5	56.5	1.0299	1.0965
1100	5.650	18.832	44.5	59.5	1.0328	1.1057
1200	6.178	20.592	47.5	62.5	1.0356	1.1148
1300	6.706	22.352	50.5	65.5	1.0384	1.1238
1400	7.234	24.112	53.5	68.5	1.0409	1.1327
1500	7.762	25.872	56.5	71.5	1.0438	1.1414

Table 3.3 Maximum and minimum hydrogen density, mass, and storage volume in the caverns at different depths.

Depth *h* (m)	**ρ (kg m^{-3})**		***m* (ton)**		**$V_{storage}$ (10^6 Nm^3)**	
	Min.	**Max.**	**Min.**	**Max.**	**Min.**	**Max.**
1000	3.837	11.464	2302	6879	25.61	76.51
1100	4.181	12.428	2508	7457	27.90	82.94
1200	4.516	13.358	2710	8015	30.14	89.15
1300	4.844	14.256	2906	8554	32.33	95.15
1400	5.143	15.124	3086	9075	34.32	100.94
1500	5.476	15.964	3285	9579	36.55	106.55

where $\rho_N\left(=0.0899\frac{\text{kg}}{\text{m}^3}\right)$ is the density [kg m^{-3}] of the hydrogen at the Normal Condition (standard condition; 1 atm, 0°C.

Table 3.2 shows the minimum-maximum values of the pressure and temperature values that will occur in the cavern at different depths and the hydrogen compressibility factor values calculated based on these parameters.

Table 3.3 shows the density of the hydrogen gas that can be stored in the cavern at the maximum and minimum storage conditions and the storage mass and volume values depending on these densities.

3.3 Results and discussion

For the safe and efficient storage of hydrogen in very large quantities, the determination of the structural properties, as well as the storage parameters under pressure, is key. These parameters are temperature, pressure, compressibility factor (z), and cushion gas/working gas ratio (R). Gases have different ductility factors in terms of their structural properties and this parameter is important in determining the storage rates of gases. For safe gas storage in salt caverns, parameter *z* should also be set.

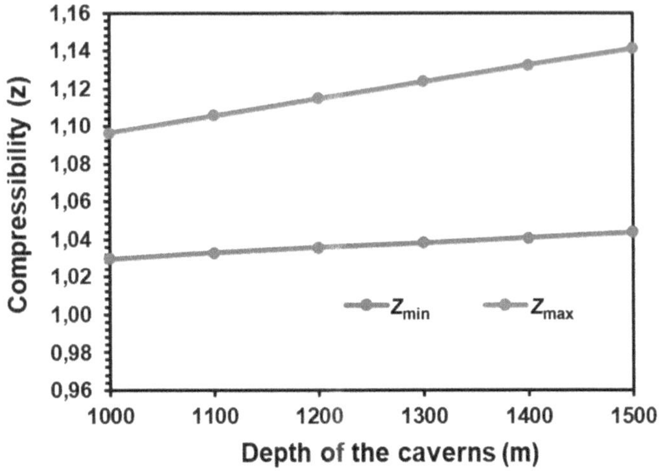

Figure 3.2 Compressibility factor of hydrogen gas for minimum and maximum storage conditions in the caverns at the different depths.

Fig. 3.2 shows the variation of the parameter z with respect to depth. The minimum and maximum compressibility factor values (z_{max} and z_{min}) of hydrogen gas are 1.03 for the depth of 1000 m and 1.14 for the depth of 1500 m, respectively. This situation shows that the compressibility factors of hydrogen will increase linearly as the depth increase. However, by comparison, it is seen that the increase in the maximum compressibility factor is higher than the other.

Fig. 3.3 shows the variation of the amount of stored hydrogen in the caverns built at six different depths. It is seen that there is an increase in the amount of stored hydrogen with the increase in depth. While a minimum of 25.61×10^6 Nm^3 of hydrogen is stored in the 1000 m deep cavern, a maximum of 76.51×10^6 Nm^3 of hydrogen can be stored. While a minimum of 36.55×10^6 Nm^3 of hydrogen is stored in the cavern at a depth of 1500 m, a maximum of 106.55×10^6 Nm^3 of hydrogen can be stored.

Fig. 3.4 shows that the cushion gas increases volumetrically with increasing depths according to the calculations made using the determined parameters in underground salt caverns. For each cavern with a constant volumetric size of 6.3×105 m^3, the percentages of cushion gases of six different caverns with central depths of 1000–1100–1200–1300–1400 and 1500 m are 33.5%, 33.7%, 33.85%, 33.95%, 34%, and 34.25%, respectively. To maintain the stability of the caverns, with the effect of the new temperature and pressure values that occur depending on the depth, it is necessary to provide the internal-external pressure balance by increasing the cushion gas ratio.

Fig. 3.5 shows the change in the ratio of cushion gas to working gas for storage of hydrogen gas depending on depth. For hydrogen gas, this ratio decreases linearly with depth. This reduction depends on the physical and chemical properties of hydrogen gas as well as on maintaining cavern stability. At the depths of 1000 and

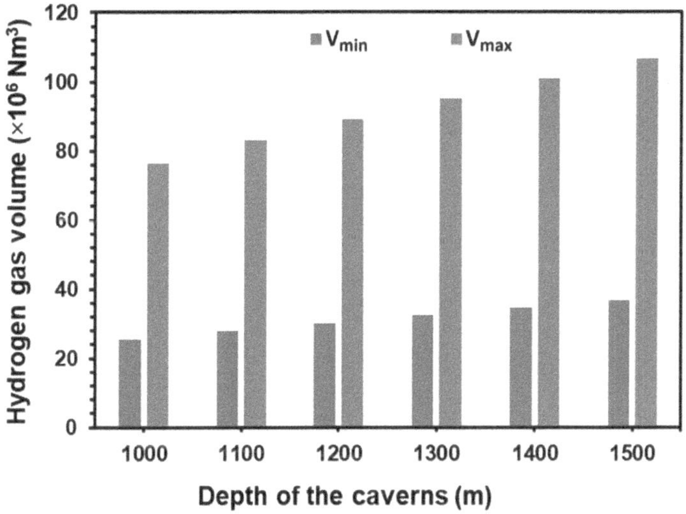

Figure 3.3 Minimum and maximum volumes of hydrogen stored in the caverns at the different depths.

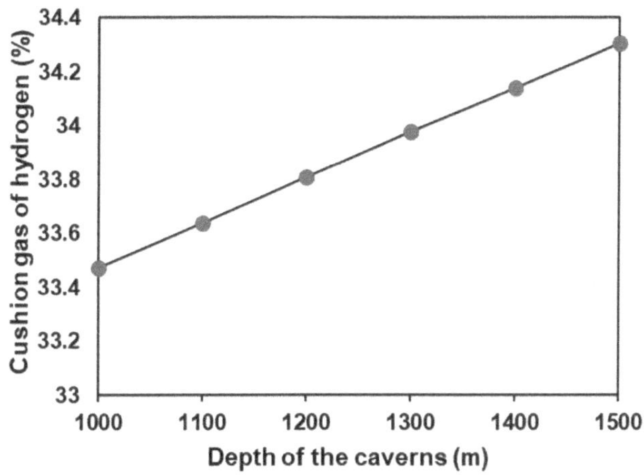

Figure 3.4 Change with a depth of cushion gas of hydrogen gas at the different depths.

1500 m, this ratio is determined as a maximum of 1.985 and a minimum of 1.915, respectively.

Fig. 3.6 gives the energy value depending on the depth of the cushion gas-related to the hydrogen gas storage. Because cushion gas is an important cost item, it is extremely important for a business. While minimum energy of 90.097 GWh of cushion gas is stored in the 1000 m deep cavern, the maximum energy of 128.57 GWh of cushion gas can be stored.

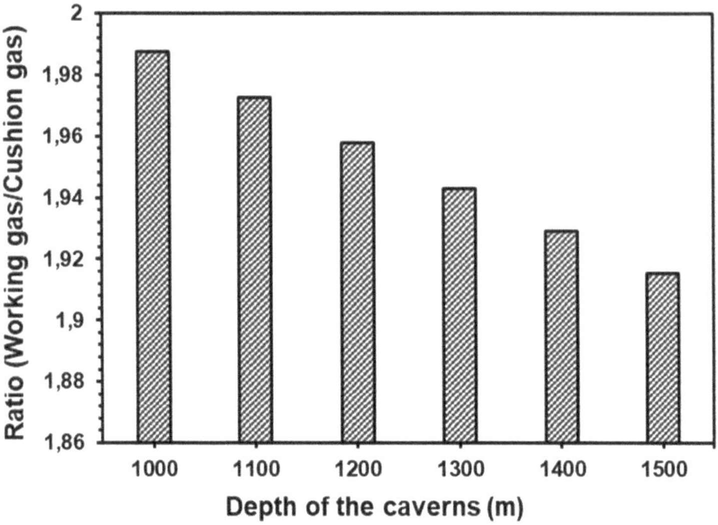

Figure 3.5 Variation of the ratio of working gas to cushion gas at different depths.

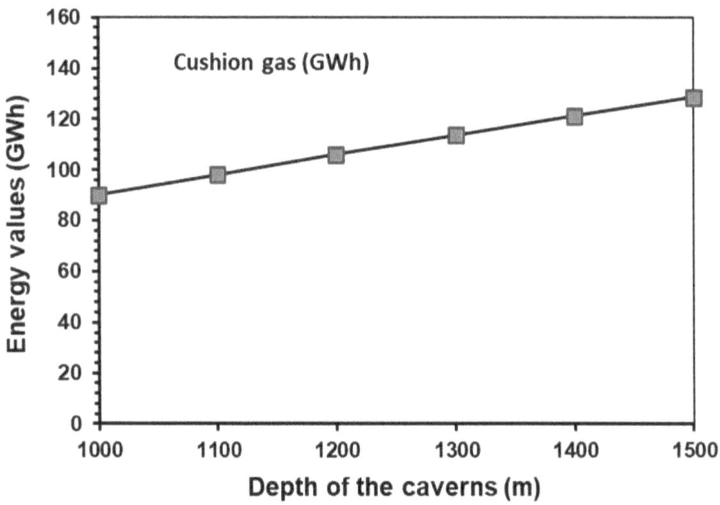

Figure 3.6 Energy values of hydrogen cushion gas for different depths.

Fig. 3.7 gives the usable energy value of stored hydrogen gas for six caverns at different depths. This value refers to the part that can be used in injection and withdrawal. While minimum energy of 181.69 GWh of hydrogen is stored in the 1000 m deep cavern, the maximum energy of 249.84 GWh of hydrogen can be stored.

As a result of the calculations made according to these data, it is determined that the working gas that has the energy of approximately 217 GWh (upper thermal

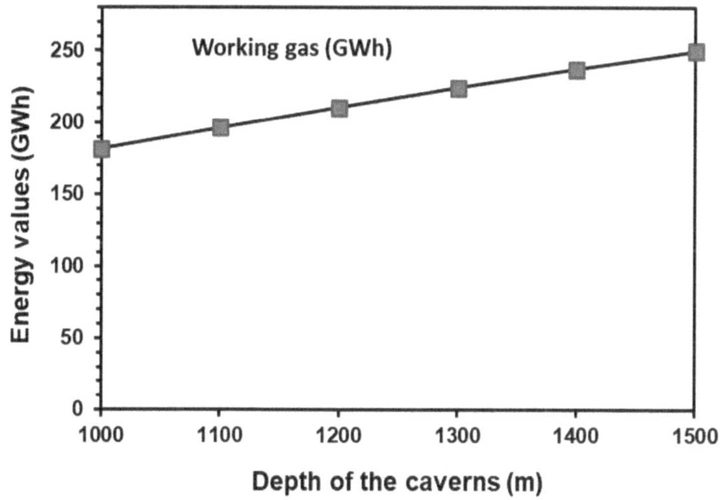

Figure 3.7 Energy values of hydrogen working gas for different depths.

energy value) could be stored for each cavern with a volume of 6.3×10^5 m^3. This value means that 1300 GWh of hydrogen energy can be stored as a working gas for six caverns. The results show that hydrogen gas storage systems (UHGS) in salt domes have great potential in terms of large-scale storage of hydrogen. However, the cushion gas rate stands out as an important parameter to be considered in terms of cost.

3.4 Conclusions

In this study, to determine the storage potential of hydrogen gas in underground salt caverns, the changes of some basic parameters with depth are discussed. Based on these parameters, the results obtained by determining the minimum and maximum values of hydrogen gas that can be stored for the same type of caverns at different depths are examined.

It is observed that the increase in temperature and pressure parameters with the increase in depth also affects the compressibility factor value of the hydrogen gas to be stored linearly. This increase in the Z value of hydrogen gas indicates that more hydrogen gas can be stored in deeper designed caverns of the same volume.

Based on the energy value of the stored working gas, it is determined that the difference between hydrogen energies that can be stored between 1000 and 1500 m depths will be approximately 70 million GWh. This difference highlights the importance of storing hydrogen deeper in underground salt domes in terms of energy value.

As the depth increases in underground salt caverns, the amount of cushion gas increases with the hydrogen stored. The rate of cushion gas that should be left in the caverns without being used is an important issue that should be evaluated as an investment cost.

It is determined that the working gas/cushion gas ratio decreases with increasing depth. This means that although the working gas amount increases in deeper storage, the increase of cushion gas will be proportionally higher. Therefore, this situation should be taken into consideration in terms of determining the optimum storage depth.

Although the increase in the amount of hydrogen gas that can be stored with the increase in depth in the salt domes is an advantage, the increasing rate of cushion gas will be an important factor in terms of investment.

Finally, the determination of the optimal cushion gas/working gas ratio was determined not only to protect the caverns but also an important factor in the energy efficiency during the storage and withdrawal of hydrogen gas in the caverns.

References

[1] J. Andersson, S. Gronkvist, Large-scale storage of hydrogen, Int. J. Hydrog. Energy 44 (2019) 11901–11919.

[2] A. Ho, K. Mohammadi, M. Memmott, J. Hedengren, K.M. Powell, Dynamic simulation of a novel nuclear hybrid energy system with large-scale hydrogen storage in an underground salt cavern, Int. J. Hydrog. Energy 46 (2021) 31143–33115.

[3] San, S., Investigation of hydrogen storage method in salt domes. (MSc thesis), Institute of Natural and Applied Sciences, University of Cukurova, Adana, Turkey, 2019.

[4] Olaf Kruck, Fritz Crotogino, Ruth Prelicz, Tobias Rudolph, Overview on all known underground storage technologies for hydrogen, Grant Agreement No: 2/93, 303417, 2013.

[5] X. Liu, X. Shi, Y. Li, P. Li, K. Zhao, H. Ma, et al., Maximum gas production rate for salt cavern gas storages, Energy 234 (2021) 121211.

[6] B. Pan, X. Yin, Y. Ju, S. Iglauer, Underground hydrogen storage: influencing parameters and future outlook, Adv. Colloid Interface Sci. 294 (2021) 102473.

[7] M. Lysyy, M. Fernø, G. Ersland, Seasonal hydrogen storage in a depleted oil and gas field, Int. J. Hydrog. Energy 46 (2021) 25160–25174.

[8] M. Kanaani, B. Sedaee, M. Asadian-Pakfar, Role of cushion gas on underground hydrogen storage in depleted oil reservoirs, J. Energy Storage 45 (2022) 103783.

[9] X. Zhang, W. Liu, D. Jiang, W. Qiao, E. Liu, N. Zhang, et al., Investigation on the influences of interlayer contents on stability and usability of energy storage caverns in bedded rock salt, Energy 231 (2021) 120968.

[10] D.G. Caglayan, N. Weber, H.U. Heinrichs, J. Linßen, M. Robinius, P.A. Kukla, et al., Technical potential of salt caverns for hydrogen storage in Europe, Int. J. Hydrog. Energy 45 (2020) 6793–6805.

[11] Crotogino, F., Donadei, S., Bunger, U., Landinger, H., Large-scale hydrogen underground storage for securing future energy supplies, in: Proceedings of the Eighteenth World Hydrogen Energy Conference (WHEC2010), Parallel Sessions Book 4: Storage Systems/Policy Perspectives, Initiatives and Cooperation's WHEC, 16–21 May 2010, Essen, Germany, 2010.

[12] A. Ozarslan, Large-scale hydrogen energy storage in salt caverns, Int. J. Hydrog. Energy 37 (2012) 14265–14277.

[13] E.W. Lemmon, M.L. Huber, D.G. Friend, Standardized equation for hydrogen gas densities for fuel consumption applications, Environ. Prot. Agency (2006) 01–0434.

Energy storage in capacitor banks

4

Satyaranjan Jena[1], *Pradeep Kumar Sahu*[1] *and Madan Mohan Sahu*[2]
[1]School of Electrical Engineering, KIIT, Deemed to be University, Bhubaneswar, Odisha, India, [2]Department of Electrical Engineering, National Institute of Technology Rourkela, Rourkela, Odisha, India

Chapter Outline

Emerging Trends in Energy Storage Systems and Industrial Applications. DOI: https://doi.org/10.1016/B978-0-323-90521-3.00015-6

Highlights of the chapter

1. Among all energy storage devices, the capacitor banks are the most common devices used for energy storage. The capacitor bank has advantages that can provide a very high current for short period.
2. Energy storage capacitor banks are widely used in pulsed power for high-current applications, including exploding wire phenomena, shock-less compression, and the generation, heating, and confinement of high-temperature, high-density plasmas, and their many uses in this chapter.
3. Highlights various types of capacitor storage systems and their brief comparisons.
4. Mathematical modeling of energy storage capacitors, as well as their design and characteristics, are briefly discussed in this chapter.

Nomenclature

L	life expectancy of capacitor
L_0	reference life data provided by the manufacturer
Q	quality factor of the capacitor
q	exponent which is determined experimentally
K_v	voltage coefficient
U_a	actual operating voltage across a capacitor
U_R	rated capacitor voltage
$T1$	actual operating temperature
$T2$	maximum Operating temperature
Δt	ripple current rise
C	capacitance of charging capacitor
Lc	inductance of transmission line
$R_{\min}$	minimum charging resistor
V_{max}	variable maximum supply voltage
$I_{\max}$	maximum charging current
v_c	voltage across a capacitor
V	DC supply voltage
E_R	energy dissipated in series resistor R
V_R	voltage across the series resistor
η	efficiency of charging during constant current
$I_C(t)$	capacitor charging current
T_{ch}	charging time in sec.
ω_0	resonant frequency

Index

4.1 Introduction

For the overall economic development of any country, energy plays a very vital role. Till now mostly we are relying on fossil fuels for the extraction of energy. The excessive use of fossil fuel sources can create serious pollution in the environment as well as depletion of natural resources. Worldwide the power sector contributes 75% of total CO_2 emissions [1]. Because of that the power producer are searching for environment-friendly energy sources which helps for sustainable development as well as increase the living standards. For sustainable development, the government of different countries is emphasizing energy conservation and the generation of energy from renewables [2,3].

In the last few years, the renewable energy sector is growing all over the world. The main drawback of harnessing power from renewables is its intermittency. This nature of renewable energy creates a challenge for the power producer to maintain the reliability of the system. On the other hand when the energy production is in excess during off-peak hours then it must be stored to increase the efficiency as well as reliability of the system. Sometimes the storage of energy can bridge the gap between supply and demand, provide energy security, and reduce greenhouse gas emissions. Therefore, renewable generation must be paired with an efficient storage system. As it is not easy to store electricity, researchers are continuously searching for new devices and mechanisms to store electrical energy. The storage capacity is increasing continuously as per the International Energy Agency (IEA). In Fig. 4.1 a comparative analysis is given on Annual energy storage deployment for the year 2013–2019 for grid-scale and behind the meter in gigawatt. In the year 2020, the total addition is 5 kW out of which two-thirds is grid-scale [4].

Electric energy storage has multiple benefits, reduction in transmission congestion, reduce the cost and need of major infrastructure, reduction in energy bills in

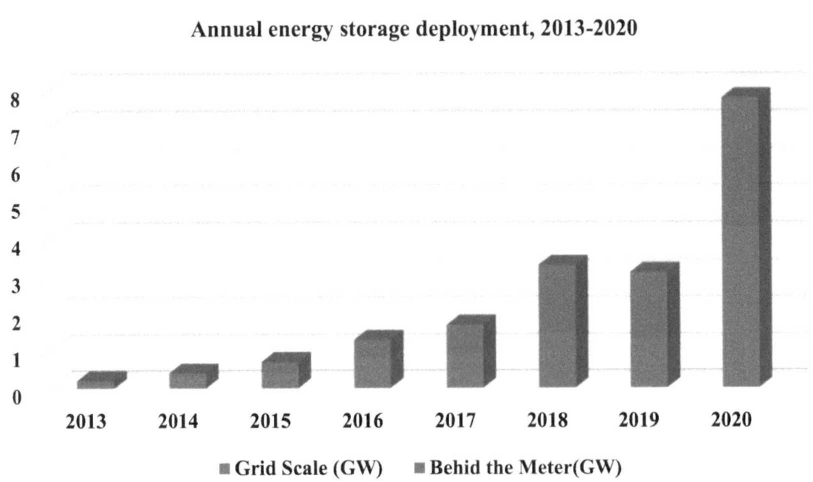

Figure 4.1 Annual energy storage deployment, 2013–2020 [4].

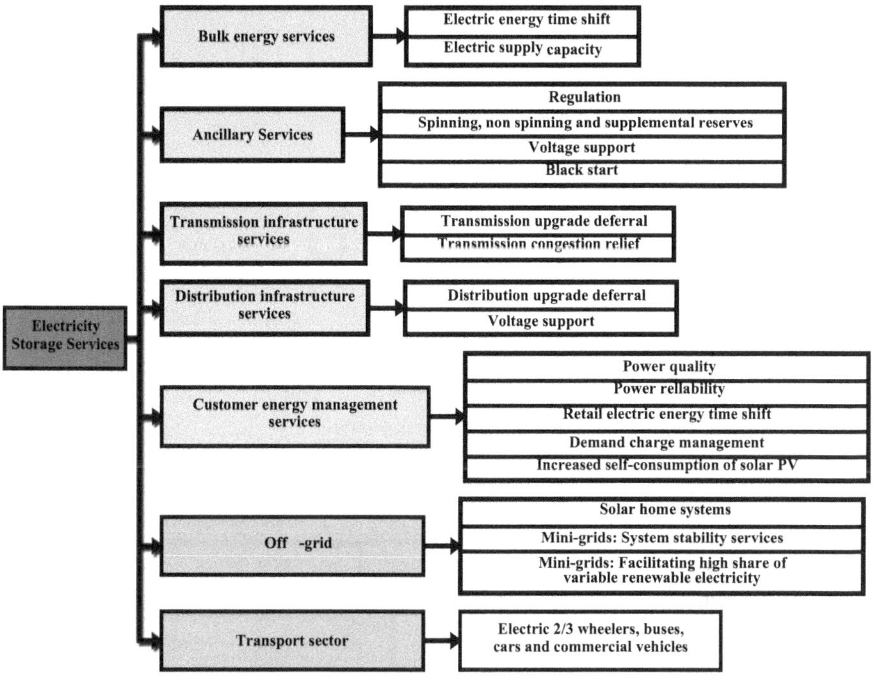

Figure 4.2 Electricity storage services [6].

case of behind-the-meter application, and peak demand reduction. In the era of the energy transition, it will provide the service from power producers to end-users. The other need for energy storage is for low-power applications like iPods, iPhones, and smartphones. It is also widely used in the transport sector, telephone industry, and health care sector [5,6]. Fig. 4.2 indicates the type of services provided by the energy storage system.

All the energy storage techniques are broadly classified into four categories as shown in Fig. 4.3[7].

1. Electrochemical storage system
2. Electromagnetic storage system
3. Thermodynamic storage system
4. Mechanical storage system

Every energy storage technology has its advantages and disadvantages. In Table 4.1 the strength and weakness of different storage technology are elaborated [7,8].

The choice of the storage technology depends on various factors like efficiency, self-discharge time, service life period, cost, and power rating in Table 4.2 a comparison based on the technical and economic parameters is highlighted for different storage technology [9].

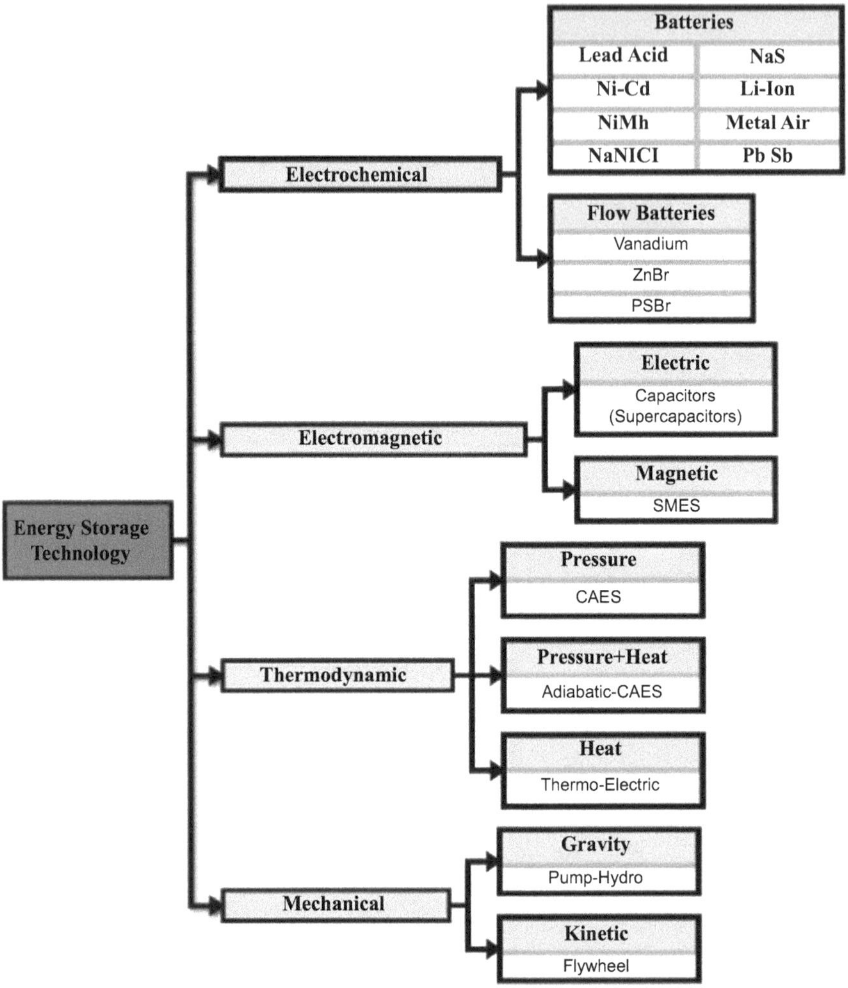

Figure 4.3 Different energy storage systems [5].

Another aspect of choosing energy storage technology mainly depends on the power requirement and the period of discharge at the rated power. In Fig. 4.4 below a comparison between system power rating and the discharge time at rated power is shown.

In the current scenario, electrochemical energy storage technologies are in demand due to their compactness and reliability. The different electrochemical storage technologies available in the market are fuel cells, batteries, and Electrochemical supercapacitors. As compared to other storage systems, capacitive energy is more compact and highly reliable. This compactness is achieved due to increase energy density. In the field of capacitive energy technology, ESs are more popular as compared to the conventional capacitor. In the next section, ESs energy storage technology is described briefly [10–13].

Table 4.1 Strength and weakness of different energy storage systems.

Technology	Strength	Weakness
Electrochemical storage system		
Lead acid batteries	25–65	74–89
Nickel battery	60–105	71
Lithium battery	190–375	90–97
Sodium battery	155–225	72–81
Fuel cell	–	45–53
Electromagnetic storage system		
Capacitors	– Very high efficiency – High power capacity	90–95
SMES	10	92
Thermodynamic storage system		
Compress air energy storage (CAES)	2–7	40–55
Adiabatic CAES	2–8	60–68
Thermo electric	100–300	45–75
Mechanical storage system		
Flywheel	210	83–93
Pump storage plant	0.35–1.1	70–82

Table 4.2 Technical and economic parameter of energy storage systems.

Technology	**KWh/m^3**	**Efficiency (%)**	**Self-discharge (%/ day)**	**Service life (in year)**	**Cost (€/kWh)**
Electrochemical storage system					
Lead acid batteries	25–65	74–89	0.17	10	90–335
Nickel battery	60–105	71	–	–	385–1100
Lithium battery	190–375	90–97	0.008–0.041	15	140–180
Sodium battery	155–225	72–81	–	17	265–645
Fuel cell	–	45–53	–	–	2305
Electromagnetic storage system					
Capacitors	10	90–95	0.004–0.013	10	5150
SMES	10	92	10–12	30	13,570–75,670
Thermodynamic storage system					
Compress air energy storage (CAES)	2–7	40–55	0–10	40	40–80
Adiabatic CAES	2–8	60–68	–	–	600–800
Thermo electric	100–300	45–75	–	–	0.2–8
Mechanical storage system					
Flywheel	210	83–93	72–100	–	650–2625
Pump storage plant	0.35–1.1	70–82	0–0.5	40–100	40–180

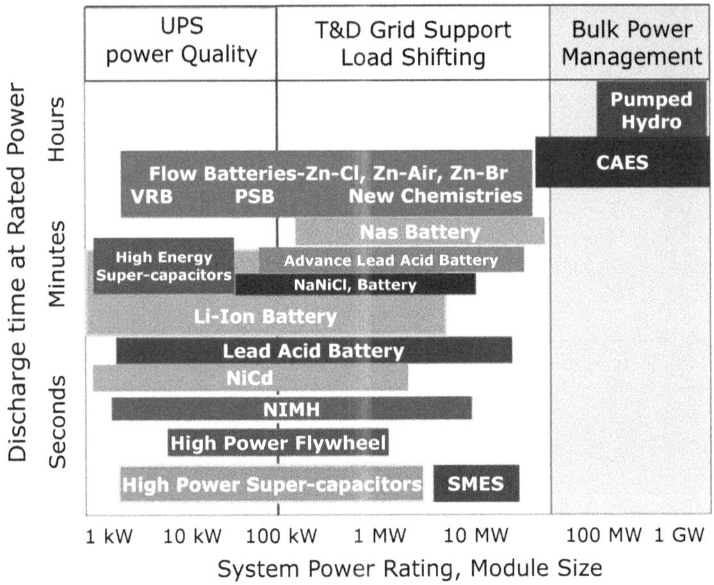

Figure 4.4 Power rating of the different storage systems.

4.2 Energy storage capacitor

The capacitor bank used for bulk energy storage has the properties like large peak current, low inductance, high di/dt rating, better reliability, long life period, and improved fault tolerance capability. The selection of materials for capacitive storage generates much scientific interest. The materials of each component of the capacitor bank are selected to perform multiple functions. Both surface and chemical charge storage can be incorporated by calculating the optimum mixture of each material. The electrolyte materials used in the capacitor bank can be determined based on optimum pore-size distribution. The electrolyte should supply or omits the ions at a higher rate. The behavior of charging ions solvation is incorporated which changes with the size of the pore in the range of macropores to the extreme size. Here an intercalation reaction takes place between neighboring sheets of the material. The optimum electrolyte and salt mixture for high charge transfer and conductivity can be designed that focus on the cost feasibility. The mathematical modeling and simulation of these complex structure capacitor banks can be briefly understood by advanced computational tools. It is also challenging for anyone to provide a contentious electric path in the electrode for any energy storage unit. Then, it is easy for the electron to remove or insert very rapidly. The design of capacitors also plays an important role in high-power devices. Various nanosized porous materials have been investigated nowadays for designing capacitors [14,15]. These materials have inherent properties like improved chemical redox nature and extreme surface areas.

In general, the capacitor is divided into three types dielectric, electrolytic, and Electrochemical Capacitors (ECs). The dielectric capacitor and electrolyte capacitors are treated as conventional capacitors and ECs are known as supercapacitors (SCs) of ultracapacitors

4.2.1 Conventional capacitor

A conventional capacitor or dielectric capacitor consists of two conducting electrodes separated by a dielectric insulating medium. When a voltage is applied to a capacitor, opposite charges accumulated at the surface of each electrode. The charges are kept separated by the insulator, thereby the electric field produced lets the capacitor store the energy. Different types of capacitors are reported in the literature. Capacitors connected to electronic and power circuits are of electrolyte type or dielectric type and are usually solid-state devices. These capacitors have properties like rapid response time and a long-life period. However, these devices store less energy mostly lower than 0.1 Wh/kg. Therefore, these are not commonly used for bulk energy storage.

4.2.2 Electrochemical capacitors

Another type of capacitor is the electrochemical capacitor [16] or supercapacitors (SCs). Here electrodes of high-surface-area and electrolytes of acetonitrile or sulfuric acid are commonly used for bulk energy storage. According to current R&D activities, ECs are categorized as electrostatic double-layer capacitors, pseudocapacitors, and hybrid capacitors as shown in Fig. 4.5.

4.2.2.1 Electrostatic double-layer capacitors

The energy is stored at the surface of the electrostatic double-layer capacitors (EDLCs) (Fig. 4.6). As compared to solid-state capacitors, bulk power can be stored across these capacitors and can store bulk energy. However, the stored energy is still less than 10 Wh/kg. The cell voltages are limited to prevent the decomposition

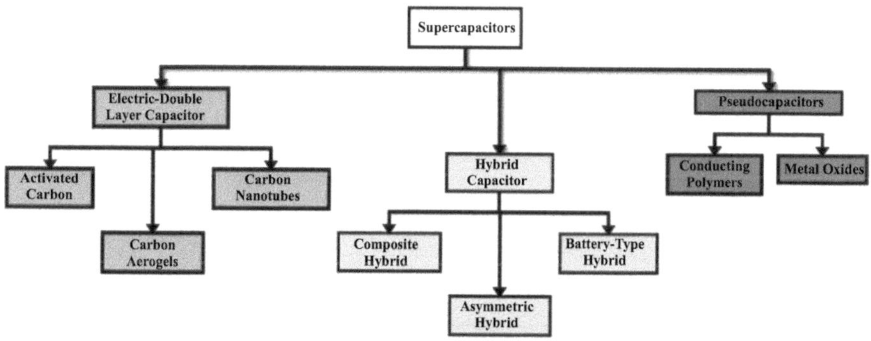

Figure 4.5 Classification of supercapacitor [17].

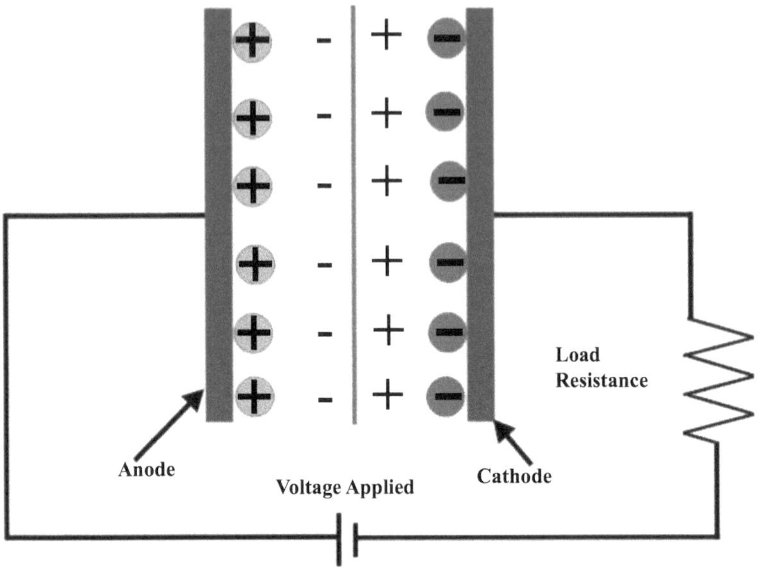

Figure 4.6 Electrostatic double-layer capacitors [18].

of the liquid electrolytes: less than 1 V for water-based and around 3 V for non-aqueous electrolytes. These capacitors are commonly used for short periods of the energy demand of the range from 0.01 to 100 seconds. These are more reliable and less expensive and used in trucks and cranes.

4.2.2.2 Pseudocapacitors

Another capacitor bank called pseudo capacitors (Fig. 4.7) can be designed by hybridization of double-layer capacitors and batteries where the materials' surface rials play a major role [19]. Therefore, these capacitor banks can handle a large amount of energy as compared to the surface capacitors. However, these banks face several challenges like reliability and technical issues like the selection of dielectric medium [20]. The materials especially transition-metal oxides like manganese or vanadium oxides are commonly used in the construction of pseudocapacitors [21]. Generally, many crystalline materials with flat discharge levels like vanadium pentoxide react with lithium and are converted into amorphous substances [22]. During this reaction, a lattice structure is formed by randomized the vanadium and lithium ions. It exhibits a similar sloping discharge level as that of a capacitor.

Nowadays hydrated ruthenium oxide, another compound material has gained scientific interest [21,23]. These capacitors store the highest energy of approximately 200 Wh/kg. The chemical and physical properties of these unique compound materials should be understood briefly. Mainly metal oxides and conducting polymer are used in pseudocapacitors to store energy.

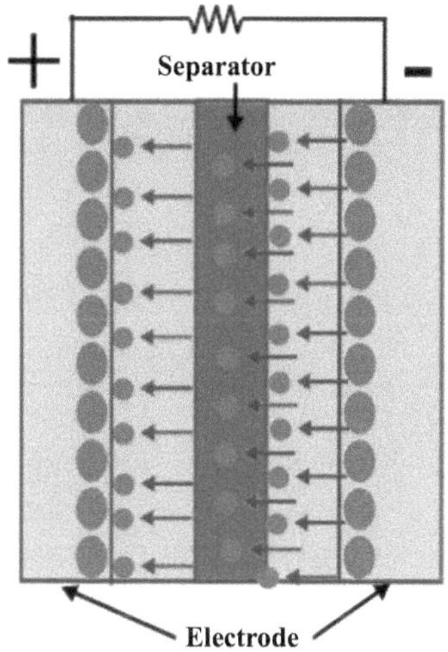

Figure 4.7 Pseudocapacitors [18].

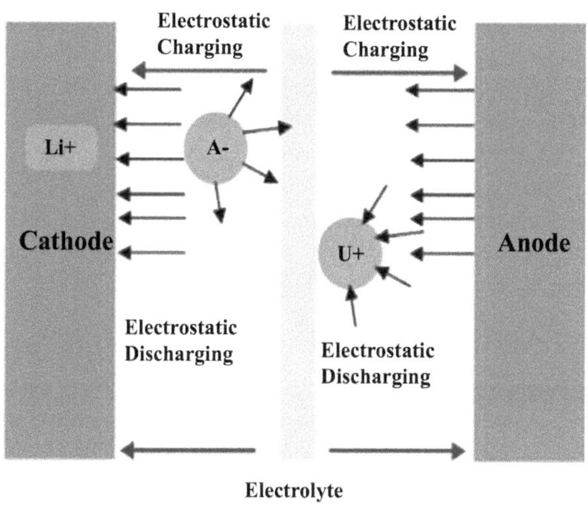

Figure 4.8 Hybrid supercapacitors [18]

4.2.2.3 Hybrid supercapacitors

The combination of EDLCs and pseudo capacitors is known as a hybrid (Fig. 4.8). This can be used for faradaic as well as non-faradaic processes. The energy and power density of hybrids are more as compared to EDLCs. Irregular electrodes are used for the construction of these capacitors. For example, in Lithium-ion capacitors A^- and Li^+ are the ions present in the electrolyte for necessary reaction as shown in Fig. 4.8. In most cases for hybrid capacitors, the charging and discharging

Table 4.3 Comparison between conventional capacitor, battery, and supercapacitor [13].

Storage device characteristics	Battery	Capacitor	Supercapacitor
Charging time	$1 < t < 5$ h	$10^{-3} < t < 10^{-6}$S	1–30 s
Discharging time	$T > 0.3$ h	$10^{-3} < t < 10^{-6}$S	1–30 s
Power density (W/kg)	< 1000	> 1,000,000	10,000
Energy density (Wh/kg)	10–100	< 0.1	1–10
Life time (Cycle number)	1000	10^6	10^6
Charge/discharge efficiency	0.7–0.85	> 0.95	0.85–0.98

time is around 10 minutes–20 minutes with 300 w/kg peak power density. The value will differ according to the material used for the electrode in the case of a hybrid capacitor. According to the alignment of Electrode, a hybrid capacitor is dived into, battery type, composite and asymmetric hybrids. Hybrid capacitors are used in security cameras, computer servers, and backup devices [18].

4.2.3 Comparison of supercapacitor and other storage devices

As compared to battery or electrolyte capacitors, the volume and weight SCs are less. Also, SCs have higher power and energy densities. The SCs have a 5000-time longer life cycle as compared to the battery [24–26]. The performance of battery band SCs is complementary to each other because SCs can support various power but the battery can support a power level up to a limited value. The energy-to-weight ratio of batteries is much higher than SCs [27–29]. To meet high power requirements hybrid energy system is an emerging option [18,30]. A comparison between conventional capacitor, battery, and supercapacitor is given in Table 4.3 below. The modeling of the capacitor is highlighted in the next section.

4.3 Capacitor model

Energy storage capacitors are commonly modeled as lumped RLC (resistor-inductor-capacitor) circuits. Here, equivalent series resistance (ESR) represents the resistive and dielectric losses in the capacitor, and equivalent series inductance (ESL) represents the inductance of the capacitor lead and current path through the capacitor.

4.3.1 Capacitor parameters

The exact mathematical capacitor model is more complex which is shown in Fig. 4.9. The simplified model of the capacitor is given in Fig. 4.10. Here, ESR and ESL are the series resistance and inductance respectively. These design parameters are more important while the capacitor is used at high voltage or high frequency. The losses due to ESL can be reduced by taking low loss factor dielectric film or

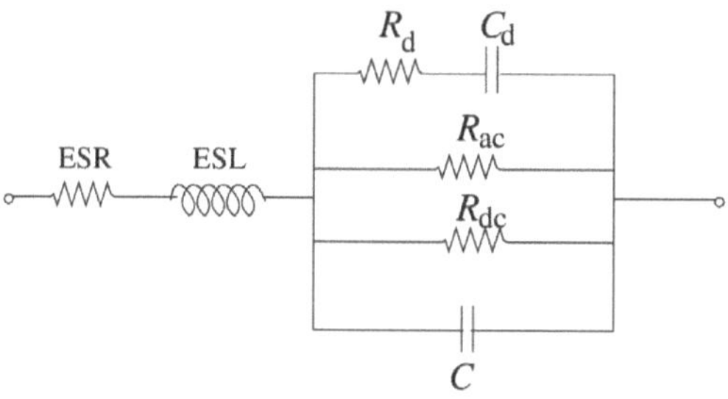

Figure 4.9 Capacitor models: a comprehensive equivalent circuit.

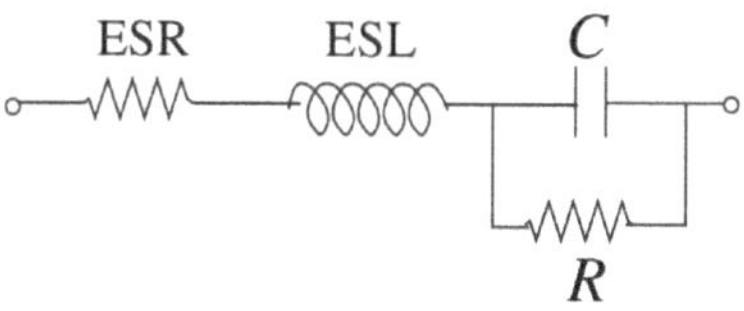

Figure 4.10 Capacitor models: a simplified version.

fluid used as a dielectric medium. This loss can also be minimized by using proper welding and soldering approach for bushes and leads. However, the ESR losses increase at a high frequency due to the skin effect. The capacitor di/dt rating is degraded due to the presence of ESL. This can be minimized by using strategies like the use of high width-to-length ratio metal foil, extended foil design, radial bushing, and the use of parallel pads. Here, R_{dc} and R_{ac} are the equivalent resistance for dc and ac dielectric losses respectively. Another important parameter of the capacitor is dielectric absorption in which the dielectric of the capacitor cannot be polarized immediately. Due to this property, the capacitor cannot be fully discharged even after being short-circuited for a short period. Due to the presence of parasitic resistance Rd and capacitance Cd, the capacitor recovers a small amount of dc voltage at its terminal even after fully discharged.

4.3.2 Shot life of capacitor

The capacitor can sustain the number of charging-discharging cycles before failure is called shot time. The shot time is commonly represented by its survival possibility rate. The survival rate versus no. of the shot curve of the capacitor is provided by the manufacturers. The capacitor life period depends on several factors like; operating frequency, the magnitude of voltage reversal, temperature, and operating voltage. The capacitor life period curve is given in Fig. 4.11.

The size and life of the capacitor are impacted by the significant voltage reversal. The capacitor life can be extended by minimizing the degree of voltage reversal in the normal mode of operation. The reversal of the voltage creates stress on the capacitor which also comprises enhancement of the electric field in the dielectric medium, more power losses, and more heating on the conductor. The capacitor life

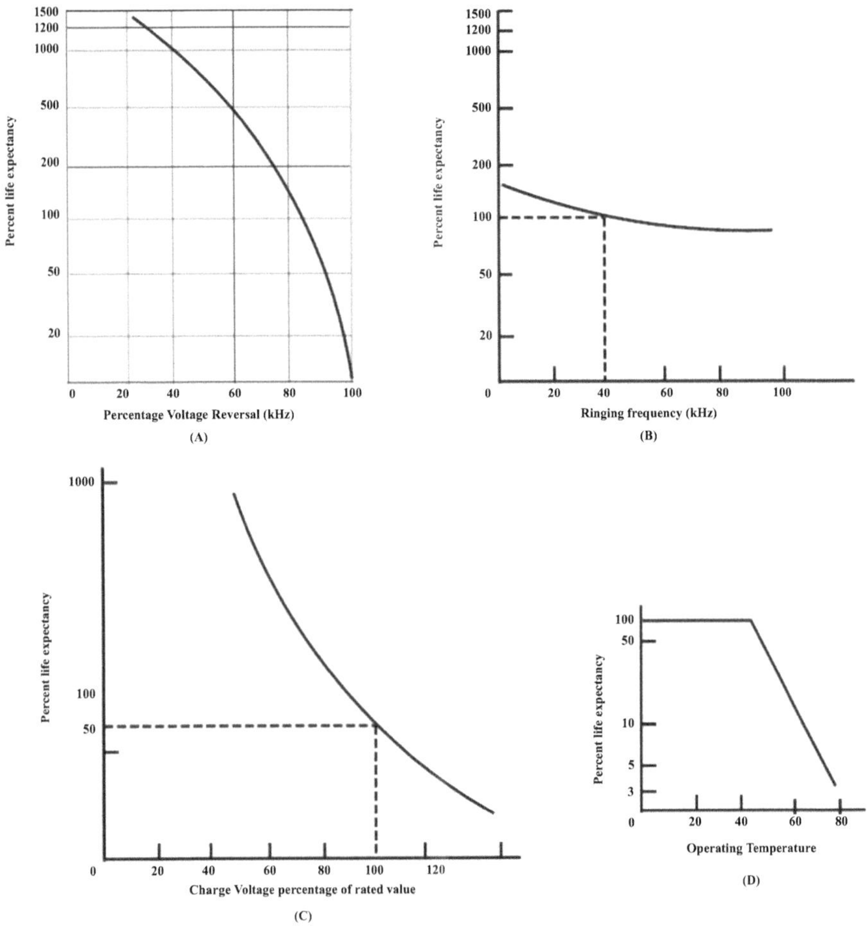

Figure 4.11 Curves of capacitor life expectancy versus (A) voltage reversal (B) ringing frequency (C) charge voltage (D) operating temperature.

test data was provided by the manufacturer under different conditions and various types of dielectric and electrode systems. The test data integrated with the empirical equation of life-scaling is applied for the prediction of the capacitor life under different environments. The voltage reversal factor is given by (4.1) and plotted in Fig. 4.11A.

$$\frac{L}{L_0} = \left[\frac{Q}{Q_0}\right]^q \tag{4.1}$$

where L and L_0 are the life of the capacitor and reference life data provided by the manufacturer whereas Q and Q_0 are the quality factor of the capacitor and its reference value. Here, q is the exponent which is determined experimentally.

Generally, the energy storage capacitor used for fast discharge applications is classified as slow, medium, and fast which regulates the discharging rate by the internal inductance of the capacitor called ringing frequency. In a discharge circuit, the upper current and the frequency limit are set by the ringing frequency. The typical value of the ringing frequency is in the order of 100–200 kHz. It is experimentally validated that a capacitor operated at 30% of ringing frequency, the capacitor life is increased by half. The relationship between capacitor life expectancy and ringing frequency is almost linear as shown in Fig. 4.11B.

The operating voltage creates stress on the capacitor dielectric medium which also affects life expectancy. If the applied voltage approaches the rated voltage, then the electrolyte consumption is more due to self-heating in the dielectric layer. So, the low operating voltage enhances the life expectancy of the capacitor. The operating voltage factor is given as follows;

$$Kv = \left[\frac{U_a}{U_R}\right]^{-n} \tag{4.2}$$

if

$$0.5 \le \frac{U_a}{U_R} \le 0.8 \rightarrow n = 3$$

$$0.8 \le \frac{U_a}{U_R} \le 1 \rightarrow n = 5$$

where K_v is the voltage coefficient and n is the next valid condition. U_a and U_R are the actual operating and rated voltage. The relationship between charge operating voltage and life of capacitor is plotted in Fig. 4.11C.

The life expectancy of the capacitor generally obeys "the doubling 10°C rule." Here, the life period of the capacitor is doubled for every 10°C operating temperature reduction. The capacitor life expectancy can be determined by the following equation and its plot is shown in Fig. 4.11D.

$$L_2 = L_1 \times 2\frac{T_1 - (T_2 + \Delta T)}{10} \tag{4.3}$$

where L and L_0 are expected life (hours) and specified life (hours) at actual operating temperature (T_2 in °C) and maximum operating temperature (T_1 in °C) respectively with ripple current rise Δt °C.

4.3.3 Test methods

A simple and direct method for the calculation of ESR is reported in [31]. Here, the ratio of ripple voltage to ripple current of the capacitor determines the ESR parameter. But this method for calculating the exact value is more expensive and complex

Table 4.4 Comparison of various test methods for equivalent series resistance.

Methods	Design	Cost	precision	Noise effect	Elementary circuit requirement
Direct method [31]	Complex	High	High	Minimum	Minimum
Voltage measurement method [32]	Moderate	Low	Low	Moderate	Minimum
Experimental method [33]	Simple	High	High	High	Maximum

in design for the small ripple current. ESR can also be calculated by using voltage measurement only [32]. Under some conditions, assuming a constant inductor ripple current, ESR can be calculated from output ripple voltage. However, this method does not provide high precision. The ESR of the capacitor storage unit can also be determined by a laboratory method which is reported in [33,34]. But, the realization of this method is expensive as it requires a power amplifier, a signal generator, an oscilloscope, and an RC circuit. Another simplified method to calculate this parameter of the capacitor is briefly presented in [35]. In this method, the ESR is determined by coupling a square waveform voltage signal with a reference resistor. Here, a reference resistor is connected in series with a capacitor and the step response of output capacitor voltage determines the ESR. The comparison among all the methods is highlighted in Table 4.4.

The ESL parameter of the capacitor is ignored in many applications because of its small value. ESL can be determined by various experimental methods like; (1) a low voltage is supplied for charging the capacitor then-current waveform is observed with short-circuited capacitor [36,37], (2) the differential method in which the capacitor is connected in series with the additional known capacitor [38], (3) variable inductance method [39].

4.3.4 Switch/triggering pulse generator

The switch used for the capacitor bank should hold off the voltage at the time of the charging period, close it with small jitter, and allow the high current. Generally, switching to a capacitor bank uses spark gaps. A sole 2-ball spark gap is placed in series with the load as a result the capacitor charges as long as the gap is filled. Nowadays, a 2-ball spark gap consists of a triggering pin. The 3- pin triggertron is also used in some applications due to its low cost and reliable switch for capacitor bank. Triggertron is working in surrounding pressure and working voltage up to 40 kV. Vacuum spark gaps are also commonly used as capacitor switching with an operating voltage of 100 v to 20 kV. However, constant pumping and vacuum machinery are required for the vacuum gap. This technique is suitable for low-inductance capacitor banks due to its wide voltage range and reliability.

A four-element spark gap can also be designed for the high voltage range of 50–100 kV with a 10–8 seconds jitter. It is similar to the 3-electrode gap with the separate triggering element. The electrode placed at the center has a voltage in between the lower and upper electrodes. The fourth electrode is used as the spark plug. The central electrode drives by triggering the gap to crack the voltage of any one gap. The spark plug gap can be broken down by applying these trigger pulses. When one-half of the gap cracks then the other half is also overfilled and cracked. A large capacitor bank is designed to provide a peak current of range 10-15MA at the rate of 10^{12} A/s. A single spark gap cannot supply such a high current. Spark gaps with low jitter are used in parallel for high current applications. Generally, the design of a capacitor bank needs a separate spark gap. These designs may not be available in the commercial market.

4.3.5 Transmission system

Based on the flexibility and inductance of the capacitor bank, the bus-bar, parallel plate transmission line, or array of coaxial cable are commonly used as the transmission system. The open bus system is suitable for the power transmission of the low inductance capacitor bank. The bus system for the bank is designed to provide the maximum short-circuited current. The designer should have knowledge that opposite current repels whereas parallel currents attract. This concept should be considered for designing the bus system of the bank. The large current in the kA range will produce a magnetic field which may damage the bus.

The coaxial cables are also suitable for the flexible load and low inductance transmission line [40]. The standard cables used for the voltage range of 10 to 100 kV are RG-8/U, RG-14/U, RG-17/U, and RG-19/U. These cables can handle a current amplitude of 30 kA with a repetition rate of 2 shots/min. The typical high-frequency inductance of the coaxial cable is 0.08 μH per foot. The capacitor bank also uses some special low-inductance coaxial cable [41]. The cables available commercially are expressed by their diameter. The cable-type 17/1 has an outer conductor of RG-17/U and an inner conductor of RG-14/U. Polyethylene foam is used as the central core of the coaxial cable. It behaves as the dielectric medium for the conductors of the cable. RG-17/U vinyl jackets are used as the outer cover for the cable. These cables have an inductance of 0.036 μH/foot at low frequency and resistance of roughly 1 mΩ/foot.

Parallel plate transmission lines for the desired inductance are also well suited for capacitor banks. It has the inductance which is represented as;

$$L = 4\pi \times 10^{-7}\frac{t}{w}H/m \tag{4.4}$$

where w and t are the width and separation distance in m respectively. Two opposite direction currents which flow in the parallel plate produced a magnetic force that may displace the plates. The magnetic force produced at the plates is given by;

$$P = 2\pi\left(\frac{I}{w}\right)\text{atm} \tag{4.5}$$

where I am the current flows through the parallel plates in A. The amplitude of the magnetic force is very large even though it lasts for a small duration. It is challenging for anyone to hold the parallel plates together.

4.3.6 Power feed

The power feed for the capacitor bank is used to collect the discharge current at one end through a coaxial cable and supply this current on the other end to the load. It is mechanically arranged for tightening many coaxial cables. The efficient delivery of energy can be possible from the bank when it has low inductance. Generally, the parallel plate transmission line filled with dielectric layers requires a collector design (Fig. 4.12).

The cable-receiving end can be designed in such a way that the surface of the cable is free from any breakdown. Again, there should not be occurred any enhancement in stress on the dielectric medium and the insulation of the cable is stripped [42]. By using special type electrodes, stress grading, and high dielectric strength insulating liquid, the above problems can be overcome. The insulation grading of the collector-cable junction is suitable for a capacitor bank which is shown in [42].

A single collector plate is used to collect the large current which is carried out by the various coaxial cables. The current density of the collector plate is quite high, normally in the range of 10 MA/m. The pressure developed on the plate [43,44] at this high current density can be determined by (4.5) and found as 10,000 psi. This high pressure can be handled by using strong clamping. Thick aluminum plates are commonly used as collector plates in the design of a megajoule capacitor bank [45]. These collector plates are fitted with the bank through steel tie bolts and backed by rubber sheets.

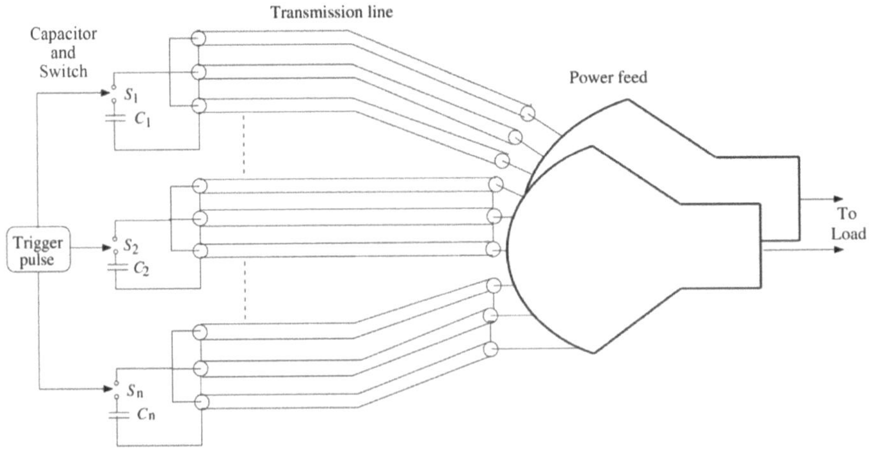

Figure 4.12 Large energy storage capacitor bank configuration.

4.4 Topology of capacitor bank circuit

The voltage is supplied to charge the high-energy storage capacitor bank. Similarly, the discharge operation of the bank is initiated by applying a command trigger communicated to the start switch. The transmission line is used to carry the discharging current which is sent to the load by a power feed. Fig. 4.13 given below shows the layout of a typical high-energy storage capacitor bank. The crowbar switch is placed in the capacitor bank circuit based on its function. The excessive voltage reversal may cause failure on the high-energy-density capacitor bank. In this scenario, both the crowbar switch and capacitor bank are located nearer to each other. This switch may also be located in between the start switch and the capacitor bank. The desired current is allowed by the crowbar switch and passes through the start switch to initiate the discharging operation. In this way, the capacitor life can be prolonged by reducing erosion. In other operations like producing a high magnetic field or plasma confinement, the unidirectional current is allowed to flow in an under-damped inductive load through the crowbar switch. The externally triggered pulses are used for the operation of these switches. Another operation like flux compression generators is designed in such a way that the crowbar switch is operated to short the load.

It is difficult to connect the crowbar switches in some capacitor energy bank circuit topology due to their continuous on and off operation. These switches are remained open during the discharging operation to the load and closed by applying a triggered pulse during low voltage across the switch. In various topologies, these switches are triggered accurately at their self-breaking voltage.

4.4.1 Equivalent circuit of an energy storage capacitor bank

A typical low-energy storage capacitor bank schematic diagram is illustrated in Fig. 4.14. The bank consists of a capacitor bank of capacitance C_s, a charging resistor R_c, a start switches S_1, transmission line T_l, a crowbar switch S_2, and a dissipating resistor R_d with an inductive load of L_0. Here the capacitor Cs starts charging to

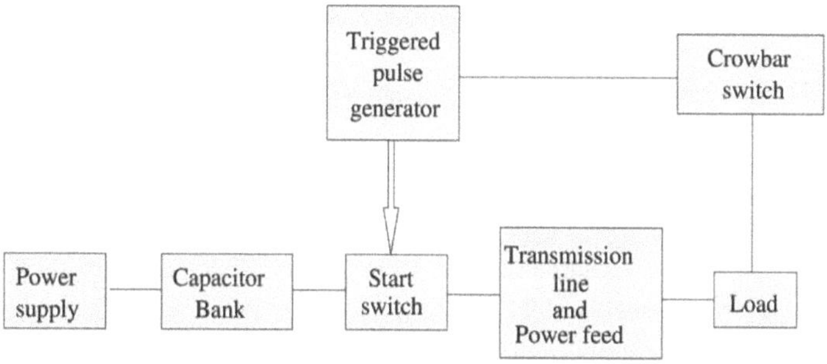

Figure 4.13 Energy storage capacitor bank circuit topology with crowbar switch.

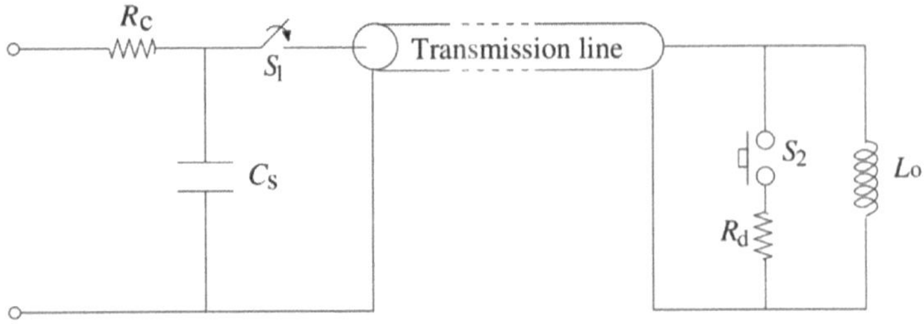

Figure 4.14 Discharging circuit for energy storage capacitor bank.

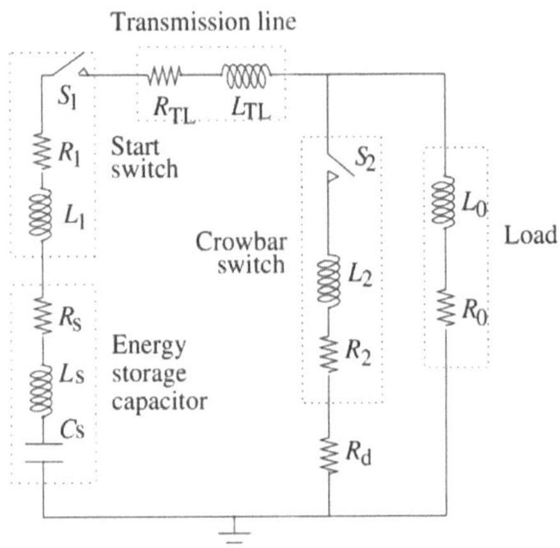

Figure 4.15 Large capacitor bank equivalent circuit.

a voltage V_0 through the charging resistor R and the start switch S_1 is used to discharge the bank. The current from the bank supplies the inductive load of L_0 through a transmission line T_1. The protection of the load can be achieved through a dissipating resistor R_d and a crowbar switch S_2.

The equivalent circuit of a low-energy capacitor bank has the following symbols. The operation of the capacitor bank can be briefly studied by its equivalent circuit. Anyone can calculate the peak current of the bank and the time to attain this current. A typical 25 kJ energy storage capacitor bank is shown in Fig. 4.15 [46]. The parameter details of the equivalent circuit are given below.

1. C_s, L_s, and R_s—capacitance, equivalent self-inductance, and equivalent resistance of capacitor bank.
2. L_1 and R_1—self-inductance and resistance of start switch S_1
3. L_{TL} and R_{TL}—self-inductance and resistance of transmission line
4. L_2 and R_2—self-inductance and resistance of crowbar switch S_2
5. L_0 and R_0—self-inductance and resistance of the connected load
6. R_d—damping resistor

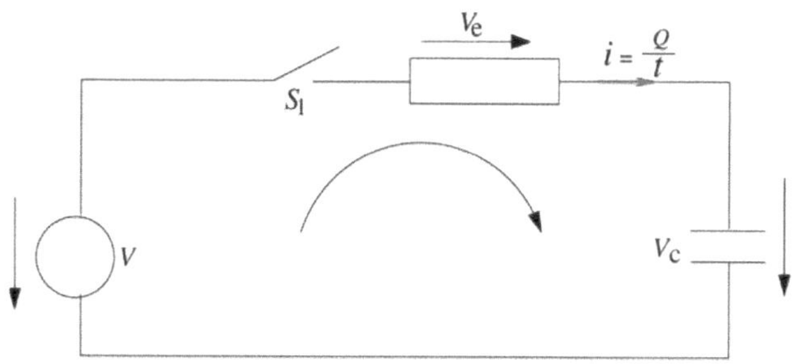

Figure 4.16 Charging circuit for a capacitor bank.

4.5 Charging and discharging operation

The basic need of an energy storage system is to charge as quickly as possible, store maximum energy, and discharge as per the load demand. The charging of the capacitor bank depends on the amount of supply voltage and the duration of charging and discharging operation depends on the type of the load connected across it as shown in Fig. 4.16. Depending on the charging methods, the chargers are classified as; Constant voltage, constant current, constant power, and resonant charging.

4.5.1 Constant voltage charging

In constant voltage charging, the voltage of the capacitor is kept constant by applying variable voltage with a maximum voltage V_{max}. The voltage is applied to the capacitor through a series resistor "R" as shown in Fig. 4.4. When the switch S_1 is closed, the current flows through the capacitor. Initially, the capacitor stores the charges with higher rates. When it gets almost fully charge the current through it becomes negligible.

Let, for a power rating of P, the maximum current flow through the capacitor is I_{max}. Then the minimum value of the resistance to charge the capacitor is obtained by,

$$R_{min} = \frac{V_{max}}{I_{max}} \tag{4.6}$$

During Charging, the voltage across the capacitor is exponentially rising, and expressed by,

$$V_C = V \quad (1 - e^{-t/RC}) \tag{4.7}$$

where

1. V_C = voltage across the capacitor.
2. V = DC supply voltage.
3. T = duration of charge

Similarly, the charge stored through the capacitor is expressed as

$$Q_C = CV \quad (1 - e^{-t/RC}) \tag{4.8}$$

As per Eq. (4.8), the charge of the capacitor (Q_C) also increases exponentially and the charging time of the capacitor depends on the value of RC that is the time constant of the circuit. The capacitor charges up to 99% of the full value in a five-time constant. So, we can say that the charging time of the capacitor is five times the time constant.

$$T_{ch} = 5RC \tag{4.9}$$

During this period the amount of energy stored in the capacitor is,

$$E_c = ½CV^2 \tag{4.10}$$

Also, Energy dissipated in the series resistor is given by,

$$E_R = \int_0^{\infty} \frac{V_R^2}{R} dt = \int_0^{\infty} \frac{(V.e^{-t/RC})^2}{R} \quad dt = \frac{1}{2} \quad CV^2 \tag{4.11}$$

And Energy comes from a source that is

$$E_S = E_C + E_R = CV^2 \tag{4.12}$$

So, the Efficiency of charging of capacitor in the constant voltage method is

$$\eta = \frac{E_C}{E_S} = \frac{\frac{1}{2}CV^2}{CV^2} = 50\% \tag{4.13}$$

4.5.2 Constant current charging

In the constant current charging method, a constant value of current is flown through the capacitor till the capacitor gets fully charged. As a result, the capacitor charges at a constant rate as shown in Fig. 4.17. This method gives faster charging than the constant voltage method. As the current is constant during the charging time, it is mostly used to charge a small value of the capacitor.

The capacitor charging current is given by,

$$I_C(t) = C\frac{dV}{dt} \tag{4.14}$$

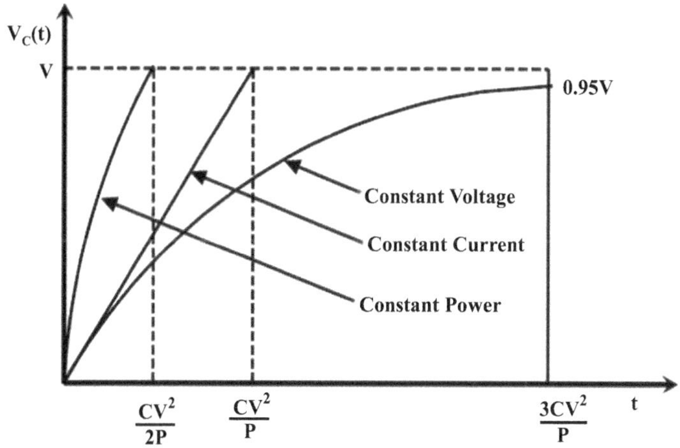

Figure 4.17 Capacitor voltages during charge.

By integrating both sides, we get

$$\int_{0}^{V_C(t)} dV = \int_{0}^{t} \frac{I_C}{C} \, dt \tag{4.15}$$

Hence, the charging voltage is given by,

$$V_C(t) = \frac{I_C}{C} t \tag{4.16}$$

$$Q_C(t) = I_C t \tag{4.17}$$

Eq. (4.13) shows that the voltage across the capacitor is increased with a constant slope with respect to time. The charging time T_{ch} depends on the current flow through the capacitor.

$$T_{ch} = \frac{CV}{I_C} = \frac{CV^2}{P} \tag{4.18}$$

As the charging current increases, the charging time decreases accordingly. For fast charging, the capacitor current is maintained to maximum value that is I_{max}. The efficiency is higher than the constant voltage method of charging.

4.5.3 Constant power charging

In constant power charging, the current and voltage both are variable. Due to the absence of the series resistance, it gives high efficiency than the other two methods.

For constant power supply,

$$P = I_C(t).V_C(t) \tag{4.19}$$

The capacitor current in terms of the capacitor voltage is given by,

$$I_C(t) = C\frac{dV_C(t)}{dt} \tag{4.20}$$

Multiplying both sides by $V_C(t)$,

$$C.V_C(t).dV_C = I_C(t).V_C(t).dt = P.dt \tag{4.21}$$

By integrating we get,

$$\frac{1}{2}C.V_C^2(t) = P.t \tag{4.22}$$

The charging voltage is defined as;

$$V_C(t) = \sqrt{\frac{2P.t}{C}} \tag{4.23}$$

$$Q_C(t) = \sqrt{2CP.t} \tag{4.24}$$

From the above expression, it is seen that the charging of the capacitor depends on the square root of time, as sown in Fig. 4.17.

The time required to charge the capacitor to its full value is,

$$T_{ch} = CV^2/2P \tag{4.25}$$

From the charging time expression of the above three methods, it can be easily observed that the constant power charging is faster than the other charging methods. Also, it gives high efficiency than the other two methods due to the absence of the series resistance.

4.5.4 Resonant charging

In this method, a thyristor and a diode in series with an inductor are connected to the capacitor as shown in Fig. 4.18. A high voltage DC output "V" produces at the rectifier output terminal, and it is filtered through a high value of capacitor C_2. When the thyristor triggers, the energy of the capacitor C_2 transfer to the capacitor C_1 through the inductor L and diode D_1. The circuit behaves as a resonant circuit and the inductor continue to flow the current till the voltage across C_1 becomes

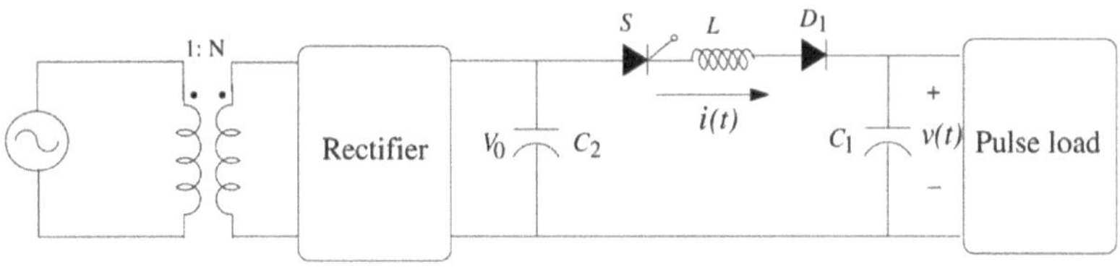

Figure 4.18 Circuit for resonant charging.

2 V. When the thyristor gets off the energy at capacitor C_1, flow to the capacitor bank.

The voltage across the capacitor C_2 and the current is defined as,

$$V_C(t) = V(1 - \cos\omega_0 t) \tag{4.26}$$

$$I_C(t) = V\sqrt{\frac{C_1}{L}} \quad \sin\omega_0 t \tag{4.27}$$

Due to the circuit resonance, the voltage and current oscillate sinusoidally with an oscillating frequency of ω_0.

where,

$$\omega_0 = 1/\sqrt{LC_1} \tag{4.28}$$

The charging time T_{ch} can be calculated from Eq. (4.27) when the current reaches zero.

$$T_{ch} = \pi\sqrt{LC_1} = \pi/\omega_0 \tag{4.29}$$

This charging method is efficient and the charging time depends on the value of L & C_1. Due to the presence of a high voltage capacitor of large value that is C_2 it is more expensive than other methods of charging.

4.6 Application of capacitor bank storage system

In recent days all electrical systems are looks towards energy efficiency. One of the major loss components is the presence of reactive power and unbalancing systems in the system, mainly on the distribution side. Also, presence of the power electronics devices the harmonics are injected into the power system. So, capacitors are widely applied in most electrical power sectors to improve power quality. Due to the compactness of the capacitor, it is sometimes used as an energy storage device instead of a battery; like in hybrid electric vehicles, UPS, etc.

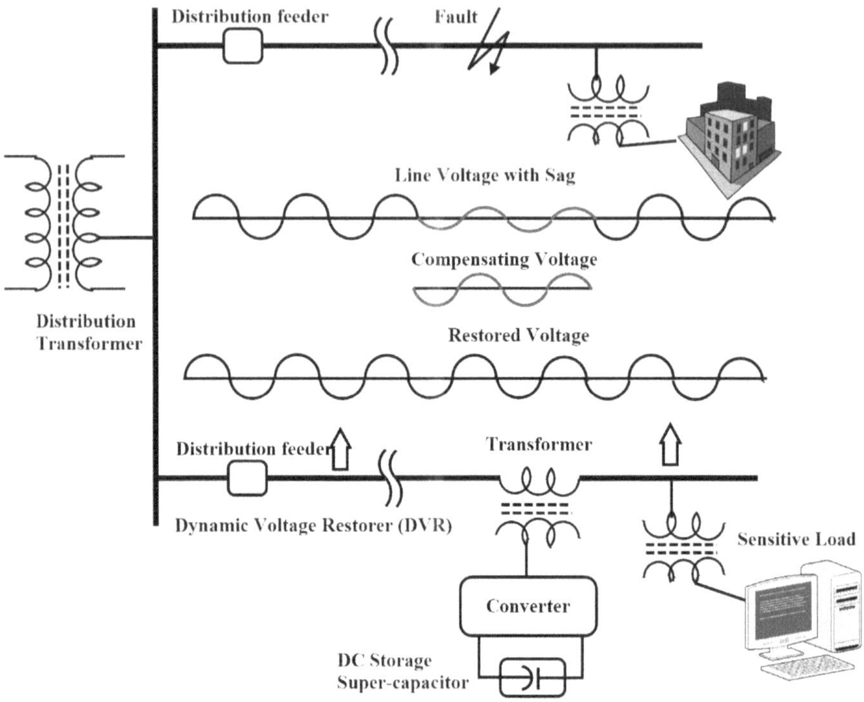

Figure 4.19 Power quality improvement.

4.6.1 Power quality improvement

Any deviation or fluctuations of voltage, the frequency that results in misoperation of electronics and electrical devices is called power quality problems. Voltage sag swell, voltage dip, over-voltage, and injection of harmonics low power factor are some examples of power quality issues [17,47]. These problems mainly arise on the distribution side due to the presence of nonlinear loads, unbalanced loads, and power electronics converters. The capacitor can act as an energy storage device for the system to improve power quality and reliability (Fig. 4.19) [47]. Capacitor banks are widely used in power distribution networks as power factor improvement devices. Also, in the flexible power transmission system, high voltage capacitors are used with power electronics converters to improve the reliability of power. Such as; Static Condenser, Static shunt Compensate or the (SVC, STATCOM), static series compensator (SSSC, TCSC), UPFC, Voltage regulator, voltage stabilizer, etc.

4.6.2 Power factor improvement

Considering the electrical loads, almost all loads are inductive and absorb reactive power from the grid. Total power at the generator end consists of real power P and

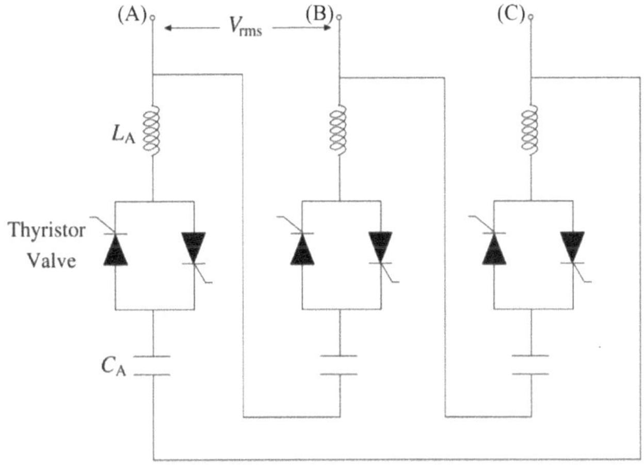

Figure 4.20 Switched capacitor bank circuit.

reactive power Q, which support the line as well as load impedance. It produces a low power factor, which gives rise to the current, voltage drop, and loss in the system. To improve the power factor, capacitor banks are connected at the load end to generate the same amount of reactive power required by the load. The amount of the reactive power can be varied as per the load, by using power electronics devices as shown in Fig. 4.20, called switched capacitor bank. These are mostly preferable where inductive loads fluctuate largely. The capacitor banks are automatically turned on or off depending upon the voltage profile or VAR required of the system. The alternate method to improve the power factor is by using a static compensator (STATCOM) and Static var compensator (SVC). The STATCOM and SVC can either generate or absorb the reactive power in synchronization with the supply voltage. These devices at the load end neutralize the burden of reactive power of the source.

The capacitor bank is important for improving the power factor for any electrical system. The economic impact of electrical power consumed at a high power factor is briefly analyzed by taking a case study of Lebanon [47]. It is found that the electrical energy cost is significantly increased for an unaddressed power factor. The decrease in the cost of energy indicates that there is a reduction in the amount of electrical energy. The cost of electrical energy can be reduced in this manner. Therefore, capacitors are used to improve the power factor which makes the system more efficient in energy usage for transmission and distribution systems. Currently, Lebanon has a capacity of power generation of 2500 MW whereas the peak load demand in the year is 3800 MW which is quite low compared to the generation capacity. This deficiency of power between the generation and demand can be fulfilled by operating the power factor of the system close to 0.95. Penalty may be implemented for that consumers who are operated their electrical load below 0.8 power factor. The performance of the power system can be significantly improved from both an economical and technical point of view by operating different types of equipment like generators, transformers, and transmission lines close to the unity power factor. The installation of auxiliary equipment in the power system like switchgear, bus bar, etc., may also consume more reactive power and an appropriate size of the capacitor is connected to counter this inductive power.

4.6.3 voltage stabilizer

Energy storage devices are a more important role in utility grid stabilization. When more numbers of renewable energy sources are introduced into the grid, the need for voltage stabilization becomes essential [17].

The voltage profile gets disturbed like; voltage fluctuations, surge, and spikes, voltage sag, and swell, etc. These Short term or long-term voltage fluctuations can be avoided by injecting or absorbing the required amount of reactive power into the system. With the amount of reactive power injection to the grid, the voltage profile becomes higher and vice versa.

STATCOMs and SVCs models are widely used to stabilize the system voltage within a specified range. As the voltage fluctuation in the distribution system is short-lived, less than 10 cycles, Supercapacitor is an effective option for energy storage and release. [48] In another way, we can say that during the low voltage profile some amount of energy from supercapacitor storage fed the system to balance the system voltage. In this condition, the STATCOM behaves as an inverter and the supercapacitor is an energy storage element (Fig. 4.21).

4.6.4 Hybrid electric vehicle

Most companies practice batteries as the power storage element in hybrid vehicles. But the battery-powered vehicles have some limitations like; low power density, high temperature, long charging time, and limited charge and discharge time. The electrochemical capacitor can overcome these limitations.

But the problem with it, having a high cost and low energy density [17,48]. The supercapacitor cannot meet the power demand for the propulsion of the vehicle, but it can give the support peak load demand like; acceleration, frequent start and stop of the engine, and high torque [48].

1. In addition to the application in electric vehicles, a supercapacitor can maximize the performance of an internal combustion engine (ICE). During the normal running conditions of the vehicle, the engine fed power to the wheel as shown in Fig. 4.22 and during the operation of the clutch the engine power is stored at the ultra-capacitor. These powers are

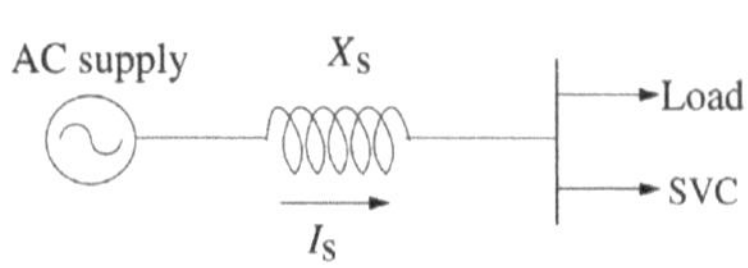

Figure 4.21 Utility voltage stabilization by using static var compensator.

further used to peak up the vehicle in the next moment. The following functions are added to hybrid electric vehicles by using a supercapacitor [49,50].

2. Offset fuel consumption during the ideal stop of the engine.
3. Maximum regenerative power recovered during the braking operation.
4. Optimal use of onboard energy storage system and the engine with energy management.
5. Support of high torque pick up, and significant boosting during acceleration.
6. Emission-less vehicle operation is possible if a sufficient Energy storage system is available.

4.6.5 Uninterrupted power supply

An uninterrupted power supply (UPS) system, store the energy and gives backup consisting of power converters with energy storage units (Fig. 4.23). These are

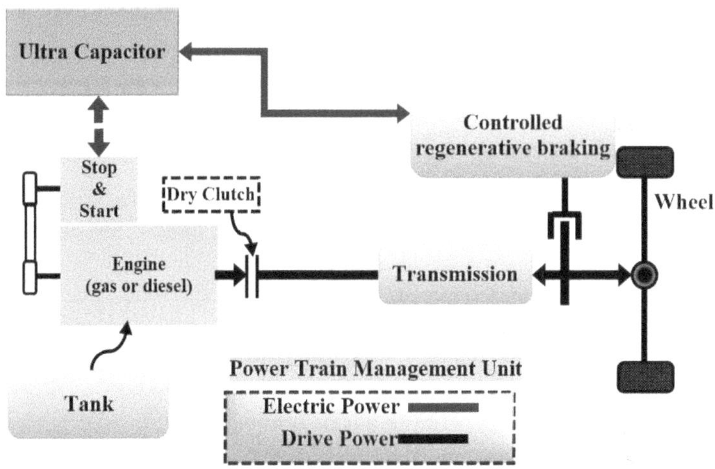

Figure 4.22 Hybrid electric vehicle.

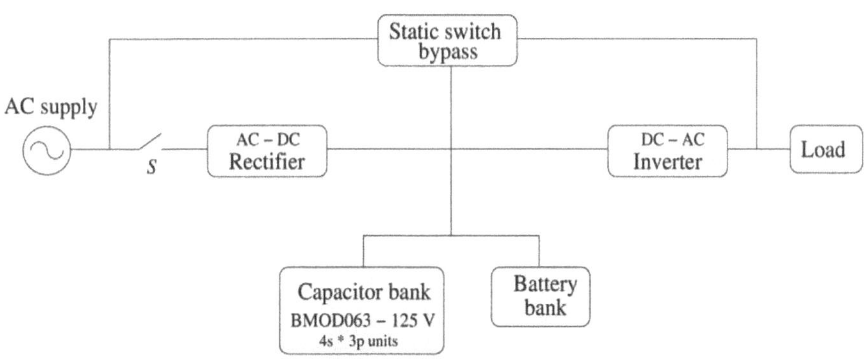

Figure 4.23 Capacitor bank used in uninterrupted power supply.

mostly used in highly sensitive areas like; hospitals, data centers, data server offices, etc. [49]. In case of any abnormality occurs the UPS transfers the power to critical loads also during the time of nonavailability of power, the UPS acts as a backup energy source. UPS normally gives the backup for a small period depending upon the types of energy storage device used.

During the operation of the UPS system, the DC voltage that appears at the terminal of the battery is not purely DC. Some high-frequency harmonics are present with the DC voltage, due to the power electronics converter operation. This fluctuation of voltage reduces the performance of the battery. The use of a high voltage capacitor bank at the terminal of the battery will smoothen the DC voltage. As a result, the performance, and life of the battery increase.

Supercapacitors are also used on the AC side of the UPS system to prevent the momentary fluctuation of the output voltage [50]. Output voltage at the inverter side never is purely sinusoidal without using a capacitor. Supercapacitors are the best option for smoothing the output voltage as well as mitigating the short-term voltage perturbations.

4.6.6 Renewable energy application

In Solar photovoltaic applications, the amount of energy generated from solar radiation is stored in the battery storage system. But the life span of the battery is quite less (3 to 5 years) as compared to the supercapacitor (about 20years). [17] Hence use of a supercapacitor will reduce the life cycle cost of a total solar photovoltaic system.

Considering the energy efficiency, the supercapacitor has higher energy efficiency compared to the battery. It saves almost 20% extra power during charging and discharging operations compared to a lead-acid battery [46]. Another advantage of the supercapacitor is to operate effectively in a wide range of temperatures. So, it can be used in remote areas of solar or wind farms.

For position control of solar panels and pitch control of wind turbine blades to track maximum power, additional power sources are needed. These are normally taken from battery-driven energy storage systems. The requirement of power is for a small duration. So, a supercapacitor can be the best option for this purpose.

4.6.7 Portable power supply

Supercapacitors may use as a rechargeable stand-alone power source for some portable electronic devices. Supercapacitor offers quick charging and discharging time with high efficiency. For example, the power light-emitting diode of instant rechargeable safety light. The supercapacitor can be used in a "no-battery power supply for the mini frequency-domain photon migration (FDPM)"—a handheld, noninvasive breast cancer detector based on the principle of FDPM [51,52].

4.6.8 *Adjustable speed drive*

Adjustable speed drives (ASD) are commonly found in industrial applications. Due to the power fluctuations like transient, under-voltage, over-voltage, etc., the drives are not performed satisfactorily and the continuous running process may interrupt. An energy storage system should connect to the ASD to ride through the disturbance at full power. The supercapacitor has the unique advantage of quick response to disturbances.

4.7 Conclusion

The design and construction of a large capacitor bank for bulk energy storage are briefly discussed in this chapter. The complexity of the capacitor storage can be reduced into a simplified equivalent circuit comprising R, L, and C. Methods are also highlighted to find out the capacitor different parameters like capacitance, ESR, and EPR from the equivalent circuit. The various components like capacitor units, switches, transmission lines, and power feed for the construction of a large capacitor bank are discussed here. These components are designed for the proper power flow from the capacitor bank to the connected load. The operation of a typical large energy storage bank of 25 MJ is discussed by taking the equivalent circuit. The merits and demerits of energy storage capacitors are compared with the other energy storage units. The basic need of an energy storage system is to charge as quickly as possible, store maximum energy, and discharge as per the load demand. The charging and discharging operation of the capacitor bank depends on the amount of supply voltage and the duration of charging. Different charging methods like the constant voltage, constant current, constant power, and resonant charging are discussed briefly in this chapter.

The energy storage capacitor bank is commonly used in different fields like power electronics, battery enhancements, memory protection, power quality improvement, portable energy sources, high power actuators, ASDs, hybrid electric vehicles, high power actuators, off-peak energy storage, and military and aerospace applications. The capacitor banks have high power density, and low ESR, are compact and have long-life cycles. It is connected with storage batteries to enhance the life cycle of the battery. The power quality can be enhanced in the power system. The ride-through capability can also be provided by the capacitor bank in the regulated speed drives. It will compensate for the short voltage fluctuation and the large voltage sags during the transient conditions.

References

[1] S. Manish, I.R. Pillai, R. Banerjee, Sustainability analysis of renewables for climate change mitigation, Energy Sustain. Dev. 10 (4) (2006) 25–36.

[2] Y. Lu, Z.A. Khan, M.S. Alvarez-Alvarado, Y. Zhang, Z. Huang, M. Imran, A critical review of sustainable energy policies for the promotion of renewable energy sources, Sustainability 12 (12) (2020) 5078.
[3] R.M. Elavarasan, G.M. Shafiullah, S. Padmanaban, N.M. Kumar, A. Annam, A.M. Vetrichelvan, et al., A comprehensive review on renewable energy development, challenges, and policies of leading Indian states with an international perspective, IEEE Access. 8 (2020) 74432–74457.
[4] IEA, Energy Storage, IEA, Paris, 2021. Available from: https://www.iea.org/reports/energy-storage.
[5] A.Z.A. Shaqsi, K. Sopian, A. Al-Hinai, Review of energy storage services, applications, limitations, and benefits, Energy Rep. (2020).
[6] E. IRENA, Storage and Renewables: Costs and Markets to 2030, Int Renew Energy Agency, Abu Dhabi, 2017.
[7] M. Sterner, I. Stadler (Eds.), Handbook of Energy Storage: Demand, Technologies, Integration, Springer, 2019.
[8] S. Hajiaghasi, A. Salemnia, M. Hamzeh, Hybrid energy storage system for microgrids applications: a review, J. Energy Storage 21 (2019) 543–570.
[9] D.O. Akinyele, R.K. Rayudu, Review of energy storage technologies for sustainable power networks, Sustain. Energy Technol. Assess. 8 (2014) 74–91.
[10] R.F. Ramazanov, B.E. Fridman, K.S. Kharcheva, O.V. Komarov, R.A. Serebrov, Conceptual design of 2 MJ capacitive energy storage, Defence Technol. 14 (5) (2018) 622–627.
[11] Y.S. Jin, Y.B. Kim, J.S. Kim, C. Cho, K.S. Yang, S.H. Kim, et al., A 4.8-MJ pulsed-power system for electromagnetic launcher experiment, IEEE Trans. Plasma Sci. 43 (10) (2015) 3369–3373.
[12] B.E. Fridman, R.S. Enikeev, S.V. Korotkov, N.A. Kovrizhnykh, K.M. Lobanov, R.A. Serebrov, A 0.5-MJ 18-kV module of capacitive energy storage, IEEE Trans. Plasma Sci. 39 (2) (2011) 769–774.
[13] B.E. Fridman, B. Li, V.A. Belyakov, R.S. Enikeev, N.A. Kovrizhnykh, Y.L. Kryukov, et al., A 1-MJ capacitive energy storage, Instrum. Exp. Tech. 54 (5) (2011) 695–698.
[14] M. Khairy, S.A. El-Safty, Nanosized rambutan-like nickel oxides as electrochemical sensor and pseudocapacitor, Sens. Actuators B Chem. 193 (2014) 644–652.
[15] J. Jiang, J. Liu, W. Zhou, J. Zhu, X. Huang, X. Qi, et al., CNT/Ni hybrid nanostructured arrays: synthesis and application as high-performance electrode materials for pseudocapacitors, Energy Environ. Sci. 4 (12) (2011) 5000–5007.
[16] N. Kularatna, Electronic Circuit Design: From Concept to Implementation, CRC Press, 2017.
[17] A.K. Shukla, A. Banerjee, M.K. Ravikumar, A. Jalajakshi, Electrochemicalcapacitors: technical challenges and prognosis for future markets, Electrochim. Acta 84 (2012) 165–173.
[18] P. Sharma, V. Kumar, Current technology of supercapacitors: a review, J. Electron. Mater. (2020) 1–13.
[19] Y. Zhang, H. Feng, X. Wu, L. Wang, A. Zhang, T. Xia, et al., Progress of electrochemical capacitor electrode materials: a review, Int. J. Hydrog. Energy 34 (11) (2009) 4889–4899.
[20] C. Zhong, Y. Deng, W. Hu, J. Qiao, L. Zhang, J. Zhang, A review of electrolyte materials and compositions for electrochemical supercapacitors, Chem. Soc. Rev. 44 (21) (2015) 7484–7539.

[21] C.Q. Yi, J.P. Zou, H.Z. Yang, L.E.N.G. Xian, Recent advances in pseudocapacitor electrode materials: transition metal oxides and nitrides, Trans. Nonferrous Met. Soc. China 28 (10) (2018) 1980–2001.
[22] G. Wang, X. Lu, Y. Ling, T. Zhai, H. Wang, Y. Tong, et al., LiCl/PVA gel electrolyte stabilizes vanadium oxide nanowire electrodes for pseudocapacitors, ACS Nano 6 (11) (2012) 10296–10302.
[23] M.T. Brumbach, T.M. Alam, R.H. Nilson, P.G. Kotula, B.B. McKenzie, R.G. Tissot, et al., Ruthenium oxide–niobium hydroxide composites for pseudocapacitor electrodes, Mater. Chem. Phys. 124 (1) (2010) 359–370.
[24] Castiglia, V., Campagna, N., Di Tommaso, A. O., Miceli, R., Nevoloso, C., Pellitteri, F., et al. Modelling, simulation and characterization of a supercapacitor in automotive applications. IEEE Trans. Indust. Appl., 2022.
[25] Prasad, G. G., Shetty, N., Thakur, S., &Bommegowda, K. B., Supercapacitor technology and its applications: a review, in: IOP Conference Series: Materials Science and Engineering vol. 561, no. 1, IOP Publishing, 2019, October, p. 012105.
[26] Zhai, N. S., Yao, Y. Y., Zhang, D. L., & Xu, D. G., Design and optimization for a supercapacitor application system, in: Proceedings of the International Conference on Power System Technology, IEEE, 2006, October, pp. 1–4.
[27] C.M. Krishna, Managing battery and supercapacitor resources for real-time sporadic workloads, IEEE Embedded Syst. Lett. 3 (1) (2010) 32–36.
[28] L.C. Silva, J.J. Eckert, M.A. Lourenço, F.L. Silva, F.C. Corrêa, F.G. Dedini, Electric vehicle battery-ultracapacitor hybrid energy storage system and drivetrain optimization for a real-world urban driving scenario, J. Braz. Soc. Mech. Sci. Eng. 43 (5) (2021) 1–15.
[29] Jayasinghe, S. D. G., Vilathgamuwa, D. M., &Madawala, U. K., A new method of interfacing battery/supercapacitor energy storage systems for distributed energy sources, in: Proceedings of the Conference Proceedings IPEC, IEEE, 2010, October, pp. 1211–1216.
[30] B.K. Kim, S. Sy, A. Yu, J. Zhang, Electrochemical supercapacitors for energy storage and conversion, Handb. Clean. Energy Syst. (2015) 1–25.
[31] P. Venet, F. Perisse, M.H. El-Husseini, G. Rojat, Realization of a smart electrolytic capacitor circuit, IEEE Ind. Appl. Mag. 8 (1) (2002) 16–20.
[32] Chen, Y. M., Chou, M. W., & Wu, H. C., Electrolytic capacitor failure prediction of LC filter for switching-mode power converters, in: Proceedings of the Fourtieth IAS Annual Meeting. Conference Record of the 2005 Industry Applications Conference, 2005, vol. 2, IEEE, 2005, October, pp. 1464–1469.
[33] Amaral, A. M., & Cardoso, A. M., An ESR meter for high frequencies, in: Proceedings of the International Conference on Power Electronics and Drives Systems, vol. 2, IEEE, 2005, November, pp. 1628–1633.
[34] Amaral, A. M., & Cardoso, A. M., An experimental technique for estimating the ESR and reactance intrinsic values of aluminum electrolytic capacitors, in: Proceedings of the IEEE Instrumentation and Measurement Technology Conference Proceedings. IEEE, 2006, April, pp. 1820–1825.
[35] K.M. Tsang, W.L. Chan, Simple method for measuring the equivalent series inductance and resistance of electrolytic capacitors, IET Power Electron. 3 (4) (2010) 465–471.
[36] H.J. Park, J.C. Kim, S. Kwak, Deep learning-based estimation technique for capacitance and ESR of input capacitors in single-phase DC/AC converters, J. Power Electron. (2021) 1–9.

[37] Ennis, J. B., MacDougall, F. W., Cooper, R. A., & Bates, J., Repetitive pulse application of self-healing high voltage capacitors, in: Proceedings of the Conference Record of the Twenty-Fifth International Power Modulator Symposium, 2002 and 2002 High-Voltage Workshop, IEEE, 2002, June, pp. 634–638.
[38] W.C. Nunnally, M. Kristiansen, M.O. Hagler, Differential Measurement of fast energy discharge capacitor, inductance, and resistance, IEEE Trans. Instrum. Meas. 24 (2) (1975) 112–114.
[39] B.R. Hayworth, How to tell a nanohenry from a microfarad. EID–Electronic, Instrumentation (1972) 36–39.
[40] Giesch, M., Specifications for a low inductance coaxial cable for high current pulses (No. CERN-NPA-Int-61-22).Young, M. P., & Ford, R. D. (1971), 1961.
[41] A Low Inductance 120 kV energy storage module. IEEE Trans. Nucl. Sci. 18(4), 240–242.
[42] T.E. James, Field sketching as applied to electric field design problems, High. Volt. Pulse Technol. (1968).
[43] K. Takaki, High-voltage and pulsed power technologies, Agritech: Innovative Agriculture Using Microwaves and Plasmas, Springer, Singapore, 2022, pp. 25–48.
[44] G. Wang, Z. Lu, Y. Li, L. Li, H. Ji, A. Feteira, et al., Electroceramics for high-energy density capacitors: current status and future perspectives, Chem. Rev. 121 (10) (2021) 6124–6172.
[45] W. Bostick (Ed.), Energy Storage, Compression, and Switching, Springer Science & Business Media, 2013.
[46] B. Maurya, Global overview of large-scale photovoltaic system and its electrical energy storage implementation. computational and experimental methods in, Mech. Eng. (2022) 143–154.
[47] K. Wahab, M. Rahal, R. Achkar, Economic improvement of power factor correction: a case study, J. Power Energy Eng. 9 (6) (2021) 1–11.
[48] J.R. Miller, P. Simon, Electrochemical capacitors for energy management, Sci. Mag. 321 (5889) (2008) 651–652.
[49] A.P. Yu, V. Chabot, J.J. Zhang, Applications of electrochemical supercapacitors, Electrochem. Supercapacit. Energy Storage Deliv.: Fundament. Appl. (2013) 317–334.
[50] B.E. Conway, Electrochemical Supercapacitors: Scientific Fundamentals and Technological Applications, Springer Science & Business Media, 2013.
[51] Y. Yang, Y. Han, W. Jiang, Y. Zhang, Y. Xu, A.M. Ahmed, Application of the supercapacitor for energy storage in China: role and strategy, Appl. Sci. 12 (1) (2022) 354.
[52] A. Lahyani, P. Venet, A. Guermazi, A. Troudi, Battery/supercapacitors combination in uninterruptible power supply (UPS), IEEE Trans. power Electron. 28 (4) (2012) 1509–1522.

Energy management systems for battery electric vehicles

5

Metha Islameka[1,2], *Bentang Arief Budiman*[1,3], *Firman Bagja Juangsa*[3] *and Muhammad Aziz*[4]
[1]National Center for Sustainable Transportation Technology, Bandung, Indonesia, [2]Department of Design Engineering, Politeknik Manufaktur Bandung, Bandung, Indonesia, [3]Faculty of Mechanical and Aerospace Engineering, Institut Teknologi Bandung, Bandung, Indonesia, [4]Institute of Industrial Science, The University of Tokyo, Tokyo, Japan

Chapter Outline

Highlights

- Electric motors and batteries as key components of electric vehicles are explained.
- Configurations of the propulsion system for electric vehicles are discussed, including power flow diagrams.
- Strategies for saving energy in an electric vehicle such as regenerative braking, range extender implementation, and fast charging system are introduced.

Emerging Trends in Energy Storage Systems and Industrial Applications. DOI: https://doi.org/10.1016/B978-0-323-90521-3.00006-5

Nomenclature

A_f	The surface area of the front of the vehicle (m^2)
a	Acceleration or deceleration of the vehicle (m/s^2)
C_d	Aerodynamic drag coefficient
E	The capacity of the battery (Wh)
E_0	The initial energy of the battery (Wh)
E_{int}	Battery energy used (Wh)
F_R	Resistance force that the traction force needs to be overcome (N)
F_{bf}	Braking force on the front axle (N)
F_{bfmax}	Maximum braking force for the front axle (N)
F_{br}	Braking force on the rear axle (N)
F_{brmax}	Maximum braking force for the rear axle (N)
F_i	Inertia force (N)
F_t	Traction force acting on the object (N)
f	Voltage frequency of the supply (Hz)
f_{rl}	Resistance coefficient of wheel rotation
g	Acceleration due to gravity (m/s^2)
h	Distance between center of mass of the vehicle and surface/road (m)
I	Current (A)
K_{bf}	The proportion of total braking force for the front axle
K_{br}	The proportion of total braking force for the rear axle
L	Wheelbase (m)
l_r	Distance between center of mass of the vehicle and rear axle (m)
m	Mass of the vehicle (kg)
n_r	Rotor speed (rpm)
n_s	Synchronous speed (rpm)
P	Number of poles
P_{int}	The internal power of the battery (W)
$P_{required}$	Power requirement of the electric vehicle (W)
R_a	Aerodynamic resistance (N)
R_g	Resistance of the slope of the road (N)
R_{int}	Resistance in the battery (Ω)
R_{rl}	Resistance of the wheel rotation (N)
SOC	Battery state of charge (%)
$SOC_{initial}$	Initial state of charge (%)
s	Position (m)
t	Time (s)
t_i	Time of the drive cycle (s)
V_{oc}	Battery open-circuit voltage (V)
v	Speed (m/s)
W	Weight of the vehicle (N)
x	Time of the new speed (s)
x_1	Time at the first point (s)
x_2	Time at the second point (s)
y	Value of the new speed data (m/s)
y_1	Speed value at the first point (m/s)
y_2	Speed value at the second point (m/s)
z	Braking rate of the vehicle

Greek letters

θ The slope of the road (°)
μ Friction coefficient
ρ The density of air (kg/m^3)

5.1 Introduction

Recent automotive technology development puts a great effort into environmental considerations, including reducing the amount of CO_2 emission due to the increase in CO_2 and other air pollutants produced by internal combustion engine (ICE) vehicles. The amount of CO_2 and the pollutants produced is directly proportional to the number of vehicles using ICE, and continuous air pollution can cause various diseases and even death [1]. The WHO reported that air pollution kills about seven million people worldwide each year [2]. To reduce the amount of air pollution produced by vehicles, governments in many countries have begun to tighten regulations on emissions. Finding the technology for ICE vehicles to significantly reduce pollutants becomes more challenging since the ICE technology has reached saturated conditions. The most widely used techniques to reduce air pollution for ICE include adding advanced filters to the exhaust engine [3], using cleaner energy sources like biofuel [4], or optimizing the vehicle body shape to minimize drag forces significantly [5].

Several alternative energy sources can be utilized to replace gasoline as the primary energy source of vehicles, such as biodiesel, ethanol, hydrogen, natural gas, and electricity [6]. From all these alternatives, electricity as the vehicle energy source is becoming increasingly popular with existing technological developments. A battery electric vehicle (BEV) is a vehicle that uses electrical energy to drive and store this energy in batteries. Some vehicle manufacturers can warrant the battery service for up to eight years or 100,000-mile, which is very promising to get rid of ICE vehicles [7]. The National Renewable Energy Laboratory reports that current batteries can operate up to 12–15 years in temperate climates or 8–12 years in extreme environments [7]. By replacing ICE vehicles with BEVs, the vehicle does not produce as much carbon as ICE with fuel, making it more environmentally friendly. Vehicle noise can also be reduced due to the application of electric motors as powertrain, which is much quieter than ICE. Because it only uses fewer major components, BEV maintenance is simpler and cheaper than vehicles with ICE. Besides the more affordable maintenance, fewer components also expand the vehicle space for other features or larger cargo space.

In recent decades, BEVs have been massively developed and introduced to the public to reduce the distributed CO_2 emission from the vehicle and reduce the overall CO_2 emission by integrating renewable energy utilization. BEVs can also be integrated by grid services to assure power supply in a specific region or to work as an energy carrier during disaster mitigation [8]. These can be realized by regulating electric prices based on time and location during charging and discharging, and thus, the BEV owners can choose to charge based on current prices. The electric price regulation is expected to cut peak electric load, which directly reduces the operating cost of grid services.

The integration of BEV to the grid, known as vehicle-to-grid (V2G), has been developed and demonstrated intensively. Charging and discharging behaviors of the battery in the BEV is entirely controllable, leading to the possibility to utilize BEV as energy storage, in addition to its primary function as a vehicle. Moreover, as both charging and discharging can be performed responsively in a concise time, with a wide distribution and utilization of BEVs across the region, V2G is believed to have a high opportunity to be promising large-scale energy storage in the future. The BEVs parked and connected to the charger connected to the grid can be utilized as a large-scale battery. If there is an increase in demand from the grid, BEVs can be controlled to supply the electricity (discharging). On the other hand, when there is surplus electricity in the grid, due to lower demand or higher supply of renewable energy, these aggregated BEVs can absorb this surplus electricity [9]. Three important points must be fulfilled to facilitate the V2G system: (1) the connection between the electricity grid and BEVs, (2) the communication system which can facilitate bi-directional communication among the parties, including BEVs, owner/driver, and operator, and (3) the objective measurement system which can guarantee the measured amount of electricity which is absorbed or discharged by BEVs [10].

Currently, BEVs are being commercially and intensively manufactured in mass production but cannot compete with the energy generated from using the ICE. The storage density of electric energy in batteries is currently much lower than in gasoline, so those electric vehicles can only carry a limited amount of electrical energy [11]. This problem raises concerns for potential users of BEVs, known as range anxiety, which is the concern of potential users of electric vehicles about the distance that BEVs can cover. Therefore, BEV developers' main goal is to reduce the range of anxiety of potential BEV users by increasing BEVs' energy efficiency by optimizing the battery management system and by developing specific methods to increase the range that the BEV can achieve.

The development of technology for BEVs focuses on the constraints of energy storage in batteries, starting from the battery capacity and lifetime to the lack of charging station infrastructure. One constraint of a BEV is that the higher the number of batteries installed, the heavier the vehicle's mass will be, and consequently, the higher the price required and the greater the energy needed to drive the vehicle. Currently, the price of a battery can be as high as 40%–50% of the total BEV price [12]. Several types of batteries, such as lithium-ion, nickel metal, lead-acid, and ultracapacitors, have been introduced. Lithium-ion types are more widely used and developed due to their easy and cheap access, good high-temperature performance, and the high power-to-weight ratio [13]. However, this liquid electrolyte-based battery has major problems in relation to its low capacity and safety. The battery's energy storage density with current technology is only one-tenth of the energy storage density of gasoline. In terms of the safety issue, the thermal runaway might happen due to a short circuit during operation caused by dendrite generation on the electrode surface or the liquid electrolyte leakage.

Many studies have tried to solve the battery capacity problem, such as solid-state battery usage and fast charging development [14]. The solid-state battery uses all solid electrodes and electrolytes, increasing battery capacity up to four times higher than a liquid electrolyte-based battery. However, mechanical problems such as crack formation during the charging-discharging process may significantly reduce the battery performance [15]. The crack formation causes the battery to have low reliability and endurance.

Other current challenges related to batteries relate to the mass production of bat teries with large capacity and the battery charging time, which is still much longer than refilling gasoline, causing the BEVs to be less competitive than ICE vehicles. Current technology of the fastest charging requires around 20 minutes to charge the battery for approximately 300 km. Even though both solid-state batteries and fast charging technology are still not causing the BEVs to beat ICE vehicles, the exponential research progress in these fields is very promising.

The charging type of electric vehicles can be divided into three levels: slow charging, fast charging, and ultrafast charging. The difference between these three levels is in the duration and charging rate. Slow charging is carried out with rates less than 4 kW, fast charging is from 10 to 20 kW, and ultrafast charging has higher than 20 kW rates and can even reach 400 kW [16]. Slow charging is generally conducted with low-level voltage, such as 100–220 V, using the onboard converter. However, as the onboard converter's capacity is limited, fast and ultrafast charging is conducted using converters installed on the charger side. As it correlates strongly to the charging behavior, battery energy management for BEVs is considered important. It strongly influences the efficient charging and discharging of the battery, sufficient power supply for driving, and long battery lifetime.

However, it is worth noting that even though both solid-state batteries and fast charging technology are still not causing the BEVs to beat ICE vehicles, the exponential research progress in these fields is very promising. Based on the International Energy Agency (IEA), in recent years, the global electric vehicle industry has shown significant developments, and this is coupled with the start of support from individual countries' policies and technological advances [17]. Integrating electric vehicles with power systems is also implemented to make electric vehicles and power systems mutually beneficial. Governments worldwide have even begun to actively encourage the change of vehicles from ICE to BEV by issuing regulations related to vehicle emissions. Considering the fast progress in BEVs research and development, we certainly will find the BEVs as major vehicles in the future.

5.2 Propulsion system in battery electric vehicles

A propulsion system plays a vital role in determining vehicle performance during operation. It consists of a battery package containing battery cells, an electric motor including the inverter, and mechanical transmission. One of the advantages of the propulsion system in BEVs is the use of an electric motor that commonly can generate instant torque at low speed. This behavior enables the propulsion system not to use a complex transmission, and in extreme cases, the only transmission installed is a differential gear. The battery takes up the most space of the BEVs due to its low energy density. However, this can also be beneficial if the battery can be positioned below the floor to reduce the vehicle's center of gravity to be as low as possible. A lower center of gravity can put the vehicle in a very stable state, especially in an accident: the impact load might not cause a rollover of the vehicle. For pure BEVs, the battery is directly connected to plug-in charging components. Fig. 5.1 shows the schematic figure of the simplest propulsion system in BEVs.

The amount of energy needed by BEVs can be obtained by understanding their propulsion system. A BEV propulsion system model that can calculate the flow of mechanical and electrical power can be developed and simulated using driving cycles [18]. The driving cycle is speed data as a function of time which is usually used to measure ICE vehicle emissions and fuel consumption. Now, it can also be used to measure the energy consumption of electric vehicles [19]. Furthermore, the vehicle resistant forces that depend on the road profile, such as rolling resistance, aerodynamic resistance, and grade resistance, can be calculated by drawing a free body diagram [20]. Based on the series of speeds that an electric vehicle passes and the total resistant forces, the rotational speed and traction at the vehicle wheels—and subsequently the required power—can be calculated.

Since most electric vehicles only have one fixed gear [21], the electric motor's power can also be calculated directly. Finally, the state of charge (SOC) value in the battery can be determined by knowing the amount of power that works on the electric motor. This systematic analysis is typical in the automotive industry to investigate vehicle performance, especially for evaluating the performance of a new type of vehicle. The analysis can be done experimentally in which the prototype is tested in an artificial obstacle road or numerically by developing a model and simulation of the propulsion system. The detailed analysis of each of these components is explained in the following sections.

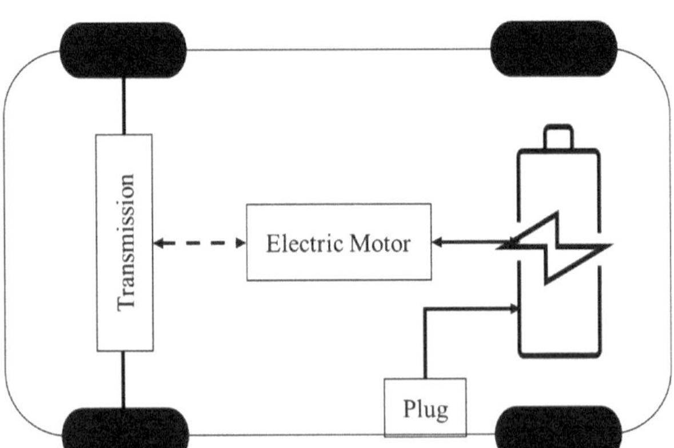

Figure 5.1 Schematic of the primary propulsion system in battery electric vehicles.

5.2.1 Driving cycles

The driving cycle is a series of vehicle speed data against time. For ICE vehicles, the driving cycle is mostly used to test the vehicle's performance and measure its emission. Due to the standard cycle implementation, the testing results can be compared directly with other vehicle data. For electric vehicles, on the other hand, the driving cycle can be used to simulate a vehicle's propulsion system to estimate the amount of energy consumption required by an electric vehicle for a certain speed and time.

During the testing, an ICE vehicle or electric vehicle is placed in a dyno testing machine. Those vehicles can be brand new from the factory or used vehicles with certain mileage services. In the testing machine, the vehicle is driven following the driving cycle standard. Traction force, heat generation from the engine, emission, fuel consumption (for an ICE vehicle), and battery SOC (for an electric vehicle) are usually measured. Furthermore, the use of the driving cycle in the simulation can give an insight into the comparison of different types of electric vehicle propulsion systems with different types of components before the manufacture of electric vehicles. The driving cycle helps to estimate how an electric vehicle consumes energy at various speeds. Certain conditions, such as acceleration, deceleration, traffic conditions, constant speed, and other parameters, can be predicted. Apart from the speed, the time variation also predicts how long an electric vehicle can move with the remaining energy.

The driving cycle has many standards that are usually regulated differently in each country. The difference is usually due to different driving environments, such as traffic, road, and weather condition in each country. For example, Japan has a driving cycle of JC08 chassis dynamometer test cycle for light vehicles (<3500 kg), as shown in Fig. 5.2 [22].

Fully phased in by October 2011, JC08 driving cycles were determined to have 2 phases: cold start and hot start. The percentages of each phase are 25% JC08 cold start and 75% JC08 hot start. The total duration of the JC08 driving cycle is 1204 seconds with a total distance of 8171 km, with the average speed of the JC08 driving cycle of 24.4 km/h, and with a maximum speed of 81.6 km/h.

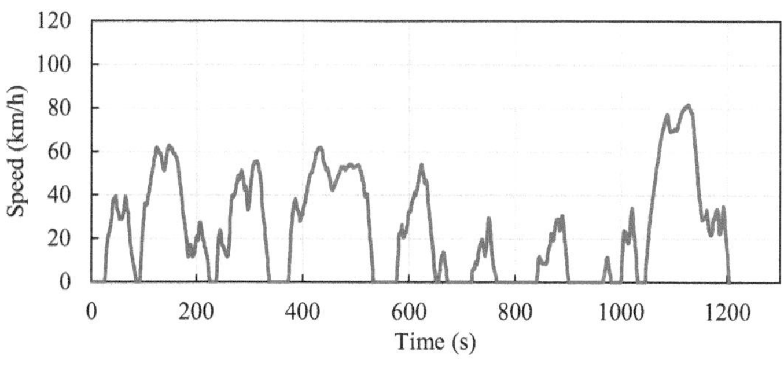

Figure 5.2 Speed part in the JC08 driving cycle.

Another example of a driving cycle standard is FTP75 from the United States for light-duty vehicles, which also has a different curve, as shown in Fig. 5.3 [23,24].

As shown in Fig. 5.3, there are three phases in the FTP75 driving cycle, namely cold start, stabilized, and hot start. However, before starting the third phase, hot start, the engine is stopped for 10 minutes. The total duration of the FTP75 driving cycle is 2474 seconds, of which cold start and hot start have a duration of 505 seconds each. The average speed of the FTP75 driving cycle is 25.6 km/h with a maximum speed of 91.2 km/h.

Another example is Europe's New European Driving Cycle (NEDC) driving cycle standard for light-duty vehicles, which was last updated in 1997 to accommodate the European traffic [25], as shown in Fig. 5.4. There are two phases in the NEDC driving cycle: Phase 1, consisting of four ECE15 cycles, and Phase 2, which is EUDC (Extra Urban Driving Cycle) [26]. ECE15 is the driving cycle that is used to show the urban area, and EUDC is a driving cycle used to denote an extra-urban area. The total duration of the NEDC driving cycle is 1180 seconds, followed by phase 2, which begins after 780 seconds. The average speed of the NEDC driving cycle is 33.3 km/h with a maximum speed of 120 km/h.

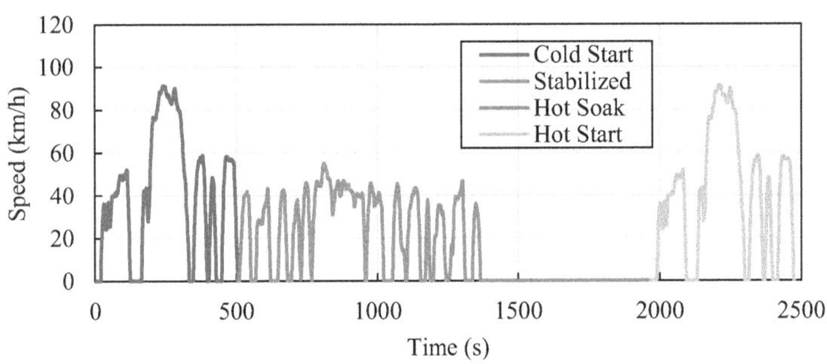

Figure 5.3 Speed part in the FTP75 driving cycle.

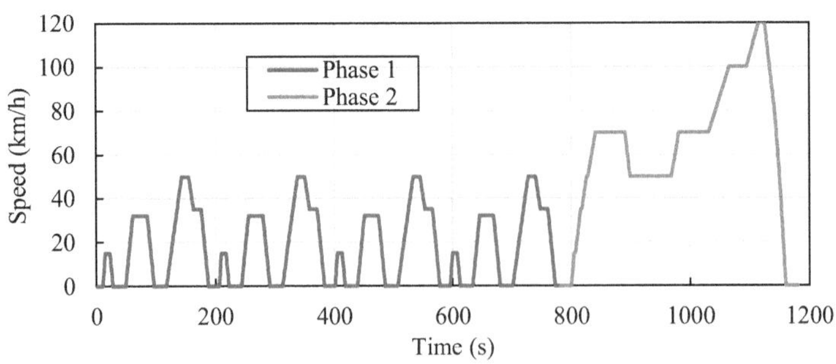

Figure 5.4 Speed part in the new European driving cycle driving cycle.

All the above-mentioned countries and regions have a strong automotive industry, so the national/regional driving cycle standard must be developed. Another main purpose of the driving cycle introduction is to ensure that the vehicle's performance testing can be easily compared and evaluated by other vehicle types. This is very important, especially for new vehicle type designs, which need to be evaluated before they appear in the regional market. However, these regional driving cycle standards might not be beneficial when the vehicles are sold in the global market. The good performance of vehicles in one region does not mean good performance in other regions. To fairly compare the testing results, the European Union has created world-standard driving cycles so that the comparison of the propulsion system for each vehicle can be more uniform, namely the Worldwide Harmonized Light Vehicle Test Procedure (WLTP).

The WLTP driving cycle, which was introduced in 2017, is a series of speed data against time adjusted to different countries' driving cycles [27]. Fig. 5.5 shows a given part of the speed of the WLTP driving cycle.

The difference in the WLTP driving cycle compared to other driving cycles is that the WLTP driving cycle has a class that is adjusted to the type of vehicle and the maximum speed of the vehicle. Table 5.1 shows an explanation of each class to determine the type of WLTP drive cycle to be used [28].

Table 5.1 shows the classes in WLTP based on the specifications of the vehicle. Class selection is made based on the value of the power-to-weight ratio (PWR) of the vehicle to be tested. The unit power is in Watt, while the weight refers to the curb weight, which is the vehicle weight excluding the driver, passenger, and luggage, and has a unit of kg. In practice, although the obtained PWR value exceeds 22, if the value is close to 22, the vehicle can still be classified as Class 1. This classification accommodates a wide range of vehicle types; therefore, the performance testing can be compared fairly.

By applying the driving cycle standard during the dyno test, the vehicle performance may be analyzed, but it cannot capture the actual performance of the vehicle during its services. The vehicle's performance mostly depends on the driving style, traffic density, road profile, vehicle's load, and external conditions such as the weather. For some automotive producer countries, the national driving cycle

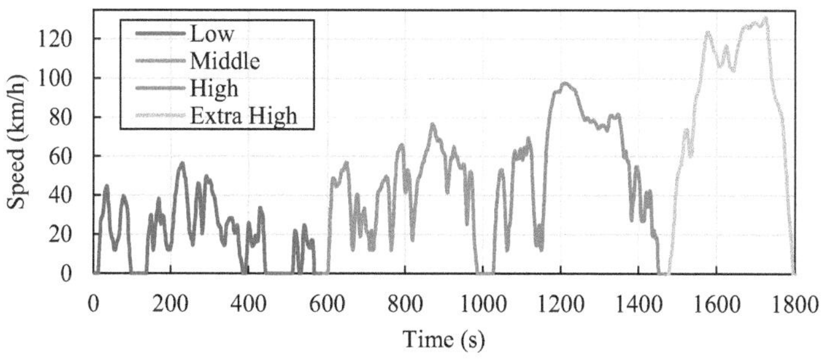

Figure 5.5 Speed part in the worldwide harmonized light vehicle test procedure driving cycle.

Table 5.1 Class for each vehicle specification.

Class	Description
Class 1	PWR less than 22 This includes a low and a medium speed part. If the maximum speed is more than 70 km/h, phase low is repeated after phase medium. If the maximum speed is less than 70 km/h, the phase medium is replaced by a repetition of phase low.
Class 2	PWR between 22 to 34 This includes a low, medium, and high-speed part. If the maximum speed is less than 90 km/h, phase high is replaced by a repetition of phase low.
Class 3	PWR more than 34 This includes a low, a medium, a high, and an extra high-speed part. If the maximum speed is less than 135 km/h, phase high is replaced by a repetition of phase low.

standard may accommodate this issue since the standard is made considering the traffic uniqueness of their countries. However, most countries do not have such standards. Alongside existing current driving cycle standards, both from the automotive producer countries and world standards, driving cycle data can also be retrieved by adjusting needs. This can assist the vehicle designer in estimating the total energy consumption more accurately for vehicles under development.

Numerous research has been done to find a way to retrieve speed versus time data for the driving cycle [29–31]. The most common method is using a global positioning system (GPS)-based application and a gyroscope-based application [32,33]. The advantage of utilizing this driving cycle is that the performance can be predicted better based on the actual condition. For example, an electric bus used for interring region transportation must have different characteristics from an electric bus used for rapid transit. The bus for interring region usually requires high top speed and high energy efficiency during constant speed because most of them use the highway or county roads which have few stopping points. In contrast, bus rapid transit requires frequent stopping during services, so the bus features must accommodate this. The regenerative braking technology becomes more critical here to absorb kinetic energy efficiently and as much as possible when the bus is braking as it approaches the bus station. These characteristics can be investigated by developing the driving cycle depending on their application. Furthermore, the component selection to acquire the most efficient energy consumption for a specific service can be conducted by implementing an optimization method.

From the explanation of the driving cycle, the actual driving cycle is still required to be developed according to the needs. The speed versus time curve data is usually collected several times to obtain more accurate data representing actual traffic conditions. After obtaining the collected driving cycle data, several treatments need to be conducted for the anomaly data that always exist at the time of data collection. Note that these treatments should not alter the data interpretation, which may not represent the actual traffic condition.

5.2.1.1 Comparison of speed values based on calculations and application usage

In a general GPS-based application, the speed of a vehicle can be obtained directly from the application. However, these speed data are calculated values from the position data obtained by the GPS, which works to record every change in position within a certain time; thus, the automatic calculation provided by the application might not be accurate.

Fig. 5.6 shows how different data rates are obtained from applications and manual calculations. It can be seen there are different recorded data. Furthermore, at certain times, the GPS data is failed to be recorded, which might happen due to a lost connection while recording the GPS data. To assure accurate data are obtained for driving cycle development, the speed data from the manual calculation, which is based on recorded position data, are used. By knowing the recorded changes in position and time, the vehicle speed at one time can be calculated using Eq. (5.1).

$$v(t) = \frac{\Delta s(t)}{\Delta t} \tag{5.1}$$

where $v(t)$ is the vehicle speed at a certain time, $\Delta s(t)$ is the vehicle displacement at a certain time, and Δt is the change in the time it takes the vehicle to change positions.

5.2.1.2 Change the speed value to 0 m/s for the actual stop condition

The recorded position using the GPS might produce an error that is caused by connectivity loss or other issues. To ensure the quality of the speed curve is high, the recorded data can be validated with an actual condition, especially when the vehicle

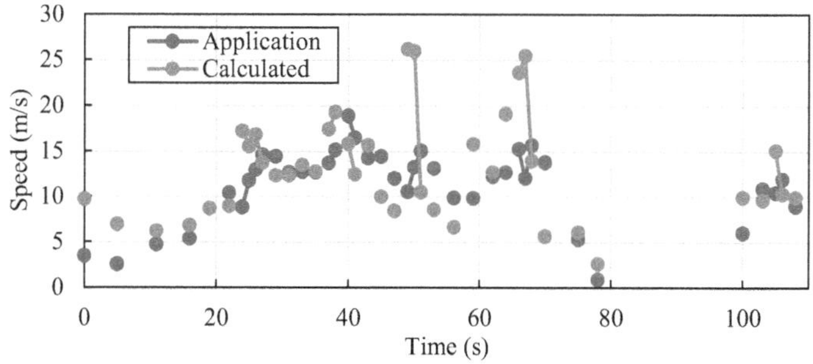

Figure 5.6 Comparison of speed values based on calculations and application usage.

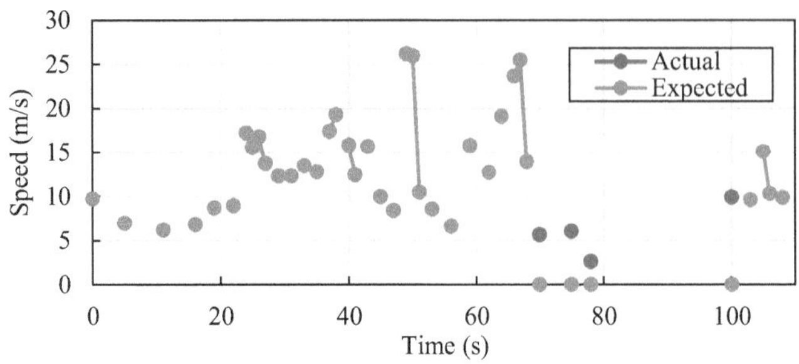

Figure 5.7 Change the speed value to 0 m/s at the stop.

does not move (in rest condition). For example, at the bus stop, it is possible the speed data is not calculated as 0 m/s, which is not valid. To conduct such validation, the driver must record the time and duration of the vehicle when it is not moving. Fig. 5.7 shows the example change in data rate at 70 seconds–100 seconds. This is conducted because when verifying the latitude and longitude at that time, it shows that the vehicle is at the rest condition.

5.2.1.3 Filling in blank data with linear regression

It is quite common during recording the GPS data that some position data are not recorded every second due to connectivity lost. Consequently, the speed calculated based on the position data will be blank, as shown in Fig. 5.7. To overcome this issue, linear regression can be performed to fill in the speed data using Eq. (5.2).

$$\frac{(y - y_1)}{(y_2 - y_1)} = \frac{(x - x_1)}{(x_2 - x_1)} \tag{5.2}$$

where y is the value of the new speed data, y_1 is the speed value at the first point, y_2 is the speed value at the second point, x is the time of the new speed, x_1 is the time at the first point, and x_2 is the time at the second point. Fig. 5.8 shows the changes in the speed data series using linear regression on blank data.

5.2.1.4 Changing data spikes with linear regression

Apart from the blank data, some spike data can also be transformed by linear regression. The Economic Commission for Europe (ECE) establishes safe braking rate regulations for vehicles, and this braking rate is calculated by dividing the acceleration or deceleration of a vehicle by the acceleration due to gravity [34].

$$z(t) = \frac{a(t)}{g} \tag{5.3}$$

where $z(t)$ is the braking rate of the vehicle at a certain time, $a(t)$ is the vehicle acceleration or deceleration at a certain time, and g is the acceleration of gravity.

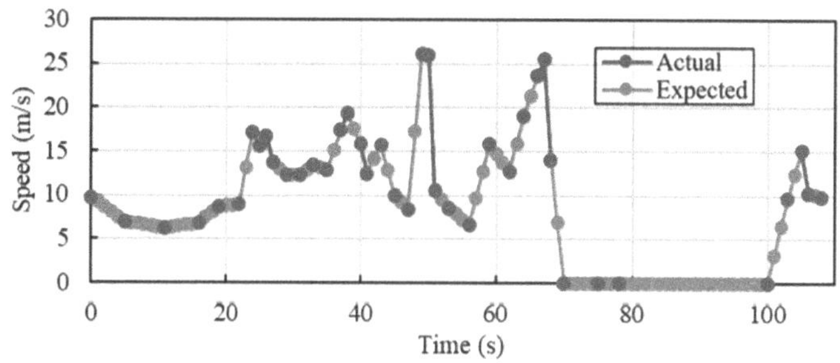

Figure 5.8 Filling in blank data with linear regression.

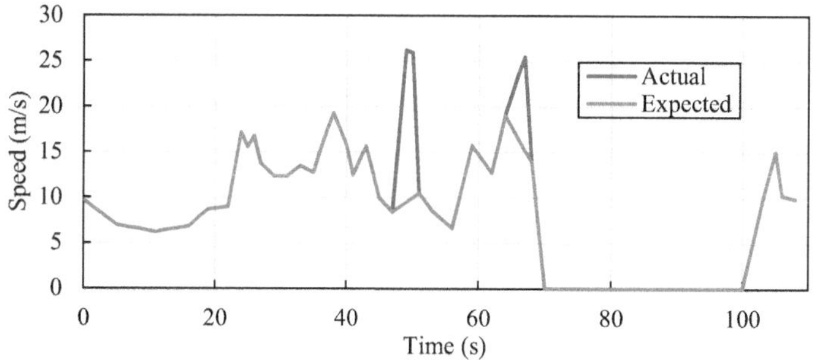

Figure 5.9 Changing data spikes with linear regression.

Several data spikes have acceleration and deceleration values exceeding 1 g. For example, in this study, data improvement continues until the value of vehicle acceleration and deceleration is below 0.8 g. Fig. 5.9 shows the change in the speed data series after linear regression on the data with spike data. This simple data filtering method is still considered to be effective in generating a driving cycle.

5.2.2 Free body diagram of a vehicle

In designing a vehicle, the first step that must be done is to make sure the designed propulsion system can generate traction in drive wheels that oppose any resistant force and provide sufficient acceleration of the vehicle. Fig. 5.10 shows a free body diagram of a moving vehicle with acceleration with all possible forces imposed on the vehicle shown.

Generally, three major resistant forces appear rolling resistance, aerodynamic resistance, and grade resistance. Considering the d'Alembert principle, the fictitious force or inertia force can be added to simplify the dynamic systems into static systems. This fictitious force's value equals the resultant force acting on the center of mass of an accelerating object but in the opposite direction.

$$F_t(t) - F_R(t) = 0 \tag{5.4}$$

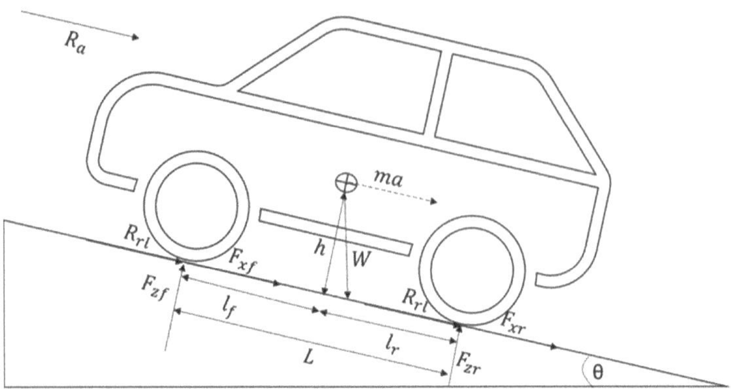

Figure 5.10 Forces acting on the vehicle.

$$F_t(t) = F_i(\mathrm{t}) + R_a(\mathrm{t}) + R_{rl}(\mathrm{t}) + R_g(t) \tag{5.5}$$

where $F_t(t)$ is the traction force acting on the object at a certain time and $F_R(t)$ is the resistance force that needs to be overcome by the traction force at a certain time. The resistance force on the vehicle in the form of $R_a(t)$ is the aerodynamic resistance at a certain time; $R_{rl}(t)$ is the resistance of the wheel rotation at a certain time; $R_g(t)$ is the resistance of the slope of the road at a certain time; and $F_i(t)$ is inertia force at any given time. The fictitious force is then calculated as follows,

$$F_i(t) = m \quad a(t) \tag{5.6}$$

where m is vehicle mass and $a(t)$ is vehicle acceleration at a certain time. The calculation of aerodynamic resistance, rolling resistance, and grade resistance occurring in the vehicle is formulated in Eqs. (5.7–5.9).

$$R_a(\mathrm{t}) = \frac{1}{2}\rho C_d A_f v(t)^2 \tag{5.7}$$

$$R_{rl}(t) = f_{rl}(t)W \tag{5.8}$$

$$R_g(t) = W\sin\theta \tag{5.9}$$

where ρ is the density of air, C_d is the aerodynamic coefficient, A_f is vehicle frontal area, $f_{rl}(t)$ is the wheel rotation resistance coefficient at a certain time, W is vehicle weight, and θ refers to the slope of the road the vehicle is traveling on.

The resistance coefficient of wheel rotation can be determined using Eq. (5.10) [35].

$$f_{rl}(t) = 0.01\left(1 + \frac{v(t)}{147}\right) \tag{5.10}$$

All resistant forces described here always change at any time, depending on road conditions. These three resistant forces have different characteristics. Aerodynamic force behaves exponentially to vehicle speed, whereas grade resistance only depends on an elevated angle. Both forces are expected to be minimized to increase energy consumption efficiency. In contrast, the appearance of rolling resistance, which depends slightly on vehicle speed, is required to prevent the wheels from a slip. The resistance depends on the tire material properties, tire pattern, tire pressure, and road type. Fig. 5.11 shows an example calculation of resistant forces plotted with speed for a city car.

The current BEV development related to these resistant forces is still an interesting topic. In general, the BEV body's shape is designed and optimized to minimize drag forces significantly while simultaneously increasing downforce to avoid slip. Drag force reduction in the BEV body can minimize the energy required to oppose aerodynamic resistance, which will reduce electric energy consumption directly. This reduction will assist the vehicle to operate at a longer range with limited battery capacity. The BEV platform is designed to be as light as possible to reduce inertia forces by using composite materials and monocoque structures [36]. In this case, the downforce plays an important role in assuring the slip in wheels does not occur. The slip may happen especially for BEVs since the electric motor can generate instant torque for low rotational speed. A BEV relatively has higher acceleration than ICE vehicles due to the electric motor application. Furthermore, the electric motor behavior also brings advantages of BEV to oppose grade resistance, especially for a big elevation angle.

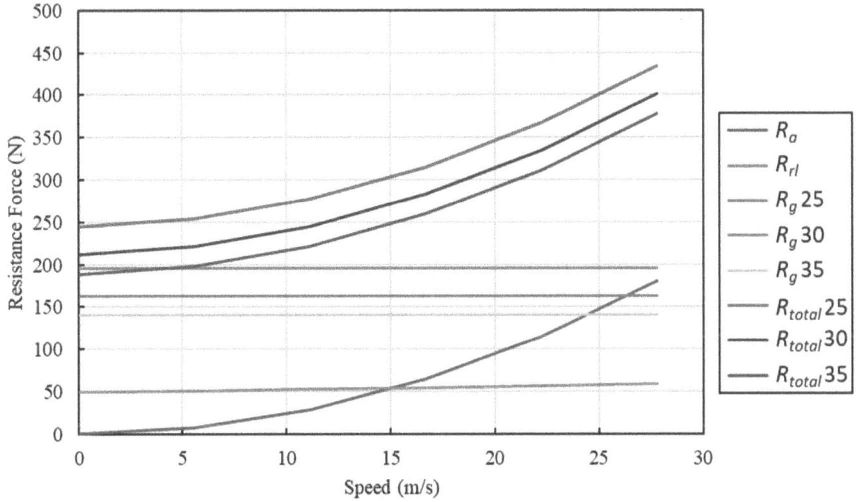

Figure 5.11 Resistant forces versus vehicle speed curves.

5.2.3 Key drivetrain components for battery electric vehicle

There are fewer components in electric vehicles compared to ICE vehicles which can potentially reduce the maintenance costs of the former. The following section describes the key drivetrain components for BEV, including their types and characteristics.

5.2.3.1 Electric motor

An electric motor is a component used to convert electrical energy into kinetic energy and vice versa. The electric motor function can be turned into a generator that converts kinetic energy to electrical energy. As a generator, the kinetic energy is converted to electric energy during the braking process called regenerative braking. The electric energy is then used to charge the battery. A further detailed explanation regarding this process is discussed in the section on regenerative braking.

Electrical energy can be converted to mechanical energy by utilizing the magnetic field's interaction and current winding that generates forces. Compared to ICE, the electric motor has a much higher efficiency in converting the electrical to mechanical energy, reaching 70%–95%. Generally, electric motors have a large torque and power density, and the efficiency of most electric motors reaches the maximum at high torque and low speed. Electric motors are divided into two types based on the current sources, namely alternating current (AC) electric motors and direct current (DC) electric motors. Electric vehicles can use both AC and DC motors.

5.2.3.1.1 DC electric motor

There are two types of DC electric motors: brushed and brushless DC motors. Brushed DC motors are the simplest type of DC motor and are mostly used to illustrate how electric motors work. As in the term, the difference between a brushed DC motor and a brushless DC motor lies in the use of a brush on the electric motor. Fig. 5.12 shows the schematic of the brushed DC electric motor.

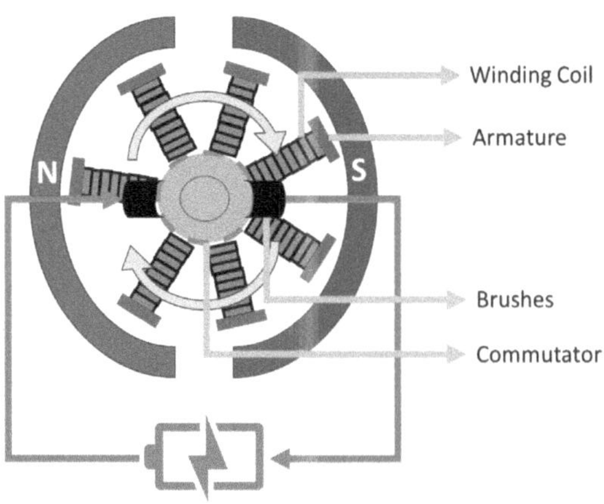

Figure 5.12 Schematic of brushed DC electric motor.

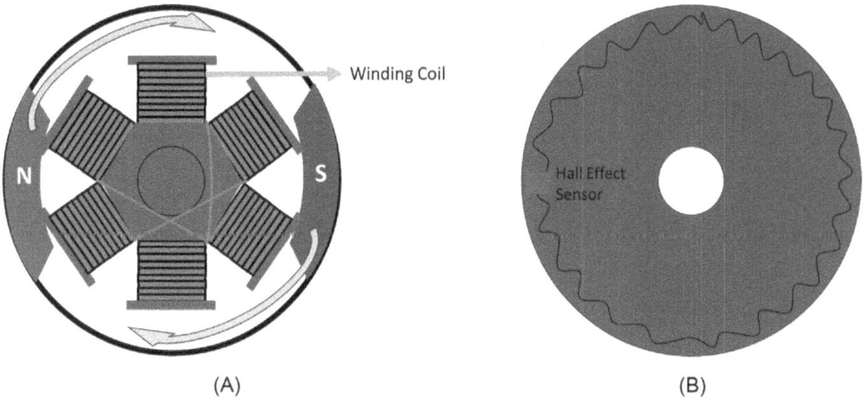

Figure 5.13 Schematic of (A) brushless DC motor and (B) hall effect sensor.

As shown in Fig. 5.12, the brushed electric motor is like other types of electric motors. The main components of an electric motor are the rotor and stator. The rotor is a component that rotates in the electric motor, while the stator is a stationary component in the electric motor. The rotor can rotate due to the current flowing from the source, in this case, the battery, to the brushed one, which is in direct contact with the commutator. Because it consists of several segments, the commutator will periodically reverse its current direction. The reverse current direction in the stator's magnetic field, which is a permanent magnet, produces rotation or torque in the armature, which is part of the rotor. Because there is a direct contact between the brush and commutator on a brushed DC motor which causes a lot of friction to occur and the material to wear down and subsequently causes the need for periodic checks and power losses which are also related to the efficiency of the electric motor itself. Friction between the brush and commutator also produces noise which can cause interference to the surrounding microelectronics. The repeated switching of the current also can cause sparks at the commutator contacts [37]. Because of these limitations, brushless DC motor (BLDC) is more widely used for electric vehicles than brushed DC motors. The schematic of BLDC is in Fig. 5.13.

Fig. 5.13 shows the schematic of the brushless DC electric motor and the Hall effect sensor. BLDC is the type that is widely used for electric vehicles, for instance, by these three-wheeler manufacturers: Goenka Electric Motors, Speego Vehicles, and Kinetic Green [38]. There are two types of BLDC based on the position of the stator and rotor, namely the internal stator and the external stator. The stator in the BLDC is in the form of three pairs of coil winding that will be energized by direct current, while the rotor in BLDC is a permanent magnet. To produce rotation in the rotor, three pairs of coils on the stator are not energized simultaneously. However, an excitation sequence is carried out from each of these coil windings, or usually, every two-coil winding.

In the absence of a brush and commutator, the DC flow is given with the help of an electronic controller, which consists of a position sensor that will indicate which coil needs to be energized. The sensor that is commonly used is the hall effect sensor, and this sensor will detect magnetic information on the rotor. Then, the electronic controller can know which coil winding on the rotor will be energized next so that it follows the right excitation sequence. By varying the current flowing through the stator, the speed of the motor can be varied. Hall effect sensors will help replace the commutator function on a brushed DC motor that helps to switch the coil on and off. One of the advantages of BLDC motors for electric vehicles is the regenerative braking feature, which is discussed in 5.3.1.

5.2.3.1.2 AC electric motor

AC electric motors refer to the sinusoidal-powered electric motors, divided into two groups: synchronous AC electric motors and asynchronous or induction AC electric motors.

The naming of the synchronous AC electric motor is because the rotational speed of the rotor is the same as the speed of the AC electric waveform. Currently available synchronous AC motors consist of a switched reluctance motor, stepper motor, and permanent magnet synchronous motor. One type of synchronous AC often used for electric vehicles is a permanent magnet synchronous motor (PMSM). The stator and rotor components at PMSM are not much different from BLDC. The factor distinguishing them is the way the coil is wound.

Stator winding at BLDC is concentrated, whereas, in PMSM, it is distributed (see Fig. 5.14). Due to differences in the stator winding, the resulting back emf waveform is different. In BLDC, the resulting back emf wave is trapezoidal, while in PMSM, the back emf wave is sinusoidal. Since electric vehicles use batteries to store energy, an inverter is needed to convert DC to AC for PMSM. The advantages of PMSM are that it is small in size, lightweight has high power density, and has

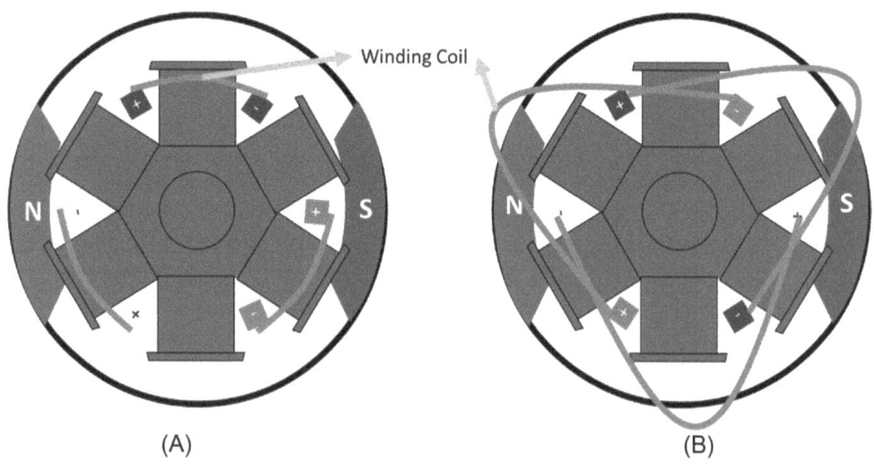

Figure 5.14 The difference in stator winding in (A) brushless DC and (B) permanent magnet synchronous motor.

excellent dynamic response characteristics, which make it very suitable for high-speed applications [39].

One of the characteristics of an asynchronous or induction AC electric motor is that it does not use a permanent magnet. The rotor on an induction motor is usually a squirrel cage, and the stator is a 3-phase coil winding. The way to make the rotor move in an induction motor is to provide AC to the coil on the stator. Because the direction of electrons flows in AC switches back and forth, the stator finally produces a magnetic field rotation. Magnetic field rotation cannot be generated if DC is applied. So, if an induction motor is to be used for electric vehicles, an inverter is needed to convert the DC from the battery into AC. The rotational speed of the magnetic field generated in the stator is known as the synchronous speed. Eq. (5.11) shows how to calculate the synchronous speed.

$$n_s = \frac{120 \times f}{P} \tag{5.11}$$

where n_s is synchronous speed, f is the voltage frequency of the supply, and P is the number of poles.

The rotation of the magnetic field produced by the stator will rotate the rotor. The frequency of rotation of the magnetic field is equal to the difference between the electric frequency applied to the electric motor and the frequency of the rotor. Because the resulting motor motion is obtained from magnetic field induction, an asynchronous AC electric motor is also called an induction motor. Because it depends on the speed difference between the rotor and the rotational speed of the magnetic field, a speed sensor is needed when using this type of electric motor. The difference in the rotational speed of the magnetic field produced by the stator and the rotational speed of the rotor is called slip. If the slip value is "0", the rotor speed is equivalent to the rotating magnetic flux, and when this happens, there is no flux cutting acting on the rotor windings. Therefore, emf will not be generated in the rotor coil so that the motor does not produce torque. This is why this type of motor is referred to as an asynchronous motor because it requires an asynchronous speed between the rotor speed and the rotational speed of the magnetic flux. If the slip is positive, then the electric motor will function as a motor, and when the slip is negative, the electric motor changes its function to a generator. Eq. (5.12) shows how to calculate the slip-on of an electric induction motor.

$$\text{Slip} = \frac{n_s - n_r}{n_s} \tag{5.12}$$

where n_r is rotor speed.

One of the biggest electric vehicles producer, Tesla, decided to change the electric motor type from an induction motor to a permanent magnet motor, and this change increased the efficiency by 4%, and the increased efficiency allows for an increase in the driving distance [40]. The overall drivetrain efficiency of the Tesla EV increased by an average of 10%, while range driving distances increased by about 5%, and the largest increase was city driving distances by 14%. Over 90% of current EV models use neodymium (NdFeB)-based permanent magnet motor as part of the EV drivetrain.

5.2.3.2 Batteries

Battery is a component that stores electrical energy, which becomes a power source for electric vehicles to move. The types of batteries used for electric vehicles are continually evolving. Table 5.2 shows the most frequently used battery types for electric vehicles.

Comparing the parameters shown in Table 5.2, lithium-ion batteries have better specifications than other types. Starting from the life cycle, energy density, charging efficiency, and self-discharging rate. Lead acid has a lower life cycle compared to different types of batteries. Likewise, the energy density value of lead acid is the lowest compared to other types. To increase the battery energy density, current research on lithium batteries is to combine lithium with other chemicals such as lithium nickel manganese cobalt aluminum oxide (NCA) and lithium nickel manganese cobalt oxide. However, for the battery price, the lithium iron phosphate battery type has a lower price of 80 USD/kWh [41].

Solid-state batteries are expected to be the next generation of battery technology due to their superiorities. Solid-state batteries use solid electrolytes, which enables them to increase energy capacity and simultaneously improve safety during operation. The solid phase avoids the electrolyte leakage as well as a short circuit which is still a challenging issue for liquid-based battery applications. However, the solid-state battery remains a problem in durability and large-scale manufacturing. During the charging-discharging process, the electrodes experience expansion and shrinkage, giving the pressure load to the solid electrolyte. This loading can generate crack initiation and propagation, especially in regions near the interface between the electrode and electrolyte. The crack generation can increase internal electric resistance, which may cause performance degradation rapidly. The interface region between electrode and electrolyte also still becomes a big problem due to the gap appearing between two surfaces. This gap can significantly reduce the power density of the battery. Fig. 5.15 shows the difference between liquid-based batteries and solid-state batteries.

Table 5.2 Popular battery specifications based on the material are used for electric vehicles [42,43].

Battery type	Lead-acid battery	Nickel battery	Li-ion battery
Nominal voltage (V)	2	1.2	3.2–3.7
Energy density (Wh/kg)	30–50	50–120	100–270
Power density (W/kg)	180	150–1000	250–680
Charging efficiency (%)	50–95	70–90	80–90
Self-discharge rate (%/month)	5	20–30	3–10
Service life (cycle)	200–300	300–1000	600–3000

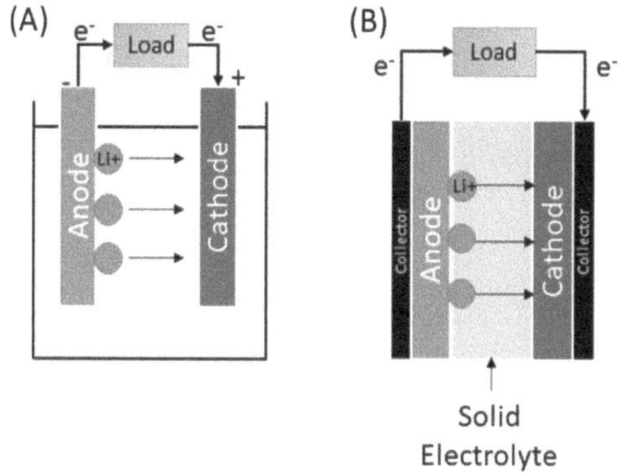

Figure 5.15 Schematic figures of (A) liquid electrolyte battery and (B) solid-state battery.

Some of the common parameters used to assess the condition of a battery are SOC, State of Health (SOH), and Depth of Discharge (DoD) [44]. SOC represents how much energy capacity is left in the battery, usually expressed as a percentage. SOH means the battery's ability to deliver and store electrical energy compared to its ideal conditions, and DoD, which is the opposite of SOC, represents how much energy capacity the battery has used. Same as SOC, DoD is expressed in a percentage. To find the SOC value in the battery, the DoD of the battery can be calculated first by calculating the power used by the electric motor to drive the electric vehicle. Eq. (5.13) shows the calculation of the current amount that enters the battery based on the power from the electric motor, additional load power, voltage, and resistance in the battery. Additional load power may consist of the use of lights, air conditioning, and fans.

$$I(t) = \frac{V_{oc} - \sqrt{V_{oc}^2 - 4R_{int}P_{required}(t)}}{2R_{int}} \quad (5.13)$$

where $I(t)$ is the current entering the battery at a certain time, V_{oc} is an open-circuit voltage of the battery, R_{int} is the internal resistance of the battery, and $P_{required}(t)$ is the total power requirement of the electric vehicle at a certain time.

Eq. (5.14) shows the calculation of the internal power of the battery based on the current value from Eq. (5.13).

$$P_{int}(t) = V_{oc} \bullet I(t) \quad (5.14)$$

where P_{int} is the internal power of the battery at a certain time.

The next step is to determine the initial battery energy. Eq. (5.15) shows the calculation for the unit of the battery capacity is in kWh.

$$E_0 = 3600 \times 1000 \times E \times SOC_{initial} \quad (5.15)$$

where E_0 is the initial battery energy, E is the capacity of the battery, and $SOC_{initial}$ is the amount of the initial SOC.

From the internal power and the initial energy of the battery data, the reduction of the electric vehicle energy during the trip can be calculated by using Eq. (5.16) after multiplying the $P_{int}(t)$ value by -1.

$$E_{int}(t) = E_0 + \int_{t_i}^{t_{i+1}} (-P_{int}(t))dt \tag{5.16}$$

where $E_{int}(t)$ is the amount of battery energy used at a certain time and t_i is the i second time based on the driving cycle.

Eq. (5.17) shows how to find out the amount of SOC based on the amount of energy.

$$SOC(t) = \frac{E_{int}(t)}{E} \times 100 \tag{5.17}$$

where SOC(t) is the state of charge of the battery at a certain time.

A BEV uses battery cells integrated into the battery package to fulfill the electric energy running the electric motor and other auxiliary components. Three common types of battery cells are pouch battery, cylindrical battery, and prismatic battery. The pouch battery has the advantage of being lightweight while the cylindrical battery has a strong structure so that it can also be used to increase the structural rigidity of the BEV, and the prismatic battery has the superiority due to the high energy capacity for each cell, making it easy to control the balancing of the battery.

In integrating battery cells, the battery pack is equipped with a battery management system (BMS) that balances the battery cell capacity during charging and discharging either actively or passively. During operation, each battery cell in the package might have different SOC due to various factors such as temperature difference, internal resistance, or manufacturing and assembling quality. This difference can cause one cell to get overcharged while others are undercharged. In the long run, the SOC difference can decrease the battery's performance and reliability. To avoid this, the BMS rebalances the SOC of the cells. BMS also monitors the temperature and SOH of the battery cells to accurately determine the inspection and repair schedule.

5.2.4 Configurations for the propulsion system

The key components must be selected and assembled properly to be the propulsion systems during the design process to obtain high-performance electric vehicles. The main objective of the design is to find an optimized configuration that can consume electric energy stored in the battery effectively. These include adding ICE in the configuration to assure long-range travel and simultaneously keep the price affordable. The are many configurations that have been proposed to achieve the goal [45]; however, those can be divided into two major configurations of parallel hybrid and series hybrid configurations [45], as shown in Figs. 5.16 and 5.17, respectively.

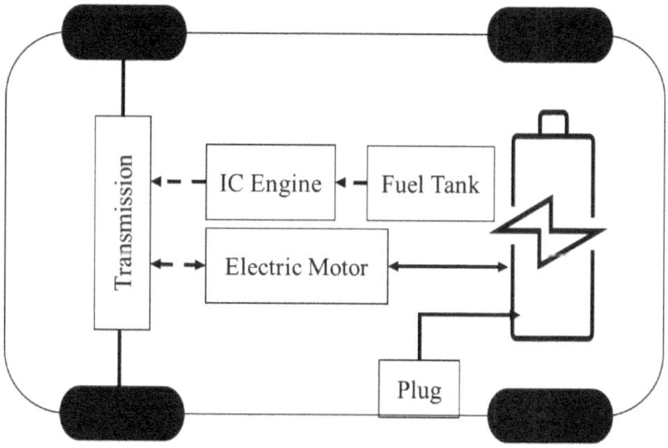

Figure 5.16 Propulsion system in parallel hybrid electric vehicles.

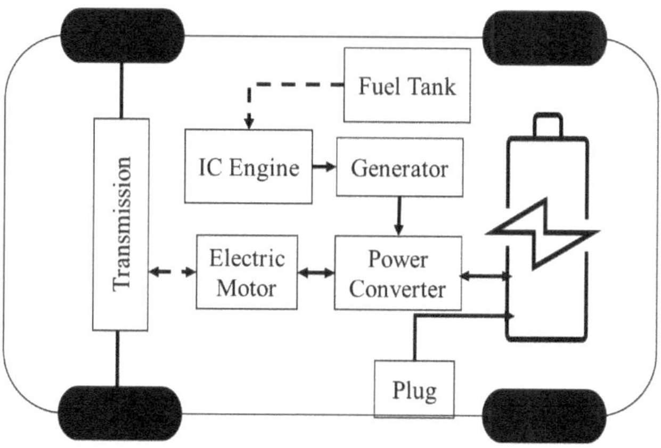

Figure 5.17 Propulsion system in series hybrid electric vehicles.

As shown in Fig. 5.16, a parallel hybrid electric vehicle is a type of electric vehicle that has two propulsion components to operate: an electric motor and an ICE. This vehicle can use either an electric motor and ICE or use both to drive the vehicle. The electric motor gets energy stored in the battery so the vehicles can charge the battery. Because of that, this type of electric vehicle can also be referred to as a plug-in hybrid electric vehicle (PHEV). Furthermore, because it has two powertrains, PHEV control will be more complicated than other types of electric vehicles, but with the two types of powertrains, drivers can manage the energy of the vehicle more freely so that they can use energy optimally [46]. One example of a PHEV is the Kia Niro Plug-In Hybrid.

As shown in Fig. 5.17, the series hybrid electric vehicle is simpler than the parallel hybrid electric vehicle because only the electric motor is used as the propulsion component. ICE is also still used in this type of vehicle, but its function is different from the parallel type. ICE in this type functions to charge the battery while running. Because it uses ICE to increase the range of vehicles, this type of vehicle can also be referred to as an extended-range electric vehicle (EREV). One example of EREV is the BMW i3.

The appearance of ICE in the propulsion system is intended to assure the driver has no range anxiety. Range anxiety is defined as the inconvenience of driving electric vehicles due to the possibility of running out of electric energy before reaching the destination. Note that range anxiety is not only related to mileage today but electric vehicles can also be used for long-range travel. For example, Tesla claims the tesla model S can run up to 650 km for one charging time. However, the range anxiety also occurs due to the slow charging time and the lack of charging infrastructures. Those reasons cause the hybrid configuration to still be relevant. After fast charging technology is found and a massive charging infrastructure is built, the ICE in electric vehicles propulsion system may not be required anymore. The massive deployment of electric vehicles happens proportionally to the increase of battery capacity and the reduction of battery manufacturing costs.

5.3 Strategies for energy management systems in electric vehicle

Apart from several configurations for propulsion systems for electric vehicles, several strategies can also be carried out using key components in electric vehicles, some of which are the regenerative braking system, range extender, and charging strategy. Fig. 5.18 shows an example of the schematic figures of electrical and mechanical power flows in an electric bus for rapid transit.

The mechanical power flows appear from the electric motor to the wheel through the transmission. During regenerative braking implementation, the power flows also occur in opposite directions. Other mechanical power flows occur to pump liquid cooling and hydraulic oil pump in the conventional steering system. In general, the mechanical power source comes from an auxiliary electric motor. The two main components, the battery cell, and the electric motor must be cooled continuously. Both components play an essential role in generating traction for the electric vehicle.

To minimize the range anxiety of the driver, the range extender component is installed. It consists of an ICE engine and a generator that is connected to an onboard charger. The range extender turns on if the SOC of the battery is at a low level, for example, 20% or 10%, depending on the controller set up during installation. This type of electric bus is a series of hybrid vehicles. In the future, when the fast-charging stations are deployed, and the technology for high battery density matures, the range extender can be removed.

Electrical power flows in an electric vehicle appear to be two voltage values: low voltage (LV) and high voltage (HV). The high voltage is used to turn on the electric motor so that it can generate the traction required for electric vehicles,

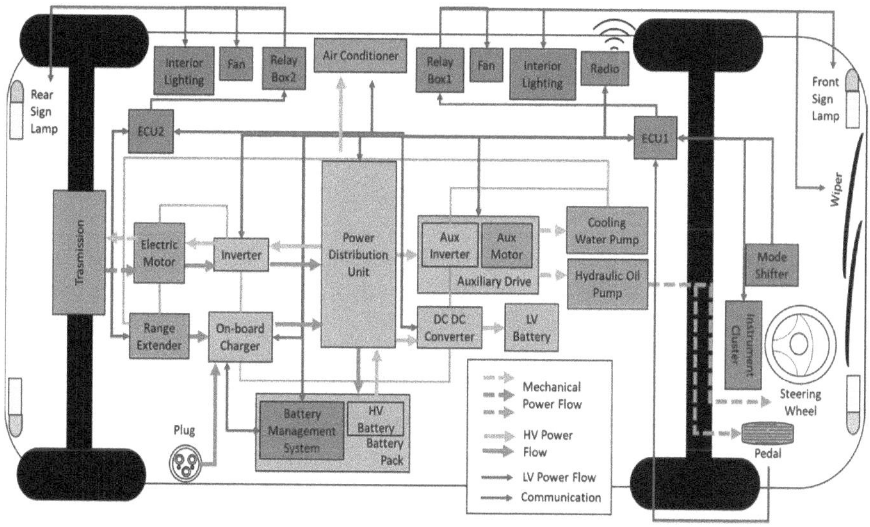

Figure 5.18 Schematic figure of electrical and mechanical power flows in an electric vehicle.

while the low voltage is used to turn on the auxiliary component such as a lamp, fan, air conditioner, and radio. Besides those two wiring systems, the wiring data communication is also required to ensure that the vehicle's performance can be observed on the dashboard by the driver. Furthermore, the data communication wire is also required to give a command to both main and auxiliary inverters in controlling electric motor rotational speed.

Fig. 5.19 shows an example model of an energy management system for electric buses created in Simulink. From this model, the energy consumption required to execute standard driving cycles can be obtained. These results can then be compared with other types of energy management systems. In designing new electric vehicles, an energy management system must be developed at an early stage of the design process. Thus, the system implemented in the vehicle can be assured to work well with high performance. Several key components can be modified to have an optimized design system, such as regenerative braking system, range extender application, and charging strategy.

5.3.1 Regenerative braking

Regenerative braking is a way to harvest electrical energy from the braking mechanism of electric vehicles. Unlike mechanical braking, which converts vehicle motion energy into friction and heat energy, regenerative braking allows an electric motor to operate as a generator capable of absorbing vehicle motion energy into electrical energy when the vehicle is slowing down. Regenerative braking can be designed for all-electric vehicles as it does not require adding any components to the vehicle. The period of recharging the electrical energy to the battery during the regenerative braking process does not increase battery degradation even when the battery temperature is 10°C [47].

However, the regenerative braking function has certain limitations to maximize its use. The amount of regenerative braking is limited by the effective capacity of the powertrain, traction, and stability of the electric vehicles [48]. Regenerative braking is limited by vehicle speed, accelerator pedal and braking signals, and battery voltage [49]. Regenerative braking is not capable of harvesting energy below a certain speed and can harm the overall efficiency of the vehicle [50]. The SOC of the battery should not be more than 90% when using the regenerative braking system to avoid the battery from becoming overcharged. The SOC of the battery also should not be less than 10% to avoid the battery becoming over-discharged. The SOC of the battery should not be too low because the resistance in the battery will be very high [51].

The regenerative braking system works by changing the function of an electric motor into a generator. The braking pedal can be used as a trigger for changing the function of an electric motor into a generator so that during the deceleration, the kinetic energy of the vehicle wheels will be converted into electrical energy by the generator. This electrical energy can then be stored in the battery to increase the energy in the battery and the range of which the electric vehicles can go.

Since the amount of electrical energy that can be recovered is limited to the specifications of the electric motor used, the braking system of an electric vehicle needs to combine regenerative braking and friction braking. The braking control system can be implemented to maximize the use of the regenerative braking system. It is expected to maintain the stability of the braking system and still be able to recover a lot of energy. Strategy for battery saving in BEV with regenerative braking can be done in serial, parallel, or other strategies that have been modified by the researchers. The braking control system in a serial strategy will use regenerative braking up to the

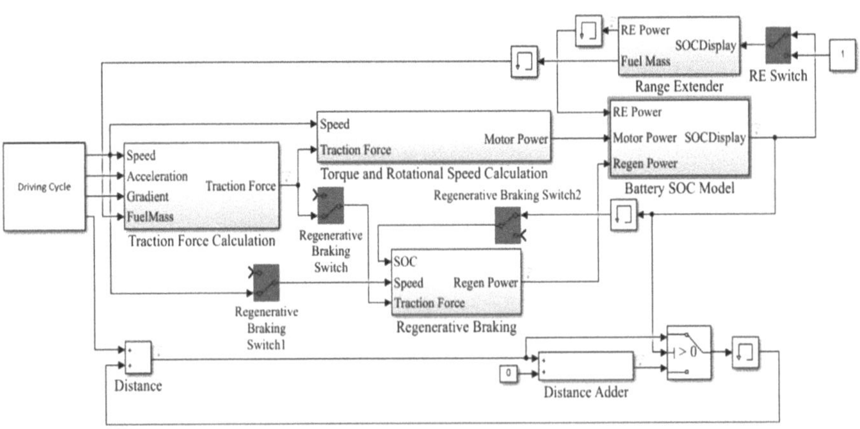

Figure 5.19 Example of Simulink model for analyzing the performance of energy management system in an electric bus.

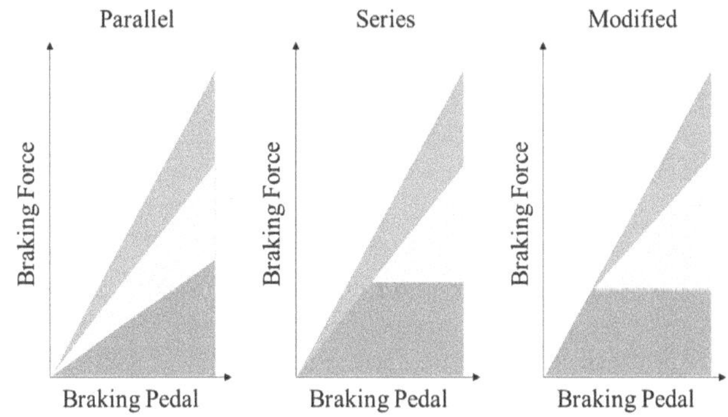

Figure 5.20 The braking control system for a rear-wheeled vehicle consists of the front axle hydraulic braking (orange), the rear axle hydraulic braking (yellow), and the regenerative braking (green) [52–54].

maximum value of torque that the tires and the road surface can receive, and then the rest uses mechanical braking [55]. Meanwhile, the parallel braking control system is braking that uses regenerative braking and mechanical braking simultaneously [56]. Fig. 5.20 shows the differences between the three braking control systems, namely the serial, parallel braking control system, and the modification of the serial system.

As shown in Fig. 5.20, orange indicates the hydraulic braking for the front axle, yellow indicates the hydraulic braking for the rear axle, and green indicates regenerative braking. The difference can be seen in how each strategy divides the braking force based on the size of the braking pedal position. The parallel braking system will evenly divide the braking force required by the vehicle to reduce its speed to the three types of braking available on the electric vehicle. In contrast, the serial braking system will maximize the use of the regenerative braking function together with hydraulic braking on the front wheels. If the required braking force exceeds the maximum regenerative braking capability, the two hydraulic brakes for the front or rear axle are used together.

The modified braking system is an extension of the serial strategy. This is because the serial strategy can recover more energy than the parallel strategy [57]. The difference is that the modified braking will fully use regenerative braking to the maximum limit of the motor capability used. If the value of the required braking force has exceeded the maximum limit of the electric motor's capability, braking will be carried out with the help of hydraulic braking on the front and rear axles.

The braking force distribution is based on the braking ratio of the front and rear axles, whose value is obtained from the maximum value of the braking force of the front and rear axles of the vehicle [20]. Eq. (5.18) shows the braking force ratio for the front and rear axles of the vehicle.

$$\frac{K_{\mathrm{bf}}}{K_{\mathrm{br}}} = \frac{F_{\mathrm{bfmax}}}{F_{\mathrm{brmax}}} \tag{5.18}$$

where K_{bf} and K_{br} are respectively the proportion of the total braking force for the front axle and rear axle and F_{bfmax} and F_{brmax} are the maximum braking force for the front axle and rear axle, respectively.

The determination of the maximum amount of braking force on the front and rear axles is limited by the ideal braking distribution curve. This value limits when the front and rear wheels will lock. If the braking force ratio is above the ideal braking distribution curve, the rear wheels will lock earlier than the front wheels. Conversely, if the operating braking force is below the ideal curve of braking distribution, the front wheels will lock before the rear wheels [50]. The braking forces are also limited by the wheel and road adhesion [58]. Eqs. (5.19) and (5.20) show the ideal braking distribution curve formula [59],

$$F_{bf} = \mu \times \left(\frac{W}{L} \times (l_r + (\mu \times h)) \right) \tag{5.19}$$

$$F_{br} = \frac{1}{2} \left[\frac{W}{h} \sqrt{l_r^2 + \frac{4hL}{W} F_{bf}} - \left(\frac{W \times l_r}{h} + 2F_{bf} \right) \right] \tag{5.20}$$

where F_{bf} and F_{br} are the front axle and rear axle braking force, respectively, μ is the friction coefficient between the road and the wheels, L is the wheelbase, l_r is the distance between the vehicle's center of mass and rear axle, and h is the distance between the center of mass of the vehicle and the surface/road.

Fig. 5.21 shows a schematic figure of the braking control system applicable for an electric vehicle. There is mechanical braking in the front and rear wheel axles which work simultaneously. In general, the mechanical braking system applies a

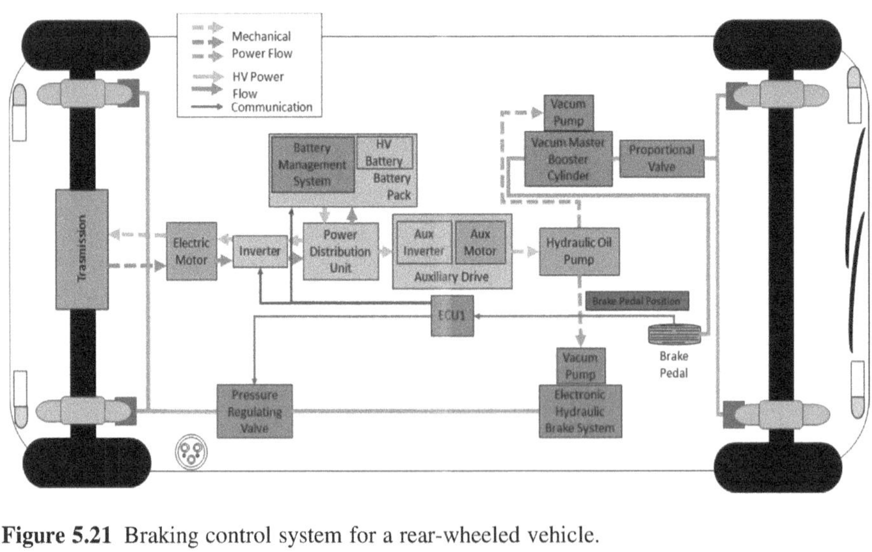

Figure 5.21 Braking control system for a rear-wheeled vehicle.

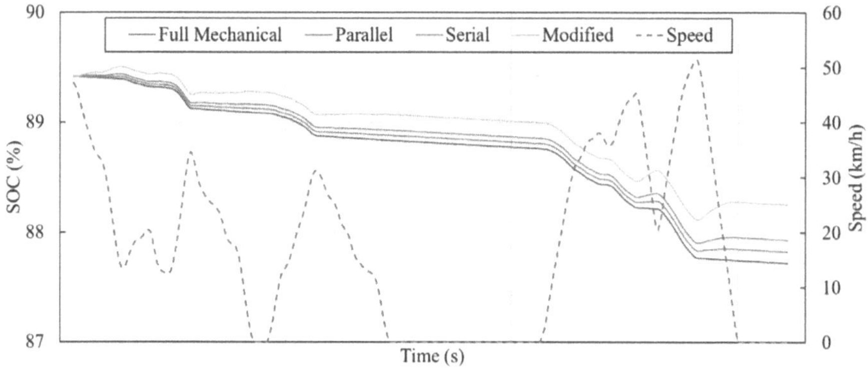

Figure 5.22 Typical SOC curves of an electric vehicle with different types of braking system strategies implementing WLTP driving cycles.

hydraulic system. The regenerative braking system is only implemented on rear wheels since the electric motor connects to them. When the driver pushes the brake pedal, depending on the control system, they will give a command to both mechanical braking and regenerative braking. Regenerative braking is usually used for light braking, whereas sudden braking or emergency braking will be handled by mechanical braking. A typical problem of regenerative braking is a slow response because there is a certain duration to switch the electric motor to be a generator. Furthermore, the ability of the electric motor as a generator to absorb kinetic energy from the rear wheel is limited. For emergency braking, the mechanical braking system must take care of the braking process. Otherwise, an accident might happen.

Fig. 5.22 shows typical SOC curves of electric vehicles implementing WLTP driving cycles. It can be seen that different regenerative braking strategies generate different curves. The energy consumption using a modified regenerative braking strategy is the smallest compared to other braking strategies. The energy consumption difference becomes larger with the increasing operating time of the vehicle. The regenerative braking role becomes more significant when the electric vehicles pass through hilly terrain. In electric buses for rapid transit having a dedicated line, the regenerative braking can be maximized by absorbing as much kinetic energy as possible when the bus brakes near a bus stop.

5.3.2 *Range extender*

A range extender is an additional power unit (APU) that is used to recharge batteries in electric vehicles, thereby increasing the travel range for electric vehicles. As previously described, the range extender is part of a series of hybrid vehicles or EREVs. The range extender consists of an internal combustion engine to rotate a small generator that converts rotational kinetic energy to electric energy. This electric energy is then used to charge the battery. The specification of the internal combustion engine used for the range extender is different from the ICE for conventional vehicles. In a range extender, the rotational speed is more desired to assure the generator can charge batteries, whereas the torsion remains constant and relatively small. The use of a range extender aims to ensure that whenever the SOC of the battery becomes very low, the drivers do not need to be anxious because they can start the engine and charge the battery while still traveling to the destination. The distribution of gas station infrastructure to fill range extender fuel is still prevalent in many regions compared to charging stations.

To maximize the use of the range extender, several studies were conducted, such as the active time of the range extender, the proper battery SOC limit for charging using the range extender, and the mass variation of the range extender used. The mass variation of the range extender is sufficient to affect the efficiency of the use of the range extender in electric vehicles due to the mass factor that affects the energy needed by the vehicle to move.

The variety of energy management systems ranges from the battery types, the braking strategies, and the use of different range extenders and masses. Reksowardojo et al. [18] used this combination of variables for e-trike and showed that, in terms of efficiency and effectiveness, the NCA battery with serial regenerative braking and no RE needed was the best combination. Kusuma et al. [60] also simulate the different energy management systems applied in the electric bus with a range of extender usages and regenerative braking strategies. Results show that the energy management system using a series regenerative braking strategy has the optimum ability to reduce total energy consumption. Apart from using several variables, Song et al. [61] created a control strategy based on the charging management for EREV. By determining the best turn-off time of the range extender, the driver can maximize the available battery energy before it is recharged at the charging station. Lee et al. [62] also developed a controller which could improve average fuel consumption by 9.16% in 10 randomly mixed cycles by balancing the SOC. Fig. 5.23 shows a typical SOC with and without the range extender.

As shown in Fig. 5.23, without a range extender, the electric vehicle might not reach its destination (red line). However, by setting the control to turn on the range extender at 30%–40% SOC, the electric vehicle can complete the intended route (blue line). Furthermore, that additional regenerative braking feature can cause the duration of range extender utilization to become shorter since some kinetic energy is absorbed to charge the battery during the braking process.

5.3.3 Charging system

Charging electric vehicles can be done alone at home as most electric vehicle owners have their chargers or at charging stations. For the latter, the infrastructure for these is still being expanded. In addition to expanding the charging station infrastructure, fast charging is also needed to reduce charging time. Of the 47% stock of light-duty electric vehicles, China owns 80% of fast chargers that can be accessed by the public [17]. EV charging is like charging a cell phone or computer in the sense that the battery in an electric vehicle is connected to a power source that can charge it. Most sockets use AC or alternating current. Batteries typically use DC

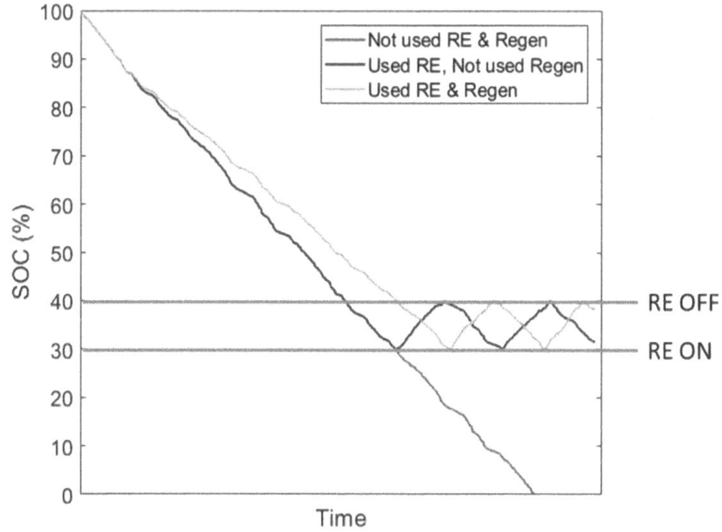

Figure 5.23 Typical state of charge curves of an electric vehicle with and without the range extender.

electricity to charge. Therefore, the part of the charging process involves converting AC to DC.

There are many standards applied for EV charging, including the onboard charger, charging connector, charging cable, off-board charger, communication system, and grid connection. Many EVs are currently adopting a J1772-based connector for charging, except Tesla. However, Tesla makes adapters that allow their vehicles to charge using the J1772 or CHAdeMO connector (see Table 5.3). Table 5.3 shows the different types of charging connectors that vary depending on the country and the type of current.

As previously mentioned, Tesla has a type of charging connector that can be used for AC and DC, and it is for all markets except the EU. EVs are equipped with an onboard charger. However, it is generally used only for AC charging with limited capacity. On the other hand, high-capacity charging can be conducted using an off-board charger, which is usually equipped with a converter, and the electricity is transferred in DC mode. Table 5.4 shows the comparison of different DC charging standards and specifications.

As could be seen in Table 5.4, not all the charging standards are ready for V2G except for CHAdeMO and GB/T, although both CCS1 and CCS2 are also being prepared to be V2G-ready. Furthermore, those different types of charging standards are currently distributed throughout the globe, leading to several practical problems and inconveniences, including the demand for a converter. The unification and/or flexible conversion among those charging standards are expected to be able to facilitate more flexible and convenient charging.

Apart from adding components such as a range extender or using the regenerative braking feature that can be used using an electric motor, a charging strategy for

Table 5.3 Type of charging connector.

Current	Japan	North America	EU.	
AC	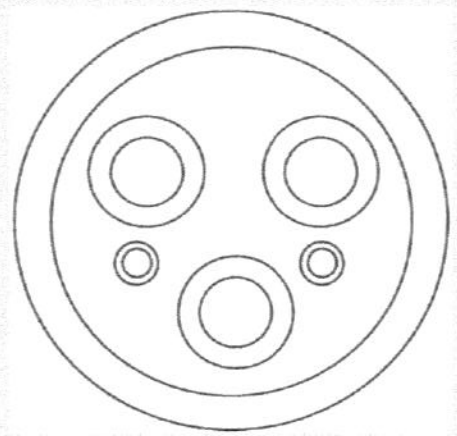 J1772 (Type 1)	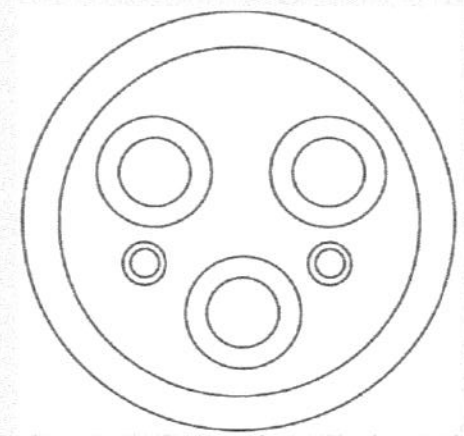 J1772 (Type 1)	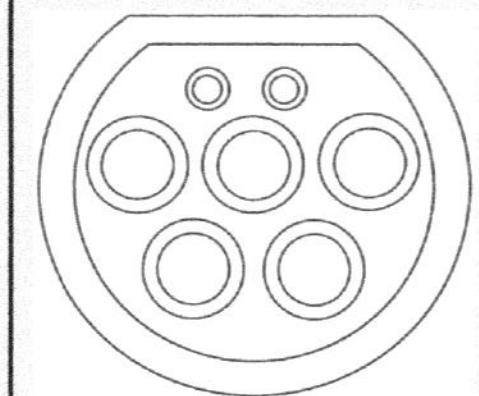 Mennekes (Type 2)	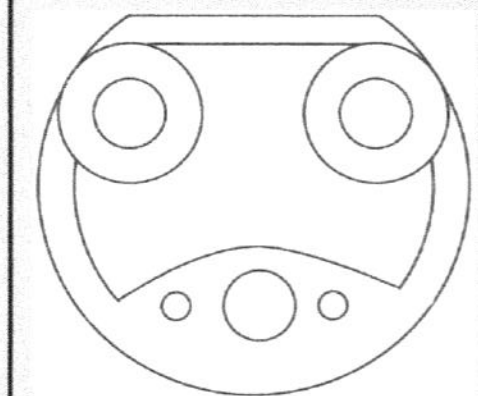 Tesla
DC	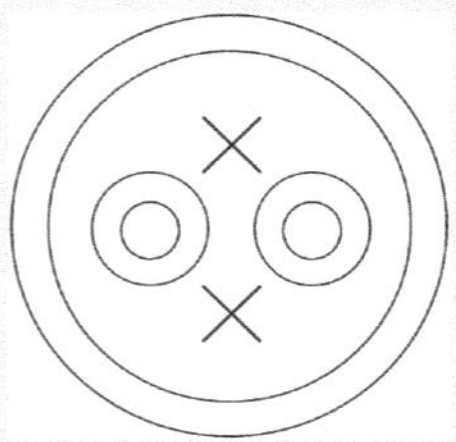 CHAdeMO	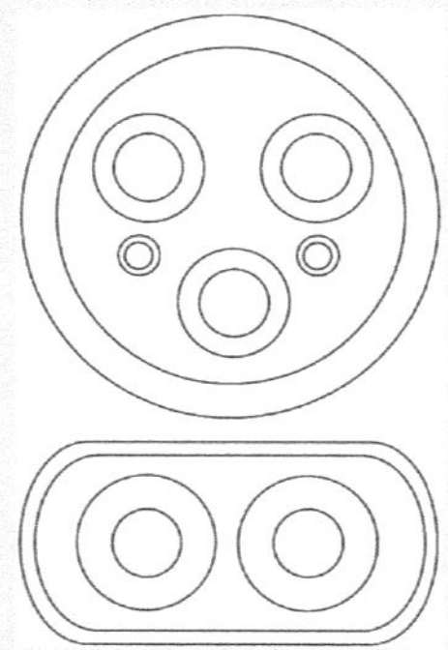 CCS1	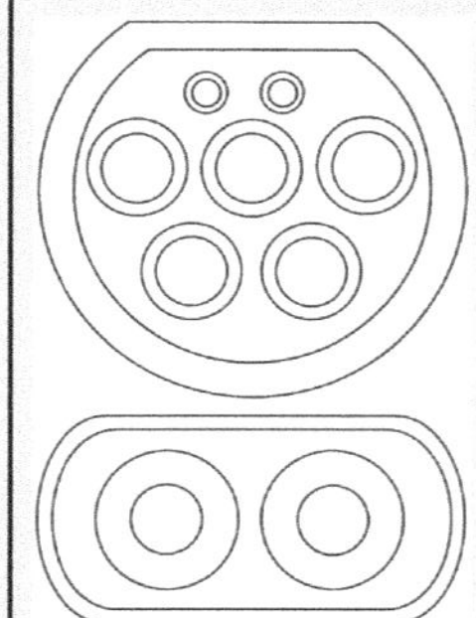 CCS2	

Table 5.4 Different DC charging standards and specifications.

Specification	Maximum power (kW)	Location	Communication protocol	Related standards	V2G readiness	Coupler lock
CCS1	500	US	PLC (ISO)	IEC 61851 SAE J1772	Being developed	Inlet
CCS2	500	EU, South Korea, Australia	PLC (ISO 15118)	IEC 61851	Being developed	Inlet
CHAdeMO	400	Global	CAN (ISO 11898)	IEC 61851 SAE J1772	Ready	Connector
Tesla	250	Global	CAN (SAE J2411)		No	Inlet
GB/T	237	China, India	CAN (SAE J1939)	IEC 61851	Being developed	Connector
New GB/T	900	China	CAN (SAE J1939)	IEC 61851	Ready	Inlet

the battery is also needed to maximize the energy stored in the battery. The charging strategy that can be done is to optimize the current combination as Li et al. [63] did so that it can reduce the charging time by 37%, or to create a control algorithm like Elma [64] did so that it can set a schedule when it is best for an electric vehicle to charge. Elma [64] proposed a control algorithm that could reduce peak demand for charging electric vehicles by 45%.

Charging of EVs also can be conducted wirelessly. This kind of wireless charging has some advantages, including convenient handling during charging, safety due to sparking avoidance, and a possibility for dynamic charging during moving or temporary stopping at the traffic light [65]. However, it suffers disadvantages of limited power density and low power transfer, especially when the distance between the transmitter and receiver increases.

To answer these limitations, there are currently two main wireless technologies available: inductive and capacitive power transfers. The former is conducted by coupling the magnetic resonance between the transmitter and receiver, while the latter is facilitated via the electric field's interaction between two different capacitors. Therefore, EV charging is currently developed based on inductive power transfer, facilitating larger air gaps and higher power transfer rates. The key components of a wireless charging system can be classified into transmitter and receiver sides. The transmitter includes a high-frequency inverter, rectifier, transmitting coil, and compensation network. On the other hand, the receiver consists of the rectifier, compensation network, and receiving coil [66].

5.4 Conclusion

Energy management systems in electric vehicles have been discussed, ranging from propulsion systems to strategies for energy management systems in electric vehicles. Before managing energy in electric vehicles, the amount of energy needed by electric vehicles is calculated. To determine the amount of energy needed by an electric vehicle, information about the driving cycle and the key component specifications of the electric vehicle is needed. Based on the conservation of energy, the power flow from the driving cycles to the battery can be determined so that the energy value needed by an electric vehicle for certain driving cycles can be obtained.

Existing types of strategies for energy management systems are parallel and serial hybrid electric vehicles. Furthermore, the addition of features to electric vehicles is carried out starting from regenerative braking, range extender, and charging strategy.

Acknowledgment

This chapter is funded by Indonesia Endowment Fund for Education (LPDP) under Research and Innovation Program (RISPRO) with contract no. PRJ-85/LPDP/2020.

References

[1] United States Environmental Protection Agency. Sources of Greenhouse Gas Emissions [Internet]. Available from: https://www.epa.gov/ghgemissions/sources-greenhouse-gas-emissions, 2021 [cited 7.12.21].

[2] World Health Organization. Air Pollution [Internet]. Available from: https://www.who.int/health-topics/air-pollution#tab = tab_1, 2021 [cited 28.01.21].

[3] J.D. Herner, S. Hu, W.H. Robertson, T. Huai, M.-C.O. Chang, P. Rieger, et al., Effect of advanced aftertreatment for PM and NO x reduction on heavy-duty diesel engine ultrafine particle emissions, Env. Sci. Technol. 45 (6) (2011) 2413–2419.

[4] V.G. Dovì, F. Friedler, D. Huisingh, J.J. Klemeš, Cleaner energy for sustainable future, J. Clean. Prod. 17 (10) (2009) 889–895.

[5] F. Muyl, L. Dumas, V. Herbert, Hybrid method for aerodynamic shape optimization in automotive industry, Comput. Fluids 33 (5–6) (2004) 849–858.

[6] US Department of Energy. Alternative fuels and advanced vehicles [Internet]. Available from: https://afdc.energy.gov/fuels/, [cited 15.01.21].

[7] US Department of Energy. Electric Vehicle Benefits and Considerations [Internet]. Available from: https://afdc.energy.gov/fuels/electricity_benefits.html, [cited 15.01. 21].

[8] M. Huda, T. Koji, M. Aziz, Techno economic analysis of vehicle to grid (V2G) integration as distributed energy resources in Indonesia power system, Energies. 13 (5) (2020) 1162.

[9] M. Aziz, T. Oda, T. Mitani, Y. Watanabe, T. Kashiwagi, Utilization of electric vehicles and their used batteries for peak-load shifting, Energies. 8 (5) (2015) 3720–3738.

[10] L. Drude, L.C.P. Junior, R. Rüther, Photovoltaics (PV) and electric vehicle-to-grid (V2G) strategies for peak demand reduction in urban regions in Brazil in a smart grid environment, Renew. Energy. 68 (2014) 443–451.

[11] Jimenez TR, Ibarra NC, Ramirez JM., Main motor controllers and regenerative braking of an electric vehicle, in: 2016 IEEE PES Transmission & Distribution Conference and Exposition-Latin America (PES T&D-LA), IEEE, 2016. pp. 1–6.

[12] Bloomberg Quint, Electric Car Batteries: How Cost Effective Are They Really? [Internet]. Available from: https://www.thequint.com/tech-and-auto/change-what-you-can/the-cost-of-electric-vehicle-batteries, 2019 [cited 29.01.21].

[13] EnergySage, How do electric car batteries work? [Internet]. Available from: https://www.energysage.com/electric-vehicles/101/how-do-electric-car-batteries-work/, 2021 [cited 17.01. 21].

[14] Rahardian S, Budiman BA, Sambegoro PL, Nurprasetio IP. Review of solid-state battery technology progress, in: 2019 6th International Conference on Electric Vehicular Technology (ICEVT). IEEE, 2019, pp. 310–315.

[15] Qi Y, Ban C, Harris SJ., A new general paradigm for understanding and preventing Li metal penetration through solid electrolytes. Joule, 2020.

[16] M. Aziz, T. Oda, M. Ito, Battery-assisted charging system for simultaneous charging of electric vehicles, Energy. 100 (2016) 82–90.

[17] International Energy Agency, Global EV Outlook 2020 [Internet]. Available from: https://www.iea.org/reports/global-ev-outlook-2020, 2020 [cited 29.01.21].

[18] I.K. Reksowardojo, R.R. Arya, B.A. Budiman, M. Islameka, S.P. Santosa, P.L. Sambegoro, et al., Energy management system design for good delivery electric trike equipped with different powertrain configurations, World Electr. Veh. J. 11 (4) (2020) 76.

[19] E.G. Giakoumis, Driving and Engine Cycles, Vol. 1, Springer, 2017.

[20] J.Y. Wong, Theory of Ground Vehicles, John Wiley & Sons, 2008.

[21] J. Larminie, J. Lowry, Electric Vehicle Technology Explained, John Wiley & Sons, 2012.
[22] DieselNet. Japanese JC08 Cycle [Internet]. Available from: https://dieselnet.com/standards/cycles/jp_jc08.php, 2007 [cited 12.02.21].
[23] DieselNet. FTP-75 [Internet]. Available from: https://dieselnet.com/standards/cycles/ftp75.php, 2014 [cited 12.02.21].
[24] United States Environmental Protection Agency. EPA Federal Test Procedure (FTP) [Internet]. Available from: https://www.epa.gov/emission-standards-reference-guide/epa-federal-test-procedure-ftp, [cited 12.02. 21].
[25] Shale-Hester T. What are the NEDC fuel economy tests? [Internet]. Available from: https://www.autoexpress.co.uk/car-news/107032/what-are-the-nedc-fuel-economy-tests, 2019 [cited 12.02.21].
[26] DieselNet. ECE 15 + EUDC/NEDC [Internet]. Available from: https://dieselnet.com/standards/cycles/ece_eudc.php, 2013 [cited 12.02.21].
[27] European Automobile Manufacturers' Association. What is WLTP and How Does It Work? [Internet]. Available from: https://www.wltpfacts.eu/what-is-wltp-how-will-it-work/, [cited 12.02.21].
[28] United Nations Economic Commission for Europe. DHC-12th session [Internet]. Available from: https://unece.org/DAM/trans/doc/2012/wp29grpe/WLTP-DHC-12-07e.xls, 2012 [cited 12.02.21].
[29] W.T. Hung, H.Y. Tong, C.P. Lee, K. Ha, L.Y. Pao, Development of a practical driving cycle construction methodology: a case study in Hong Kong, Transp. Res. Part. D. Transp Env. 12 (2) (2007) 115–128.
[30] A. Ashtari, E. Bibeau, S. Shahidinejad, Using large driving record samples and a stochastic approach for real-world driving cycle construction: winnipeg driving cycle, Transp. Sci. 48 (2) (2014) 170–183.
[31] U. Galgamuwa, L. Perera, S. Bandara, Developing a general methodology for driving cycle construction: comparison of various established driving cycles in the world to propose a general approach, J. Transp. Technol. 5 (04) (2015) 191.
[32] Islameka M, Haq IN, Leksono E, Yuliarto B., Energy consumption simulation and analysis of rear-driven electric bus with regenerative braking, in: ICEVT 2019—Proceeding: 6th International Conference on Electric Vehicular Technology 2019, 2019.
[33] Islameka M, Leksono E, Yuliarto B., Modelling of regenerative braking system for electric bus, in: Journal of Physics: Conference Series [Internet]. Available from: https://iopscience.iop.org/article/10.1088/1742-6596/1402/4/044054, 2019.
[34] United Nations, Uniform provisions concerning the approval of vehicles of categories M, N and O with regard to braking [Internet]. Available from: https://unece.org/fileadmin/DAM/trans/main/wp29/wp29regs/2015/R013r8e.pdf, 2014 [cited 22.01.21].
[35] F.L. Mannering, S.S. Washburn, Principles of Highway Engineering and Traffic Analysis, John Wiley & Sons, 2020.
[36] F. Arifurrahman, B.A. Budiman, M. Aziz, On the lightweight structural design for electric road and railway vehicles using fiber reinforced polymer composites—a review, Int. J. Sustain. Transp. Technol. 1 (1) (2018) 21–29.
[37] Arulmozhi MB, Arthi MS, Sivasankari MA. New Approach to a Hybrid Fuzzy-Sliding Mode Control to a Brushless AC Motor Scheme. 2019.
[38] Karthik SH. Types of Motors used in Electric Vehicles [Internet]. Available from: https://circuitdigest.com/article/different-types-of-motors-used-in-electric-vehicles-ev, 2019 [cited 22.01.21].

[39] Wang H, Leng J. Summary on development of permanent magnet synchronous motor, in: 2018 Chinese Control And Decision Conference (CCDC), IEEE, 2018, pp. 689–693.
[40] Bower G. Tesla Turns 4% Motor Efficiency Improvement Into 10% Range Increase [Internet]. Available from: https://insideevs.com/news/348504/tesla-improves-motor-efficiency-increase-range/, 2019 [cited 09.02.21].
[41] Henze V. Battery Pack Prices Cited Below $100/kWh for the First Time in 2020, While Market Average Sits at $137/kWh [Internet]. Available from: https://about.bnef.com/blog/battery-pack-prices-cited-below-100-kwh-for-the-first-time-in-2020-while-market-average-sits-at-137-kwh/, 2020 [cited 07.02.21].
[42] K. Liu, K. Li, Q. Peng, C. Zhang, A brief review on key technologies in the battery management system of electric vehicles, Front. Mech. Eng. 14 (1) (2019) 47–64.
[43] Battery University. Summary Table of Nickel-based Batteries [Internet]. Available from: https://batteryuniversity.com/article/bu-215-summary-table-of-nickel-based-batteries, 2021.
[44] M. Murnane, A. Ghazel, A closer look at state of charge (SOC) and state of health (SOH) estimation techniques for batteries, Analog. Dev 2 (2017) 426–436.
[45] M.R. Wahid, B.A. Budiman, E. Joelianto, M. Aziz, A review on drive train technologies for passenger electric vehicles, Energies. 14 (20) (2021) 6742.
[46] X. Zhang, C. Mi, Vehicle Power Management: Modeling, Control and Optimization, Springer Science & Business Media, 2011.
[47] P. Keil, A. Jossen, Impact of dynamic driving loads and regenerative braking on the aging of lithium-ion batteries in electric vehicles, J. Electrochem. Soc. 164 (13) (2017) A3081.
[48] Halmeaho T, Rahkola P, Pellikka A-P, Ruotsalainen S, Tammi K. Electric city bus energy flow model and its validation by dynamometer test, in: 2015 IEEE Vehicle Power and Propulsion Conference (VPPC), IEEE, 2015, pp. 1–6.
[49] J. Choi, J. Jeong, Y. Park, S.W. Cha, Evaluation of regenerative braking effect for E-REV bus according to characteristic of driving cycle, Int. J. Precis. Eng. Manuf. Technol. 2 (2) (2015) 149–155.
[50] P. Fajri, S. Lee, V.A.K. Prabhala, M. Ferdowsi, Modeling and integration of electric vehicle regenerative and friction braking for motor/dynamometer test bench emulation, IEEE Trans. Veh. Technol. 65 (6) (2015) 4264–4273.
[51] X. Cao, T. Ishikawa, Optimum design of a regenerative braking system for electric vehicles based on fuzzy control strategy, IEEJ Trans. Electr. Electron. Eng. 11 (2016) S186–S187.
[52] M. Islameka, C. Kusuma, B. Budiman, Influence of braking strategies for electric trike energy consumption, Int. J. Sustain. Transp. Technol. [Internet] 3 (1) (2020) 20–25. Available from. Available from: https://unijourn.com/article/5eb61fadabefcd1faa650cff/5ae99ad07348a8567766abe2/4.html.
[53] B. Xiao, H. Lu, H. Wang, J. Ruan, N. Zhang, Enhanced regenerative braking strategies for electric vehicles: dynamic performance and potential analysis, Energies. 10 (11) (2017) 1875.
[54] C. Lv, J. Zhang, Y. Li, Y. Yuan, Mechanism analysis and evaluation methodology of regenerative braking contribution to energy efficiency improvement of electrified vehicles, Energy Convers. Manag. 92 (2015) 469–482.
[55] Tousi SMR, Golpayegani SO, Sharifian E. Anti-lock regenerative braking torque control strategy for electric vehicle, in: 2016 IEEE International Conference on Industrial Technology (ICIT). IEEE, 2016. pp. 1418–1423.

[56] W. Xu, H. Chen, H. Zhao, B. Ren, Torque optimization control for electric vehicles with four in-wheel motors equipped with regenerative braking system, Mechatronics. 57 (2019) 95–108.
[57] C. Qiu, G. Wang, M. Meng, Y. Shen, A novel control strategy of regenerative braking system for electric vehicles under safety critical driving situations, Energy 149 (2018) 329–340.
[58] Guo J, Wang J, Cao B., Study on braking force distribution of electric vehicles, in: 2009 Asia-Pacific Power and Energy Engineering Conference, IEEE, 2009, pp. 1–4.
[59] Z. Zhang, Y. Dong, Y. Han, Dynamic and control of electric vehicle in regenerative braking for driving safety and energy conservation, J. Vib. Eng. Technol. 8 (1) (2020) 179–197.
[60] C.F. Kusuma, B.A. Budiman, I.P. Nurprasetio, M. Islameka, A.H. Masyhur, M. Aziz, et al., Energy management system of electric bus equipped with regenerative braking and range extender, Int. J. Automot. Technol. 22 (6) (2021) 1651–1664.
[61] W. Song, X. Zhang, Y. Tian, L. Xi, A charging management-based intelligent control strategy for extended-range electric vehicles, J. Zhejiang Univ. A. 17 (11) (2016) 903–910.
[62] Lee W, Jeoung H, Park D, Kim N., An adaptive energy management strategy for extended-range electric vehicles based on Pontryagin's minimum principle, in: 2018 IEEE Vehicle Power and Propulsion Conference (VPPC), IEEE, 2018, pp. 1–4.
[63] Y. Li, K. Li, Y. Xie, J. Liu, C. Fu, B. Liu, Optimized charging of lithium-ion battery for electric vehicles: adaptive multistage constant current–constant voltage charging strategy, Renew. Energy. 146 (2020) 2688–2699.
[64] O. Elma, A dynamic charging strategy with hybrid fast charging station for electric vehicles, Energy. 202 (2020) 117680.
[65] P. Machura, Q. Li, A critical review on wireless charging for electric vehicles, Renew. Sustain. Energy Rev. 104 (2019) 209–234.
[66] C. Panchal, S. Stegen, J. Lu, Review of static and dynamic wireless electric vehicle charging system, Eng. Sci. Technol. Int. J. 21 (5) (2018) 922–937.

Electrochemical energy storage part I: development, basic principle and conventional systems

6

Udita Bhattacharjee, Shuvajit Ghosh*, Madhushri Bhar* and Surendra Kumar Martha*
Department of Chemistry, Indian Institute of Technology Hyderabad, Sangareddy, Telangana, India

Chapter Outline

Highlights

- An overview of EES history and prognosis
- An outline of conventional EES batteries, supercapacitors, fuel cells
- A brief insight into the energy storage mechanism, and electro-active materials
- Electrochemical performances of various EES and their applications

* Equal contributing authors.

Emerging Trends in Energy Storage Systems and Industrial Applications. DOI: https://doi.org/10.1016/B978-0-323-90521-3.00001-6

Nomenclature

1D	one dimensional
2D	two dimensional
3D	three dimensional
AC	activated carbon
AFCs	alkaline fuel cells
ATR	autothermal reforming
CIB	chloride ion battery
CNT	carbon nanotube
COF	covalent organic framework
CSR	catalytic steam reforming
CV	cyclic voltammetry
DCB	dual carbon battery
DEC	diethyl carbonate
DIB	dual ion battery
DMC	dimethyl carbonate
DME	dimethoxy ethane
DOL	1,3-dioxolane
ECs	electrochemical capacitors
EDLCs	electric double layer capacitors
EE	electrochemical efficiency
EES	electrochemical energy storage
EIS	electrochemical impedance spectroscopy
EV	electric vehicle
FCs	fuel cells
FEC	fluoroethylene carbonate
FIBs	fluoride ion batteries
HTSSFIB	high-temperature solid state FIB
HUCs	hybrid ultracapacitors
ICE	internal combustion engines
IHP	inner helmholtz plane
IR	internal resistance
KIBs	potassium-ion batteries
LAB	lead acid battery
LCO	lithium cobalt oxide
LHCE	localized high concentrated electrolyte
LIBs	lithium-ion batteries
Li-NCA	lithium nickel cobalt aluminum oxide
Li-NMC	lithium nickel manganese cobalt oxide
LIPON	lithium phosphorous oxynitride
LISICON	lithium superionic conductors
LTO	lithium titanate
MCFC	molten carbonate fuel cell
MF	metal fluoride
MIB	Mg-ion batteries
MOF	metal organic framework
Mr	membrane reforming

NASICON	sodium superionic conductors
Ni-MH	nickel metal hydride
OCV	open circuit voltage
OER	oxygen evolution reaction
OHP	outer helmholtz plane
ORR	oxygen reduction reaction
P2VP	poly-2-vinyl pyridine
PAFCs	phosphoric acid fuel cells
PAN	polyacrylonitrile
PANI	polyaniline
PC	propylene carbonate
PE	polyethylene
PEG	polyethylene glycol
PEMFCs	polymer electrolyte membrane fuel cells
PEO	polyethylene oxide
PMMA	poly methyl methacrylate
POC	poly octamethylene citrate
POX	partial oxidation
PPy	polypyrrole
PVC	polyvinyl chloride
PVDF	polyvinylidene fluoride
PVDF-HFP	polyvinylidene fluoride-*co*-hexafluoropropylene
RTFIB	room temperature FIB
RTILs	room temperature ionic liquids
SEI	solid electrolyte interface
SHE	standard hydrogen electrode
SIBs	sodium-ion batteries
SOFC	solid oxide fuel cell
SSE	solid state electrolyte
SSWB	sodium sea-water battery
STP	standard temperature and pressure
STR	steam reforming
TEGDME	tetraethylene glycol dimethyl ether
VC	vinylene carbonate
VRLAs	valve regulated lead acid batteries
WiBS	water in bisalt
WiS	water in salt
XRD	X-ray diffraction

6.1 General introduction

The storage of energy and its conversion for various applications has been a prevalent question throughout human civilization's growth and development. The depletion of non-renewable sources like coal, oil, and other petroleum products has led to significant research on renewable sources for energy storage and conversion technologies during the last few decades. Various technological advancements resulted in the utilization of naturally abundant sources like solar, wind, and water through their conversion into usable forms of energy. But the availability and conversion feasibility of these sources of energy are primarily dependent on factors like geographical location and climate. Thus, to find a route to make energy accessible for human needs worldwide, energy storage devices involving various fundamental forms of energy like mechanical, thermal, electrochemical, magnetic, etc., have been explored. Mechanical energy storage devices store energy in the form of potential or kinetic energy. Prominent mechanical energy storage technologies include hydroelectric storage (potential energy of water), compressed air storage (kinetic energy), and flywheel storage (kinetic energy of the highly accelerated rotor wheel). The mechanical energy storage devices have storage capacities up to megawatts, but the infrastructural and location issues limit their applicability. Thermal energy storage devices store energy in the form of heat by heating water like a medium, but similar infrastructural shortcomings are associated with these devices. Superconducting magnetic energy storage utilizes the energy in the magnetism of a coil. Extreme temperature conditions are required to generate this form of energy, thus limiting its utility [1].

Electrochemical energy storage systems (EES) utilize the energy stored in the redox chemical bond through storage and conversion for various applications. The phenomenon of EES can be categorized into two broad ways: One is a voltaic cell in which the energy released in the redox reaction spontaneously is used to generate electricity, and the other is an electrolytic cell in which the electrical energy is used to undergo the redox reactions at the electrode. Due to the lesser extremity in the chemical redox reaction conditions, EES devices can be utilized for energy consumption in portable electronic devices, grids, start-stop applications, electric vehicles, etc., in a wide span of geographical locations and climatic conditions. Thus, EES is being explored as a sustainable form of energy. Based on the reversibility of the chemical reaction involved in the electrochemical cell, they are classified as rechargeable (reversible redox reaction) and non-rechargeable (irreversible redox reaction) systems. Theoretically, all electrochemical reactions appear to be reversible, but due to the necessity of some extreme conditions, not all redox reactions can be reversed. In the case of a rechargeable energy storage device, during the process of discharging, the cell acts like a voltaic cell, and during charge, the cell functions like that an electrolytic one [2]. The most prominent types of EES systems which have dominated the market since the late 20th century are batteries, fuel cells, and supercapacitors. Batteries are the combination of electrochemical cells in which the energy is stored in the bulk of the active electrode material, that is the electrodes are the redox-active material as well as the charge transfer medium. In contrast, in fuel cells, the redox-active material (fuel, i.e., hydrogen, oxygen, methanol, etc.) is supplied from outside, and the electrodes act as a facilitator for the redox reaction to take place. Supercapacitors, on the other hand, store energy on the electrode surface either by double-layer capacitive phenomenon or by redox-active surface reactions. Various types of these energy storage systems and their hybrid combinations, which have been explored, are represented in Scheme 6.1

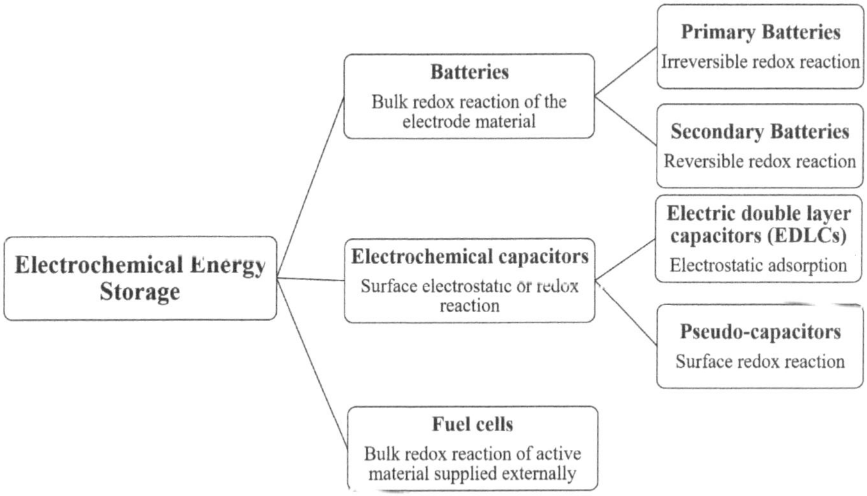

Scheme 6.1 Schematic representation of the broad categories of electrochemical energy storage.

and also discussed in the later section of this chapter. This chapter mostly emphasizes the following aspects:

- Periodical development of different EES systems from their inception till the present day
- A brief overview of the various EES systems from mechanistic as well as materials perspective, their shortcomings, and their future possibilities for improvement in performance
- Future of the EES systems in terms of newer electrochemical mechanisms as well as sustainability
- Comparative outlook on the performance of the various EES systems in terms of energy and power outputs.

6.2 History

The development of the three prominent EES systems through their electrochemistry and electrode-electrolyte components is elaborated in Table 6.1. through a timeline since the inception of the first known EES device [3–10].

6.3 Thermodynamics and basic principle

The energy involved in the bond breaking and bond making of redox-active chemical compounds is utilized in these systems. In the case of batteries and fuel cells, the maximum energy that can be generated or stored by the system in an open circuit condition under standard temperature and pressure (STP) is dependent on the

Table 6.1 Timeline of the development of prominent electrochemical energy storage.

Time scale	Batteries	Fuel cells	Electrochemical capacitors
1800–50	1800: Volta pile 1836: Daniel cell	1800s: Electrolysis of water 1838: First hydrogen fuel cell (gas battery)	–
1850–1900	1859: Lead-acid battery 1866: Leclanche cell 1899: Nickel cadmium battery	1859: Mond gas-powered fuel cell 1893: Working principle of fuel cell 1896: First fuel cell with practical utility 1900: Zirconia fuel cell	1879: Description of double-layer formation at solid/electrolyte interface
1900–50	1901: Nickel iron battery 1942: Mercury oxide dry cell 1950s: Urry battery	Development by William W. Jacques and Emily Boar 1933: First practical fuel cell with air and hydrogen	–
1950–2000	1960s: Ni-MH batteries 1966: Na-S batteries 1960s: ZEBRA batteries 1970s: Primary Lithium batteries 1991: Commercialization of LIB	1955: Ion exchange polymer-based fuel cell 1959: KOH electrolyte-based fuel cell 1960: molten salt fuel cell 1961: Phosphoric acid fuel cell 1990: Direct methanol-based fuel cell	1957: Patented first carbon-based electrolytic capacitor 1969: Non-aqueous electrolyte-based supercapacitor 1978: Mercury based supercapacitor used for computer backup 1991: Pseudocapacitive RuO_2
2000–present	2000s: Na ion batteries, Dual carbon batteries, Potassium, magnesium, zinc ion batteries	Combination of various fuels and electrolytes and their commercialization	Various double-layer capacitive and pseudocapacitive materials and hybrid systems were studied and commercialized

individual redox potentials of the reaction involved in the electrode ($E^0{}_{cathode/anode}$ vs SHE). The energy stored in pseudo-capacitors is also partially dependent on the potential redox reaction involved, but the only difference is that it depends solely on the surface redox activities. The cell voltage $E^0{}_{cell}$ can be represented by Eq. (6.1)

$$E^0_{cell} = E^0_{cathode} - E^0_{anode} \quad (6.1)$$

For a reaction, $aA + bB \leftrightarrow cC + dD$ which is not in STP condition, the E_{cell} is given by Eq. (6.2)

$$E_{cell} = E^0_{cell} - \frac{RT}{nF} \frac{[C]^c[D]^d}{[A]^a[B]^b} \quad (6.2)$$

where n, number of electrons involved, F (Faraday's constant) = 96,500 C, R, Universal gas constant, T is the absolute temperature in Kelvin, $[A]$, $[B]$, $[C]$, $[D]$ are the concentration of the reactive species and a,b,c,d are their stoichiometric amounts, respectively. The maximum energy that can be attained is given by the Gibbs free energy change (ΔG), which can be expressed in terms of cell potential represented in Eq. (6.3)

$$\Delta G = -nFE_{cell} \quad (6.3)$$

Similar to all practical thermodynamic processes, energy conversion is also not completely elastic. Thus, on connecting to the external load, losses in the open-circuit voltage occur due to various processes, including polarization in the electrode/ electrolyte interface and internal resistances. Loss due to polarization may be of two types activation polarization and concentration polarization. Activation polarization develops due to the activation of the electrode/electrolyte interface towards the redox reaction. In contrast, concentration polarization occurs due to the difference in concentration of electrode active material at the electrode-electrolyte interface and in bulk. Another factor that incurs a loss in the energy conversion process while being connected to an external load is the internal impedance of the cell, which is a combined effect of the resistance of the electrolyte, active material—current collector interface, electronic resistance of the active material, etc. This loss is also called ohmic polarization or IR loss. Thus, the available cell potential considering all the losses can be represented by Eq. (6.4)

$$E = E^0 - \left[(\eta_{ct})_a + (\eta_c)_a\right] - \left[(\eta_{ct})_c + (\eta_c)_c\right] - \text{IR} \quad (6.4)$$

where E^0 = electromotive force or open-circuit voltage of the cell, $(\eta_{ct})_a$, $(\eta_{ct})_c$ = activation polarization at anode and cathode, $(\eta_c)_a$, $(\eta_c)_c$ = concentration polarization at anode and cathode, i = operating current of the cell, R = internal

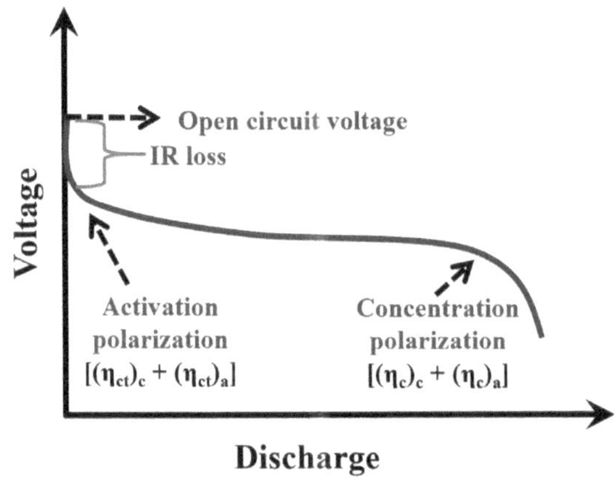

Figure 6.1 A schematic representation of the polarization in a redox-active electrode [2].

resistance of the cell. The maximum theoretically available energy can be obtained only when the cell operates close to open circuit potential, that is the polarization effects and IR drop are negligible (Fig. 6.1).

On the other hand, in the case of double-layer capacitors, the maximum energy stored is dependent on the voltage applied across the electrodes so that maximum charge storage takes place at the electrode/ electrolyte interface. In the case of a double-layer capacitor, when a suitable potential is applied across the electrodes, the electrodes become oppositely charged, and hence oppositely charged ions from the electrolyte are adsorbed onto the electrodes through electrostatic force, and an electrical double layer is formed at the interface. According to the Helmholtz model, the charges are concentrated on both sides of the electrode surface. The closest distance for the approach of non-specifically adsorbed ions near the charged surface is the outer Helmholtz plane, and the closest approach of specifically adsorbed ions to the electrode surface is termed as inner Helmholtz layer (Fig. 6.2)

Each electrode with opposite charges on both surfaces acts as a physical capacitor separated by a distance in molecular dimensions. The anode and cathode together signify two physical capacitors connected in series, and their equivalent capacitance can be represented by Eq. (6.5).

$$\frac{1}{C_{\text{equivalent}}} = \frac{1}{C_{\text{anode}}} + \frac{1}{C_{\text{cathode}}} \tag{6.5}$$

Considering symmetrical capacitors, $C_{\text{anode}} = C_{\text{cathode}} = C$, the equation becomes, $C_{\text{equivalent}} = \frac{1}{2}C$

The maximum number of ions adsorbed onto the electrodes depends on factors like the number of surface-active sites on the electrode, the dependence of electrostatic force in the electrode to the distance from the electrode, and the maximum

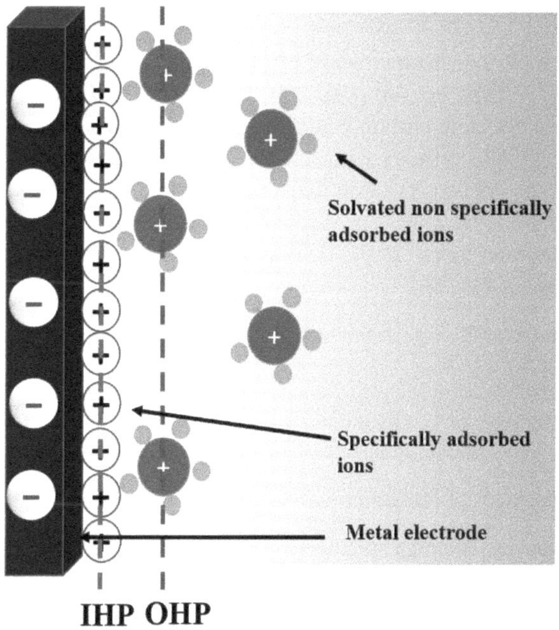

Figure 6.2 Electric double-layer formation (Helmholtz model) [2].

voltage achievable with respect to the electrolyte. The capacitive electrodes have a significantly less response time for changing potentials than batteries and fuel cells as there is no involvement of bulk electrode redox reactions. The maximum energy stored in an electrochemical capacitor is represented in Eq. (6.6).

$$E = \frac{1}{2}CV^2 \tag{6.6}$$

E is the maximum energy stored in a capacitor, C is the capacitance in Farad and V is the voltage applied across the electrodes. The maximum energy is not the deliverable energy due to internal resistance losses when connected to a load similar to batteries and fuel cells [2,11,12].

6.4 Batteries

Batteries are a combination of electrochemical cells connected in series or parallel. Each electrochemical cell stores energy in the form of chemical bonds and the conversion of energy to usable forms occurs through redox reactions. The redox reactions can be reversible or irreversible depending on the spontaneity of the reaction. Thus, batteries can be broadly classified into primary and secondary batteries and are briefly discussed in the following section.

6.4.1 Primary batteries

Primary batteries are non-rechargeable cells that are discarded after one-time use. These batteries are existing for over 100 years. The energy density varies from <50 Wh kg^{-1} for the zinc-carbon batteries to >400 Wh kg^{-1} for lithium batteries. The main advantages of primary batteries are (1) good shelf life (e.g., 3–10 years), (2) simple and easy to use, (3) low maintenance, and (4) good energy and power density. These batteries are used in various applications based on the configuration of different shapes and sizes [2,13,14]. Depending on the cell components, primary batteries can be classified as

1. Liquid cathode batteries (e.g., Lithium/sulfur dioxide (Li/SO_2), Lithium/thionyl chloride ($Li/SOCl_2$) batteries)
2. Solid-state electrolyte batteries (e.g., lithium-iodine cells)
3. Solid cathode batteries (e.g., lithium manganese Oxide (Li/MnO_2) batteries, lithium polycarbon fluoride cells, lithium iron, zinc-manganese di-oxide, and alkaline batteries).

The major advantages of lithium primary batteries over other conventional batteries are

1. Lithium metal possesses a high specific capacity (i.e., 3860 mAh g^{-1}) and specific energy (200 Wh kg^{-1}) with a nominal voltage up to 3.7 V depending on the cathode. In contrast, in most of the other primary batteries, the voltage is limited to 1.5 V.
2. Lithium batteries can work over a wide temperature range from 70°C to −40°C, even up to 150°C or as low as −80°C.
3. Lithium batteries have a superior shelf life of up to 10 years at room temperature and one year at 70°C.
4. Typically, a flat discharge curve for most lithium batteries.

6.4.1.1 Liquid cathode batteries

6.4.1.1.1 Lithium/sulfur dioxide (Li/SO_2) batteries

Li/SO_2 cells consist of the metallic lithium anode and liquid cathode with sulfur dioxide as active material on porous carbon current collector. Acetonitrile with dissolved lithium bromide and sulfur dioxide is used as the electrolyte. The internal cell pressure is very high due to liquid SO_2 (vapor pressure 4×10^5 Pa at 20°C). During discharge, Li reacts with SO_2 to form lithium dithionite, and the cell pressure is reduced. The electrochemical cell reaction is as follows.

$$2Li + 2SO_2 \rightarrow Li_2S_2O_4 \tag{6.7}$$

The nominal voltage of these cells is 2.95 V, and the specific energy is 260 Wh kg^{-1}. The cell can operate between the temperature range of −55°C and 70°C. It has a low self-discharge rate (<2% per year) and long shelf life so that it can be stored for up to 10 years at room temperature. These batteries are mainly used for defense applications and emergency power units in aircraft.

6.4.1.1.2 Lithium/thionyl chloride ($Li/SOCl_2$) batteries

$Li/SOCl_2$ cells comprise Li-metal anode and liquid thionyl chloride ($SOCl_2$) as cathode active material on the porous carbon electrode. $SOCl_2$-$LiAlCl_4$ is used as the electrolyte. The electrochemical cell reaction is shown below.

$$4Li + 2SOCl_2 \rightarrow 4LiCl + S + SO_2 \quad (6.8)$$

The cells have a high operating voltage of 3.5 V. Due to the formation of SO_2 during discharge, the internal cell pressure is increased. Moreover, elemental sulfur blocks the cathode and retards the cell performance. In the presence of BrCl additives, the formation of hazardous sulfur or intermediates is minimized. The discharged product that is LiCl, forms a passivating layer on the anode surface that improves the shelf life and lowers the self-discharge rate. These batteries are used in medical equipment, military GPS systems, sensors, etc.

Further, lithium/sulfuryl Chloride (Li/SO_2Cl_2) batteries with an operating voltage of 3.7 V follow similar mechanisms Eq. (6.8) to $Li/SOCl_2$ cells.

$$2Li + SO_2Cl_2 \rightarrow 2LiCl + SO_2 \quad (6.9)$$

Sulfur is not produced during discharge. Thus, this system eliminates hazardous side reactions and is comparatively safer than $Li/SOCl_2$ cells. However, these cells are not commercialized due to storage limitations.

6.4.1.2 Solid-state electrolyte batteries

6.4.1.2.1 Lithium-iodine cells

These cells were first invented in 1972 for application in cardiac pacemakers. The cathode is a solid organic compound of iodine in poly-2-vinyl pyridine (P2VP), and the anode is Li-metal. The cathode acts as an electronic conductor. I_2 typically contributes 90% of the weight in the cathode. The electrolyte is lithium iodide which conducts Li-ion only. The electrochemical cell reaction is,

$$2Li + nI_2 \rightarrow (n-1)I_2 + 2LiI \quad (6.10)$$

The nominal cell voltage is 2.8 V. Lithium-iodine batteries possess capacities of 120–250 mAh with a cell life of 10 years. The advantage of these cells for pacemaker applications is that the voltage falls slowly at the end of life, and it can be easily detected in routine follow-up for replacement.

Other solid-state electrolyte lithium primary batteries include lithium bromine trifluoride Battery (Li-BrF_3), Li/LiI-Al_2O_3/PbI_2 cells, carbon tetramethylammonium penta-iodide batteries $Li/LiI/Me_4NI_5 + C$, etc.

6.4.1.3 Solid cathode batteries

6.4.1.3.1 Lithium manganese oxide (Li-MnO_2) batteries

Lithium Manganese Oxide batteries are among the most common commercial primary batteries and grab 80% of the lithium battery market. The cells consist of Li-metal as the anode, heat-treated MnO_2 as the cathode, and $LiClO_4$ in propylene carbonate and dimethoxyethane organic solvent as the electrolyte. During lithiation, Mn^{IV} is reduced to Mn^{III} due to the formation of an intercalation compound, Li_xMnO_2. The overall cell reaction is as follows.

$$xLi + Mn^{IV}O_2 \rightarrow Li_xMn^{III}O_2 \quad (6.11)$$

The nominal cell voltage is 3 V, and the energy density is 250 Wh kg^{-1}. In 1976, Sanyo first manufactured Li-MnO_2 cells. Different configured Li-MnO_2 cells are available in the market depending on size, weight, and power source applicability. These cells are mainly used in watches, cameras, calculators, memory backups, RFID tags, etc. Duracell and Panasonic produce various cells, from CR1025 coin cells (30 mAh) to cylindrical cells (2.4 Ah).

6.4.1.3.2 Lithium polycarbon fluoride cells

The lithium–carbon monofluoride system is one of the most successful early lithium primary cells developed in 1973. It has a very high theoretical specific energy density of 2600 Wh kg^{-1} and an open circuit voltage range of 2.8–3.3 V with respect to cathode composition. The cell uses Li-metal as anode and polycarbon monofluoride $(CF)_n$ as a cathode, an intermediate formed by the fluorination of carbon. The electrolyte used is $LiAsF_6$ in butyrolactone or $LiBr_4$ in propylene carbonate (PC) and dimethoxyethane. A generalized cell reaction is as follows.

$$nxLi + (CF_x)_n \rightarrow nC + nxLiF \quad (6.12)$$

Touhara et al. proposed the reaction mechanism in graphite fluorides with $x > 0.5$ that (1) initially fluorine is intercalated in each alternate layer as –CFCF–, later (2) the stacking sequence changes into –CCFCCF–. The hexagonal symmetry is maintained in both the two $(CF_{0.5})_n$ stages. The major drawback is the decrease in electrical conductivity to 10^{-14} S cm^{-1} due to the C–F bond relative to graphite (10^4 S cm^{-1}).

Several companies manufacture these cells. Nippon Steel Co. produces rechargeable cells with carbon fiber electrodes. 5 Ah cylindrical cells are used in military applications. These cells are also used in artificial cardiac pacemakers, aerospace applications, clocks, and memory backup batteries. Ultralife produced Li//CF_x-MnO_2 hybrid cells that can be stored for 15 years and can deliver 14% more capacity than Li/MnO_2 cells.

6.4.1.3.3 Lithium iron batteries

These batteries can be used as an alternative to alkaline batteries as the working potential is 1.5 V. It possesses a flat discharge profile in comparison to Zn-MnO_2

cells. Li-FeS_2 cells consist of a Li-metal anode, iron disulfide as the cathode, and LiI dissolved in propylene carbonate or dioxolane or dimethoxyethane as an electrolyte. The cell reaction is as follows,

$$Li + FeS_2 \rightarrow Fe + Li_2S \tag{6.13}$$

A two-step voltage plateau is observed in the voltage versus discharge profile. At 1.7 V, $Fe_{1-x}S$ $(0<x<0.2)$ and $Li_{2+x}Fe_{1-x}S_2$ $(0<x<0.33)$ are formed and at 1.5 V, Li_2S and Fe are formed as final products. The electrochemical performance strongly depends on the cathode particle size. The lithiation process in FeS_2 is enhanced by reducing the particle size from 10 μm to 100 nm with an increase in surface area per unit volume. AA-size Li-FeS_2 batteries are commercially available in markets. These cells are used in high current drain applications like digital cameras, cellular telephones, and motor-driven toys.

6.4.1.3.4 Zinc-manganese di-oxide batteries (Leclanché, Zinc Chloride Cell, and alkaline batteries)

Zinc carbon batteries are primary "dry cells" that have existed for over 100 years. It consists of zinc as an anode (i.e., the cell container) and carbon blended manganese dioxide as a cathode. The cathode material is placed around a carbon collector rod that collects current from MnO_2. An aqueous paste of NH_4Cl is used as the electrolyte. During discharge, zinc is oxidized to $ZnCl_2$ by Cl- charge carrier, and MnO_2 is reduced to Mn_2O_3. The overall cell reaction is as follows

$$Zn + 2MnO_2 + 2NH_4Cl + H_2O \rightarrow ZnCl_2 + Mn_2O_3 + 2NH_4OH \tag{6.14}$$

The nominal cell voltage is 1.5 V. These cells have a short life span. When the cell is in use, the zinc container gradually corrodes with the reaction of NH_4Cl, and leakage occurs. Zinc carbon batteries are used in transistor radios, toys, flashlights, remote controls, etc. Instead of NH_4Cl, $ZnCl_2$ paste is often used in heavy-duty type *zinc chloride cells* for industrial applications. These cells have comparatively low leakage issues. The overall cell reaction is

$$Zn + 2MnO_2 + H_2O \rightarrow Zn(OH)_2 + Mn_2O_3 \tag{6.15}$$

Alkaline batteries have a higher shelf life, and energy density compared to zinc-carbon and zinc chloride cells. Potassium hydroxide is used as an electrolyte in alkaline batteries in place of NH_4Cl or $ZnCl_2$, and the electrode components are unchanged. As the amount of OH− consumed is equal to OH− production, the electrolyte remains the same. The electrochemical cell reactions involved in alkaline batteries are,

$$Zn + 2MnO_2 \rightarrow ZnO + Mn_2O_3 \tag{6.16}$$

These cells are commercially available since the 1950s in standardized cylindrical forms for household battery applications. Alkaline batteries also have leakage problems due to the side reactions and evolution of H_2 gases that rupture the case sealing. KOH absorbs CO_2 from the air through the leak in the steel shell and forms potassium carbonate that comes out from the battery leading to cell death.

6.4.1.3.5 Zinc–mercuric oxide battery

Zinc–mercuric oxide battery is a primary alkaline battery with mercuric oxide as cathode and zinc as an anode in an alkaline electrolyte (e.g., KOH or NaOH). Graphite is mixed with mercuric oxide to increase the conductivity and reduce the formation of large Hg particles. These cells have higher energy density and long shelf life (e.g., up to 10 years) than zinc-manganese di-oxide batteries. The nominal cell voltage is 1.35 V. The redox electrochemical cell reaction is shown below, where Zn is oxidized to ZnO, and HgO is reduced to elemental Hg.

$$Zn + HgO \rightarrow ZnO + Hg \tag{6.17}$$

During World War II, the cells were used in military applications, for example, walkie-talkies and metal detectors. In post-war, button and cylindrical cells are manufactured for hearing aids, transistor radios, electric wristwatches, etc. Duracell is one of the manufacturers of zinc–mercuric oxide batteries. These cells are banned in many countries because of toxic mercury. The potential alternative of zinc–mercuric oxide battery is manufactured as alkaline batteries and silver oxide batteries.

6.4.1.3.6 Zinc-silver oxide battery

Zinc-silver oxide battery is another type of primary alkaline battery consisting of zinc as an anode, silver oxide (Ag_2O) blended with graphite as a cathode, and an alkaline electrolyte. During the cells' operation, Ag_2O is reduced to elemental Ag, and Zn is oxidized to Zn(II). The overall cell reaction is,

$$Zn + Ag_2O \rightarrow ZnO + 2Ag\downarrow \tag{6.18}$$

The specific energy density of the cell is 130 Wh kg^{-1}, and the open-circuit voltage is 1.6 V. These cells are commonly used in hearing aids, cameras, and watches as button cells. It delivers a wide range of capacities from 5 to 250 mAh. These batteries can tolerate heavy current load pulses and have military applications as torpedo batteries. Though divalent silver oxide (AgO) exhibits a 40% higher capacity than monovalent silver oxide (Ag_2O) for the same size battery, it is unstable in the alkaline medium as AgO decomposes into Ag_2O with the evolution of O_2 gas. Further, divalent silver oxide shows a two-step discharge profile which is not desirable for tight regulated electronic applications. Zinc-silver oxide battery was used in Apollo lunar spacecraft.

6.4.2 *Secondary batteries*

Secondary batteries are widely used rechargeable batteries for over 100 years. These batteries can be recharged after discharge without disposing of them. Different types of rechargeable systems are briefly discussed in the following section.

6.4.2.1 Lead-acid batteries

Lead-acid batteries (LAB) are the first discovered rechargeable batteries and still occupy almost 50% of the commercially available batteries' market share. The LAB market continues to expand to date because of its newer applications, including electric vehicles. LABs comprise a spongy lead (Pb) anode, lead dioxide (PbO_2) cathode, and ~4.5 M H_2SO_4 as the electrolyte. The separator is polymeric (Polyethylene (PE), polyvinyl chloride (PVC)), or absorbent glass mat or gelled electrolyte with PE/PVC separator. During charge, the lead dioxide (PbO_2) at the cathode and lead (Pb) at the anode are converted to lead sulfate ($PbSO_4$) through the consumption of H_2SO_4 in the electrolyte. During discharge, the processes are reversed at both electrodes. The cell voltage at open circuit conditions is 2.06 V [2].

$$\mathrm{PbO_2}(s) + \mathrm{H_2SO_4}(aq) + 2H^+(aq) + 2e^- \underset{\leftarrow \text{Charge}}{\overset{\text{Discharge}}{\longrightarrow}} \mathrm{PbSO_4}(s) + 2\mathrm{H_2O}(l), E^0 = 1.7\ V \text{ versus SHE} \tag{6.19}$$

$$\mathrm{Pb}(s) + \mathrm{H_2SO_4}(aq) \underset{\leftarrow \text{Charge}}{\overset{\text{Discharge}}{\longrightarrow}} \mathrm{PbSO_4}(s) + 2H^+(aq) + 2e^-, E^0 = -0.36\ \mathrm{V} \text{ versus SHE} \tag{6.20}$$

$$\mathrm{PbO_2}(s) + \mathrm{Pb}(s) + 2\mathrm{H_2SO_4} \underset{\leftarrow \text{Charge}}{\overset{\text{Discharge}}{\longrightarrow}} \mathrm{PbSO_4}(s) + 2\mathrm{H_2O}(l), E^0_{\text{cell}} = 2.06\ \mathrm{V} \tag{6.21}$$

The fabrication of a lead-acid cell requires the stepwise formation of a grid (current collector), the active material paste mixing, pasting, curing, soaking in the electrolyte, and formation cycling. The grid is composed of lead, Ca, and Sn alloy, which has optimum corrosion resistance. The active material paste containing leady oxide (a mixture of Pb and PbO), lignin, barium sulfate, sulfuric acid, water, other expanders, and additives are mechanically mixed. Additionally, 25% Pb_3O_4 is added to the positive active material paste to improve the formation efficiency. This mixture is then pasted onto the grid and washed in 1.1 M H_2SO_4 (prickling). Subsequently, curing of the electrodes at 60°C, >70% humidity condition. During curing, active materials strongly bind with the grid, and basic sulfates, corrosion layer, and pores are formed. The electrodes are assembled with a separator between and then soaked in 4.5 M H_2SO_4 followed by formation cycling. During this process, lead and lead dioxide are formed at the anode and cathode, respectively [15,16].

Though the LAB chemistry is an age-old established one various secondary processes like irreversible sulfation at the negative plate, gassing, grid corrosion at the positive electrode, water loss, and acid stratification cause their degradation and failure. Various strategies have been applied to minimize the degradation process, like introducing valve-regulated lead-acid batteries, which take care of gassing reaction through an internal oxygen cycle, and the emergence of absorptive glass mat separators, which solved the issue of acid stratification. Similarly, to account for the irreversible sulfation, various nanocarbon along with metal oxides additives were used to decrease the size of lead sulfate formed and increase the electronic conductivity, which in turn resulted in the higher reversibility of the process [17,18] and also various conducting polymer coatings and anti-corrosion coatings [19,20] were studied to mitigate the issue of grid corrosion at the positive electrode. Despite the lower energy density ($\sim$30 Wh kg^{-1}), LAB systems still occupy a significant battery market share.

6.4.2.2 Nickel metal hydride (Ni-MH) batteries

Ni-MH battery consists of nickel oxyhydroxide (NiOOH) as a cathode, hydrogen-absorbing alloy (MH) as an anode in the charged state, and aqueous KOH as the electrolyte. The anode is of two types: AB_5 type where A = La, Ce, Mo; B = Ni, Co, Al, Mn and AB_2 type where A = Ti, V; B = Zr, Ni, Cr. AB_2 alloys have a higher capacity (360–400 Ah kg^{-1}) than AB_5 alloy (270–290 Ah kg^{-1}). In contrast, AB_5 alloy exhibits good electrochemical performance at a high discharge rate and high-temperature stability. The self-discharge rate of Ni-MH cells is quite high depending on temperature (20% on 1st day and 10% per month at RT and thrice at 45°C). This issue can be solved by improved positive electrode designing and separator engineering. NiOOH is reduced to $Ni(OH)_2$, and MH is oxidized to a metal alloy during discharge. The overall cell voltage is 1.35 V. The cell reaction is as follows,

$$NiOOH + MH \underset{\leftarrow \text{Charge}}{\overset{\text{Discharge}}{\longrightarrow}} Ni(OH)_2 + M \quad (6.22)$$

The currently available Ni-MH cells are 30 mAh to 250 Ah capacity based on the cell design for industrial or consumer applications. Ni-MH cells have been potentially used by Toyota, Honda, and Ford companies for their EVs and HEVs. At present, the Li-ion battery is gradually replacing Ni-MH batteries due to their superior electrochemical performance [2,21,22].

6.4.2.3 Lithium-ion batteries

Lithium-ion batteries have conquered the global market of EES devices since its commercialization by SONY in 1991. It has wide applications in portable electronics consumer appliances (e.g., smartphones, laptops, cameras, etc.), grid storage, transportation (e.g., EVs, E-buses, E-bicycles), etc. LIB has a high energy density (150–200 Wh kg^{-1}) and high power density with >1000 cycles. A conventional Li-ion cell consists of layered transition metal (M = Mn, Fe, Co, Ni, Cr, etc.) oxide or phosphate-based cathodes and graphite as an anode. The electrolyte is $LiPF_6$ salt dissolved in carbonate-based solvents like ethylene carbonate (EC), diethyl carbonate (DEC), dimethyl carbonate (DMC), etc. During charge, Li^+ from the cathode moves towards the anode via the electrolyte medium and forms an intercalation compound (LiC_6) with graphite. During discharging, opposite phenomena happen. Al and Cu are used for cathode and anode current collectors. The net cell reaction during the charge-discharge process is as follows.

$$LiC_6 + Li_{1-x}MO_2 \underset{\leftarrow \text{charge}}{\overset{\text{Discharge}}{\longrightarrow}} Li_{1-x}C_6 + LiMO_2 \tag{6.23}$$

A brief description of the recent progress on LIB electrode materials is given below.

LIB cathode materials—the cathode is an oxidizing electrode having high working potential (i.e., >4 V). It should contain a redox-active transition metal. LIB cathodes can be broadly classified into (1) insertion-type and (2) conversion-type materials based on their lithiation and delithiation phenomena.

6.4.2.3.1 Insertion types cathodes

Layered (2D), spinel (3D), and olivine (1D) structured cathodes fall under this category. Insertion cathodes have theoretical capacities of 274–280 mAh g^{-1}. $LiCoO_2$ (LCO) is the most used cathode material introduced by Goodenough [23]. The material has an easy synthesis method, low self-discharge rate, and good electrochemical performance (> 500 cycles). However, the structural phase change above 4.2 V limits the achievable capacity of 140 mAh g^{-1} due to the extraction of ½ of the Li at an upper cut-off potential of 4.2 V [24]. LCO batteries are thermally unstable at >200°C because of the reaction with organic electrolytes and evolved O_2 (Eq. 6.24) [25].

$$Li_{0.5}CoO_2 = \frac{1}{2}LiCoO_2 + \frac{1}{6}Co_3O_4 + \frac{1}{6}O_2 \tag{6.24}$$

For this reason, all the batteries from Boeing airlines were called off in 2003 after several accidents were reported. Surface coating with $FePO_4$, ZrO_2, Al_2O_3, etc., increases the chemical stability of LCO and reduces the thermal runaway problem. LIB research is moving towards low to non-cobalt-based electrode materials considering the issues of limited availability, cost-effectiveness, and environmental hazards of cobalt.

$LiNiO_2$ and $LiMnO_2$ are cheaper and eco-friendlier cathodes and isostructural to LCO. Phase pure $LiNiO_2$ is difficult to synthesize due to the cation mixing of Ni^{+2}/Li^+, and the actual structure becomes $Li_{1-y}Ni_{1+y}O_2$ with Li deficiency. It delivers 20%–30% more capacity than LCO, resulting in high energy density. However, Ni-O's poor thermal stability (bond energy is even worse than Co-O) hinders its commercial application [26]. $LiMnO_2$ has improved thermal stability. Mn-based electrodes face some common problems, that is (1) electrolyte instability in high voltage region, (2) Jahn-Teller distortion and structural phase change, (3) disproportion of Mn^{+3} to Mn^{+2}, and Mn^{+4} resulting in Mn^{+2} dissolution in electrolyte and formation of cathode electrolyte interface. Therefore, mixed transition-metal oxides

$LiNi_{1-y-z}Mn_yCo_zO_2$ have emerged to overcome the issues of individual metal oxides cathodes. It delivers a capacity of 160–180 mAh g^{-1} in the same operational condition. The synergistic effects of Ni offering high capacity, Mn providing good cycle life, and Co increasing the electronic conductivity result in better electrochemical performance. Further, Co decreases Ni^{+2}/Li^{+} cation disordering. $LiNi_{1/3}Mn_{1/3}Co_{1/3}O_2$ is one of the promising cathodes for HEV and EV applications. It delivers a high capacity of 180 mAh g^{-1} in 2.8–4.4 V [27]. Incorporation of Al in $LiNi_xCo_yAl_zO_2$ (NCA, $x + y + z = 1$) improves the cycling stability with 200 mAh g^{-1} discharge capacity. It has wide commercial applications, for example, Tesla EVs. Further, Li, Mn-rich – NMC cathode has a general formula $xLi_2MnO_3.(1-x)LiMO_2$ (M = Ni, Co, Mn) delivers a capacity of >250 mAh g^{-1} in the wide voltage range of 2.5–4.8 V with an energy density of 1000 Wh kg^{-1} [28]. However, Li_2MnO_3 is activated above 4.4 V, and the structural non-stoichiometry results layered to spinel transformation with a voltage decay below 3 V resulting in loss of energy. A spinel-type cathode, for example, $LiMn_2O_4$ provides a unique three-dimensional diffusion pathway and enhances Li^+ kinetics [29]. Various dopants like Ti, Mg, Zn, Fe, Ni, etc., can suppress the previously mentioned challenges arising in Mn-based electrodes. Polyanionic type phospho-olivines ($LiMPO_4$, M = Fe, Mn, Ni, Co) are the emerging cathode materials for LIBs due to high structural stability (strong M–P–O bond), high abusive tolerance, flat voltage curve in discharge profile, and good electrochemical performance at high current density [30]. $LiFePO_4$ is isostructural in both lithiated and delithiated states ($FePO_4$). Thus the practical obtained capacity is similar to the theoretical capacity of 170 mAh g^{-1}. Among all olivine cathodes, $LiFePO_4$ is commercialized. However, because of low working voltage (3.3 V), low Li^+ diffusion coefficient (10^{-12} $cm^{-2}s^{-1}$), and low conductivity (10^{-9} S cm^{-1}), these cathodes cannot be used for high energy density applications. Further, fluorophosphate and fluorosulphates type cathode materials such as $LiFeSO_4F$, and $LiVPO_4F$ in the tavorite category possess high electronic/ionic conductivity [31]. $LiVPO_4F$ delivers 130 mAh g^{-1} capacity at 4.2 V. Lithium transition metal orthosilicates, that is Li_2MSiO_4 where M = Co, Mn, Fe has a high theoretical capacity (333 mAh g^{-1} considering $2e^-$ process), high thermal stability because of strong Si–O bond, and abundant in nature [32]. The low electronic conductivity of silicates and structural instability can be improved by doping and surface coatings. Further, lithium excess disordered rock salt cathodes are an emerging research topic in recent years [33]. It is a combination of redox-active low valence transition metal, for example, Mn^{+2}/Mn^{+4}, and high valence electrochemically inactive charge compensator, for example, V^{+4}/V^{+5}. Partial substitution of O by F reduces oxygen loss during cycling. These types of cathodes deliver a high discharge capacity of 350–400 mAh g^{-1}. Table 6.2 summarizes the properties of insertion cathodes.

6.4.2.3.2 Lithium-ion batteries anode materials

Pure Li-metal was used as an anode in the past [34]. However, dendrite formation during long cycling and low-temperature operation resulting in a short circuit restricted its further application. LIB research was revitalized during the 1980s

Table 6.2 Different types of cathode materials used in lithium-ion batteries.

Crystal structure	Cathode materials	Theoretical capacity (mAh g^{-1})	Working potential (V)	Practical capacity (mAh g^{-1})
Layered	$LiCoO_2$	274	3.8	148
	$LiNiO_2$	275	3.8	150
	$LiMnO_2$	285	3.3	140
	$LiNi_{0.33}Mn_{0.33}Co_{0.33}O_2$	280	3.7	160
	$LiNi_{0.8}Co_{0.15}Al_{0.05}O_2$	279	3.7	199
Spinel	$LiMn_2O_4$	148	4.1	120
Olivine	$LiFePO_4$	170	3.4	165
	$LiMnPO_4$	171	3.8	168
	$LiCoPO_4$	167	4.8	125

when a layered structured graphite anode was invented (specific capacity 372 mAh g^{-1}, working potential 0.1 V) [35]. Graphite forms an intercalation type compound LiC_6 on lithiation in a reversible way. It shows stable capacity retention over 1000 cycles with >99% coulombic efficiency and minimum volume change (~10%) during cycling. Extensive research has been done on carbonaceous materials such as LIB anode. Apart from graphite, hard carbon, graphene oxide, graphene, and carbon nanotubes are reported as LIB anode materials. A passivation layer of solid electrolyte interface (SEI) is formed on the carbon electrode surface when the electrolyte solvents are reduced below 100 mV. An ideal SEI is typically 30–50 nm and provides structural integrity and high ionic conductivity with no electronic conductivity. The composition of SEI is Li_2O, Li_2CO_3, LiF, ROLi, $ROCO_2Li$, Li-alkoxides, etc [36]. Lithium titanate ($Li_4Ti_5O_{12}$ or LTO theoretical capacity 175 mAh g^{-1}) is an intercalation type metal oxide anode with a flat voltage plateau at 1.5 V [37]. It does not form SEI during the lithiation process. It is a "zero strain" material as it undergoes a slight volume change (~0.2%) during cycling. LTO is the right choice for high power and high cycle life applications.

As graphite has limited capacity, LIB research has moved towards a high specific capacity alloy type (e.g., Sn, Sb) and conversion type anodes (metal oxides, e.g., Fe_2O_3, Co_3O_4, MnO_2) [38,39]. Silicon has gathered the most attractive alloy anode due to the high theoretical capacity of 4200 mAh g^{-1} corresponding to $Li_{4.4}Si$. Unlike graphite, SEI is also formed on Si anode. Sn also has a high theoretical capacity of 960 mAh g^{-1} in its lithiated state of $Li_{17}Sn_4$. Binary alloy anode is of two types, that is (1) one metal is inactive and acts as a buffer during structural changes, and the other metal participates in alloy- dealloying reaction, for example, Fe-Sn. The inactive metal should be permeable to Li^+. In another type, (2) both the metals actively participate in alloying-dealloying reactions, for example, Sb-Sn. In conversion type reaction (e.g., Metal oxides and sulfides), Li_2O is formed in the initial cycle with low electrical conductivity. It traps the metallic nanoparticles into the matrix resulting in high irreversible capacity loss and larger hysteresis. However, these materials suffer from huge volume expansion (>300%) and electrical contact loss during the prolonged charge-discharge process. Conductive carbon composite enhances mechanical stability, increases conductivity, and accommodates volume change. Research and development are going on several types of anode materials. Graphite remains the "state of art" anode in LIBs for easy availability, low redox potential, high cycling stability, and excellent electrochemical performance.

LIB electrolyte—the electrolyte should have high ionic conductivity (10^{-2} S cm^{-1}), high operating voltage stability (>4.5 V), no chemical reactivity with the electrodes, and less flammable. Typically used LIB electrolytes are Li^+ containing salts like $LiPF_6$, $LiBr_4$, LiTFSI, etc., dissolved in carbonate-based solvents like diethyl or dimethyl or ethylene carbonate (DEC/DMC/EC) [40]. Additives like fluoroethylene carbonate (FEC) or vinylene carbonate (VC) help compact and stable SEI formation that gives a better result to the electrochemical cell performance. Room temperature ionic liquid (RTILs) based electrolytes overcome the problem of the high flammability of carbonate-based electrolytes and are stable up to 6 V. Imidazolium cation and bis-(trifluoromethylsulfonyl) imide anion are mostly reported as RTILs with structural modifications [41]. Further, polymer-based electrolytes are intermediate between liquid and solid electrolytes that replace both liquid electrolytes and separators. Polyethylene oxide (PEO), and polyethylene glycol (PEG) in Li-salts are examples of polymer electrolytes. However, solid-state electrolytes mitigate the issues of operating voltage (Stable up to 5.5 V) and flammability [42]. Various forms of solid-state electrolytes are NASICON type, that is $NaM_2(PO_4)_3$ M = Zr, Ti; LISICON type $Li_{14}Zn(GeO_4)_4$, that is garnet type, that is $Li_3Ln_3M_2O_{12}$ M = Te, W, Ln = Pr, Nd, Y, Gd, Lu, etc. LIPON (lithium phosphorous oxynitride) is another reported solid-state electrolyte having a wide voltage stability window between 0–5.5 V. However, the liquid electrolytes are used in commercial LIBs for high ionic conductivity, chemical stability, cost-effectiveness, and easy availability.

6.4.2.4 Sodium-ion batteries

Although LIB is leading the rechargeable battery market, experts forecast that LIB manufacturing will face future challenges. The uneven distribution and limited lithium and cobalt resources cannot fulfill high consumer demand. Depending on abundant availability, cost, and reduction potential (−2.71 vs SHE), Sodium-ion batterie (SIB) could be an ideal alternative. SIB consists of hard carbon as anode and insertion type transition metal oxide (Na_xMO_2) as a cathode. The electrolyte is $NaPF_6$ or $NaClO_4$ salt dissolved in carbonate-based organic electrolyte like EC/PC/DEC. SIB system follows similar electrochemistry to LIB during the sodiation and dissociation process. SIB research is in progress by tuning the LIB electrodes in novel structural, architectural design, and sodium host construction.

SIB cathodes can be broadly classified into five different categories.

1. Layered transition metal oxides that is $NaMO_2$, M = Co, Fe, Mn, Ni, Ti.
2. Phosphate polyanions
3. NASICON types

4. Sodium fluorophosphate
5. Prussian blue analog

In layered transition metal oxides, Na–ions are sandwiched into the repeated MO_6 layers. Based on the structural arrangement, Delmas classified the oxide cathodes into O3 (ABCABC stacking), P2 (ABBA stacking), and P3 (ABBCCA stacking) where, O and P signifies octahedral and trigonal prismatic sites for Na^+. Na_xCoO_2 was one of the most studied materials for the great success of the $LiCoO_2$ analog. The effect of oxygen non-stoichiometry Na_xCoO_{2-y} and multiphase voltage profile deteriorate the electrochemical performance. Layered $NaMnO_2$ has versatile tunnel-like structures, but it suffers from Mn dissolution and structural distortion. $NaFeO_2$ exhibits high reversible capacity (e.g., 80–100 mAh g^{-1} capacity at 3.4 V cut-off voltage) for $Fe^{+3/+4}$ redox. Doped layered oxides, for example, P2-$Na_{2/3}Fe_{1/2}Mn_{1/2}O_2$, $NaFe_{1/2}Co_{1/2}O_2$, P2-$Na_{2/3}Mn_{1/2}Fe_{1/2}O_2$ shows synergistic effects of good ionic conductivity of Na_xCoO_2, the high redox potential of Na_xFeO_2, and high specific capacity of Na_xMnO_2. Polyanionic cathodes for SIBs have been extensively studied for high operating voltage and high structural stability on cycling but suffer from poor conductivity. Olivine structured $NaFePO_4$ shows 150 mAh g^{-1} discharge capacity at 2.9 V but has inferior electrochemical performance due to the phase change during sodiation and desodiation process. 3D structured NASICON type cathodes $Na_3V_2(PO_4)_3$ show enhance Na^+ kinetics with a reversible capacity of 117 mAh g^{-1} and a potential plateau of 3.4 V. Introduction of fluorine to the polyanion cathodes for example, $Na_3V_2O_{2x}(PO_4)_2F_{3-2x}$ enhances the capacity by boosting the activity of redox couple. Prussian blue analogs possess cage-like unique metal-organic framework (MOF) structures with a general formula of $Na_xMFe(CN)_6$ ($0 \le x \le 2$; M = Fe, Zn, Co, Ni, Mn, Cu). It shows a high capacity for the two-electron redox process of M and Fe. However, the release of HCN at high temperatures (>200°C) is still challenging [43–46]. A comparison table of some of best performing SIB cathodes in half-cell is shown below (Table 6.3).

SIB anodes are mainly of three types, that is (1) intercalation, (2) alloy, and (3) conversion-based materials, unlike LIBs. Non-graphitized hard carbons (sodiation potential <100 mV and high reversible capacity of 300 mAh g^{-1}) with disordered graphene sheets are the most attractive anode materials [47]. The electrochemical performance of hard carbon depends on defects, the interlayer spacing, surface functional groups, porosity, and electrolyte nature. The largely irreversible capacity loss during the initial cycles due to SEI formation and undesirable side reactions can be mitigated by optimum calcination temperature (700°C–1500°C), proper precursors (e.g., biowastes, sucrose, glucose, pitch, etc.), and pretreatment process (e.g., surface activation with base treatment). Various titanium oxides-based intercalation anodes have been reported for SIBs. CNT-TiO_2 hybrid delivers a stable capacity of 100 mAh g^{-1} at 1 Ag^{-1} over 1000 cycles [48]. Structurally stable and flat voltage profile sodium titanates like tunnel structured $Na_2Ti_6O_{13}$ and layered structured $Na_2Ti_3O_7$ have been reported. Alloy type SIB anodes include $Na_{15}Sn_4$, Na_3Sb, Na_3P etc. Transition metal oxides (e.g., Fe_2O_3, Fe_3O_4, SnO_2, Co_3O_4) and sulfides (CoS_2, MoS_2, SnS) are examples of conversion type anodes

Table 6.3 Different cathode materials for sodium-ion batteries.

Sodium-ion batterie cathode material	**Avg. voltage in V (Voltage range)**	**Capacity ($mAh\ g^{-1}$) @ Current rate**	**Capacity retention @ number of cycles**
O3-$NaNi_{0.5}Mn_{0.2}Ti_{0.3}O_2$	~3 (2.0–4.0)	135 @ 1 C	85% @ 200
O3- $NaNi_{0.25}Co_{0.25}Fe_{0.25}Ti_{0.25}O_2$	~3.1 (2.0–3.9)	116 @ 1 C	93% @ 100
O3-$NaNi_{0.5}Mn_{0.3}Co_{0.2}O_2$	~2.8 (2.0–4.0)	136 @ C/10	64% @ 200
P2-$Na_{0.6}Li_{0.2}Mn_{0.8}O_2$	~2.6 (2.0–4.6)	190 @ C/15	100% @ 100
P2-$Na_{0.66}Fe_{0.5}Mn_{0.5}O_2$	~2.75 (1.5–4.3)	190 @ C/10	79% @ 30
$Na_3V_2(PO_4)_3$	~3.4 (2.7–3.7)	107 @ C/10	93% @ 80
$Na_3V_2(PO_4)_2F_3$	~3.9 (2.5–4.3)	117 @ 1 C	92.4% @ 100
$Na_4Fe(CN)_6$	~3.4 (2.0–3.9)	87 @ 1 C	88% @ 500

[49]. Both these alloy and conversion type SIB anodes have similar advantages and disadvantages to LIB anodes.

6.4.2.5 Potassium-ion batteries (KIBs)

Potassium-ion batteries (KIB) is promising alternatives to LIB for its comparable standard electrode potential (−2.93 V vs SHE) and a theoretical capacity of 591 mAh g^{-1}. Layered potassium transition metal oxides ($K_{0.3}MnO_2$, $K_{0.55}CoO_2$) and Prussian blue analogs are used as cathodes. Besides potassium metal, graphite is used as an anode, and it forms different phases of K-graphite intercalation compounds with a reversible capacity of 208 mAh g^{-1} at 5 mA g^{-1} current density. Due to the higher reactivity of potassium, ionic liquids and solid polymer electrolytes [e.g., poly (propylene carbonate)-KFSI] are preferred use over carbonate-based solvents. The research and development of KIB are still in the infancy stage, and it needs to monitor the electrode design to overcome the sluggish reaction kinetics and poor electrochemical performance [50,51].

6.4.2.6 Multivalent rechargeable batteries

The electrochemical cell reaction in multivalent rechargeable batteries (Mg^{+2}, Ca^{+2}, Al^{+3}, Zn^{2+}, etc.) involves two or more electrons in the redox process resulting in high capacity (i.e., 2−4 times higher than LIBs) and high volumetric energy density. The electrode materials are less sensitive to moisture than LIBs. Mg-ion, Ca-ion, and Al-ion batteries' standard reduction potentials are −2.37, −2.87, and −1.67 V, respectively. Pure metals serve as an anode. Rocking chair transition metal oxides (e.g., V_2O_5, MnO_2) and hybridized with carbonaceous materials (e.g., graphite, graphene, CNT),

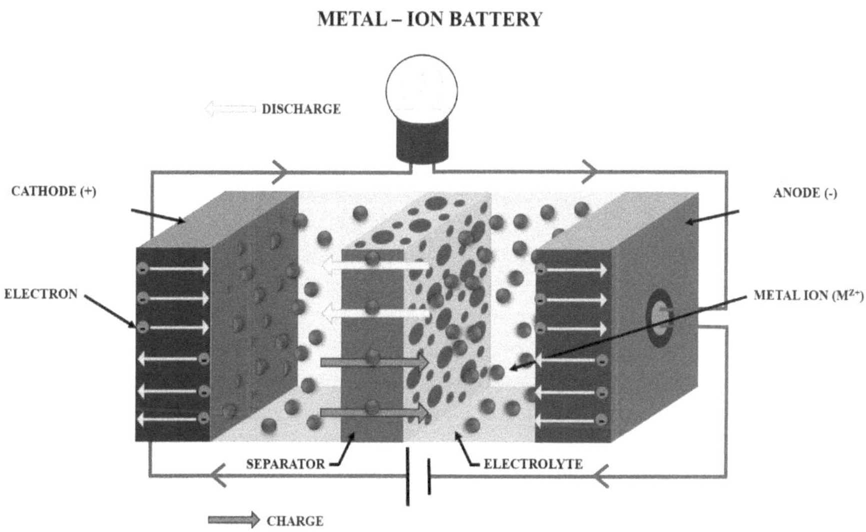

Figure 6.3 Schematic representation of the working principle of metal-ion batteries.

conductive polymers (e.g., PANI, PPy) are mostly used for cathodes. However, the significant challenges of multivalent rechargeable batteries are (1) lack of cathode materials with high diffusivity of metal-ion, (2) high polarization of multivalent cations causing largely irreversible capacity loss, and (3) formation of a surface passivation layer on metal electrode due to the electrolyte decomposition blocking the ionic and electronic transport. Therefore, the proper choice of electrode materials and compatible electrolytes for ion diffusion is of great importance to overcome the technological and fundamental issues of these emerging rechargeable systems [52,53]. (Fig. 6.3)

6.5 Electrochemical capacitors

Electrochemical capacitors (ECs) store energy through surface reactions and hence, are fast processes. They are also referred to as supercapacitors or ultracapacitors. ECs consist of two electrodes immersed in an electrolyte. The electrodes' designation, specifically as anode and cathode, either depends on the polarity of the potential applied across the electrodes or on the surface redox activity involved in any of the electrodes. The electrolytes used can be aqueous or non-aqueous. The non-aqueous electrolytes have higher resistivity and hence, possess limited ionic mobility as compared to aqueous electrolytes. But the evolution of non-aqueous systems can be attributed to the potential limitation of the aqueous systems as the electrolysis of water occurs beyond 1.23 V. As the energy of an electrochemical capacitor is directly proportional to the square of the voltage, the broader voltage range overshadows the decrease in the ionic mobility and resulting capacitance till > 4.5 V. The aqueous electrolytes can be either acidic (low concentration ~ 1 M H_2SO_4), basic (low concentration ~ 1 M KOH) or salt (~ 1 M Na_2SO_4). In contrast, the non-aqueous electrolytes contain salts dissolved in organic (alkyl carbonates) based solvents or room temperature ionic liquids. The ionic conductivity of aqueous electrolytes is higher than the non-aqueous ones by a factor of 10^3, and the ionic conductivity of the ionic liquid is even lower in the order of 10^{-4} S cm^{-1}. But the trend of higher potential stability for the electrolytes is reversed as compared to the ionic conductivity, that is aqueous electrolyte (<1.5 V) $<$ carbonate-based electrolyte (<4.5 V) $<$ Ionic liquids (4.0–6.0 V). However, electrochemical capacitors' charge storage property depends on the interactions at the electrode-electrolyte interface. The surface of the electrode active material can be differentiated based on its redox activity. If the active material undergoes a redox reaction or faradaic reaction and high surface adsorption, these are considered pseudocapacitive. In contrast, if the active material works on the principle of conventional adsorption double layer formation of oppositely charged ions, then these are classified as electric double layer type materials. Thus, depending on the property of the electrode active material, the electrochemical supercapacitors can be classified into two broad categories, which are briefly discussed here: electric double-layer capacitors (EDLCs) and pseudo-capacitors. ECs differ from batteries in the charge storage mechanism. The voltage profile for ECs is sloping in nature with an ideally constant slope, and the cyclic voltammetry shows a rectangular profile as the current response remains almost unchanged during the change in potential due to the absence of faradaic reaction. Pseudo-capacitors undergo faradaic reactions, but the reactions are surface-bound and thus, provide a mid-way between surface adsorbing EDLCs and bulk faradaic reactions observed in batteries [2,12,54].

6.5.1 Electric double-layer capacitors

The working principle of EDLCs essentially involves the formation of an oppositely charged ionic double layer at each electrode-electrolyte interface. The thickness of the double layer lies in the range of 5−10 A° depending on the physical properties of the electrolyte [2]. On applying a potential difference across the electrodes, that is while charging, the ions in the electrolyte get adsorbed onto the oppositely charged electrode surface through electrostatic force of attraction (physisorption) ideally. The process is reversed on discharging, and the electrodes and the electrolyte ions revert to their original state. Apart from these, reversible chemisorption of ions can also take place at the electrode surfaces. The amount of ions stored on the electrode surface depends on various factors like high surface area, porosity, and surface functionalities. The most commonly used electrode material for EDLCs is activated carbon (AC), that is carbon with high surface area and surface functionality [12,55,56]. Conventionally EDLCs are comprised of two symmetrical activated carbon electrodes immersed in aqueous or non-aqueous electrolytes. The EDLCs can deliver power density in kWh kg^{-1}, but the energy density is limited to a range of 5−10 Wh kg^{-1}.

6.5.2 Pseudo-capacitors

Pseudo-capacitors involve redox-active electrode surfaces. The energy is stored in the pseudo-capacitor in the form of a redox chemical reaction. The application of external potential facilitates the activation of the redox couple on the electrode surface during charge. During the discharge, this process is reversed, and the redox-active surface groups redirect to the original state. The electrolyte provides the internal ionic circuit for the redox reaction to occur. The energy stored in pseudo-capacitors is attributed to the redox chemical reaction taking place at the electrode surface. Thus, pseudo-capacitors have a higher energy density compared to the EDLCs. The pseudocapacitive reaction can follow three mechanisms [57]:

1. Underpotential deposition: This phenomenon occurs when ions deposit on the two-dimensional metal electrolyte interface at potentials lower than their reversible reduction potential. (e.g., H^+ on Pt, Pd^{2+} on Au). The capacitance observed through this phenomenon is high (~2200 μF cm^{-2}), but the potential window is tiny (0.3−0.6 V), and hence energy density is limited.
2. Redox pseudocapacitance: This is the most common process observed in pseudo-capacitors where faradaic reactions occur on the surface of the active material across which potential is being applied. (e.g., metal oxides like RuO_2, conductive polymers).

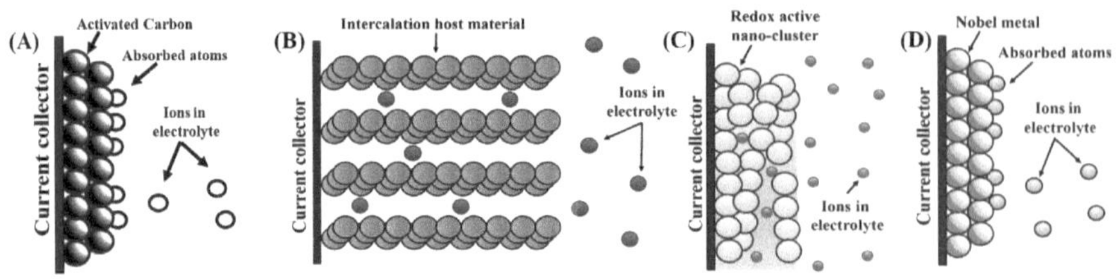

Figure 6.4 Schematic representation of charge storage phenomena of (A) double-layer capacitive electrode (B) Intercalation pseudocapacitive electrode (C) redox-active pseudocapacitive electrode and (D) underpotential pseudocapacitive material [57].

This process can be described as reversible electrochemical adsorption. A capacitance of ~5000 F cm^{-3} can be achieved depending on the surface redox-active couple.

3. Intercalation pseudocapacitance: This occurs through the redox-active intercalation chemistry of electrode materials which has a faster time scale compared to the conventional battery-type intercalation chemistry (e.g., Nb_2O_5, nanosized LTO, V_2O_5). The intercalation chemistry is fundamentally similar to that of lithium-ion batteries. In this process, the contribution from surface redox pseudocapacitance and electrostatic pseudocapacitance is also observed in addition to the intercalation phenomenon. This involves a wider potential window, and hence higher energy density is achieved compared to EDLCs.

The mechanism of EDLC and pseudocapacitive reactions is depicted in Fig. 6.4, and some of the materials reported for the different types of ECs are given in Table 6.4.

Asymmetric configurations of ECs have also been studied. These systems are composed of two different electrode materials. The two different electrode materials can either have a similar capacitive phenomenon of charge storage or have two different charge storage phenomena [54]. The latter one is called a hybrid capacitor which is discussed in Chapter 21. The asymmetric electrochemical capacitors are fabricated to utilize wider potential windows and varied charge storage capacities striving toward higher energy densities without sacrificing the power density capabilities.

6.6 Fuel cells

Fuel cells (FCs) are the oldest EES system. Unlike a battery, FCs are not self-sufficient as fuels must be supplied continuously to keep the cell functioning. A fuel cell converts the energy stored inside chemical bonds of fuel into electrical energy. All kinds of FCs are made of anode, cathode, electrolyte, and catalysts. Catalyst at anode dissociates hydrogen into cations and electrons, and catalyst at cathode recombines hydrogen and oxygen to form water. A membrane is used that suitably allows only specific ions. Since its first inception in the 19th century, a lot of variants of FCs are reported in the literature and upscaled at a larger scale. FCs can be classified on multiple grounds [58]. The first classification is based on

Table 6.4 Some examples of active materials reported for electrochemical capacitors in literature [54].

Types of capacitors	Active materials	Electrolyte	Capacitance (Fg^{-1})
EDLC	Activated Carbon (~2000 m^2g^{-1}) KOH activated Carbon from resol (~2700 m^2g^{-1}) Carbon nanosheet from sodium gluconate SWCNT based macroscopic bulk Porous graphene oxide Mesocarbon/ graphene Nitrogen enriched mesoporous carbon nanofiber array	Aqueous electrolyte Organic electrolyte 6 M KOH 1 M H_2SO_4 1 M $TEABF_4$/ AN Organic electrolytes Aqueous 6 M KOH Aqueous	200 100 307 140 @150 Ag^{-1} 100 @ 120 Ag^{-1} 80 160 197 264, 325
Pseudo capacitor	Pseudocapacitive metal oxides: Hydrous RuO_2 RuO_x. nH_2O MnO_2/ CNT V_2O_5/ C-core shell nanowire Conducting polymer-based PANI hydrogel PANI/Co-based MOF/Carbon cloth PPY/ 3D GO paper	Aqueous Aqueous Aqueous Aqueous Aqueous Aqueous Aqueous	720 1300 @ 25 mV s^{-1} 1230 128.5 Fcm^{-2} @10 mVs^{-1} 750 @ 1 A g^{-1} 2146 mF cm^{-2} 198 Fcm^{-3}

electrolytes, and the second one is based on the temperature of operation. (Figs. 6.5 and 6.6)

The comparison of various FCs is presented in Table 6.5 [59–61].

The generalized energy conversion steps of alkaline fuel cells (AFCs) are as follows: (1) Hydrogen enters from the anode side. Catalyst at anode causes oxidation of Hydrogen and splits it into protons (H^+) and electrons. Oxidation occurs at the anode side. (2) The electron moves from the anode to the cathode side via an external circuit. This movement of electrons produces DC electricity. Finally, the electron reached the cathode side. (3) Oxygen is introduced from the cathode side. Catalyst at the cathode, with the help of electrons from the anode side, reduces Oxygen to form hydroxide ions. Subsequently, the reduction occurs at the cathode side. These hydroxide ions move towards the anode via an internal circuit through the electrolyte. (4) At the anode, ions and electrons combine to form water. The basic mechanism of all types of FCs is almost the same. Hydrogen and Oxygen finally produce water. However, it varies slightly depending upon the type of cell. In AFC, OH^- ions move from the cathode to the anode, and recombination

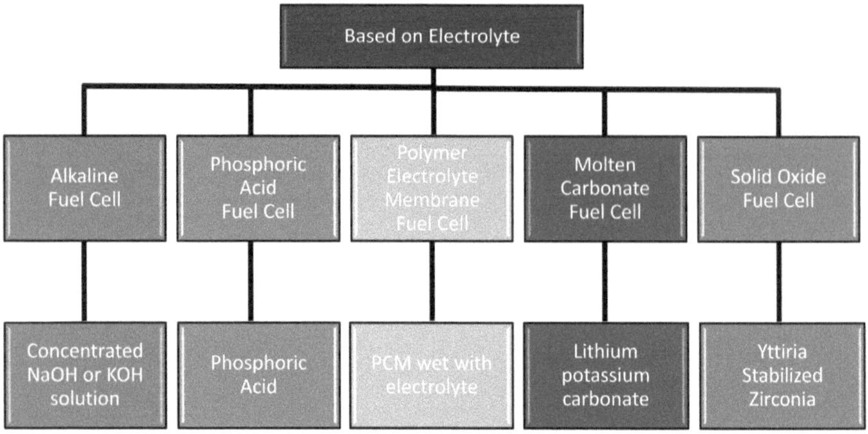

Figure 6.5 Classification of fuel cells based on types of electrolytes.

reactions occur at the anode. In Polymer Electrolyte Membrane Fuel Cell (PEMFC), H^+ produced on the anode side move to the cathode and is combined with oxygen to form water. Water forms at the cathode side. Proton exchange membrane in between cathode and anode is permeable to only protons. It does not allow OH^- to move across. Therefore, depending on the medium and electrolyte pH, ion movements differ from one variant to another. However, as long as fuel is supplied to the cell, it will continue to generate electricity irrespective of the difference in mechanism among the variants [59].

Chemical reactions that occur at different pH ranges are as follows –

In acidic electrolyte:

$$\text{Cathode reaction: } O_2 + 4H^+ + 4e^- \rightarrow 2H_2O; E^0 = 1.229\text{V} \tag{6.25}$$

$$\text{Anode reaction: } H_2 \rightarrow 2H^+ + 2e^-; E^0 = 0.000\text{V} \tag{6.26}$$

$$\text{Overall reaction: } 2H_2 + O_2 \rightarrow 2H_2O; E^0 = 1.229\text{V} \tag{6.27}$$

In alkaline electrolyte:

$$\text{Cathode reaction: } O_2 + 2H_2O + 4e^- \rightarrow 4OH^-; E^0 = 0.401\text{V} \tag{6.28}$$

$$\text{Anode reaction: } H_2 + 2OH^- \rightarrow 2H_2O + 2e^-; E^0 = -0.828\text{V} \tag{6.29}$$

$$\text{Overall reaction: } 2H_2 + O_2 \rightarrow 2H_2O; E^0 = 1.229\text{V} \tag{6.30}$$

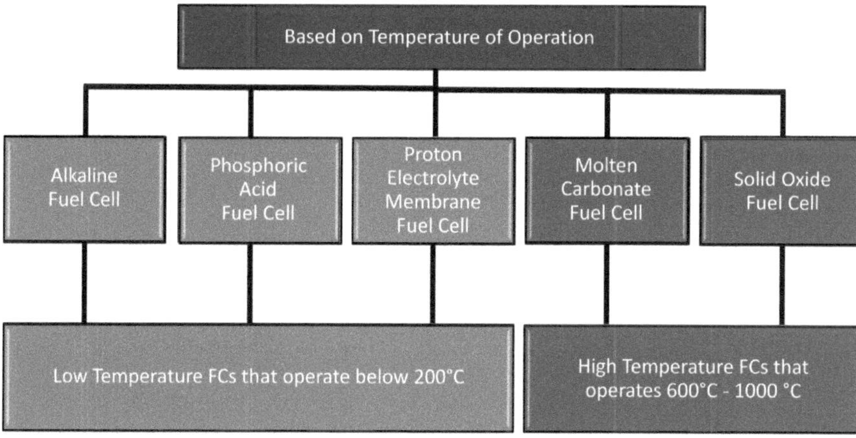

Figure 6.6 Classification of fuel cells based on temperature of operation.

Oxidation of Hydrogen occurs readily on Pt-based catalysts. The kinetics of hydrogen oxidation reaction at the anode is very fast. First, adsorption of hydrogen gas on the catalyst bed occurs, followed by the dissociation of the molecule according to

$$2\mathrm{Pt}(s) + \mathrm{H}_2 \rightarrow 2\mathrm{Pt} - H_{\mathrm{ads}} \tag{6.31}$$

$$\mathrm{Pt} - H_{\mathrm{ads}} \rightarrow \mathrm{H}^+ + e^- + \mathrm{Pt}(s) \tag{6.32}$$

[Pt(s) is free Pt surface and Pt-H_{ads} is H_2 adsorbed on Pt surface]

$$\text{Theoverallreaction is: } \mathrm{H}_2 \rightarrow 2\mathrm{H}^+ + 2e^- \tag{6.33}$$

Although the reaction is fast on the Pt surface, operational issues may occur if impure hydrogen is used. Pure hydrogen provides the best electrochemical performance but is expensive and difficult to store. The alternative to pure hydrogen is natural gas, propane, or alcohol. These substances must be reformed into hydrogen. But always, some contaminants remain after gas clean up. As with methanol oxidation catalysts, CO poisons the catalyst by blocking the active site of the surface. Therefore, alloys are used in those cases instead of pure Pt catalyst to improve catalytic activity with contaminated hydrogen [60].

O_2 is a very stable molecule. It is challenging to break the oxygen molecule due to high bonding forces. The exact mechanistic path of Oxygen Reduction Reaction (ORR) is a question of long debate due to several parallels and consecutive reactions occurring simultaneously. There are mainly two pathways, *viz.* $4e^-$ pathway

Table 6.5 Comparison of various fuel cells.

Properties	Alkaline fuel cell	Polymer electrolyte membrane fuel cell	Phosphoric acid fuel cell	Molten carbonate fuel cell	Solid oxide fuel cell
Anode reaction	$H_2 + 2OH^- \rightarrow 2\,H_2O + 2e^-$	$H_2 \rightarrow 2\,H^+ + 2e^-$	$H_2 \rightarrow 2H^+ + 2e^-$	$H_2 + CO_3^{2-} \rightarrow H_2O + CO_2 + 2e^-$	$H_2 + O^{2-} \rightarrow H_2O + 2e^-$
Cathode reaction	$O_2 + 2H_2O + 4e^- \rightarrow 4OH^-$	$O_2 + 4\,H^+ + 4e^- \rightarrow 2H_2O$	$O_2 + 4\,H^+ + 4e^- \rightarrow 2H_2O$	$O_2 + 2CO_2 + 4e^- \rightarrow 2\,CO_3^{2-}$	$O_2 + 4e^- \rightarrow 2O^{2-}$
Operating temperature (°C)	60–90	50–120	160–220	600–800	800–1000 (500–600 also possible)
Electrolyte	35%–50% KOH	Polymer membrane (Nafion/ Dow)	Concentrated Phosphoric acid	Molten carbonates (Li_2CO_3)	Yttiria stabilised zirconia (Y_2O_3 –ZrO_2)
Charge carrier	OH^-	H^+	H^+	CO_3^{2-}	O^{2-}
Cell efficiency (%)	50–60	50–60	53–55	60–65	55–65
Application	Space, traction, military	Space, traction, military	Stationary power systems	Power generation	Power generation

and $2e^-$ pathway. The $2e^-$ pathway generates hydrogen peroxide, one of the most vital intermediates in ORR. The number of steps is more in the case of ORR than hydrogen oxidation at the anode. ORR occurs at 100–140 mV lower than theoretical oxygen potential due to sluggishness in the adjustment of potentials. Catalyst structure, crystallinity, geometry, and d-band valency of metals play an essential role in ORR reaction. ORR is highly sensitive to catalyst structure. The reaction only happens at a specific active site of the catalyst. When the particle size is smaller (e.g., ~5 nm), the relative surface area increases, but the number of active sites decreases. Pretreatment of the catalyst or electrochemical activation of the electrodes is needed to improve the ORR performance [60].

FC is an electrochemical energy converter. Its efficiency can be measured in several terms – thermodynamic efficiency, electrochemical efficiency, electrochemical efficiency, faradic efficiency, and total efficiency. The free energy of a chemical reaction is converted into electrical energy. The thermodynamic efficiency of an electrochemical energy converter is

$$\eta = \frac{\Delta G}{\Delta H} = 1 - \frac{T\Delta S}{\Delta H} \tag{6.34}$$

The thermodynamic efficiency of all FCs is >90%, which is much higher than the most efficient thermal engine (only 50%). An interesting thing to note here is that the efficiency of the formic acid cell and hydrocarbon air-fuel cell is more than 100%. Because below 100°C, the product water is in liquid form, ΔS is negative, and efficiency is 100%. But at temperatures >100°C, water is liberated as vapor, ΔS becomes positive, and efficiency surpasses 100%. However, practical efficiency is much lesser due to overpotential losses. The thermodynamic efficiency of FCs decreases with increasing temp, which is in contrast with internal combustion engines (ICEs). With increasing the temperature of ICEs, the temperature difference between heat sinks and heat sources increases, η increases. Electrochemical efficiency (EE) measures the change in efficiency by the losses due to electrode overpotential and the cell's electrolyte resistance. EE extracts more information from FC than thermodynamic efficiency because it relates directly to the electrode kinetics of inner processes. That is why EE is more relevant than thermal efficiency. Apart from operation life, cost, and other parameters, EE can compare two different types of FCs.

$$EE(\eta_{el}) = \frac{-\mathrm{nFE_k}}{-\mathrm{nFE^0}} = \frac{-\mathrm{nFE_k}}{\Delta G} = \frac{\mathrm{E_k}}{\mathrm{E^0}} \tag{6.35}$$

$$\left(E_k = \text{Operating potential and } E^0 = \text{Standard electrode potential}\right)$$

EE of H_2-O_2 FC is as high as 0.9 at low current densities and decreases slowly when a larger current is drawn from the cell until a limiting value is reached. E_k is always lesser than E^0 due to polarization effects and voltage drop owing to ohmic resistance. Therefore, practical efficiency (η_P) under load conditions can be expressed as

$$\eta_p = \frac{-\,nFE_k}{\Delta H} = \frac{E_k}{\left(E_0 - T\left(\frac{dE_0}{dT}\right)\right)} \tag{6.36}$$

Faradic efficiency (η_f) is defined as the ratio of experimental current observed in an FC (I_{exp}) and maximum possible theoretical current (I_m) based on amounts of reactants consumed. It is analogous to current efficiency in conventional electrochemical cells. In most FCs, faradic efficiency is <1 due to parallel electrochemical reactions yielding fewer electrons per mole of reactant consumed, chemical reactions of electrode catalyzed by the electrode, and direct reaction between two electrodes. Total efficiency includes all types of efficiencies, including the pumping of fuel and oxidation material, the necessary energy for heating, cooling, electrolyte pipes, compression, auxiliary installations, etc. Mathematically, the total efficiency can be calculated as

$$\eta_{\text{total}} = \eta_t + \eta_f + \eta_{\text{el}} + \eta_p \tag{6.37}$$

The total efficiency of an FCs, in general, is greater than gas-electric, diesel-electric, microturbines, that is most thermal engines. Some heat is always produced during the conversion of chemical to electrical energy by FCs. Utilization of that heat into usable energy is called "Cogeneration." Cogenerative FC systems simultaneously use electricity and heat, and it increases the total efficiency of the system [61].

$$\eta_{\text{overall}} = \frac{Q + E}{Q_0} \tag{6.38}$$

(Q is usable energy obtained in heat, E is electrical energy, Q_0 is the heat content of fuel)

FCs also bring several environmental benefits as well. Conventional generation of electricity produces CO_x, SO_x, NO_x, particulates, etc. Fuel cell power plant emissions are at least 10 times lower than that of traditional power plants. Substitution of FCs in power plants will improve air quality by lowering the volume of greenhouse gas emissions. By product of FCs reactions is water. To run an FC power plant, hardly any external source of water is required. But large steam power plants need a massive amount of water at each cycle. The quality of discharged wastewater is better and less in quantity in FCs than the traditional power plants. Therefore, FCs can eliminate the water quality problem associated with thermal discharge. Quiet electrochemical nature FCs eliminate many sources of noise associated with conventional steam-powered plants [62].

Performance-wise, FCs are closer to ICEs. Electric vehicles run by FCs are an attractive economic option and can be a better choice for long-range heavy vehicles than battery-run EVs. The main fuel of FCs is Hydrogen. Pure hydrogen is very difficult to store. H_2 can be stored in CNTs, Metal hydrides, Ammonia (NH_3), Hydrazine (N_2H_4), etc. H_2 storage in alcohols or hydrocarbons can also be an option. Such hydrocarbons and alcohols can be reformed into hydrogen-rich synthesis gases by several methods. The main processes are Partial Oxidation (POX), Catalytic Steam Reforming (CSR), and Autothermal Reforming (ATR). Electrolysis of water is one of the fewest processes that does not require fossil fuel burning and one of the cleanest methods to produce pure hydrogen along with pure oxygen. During the reforming of CSR and ATR process, CO is formed, which poisons the catalyst. Thus, the presence of CO is one of the major issues for FCs operating at <250°C. Favorable routes for hydrogen production from different substrates can be categorized in this way.

1. H_2 from alcohols can be produced by POX, STR, ATR, and Mr processes.
2. H_2 from hydrocarbons can be generated by catalytic steam reforming, POX, ATR, membrane reactors, propane cracking, and coal gasification.

Reforming hydrocarbons into hydrogen always produces small amounts of CO in the gas stream. CO is not a problem for high-temp FCs that can use CO as fuel for anode reaction. But low-temperature FCs must eliminate CO. The gas clean-up is done in two steps as the level of CO should be reduced to a ppm level for the smooth functioning of FCs as given below:

Step 1: Water gas-shift conversion of CO into CO_2; ($CO + H_2O \rightarrow CO_2 + H_2$). This step not only reduces CO to the equilibrium value (in the range of 1% CO) but also produces more hydrogen.

Step 2: This step is essential as it lowers CO down to the 5–100 ppm range. The three most common techniques are—selective oxidation, methanation, and the use of a selective membrane [63,64].

Hydrogen can also be stored in physical ways other than the chemical pathways discussed above. Pure hydrogen is one of the problematic gas to store. The hydrogen/air mixture is explosive. Many gases are far more explosive than hydrogen. But in case of a leak in hydrogen, it will disperse so quickly that it is near impossible to resist it from the ignition. Therefore, safety measure is one of the major concerns for H_2-based FCs. Storing gaseous hydrogen directly on board the vehicle or next to the stationary application can avoid contamination problems. Gaseous hydrogen has a low energy density. This is one of the disadvantages of using pure H_2 gas. Compressing hydrogen into a tank can be achieved, but the tank's size and volume may pose a restriction. Gaseous hydrogen is suited in fuel buses and fleet operations where the driving range is of little interest. Liquid hydrogen can be an alternative. It can be stored at −253 °C. However, liquid hydrogen tanks must be well insulated, and even 1%–2% of loss by evaporation must be monitored from an economic point of view. However, the disadvantage of using liquid hydrogen is the energy required to convert hydrogen gas into a liquid is not economically viable. The hydrogen content of solid-state materials can be leveraged to use as hydrogen storing media. Hydrogen is stored in metal as bound hydrogen or as a gas adsorbed into the structure. Metal hydrides are commonly used for hydrogen storage.

The method of releasing hydrogen is heating the metal hydride. Graphene, CNTs, and other carbon nano frameworks can also store hydrogen [65–67].

6.7 Conclusions

The EES systems are sought to provide for the ever-increasing energy demand across the globe. The basis of EES systems from thermodynamic as well as reactivity perspectives along with their development timeline are elaborated in this chapter. The prominent types of energy storage systems have been discussed briefly in this chapter.

- Rechargeable batteries have found their utility in various applications like electric vehicles, grid storage, portable electronics, etc. LIBs have dominance in the battery market with energy densities >200 Wh kg^{-1} along with other systems like NiMH and lead-acid batteries. Potential analogs of LIBs like other alkalies (Na, K, Mg) metal ion batteries are in the research and development stage.
- Fuel cells are considered for energy conversion applications in the case of telecommunication with 99.99% efficiency, landfills, and water plants, and also many fuel cell cars have been launched for transport purposes.
- Electrochemical supercapacitors are being considered in conventional as well as hybrid forms with batteries and fuel cells for large-scale applications. Conventional supercapacitors are believed to be of use in case regenerative systems like elevators, cranes, construction machines, and battery backups.

The future of the discussed conventional EES devices through hybridization and the future of EES through sustainable alternatives hand higher energy output are discussed in the following chapter.

Acknowledgments

Udita and S. K. Martha acknowledge DST-SERB (Sanction Order: CRG/2018/003543), Shuvajit acknowledges CSIR (letter no. 09/1001(0067)/2019-EMR-I), and Madhushri acknowledges DST-INSPIRE (code: IF180708) for fellowship and financial support to this work.

References

[1] B. Kale, S. Chatterjee, Electrochemical energy storage systems: India perspective, Bull. Mater. Sci. 43 (1) (2020) 96.
[2] T. Reddy, Linden's Handbook of Batteries, fourth ed., McGraw Hill Professional, 2010, p. 1457.
[3] J.M. Andújar, F. Segura, Fuel cells: history and updating. A walk along two centuries, Renew. Sustain. Energy Rev. 13 (9) (2009) 2309–2322.

[4] Electrochemical Energy Storage - Energy Storage Options and Their Environmental Impact (RSC Publishing) [Internet]. [cited 2021 Mar 13]. Available from: https://pubs.rsc.org/en/content/chapter/9781788015530-00115/978-1-78801-553-0.
[5] Energy Storage Systems: An Introduction – Nova Science Publishers [Internet]. [cited 2021 Mar 12]. Available from: https://novapublishers.com/shop/energy-storage-systems-an-introduction/.
[6] M. Mounika, G.D. Kumar, D. Danalakshmi, Evolution of hybrid super capacitors and its future, Pathway 4 (6) (2019) 10.
[7] N. Nitta, F. Wu, J.T. Lee, G. Yushin, Li-ion battery materials: present and future, Mater. Today 18 (5) (2015) 252–264.
[8] A.K. Samantara, S. Ratha, Historical background and present status of the supercapacitors, in: A.K. Samantara, S. Ratha (Eds.), Materials Development for Active/Passive Components of a Supercapacitor: Background, Present Status and Future Perspective [Internet], Springer, Singapore, 2018, pp. 9–10. Available from: https://doi.org/10.1007/978-981-10-7263-5_2.
[9] D.D. Sarma, A.K. Shukla, Building better batteries: a travel back in time, ACS Energy Lett 3 (11) (2018) 2841–2845.
[10] M.S. Whittingham, History, evolution, and future status of energy storage, Proc. IEEE 100 (Special Centennial Issue) (2012) 1518–1534.
[11] Chapter 6. Electrochemical energy storage—Annual Reports Section 'C' (Physical Chemistry) (RSC Publishing) [Internet]. [cited 2021 Mar 13]. Available from: https://pubs.rsc.org/en/content/articlelanding/1999/PC/pc095163#!divAbstract.
[12] M. Winter, R.J. Brodd, What are batteries, fuel cells, and supercapacitors? Chem. Rev. 104 (10) (2004) 4245–4270.
[13] C. Julien, A. Mauger, A. Vijh, K. Zaghib, Lithium batteries, in: C. Julien, A. Mauger, A. Vijh, K. Zaghib (Eds.), Lithium Batteries: Science and Technology [Internet], Springer International Publishing, Cham, 2016, pp. 29–68 [cited 2021 Mar 13]. Available from: https://doi.org/10.1007/978-3-319-19108-9_2.
[14] T. Sarakonsri, R.V. Kumar, Primary batteries, High Energy Density Lithium Batteries [Internet], John Wiley & Sons, Ltd, 2010, pp. 27–52 [cited 2021 Mar 13]. Available from: https://onlinelibrary.wiley.com/doi/abs/10.1002/9783527630011.ch2.
[15] D. Pavlov, Lead-Acid Batteries: Science and Technology: A Handbook of Lead-Acid Battery Technology and Its Influence on the Product, Elsevier, 2017, p. 722.
[16] J. Jung, L. Zhang, J. Zhang, Lead-Acid Battery Technologies: Fundamentals, Materials, and Applications, Vol. 8, Crc Press, 2015.
[17] N. Vangapally, S. Jindal, S.A. Gaffoor, S.K. Martha, Titanium dioxide-reduced graphene oxide hybrid as negative electrode additive for high performance lead-acid batteries, Journal of Energy Storage 20 (2018) 204–212.
[18] V. Naresh, U. Bhattacharjee, S.K. Martha, Boron doped graphene nanosheets as negative electrode additive for high-performance lead-acid batteries and ultracapacitors, Journal of Alloy Compound. 797 (2019) 595–605.
[19] V. Naresh, L. Elias, S.A. Gaffoor, S.K. Martha, Corrosion resistant polypyrrole coated lead-alloy positive grids for advanced lead-acid batteries, Journal of the Electrochemical Society 166 (2) (2019) A74.
[20] V. Naresh, L. Elias, S.K. Martha, Poly(3,4-ethylenedioxythiophene) coated lead negative plates for hybrid energy storage systems, Electrochimica Acta 301 (2019) 183–191.
[21] nickelmetalhydride_appman.pdf [Internet]. [cited 2021 Mar 13]. Available from: https://data.energizer.com/pdfs/nickelmetalhydride_appman.pdf.

[22] B. Hariprakash, A.K. Shukla, S. Venugoplan, Secondary batteries—nickel systems | nickel–metal hydride: overview, in: J. Garche (Ed.), Encyclopedia of Electrochemical Power Sources [Internet], Elsevier, Amsterdam, 2009, pp. 494–501 [cited 2021 Mar 13]. Available from: https://www.sciencedirect.com/science/article/pii/B9780444527455001581.

[23] K. Mizushima, P.C. Jones, P.J. Wiseman, J.B. Goodenough, $LixCoO_2$ (0. Solid State Ionics), 1981 Aug 1;3–4:171–4.

[24] J.N. Reimers, J.R. Dahn, Electrochemical and in situ X-ray diffraction studies of lithium intercalation in Li_xCoO_2, J. Electrochem. Soc. 139 (8) (1992) 2091.

[25] C.-H. Doh, D.-H. Kim, H.-S. Kim, H.-M. Shin, Y.-D. Jeong, S.-I. Moon, et al., Thermal and electrochemical behaviour of C/Li_xCoO_2 cell during safety test, J. Power Sources 175 (2) (2008) 881–885.

[26] T. Sasaki, T. Nonaka, H. Oka, C. Okuda, Y. Itou, Y. Kondo, et al., Capacity-Fading Mechanisms of $LiNiO_2$-based lithium-ion batteries: I. Analysis by electrochemical and spectroscopic examination, J. Electrochem. Soc. 156 (4) (2009) A289.

[27] K.M. Shaju, G.S. Rao, B.V.R. Chowdari, Performance of layered Li $(Ni_{1/3}Co_{1/3}Mn_{1/3})$ O_2 as cathode for Li-ion batteries, Electrochim. Acta 48 (2) (2002) 145–151.

[28] M.M. Thackeray, S.-H. Kang, C.S. Johnson, J.T. Vaughey, R. Benedek, S.A. Hackney, Li_2MnO_3-stabilized $LiMO_2$ (M = Mn, Ni, Co) electrodes for lithium-ion batteries, J. Mater. Chem. 17 (30) (2007) 3112–3125.

[29] R.J. Gummow, A. De Kock, M.M. Thackeray, Improved capacity retention in rechargeable 4 V lithium/lithium-manganese oxide (spinel) cells, Solid. State Ion. 69 (1) (1994) 59–67.

[30] A.K. Padhi, K.S. Nanjundaswamy, J.B. Goodenough, Phospho-olivines as positive-electrode materials for rechargeable lithium batteries, J. Electrochem. Soc. 144 (4) (1997) 1188.

[31] N. Recham, J.-N. Chotard, L. Dupont, C. Delacourt, W. Walker, M. Armand, et al., A 3.6 V lithium-based fluorosulphate insertion positive electrode for lithium-ion batteries, Nat. Mater. 9 (1) (2010) 68–74.

[32] A. Nytén, A. Abouimrane, M. Armand, T. Gustafsson, J.O. Thomas, Electrochemical performance of Li_2FeSiO_4 as a new Li-battery cathode material, Electrochem. Commun. 7 (2) (2005) 156–160.

[33] Z. Lun, B. Ouyang, D.A. Kitchaev, R.J. Clément, J.K. Papp, M. Balasubramanian, et al., Improved cycling performance of li-excess cation-disordered cathode materials upon fluorine substitution, Adv. Energy Mater. 9 (2) (2019) 1802959.

[34] M. Li, J. Lu, Z. Chen, K. Amine, 30 years of lithium-ion batteries, Adv. Mater. 30 (33) (2018) 1800561.

[35] R. Yazami, P. Touzain, A reversible graphite-lithium negative electrode for electrochemical generators, J. Power Sources 9 (3) (1983) 365–371.

[36] H. Azuma, H. Imoto, S. Yamada, K. Sekai, Advanced carbon anode materials for lithium ion cells, J. Power Sources 81 (1999) 1–7.

[37] C.P. Sandhya, B. John, C. Gouri, Lithium titanate as anode material for lithium-ion cells: a review, Ionics. 20 (5) (2014) 601–620.

[38] W.-J. Zhang, A review of the electrochemical performance of alloy anodes for lithium-ion batteries, J. Power Sources 196 (1) (2011) 13–24.

[39] Y. Lu, L. Yu, X.W.D. Lou, Nanostructured conversion-type anode materials for advanced lithium-ion batteries, Chem 4 (5) (2018) 972–996.

[40] A. Mauger, C.M. Julien, A. Paolella, M. Armand, K. Zaghib, A comprehensive review of lithium salts and beyond for rechargeable batteries: progress and perspectives, Mater. Sci. Eng.: R: Rep. 134 (2018) 1–21.

[41] H. Srour, L. Chancelier, E. Bolimowska, T. Gutel, S. Mailley, H. Rouault, et al., Ionic liquid-based electrolytes for lithium-ion batteries: review of performances of various electrode systems, J. Appl. Electrochem. 46 (2) (2016) 149–155.
[42] F. Zheng, M. Kotobuki, S. Song, M.O. Lai, L. Lu, Review on solid electrolytes for all-solid-state lithium-ion batteries, J. Power Sources 389 (2018) 198–213.
[43] D. Kundu, E. Talaie, V. Duffort, L.F. Nazar, The emerging chemistry of sodium ion batteries for electrochemical energy storage, Angew. Chem. Int. (Ed.) 54 (11) (2015) 3431–3448.
[44] T. Wang, D. Su, D. Shanmukaraj, T. Rojo, M. Armand, G. Wang, Electrode materials for sodium-ion batteries: considerations on crystal structures and sodium storage mechanisms, Electrochem. Energy Rev. 1 (2) (2018) 200–237.
[45] P.K. Nayak, L. Yang, W. Brehm, P. Adelhelm, From lithium-ion to sodium-ion batteries: advantages, challenges, and surprises, Angew. Chem. Int. (Ed.) 57 (1) (2018) 102–120.
[46] M. Sawicki, L.L. Shaw, Advances and challenges of sodium ion batteries as post lithium ion batteries, RSC Adv. 5 (65) (2015) 53129–53154.
[47] E. Irisarri, A. Ponrouch, M.R. Palacin, Hard carbon negative electrode materials for sodium-ion batteries, J. Electrochem. Soc. 162 (14) (2015) A2476.
[48] S. Ghosh, V.K. Kumar, S.K. Kumar, S. Biswas, S.K. Martha, An insight of sodium-ion storage, diffusivity into TiO_2 nanoparticles and practical realization to sodium-ion full cell, Electrochim. Acta 316 (2019) 69–78.
[49] L. Wang, J. Światowska, S. Dai, M. Cao, Z. Zhong, Y. Shen, et al., Promises and challenges of alloy-type and conversion-type anode materials for sodium–ion batteries, Mater. Today Energy 11 (2019) 46–60.
[50] A. Eftekhari, Z. Jian, X. Ji, Potassium secondary batteries, ACS Appl. Mater. Interfaces 9 (5) (2017) 4404–4419.
[51] C. Vaalma, D. Buchholz, S. Passerini, Non-aqueous potassium-ion batteries: a review, Curr. Opin. Electrochem. 9 (2018) 41–48.
[52] Y. Wang, R. Chen, T. Chen, H. Lv, G. Zhu, L. Ma, et al., Emerging non-lithium ion batteries, Energy Storage Mater. 4 (2016) 103–129.
[53] A. Ponrouch, J. Bitenc, R. Dominko, N. Lindahl, P. Johansson, M.R. Palacín, Multivalent rechargeable batteries, Energy Storage Mater. 20 (2019) 253–262.
[54] Y. Wang, Y. Song, Y. Xia, Electrochemical capacitors: mechanism, materials, systems, characterization and applications, Chem. Soc. Rev. 45 (21) (2016) 5925–5950.
[55] J.R. Miller, P. Simon, The chalkboard: fundamentals of electrochemical capacitor design and operation, Electrochem. Soc. Interface 17 (1) (2008) 31.
[56] P. Simon, A. Burke, Nanostructured carbons: double-layer capacitance and more, Electrochem. Soc. Interface 17 (1) (2008) 38–43.
[57] Y. Shao, M.F. El-Kady, J. Sun, Y. Li, Q. Zhang, M. Zhu, et al., Design and mechanisms of asymmetric supercapacitors, Chem. Rev. 118 (18) (2018) 9233–9280.
[58] L.J.M.J. Blomen, M.N. Mugerwa (Eds.), Fuel Cell Systems [Internet], Springer US, 1993 [cited 2021 Mar 13]. Available from. Available from: https://www.springer.com/gp/book/9780306441585.
[59] Sales10 MB, 11, Floor 1st, Mahal K, Road 192 Dr D. N., FortMumbai 400 001, et al. Fuel Cell Technology Handbook [Internet]. Routledge & CRC Press. [cited 2021 Mar 13]. Available from: https://www.routledge.com/Fuel-Cell-Technology-Handbook/Hoogers/p/book/9780367395773.
[60] K. Maheshwari, P.D.S. Sharma, D.A.K. Sharma, P.S. Verma, Fuel cell and its applications: a review. Int. J. Eng. Res. Technol. [Internet]. 2018 Jun 2 [cited 2021 Mar 13];7(6). Available from: https://www.ijert.org/research/fuel-cell-and-its-applications-a-review-IJERTV7IS060019.pdf, https://www.ijert.org/fuel-cell-and-its-applications-a-review.

[61] Y.M.A. Welaya, M.M. El Gohary, N.R. Ammar, A comparison between fuel cells and other alternatives for marine electric power generation, Int. J. Nav. Architect. Ocean. Eng. 3 (2) (2011) 141–149.
[62] K.V. Kordesch, G.R. Simader, Environmental impact of fuel cell technology, Chem. Rev. 95 (1) (1995) 191–207.
[63] J.O. Abe, A.P.I. Popoola, E. Ajenifuja, O.M. Popoola, Hydrogen energy, economy and storage: review and recommendation, Int. J. Hydrog. Energy 44 (29) (2019) 15072–15086.
[64] D.J. Durbin, C. Malardier-Jugroot, Review of hydrogen storage techniques for on board vehicle applications, Int. J. Hydrog. Energy 38 (34) (2013) 14595–14617.
[65] L. Schlapbach, A. Züttel, Hydrogen-storage materials for mobile applications, Materials for Sustainable Energy [Internet], Co-Published with Macmillan Publishers Ltd, UK, 2010, pp. 265–270 [cited 2021 Mar 13]. Available from: https://www.worldscientific.com/doi/abs/10.1142/9789814317665_0038.
[66] P. Jena, Materials for hydrogen storage: past, present, and future, J. Phys. Chem. Lett. 2 (3) (2011) 206–211.
[67] A. Züttel, Hydrogen storage methods, Naturwissenschaften 91 (4) (2004) 157–172.

Thermal energy storage systems

7

Ali Sohani[1], Mohammad Hassan Shahverdian[1], Siamak Hoseinzadeh[2] and Mamdouh El Haj Assad[3]
[1]Lab of Optimization of Thermal Systems' Installations, Faculty of Mechanical Engineering-Energy Division, K.N. Toosi University of Technology, Tehran, Iran, [2]Department of Planning, Design, Technology of Architecture, Sapienza University of Rome, Rome, Italy, [3]Sustainable and Renewable Energy Engineering Department, University of Sharjah, Sharjah, United Arab Emirates

Chapter Outline

Highlights

- The working principle of sensible and latent thermal storage systems is provided.
- Organic and inorganic thermal storage materials are introduced.
- The commercialized sensible and latent thermal storage technologies are described.
- A comprehensive literature review of the recently conducted research works is given.

Nomenclature

Symbols

c_P	The heat capacity at constant pressure (kJ per kg per K)
G	Gibbs free energy (kJ)
V	Volume (m^3)
S	Enthalpy (kJ K^{-1})
x_i	The contribution of ith material into the substance
n	Number
m	Mass (kg)

Emerging Trends in Energy Storage Systems and Industrial Applications. DOI: https://doi.org/10.1016/B978-0-323-90521-3.00019-3

h	Enthalpy (kJ kg^{-1})
Q	Heat (kJ)
R	The gas constant (kJ kg^{-1} K or kJ $kmol^{-1}$ K^{-1})
T	Temperature (K)
R_1	Reaction number 1
R_2	Reaction number 2
Ratio	Ratio
P_1	Product number 1
P_2	Product number 2

Greek symbols

ρ	Density (kg m^{-3})
Δ	Difference

Subscripts

sensible	Sensible heat
end	End of the time period
TESS	Thermal energy storage system
start	Beginning of the time period
oil	Oil
water	Water
latent	Latent heat
a	State "a"
b	State "b"
a→b	Transformation from state "a" to state "b"
phase change	Phase change material
fg	Fluid to gas transformation
m	Melting
reaction	Reaction

Abbreviations

TESS	Thermal energy storage system
PCM	Phase change material

7.1 Introduction and the working principle

A storage unit is a place in which a material or energy is gathered, kept, and used when it is needed [1–4]. Charge, storage, and discharge are the technical terms for the three mentioned processes, respectively [5–7]. A schematic description of the three indicated processes, which is called a storage cycle, is presented in Fig. 7.1.

As the name indicates, a thermal energy storage system, briefly called TESS here, is a place to store thermal energy [8–10]. Depending on the working temperature level of the thermal storage unit compared to the environment or any other places the thermal storage unit has the interaction, the stored thermal energy could be given in the form of heating or cooling [11–13]. For example, when the thermal energy is stored in the form of ice, it provides cooling if the discharging process is

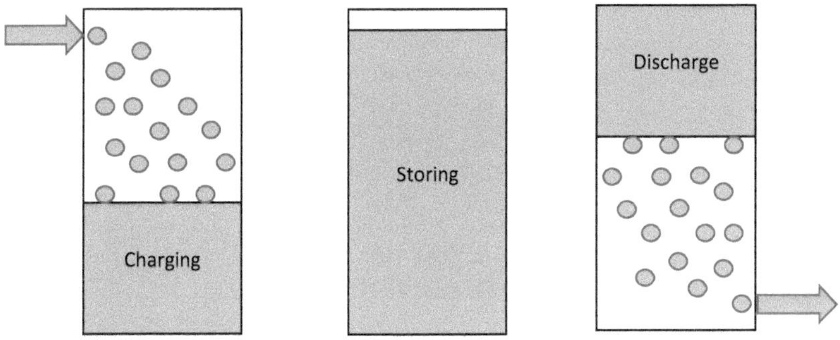

Figure 7.1 Schematic description of a storage cycle.

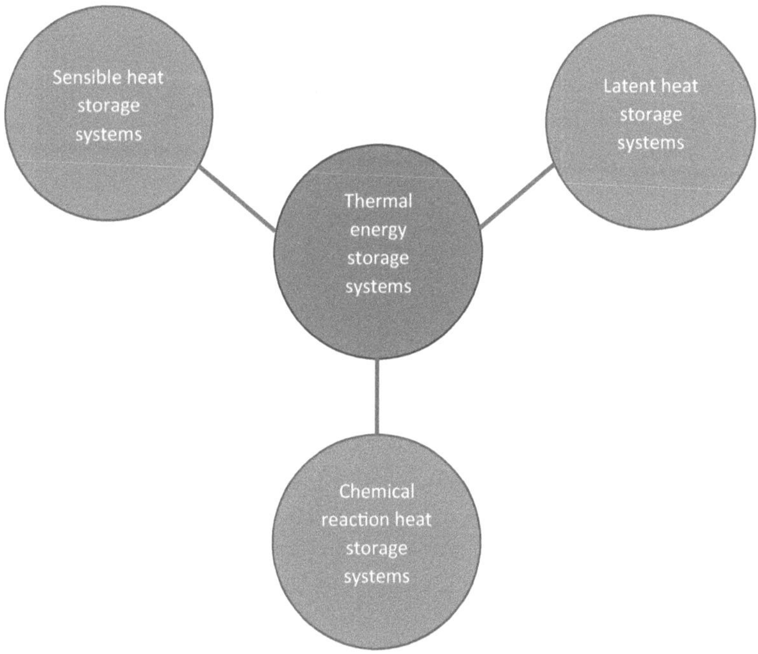

Figure 7.2 Different classifications of thermal storage systems.

done in the interaction with water or the ambient. On the other hand, if the thermal energy is stored in the form of high-pressure vapor, discharging the stored energy to a medium like water or the ambient is accompanied by heating them [14–16].

Thermal storage systems could be classified based on different criteria, including [17–19]:

- The storing mechanism (sensible heat storage, latent heat storage, or chemical reaction heat storage, as per Fig. 7.2),
- The state of the medium heat is stored in (solid, liquid, or gas),

- The period between charge and discharge processes, or in another word, the storage period (hourly, daily, weekly, monthly, seasonal, annual, and so on)
- Usage (heating or cooling, which has been shortly discussed).

7.1.1 Sensible thermal storage systems

In a sensible TESS, the stored heat leads to changing the temperature of the medium which absorbs that, and it does not change the phase of the material [20–22]. In other words, the charge and discharge processes in a sensible TESS result in "sensible" heating or cooling of the material that keeps energy. Therefore, the amount of the stored energy could be calculated based on Eq. (7.1) [23]:

$$Q_{\text{sensible TESS}} = \int_{T_{\text{start}}}^{T_{\text{end}}} m\ c_P\ dT = \int_{T_{\text{start}}}^{T_{\text{end}}} \rho\ V\ c_P\ dT \tag{7.1}$$

Where Q and T stand for the stored heat, and temperature, respectively. Furthermore m, ρ , V and c_P, are mass, density, volume, and the specific isobaric heat capacity of the absorbing substance, while subscripts sensible, TESS, start, and end represent sensible, TESS, and starting and ending time of charging process, respectively. If the mass (density) is constant and changes in c_P the variation of temperature are neglected, that is it is considered constant, Eq. (7.1) could be written in the form of Eq. (7.2) [23].

$$Q_{\text{sensible TESS}} = m\ c_P \int_{T_{\text{start}}}^{T_{\text{end}}} dT = m\ c_P(T_{\text{end}} - T_{\text{start}}) = \rho\ V\ c_P(T_{\text{end}} - T_{\text{start}}) \tag{7.2}$$

According to Eq. (7.1), the heat absorption capacity of sensible heat storage is a function of three parameters, the mass of the thermal storage absorbing material, the heat capacity of the substance that keeps energy, and the temperature difference occurs during the sensible charging of the heat storage material. The values of c_p for different materials are reported in Table 7.1.

Example 7.1: Water in a tank is going to be utilized as thermal storage material. The tank is a cylinder, and full of water at the atmospheric pressure, with a diameter of 3 m and a height of 4 m. In the beginning, water is in the ambient temperature, which is 25°C, and at the end, the temperature of that is equal to 75°C.

(a) If the specific isobaric heat capacity and density of water as a function of temperature have variation trends like Fig. 7.3, find the heat absorption capacity of water.
(b) If instead of the function considered in part (a), a constant specific heat capacity is assumed for water, how much does the calculated capacity change?
(c) If instead of water, oil with the heat capacity of 1.88 $kJ\ kg^{-1}\ K^{-1}$ is employed, how much does the capacity of TESS change?

Table 7.1 The values of c_p for different materials. Although the values were extracted from [24], they are available in some other textbooks like [25–27].

No.	Material	Phase*	c_p (kJ kg^{-1} K^{-1})
1	Air (typical room conditions)	(G)	1.012
2	Air [Sea level, dry, 0°C (273.15K)]	(G)	1.005
3	Aluminum	(S)	0.897
4	Ammonia	(L)	4.700
5	Animal tissue (incl. human)	Mixture	3.500
6	Antimony	(S)	0.207
7	Argon	(G)	0.520
8	Arsenic	(S)	0.328
9	Asphalt	(S)	0.920
10	Beryllium	(S)	1.820
11	Bismuth	(S)	0.123
12	Brick	(S)	0.840
13	Cadmium	(S)	0.231
14	Carbon dioxide CO_2	(G)	0.839
15	Chromium	(S)	0.449
16	Concrete	(S)	0.880
17	Copper	(S)	0.385
18	Diamond	(S)	0.509
19	Ethanol	(L)	2.440
20	Gasoline (octane)	(L)	2.220
21	Glass	(S)	0.840
22	Glass, crown	(S)	0.670
23	Glass, flint	(S)	0.503
24	Glass, pyrex	(S)	0.753
25	Glass, silica	(S)	0.840
26	Gold	(S)	0.129
27	Granite	(S)	0.790
28	Granite	(S)	0.790
29	Graphite	(S)	0.710
30	Gypsum	(S)	1.090
31	H_2O @ −10°C (ice)	(S)	2.050
32	H_2O @ 100°C (saturated liquid)	(L)	4.181
33	H_2O @ 100°C (saturated vapor or steam)	(G)	2.080
34	H_2O @ 25°C	(L)	4.181
35	Helium	(G)	5.193
36	Hydrogen	(G)	14.300
37	Hydrogen sulfide H2S	(G)	1.015
38	Iron	(S)	0.412
39	Lead	(S)	0.129
40	Lithium	(S)	3.580
41	Lithium @ 181°C	(L)	4.379
42	Magnesium	(S)	1.020
43	Marble, mica	(S)	0.880

(*Continued*)

Table 7.1 (Continued)

No.	Material	Phase*	c_p (kJ kg^{-1} K^{-1})
44	Mercury	(L)	0.140
45	Methane at 2°C	(G)	2.191
46	Methanol	(L)	2.140
47	Molten salt (142°C–540°C)	(L)	1.560
48	Neon	(G)	1.030
49	Nitrogen	(G)	1.040
50	Oxygen	(G)	0.918
51	Paraffin wax $C_{25}H_{52}$	(S)	2.500
52	Polyethylene (rotomolding grade)	(S)	2.303
53	Sand	(S)	0.835
54	Silica (fused)	(S)	0.703
55	Silver	(S)	0.233
56	Sodium	(S)	1.230
57	Soil	(S)	0.800
58	Steel	(S)	0.466
59	Tin	(S)	0.227
60	Titanium	(S)	0.523
61	Tungsten	(S)	0.134
62	Uranium	(S)	0.116
63	Water	(L)	4.181
64	Wood	(S)	1.700 (from 1.200 to 2.900)
65	Zinc	(S)	0.387

*(*S*): solid; (*L*): liquid; (*G*): gas.

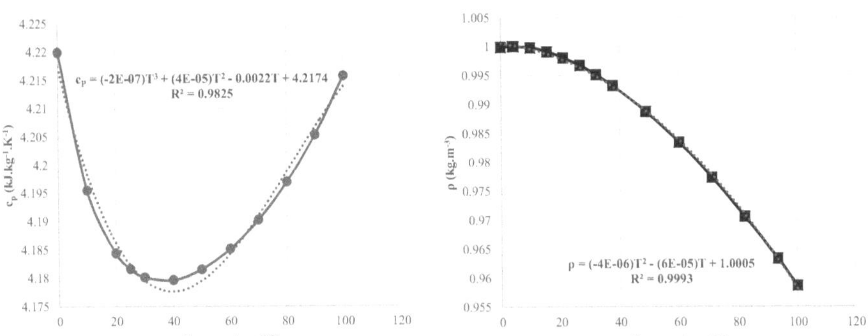

(A) **(B)**

Figure 7.3 Variation of the properties of the pure water liquid at the atmospheric pressure (A) isobaric specific heat capacity (c_P); (B) density (ρ).

Solution:

(a) Using Eq. (7.1)

$$Q_{\text{sensible TESS}} = \int_{T_{\text{start}}}^{T_{\text{end}}} m \; c_P \; dT = \int_{T_{\text{start}}}^{T_{\text{end}}} \rho \; V \; c_P \; dT = V \int_{T_{\text{start}}}^{T_{\text{end}}} \rho \quad c_P \; dT$$

$$= V \int_{T_{\text{start}}}^{T_{\text{end}}} \{(-4E-06)T^2 - (6E-05)T + 1.0005\}$$
$$\times \{(-2E-07)T^3 + (4E-05)T^2 - 0.0022T + 4.2174\} \; dT$$

$$= V \int_{T_{\text{start}}}^{T_{\text{end}}} \{(-4E-06)T^2 - (6E-05)T + 1.0005\}$$
$$\times \{(-2E-07)T^3 + (4E-05)T^2 - 0.0022T + 4.2174\} \; dT$$
$$= V\{(1.33E-13)T^6 - (2.96E-11)T^5 - (5.3E-08)T^4 + (7.83E-06)T^3 + (1.26E-04)T^2 + 4.22T\}_{25}^{75}$$
$$= 6025.5 \text{ J} = 6.0255 \text{ kJ}$$

(b) Eq. (7.2) could be used:

$$Q_{\text{sensible TESS}} = \rho \; V \; c_P(T_{\text{end}} - T_{\text{start}}) = (1.000)(\frac{\pi}{4} \times (3)^2 \times 4)(4.18)(75-25) = 5909.3 \text{ J} = 5.9093 \text{ kJ}$$

Therefore, the change in the calculated capacity in percentage is:

$$\frac{|6.0255 - 5.9093|}{6.0255} \times 100 = 1.93$$

As observed, only below 2% changes in the heat absorption capacity are observed, which is negligible in the engineering application. Consequently, it is concluded that using constant values c_P ρ could bring accurate results.

(c) The capacity for both liquids could be computed by Eq. (7.2). Since all the parameters except for ρ are the same, we have:

$$\frac{Q_{\text{oil}}}{Q_{\text{water}}} = \frac{\rho_{\text{oil}}}{\rho_{\text{water}}} = \frac{1.88}{4.18} = 0.4498$$

It means a 45.02% decrement.

7.1.2 Latent thermal energy storage systems

In a latent TESS, the heat given from the source to the absorbing material during the charging process leads to changing the phase of the material [28–30]. It means that before charging, the storing material is in one and after that, it is in another phase. Therefore, the heat transfer between the source and the TESS is in the form of "latent" heating or cooling, which results in the phase transformation [31–33] (Fig. 7.4).

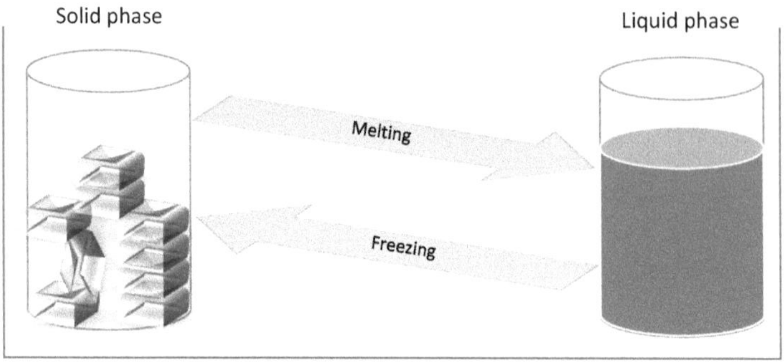

Figure 7.4 Schematic description of a latent heat storage system.

Generally, in addition to the phase change, a small fraction of the given heat might make sensible heating or cool in the absorbing material, as well. Therefore, if initially, the material is in the phrase "a", and then, it goes to the phase "b", Eq. (7.3) provides the amount of the stored thermal energy [23]:

$$Q_{\text{latent, TESS}} = Q_{a,\ \text{sensible}} + Q_{a\to b} + Q_{b,\ \text{sensible}} \tag{7.3}$$

Which $Q_{a,\ \text{sensible}}$ is the sensible heat storage, which leads to moving the state of the absorbing material from the temperature level of the initial condition (T_{initial}) to the phase-changing ($T_{\text{phase change}}$) condition. $Q_{a,\ \text{sensible}}$ is obtained from Eq. (7.4) [23]:

$$Q_{a,\ \text{sensible}} = m\ c_{p,a}(T_{\text{initial}} - T_{\text{phase change}}) \tag{7.4}$$

Similarly, $Q_{b,\ \text{sensible}}$ stands for the sensible heat gained by changing the temperature level from the phase change to the final condition, which is determined by Eq. (7.5) [23]:

$$Q_{b,\ \text{sensible}} = m\ c_{p,b}(T_{\text{phase change}} - T_{\text{final}}) \tag{7.5}$$

The amount of thermal energy storage through the phase change "a" to "b" ($Q_{a\to b}$) is computed via Eq. (7.6) [23]:

$$Q_{a\to b} = m\ h_{a\to b} \tag{7.6}$$

where $h_{a\to b}$ is the heat of phase change per unit mass of the absorbing material. The values of h_{fg} for different materials are given in Table 7.2.

Example 7.2: Consider 1 kg of water. If it converts from liquid to gas, how much bigger it could absorb the heat compared to the condition that goes from the melting point to the boiling point (0°C to 100°C)?

Table 7.2 Table of h_{fg} for different materials. Although the values were extracted from [34], they are available in some other textbooks like [35].

No.	Substance	Organic	Melting point, T_m (°C)	h_{fg} (kJ kg^{-1})
1	1-Cyclohexylooctadecane	✔	41	218
2	2-Heptadecanone	✔	48	218
3	3-Heptadecanone	✔	48	218
4	4-Heptadacanone	✔	41	197
5	9-Heptadecanone	✔	51	213
6	Acetamide	✔	81	241
7	Acetic acid	✔	17	184
8	a-Chloroacetic acid	✔	61	130
9	Acrylic acid	✔	68	115
10	Actanilide	✔	119	222
11	Alpha napthol	✔	96	163
12	Aluminum	✘	660	397
13	a-Nepthylamine	✔	59	93
14	Azobenzene	✔	67	121
15	Bees wax	✔	62	177
16	Benzamide	✔	127	169
17	Benzoic acid	✔	122	143
18	Bromcamphor	✔	77	174
19	Camphene	✔	50	238
20	Camphenilone	✔	39	205
21	Capric acid	✔	36	152
22	Caprilic acid	✔	16	149
23	Caprylone	✔	40	259
24	Catechol	✔	104	207
25	Cetyl acid	✔	49	141
26	Chloroacetic acid	✔	56	130
27	Copper	✘	1085	209
28	Cyanamide	✔	44	209
29	Dinto toluent (2,4)	✔	70	111
30	Diphenyl amine	✔	53	107
31	Docasyl bromide	✔	40	201
32	Durene	✔	79	156
33	Elaidic acid	✔	47	218
34	Formic acid	✔	8	247
35	Glautaric acid	✔	98	156
36	Glycerin	✔	18	199
37	Glycolic acid	✔	63	109
38	Glyolic acid	✔	63	109
39	Gold	✘	1064	64
40	Heptadecanone	✔	41	201
41	Heptaudecanoic acid	✔	61	189
42	Hydrocinnamic acid	✔	48	118

(Continued)

Table 7.2 (Continued)

No.	Substance	Organic	Melting point, T_m (°C)	h_{fg} (kJ kg^{-1})
43	Hypophosphoric acid	✔	55	213
44	Iron	✘	1538	247
45	KNO_3	✘	337	116
46	KNO_3(10%)/$NaNO_3$	✘	290	170
47	KNO_3/KBr(4.7%) / KCl(7.3%)	✘	342	140
48	KNO_3/KCl(4.5%)	✘	320	150
49	KOH	✘	360	167
50	Lauric acid	✔	44	212
51	Lead	✘	327	23
52	Lithium	✘	181	432
53	Methyl behenate	✔	52	234
54	Methyl bromobenzoate	✔	81	126
55	Methyl eicosanate	✔	45	230
56	Methyl fumarate	✔	102	242
57	Methyl palmitate	✔	29	205
58	$Mn(NO_3)_2 \cdot 6H_2O$ / $MnCl_2 \cdot 4H_2O$ (4%)	✘	15–25	126
59	Myristic acid	✔	58	199
60	$Na_2SiO_3 \cdot 5H_2O$	✘	72	267
61	NaCl(26.8%)/NaOH	✘	370	370
62	NaCl(42.5%)/KCl(20.5%)/$MgCl_2$	✘	385–393	410
63	NaCl(5.0%)/$NaNO_3$	✘	282	212
64	NaCl(5.7%)/$NaNO_3$(85.5%) / Na2SO4	✘	287	176
65	NaCl/KCL(32.4%) /LiCl(32.8%)	✘	346	281
66	NaCl/$NaNO_3$(5.0%)	✘	284	171
67	$NaCl \cdot Na_2SO_4 \cdot 10H_2O$	✘	18	286
68	$NaNO_2$	✘	310	174
69	$NaNO_3$	✘	282	212
70	NaOH	✘	318	158
71	NaOH/Na_2CO_3(7.2%)	✘	283	340
72	Nitro naphthalene	✔	57	103
73	O-Nitroaniline	✔	50	93
74	Oxolate	✔	54	178
75	O-Xylene dichloride	✔	55	121
76	Palmitic acid	✔	55	163
77	Paraffin 14-Carbons	✔	6	228
78	Paraffin 15-Carbons	✔	10	205
79	Paraffin 16-Carbons	✔	17	237
80	Paraffin 17-Carbons	✔	22	213
81	Paraffin 18-Carbons	✔	28	244

(*Continued*)

Table 7.2 (Continued)

No.	Substance	Organic	Melting point, T_m (°C)	h_{fg} (kJ kg^{-1})
82	Paraffin 19-Carbons	✔	32	222
83	Paraffin 20-Carbons	✔	37	246
84	Paraffin 21-Carbons	✔	40	200
85	Paraffin 22-Carbons	✔	44	249
86	Paraffin 23-Carbons	✔	48	232
87	Paraffin 24-Carbons	✔	51	255
88	Paraffin 25-Carbons	✔	49	238
89	Paraffin 26-Carbons	✔	56	256
90	Paraffin 27-Carbons	✔	59	236
91	Paraffin 28-Carbons	✔	62	253
92	Paraffin 29-Carbons	✔	63	240
93	Paraffin 30-Carbons	✔	65	251
94	Paraffin 31-Carbons	✔	68	242
95	Paraffin 32-Carbons	✔	70	170
96	Paraffin 33-Carbons	✔	74	268
97	Paraffin 34-Carbons	✔	76	269
98	p-Bromophenol	✔	64	86
99	p-Dichlorobenzene	✔	53	121
100	Pentadecanoic acid	✔	53	178
101	Phenol	✔	41	120
102	Phenylacetic acid	✔	77	102
103	p-Joluidine	✔	43	167
104	p-Lattic acid	✔	26	184
105	Polyethylene glycol 600	✔	20	146
106	p-Xylene dichloride	✔	100	139
107	Quinone	✔	115	171
108	Silver	✘	962	105
109	Sodium sulfate ($Na_2SO_4 \cdot 10H_2O$)	✘	32	252
110	ß-Chloroacetic acid	✔	56	147
111	Stearic acid	✔	69	199
112	Stibene	✔	124	167
113	Succinic anhydride	✔	119	204
114	Thiosinamine	✔	77	140
115	Thymol	✔	52	115
116	Titanium	✘	1668	296
117	TME (63%) /H_2O(37%)	✔	30	218
118	Trimyristin	✔	33	201
119	Tristearin	✔	56	191
120	Water	✘	0	334
121	Zinc	✘	420	112

Solution:

The heat absorbed during the sensible cooling could be computed from Eq. (7.2). In addition, the amount of thermal energy storage through phase change is determined via Eq. (7.6). Since both conditions have the same mass:

$$\text{Ratio} = \frac{h_{fg}}{c_p \Delta T} = \frac{2256.4}{(4.18)(100)} = 5.40$$

This example demonstrates the much bigger (5.40 times greater) thermal storage capacity of the phase change process of water compared to the condition it experiences sensible heating in a wide range, from 0°C to 100°C.

Latent thermal storage systems could be divided into three groups, which are organic, inorganic, and eutectic materials.

7.1.2.1 Organic thermal storage materials

Organic latent thermal storage materials are the substances in which carbon atoms are found, and if they are not produced at the lab scale, originate from the living micro- or macro-organism. Organic latent thermal storage materials are themselves classified into paraffin (wax) and non-paraffin substances [36–38]. They enjoy the advantages like not having overcooling, as well as high stability and high phase change heat per unit mass of the substance. Nonetheless, they usually have low thermal conductivity levels and high corrosion probability, while using them imposes more expensive costs than inorganic materials [39–41].

Paraffin substances are hydrocarbons that have a linear structure Fig. 7.5, and the chemical formula of C_nH_{2n+2}. In a paraffin substance, there is a single-bound connection between each atom and another one. In the practical applications, paraffin with a number of carbon atoms between 14 and 34 is employed, which offers the phase change temperature values from 5.5°C to 75.9°C. Paraffin is also known as alkanes.

Non-paraffin latent thermal storage materials cover a wide range of compounds, including fatty acids and esters. Lauric acid, with a melting point of 44.2°C is one of the most famous phase change materials (PCMs), which is found in palm trees, coconut oil, astrocaryum murumuru, attalea cohune, and so on. The ester obtained

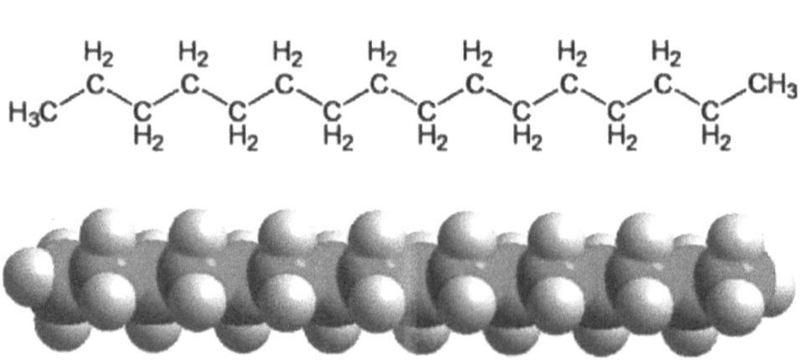

Figure 7.5 Schematic description of a paraffin substance that has a linear structure.

from methyl ester, which is vegetable oil, could be also given as an example of an ester PCM. It has a melting point equal to 27.6°C [42].

7.1.2.2 Inorganic latent thermal storage materials

On the other hand, the substances which do not have carbon atoms are put into the inorganic latent thermal storage materials. Inorganic materials are divided into two sub-categories, which are metallic and salt hydrates compounds [43–45]. They have the advantages like having a broader range of operation, lower cost, higher thermal conductivity, and heat of phase change per unit of mass compared to the organic materials, whereas they have shortcomings like higher degradation rate and corrosion possibility, as well as lower stability compared to organic substances [46–48].

As the name indicates, metallic compounds are the ones that contain metals. Gallium is an example of metallic latent heat storage material whose melting point is 29.7°C. Indium and Zinc are two other metallic PCMs, which have melting points of 157°C and 419.5°C, respectively. Salt hydrates are another member of inorganic latent thermal storage material. Sodium acetate with the chemical formula of $C_2H_3NaO_2$ is one example of such a substance while $CaCl_2$ is another. Salt hydrates can form crystals with a number of water molecules.

7.1.2.3 Eutectic thermal storage materials

In addition to the pure substances, including the two groups which have been already introduced, the mixtures could be also utilized as TESS. In this case, since the goal is storing and releasing energy at a constant temperature level, a condition in which the mixture turns from the liquid to the solid phase and vice versa is chosen, which is called the eutectic point. In general, a eutectic system refers to a mixture of substances, which is homogenous and has a lower melting point than each substance individually. The word is made by a combination of the two Greek phrases: "eu" and "tactic," which mean "well" and "melting," respectively [49–51].

Different kinds of eutectic systems are employed in different applications. For instance, nitrate-based molten salts are utilized as TESS in solar power plants. One of the most common mixtures is the binary combination of $NaNO_3$ and KNO_3, which have the melting temperature value of 307°C and 337°C at the atmospheric pressure. When a homogenous mixture with 47.5% mol fraction of the former and 52.5% mol fraction of the latter is created, a eutectic system is achieved. The outcome, that is obtained binary mixture with the given mol fractions, could offer the thermal storage at its melting level, which is 222°C at the atmospheric pressure.

To determine the eutectic point and the corresponding mol or weight fractions, two approaches could be employed:

- Drawing the phase diagrams and determining the point based on that,
- Using the governing equations and obtaining the point using it,

Both methods are briefly introduced, and one example for each one is solved for better understanding.

7.1.2.3.1 Determining the eutectic point using the phase diagram

Fig. 7.6 depicts the phase change diagram for a binary mixture. When there is the solution in the liquid phase, it is called L, which is composed of the substances A and B. The pure A in the solid form is indicated by α, while β shows the solid condition of pure B. As observed the vertical axis in a phase diagram is the temperature and the horizontal one is the molar or weight composition of A and B. In general, five regions could be identified in this figure:

- I: Liquid region (L): It is the area in which A and B are combined as a binary mixture, and no solid crystals, neither α nor β exists.
- II: Liquid region with α (α + L): If the mol or weight fraction of A is in a low range, by reducing the temperature, α crystals start to form. In other words, B is solvent and A is solute, and by decreasing the temperature, the solubility of A in B goes down, which leads to solid parts of A (α) in the liquid.
- III: Liquid region with β (β + L): It is similar to region II. The only difference is A is dominant in this region instead of B. Therefore, β solid particles are seen in the mixture.
- IV: Only α region: It is the part of the phase diagram where the only solid-state of A, that is α exists. In other words, if the temperature is low enough all A turns into α.
- V: Only β region: solid particles could be found in that region.
- VI: α + β region: In that part, the solid forms of α and β could be seen.

Since the eutectic point is the point in which all the liquid (L) turns into the solid (α + β), by drawing the phase diagram, the eutectic point could be found, as it is indicated in Fig. 7.6. Example 7.3 investigates a real case by this method.

Example 7.3: The phase diagram for the binary solution of $NaNO_3$ and KNO_3 is illustrated in Fig. 7.7. Find both temperature and the molar fraction of each substance at eutectic point for this mixture.

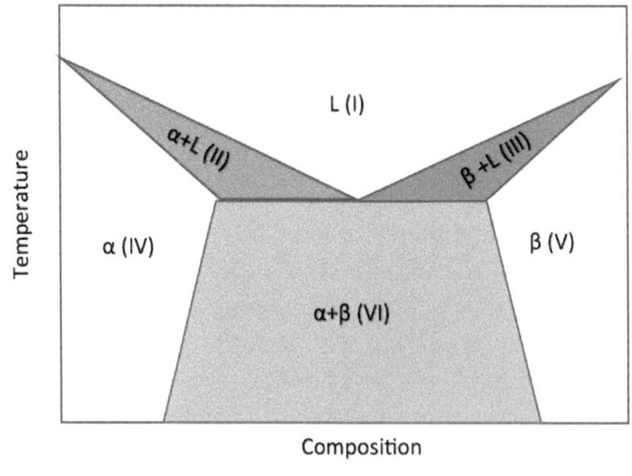

Figure 7.6 A phase diagram for the binary solution of L that is composed of A and B. A and B have the solid forms of α and β, respectively.

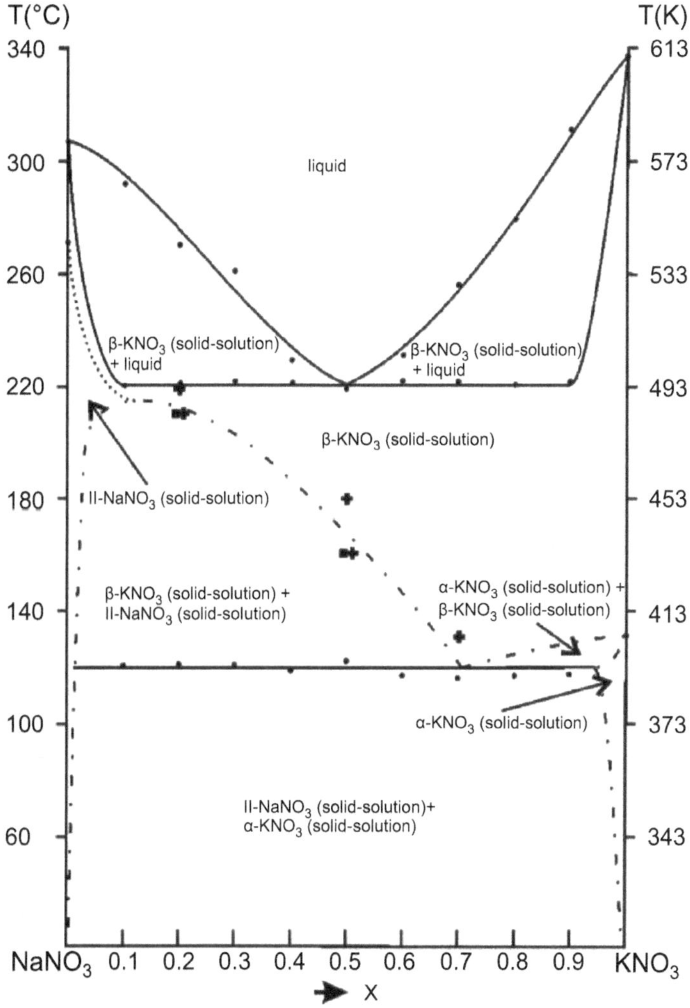

Figure 7.7 The phase diagram for the binary solution of $NaNO_3$ and KNO_3 [52].

Solution:

Based on Fig. 7.7 and by employing the drawing approach, the temperature at the eutectic point for the investigated mixture is obtained at 222°C, which has been introduced previously in this part. Based on Fig. 7.7 the molar fractions of $NaNO_3$ and KNO_3 are approximately 52.5% and 47.5%, respectively, which are in the agreement with the previously indicated data.

7.1.2.3.2 Using the governing equations

The characteristics of the eutectic point could be also determined using the governing equations. Although the drawing method gives approximate values and using that is much easier than writing and solving the governing equations, the mathematical way can offer the exact parameters.

In the mathematical approach, by following the stages indicated in the continue, the values are determined [53]:

1. Initially, Eq. (7.7) is written by knowing the values for the fusion process:

$$G = H - TS \Rightarrow \begin{cases} H = G + TS \\ \left(\frac{\partial G}{\partial T}\right)_P = -S \end{cases} \Rightarrow H = G - T\left(\frac{\partial G}{\partial T}\right)_P \tag{7.7}$$

2. Next, the partial derivative of the Gibbs's energy divided by temperature is found with respect to the temperature at constant pressure:

$$\frac{\left(\frac{\partial G}{T}\right)_P}{\partial T} = \frac{1}{T}\left(\frac{\partial G}{\partial T}\right)_P - \frac{1}{T^2}G = -\frac{1}{T^2}\left\{G - T\left(\frac{\partial G}{\partial T}\right)_P\right\} = -\frac{H}{T^2} \tag{7.8}$$

3. Then, the relation for the chemical potential is written, and after a simplification [Eq. (7.9)], for the equilibrium condition [Eq. (7.10)], finding a similar derivative compared to the free Gibbs' energy is done [Eq. (7.11)]:

$$\mu_i = \mu_i^0 + RT\ln\frac{a_i}{a} \approx \mu_i^0 + RT\ln x_i \tag{7.9}$$

$$\mu_i = \mu_i^0 + RT\ln x_i \Rightarrow \mu_i^0 = -RT\ln x_i \tag{7.10}$$

$$\left(\frac{\frac{\partial \mu_i}{T}}{\partial T}\right)_P = \frac{\partial}{\partial T}(R\ln x_i) \Rightarrow R\ln x_i = -\frac{H_i^0}{T} + K \tag{7.11}$$

4. For a pure substance that has the melting temperature of T^0 and phase-change enthalpy of H^0, the integration coefficient is obtained from Eq. (7.12):

$$x_i = 1 \Rightarrow T = T_i^0 \Rightarrow K = \frac{H_i^0}{T_i^0} \tag{7.12}$$

5. In addition:

$$R\ln x_i = -\frac{H_i^0}{T} + \frac{H_i^0}{T_i^0} \tag{7.13}$$

6. Therefore, for a mixture that is the combination of n substances, a system of equations could be found:

$$\begin{cases} \ln x_i + \frac{H_i^0}{RT} - \frac{H_i^0}{RT_i^0} = 0 \\ \sum_{i=1}^{n} x_i = 1 \end{cases} \tag{7.14}$$

$$\begin{cases} \forall i < n \Rightarrow \ln x_i + \dfrac{H_i^0}{RT} - \dfrac{H_i^0}{RT_i^0} = 0 \\ \ln\left(1 - \sum_{i=1}^{n-1} x_i\right) + \dfrac{H_n^0}{RT} - \dfrac{H_n^0}{RT_n^0} = 0 \end{cases}$$

7. By solving the system of equations presented in Eq. (7.14), all the values for the eutectic point are determined:

$$\begin{bmatrix} \Delta x_1 \\ \Delta x_2 \\ \Delta x_3 \\ \vdots \\ \Delta x_{n-1} \\ \Delta T \end{bmatrix} = \begin{bmatrix} \frac{1}{x_1} & 0 & 0 & \cdots & 0 & -\frac{H_1^0}{RT^2} \\ 0 & \frac{1}{x_2} & 0 & \cdots & 0 & -\frac{H_2^0}{RT^2} \\ 0 & 0 & \frac{1}{x_3} & \cdots & 0 & -\frac{H_3^0}{RT^2} \\ \vdots & \vdots & \vdots & \ddots & 0 & \vdots \\ 0 & 0 & 0 & \ddots & \frac{1}{x_{n-1}} & -\frac{H_{n-1}^0}{RT^2} \\ \frac{-1}{1-\sum_{i=1}^{n-1} x_i} & \frac{-1}{1-\sum_{i=1}^{n-1} x_i} & \frac{-1}{1-\sum_{i=1}^{n-1} x_i} & \cdots & \frac{-1}{1-\sum_{i=1}^{n-1} x_i} & -\frac{H_n^0}{RT^2} \end{bmatrix}^{-1} \times \begin{bmatrix} \ln x_1 + \frac{H_1^0}{RT} - \frac{H_1^0}{RT_1^0} \\ \ln x_2 + \frac{H_2^0}{RT} - \frac{H_2^0}{RT_2^0} \\ \ln x_3 + \frac{H_3^0}{RT} - \frac{H_3^0}{RT_3^0} \\ \vdots \\ \ln x_{n-1} + \frac{H_{n-1}^0}{RT} - \frac{H_{n-1}^0}{RT_{n-1}^0} \\ \ln\left(1 - \sum_{i=1}^{n-1} x_i\right) + \frac{H_n^0}{RT} - \frac{H_n^0}{RT_n^0} \end{bmatrix} \tag{7.15}$$

As the final point in this section, it is worth mentioning that using a eutectic system for TESS is accompanied by the advantages like high conductivity and heat storage capacity, while they are widely available in the market. However, they have also the disadvantages like corrosion and phase segregation.

7.1.3 Chemical reaction thermal (thermochemical) storage systems

When a number of substances take part in a chemical reaction, based on the chemical energy of the products compared to the reactants, heat is dissipated or absorbed. For instance, if the reaction shown in Eq. (7.16) is considered:

$$R_1 + R_2 + Q_{\text{reaction}} \rightarrow P_1 + P_2 \tag{7.16}$$

Eq. (7.16) demonstrates that when R_1 is added R_2, a chemical reaction happens in which P_1 and P_2 are obtained. This process needs the amount of Q_{reaction}. Therefore, heat could be given to the chemical reaction to produce and P_2. Similarly, when two substances P_1 and P_2 are mixed together, the reaction occurs on the reverse side, which leads to forming R_1 and R_2, and discharging the thermal energy. Consequently, a system of chemical reactions could be employed as a thermal storage system, which is known as a chemical reaction thermal storage system.

Chemical reaction thermal storage systems enjoy high-efficiency levels while they are environmentally friendly. Nonetheless, because of having more equipment, and more complexity they impose more initial and operating costs. Chemical reaction thermal storage systems are also known as thermochemical storage systems, as well [54–56].

The list of good candidates to be used as a thermochemical storage system is presented in Table 7.3.

Table 7.3 The list of good candidates to be used as a thermochemical storage system in addition to their characteristics of them. The data was found in the reference [57], and data in that reference had been itself from [58–62].

No.	Reaction	Turning temperature (K)	Group	Enthalpy change (kJ mol^{-1})
1	$2NH_3(g) \Leftrightarrow N_2(g) + 3H_2(g)$	468	One	59.0
2	$SO_3(g) \Leftrightarrow SO_2(g) + 1/2O_2(g)$	1040	One	98.9
3	$NH_4HSO_4(l) \Leftrightarrow NH_3(g) + H_2O(g) + SO_3(g)$	740	One	337.0
4	$CH_3OH(g) \Leftrightarrow 2H_2(g) + CO(g)$	420	One	90.2
5	$CH_4(g) + H_2O(g) \Leftrightarrow CO(g) + 3H_2(g)$	961	One	205.0
6	$CH_4(g) + CO_2(g) \Leftrightarrow 2CO(g) + 2H_2(g)$	959	One	247.0
7	$C_2H_6(g) \Leftrightarrow C_2H_4(g) + H_2(g)$	1136	One	138.0
8	$C_6H_{12}(g) \Leftrightarrow C_6H_6(g) + 3H_2(g)$	568	One	205.0
9	$C_7H_{14}(g) \Leftrightarrow C_7H_8(g) + 3H_2(g)$	549	One	205.0
10	$CoCl_2(g) \Leftrightarrow Co(g) + 3H_2(g)$	801	One	108.0
11	$NH_4I(s) \Leftrightarrow NH_3(g) + HI(g)$	649	Two	182.0
12	$NH_4F(s) \Leftrightarrow NH_3(g) + HF(g)$	499	Three	149.3
13	$NH_4Cl(s) \Leftrightarrow NH_3(g) + HCl(g)$	618	Three	176.0
14	$NH_4Br(s) \Leftrightarrow NH_3(g) + HBr(g)$	677	Three	188.0

(Continued)

Table 7.3 (Continued)

No.	Reaction	Turning temperature (K)	Group	Enthalpy change (kJ mol^{-1})
15	$BaO_2(s) \Leftrightarrow BaO(s) + 1/2O_2(g)$	1011	Three	80.9
16	$PbO_2(s) \Leftrightarrow PbO(s) + 1/2O_2(g)$	565	Three	55.1
17	$2CuO(s) \Leftrightarrow Cu_2O(s) + 1/2O_2(g)$	1298	Three	143.1
18	$MnO_2(s) \Leftrightarrow MnO(s) + 1/2O_2(g)$	1226	Three	133.9
19	$Li_2O_2(s) \Leftrightarrow Li_2O(s) + 1/2O_2(g)$	460	Three	38.4
20	$Na_2O_2(s) \Leftrightarrow Na_2O(s) + 1/2O_2(g)$	1200	Three	99.6
21	$SrO_2(s) \Leftrightarrow SrO(s) + 1/2O_2(g)$	416	Three	48.1
22	$MgO(s) \Leftrightarrow Mg(g) + 1/2O_2(g)$	3360	Three	752.0
23	$CaO(s) \Leftrightarrow Ca(g) + 1/2O_2(g)$	3800	Three	828.0
24	$SrO(s) \Leftrightarrow Sr(g) + 1/2O_2(g)$	355	Three	755.0
25	$BaO(s) \Leftrightarrow Ba(g) + 1/2O_2(g)$	3600	Three	729.0
26	$Li_2O(s) \Leftrightarrow 2Li(g) + 1/2O_2(g)$	2650	Three	906.0
27	$Na_2O(s) \Leftrightarrow 2Na(g) + 1/2O_2(g)$	1880	Three	634.0
28	$K_2O(s) \Leftrightarrow 2K(g) + 1/2O_2(g)$	1652	Three	544.0
29	$Rb_2O(s) \Leftrightarrow 2Rb(g) + 1/2O_2(g)$	1494	Three	502.0
30	$Cs_2O(s) \Leftrightarrow 2Cs(g) + 1/2O_2(g)$	1432	Three	481.0
31	$ZnCO_3(s) \Leftrightarrow ZnO(s) + CO_2(g)$	406	Three	67.0
32	$MgCO_3(s) \Leftrightarrow MgO(s) + CO_2(g)$	670	Three	125.0
33	$CaCO_3(s) \Leftrightarrow CaO(s) + CO_2(g)$	1110	Three	176.0
34	$SrCO_3(s) \Leftrightarrow SrO(s) + CO_2(g)$	1381	Three	234.0
35	$BaCO_3(s) \Leftrightarrow BaO(s) + CO_2(g)$	1570	Three	269.0
36	$Li_2CO_3(s) \Leftrightarrow Li_2O(s) + CO_2(g)$	1390	Three	224.0
37	$Na_2CO_3(s) \Leftrightarrow Na_2O(s) + CO_2(g)$	2710	Three	322.0
38	$K_2CO_3(s) \Leftrightarrow K_2O(s) + CO_2(g)$	2585	Three	394.0
39	$Rb_2CO_3(s) \Leftrightarrow Rb_2O(s) + CO_2(g)$	2634	Three	405.0
40	$Cs_2CO_3(s) \Leftrightarrow Cs_2O(s) + CO_2(g)$	2655	Three	408.0
41	$Mg(OH)_2(s) \Leftrightarrow MgO(s) + H_2O(g)$	531	Three	84.0
42	$Ca(OH)_2(s) \Leftrightarrow CaO(s) + H_2O(g)$	752	Three	109.0
43	$Sr(OH)_2(s) \Leftrightarrow SrO(s) + H_2O(g)$	852	Three	126.0
44	$KBF_4(s) \Leftrightarrow KF(s) + BF_3(g)$	978	Three	109.0
45	$2NF_3(s) \Leftrightarrow N_2(s) + 3F_2(g)$	945	Three	130.0
46	$MgSO_4(s) \Leftrightarrow MgO(s) + SO_3(g)$	1470	Three	280.0
47	$CaSO_4(s) \Leftrightarrow CaO(s) + SO_3(g)$	2145	Three	402.0
48	$SrSO_4(s) \Leftrightarrow SrO(s) + SO_3(g)$	2450	Three	459.0
49	$BaSO_4(s) \Leftrightarrow BaO(s) + SO_3(g)$	2660	Three	517.0
50	$Li_2SO_4(s) \Leftrightarrow Li_2O(s) + SO_3(g)$	2470	Three	444.0
51	$Na_2SO_4(s) \Leftrightarrow Na_2O(s) + SO_3(g)$	3190	Three	574.0
52	$K_2SO_4(s) \Leftrightarrow K_2O(s) + SO_3(g)$	3867	Three	676.0
53	$Rb_2SO_4(s) \Leftrightarrow Rb_2O(s) + SO_3(g)$	3900	Three	700.0
54	$Cs_2SO_4(s) \Leftrightarrow Cs_2O(s) + SO_3(g)$	3940	Three	708.0
55	$Ca(NO_3)_2(s) \Leftrightarrow CaO(s) + 2NO_2(g) + 1/2O_2$	740	Three	109.0
56	$LiNO_2(s) + LiNO_3(s) \Leftrightarrow 2LiO(s) + 2NO_2(g)$	1089	Three	109.0

(Continued)

Table 7.3 (Continued)

No.	Reaction	Turning temperature (K)	Group	Enthalpy change (kJ mol^{-1})
57	$CaCl_2.8NH_3(s) \Leftrightarrow CaCl_2.4NH_3(s) + 4NH_3(g)$	305	Four	41.0
58	$NH_4Cl.3NH_3(l) \Leftrightarrow NH_3Cl(s) + 3NH_3(g)$	2	Four	29.4
59	$MnCl_2.6NH_3(s) \Leftrightarrow MnCl_2.2NH_3(s) + 4NH_3(g)$	364	Four	47.4
60	$MgCl_2.6NH_3(s) \Leftrightarrow MgCl_2.2NH_3(s) + 4NH_3(g)$	408	Four	55.7
61	$LiH(s) \Leftrightarrow Li(s) + 1/2H_2(g)$	1220	Five	88.0
62	$NaH(s) \Leftrightarrow Na(s) + 1/2H_2(g)$	735	Five	88.0

7.2 Different employed technologies for thermal energy storage

After introducing different type of TESSs and describing the working principle of each of them, in this part, details about the employed technologies for thermal energy storage is provided.

Having the characteristics like being widely available, environmentally friendly, and having a high heat absorption capacity have made water the most common thermal energy storage material. As one of the favorite uses of water as the energy storage media, utilizing the water tank heat storage stage could be given as an example. Water tank heat storage systems are commonly used in domestic applications, where the hot water for domestic and space heating demands are provided. A general schematic of a water tank storage unit is depicted in Fig. 7.8. As this figure shows, the hot fluid from a source makes the water hot to the T_s level, which is sent for either domestic or space heating applications. The hot fluid which makes the water in the tank hot could be supplied from a fossil fuel burning source like a boiler, or a renewable energy source like a solar thermal collector. In addition, the heat absorbed from the back of a photovoltaic system could be also utilized [63].

When a renewable energy source is employed as the primary supplier for making the water in the tank hot, having a secondary heating provider is necessary because the sun is not available at night and during cloudy hours [65–67]. A schematic of such a system, in which the solar thermal collector and a fossil fuel burning boiler are used as the primary and auxiliary units to make the water hot is illustrated in Fig. 7.9.

Another common way to store thermal energy is an underground TESS. As the name suggests, in an underground energy storage system, thermal energy is stored underground. The medium could be solid, that is the soil, clay, sand, or rock, or the liquid, that is the groundwater.

In such a system, usually, a working fluid is sent to the ground, by which the charging or discharging process is done. Water is the most common fluid for this purpose because it is non-toxic and has an extremely hot capacity. As Fig. 7.10

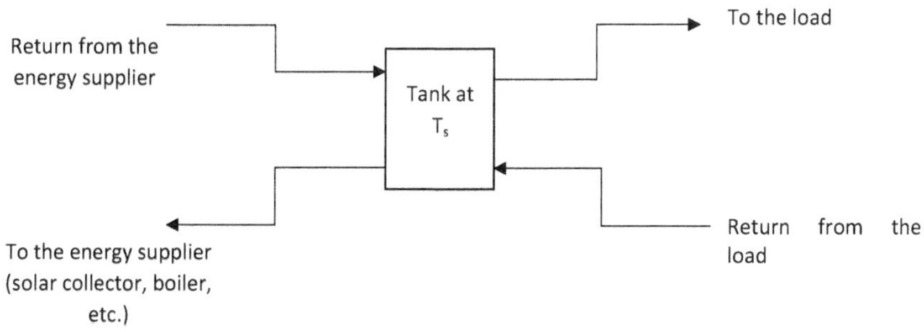

Figure 7.8 Schematic of a simple water tank heat storage system. The picture is from the reference [64], but it has been redrawn.

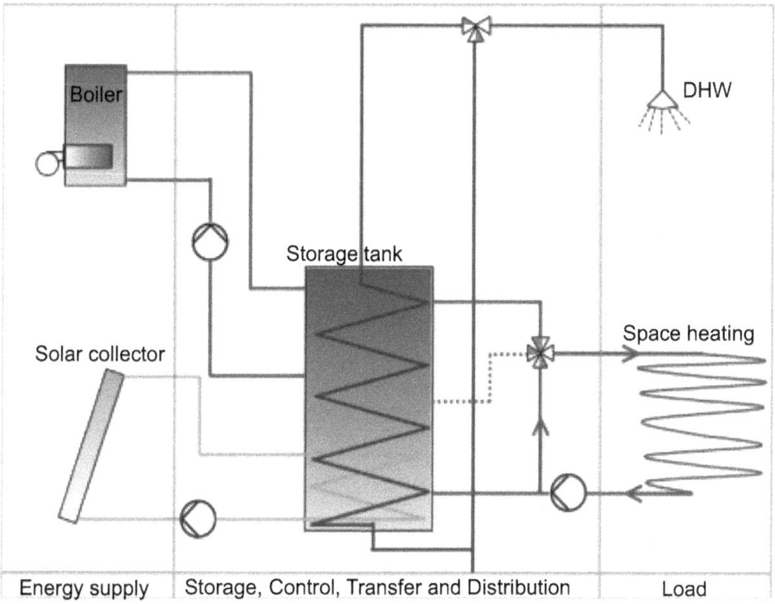

Figure 7.9 Schematic of a solar thermal collector-based water tank heat storage system with an auxiliary boiler [68].

shows, that the pipes, which are frequently made of steel or plastic, could be put in various conditions, including a vertical or horizontal position under the ground [69–71].

Underground thermal storage units are vastly utilized in the heat pumps, where the heat dissipation in the condenser in the cooling mode or the heat absorption in the evaporator in the heating mode is done by the interaction with the ground, and the mentioned process. Such heat pumps are known as ground-source types [72–74].

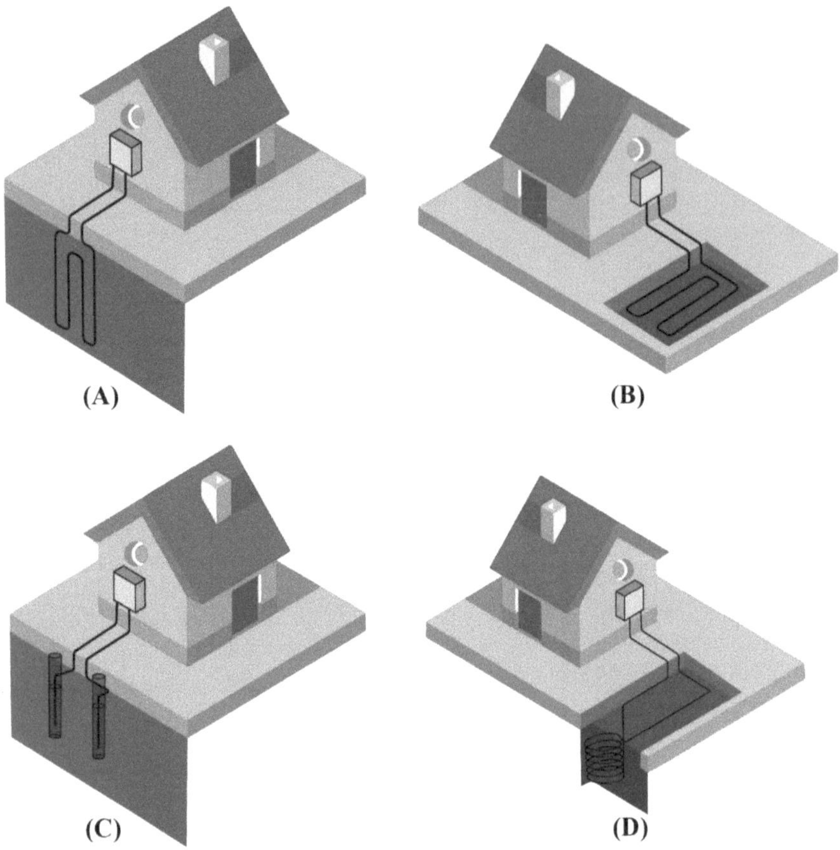

Figure 7.10 Introducing the underground thermal energy storage systems and different utilized configurations; (A) vertical loop; (B) horizontal (slinky ground) loop; (C) helical loop; (D) vertical loop inside wells.

An aquifer is another kind of underground thermal storage system. As it could be guessed from the name "Aqua," means water, in this system the storage medium is water. An aquifer is made of two basins, both located underground as shown in Fig. 7.11. One of the basins is full while another is empty. During the hot season, the heat is stored in the full basin, and in the cold season, it is used as the working fluid for space heating or other heating applications (Fig. 7.11A). Then, water is sent to the basin which was empty before, and through heat transfer with the weather which is now cold, its temperature decreases more. Afterward, during the hot season, the cold stored water in the basin is pumped into the building and it is utilized for cooling purposes (Fig. 7.11B). If the stored water was not hot or cold enough during the cold or hot seasons, auxiliary units could be also added. Aquifers have a high potential to be employed in district heating or cooling applications [75–77].

Another efficient and commercially developed thermal storage system is the packed bed thermal storage system, which is schematically shown in Fig. 7.12. A

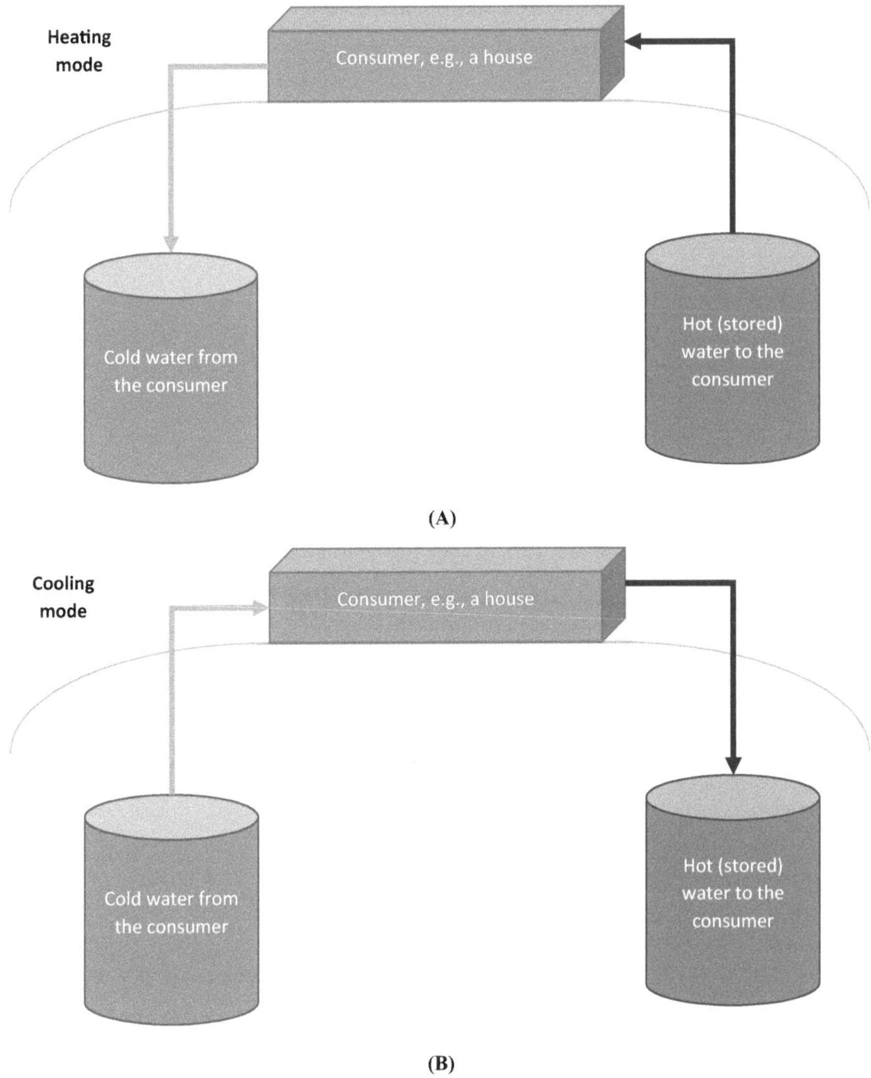

Figure 7.11 The working principle of an aquifer (A) utilizing at the heating mode; (B) using at the cooling application.

packed bed thermal storage system, which is also known as a pebble-bed type, is made of a number of solid matters (like stone or pebble, or similar materials), which are put together and form a packed bed. During the charging process, the heat or cold is stored in the solid matter, and the end of the charging period is the time when the temperature of all the solid matters is equal. In the discharge process, a fluid, like water or air passes through the bed and absorbs heat or cool for the designed purpose [78–80].

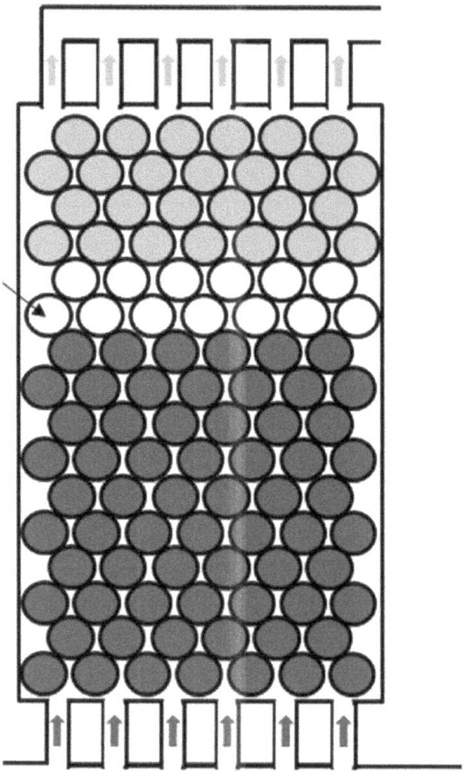

Figure 7.12 Schematic of a packed (pebble) bed thermal storage system [81].

One of the most common applications of pebble bed thermal storage systems is in the solar thermal systems for space heating. When the sun is in the sky, the thermal collector works and meets the demands. Then, and during the night, the heat absorbed in the pebbles or stones during the day is transferred to the air that passes through them. Then, it enters the space and provides heating [64].

Another thermal energy storage technology that is widely used for space heating is the Trombe wall. A schematic of a Trombe wall is illustrated in Fig. 7.13. As Fig. 7.13 depicts, in a Trombe wall system a glass is placed at the external side of the wall, which makes a rectangular channel. During the cold season, the weather from the room enters the Trombe wall, and by absorbing the energy from the sun and hot wall through the radiation and convection, it makes hot and goes to the room again for space heating [82–84].

Trombe wall could be helpful during the hot season, as well. In the hot season, it works in a way that the air which removed the cooling load from the space enters the channel and it makes the wall, which gets hot from radiation by the sun, colder. Then, instead of going again to the room, it is discharged into the atmosphere.

PCMs could be also used as masonry units. They have a much greater heat absorption capacity compared to the other materials which are widely used in the building, as reported in Fig. 7.14.

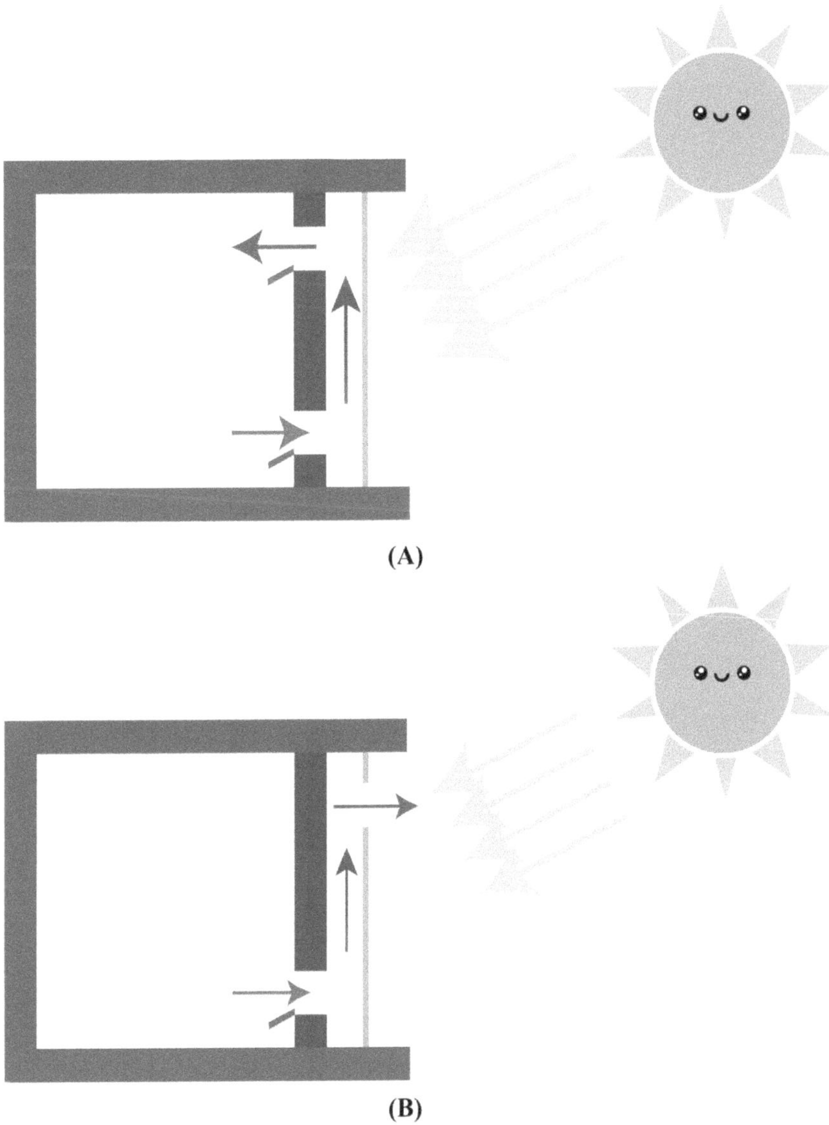

Figure 7.13 Schematic of a Trombe wall in which the operation in (A) heating mode; (B) cooling modes.

One of the applications of PCM, which is similar to the Trombe walls, is using them as the building façade. The outcome is a system that is called a solar façade thermal storage system. This system is shown in Fig. 7.15. During the charging period, that is when the sun is in the sky, the absorbed heat leads to the melting PCM. Then, by dissipating the gained energy during the sun's absence period, the thermal energy is supplied for the considered usage, for example, space heating or domestic hot water provision [86–88].

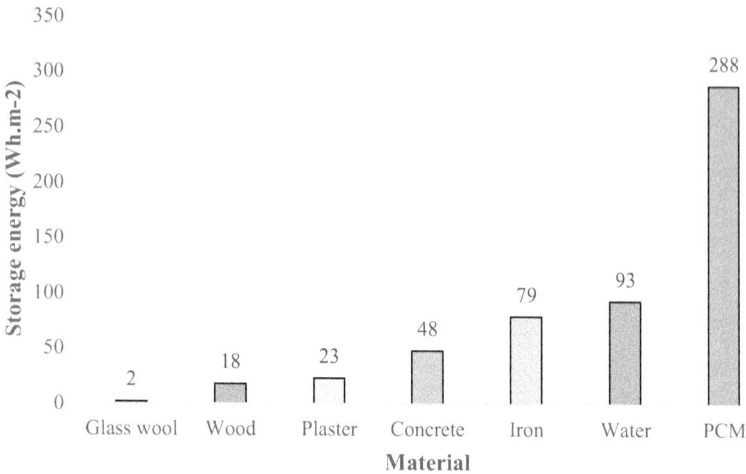

Figure 7.14 Comparison of the heat absorption capacity of a conventional PCM with other typical masonry units [85].

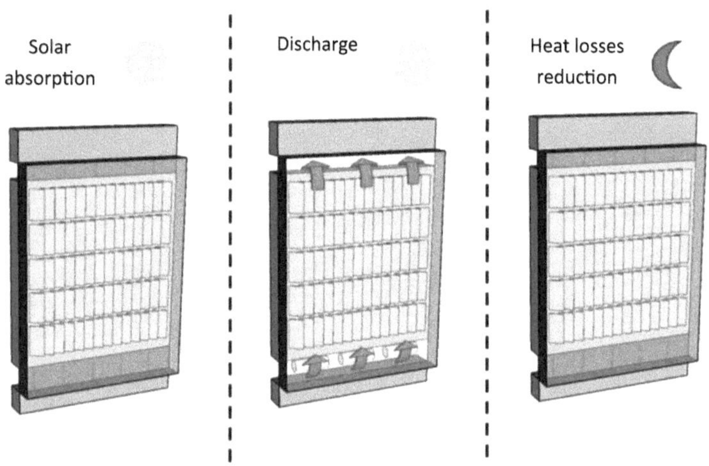

Figure 7.15 A solar façade thermal storage system [89].

An ice storage system is also taken into account as another commercially developed heat storage system. As the name suggests, it makes ice during the charging time, and then, the made ice is converted into the water during the discharge period, which provides cooling. An ice storage unit is depicted in Fig. 7.16. When electrical demand is lower than the production, surplus electricity is utilized in a mechanical chiller to make ice. The make ice is then used during the peak electricity consumption period. It is worth mentioning that because a large part of the electrical energy is required to provide cooling in the summers, using the ice storage system helps to diminish the problems in the peak electrical demands in that season [90–92].

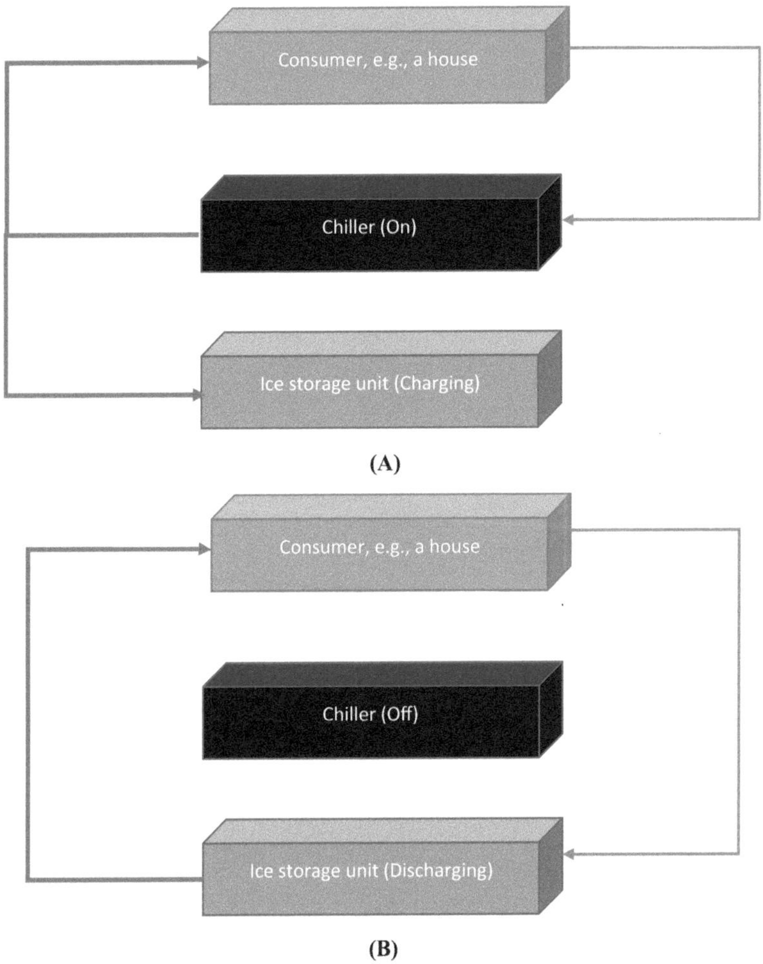

Figure 7.16 An ice storage system in (A) charge and (B) discharge modes.

As another item, using latent heat storage materials in power generation applications could be investigated. Fig. 7.17 introduce this system, where the fluid is sent from a low-temperature tank to the pipeline. In the pipeline, through receiving a high level of solar radiation, the fluid gets hot enough to either produce power itself or gets its heat to the main working fluid for that purpose. The molten salt is also prepared in the same way as the fluid in the pipeline is prepared, and it is utilized in the cases sun radiation is not high enough or the sun is not in the sky [93].

A potential design for integration of a parabolic trough solar power generation unit with a thermochemical (chemical reaction) heat storage system is shown in Fig. 7.18. When the sun is in the sky, the main working fluid absorbs the sun's radiation and produces power in a turbine. At the same time, in another circuit, another working fluid, which is called the secondary working fluid, absorbs another part of

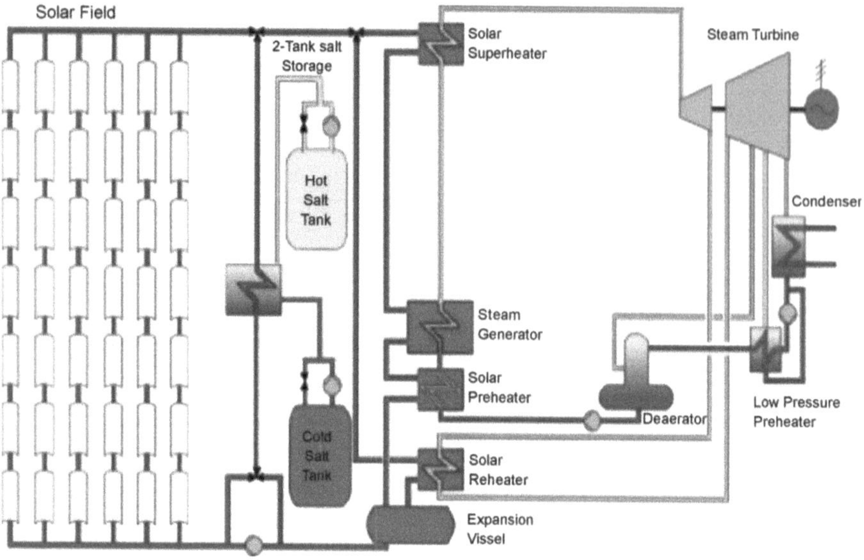

Figure 7.17 A parabolic trough solar power generation unit with molten salt as the thermal energy storage material [93].

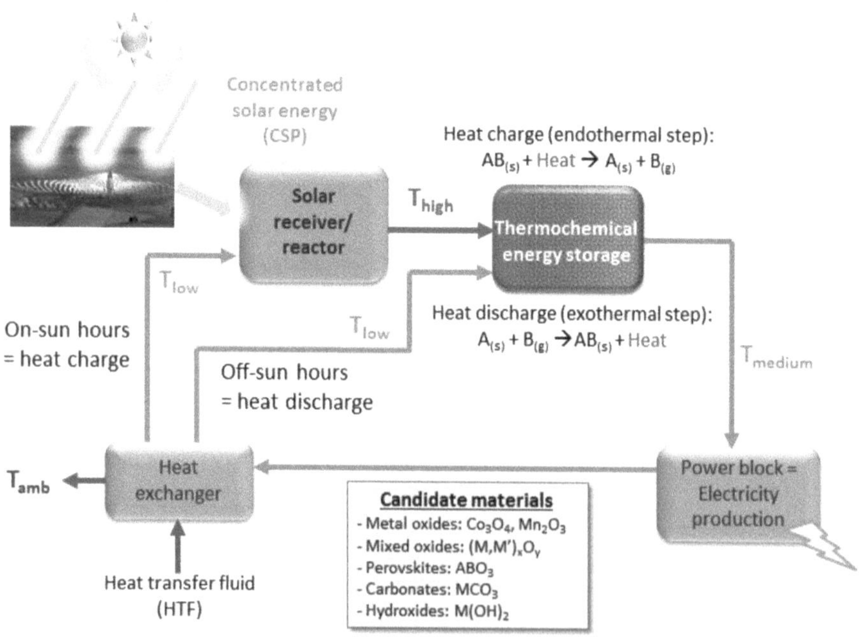

Figure 7.18 A potential design for integration of a parabolic trough solar power generation unit with thermochemical (chemical reaction) heat storage system [94].

the radiation and provides the heat for an endothermic chemical reaction. When the sun is not in the sky, the primary working fluid, which is now cold, has heat transfer with the reactants of the thermochemical energy storage charging process and gets energy from them. Consequently, the chemical reaction goes in the reverse side, which leads to making the main working fluid hot, and the hot primary working fluid goes to the turbine then, a produces power [94].

7.3 Conclusion

A review of TESSs was presented here, where sensible, latent, and chemical reaction (thermochemical) types of that were introduced. Each type was introduced by giving the working principle of that and some examples. Then, different most common thermal storage technologies, which have been commercially developed, were introduced. It was observed that a wide range of technologies, including water tank storage, ground source heat pump, aquifers, packed (pebble) bed, Trombe wall, solar façade, ice storage, molten salt, and solar thermochemical TESSs were employed as the heat storage units in a variety of applications from buildings to power plants, and so on.

References

[1] A.R. Razmi, M. Soltani, A. Ardehali, K. Gharali, M.B. Dusseault, J. Nathwani, Design, thermodynamic, and wind assessments of a compressed air energy storage (CAES) integrated with two adjacent wind farms: a case study at Abhar and Kahak sites, Iran, Energy. 221 (2021) 119902.

[2] A. Sohani, S. Hoseinzadeh, K. Berenjkar, Experimental analysis of innovative designs for solar still desalination technologies; an in-depth technical and economic assessment, J. Energy Storage. 33 (2021) 101862.

[3] R. Roushenas, A.R. Razmi, M. Soltani, M. Torabi, M.B. Dusseault, J. Nathwani, Thermo-environmental analysis of a novel cogeneration system based on solid oxide fuel cell (SOFC) and compressed air energy storage (CAES) coupled with turbocharger, Appl. Ther. Eng. 181 (2020) 115978.

[4] H. Yağlı, Y. Koç, H. Kalay, Optimisation and exergy analysis of an organic Rankine cycle (ORC) used as a bottoming cycle in a cogeneration system producing steam and power, Sust. Energy Technol. Assess. 44 (2021) 100985.

[5] Q. Meng, Y. Li, X. Ren, C. Xiong, W. Wang, J. You, A demand-response method to balance electric power-grids via HVAC systems using active energy-storage: simulation and on-site experiment, Energy Reports. 7 (2021) 762–777.

[6] J.Z. Alvi, Y. Feng, Q. Wang, M. Imran, G. Pei, Effect of working fluids on the performance of phase change material storage based direct vapor generation solar organic Rankine cycle system, Energy Reports. 7 (2021) 348–361.

[7] A. Bolzoni, A. Parisio, R. Todd, A. Forsyth, Model predictive control for optimizing the flexibility of sustainable energy assets: an experimental case study, Int. J. Electr. Power Energy Syst. 129 (2021) 106822.

[8] A. Sohani, A. Dehnavi, H. Sayyaadi, S. Hoseinzadeh, E. Goodarzi, D.A. Garcia, et al., The real-time dynamic multi-objective optimization of a building integrated photovoltaic thermal (BIPV/T) system enhanced by phase change materials, J. Energy Storage. 46 (2022) 103777.

[9] Y. Cui, J. Zhu, F. Zhang, Y. Shao, Y. Xue, Current status and future development of hybrid PV/T system with PCM module: 4E (energy, exergy, economic and environmental) assessments, Renew. Sust. Energy Rev. 158 (2022) 112147.

[10] H.A. Nasef, H. Hassan, S. Mori, S.A. Nada, Experimental investigation on the performance of parallel and staggered arrays of PCM energy storage system for PV thermal regulation, Energy Convers. Manag. 254 (2022) 115143.

[11] W.-B. Li, D. Zhou, W.-F. Liu, J.-Z. Su, F. Hussain, D.-W. Wang, et al., High-temperature BaTiO3-based ternary dielectric multilayers for energy storage applications with high efficiency, Chem. Eng. J. 414 (2021) 128760.

[12] E.C. Santos, E.N. Macêdo, M.A.B. Galhardo, T.O. Costa, A.F.P. Costa, A.U. Brito, et al., Analysis of the behavior of an air conditioning system based on the solar energy natural regulation in amazonian climatic conditions, J. Sol. Energy Eng. 143 (3) (2021).

[13] H. Selvnes, Y. Allouche, R.I. Manescu, A. Hafner, Review on cold thermal energy storage applied to refrigeration systems using phase change materials, Therm. Sci. Eng. Prog. 22 (2021) 100807.

[14] H. Ono, Y. Ohtani, M. Matsuo, T. Yamaguchi, R. Yokoyama, Optimal operation of heat source and air conditioning system with thermal storage tank using nonlinear programming, Energy. 222 (2021) 119936.

[15] X. Zhang, G. Su, J. Lin, A. Liu, C. Wang, Y. Zhuang, Three-dimensional numerical investigation on melting performance of phase change material composited with copper foam in local thermal non-equilibrium containing an internal heater, Int. J. Heat Mass Transfer. 170 (2021) 121021.

[16] R.K. De, A. Ganguly, Modeling and analysis of a solar thermal-photovoltaic-hydrogen-based hybrid power system for running a standalone cold storage, J. Clean. Prod. 293 (2021) 126202.

[17] P.K.S. Rathore, S.K. Shukla, Potential of macroencapsulated PCM for thermal energy storage in buildings: a comprehensive review, Constr. Build. Mater. 225 (2019) 723–744.

[18] A. Mehari, Z.Y. Xu, R.Z. Wang, Thermal energy storage using absorption cycle and system: a comprehensive review, Energy Convers. Manag. 206 (2020) 112482.

[19] Y. Lin, Y. Jia, G. Alva, G. Fang, Review on thermal conductivity enhancement, thermal properties and applications of phase change materials in thermal energy storage, Renew. Sust. Energy Rev. 82 (2018) 2730–2742.

[20] D. Thanganadar, F. Fornarelli, S. Camporeale, F. Asfand, J. Gillard, K. Patchigolla, Thermo-economic analysis, optimisation and systematic integration of supercritical carbon dioxide cycle with sensible heat thermal energy storage for CSP application, Energy. 238 (2022) 121755.

[21] J.-Y. Ryu, A. Alford, G. Lewis, Y. Ding, Y. Li, A. Ahmad, et al., A novel liquid air energy storage system using a combination of sensible and latent heat storage, Appl. Ther. Eng. 203 (2022) 117890.

[22] D.R. Parida, S. Advaith, N. Dani, S. Basu, Assessing the impact of a novel hemispherical diffuser on a single-tank sensible thermal energy storage system, Renew. Energy. 183 (2022) 202–218.

[23] Y.A. Cengel, M.A. Boles, Thermodynamics: An Engineering Approach 6th Editon (SI Units), The McGraw-Hill Companies, Inc., New York, 2007.

[24] Wikipedia, Values of isobaric heat capacity for different materials. <https://en.wikipedia.org/wiki/Table_of_specific_heat_capacities>, 2021 (accessed 18.02.21).
[25] D.W. Waples, J.S. Waples, A review and evaluation of specific heat capacities of rocks, minerals, and subsurface fluids. Part 1: minerals and nonporous rocks, Nat. Resour. Res. 13 (2) (2004) 97–122.
[26] D.W. Waples, J.S. Waples, A review and evaluation of specific heat capacities of rocks, minerals, and subsurface fluids. Part 2: fluids and porous rocks, Nat. Resour. Res. 13 (2) (2004) 123–130.
[27] J.C.O. Santos, M.G.O. Santos, J.P. Dantas, M.M. Conceição, P.F. Athaide-Filho, A.G. Souza, Comparative study of specific heat capacities of some vegetable oils obtained by DSC and microwave oven, J. Ther. Anal. Calorim. 79 (2) (2005) 283–287.
[28] H. Koide, A. Kurniawan, T. Takahashi, T. Kawaguchi, H. Sakai, Y. Sato, et al., Performance analysis of packed bed latent heat storage system for high-temperature thermal energy storage using pellets composed of micro-encapsulated phase change material, Energy. 238 (2022) 121746.
[29] B.D. Mselle, G. Zsembinszki, D. Vérez, E. Borri, L.F. Cabeza, A detailed energy analysis of a novel evaporator with latent thermal energy storage ability, Appl. Therm. Eng. 201 (2022) 117844.
[30] L. Coelho, M.K. Koukou, G. Dogkas, J. Konstantaras, M.G. Vrachopoulos, A. Rebola, et al., Latent thermal energy storage application in a residential building at a Mediterranean climate, Energies. 15 (3) (2022) 1008.
[31] C. Suresh, R.P. Saini, Thermal performance of sensible and latent heat thermal energy storage systems, Int. J. Energy Res. 44 (6) (2020) 4743–4758.
[32] B. Buonomo, H. Celik, D. Ercole, O. Manca, M. Mobedi, Numerical study on latent thermal energy storage systems with aluminum foam in local thermal equilibrium, Appl. Therm. Eng. 159 (2019) 113980.
[33] H. Jouhara, A. Żabnieńska-Góra, N. Khordehgah, D. Ahmad, T. Lipinski, Latent thermal energy storage technologies and applications: a review, Int. J. Thermofluid. 5–6 (2020) 100039.
[34] Wikipedia, Phase change enthalpy for a number of selected materials. <https://en.wikipedia.org/wiki/Phase-change_material>, 2021 (accessed 18.02.21).
[35] UCLA, Thermal properties of phase change materials (PCMs). <http://www.seas.ucla.edu/~pilon/downloads.htm#section4>, 2021 (accessed 18.02.20).
[36] S. Daniarta, M. Nemś, P. Kolasiński, M. Pomorski, Sizing the thermal energy storage device utilizing phase change material (PCM) for low-temperature organic Rankine cycle systems employing selected hydrocarbons, Energies. 15 (3) (2022) 956.
[37] S. Venettacci, R. Cozzolino, G.S. Ponticelli, S. Guarino, Environmental and economic life cycle assessment of thermal energy storage based on organic phase change material embedded in open-cell copper foams, Sust. Prod. Consumpt. 29 (2022) 387–405.
[38] I. Delgado-Ferreiro, J. Garcia-Ben, S. Castro-Garcia, M. Sánchez-Andújar, M.A. Señarís-Rodríguez, J.M. Bermudez-Garcia, New organic-inorganic halometallates [A][FeCl4] plastic crystals for electric and thermal energy storage. https://qies22.icms.us-csic.es/files/2022/01/F-A11.pdf
[39] S.Y. Kee, Y. Munusamy, K.S. Ong, Review of solar water heaters incorporating solid-liquid organic phase change materials as thermal storage, Appl. Therm. Eng. 131 (2018) 455–471.
[40] S.S. Magendran, F.S.A. Khan, N.M. Mubarak, M. Vaka, R. Walvekar, M. Khalid, et al., Synthesis of organic phase change materials (PCM) for energy storage applications: a review, Nano-Struct. Nano-Obj. 20 (2019) 100399.

[41] D.G. Atinafu, Y.S. Ok, H.W. Kua, S. Kim, Thermal properties of composite organic phase change materials (PCMs): a critical review on their engineering chemistry, Appl. Therm. Eng. 181 (2020) 115960.

[42] S. Serrano, C. Barreneche, A. Inés Fernández, M.M. Farid, L.F. Cabeza, Composite gypsum containing fatty-ester PCM to be used as constructive system: thermophysical characterization of two shape-stabilized formulations, Energy Build. 86 (2015) 190−193.

[43] J. Zhao, J. Sun, Y. Li, R. Xia, W. Zhang, B. Wang, et al., Wood-plastic materials with organic−inorganic hybrid phase change thermal storage as novel green energy storage composites for building energy conservation, J. Mater. Sci. (2022) 1−16.

[44] S. Khanna, S. Paneliya, P. Marathey, K. Shah, P. Prajapati, R. Chaudhari, et al., Investigation of thermophysical properties of synthesized N-hexacosane-encapsulated titania phase change material for enhanced thermal storage application, Recent Adv. Mech. Infrastruct.: Springer; (2022) 107−118.

[45] G. Sadeghi, Energy storage on demand: thermal energy storage development, materials, design, and integration challenges, Energy Storage Mater. (2022).

[46] Y. Lin, G. Alva, G. Fang, Review on thermal performances and applications of thermal energy storage systems with inorganic phase change materials, Energy. 165 (2018) 685−708.

[47] S.A. Mohamed, F.A. Al-Sulaiman, N.I. Ibrahim, M.H. Zahir, A. Al-Ahmed, R. Saidur, et al., A review on current status and challenges of inorganic phase change materials for thermal energy storage systems, Renew. Sust. Energy Rev. 70 (2017) 1072−1089.

[48] Y.E. Milián, A. Gutiérrez, M. Grágeda, S. Ushak, A review on encapsulation techniques for inorganic phase change materials and the influence on their thermophysical properties, Renew. Sust. Energy Rev. 73 (2017) 983−999.

[49] D.K. Döğüşcü, A. Altıntaş, A. Sarı, C. Alkan, Polystyrene microcapsules with palmitic-capric acid eutectic mixture as building thermal energy storage materials, Energy Build. 150 (2017) 376−382.

[50] R.P. Singh, S.C. Kaushik, D. Rakshit, Solidification behavior of binary eutectic phase change material in a vertical finned thermal storage system dispersed with graphene nano-plates, Energy Convers. Manag. 171 (2018) 825−838.

[51] S. Ushak, M. Vega, J.A. Lovera-Copa, S. Pablo, M. Lujan, M. Grageda, Thermodynamic modeling and experimental verification of new eutectic salt mixtures as thermal energy storage materials, Solar Energy Mater. Solar Cells. 209 (2020) 110475.

[52] R. Benages-Vilau, T. Calvet, M.A. Cuevas-Diarte, H.A.J. Oonk, The $NaNO_3-KNO_3$ phase diagram, Phase Trans. 89 (1) (2016) 1−20.

[53] L. Brunet, J. Caillard, P. André, Thermodynamic calculation of n-component eutectic mixtures, Int. J. Modern Phys. C. 15 (05) (2004) 675−687.

[54] L. Dai, X.-F. Long, B. Lou, J. Wu, Thermal cycling stability of thermochemical energy storage system Ca(OH)2/CaO, Appl. Therm. Eng. 133 (2018) 261−268.

[55] A. Mukherjee, R. Majumdar, S.K. Saha, C. Subramaniam, L. Kumar, Performance evaluation of an open thermochemical energy storage system integrated with flat plate solar collector, Appl. Therm. Eng. 173 (2020) 115218.

[56] J. Sunku Prasad, P. Muthukumar, F. Desai, D.N. Basu, M.M. Rahman, A critical review of high-temperature reversible thermochemical energy storage systems, Appl. Energy. 254 (2019) 113733.

[57] T. Yan, R.Z. Wang, T.X. Li, L.W. Wang, I.T. Fred, A review of promising candidate reactions for chemical heat storage, Renew. Sust. Energy Rev. 43 (2015) 13−31.

[58] W.E. Wentworth, E. Chen, Simple thermal decomposition reactions for storage of solar thermal energy, Solar Energy. 18 (3) (1976) 205–214.

[59] K. Kyaw, H. Matsuda, M. Hasatani, Applicability of carbonation/decarbonation reactions to high-temperature thermal energy storage and temperature upgrading, J. Chem. Eng. Jpn. 29 (1) (1996) 119–125.

[60] C. Wyman, J. Castle, F. Kreith, A review of collector and energy storage technology for intermediate temperature applications, Solar Energy. 24 (6) (1980) 517–540.

[61] G. Cacciola, N. Giordano, Chemical processes for energy storage and transmission, Appl. Energy. 25 (4) (1986) 315–337.

[62] M.A. Fahim, J.D. Ford, Energy storage using the BaO_2-BaO reaction cycle, Chem. Eng. J. 27 (1) (1983) 21–28.

[63] A. Sohani, H. Sayyaadi, Employing genetic programming to find the best correlation to predict temperature of solar photovoltaic panels, Energy Convers. Manag. 224 (2020) 113291.

[64] I. Sarbu, C. Sebarchievici, Solar Heating and Cooling Systems: Fundamentals, Experiments and Applications, Academic Press;, 2016.

[65] A. Sohani, M.H. Shahverdian, H. Sayyaadi, D.A. Garcia, Impact of absolute and relative humidity on the performance of mono and poly crystalline silicon photovoltaics; applying artificial neural network, J. Cleaner Prod. 276 (2020) 123016.

[66] A. Sohani, H. Sayyaadi, Providing an accurate method for obtaining the efficiency of a photovoltaic solar module, Renew. Energy. 156 (2020) 395–406.

[67] M.H. Shahverdian, A. Sohani, H. Sayyaadi, S. Samiezadeh, M.H. Doranehgard, N. Karimi, et al., A dynamic multi-objective optimization procedure for water cooling of a photovoltaic module, Sust. Energy Technol. Assess. (2021) 101111.

[68] J. Heier, C. Bales, V. Martin, Combining thermal energy storage with buildings—a review, Renew. Sust. Energy Rev. 42 (2015) 1305–1325.

[69] A. Widiatmojo, S. Gaurav, T. Ishihara, K. Yasukawa, Y. Uchida, M. Yoshioka, et al., Using a capillary mat as a shallow heat exchanger for a ground source heat pump system, Energy Build. 209 (2020) 109684.

[70] J. Warner, X. Liu, L. Shi, M. Qu, M. Zhang, A novel shallow bore ground heat exchanger for ground source heat pump applications—Model development and validation, Appl. Therm. Eng. 164 (2020) 114460.

[71] Y. Chen, H. Hua, J. Wang, P.D. Lund, Integrated performance analysis of a space heating system assisted by photovoltaic/thermal collectors and ground source heat pump for hotel and office building types, Renew. Energy. 169 (2021) 925–934.

[72] J. Han, M. Cui, J. Chen, W. Lv, Analysis of thermal performance and economy of ground source heat pump system: a case study of the large building, Geothermics 89 (2021) 101929.

[73] H. Zhang, Z. Han, X. Li, M. Ji, X. Zhang, G. Li, et al., Study on the influence of borehole spacing considering groundwater flow and freezing factors on the annual performance of the ground source heat pump, Appl. Therm. Eng. 182 (2021) 116042.

[74] H. Wang, B. Liu, F. Yang, F. Liu, Test investigation of operation performance of novel split-type ground source heat pump systems for clean heating of rural households in North China, Renew. Energy. 163 (2021) 188–197.

[75] P. Fleuchaus, B. Godschalk, I. Stober, P. Blum, Worldwide application of aquifer thermal energy storage—a review, Renew. Sust. Energy Rev. 94 (2018) 861–876.

[76] P.A. Fokker, E. Salina Borello, D. Viberti, F. Verga, J.-D. van Wees, Pulse testing for monitoring the thermal front in aquifer thermal energy storage, Geothermics 89 (2021) 101942.

[77] B. Nordell, A. Snijders, L. Stiles, 5 – The use of aquifers as thermal energy storage systems, in: L.F. Cabeza (Ed.), Advances in Thermal Energy Storage Systems, Second Edition, Woodhead Publishing;, 2021, pp. 111–138.
[78] M.-J. Li, M.-J. Li, Z.-X. Tong, D. Li, Optimization of the packed-bed thermal energy storage with cascaded PCM capsules under the constraint of outlet threshold temperature, Appl. Therm. Eng. 186 (2021) 116473.
[79] F.G. Battisti, L.A. de Araujo Passos, A.K. da Silva, Performance mapping of packed-bed thermal energy storage systems for concentrating solar-powered plants using super-critical carbon dioxide, Appl. Therm. Eng. 183 (2021) 116032.
[80] A. Gautam, R.P. Saini, Development of correlations for Nusselt number and friction factor of packed bed solar thermal energy storage system having spheres with pores as packing elements, J. Energy Storage. 36 (2021) 102362.
[81] T.R. Davenne, S.D. Garvey, B. Cardenas, M.C. Simpson, The cold store for a pumped thermal energy storage system, J. Energy Storage. 14 (2017) 295–310.
[82] L. Zhang, J. Dong, S. Sun, Z. Chen, Numerical simulation and sensitivity analysis on an improved Trombe wall, Sust. Energy Technol. Assess. 43 (2021) 100941.
[83] L. Hou, Y. Liu, T. Liu, L. Yang, Y. Feng, Q. Gao, Dynamic heat preservation at night for a Trombe wall with a built-in panel curtain in Western China, Solar Energy. 213 (2021) 284–299.
[84] N. Islam, K. Irshad, M.H. Zahir, S. Islam, Numerical and experimental study on the performance of a Photovoltaic Trombe wall system with Venetian blinds, Energy. 218 (2021) 119542.
[85] A. Hassan, M. Shakeel Laghari, Y. Rashid, Micro-encapsulated phase change materials: a review of encapsulation, safety and thermal characteristics, Sustainability. 8 (10) (2016) 1046.
[86] J. Čurpek, M. Čekon, Climate response of a BiPV façade system enhanced with latent PCM-based thermal energy storage, Renew. Energy. 152 (2020) 368–384.
[87] E. Bigaila, A.K. Athienitis, Modeling and simulation of a photovoltaic/thermal air collector assisting a façade integrated small scale heat pump with radiant PCM panel, Energy Build. 149 (2017) 298–309.
[88] J. Curpek, M. Cekon, J. Hraska, editors PCM Integrated in BiPV Ventilated Façade Concepts: Experimental Test Cell Platform and Initial Full-Scale Measurements, IOP Publishing., 2019.
[89] A. de Gracia, L. Navarro, A. Castell, Á. Ruiz-Pardo, S. Álvarez, L.F. Cabeza, Solar absorption in a ventilated facade with PCM. Experimental results, Energy Procedia. 30 (2012) 986–994.
[90] S.S.M. Ajarostaghi, K. Sedighi, M.A. Delavar, S. Poncet, Influence of geometrical parameters arrangement on solidification process of ice-on-coil storage system, SN Appl. Sci. 2 (1) (2020) 1–12.
[91] Q. Wei, Z. Liao, R. Song, P. Zhang, Z. Wang, J. Xiao, Self-learning optimal control for ice-storage air conditioning systems via data-based adaptive dynamic programming, IEEE Trans. Indust. Electr. 68 (4) (2020) 3599–3608.
[92] C. Yan, F. Wang, Y. Pan, K. Shan, R. Kosonen, A multi-timescale cold storage system within energy flexible buildings for power balance management of smart grids, Renew. Energy. 161 (2020) 626–634.
[93] U. Herrmann, B. Kelly, H. Price, Two-tank molten salt storage for parabolic trough solar power plants, Energy. 29 (5) (2004) 883–893.
[94] L. André, S. Abanades, Recent advances in thermochemical energy storage via solid–gas reversible reactions at high temperature, Energies. 13 (22) (2020) 5859.

Hybrid energy storage devices: Li-ion and Na-ion capacitors

8

Manoj Goswami[1,2], *Satendra Kumar*[1,2], *Hafsa Siddiqui*[1], *Viplov Chauhan*[1], *Netrapal Singh*[1,2], *N. Sathish*[1,2], *Mohammad Ashiq*[1,2] *and Surender Kumar*[1,2]
[1]CSIR—Advanced Materials and Processes Research Institute (AMPRI), Bhopal, Madhya Pradesh, India, [2]Academy of Scientific and Innovative Research (AcSIR), Ghaziabad, Uttar Pradesh, India

Chapter Outline

Highlights

In this chapter, we discussed the basics of hybrid energy storage devices where we have discussed the basic principle of Li-ion and Na-ion batteries, their working mechanism, and many more factors (Section 8.2). In Section 8.3, we discussed the basics of electrochemical capacitors in which, electric double-layer capacitors and pseudocapacitors are involved. In Section 8.4, we go through a brief exploration of hybrid energy storage devices in which, lithium-ion and sodium-ion capacitors are involved. We discussed the mechanism of hybrid capacitors, their working principle, used electrode material for anode and cathode, and many more.

Emerging Trends in Energy Storage Systems and Industrial Applications. DOI: https://doi.org/10.1016/B978-0-323-90521-3.00016-8

Novelty

In this chapter both hybrid ion capacitors (Lithium and Sodium-ion capacitor) are extensively discussed with their evaluation from former to latter. For industrial applications (e-vehicles, bikes, power tools, etc.), the secondary batteries could not utilize effectively due to low energy density (30–200 Wh kg^{-1}), power density (<1 kW kg^{-1}), and poor cyclic stability. The low power density of secondary batteries is the main drawback for the application in automobile industries. To accelerate any electric vehicle a high power with high energy density-based energy storage device is necessary. Secondary batteries are unable to fulfill the requirement of high energy density and high power density in a single system. To overcome this, hybrid energy storage devices were introduced by which both high energy density and power density can be achieved in a single device. HESDs are the combination of battery and supercapacitor types of the electrode in a single system. The use of pseudocapacitive anode with carbon-based cathode material will be a good strategy for high-performance hybrid capacitors. Currently, SIC is the mainly focused and low-cost HESDs as compared to LICs. The development of SICs is in the initial stages due to the lack of insertion host materials. This is a new and novel technique for the future development of the electric vehicle, regenerative braking, and many more applications.

Nomenclatures

AC	activated carbon
C_6O_6@graphhene	oxocarbon-functionalized graphene
CNT	carbon nanotube
EC	electrochemical capacitor
EDLC	electric double-layer capacitors
EDLCC	electric double-layer conventional capacitor
EDLCs	electric double-layer capacitors
EESDs	electrochemical energy storage devices
HC	hard carbons
HESDs	hybrid energy storage devices
HESDs	the hybrid energy storage devices
HIC	hybrid ion capacitor
IHP	inner Helmholtz plane
LIBs	lithium-ion batteries
LIC	lithium-ion capacitor
MWCNTs	multiwall carbon nanotubes
NICs/SICs	sodium-ion capacitors
OHP	outer Helmholtz plane
PC	pseudocapacitors
SEI	surface electrolyte interface
SWCNTs	single-wall carbon nanotubes

8.1 Introduction

In the present scenario, the world is facing the problem of a rapid increase in carbon dioxide (CO_2) amount in the environment/atmosphere. Every year tons of CO_2 and greenhouse gases are generated from vehicles, machinery, industries, and other means. The CO_2 and greenhouse gases are responsible for the earth's surface temperature increment and depletion of the ozone layer. It is a major concern to resolve this problem as soon as possible. One of the important things to solve this problem is to transfer the fossil-based energy system to sustainable and renewable systems. Systematic sustainable and renewable energy storage systems are the main focusing point for electric automobile and commercial industries. For industrial applications (e-vehicles, bikes, power tools, etc.), the secondary batteries could not utilize effectively due to low energy density (30–200 Wh Kg^{-1}), power density (< 1 kW kg^{-1}), and poor cyclic stability. For commercial applications (laptops, smartphones, tablets, power backup systems, etc.) the secondary batteries are suitable and acceptable. The low power density of secondary batteries is the main drawback for the application in automobile industries. To accelerate any electric vehicle a high power with high energy density-based energy storage device is necessary. Secondary batteries are unable to fulfill the requirement of high energy density and high power density in a single system. To overcome this, researchers look forward to making a new hybrid energy storage device by which both high energy density and power density can be achieved in a single device. A hybrid energy storage device (HESDs) is a combination of battery and capacitor type of electrodes. In the case of battery-type electrodes, the redox reaction occurs, which results in high energy density. On the other hand, capacitor-type electrodes show high power density due to the double layer formation on the surface of the electrode. Lithium-ion capacitor (LICs: also known as hybrid lithium-ion capacitors) is the most promising HESDs in which lithium-ion battery (LIB) and electrochemical capacitor (EC) was used as an anode and cathode electrode respectively, with Li^+ containing electrolyte. According to previous studies, the HESDs never replace Li-ion and Na-ion batteries to serve high energy density or energy storage per unit volume. But in the case of Li-ion and Na-ion capacitors (hybrid capacitors), the energy density can be achieved 4 to 5 times higher than the commercially available electric double-layer capacitors (EDLCs). The research on hybrid capacitors (LICs) starts in 2001 by Ammatacci and their co-workers. Firstly, they fabricated a hybrid LIC device by using nanosized $Li_4Ti_3O_{12}$ as anode and activated carbon (AC) as the cathode. They observed 3 times larger energy density than of those conventional carbon-based supercapacitors. Following this research, the working researcher looks forward to working and enhancing the properties of hybrid LICs. Till now different types of anode materials were used to enhance the performance of hybrid LICs. Due to the low abundance in nature and high cost of lithium elements researchers look forward to working on sodium-ion-based energy storage devices which possess the same intercalation and deintercalation/adsorption and desorption properties as Li-ion. As we know that Na-ion is a highly abundant and low-cost element in nature so the researcher takes an interest to explore Na-ion for batteries and capacitors.

8.2 Electrochemical energy storage devices

The basic definition of "energy storage" is to capture/store the produced energy for the first time and used it for a later time. Currently, bulk energy storage is dominated by hydraulic dams, grid energy storage, solar energy storage, fossil fuels, and many more. A rechargeable/secondary battery, which converts chemical energy to electrical energy is mostly considered an energy storage device for the daily life of human beings. Rechargeable batteries are widely used in cell phones, laptops, power tools, e-vehicles, and many more. Li-ion batteries have attracted much attention over the last few decades, as one of the most desirable rechargeable batteries. They currently represent the dominant extensive use in mobiles, laptops, and portable electronic devices. In our daily life, we noticed that the increasing functionality and performance of mobiles and electronic equipment, demands better LIBs with good stability and capacity. USA government set an aim to get 1 M of plug-in hybrid vehicles on the road in the USA with a positive intention to decrease the air pollution from fossil fuel-based vehicles. As electric cars are demanding, the market for Li-ion batteries grows rapidly (Fig. 8.1A).

Typically, LIBs are the arrangement of two electrodes (cathode and anode) which are placed into a Li-containing electrolytic solution. These electrodes were separated by a dielectric film, known as a separator. In any electrochemical energy storage device (EESDs), the main function of the separator is to prevent the short

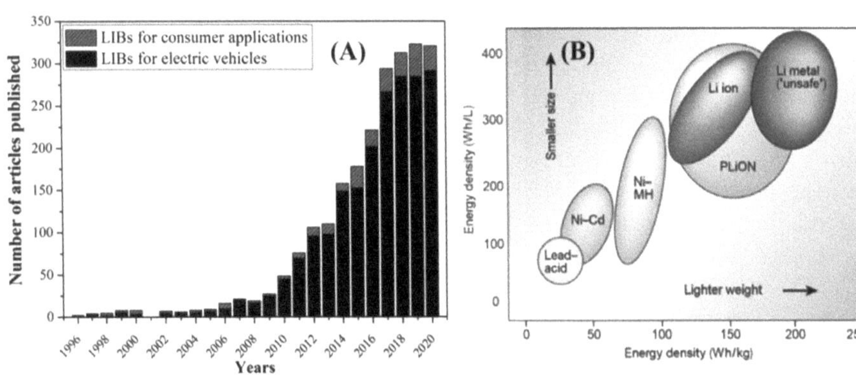

Figure 8.1 (A) Demand of lithium-ion battery (LIBs) in the last two and half decades, produced from the web of science under 'TOPIC' searched for Lithium-ion batteries powered electric vehicles and LIBs consumer use (date: 09–02–2021) (B) Comparison of the various energy storage devices with lithium-ion batteries.
Source: Reproduced with permission D. Deng, Li-ion batteries: basics, progress, and challenges, Energy Sci. Eng. 3 (2015) 385–418 [1].

circuit and flow of electrons between anode and cathode. Till now researchers use the same of those design that was commercialized by Sony about two decades ago for LIBs. During the charging process, Li-atom at the cathode releases the electrons which were towards the anode electrode through the external circuit. As the same Li-ions migrate towards the anode through the electrolytic solution and separator to combine with the electron. This same process is reversed during the discharge process. During the charging-discharging process the Li-ions shuttle between cathode and anode, this process is known as the shuttle chair mechanism. While LIBs could provide a solution to meet several requirements, tough difficulties in achieving renewable energy production via the rising demand for electric vehicles and an EES emerging market, we need to rethink the viability of Lithium which is undoubtedly the intrinsic factor of LIBs. Lithium is limited to just 20 ppm in relative abundance in the earth's crust [2]. Rechargeable sodium-ion batteries are the best alternative to LIBs because they have an abundance of materials and generic electrode potential as shown in Table 8.1.

The energy density of NIBs is low as compared to LIBs and a 0.4 V drop in potential is there at 100 mAh g^{-1} which is similar to the difference in the standard electrochemical potential for Li (3.04 V) and Na (2.71 V). The breakthrough in the work of NIBs came when Stevens and Dahn reported a large reversible capacity (300 mAh g^{-1}) with hard carbon in a sodium cell [6]. Although hard carbon was studied extensively at that time for negative electrodes but not suitable for NIBs. Due to the unique redox chemistry of Fe^{3+}/Fe^{4+} with sodium, it gives a high energy density and is so cost-effective for NIBs. Sodium-based positive electrode materials for NIBs, especially 3d transition-metal oxides with layered structures are extensively explored over the last decade. The specific capacity of LIBs and NIBs can be enhanced by developing new advanced electrode materials. Fig. 8.2 encapsulated some strategies for the enhancement of specific capacity and the selection of new electrode material, such as:

I. Dimension reduction
II. Composite formation
III. Doping and functionalization
IV. Morphology control
V. Coating and encapsulation
VI. Electrolyte modification, etc.

Table 8.1 Properties of sodium and lithium in energy storage devices [3–5].

Sr. No.	Physical properties	Na^+	Li^+
1	Relative atomic mass	23.00	6.94
2	Shannon's ionic radii/Å	1.02	0.76
3	E° (vs SHE)/V	−2.71	−3.04
4	Melting point/°C	97.7	180.5
5	Theoretical capacity of metal electrodes/mAh g^{-1}	1166	3861
6	De-solvation energy in PC/kJ mol^{-1}	157.3	218.0

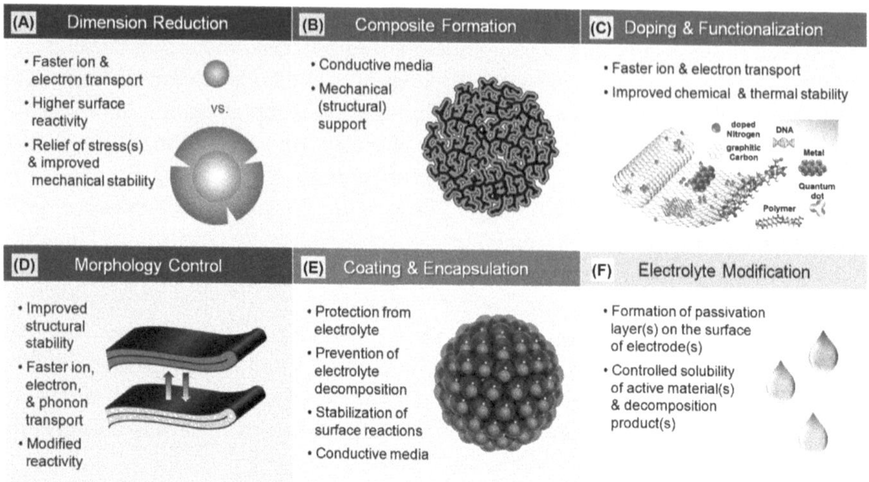

Figure 8.2 General strategies for the enhancement of properties of electrode material (A) dimension reduction, (B) composite formation, (C) doping & functionalization, (D) morphology control, (E) coating & encapsulation, and (F) electrolyte modification. Reproduced with permission Copyright (2014) Elsevier Ltd. N. Nitta, F. Wu, J.T. Lee, G. Yushin, Li-ion battery materials: present and future, Mater. Today. 18 (2015) 252–264 [7].

A porous composite may adjust the volume and also make it easier to transport Li^+ quickly. The carbon present in the composite helps for SEI formation and electronic conductivity increases. To cause aggregating particulates from 15 to 30 μm, Magasinski, etc. [8] developed a baseline material using a bottom-up technique.

8.3 Electrochemical capacitors

The main difference between a conventional capacitor and electrochemical capacitors is the absence of dielectric material. In the electrochemical capacitor, an electrolytic solution is filled with a separator between the electrodes. the electrochemical capacitor is very much efficient compared to the conventional capacitors in terms of providing power density and energy density. Fig. 8.3 shows the electrochemical capacitors which are divided into three parts that is electric double-layer capacitors (EDLC), Pseudocapacitors, and hybrid capacitors. In this section, we are going to discuss a short introduction of EDLC, Pseudocapacitors, and briefly hybrid capacitors.

8.3.1 Electric double-layer capacitors

Electric double-layer capacitors (EDLC) capacitors work on the same principle as that of the conventional capacitor in which the capacitance is a result of charge separation between two charged plates. But an EDLCC supercapacitor can store

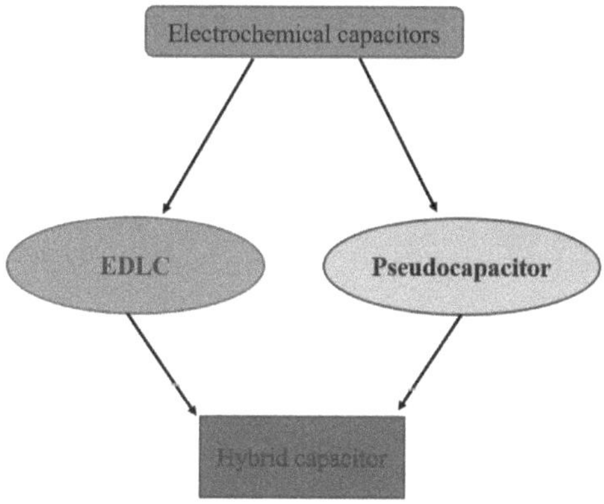

Figure 8.3 The schematic shows the type of electrochemical capacitors.

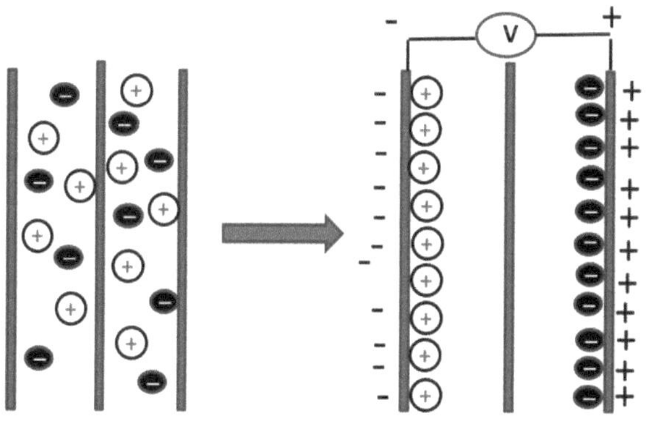

Figure 8.4 Charge storage mechanism of electric double layer capacitor in the presence of the electric field.

far more energy, ~10,000 times, when compared to the conventional capacitor due to the greater interfacial area available at the surface of the electrodes for the storage of charge. Fig. 8.4 shows how the charge is stored in an EDLC capacitor. The reversible adsorption of the cation and anions on the negatively and positively charged electrodes, respectively, forms a parallel plate-like configuration of charges at both the electrode-electrolyte interfaces. This accumulation of charge is purely electrostatic without any charge transfer at the interface and the charge generation at the surface is due to the dissociation of the surface, possible crystal defects of the material, and adsorption of ions from the solution which are physical phenomena.

The two layers of opposite charges formed at the interface are called the double layer. If there is a deficiency in or an excess of charge generation at the electrode surface, an equal amount of oppositely charged ions are created in the electrolyte near the interface to maintain the charge of electroneutrality. The double-layer is usually 5–10 Å thick and depends on the ion size and concentration of the electrolyte. Though this model gives a good idea about the charge storage in an EDLC but is not the complete picture. The above model is based on the study of Helmholtz and considers a rigid separation of opposite charges and is termed a compact layer. This model has two drawbacks—it considers neither the interactions possible in the electrolyte further beyond the compact layer nor the concentration of the electrolyte which could affect the charge concentration.

8.3.2 Pseudocapacitors

Pseudocapacitors (PC), unlike EDLC capacitors, make use of the redox reactions instead of the electrostatic attraction of the charges. These capacitors undergo faradic reactions like the conventional batteries and the capacitance arises due to the electrochemical charge transfer. In PC too accompanied by the intercalation and deintercalation of charges just like batteries but the use of the electrolyte makes it more efficient in terms of power density. As the capacitance is based on the electrochemical nature it highly depends on the available active sites. In PC the faradic charge storage is done at the electrodes. In this mechanism (Fig. 8.5) there is a charge passage from the electrolyte to the electrode and vice versa in the charging and discharging process. This redox reaction between the electrolyte and the electrode is governed by thermodynamical properties and the potential window. In a PC the capacitance arises when a potential is applied and it induces a faradaic current. This faradaic current is attributed to either the electro-sorption or the redox reaction of the electrode materials. When the electro-sorption is in the play then it is governed by the chemisorption of anions while the redox reaction is governed by charge exchange from the double layer instead of the static charge separation. Thus, when the redox reaction takes place, both the EDLC and PC charge storing mechanism take place but with the selection of the proper material, the latter can be made to dominate over the former type of supercapacitor. These materials are usually the transition metal oxides and carbon-based material of nano porosity with electro-sorption hydrogen used. In PC the electro adsorbed charge is combined with

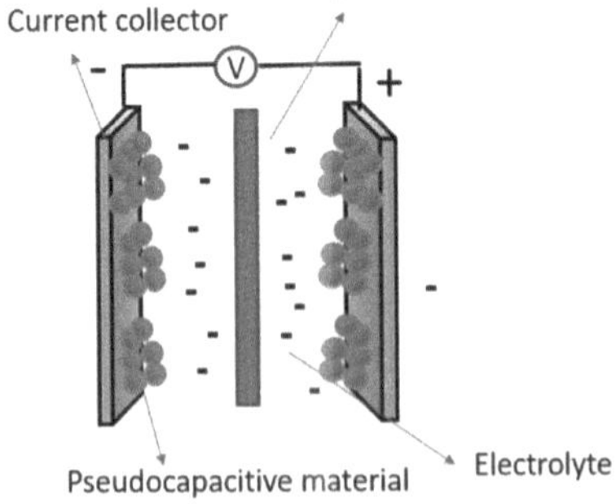

Figure 8.5 Schematic representation of pseudocapacitors.

the electrode potential controlling the extent of the charge sorption. This way the charge storage is dependent on the electrode potential and in turn the charge storing mechanism is controlled by the electron transfer instead of charge accumulation. But still, the pseudocapacitance cannot take place without the formation of the electric double layer which provides the ions from the charge transfer and is created due to the electrostatic attraction between the charged electrodes and the ions in the electrolyte.

In the metal oxides, the proton insertion and de-insertion result in pseudocapacitance. The surface layering and cationic and/or protonic diffusion can act as limiting factors in the capacitance of the PC.

8.4 Hybrid energy storage device: motivation

In the consumer electronics, automobile, aerospace, and stationery industries, electrical energy-storage devices play a key role. For example, commercial Panasonic LIBs have a specific energy of up to 200 Wh kg^{-1}, but with a maximum specific power of below 350 W kg^{-1}. Many commercial electrochemical capacitors, in comparison, have specific power values up to 10 kW kg^{-1}, but specific energies in the 5 Wh kg^{-1} band. An emerging goal is to provide high-energy and high-power in a single device with advanced electrical storage. A comparatively recent intermediate system of energy storage is a hybrid ion capacitor (HIC) in terms of power between batteries and supercapacitors, though giving supercapacitor-like cyclability qualities [9]. A major future opportunity that remains not fully operational due to shortcomings of the current secondary batteries and supercapacitor ultracapacitors technologies is the sustainable regenerative braking energy harvesting from trains, heavy automobiles, and, eventually light vehicles. Because of energy density limits, heat dissipation issues, and overall installation costs, no recent battery or supercapacitor solutions can satisfy the specifications. Regenerative braking is a significant end-use ability for HIC systems (Fig. 8.6).

Hybrid energy storage devices (HESDs) play a crucial role to bridge the gap between batteries and capacitors. It is an arrangement of two different electrodes in which a battery-type and a capacitor-type electrode are used in a single cell. In HESDs, the battery-type electrode is responsible to bring energy density and the capacitor-type electrode is responsible to bring power density at the same time and in a single device. The schematic of the arrangement of the hybrid device is shown in Fig. 8.7.

The configurations of HESDs are divided into two types that is symmetric hybrid devices and asymmetric devices. HESD with two similar types of electrodes is known as the symmetric hybrid device while with two dissimilar electrodes is known as an asymmetric hybrid device. Commercially available symmetric hybrid capacitor with AC (both electrode) and AC and MnO_2 is the primary example of symmetric and asymmetric hybrid capacitors. Firstly, a prototype of asymmetric hybrid capacitors was fabricated by Amatucci's group [10]. In this prototype, AC

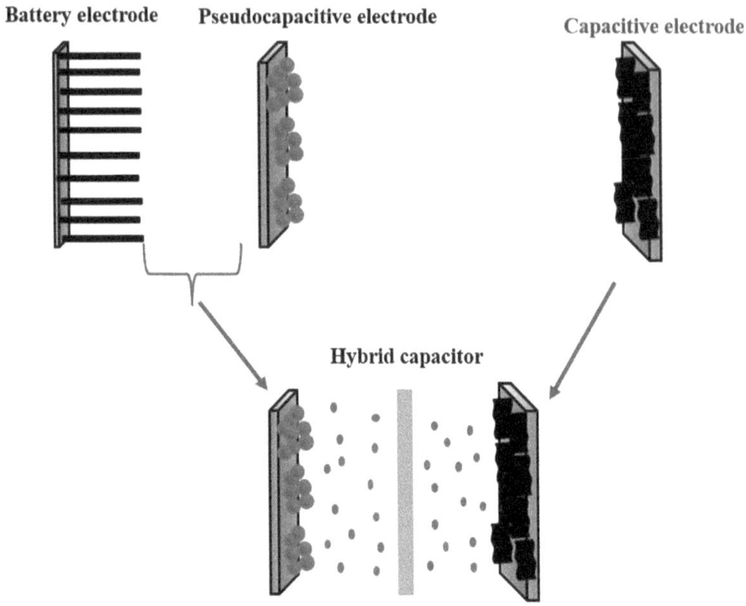

Figure 8.6 Schematic representation of a hybrid capacitor which consists of battery/pseudocapacitive type and capacitive electrode.

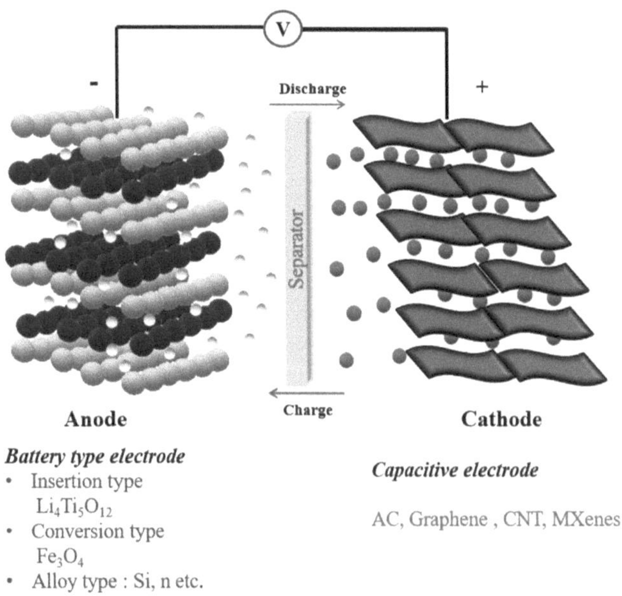

Figure 8.7 Schematic illustration of a full cell structure and electrode materials of a lithium-ion capacitor.

and $Li_4Ti_5O_{12}$ were used as capacitive and battery types of electrode material respectively. In his study, they observed that the capacitive type and battery type of the electrode are responsible for the improvement of power density and energy density of the device. In this section (8.4) of the chapter, we are going to discuss LICs and SIC in brief.

8.4.1 Hybrid lithium-ion capacitors

Lithium-ion capacitors are the most commonly known commercialized HESDs, in which graphite was used as a battery type of electrode and porous carbon was used as a capacitor type of electrode [11]. As we already discussed, the battery type electrode is responsible to bring a high energy density, and the capacitor type electrode brings a high-power density to the device. The intercalation mechanism in battery-type electrodes and double-layer formation in capacitor-type electrodes are probably considered to meet high energy density and power density for various applications that is micro-electronics, power tools, e-vehicles, and many more. High energy density as compared to EDLC, high power density as compared to batteries, high reliability, stability at large temperature ranges (−20°C to 70°C), and low self-discharge are the main advantages of LICs. However, kinetics imbalance between anode and cathode is the main drawback of LICs [12]. The basic and simple prototype of a hybrid capacitor is shown in Fig. 8.7.

8.4.2 Electrode for lithium-ion capacitors

The fast charge-discharge time and high-power capabilities of pseudocapacitive material make it a promising and attractive candidate to use in LICs. Later in the chapter, explain a strategy to improve the power conversion efficiency by introducing inorganic material in an organic solar cell is discussed. In the following section, we are going to discuss some recently used electrode materials for LICs.

8.4.2.1 Cathode

8.4.2.1.1 Carbon-based materials

In the current scenario, porous carbon is mostly employed material in LICs as cathode material. A good capacitive electrode must have three properties [13].

- High surface area
- Good conductivity
- Good electrolyte accessibility

Charge- storage in porous carbon is mainly dependent on the double layer formation at the interface between electrode and electrolyte. In easy words, the specific capacitance is generally depending on the surface area of porous carbon. Surface functionalization is also a productive method to modify the capacitance properties of carbon materials by enlarging the pseudocapacitive behavior. Activated carbon

(AC), graphene, and carbon nanotube are frequently used carbon-based materials for LICs. Here we are going to discuss some carbon-based cathode materiel in LICs.

8.4.2.1.2 Activated carbon

Activated carbon is the modified form of carbon with very small and low-volume pores. The size of pores is responsible to achieve a high surface area of AC. Due to the high porosity of AC, one gram of AC has 3000 m^2 of surface area (observed through gas absorption). Low production costs, corrosion resistance, good stability in chemicals, highly abundant in nature, and good cycling stability make AC a well-suited electrode material for LICs. AC is the firstly commercialized carbon material for the production of LICs [14]. Nevertheless, AC has some drawbacks that is low electrical conductivity and specific capacitance. To overcome these drawbacks, researchers used AC with the combination of other highly conductive materials and functionalized with beneficial groups. Ammatacci et al. [10] used AC as cathode and $Li_4Ti_5O_{12}$ as anode material for LICs. The polyacrylonitrile (PAN) was used as a precursor for the synthesis of AC. They observed 85%–90% of capacity retention after 5000 cycles, revealing a good signature of cycling stability. Recently, biomass waste pays attention due to its high abundance and recycling of raw material. Biomass-derived AC shows good physical and chemical properties, including modified functional groups on the surface, and good chemical stability. Li et al. [15] reported the synthesis procedure of nitrogen-doped activated carbon (N-AC) from corn cobs (bio-waste). They observed 2900 m^2g^{-1} of specific surface area. The prepared sample shows a specific capacitance of 185 Fg^{-1} at 0.4 Ag^{-1} of current density for half-cell configuration in organic electrolyte. The fabricated hybrid devices consisting of N-AC as cathode and Si/C as anode showed 230–141 W h kg^{-1} of energy density at 1747–30,127 w kg^{-1} of power density.

8.4.2.1.3 Graphene

Graphene is also an allotrope of carbon in which a single layer of carbon atoms is arranged in a 2-D hexagonal lattice [16]. Graphene was discovered by Andre Geim and Konstantin Novoselov in 2004. High Theoretical surface area of 2630 m^2g^{-1}, excellent electrical conductivity, charge-transfer mobility, and chemical stability make it a promising, electrode to use in energy storage devices.

Wang et al. [17] synthesized porous activated nitrogen-doped graphene sheets (A-N-GS) and nitrogen-doped graphene (N-GS) for the fabrication of dual carbon-based LIC devices. The A-N-GS//N-GS LIC device delivers 2250 w kg^{-1} of power density at 1879 wh kg^{-1} of energy density with 93.5% of capacity retention after 3000 cycles. Tu. et al. [18] reported the synthesis of ethylene glycol-assisted reduced graphene via a simple hydrothermal route. The prepared sample delivers a superior capacity of 172 mA g^{-1} at 0.1 Ag^{-1} of current density without any loss in capacity after 3000 cycles. In addition, Zhang et al. [19] used 3D graphene as cathode and Fe_3O_4/graphene nanocomposite as an anode material for LIC. Fe_3O_4/graphene//3D graphene full cell demonstrated an energy density of 204–65 W h kg^{-1} at power density ranging from 55–4600 W kg^{-1}.

8.4.2.1.4 Carbon nanotube (CNT)

A tubular rolled-like structure carbon sheet is known as a carbon nanotube. The CNT is divided into two types that is single-wall (SWCNTs) and multiwall (MWCNTs) carbon nanotubes. As the name referred to, the SWCNTs are intermediate between fullerene and a layer of graphene. However, MWCNTs consist of SWCNT in it. CNT exhibits excellent electrical conductivity, suitable pore size, hollow-like structure, and nanoscale structure making it ideal electrode material for energy storage devices.

Cai et al. [20] prepared MWCNT via the internal short circuit method. The obtained MWCNT was assembled as a cathode and AC as an anode for LIC. They obtained the energy density was four times higher than conventional EDLC. The LIC device demonstrated a specific capacitance of 59.3 Fg^{-1} at 100 $mA\ g^{-1}$ of current density. The maximum power density of 4035 $W\ kg^{-1}$ was observed at 63 $W\ h\ kg^{-1}$ of energy density with 89% of capacity retention after 1000 cycles. In addition, Zhao et al. [21] synthesize thin-film MWCNT and αFe_2O_3/MWCNT composite via hydrothermal approach and spray deposition technique respectively. The LIC device was assembled in which thin-film MWCNT was used as a cathode and αFe_2O_3/MWCNT was used as an anode. The fabricated device provides 50 $W\ h\ kg^{-1}$ of energy density at 1000 $W\ kg^{-1}$ of power density.

8.4.2.1.5 Other cathode materials used in lithium-ion capacitors

The balance of mass between two electrodes is the most important parameter to obtain high-energy density in LICs. Li^+- intercalated materials are also considered to be a well-suited cathode material for LICs. Lithium metal oxides such as Li_2MnSiO_4 [22], and $LiNi_{0.5}Mn_{1.5}O_4$ [23,24] are mainly used as Li^+ intercalated materials for LICs.

Composite materials are also used as cathode materials to enhance the electrochemical properties of LICs. The successful combination of two different materials in a composite material produced EDLC and Faradic process during the galvanometric charge-discharge process. The full cell was assembled in which, FeVO/C was used as cathode and pre-lithiated graphite was used as an anode material. The device delivers 117.9–27.3 $Wh\ kg^{-1}$ of energy density at 237.6–20,008.7 $w\ kg^{-1}$ with a capacity retention of 70.7% after 1000 cycles. Conducting polymers also pays attention to a promising pseudocapacitive material due to its flexibility, sufficient conductivity, ease of production, and low cost [25]. However, still, some drawbacks of conducting polymers prevent their practical applications. To improve, the conducting polymers are combined with metal oxide and carbon such as polyaniline (PANI)/MnO_2 [26], PANI/CNT [27], PANI/CuO [28], and many more.

8.4.2.2 Anode

In LICs, the anode plays a significant role to pull high energy density and power density with good cycling stability, and low operating potential. Anode materials with good electrical conductivity are necessary to enhance the properties of LICs.

Through this section, we are going to discuss some recently used anode material in LICs.

8.4.2.2.1 Carbon materials

In LICs, the carbon-based anode materials are divided into two types

1. Graphitized carbons
2. Non-graphitized carbons

8.4.2.2.2 Graphitized carbon

Graphite is a crystalline form of carbon in a hexagonal mesh structure. As we know, graphite is used in pencils and as lubricants. It is also a good conductor of electricity and heat. High conductivity makes it promising to use as an anode electrode in batteries and supercapacitors. In organic electrolytes, graphite can achieve up to 4.5 V due to low potential during the discharge process. However, there is still a drawback of graphite anode in LICs. During the discharge process, a chemical reaction occurs at the surface of the electrode which produces a surface electrolyte interface (SEI) layer between electrolyte and electrode. The produced SEI layer at the negative electrode is responsible for the irreversible capacity loss [29]. As we know, graphite has a layered type of structure with 0.335 nm of the interlayer gap between two layers. In this gap, Li^+ ions and solvent can easily be inserted during the intercalation reaction. The inserted organic electrolyte is reduced to generate gas, which results in the peeling of graphite layers. Therefore, the SEI layer is continuity forming and destroying [30]. In addition, graphite as anode material required some modifications to achieve better properties of LICs.

8.4.2.2.3 Non-graphitized carbon

A graphitized carbon can be transformed into crystalline graphite by heating (3000°C) but non-graphitized carbon remains in the same conditions or it can be transferred to the graphite at any temperature. Non-graphitized carbon such as chars, hard carbons (HC), charcoal, glossy carbon, amorphous carbon, and soft carbon is proposed to replace the graphite in LICs application. Amorphous carbon provides higher capacity retention and higher internal irreversible capacity loss as compared to graphite [31]. Yuan et al. [32] Investigated the performance of hard carbon, graphite, and AC as anode materials for LICs. The results conclude that the AC/HC provides better rate capability as compared to graphite. As we noticed the graphitized and non-graphitized have their advantages and disadvantages as anode material for LICs. To improve the performance of LICs, researchers have to modify the carbon material by doping other different materials. Lee et al. [33] Prepared a composite of HC and micro-crystalline graphite as anode material in LIC. The electrochemical performance of the fabricated LIC device shows the enhancement in the irreversible capacity, cycling stability, and efficiency as compared to hard carbon.

8.4.2.2.4 Titanium-based materials

For a few years, much attention had been paid to search for a well-suited anode material instead of graphite or carbon-based materials. Various titanium-based materials that is TiO_2, $Li_4Ti_5O_{12}$ (LTO), and $H_2Ti_3O_7$ have been already explored for LICs as anode materials. Lithium titanate oxide with spinal structure has been considered the most suitable anode material for LICs due to its long life and safety characteristics. However, there are some advantages and disadvantages of LTO for LICs following:

8.4.2.2.5 Advantages

- Stable a high temperature
- Lack of SEI formation, production of gas, and decomposition of electrolyte during the charge-discharge process
- Good Li^+ mobility
- Easy to synthesis
- Low cost
- Safe to use

8.4.2.2.6 Disadvantages

- Low specific capacity ($\sim$150–175 mAh g^{-1}) compared to graphite ($\sim$300 mAh g^{-1})
- Low electronic conductivity ($<10^{-13}$ S cm^{-1})

LTO shows negligible volume changes as compared to graphite, which enlarges up to 10 vol% during the charge-discharge process. The spinal crystal structure of LTO has an FD3M space group with a cubic symmetry [34,35]. In LTO, TiO_6 is the foremost body for Li^+ insertion/extraction. Basically, at the initial stage Li^+ ions inhibited the 8a site and formed a spinal structure. As the discharge process is proceed, Three Li^+ ions are transferred from 8a sites to 16c sites and the initial inhabited three Li^+ ions are moved to 16c sites from 8a sites. In the same way, the reversible process takes place during the charging process. In brief, Li^+ ions are extracted from 16c sites via 8a sites and the other ions come back to 8a sites from 16c sites. At the same time, the Faradic reaction of Ti^{4+}/Ti^{3+} allows the transition of spinal structure ($Li_4Ti_5O_{12}$) into the rock-salt structure ($Li_7Ti_5O_{12}$) [36]. Fig. 8.8 shows the crystal structure of spinal and rock-salt LTO. The transformation of spinal structure to rock-salt structure is expressed by the following electrochemical reaction Eq. 8.1.

$$Li_4Ti_5O_{12} + 3Li^+ + 3e^- \leftrightarrow Li_7Ti_5O_{12} \tag{8.1}$$

Zhang et al. [38] synthesized porous LTO nanoparticles by a precursor-directed hydrothermal approach. The LTO nanoparticles were later combined with graphene as a composite material. The LTO/graphene composite material was assembled as an anode electrode and AC was a cathode electrode. The LIC device provides a maximum power density of 7200 W h kg^{-1} with 44 W kg^{-1} of energy density. The capacity retention of 80% was observed after 10,000 cycles which is a good signature of cyclic stability. It is observed that the low energy density at high power density needs to improve. The basis on the previous studies on LIBs, ion doping and the combination of different can resolve the above-mentioned problem of LTO-based LIC devices. For instance, Lee and Yoon [39] examined that the doping of

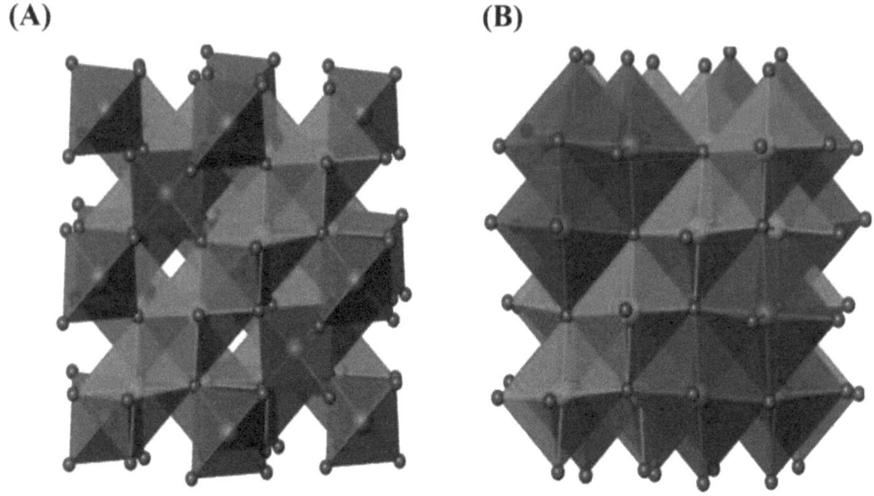

Figure 8.8 (A) $Li_4Ti_5O_{12}$ Spinel structure of $Li_4Ti_5O_{12}$ (LTO), (B) $Li_7Ti_5O_{12}$ rock-salt type of LTO structures. Blue tetrahedra show the presence of lithium, and green octahedra show the presence of disordered lithium and titanium.
Source: Reproduced with permission © 2006 American Chemical Society E.M. Sorensen, S.J. Barry, H.-K. Jung, J.R. Rondinelli, J.T. Vaughey, K.R. Poeppelmeier, Three-dimensionally ordered macroporous $Li_4Ti_5O_{12}$: effect of wall structure on electrochemical properties. Chem. Mater. 18 (2006) 1713–1714 [37].

dopants (Al^{3+}, Mg^{2+}, Cr^{3+}) does not affect the structural characteristics of LTO. The strong bond of Al-O compared to Ti-O, Al^{3+} doping enhanced the capacitance of LTO. Lee et al. [40] explain that a sufficient amount of sodium (Na) doping can improve the electrochemical performance of LTO, due to small grain size and ionic conductivity. The combination of LTO with conducting material, such as spherical LTO/CNT [41], LTO/CHNS [42], LTO/grapheme [43], etc. is also reported in reputed journals to increase the performance of LTO.

8.4.2.2.7 Pseudocapacitive materials for lithium-ion capacitor

As we previously discussed, Pseudocapacitors have some advantages over EDLCs. The Pseudocapacitance occurs due to the redox reaction or adsorption/desorption of the ions at the surface of electrode materials. In HESDs, the pseudocapacitive materials are used as an anode electrode for the fast Li^+ diffusion and enhance the rate capability of LICs as compared to carbon-based and Lithium-intercalated based electrode materials. The combination of pseudocapacitive and battery-type electrode material could provide higher energy density as well as good power density than the carbon-based HESDs. Pseudocapacitive materials such as metal hydroxide, transition metal oxides, metal nitride, and conducting polymer are recently investigated for LICs. However, low conductivity and low surface area of pseudocapacitive electrode material do meet the conditions of good electrode material. To overcome this, many strategies were implied that is formation of nanosized material, composite with highly conducting materials, doping, etc., Dunn et al. [44,45] quantitatively analyzed capacitive contribution throughout the intercalation process of Li-ions. Normally, the current versus scan rate can be indicated as Eq. (8.2).

$$i = a\nu^b \quad (8.2)$$

where I represent the measured current at a fixed voltage, ν is the sweep rate, and a, and b represents the adjustable parameter. It is observed that, if the value of b is equal to 1 then the whole process of intercalation is capacitive controlled and if $b = 0.5$ then the whole process is diffusion controlled. The value of b can be calculated by plotting the logarithm of the current versus sweep rate which can be expressed as Eq. (8.3).

$$\log i = b \log \nu + \log a \quad (8.3)$$

from Eq. (8.6), it was observed that the current can be produced from both capacitive and diffusion-controlled processes. Hence, the value of b can vary between 0.5 to 1. Now, the Eq. (8.4) can be rewritten as follows;

$$i(\nu) = k_1\nu + k_2\nu^{0.5} \quad (8.4)$$

By plotting between $\nu^{0.5}$ versus $i(\nu)/\nu^{0.5}$ the value of k_1 and k_2 can be obtained which are very useful to calculate the percentage of capacitive contribution. For example, Fig. 8.9 shows the analysis of k_1 and k_2 for $MnO/CoMn_2O_4$ and in Fig. 8.9D, the shaded part shows the capacitive contribution at a specific scan rate and the excess part shows the diffusion contribution. This method determines the nature of the pseudocapacitive electrode during the charge-discharge process.

8.4.2.2.8 Vanadium pentoxide (V_2O_5)

In recent years, the significant advances in nano sciences and nanotechnology, various nanostructured materials are used to increase the impact of Li-ion-based EESD. Among them, nanostructured V_2O_5 has been explored for the application of LICs. V_2O_5 is a low-cost material that can easily be synthesized by various techniques. Good safety, abundance, high theoretical capacity ($\sim$437 mA h g^{-1}), and multiple oxidation states make it promising anode material for LICs.

The intercalation and deintercalation mechanism of Li^+ during the reaction can be expressed as Eq. 8.5 [47].

$$V_2O_5 + xLi^+ + xe^- \leftrightarrow Li_xV_2O_5 \quad (8.5)$$

When the Li^+ ions are inserted into the layers of V_2O_5, phase transformation occurs. Fig. 8.10A shows the evolution of the phase with the increase of Li^+ intercalation. There are five types of phase that occurs during the increasing, x, of Li^+ ions [48].

(i) α-V_2O_5 (x <0.01)
(ii) ε- $Li_xV_2O_5$ (0.35< x <0.7)

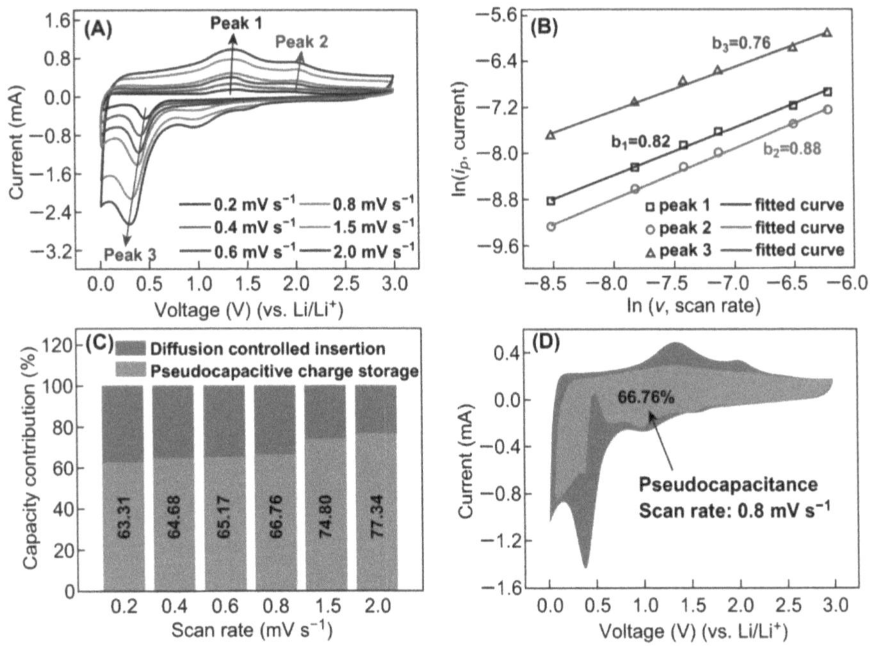

Figure 8.9 (A) CV plotes of MnO/CoMn2O4 electrode at various sweep rates in 0–3 V potential range, (B) log (i) versus log (v) plots at reduction and oxidation state, (C) capacitive and diffusion controlled capacities of $MnO/CoMn_2O_4$ electrode, (D) capacitive contribution of $MnO/CoMn_2O_4$ at 0.8 mV s^{-1}.
Source: Reproduced with permission © 2019 Springer Y. Cao, K. Geng, H. Geng, H. Ang, J. Pei, Y. Liu, et al., Metal–oleate complex-derived bimetallic oxides nanoparticles encapsulated in 3D graphene networks as anodes for efficient lithium storage with pseudocapacitance, Nano-Micro Lett. 11 (2019) 1–14 [46].

(iii) δ- $Li_xV_2O_5$ ($x = 1$)
(iv) γ - $Li_xV_2O_5$ ($x < 2$)
(v) ω - $Li_xV_2O_5$ ($x > 2$)

The Pseudocapacitance behavior of V_2O_5 was also investigated, in which the transition of V_2O_5 undergoes from capacitance to battery-like behavior while increasing the loading of V_2O_5. Sathiya et al. [50] investigated the electrochemical properties of V_2O_5-CNT composite with Li-ions. Fig. 8.10B varies with the above-described phenomenon. Divya et al. [51] used electrochemically produced γ-$Li_xV_2O_5$ as anode and AC as a cathode in LIC. The LIC provides 33.91 W h kg^{-1} of energy density at 0.22 kW kg^{-1} of power density with good capacity retention. Chen et al. [52] reported CNT/V_2O_5 and commercially available AC as an anode and cathode for LIC respectively. The fabricated device provides 40 W h kg^{-1} of energy density at 210 W kg^{-1} of power density. They observed, that at a low power density of 6.3 kW kg^{-1} the device still delivers 6.9 W h kg^{-1} of energy density.

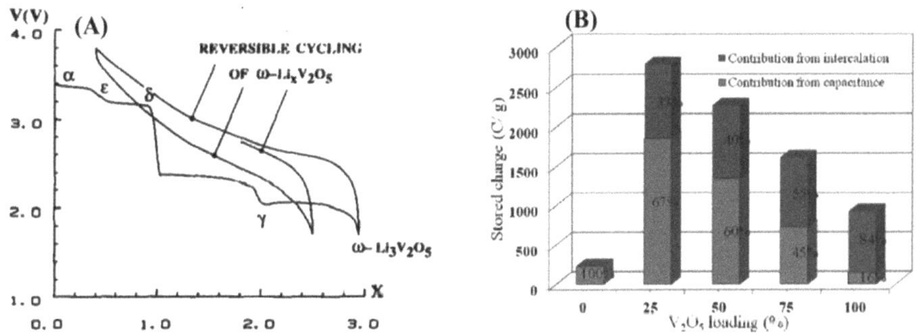

Figure 8.10 (A) Li//V_2O_5 cell showing the evolution of phase with the increasing of Li^+ intercalation. (B) Bar diagram showing the percentage of capacitive and intercalation contribution increasing the percentage of V_2O_5 loading.
Source: (A) Reproduced with permission ©Elsevier 1994 C. Delmas, H. Cognac-Auradou, J.M. Cocciantelli, M. Ménétrier, J.P. Doumerc, The LixV_2O_5 system: an overview of the structure modifications induced by the lithium intercalation, Solid. State Ion. 69 (1994) 257–264 [49]. (B) Reproduced with permission © 2011 American Chemical Society M. Sathiya, A.S. Prakash, K. Ramesha, J. Tarascon, A.K. Shukla, V_2O_5-Anchored carbon nanotubes for enhanced electrochemical energy storage, J. Am. Chem. Soc. 133 (2011) 16291–16299 [50].

The low electrical conductivity and high voltage plateau of V_2O_5 need to be overcome for the applications of LICs.

8.4.2.2.9 Niobium pentoxide (Nb_2O_5)

Nb_2O_5 is considered a favorable anode electrode for LICs due to its enhanced pseudocapacitive actions. Furthermore, Nb_2O_5 has 200 m Ah g^{-1} of theoretical capacity [53]. Dunn et al. [54] analyzed the pseudocapacitive mechanism of orthorhombic Nb_2O_5. From the results they observed, a pseudocapacitive and intercalation reaction of Li^+ ions occurs at the surface of the electrode which is responsible for the fast insertion kinetics and high power performance. Orthorhombic pseudo hexagonal and amorphous is the main crystal phase of Nb_2O_5 which generally depends on the way of synthesis conditions. The Pseudocapacitance of Nb_2O_5 is mainly dependent on the crystalline structure. According to previous studies, orthorhombic Nb_2O_5 shows higher capacitance as compared to other phases (amorphous, and pseudo hexagonal). The uses of bulk- Nb_2O_5 are found unsatisfactory for good electrochemical performance [55,56]. Nano structuring can enhance the electrochemical performance of Nb_2O_5. The uses of Nb_2O_5 can increase the active surface sites and enable a short ion transfer path. Liu et al. [57] synthesized Nb_2O_5 –nanorod via a hydrothermal approach to boost the ions-transfer rate and reduce the ion-diffusion path. It also provides a high energy density with good power density for LIC applications. However, poor electrical conductivity ($\sim 3 \times 10^{-6}$ S cm^{-1}) and high voltage plateau (1.5 V) are the main drawbacks of Nb_2O_5. To overcome these drawbacks the introduction of carbon coating or composite with highly conducting material is necessary. Wang et al. [58] fabricated a LIC device that consists of orthorhombic Nb_2O_5/graphene carbon nanosheets (GCN) as anode and GCN as a cathode. The device provides 129 W h kg^{-1} of energy density and 32 K W kg^{-1} of power density with 80% of capacity retention after 10,000 cycles.

8.4.2.2.10 Manganese oxide (MnO)

Manganese oxide has captured good attention as an anode electrode material for LICs due to its low cost, and high theoretical capacity (775 mA h g^{-1}). The high theoretical capacity and low electromotive force [1.032 V (vs Li/Li^+)] enable MnO to be the ideal choice over the graphite anode [59]. Previously various nano-micro structures of MnO that is nanotubes, nanowires, nanoparticles, micro cubes, etc. were used to enhance the performance of the device. However, the low electric conductivity and large volume expansion of pristine MnO are limited to its practical applications. To overcome the above-mentioned issues, researchers put forward two solutions firstly, the design of porous MnO structure and secondly, the introduction of a conductive agent to improve the electrical conductivity of pristine MnO. Guo et al. [60] fabricated asymmetric LICs in which, MnO@C and trisodium citrate-derived carbon (TMC) were used as anode and cathode material. The device delivers 3.9 V of operational voltage and 235 W h kg^{-1} of energy density with 25,000 W kg^{-1} of maximum power density. Lu et al. [61] synthesized micro cubes of MnO via a hydrothermal approach. The micro cubes deliver a discharge capacity of 425.5 mAh g^{-1} at 1 Ag^{-1} of current density after 495 cycles.

8.4.2.2.11 Silicon (Si) based materials

Silicon has been considered to be a most promising electrode (anode) material for LICs due to its high lithium storage capacity of 3579 mAh g^{-1} with a low discharge voltage of 0.4 V. Si provides a high theoretical capacity as compared to graphite (372 mAh g^{-1}). However, the ability of Si to tolerate numerous Li indicates that during the lithiation/delithiation process, the material undergoes major volumetric changes. Such volumetric changes of the material occurring during the full life of the anode results in a cracking phase of Si particle, the creation of immense surface stress, and, consequently adhesive problems. In addition, the continuous Si particles braking during the cycling produce a new Si surface for another electrolyte reaction. So, the braking of Si particles results in the formation of a passive layer (SEI layer). Thus, the SEI layer decreases the Coulombic efficiency of Si-based anode as compared to graphite material. To overcome this, researchers introduce nano structuring and coating of the particles to prevent/minimize the volumetric change (Table 8.2).

8.5 Hybrid Na-ion capacitor

A hybrid Na-ion capacitor is the combination of a Na-ion-based battery type anode and a capacitor type cathode. This combination can be a good replacement for a Li-ion-based hybrid device. As we have seen, to achieve high energy and high-power density with long cycle life, we need advanced hybrid ion capacitors and the use of Li-ion has proved to be an efficient option for combining the storage possibilities. But due to the scarcity of lithium, its use is constrained by both

Table 8.2 A summary of recently used anode and cathode electrode materials in lithium-ion capacitors.

Anode	Cathode	Potential (V)	The energy density (W h Kg^{-1})	Power density (W kg^{-1})	References
$LiTi_2(PO_4)_3$	AC	0–3	14	180	[62]
LTO	MOF-C	1–3	65	10,000	[63]
LTO	TrGO	1–3	45	3300	[64]
B-Si/SiO_2/C	AC	2–4.5	128	9700	[65]
TiO_2 nanowire	CNT	0–2.8	12.5	1300	[66]
TiP_2O_7	AC	0–3	13	3710	[67]
TiO_2@rGO	AC	1–3	42	8000	[12]
Nb_2O_5 − CNT	AC	0.5–3	33.5	4000	[68]
Ti_2C MXene	YP-AC	1–3.5	50	600	[69]
MoS_2 − C-RGO	PANI	0–4	188	40,000	[70]
$NiCo_2O_4$	AC	0–4.5	39.4	554	[71]
AC	$LiMn_2O_4$	0.7–3	45	800	[72]

production and cost of the product. In this scenario, Sodium ions can be a better substitute for the Li-ions. They have a $Na/Na^+ = -2.7$ V redox potential and are available in abundant amounts; 23,600 ppm while Li is only 20 ppm in the earth's crust. The research on the use of Na ion for making hybrid supercapacitors is only in its initiation period with the fabrication of the first Na-ion capacitors (NICs) in the year 2012. This NIC used V_2O_5/CNT composite as an anode with activated carbon cathode and was capable of ~40 Wh kg^{-1} energy density at a high cell voltage of 2.8 V [73]. Another cost-saving feature of the NIC is the use of inexpensive Al metal as the current collector for both electrodes as it is Na-inert. Then, when compared to Li the Na metal does not undergo metal plating on the electrodes at low potential. When non-graphitic material is used for anodes the Na atom underpotential deposition is very less into the carbon pores. Due to this reason, there is a suppression of metal dendrite formation on the electrodes which helps in making a safer device for thousands of cycles [74]. When NICs are made there is a possibility of a loss of ions in the early cycles due to the formation of a solid electrolyte interface (SEI) or maybe the Na ion can get trapped in the bulk material of the electrodes like hard carbon [74]. The limited solubility of the salts might result in the irreversible loss of ions and turn cause a degradation in the ionic conductivity. Comparative studies show that the use of non-sodiated electrodes will diminish the ion content of the electrolyte when the capacitors for hundreds of cycles deteriorate the capacity of the device [74,75]. The presodiation does not account for the ion loss during the process in the device but it has also been observed that it can increase the voltage window and extend the cyclability by controlling the voltage swings at the anode. This reduces the SEI growth, ion trapping, and volumetric expansion [75,76]. The presodiation for Na-ion-based device and Pre lithiation for

Li-ion-based device can resolve the irreversible loss of ions during the reaction. There are some following methods described below for the presodiation/prelithiation.

8.5.1 Electrochemical technique

The electrochemical presodiation/lithiation process is one of the most widely used techniques. In this process, as shown in Fig. 8.11A, the electrode is usually taken as the anode and connected to Na/Li metal in a closed circuit which can contain a simple resistor or electronic charger depending on the requirements of doping. The material is to be sodiated and the Na/Li metal is separated by an electrically insulating but ionically conducting separator. The charging process can be controlled by an external power source. When this method is used the assembly of the capacitor is made twice, once for making NICs/LICs and once before that for the sedation/lithiation process. Thus, this method is not much efficient when bulk manufacturing of the device is concerned.

8.5.2 Chemical reaction

The use of Na/Li metal to directly react with the electrode material, chemically, is a much better way when bulk manufacturing is concerned. In this method, the metal is reacted with the electrode material in the presence of an electrolyte. The reaction process can be made faster by milling the reagents for better physical contact. And using Na^+ containing compounds in place of the Na metal can be an even better method as has been done with Li. In this method the Na^+ can be incorporated into the Na ion-containing cathode, then these ions can be extracted from the cathode through an electrochemical process and can then be inserted in the anode. Fig. 8.12 shows the process of sodiation/lithiation.

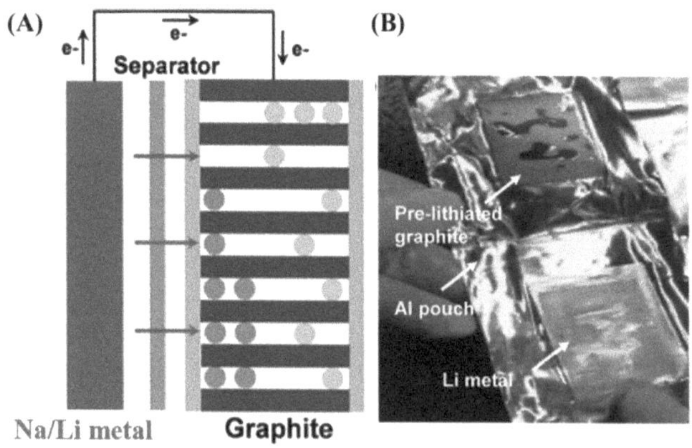

Figure 8.11 (A) schematic illustration showing the electrochemical process for presodiation/prelithiation of graphite. (B) Electrode product after pre lithiation process.
Source: Reproduced with permission © 2018 American Chemical Society J. Ding, W. Hu, E. Paek, D. Mitlin, Review of hybrid ion capacitors: from aqueous to lithium to sodium, Chem. Rev. 118 (2018) 6457–6498 [77].

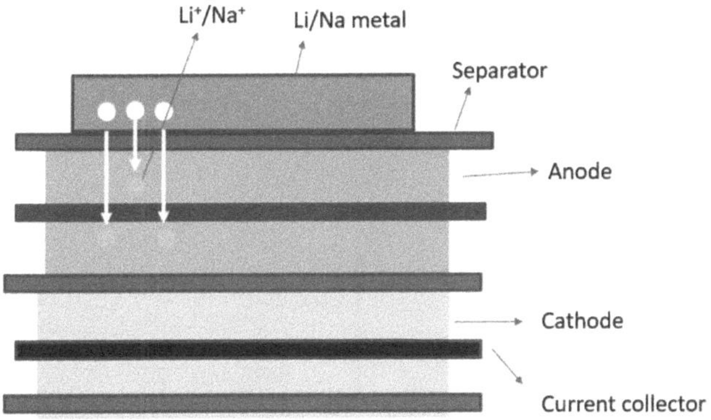

Figure 8.12 Schematic illustration shows the sodiation/lithiation process by chemical process.

Though this method is a better substitute for bulk manufacturing it suffers from the drawback of being a blind doping process, that is the amount of sodium content in the electrode can neither be monitored nor controlled.

8.5.3 Electrode materials for Na-ion capacitor

The physicochemical properties of SIC are dependent on the electrode materials which can influence the performance of the device. The electrodes required by the SIC differ from that of the LIC. This difference arises due to the difference in the ionic size of Na^+ and Li^+. The Na^+ ions being bigger can result in immense structural changes when intercalated in the electrode material thus it is more challenging to find suitable electrodes for the SIC system. To deal with this problem the electrodes with open framework electrodes are preferred so that larger ion sizes can be easily introduced into the structure. Other than this the sodiated transition metals are hygroscopic and can react even on brief exposure to the molecules in the air. This reaction will result in the formation of NaOH which can deteriorate the electrochemical properties of the material. Thus, a moisture-free environment is required to get the best results from such electrodes. One possible solution to this problem can be the doping of the material with other elements to make the electrode less responsive to its interaction with the atmosphere. Though these are some of the problems related to the electrodes that occur in the use of SIC there are some advantages also. Some researchers have reported that the intercalation and deintercalation of the Na ions are faster than that of the Li ions due to the lower Lewis acidity of the larger ion. Here we are going to discuss some recently used anode and cathode materials for NICs (Fig. 8.13).

8.5.3.1 Anode materials for Na-ion capacitor

8.5.3.1.1 Hard carbon materials

The 22.99 relative atomic mass and 1.02 Å ionic radius of Na^+ hardly allow them to intercalate in the graphitic structure but if anode material having larger tunnels in

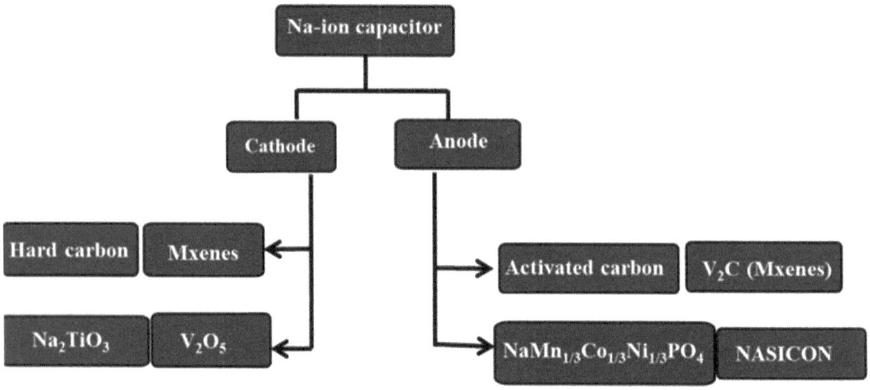

Figure 8.13 Schematic of various materials used for Na-ion capacitors as cathode and anode.

their crystal structure are created, they can be used for this purpose. To accomplish this SIC can make use of the pre-doped carbon microbead and activated carbon as the cathode. The use of hard carbon opens the possibility for the use of carbonaceous materials for making electrodes but the irreversible capacity loss is still a common problem with these anode materials and particularly in the initial cycles. The low sodiated potentials can result in the formation of dendrite structures also which results in safety hazards. Kuartani et al. were among the first ones who tried carbon-based electrodes of SIC [78]. In their work, they highlighted the fact that pre-doping of the carbon material was very important for the better performance of SIC. Using carbon microbeads and activated carbon for anode and cathode, respectively, an output potential of ~0.3 V, 70% capacity retention at 10 mA cm^{-1} current density, and a 9% decrease in cycle lifespan after about 1000 cycles. Though this is not a very impressive result but was enough to show that SICs should and could also be explored for energy storage applications. Similar to LICs, natural biomaterials were explored for SICs also. Madhvi et al. used plant-derived hard carbon as anode and mesoporous carbon from coconut shells as cathode material [79]. With an initial discharge capacity of ~216 mAh g^{-1} the SIC delivered 82 Wh kg^{-1} energy density. SICs with olive pits derived from hard carbon anodes and chemically activated hard carbon with KOH cathode have also been developed. As these SIC need to be improved in terms of cyclic stability and energy density a lot of work needs to be done in this direction. Some works have been done to achieve higher capacitance using nanopores electrodes with appropriate nitrogen content. In this Mitlin et al. [80] used polypyrrole hydrogel while Lu et al. [81] commercial pyrrole monomer for making nitrogen-doped carbon porous anodes. The polypyrrole hydrogel was treated at different temperatures and resulting in the shrinking of the granule size with interconnected granules of carbon which were interspersed with macro-porosity. This heteroatom doping with the support of high porosity shows increased reversibility of 427 mAh g^{-1} at 100 mA g^{-1}. With this material as an anode, the SIC was fabricated with the cathode of peanut shell-derived carbon to achieve energy densities of ~110 and ~40 Wh kg^{-1} at power densities of ~70 and ~14,500 W kg^{-1}, respectively. The use of this material also maintained a retentivity for about 5000 cycles. The use of a pyrrole monomer produces a rough-surfaced nitrogen-doped anode after a 1000°C thermal polymerization. This material was used for making cathode with the soft iron anode to obtain an energy density of ~245 Wh kg^{-1} at ~1620 W kg^{-1} power density for about 1000 cycles with a potential window of 0.01–4.70 V.

8.5.3.1.2 Transition metal dichalcogenides composite-based materials

The carbon-based materials can provide good specific capacitance but show dendrite formations at low sodiation potentials. Transition metal sulfides have come up as a better alternative. The MoS_2 is a two-dimensional structure in which the Mo ions are sandwiched between two sulfur atoms but due to poor conductivity and problems of agglomeration and volume changes they are used in composite forms to achieve a material with better electronic properties. But still, the MoS_2 has tremendous capabilities which can be utilized for making SICs. The Na ions can intercalate in the MoS_2 layer for termination discharge voltages greater than 0.4 V as

$$MoS_2 + xNa \leftrightarrow Na_xMoS_2 \tag{8.6}$$

and react through conversion type mechanism for voltages lower than 0.4 V, in which case we observe –

$$Na_xMoS_2 + (4 - x)Na \leftrightarrow Mo + 2Na_2S \tag{8.7}$$

Some researchers have produced a modification in the original structure of MoS_2 by achieving alloy-type anodes. Zhang et al. [82] have used monolayer carbon to make MoS_2 composite in which the interlayer modification of MoS_2 was done where an inter overlapped structure of MoS_2 and carbon was made. The layer separation in the obtained structure was around 1.02 nm which was enough to make SIC. In this SIC the intercalation of the Na ion in the MoS_2-C composite results in a phase transition of the MoS_2 from a semiconducting 2H-MoS_2 phase to a metallic IT-MoS_2 phase. Here the Na ions are electrochemically adsorbed on the MoS_2 surface helping in the rapid Na ion transport channel. Using this material as an anode and a porous carbon electrode derived from polyaniline was used to make a SIC which delivers an energy density of ~110 Wh kg^{-1} with a high power density of 12,000 W kg^{-1} for nearly 10,000 cycles. This SIC has a specific capacity of 200 mAh g^{-1} at 1 A g^{-1} with ~86% retention after 800 cycles. This has turned out an effective way to achieve better values than using hard carbon but still, there are more improvements that have been made. In a dual function sacrificing template-based strategy, the MoS_2-carbon hybrid nanotubes have been made using Sb_3S_3. Here the Sb_3S_3 plays a dual role in sulfurization and shape direction. In this method, the Sb_3S_3 acts as a template on which first the layer of Mo-containing compound is formed, and after thermal annealing MoS_2-C hybrid nanotubes are formed [83]. Using this material as anode with 0.227 and 0.27 nm interplanar distance in the composite and activated carbon as a cathode the SIC gave a maximum energy density of $\sim 107\ Wh\ kg^{-1}$. Another molybdenum-based compound, $MoSe_2$ has a comparatively narrower band gap, 1.1 eV, than MoS_2, 1.7 eV, thus making it a better electrical conductor. It has been used in SIB but it can also be used for assembling SIC as a $MoSe_2$-graphene anode with an activated carbon electrode. This material could give an energy density of $82\ Wh\ kg^{-1}$ with a retention capacity of 80% for 5000 cycles. Further Hu et al. have achieved a higher energy density of $148\ Wh\ kg^{-1}$ for a voltage range of 0–4 V with 70% retention even after 3000 cycles. They used a TiO_2 coated MoS_2 composite of N-doped carbon anode. The TiO_2 coating helps in obtaining an extended lifespan. The SnS_2 has also been explored recently for making SICs. Kim et al. achieved a specific capacitance of $600.3\ mAh\ g^{-1}$ and $304.8\ mAh\ g^{-1}$ at $10\ A\ g^{-1}$ current density to attain $108.3\ Wh\ kg^{-1}$ energy density [84]. They used SnS_2-graphene-carbon nanotube aerogel composites where the use of conductive carbon supports utilizing the pseudocapacitance of the SnS_2. They firstly coated the carbon nanotubes on GO sheets which then acted as nucleation sites for SnO_2 structures. Using hydrothermal growth of SnO_2 on the nanotubes they introduced a pressurized sulfidation method in which SnO_2 nanoparticles are transformed for making SnS_2 nanoplatelets as Eq. 8.8

$$SnO_2 + H_2S \rightarrow SnS_2 + H_2O \tag{8.8}$$

This method has the advantage over the wet chemical method in that the morphology of the carbon frame is not disturbed. This method is temperature sensitive and required to be maintained above 250°C and below 300°C. The as-obtained material can directly be used as an anode. Goswami et al. [85] used copper sulfide (CuS) as anode and Na metal as a cathode in a non-aqueous medium for a half-cell configuration (Fig. 8.14A). The use of pseudocapacitive electrodes shows capacitive controlled and diffusive controlled mechanisms at a slow scan and high scan rate respectively (Fig. 8.14B). It was observed that at $1\ mVs^{-1}$ scan rate CuS provide 74.8% of capacitive contribution (Fig. 8.14C).

8.5.3.1.3 Ti/Nb-based compounds

An even better option than transition metal dichalcogenides in terms of avoiding the dendrite formation and making the safer device is the Titanium-based materials. Other than this they are present in abundant amounts, at least enough to meet the requirements for making capacitors, undergo rare changes in a lattice structure, and are environment friendly. These materials are also capable of sustaining a longer lifespan due to the suitable Ti redox potentials ranging from half to one volt. Among the Ti-based compounds, TiO_2 has an opening structure and for the storage of Na, it has a lower average potential too. But the wide band gap of the material makes it a poor conductor and thus different methods need to be found to utilize its full potential. To achieve this goal, attempts have been made to induce oxygen vacancies and heteroatoms to produce structural defects, minimize the TiO_2 size to the nanoscale to increase Na ion diffusion efficiency, and incorporate carbon materials to increase the conductivity of the material so that better electrochemical performance could be observed.

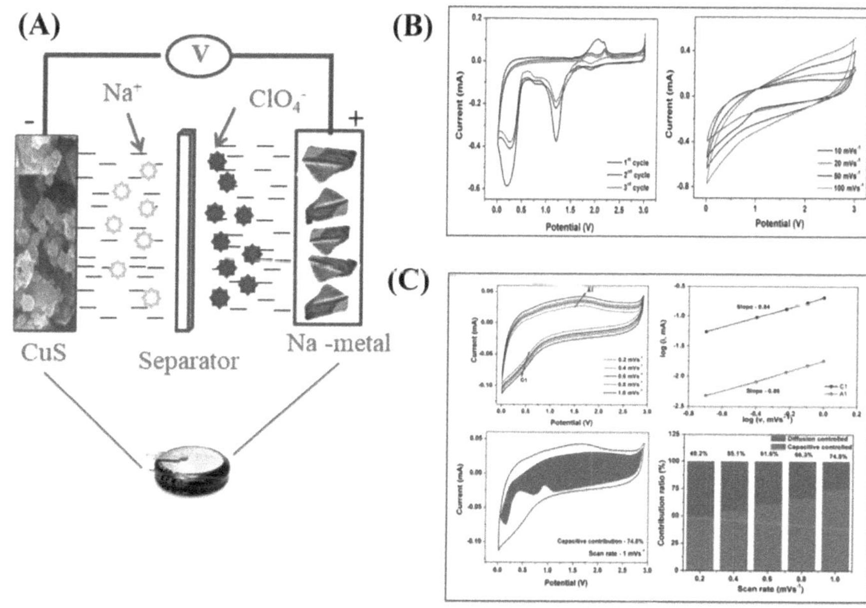

Figure 8.14 (A) Schematic representation of half-cell configuration of half-cell device. (B) CV curve at slow and fast scan rates. (C) Pseudocapacitive contribution profile of CuS at various parameters.
Source: Reproduced with permission ©The Royal Society of Chemistry and the Center National de la Recherche Scientifique 2020. M. Goswami, C. Nithya, N. Sathish, S. Kumar, N. Singh, A.K. Srivastava, S. Kumar, Electrochemical studies on crystalline CuS as an electrode material for non-aqueous Na-ion capacitors, N. J. Chem. 44 (2020) 5278–5284 [85].

An ultrafast ion transfer was achieved by Que et al. [86]. They made an oxygen-deficient TiO_{2-x}/CNT film which acts as a self-supported anode in a quasi-solid state SIC which had an activated carbon-CNT cathode system. This device showed an energy density of 109 Wh kg^{-1} with a power density of 250 W kg^{-1}. The advantage of this assembly for the anode is that the TiO_2 confined in the carbon nanofibers shortens the diffusion length for the Na ions owing to the high oxygen deficiency. Another remarkable feature of their work is the flexible film. Another work that made use of deficient sites was done by Lu et al. [87] in which they made a single crystal-like mesoporous TiO_2-graphene nanocomposite by a microwave-assisted method. The anatase TiO_2 which they used had many mesopores due to the voids between the nanocrystals building the TiO_2 structure. In this case, the voids shorten the pathway for ion diffusion. It can be seen that at the scan rates less than 5 mV s^{-1} the kinetics was controlled by the pseudocapacitive process but beyond this scan rate, the diffusion process was the dominating the kinetics of the reactions. Using activated carbon as cathode and this material as an anode the SIC shows an energy density of 64.2 Wh kg^{-1} and a long cycle life with 90% capacitive retention even after 10,000 cycles. The high energy and power density with a fast diffusion rate for Na ion can be achieved by using Nb-based compounds also. Among different structural options and oxidation states of Nb in the compounds, like amorphous phases or pseudohexagonal compounds, the Nb_2O_5 is found the most promising compound. It shows a 3.9 Å interplanar lattice spacing and a pseudocapacitive behavior. This large spacing allows the material to accept masses of Na ions and allows for a fast diffusion in the material. But again, this material too suffers from low electrical conductivity and slower reaction kinetics the electrical performance of the Nb_2O_5 suffers and restricts its use as an efficient anode for SICs. But again, employing carbon-based composites can give better performances. In this direction, Li et al. used the electrospinning technique followed by the carbonization and etching process [88]. This method produces a larger amount of mesoporous interconnected voids. The NIC based on m-Nb_2O_5/CNF anode and graphene-carbon nanofiber cathode delivered a power density of 60 kW kg^{-1} at 55 Wh kg^{-1}.

8.5.3.2 *Cathode materials for Na-ion capacitor*

8.5.3.2.1 Carbon materials

The capacitive material is usually used in SICs using the electrolyte-consuming mechanism and among different choices, commercial AC is mostly preferred [78]. Carbon materials show high thermal and chemical stability and a low redox potential which can help in obtaining an increased energy density. The AC has a very high surface area of ~3000 m^2 g^{-1} due to its micro porosity which made it one of the earliest choices for electrolyte penetration and exhibits a working potential of 1–4 V in organic electrolytes like $NaPF_6$ and $NaClO_4$ [78]. The AC utilizes the adsorption of the ions but owing to the low capacitance limit some work has been done to increase the specific surface area of the material using different structures. In this direction, Li et al. made a porous carbon nano architecture cathode and tablet-like TiO_2-C nanocomposite. Bridging the gap between the capacitive cathode and the battery-like anode they obtained a high energy density, 142.7 Wh kg^{-1} at 250 Wh kg^{-1} with cyclic retention of ~90% for 10,000 cycles. In another method for using carbon-based materials, functionalized carbon nanofibers can be used as cathode material. This increased the surface area and the interlayer spacing in the material and thus providing a higher ion diffusion efficiency. In another approach, Yang et al. have proposed using a 3D framework of AC. With an abundant number of tiny pores on the carbon-based cathode, they obtained a high energy density of ~110 Wh kg^{-1} at a power density of ~200 W kg with cyclic retention capability of about 75% over 15,000 cycles at 2 A g^{-1}.

8.5.3.2.2 MXenes

Though AC was used as the cathode material in most of the starting research in NIC, now different materials have been explored for the same purpose. Among them, MXene has been studied to a comparatively greater extent due to its capability of accommodating Na ions. These are two-dimensional inorganic compounds having atomically thin layers of transitional nitrides, metal carbides, or carbonitrides. These materials have two significant properties – metallic conductivity due to the presence of transition metals and hydrophilicity due to functional surfaces. It has a working potential ranging from 1 to 3.5 V making it a good candidate for making both anode and cathode. Simon et al. used a V2C-MXene cathode with a hard carbon anode to achieve a working potential of 3.5 V and a capacitance of $\sim 100\ F\ g^{-1}$ with a retention of 70% at a current density of $\sim 1\ A\ g^{-1}$ after 300 cycles. The lower number of cycles could be due to the change in the crystal structure due to the insertion and extraction of ions in V_2C.

8.5.3.2.3 $Na_2Fe_2(SO_4)_3$

This material, like MXene-based compounds, can also exhibit a redox potential of about 3.8 V. It can deliver a reversible capacity of $100\ mAh\ g^{-1}$ while cycling between 2 and 4.5 V for 30 cycles. Based on redox reaction charge contribution change by ferric and ferrous ions, this value is 85% of the one-electron theoretical capacity. This small drop is attributed to the decomposition of electrolytes under the 4.5 V. These electrochemical properties have been obtained for different electrolytes also. Though an acceptable specific and rate capability is seen with the use of this material as a cathode for SIC, the decrease in the capacity is still a major concern.

8.5.3.2.4 Na0.44MnO$_2$

This material has been attracting a lot of attention for use in SIB due to the presence of a large tunnel-like structure in its crystal structure which allows for the incorporation of Na ion. It also has a wide potential range of 2 to 3.8 V with a capacitive delivery of $140\ mAh\ g^{-1}$ it shows reversible Na^+ intercalation and deintercalation for the compounds in which the atom amount of Na varies from 0.25 to 0.65. in composites caused by six biphasic transitions. It can be used with MXenes as anodes but still only with a capacity drop for 60 cycles (Table 8.3).

Table 8.3 A summary of recently used anode and cathode electrode materials in Na-ion capacitors.

Anode	Cathode	Potential (V)	Energy density (W h Kg^{-1})	Power density (W kg^{-1})	References
V_2O_5-CNT	AC	0–2.8	7.5	5000	[73]
$Na_2Ti_3O_7$-CNT	AC	0–3	58.5	3000	[89]
Carbon	AC	0–4	112	67	[90]
Ti_2C-MXene	$Na_2Fe_2(SO_4)_3$	0.1–3.8	320	360	[91]
TiO_2@graphene	AC	1–3.8	64	1357	[87]
Disordered carbon	N-doped carbon microsphere	0–4.4	157	2000	[92]
Nb_2O_5@rGO	MSP-20 activated carbon	1–4.3	76	8000	[93]
$MoS_2 - C$	PANI-derived porous carbon	1–3.8	111.4	12,000	[70]
$NiCo_2O_4$	AC	0–4.5	25	300	[94]

8.6 Challenges and future perspective

In the last few years, batteries and supercapacitors are widely used for grid-connected systems and electric vehicles due to their easy accessibility. However, there are many challenges to the implementation of HESS in real-life applications:

- Increasing public awareness of the need to improve HESS technology to provide consumers with a stable power supply.
- To boost HESS's installation capability, assistance for infrastructure development, installation, and regular maintenance should be ensured.
- Because HESS technology research has remained static on a laboratory scale with just a theoretical viewpoint, programs to enhance the commercialization and industrialization of HESS technology are required.
- It is necessary to upgrade inefficient energy markets and infrastructures to entice power distribution companies to adopt HESS.
- It is necessary to upgrade inefficient energy markets and infrastructures to entice power distribution companies to adopt HESS.
- Modernization and breakthroughs in an extended lifetime, low cost, and high security are required for HESS progress. In addition, the following suggestions should be addressed shortly:
- To replace present batteries with significant improvements in lifespan and performance, new battery technologies such as aluminum, lithium-air, sodium-ion, and graphene must be developed.
- The creation of a new HESS control method based on multi-objective optimization that takes into account technological and economic constraints has yet to be targeted.
- The performance, adaptability, and dependability of HESS will be critical in the revolutionary concept of the "energy internet" for future energy supply and distribution systems.

References

[1] D. Deng, Li-ion batteries: basics, progress, and challenges, Energy Sci. Eng. 3 (2015) 385–418. Available from: https://doi.org/10.1002/ese3.95.

[2] N. Yabuuchi, K. Kubota, M. Dahbi, S. Komaba, Research development on sodium-ion batteries, Chem. Rev. 114 (2014) 11636–11682. Available from: https://doi.org/10.1021/cr500192f.

[3] K. Kuratani, N. Uemura, H. Senoh, H.T. Takeshita, T. Kiyobayashi, Conductivity, viscosity, and density of MClO 4 (M = Li and Na) dissolved in propylene carbonate and γ-butyrolactone at high concentrations, J. Power Sources. 223 (2013) 175–182. Available from: https://doi.org/10.1016/j.jpowsour.2012.09.039.

[4] M. Okoshi, Y. Yamada, A. Yamada, H. Nakai, Theoretical analysis on de-solvation of lithium, sodium, and magnesium cations to organic electrolyte solvents, J. Electrochem. Soc. 160 (2013) A2160–A2165. Available from: https://doi.org/10.1149/2.074311jes.

[5] P. Hartman, H.K. Chan, Application of the periodic bond chain (PBC) theory and attachment energy consideration to derive the crystal morphology of hexamethylmelamine, Pharm. Res. An. Off. J. Am. Assoc. Pharm. Sci. 10 (1993) 1052–1058. Available from: https://doi.org/10.1023/A:1018927109487.

[6] D.A. Stevens, J.R. Dahn, High capacity anode materials for rechargeable sodium-ion batteries, J. Electrochem. Soc. 147 (2000) 1271. Available from: https://doi.org/10.1149/1.1393348.

[7] N. Nitta, F. Wu, J.T. Lee, G. Yushin, Li-ion battery materials: present and future, Mater. Today. 18 (2015) 252–264. Available from: https://doi.org/10.1016/j.mattod.2014.10.040.

[8] A. Magasinski, P. Dixon, B. Hertzberg, A. Kvit, J. Ayala, G. Yushin, High-performance lithium-ion anodes using a hierarchical bottom-up approach, Nat. Mater. 9 (2010) 353–358. Available from: https://doi.org/10.1038/nmat2725.

[9] S. Mayrink, J.G. Oliveira, B.H. Dias, L.W. Oliveira, J.S. Ochoa, G.S. Rosseti, Regenerative braking for energy recovering in diesel-electric freight trains: a technical and economic evaluation, Energies 13 (2020). Available from: https://doi.org/10.3390/en13040963.

[10] G.G. Amatucci, F. Badway, A. Du Pasquier, T. Zheng, An asymmetric hybrid nonaqueous energy storage cell, J. Electrochem. Soc. 148 (2001) A930. Available from: https://doi.org/10.1149/1.1383553.

[11] P. Han, G. Xu, X. Han, J. Zhao, X. Zhou, G. Cui, Lithium-ion capacitors in organic electrolyte system: scientific problems, material development, and key technologies, Adv. Energy Mater. 8 (2018) 1–30. Available from: https://doi.org/10.1002/aenm.201801243.

[12] H. Kim, M.Y. Cho, M.H. Kim, K.Y. Park, H. Gwon, Y. Lee, et al., A novel high-energy hybrid supercapacitor with an anatase to 2-reduced graphene oxide anode and an activated carbon cathode, Adv. Energy Mater. 3 (2013) 1500–1506. Available from: https://doi.org/10.1002/aenm.201300467.

[13] Y. Wang, Y. Song, Y. Xia, Electrochemical capacitors: mechanism, materials, systems, characterization and applications, Chem. Soc. Rev. 45 (2016) 5925–5950. Available from: https://doi.org/10.1039/c5cs00580a.

[14] C. Sivakumar, J.N. Nian, H. Teng, Poly(o-toluidine) for carbon fabric electrode modification to enhance the electrochemical capacitance and conductivity, J. Power Sources 144 (2005) 295–301. Available from: https://doi.org/10.1016/j.jpowsour.2004.12.010.

[15] B. Li, F. Dai, Q. Xiao, L. Yang, J. Shen, C. Zhang, et al., Nitrogen-doped activated carbon for a high energy hybrid supercapacitor, Energy Environ. Sci. 9 (2016) 102–106. Available from: https://doi.org/10.1039/c5ee03149d.

[16] N.M.R. Peres, R.M. Ribeiro, Focus on graphene, N. J. Phys. 11 (2009). Available from: https://doi.org/10.1088/1367-2630/11/9/095002.

[17] R. Wang, Q. Zhao, W. Zheng, Z. Ren, X. Hu, J. Li, et al., Achieving high energy density in a 4.5 v all nitrogen-doped graphene-based lithium-ion capacitor, J. Mater. Chem. A 7 (2019) 19909–19921. Available from: https://doi.org/10.1039/c9ta06316a.

[18] F. Tu, S. Liu, T. Wu, G. Jin, C. Pan, Porous graphene as cathode material for lithium-ion capacitor with high electrochemical performance, Powder Technol. 253 (2014) 580–583. Available from: https://doi.org/10.1016/j.powtec.2013.12.008.

[19] D. Mhamane, V. Aravindan, M.S. Kim, H.K. Kim, K.C. Roh, D. Ruan, et al., Silica-assisted bottom-up synthesis of graphene-like high surface area carbon for highly efficient ultracapacitor and Li-ion hybrid capacitor applications, J. Mater. Chem. A. 4 (2016) 5578–5591. Available from: https://doi.org/10.1039/c6ta00868b.

[20] M. Cai, X. Sun, W. Chen, Z. Qiu, L. Chen, X. Li, et al., Performance of lithium-ion capacitors using pre-lithiated multiwalled carbon nanotubes/graphite composite as the negative electrode, J. Mater. Sci. 53 (2018) 749–758. Available from: https://doi.org/10.1007/s10853-017-1524-5.

[21] X. Zhao, C. Johnston, P.S. Grant, A novel hybrid supercapacitor with a carbon nanotube cathode and an iron oxide/carbon nanotube composite anode, J. Mater. Chem. 19 (2009) 8755–8760. Available from: https://doi.org/10.1039/b909779a.

[22] [40]K. Karthikeyan, S. Amaresh, J.N. Son, Y.S. Lee, Synthesis and performance of Li_2MnSiO_4 as an electrode material for hybrid supercapacitor applications, J. Electrochem. Sci. Technol. 3 (2012) 72–79. Available from: https://doi.org/10.5229/jecst.2012.3.2.72.

[23] H. Wu, C.V. Rao, B. Rambabu, Electrochemical performance of $LiNi_{0.5}Mn_{1.5}O_4$ prepared by improved solid-state method as a cathode in a hybrid supercapacitor, Mater. Chem. Phys. 116 (2009) 532–535. Available from: https://doi.org/10.1016/j.matchemphys.2009.04.028.

[24] H. Li, L. Cheng, Y. Xia, A hybrid electrochemical supercapacitor based on a 5 V Li-ion battery cathode and active carbon, Electrochem. Solid-State Lett. 8 (2005) 433–436. Available from: https://doi.org/10.1149/1.1960007.

[25] K.S. Ryu, K.M. Kim, N.G. Park, Y.J. Park, S.H. Chang, Symmetric redox supercapacitor with conducting polyaniline electrodes, J. Power Sources. 103 (2002) 305–309. Available from: https://doi.org/10.1016/S0378-7753(01)00862-X.

[26] M.U. Anu Prathap, B. Satpati, R. Srivastava, Facile preparation of polyaniline/MnO_2 nanofibers and its electrochemical application in the simultaneous determination of catechol, hydroquinone, and resorcinol, Sens. Actuators B Chem. 186 (2013) 67–77. Available from: https://doi.org/10.1016/j.snb.2013.05.076.

[27] A. Imani, G. Farzi, Facile route for multi-walled carbon nanotube coating with polyaniline: tubular morphology nanocomposites for supercapacitor applications, J. Mater. Sci. Mater. Electron. 26 (2015) 7438–7444. Available from: https://doi.org/10.1007/s10854-015-3377-5.

[28] M. Ates, M.A. Serin, I. Ekmen, Y.N. Ertas, Supercapacitor behaviors of polyaniline/CuO, polypyrrole/CuO and PEDOT/CuO nanocomposites, Polym. Bull. 72 (2015) 2573–2589. Available from: https://doi.org/10.1007/s00289-015-1422-4.

[29] B.V. Ratnakumar, M.C. Smart, S. Surampudi, Effects of SEI on the kinetics of lithium intercalation, J. Power Sources 97–98 (2001) 137–139. Available from: https://doi.org/10.1016/S0378-7753(01)00682-6.

[30] Y. Matsuo, K. Fumita, T. Fukutsuka, Y. Sugie, H. Koyama, K. Inoue, Butyrolactone derivatives as electrolyte additives for lithium-ion batteries with graphite anodes, J. Power Sources. 119–121 (2003) 373–377. Available from: https://doi.org/10.1016/S0378-7753(03)00271-4.

[31] J.-H. Kim, J.-S. Kim, Y.-G. Lim, J.-G. Lee, Y.-J. Kim, Effect of carbon types on the electrochemical properties of negative electrodes for Li-ion capacitors, J. Power Sources. 196 (2011) 10490–10495. Available from: https://doi.org/10.1016/j.jpowsour.2011.08.081.

[32] M. Yuan, W. Liu, Y. Zhu, Y. Xu, Electrochemical performance of lithium ion capacitors with different types of negative electrodes, Russ. J. Electrochem. 50 (2014) 594–598. Available from: https://doi.org/10.1134/S1023193514020074.

[33] K.J. Kim, T.S. Lee, H.G. Kim, S.H. Lim, S.M. Lee, A hard carbon/microcrystalline graphite/carbon composite with a core-shell structure as novel anode materials for lithium-ion batteries, Electrochim. Acta. 135 (2014) 27–34. Available from: https://doi.org/10.1016/j.electacta.2014.04.171.

[34] T. Ohzuku, A. Ueda, N. Yamamoto, Zero-strain insertion material of Li [Li1/3Ti5/3] O 4 for rechargeable lithium cells, J. Electrochem. Soc. 142 (1995) 1431–1435. Available from: https://doi.org/10.1149/1.2048592.

[35] W.J.H. Borghols, M. Wagemaker, U. Lafont, E.M. Kelder, F.M. Mulder, Size effects in the $Li^{4+}xTi_5O_{12}$ spinel, J. Am. Chem. Soc. 131 (2009) 17786–17792. Available from: https://doi.org/10.1021/ja902423e.

[36] X. Sun, P.V. Radovanovic, B. Cui, Advances in spinel $Li_4Ti_5O_{12}$ anode materials for lithium-ion batteries, N. J. Chem. 39 (2015) 38–63. Available from: https://doi.org/10.1039/c4nj01390e.

[37] E.M. Sorensen, S.J. Barry, H.-K. Jung, J.R. Rondinelli, J.T. Vaughey, K.R. Poeppelmeier, Three-dimensionally ordered macroporous $Li_4Ti_5O_{12}$: effect of wall structure on electrochemical properties, Chem. Mater. 18 (2006) 1713–1714. Available from: https://doi.org/10.1021/cm0604054.

[38] X. Zhang, C. Lu, H. Peng, X. Wang, Y. Zhang, Z. Wang, et al., Influence of sintering temperature and graphene additives on the electrochemical performance of porous $Li_4Ti_5O_{12}$ anode for lithium ion capacitor, Electrochim. Acta 246 (2017) 1237–1247. Available from: https://doi.org/10.1016/j.electacta.2017.07.014.

[39] B. Lee, J.R. Yoon, Preparation and characteristics of $Li_4Ti_5O_{12}$ with various dopants as anode electrode for hybrid supercapacitor, Curr. Appl. Phys. 13 (2013) 1350–1353. Available from: https://doi.org/10.1016/j.cap.2013.04.002.

[40] S.-H. Lee, H.-K. Kim, Y.-S. Yun, J.R. Yoon, S.-G. Lee, Y.-H. Lee, A novel high-performance cylindrical hybrid supercapacitor with $Li^{4-}xNaxTi_5O_{12}$/activated carbon electrodes, Int. J. Hydrog. Energy. 39 (2014) 16569–16575. Available from: https://doi.org/10.1016/j.ijhydene.2014.05.072.

[41] J.J. Yang, Y.R. Kim, M.G. Jeong, Y.J. Yuk, H.J. Kim, S.G. Park, Synthesis and electrochemical characteristics of spherical $Li_4Ti_5O_{12}$/CNT composite materials for hybrid capacitors, J. Electrochem. Sci. Technol. 6 (2015) 59–64. Available from: https://doi.org/10.5229/JECST.2015.6.2.59.

[42] H.S. Choi, T. Kim, J.H. Im, C.R. Park, Preparation and electrochemical performance of hyper-networked $Li_4Ti_5O_{12}$/carbon hybrid nanofiber sheets for a battery-supercapacitor hybrid system, Nanotechnology 22 (2011). Available from: https://doi.org/10.1088/0957-4484/22/40/405402.

[43] N. Xu, X. Sun, X. Zhang, K. Wang, Y. Ma, A two-step method for preparing $Li_4Ti_5O_{12}$-graphene as an anode material for lithium-ion hybrid capacitors, RSC Adv. 5 (2015) 94361–94368. Available from: https://doi.org/10.1039/c5ra20168c.

[44] V. Augustyn, P. Simon, B. Dunn, Pseudocapacitive oxide materials for high-rate electrochemical energy storage, Energy Environ. Sci. 7 (2014) 1597–1614. Available from: https://doi.org/10.1039/C3EE44164D.

[45] J. Wang, J. Polleux, J. Lim, B. Dunn, Pseudocapacitive contributions to electrochemical energy storage in TiO_2 (Anatase) nanoparticles, J. Phys. Chem. C. 111 (2007) 14925–14931. Available from: https://doi.org/10.1021/jp074464w.

[46] Y. Cao, K. Geng, H. Geng, H. Ang, J. Pei, Y. Liu, et al., Metal–oleate complex-derived bimetallic oxides nanoparticles encapsulated in 3D graphene networks as anodes for efficient lithium storage with pseudocapacitance, Nano-Micro Lett. 11 (2019) 1–14. Available from: https://doi.org/10.1007/s40820-019-0247-3.

[47] R. Baddour-Hadjean, C. Navone, J.P. Pereira-Ramos, In situ Raman microspectrometry investigation of electrochemical lithium intercalation into sputtered crystalline V_2O_5 thin films, Electrochim. Acta. 54 (2009) 6674–6679. Available from: https://doi.org/10.1016/j.electacta.2009.06.052.

[48] X. Liu, J. Zeng, H. Yang, K. Zhou, D. Pan, V_2O_5-based nanomaterials: synthesis and their applications, RSC Adv. 8 (2018) 4014–4031. Available from: https://doi.org/10.1039/c7ra12523b.

[49] C. Delmas, H. Cognac-Auradou, J.M. Cocciantelli, M. Ménétrier, J.P. Doumerc, The $Li_xV_2O_5$ system: an overview of the structure modifications induced by the lithium

intercalation, Solid. State Ion. 69 (1994) 257–264. Available from: https://doi.org/10.1016/0167-2738(94)90414-6.

[50] M. Sathiya, A.S. Prakash, K. Ramesha, J. Tarascon, A.K. Shukla, V_2O_5-anchored carbon nanotubes for enhanced electrochemical energy storage, J. Am. Chem. Soc. 133 (2011) 16291–16299. Available from: https://doi.org/10.1021/ja207285b.

[51] M.L. Divya, V. Aravindan, Electrochemically generated γ-$Li_xV_2O_5$ as insertion host for high-energy Li-ion capacitors, Chem. - An. Asian J. 14 (2019) 4665–4672. Available from: https://doi.org/10.1002/asia.201900946.

[52] Z. Chen, V. Augustyn, J. Wen, Y. Zhang, M. Shen, B. Dunn, et al., High-performance supercapacitors based on intertwined CNT/V_2O_5 nanowire nanocomposites, Adv. Mater. 23 (2011) 791–795. Available from: https://doi.org/10.1002/adma.201003658.

[53] S. Fu, Q. Yu, Z. Liu, P. Hu, Q. Chen, S. Feng, et al., Yolk-shell Nb_2O_5 microspheres as intercalation pseudocapacitive anode materials for high-energy Li-ion capacitors, J. Mater. Chem. A. 7 (2019) 11234–11240. Available from: https://doi.org/10.1039/c9ta02342a.

[54] J.W. Kim, V. Augustyn, B. Dunn, The effect of crystallinity on the rapid pseudocapacitive response of Nb_2O_5, Adv. Energy Mater. 2 (2012) 141–148. Available from: https://doi.org/10.1002/aenm.201100494.

[55] Y.-G. Sun, J.-Y. Piao, L.-L. Hu, D.-S. Bin, X.-J. Lin, S.-Y. Duan, et al., Controlling the reaction of nanoparticles for hollow metal oxide nanostructures, J. Am. Chem. Soc. 140 (2018) 9070–9073. Available from: https://doi.org/10.1021/jacs.8b04948.

[56] S. Lou, X. Cheng, L. Wang, J. Gao, Q. Li, Y. Ma, et al., High-rate capability of three-dimensionally ordered macroporous T-Nb_2O_5 through Li + intercalation pseudocapacitance, J. Power Sources. 361 (2017) 80–86. Available from: https://doi.org/10.1016/j.jpowsour.2017.06.023.

[57] B. Deng, T. Lei, W. Zhu, L. Xiao, J. Liu, In-plane assembled orthorhombic Nb_2O_5 nanorod films with high-rate Li + intercalation for high-performance flexible Li-ion capacitors, Adv. Funct. Mater. 28 (2018) 1704330. Available from: https://doi.org/10.1002/adfm.201704330.

[58] D. Li, J. Shi, H. Liu, C. Liu, G. Dong, H. Zhang, et al., T-Nb_2O_5 embedded carbon nanosheets with superior reversibility and rate capability as an anode for high energy Li-ion capacitors, Sustain, Energy Fuels 3 (2019) 1055–1065. Available from: https://doi.org/10.1039/c9se00052f.

[59] S. Wang, Y. Ren, G. Liu, Y. Xing, S. Zhang, Peanut-like MnO@C core-shell composites as anode electrodes for high-performance lithium-ion batteries, Nanoscale 6 (2014) 3508–3512. Available from: https://doi.org/10.1039/C3NR05916B.

[60] J. Zhang, J. Lin, Y. Zeng, Y. Zhang, H. Guo, Morphological and structural evolution of MnO@C anode and its application in lithium-ion capacitors, ACS Appl. Energy Mater. 2 (2019) 8345–8358. Available from: https://doi.org/10.1021/acsaem.9b01850.

[61] X. Fan, S. Li, L. Lu, Porous micrometer-sized MnO cubes as the anode of the lithium-ion battery, Electrochim. Acta. 200 (2016) 152–160. Available from: https://doi.org/10.1016/j.electacta.2016.03.114.

[62] V. Aravindan, W. Chuiling, M.V. Reddy, G.V.S. Rao, B.V.R. Chowdari, S. Madhavi, Carbon coated nano-$LiTi_2$(PO 4) 3 electrodes for non-aqueous hybrid supercapacitors, Phys. Chem. Chem. Phys. (2012). Available from: https://doi.org/10.1039/c2cp40603a.

[63] A. Banerjee, K.K. Upadhyay, D. Puthusseri, V. Aravindan, S. Madhavi, S. Ogale, MOF-derived crumpled-sheet-assembled perforated carbon cuboids as highly effective cathode active materials for ultra-high energy density Li-ion hybrid electrochemical

capacitors (Li-HECs, Nanoscale 6 (2014) 4387–4394. Available from: https://doi.org/10.1039/C4NR00025K.

[64] V. Aravindan, D. Mhamane, W.C. Ling, S. Ogale, S. Madhavi, Nonaqueous lithium-ion capacitors with high energy densities using trigol-reduced graphene oxide nanosheets as cathode-active material, ChemSusChem. 6 (2013) 2240–2244.

[65] R. Yi, S. Chen, J. Song, M.L. Gordin, A. Manivannan, D. Wang, High-performance hybrid supercapacitor enabled by a high-rate Si-based anode, Adv. Funct. Mater. 24 (2014) 7433–7439.

[66] Q. Wang, Z.H. Wen, J.H. Li, A hybrid supercapacitor fabricated with a carbon nanotube cathode and a TiO_2–B nanowire anode, Adv. Funct. Mater. 16 (2006) 2141–2146.

[67] V. Aravindan, M.V. Reddy, S. Madhavi, S.G. Mhaisalkar, G.V.S. Rao, B.V.R. Chowdari, Hybrid supercapacitor with nano-TiP_2O_7 as intercalation electrode, J. Power Sources. 196 (2011) 8850–8854.

[68] X. Wang, G. Li, Z. Chen, V. Augustyn, X. Ma, G. Wang, et al., High-performance supercapacitors based on nanocomposites of Nb_2O_5 nanocrystals and carbon nanotubes, Adv. Energy Mater. 1 (2011) 1089–1093.

[69] J. Come, M. Naguib, P. Rozier, M.W. Barsoum, Y. Gogotsi, P.-L. Taberna, et al., A non-aqueous asymmetric cell with a Ti2C-based two-dimensional negative electrode, J. Electrochem. Soc. 159 (2012) A1368.

[70] R. Wang, S. Wang, D. Jin, Y. Zhang, Y. Cai, J. Ma, et al., Engineering layer structure of MoS_2-graphene composites with robust and fast lithium storage for high-performance Li-ion capacitors, Energy Storage Mater. 9 (2017) 195–205.

[71] R. Ding, L. Qi, H. Wang, Porous $NiCo_2O_4$ as an anode material for 4.5 V hybrid Li-ion capacitors, 2013.

[72] D. Cericola, P. Novák, A. Wokaun, R. Kötz, Hybridization of electrochemical capacitors and rechargeable batteries: an experimental analysis of the different possible approaches utilizing activated carbon, $Li_4Ti_5O_{12}$ and $LiMn_2O_4$, J. Power Sources. 196 (2011) 10305–10313.

[73] Z. Chen, V. Augustyn, X. Jia, Q. Xiao, B. Dunn, Y. Lu, High-performance sodium-ion pseudocapacitors based on hierarchically porous nanowire composites, ACS Nano 6 (2012) 4319–4327.

[74] E.M. Lotfabad, P. Kalisvaart, A. Kohandehghan, D. Karpuzov, D. Mitlin, Origin of non-SEI related coulombic efficiency loss in carbons tested against Na and Li, J. Mater. Chem. A 2 (2014) 19685–19695.

[75] S.R. Sivakkumar, A.G. Pandolfo, Evaluation of lithium-ion capacitors assembled with pre-lithiated graphite anode and activated carbon cathode, Electrochim. Acta. 65 (2012) 280–287.

[76] K. Naoi, S. Ishimoto, J. Miyamoto, W. Naoi, Second generation 'nanohybrid supercapacitor': evolution of capacitive energy storage devices, Energy Environ. Sci. 5 (2012) 9363–9373.

[77] J. Ding, W. Hu, E. Paek, D. Mitlin, Review of hybrid ion capacitors: from aqueous to lithium to sodium, Chem. Rev. 118 (2018) 6457–6498. Available from: https://doi.org/10.1021/acs.chemrev.8b00116.

[78] K. Kuratani, M. Yao, H. Senoh, N. Takeichi, T. Sakai, T. Kiyobayashi, Na-ion capacitor using sodium pre-doped hard carbon and activated carbon, Electrochim. Acta. 76 (2012) 320–325.

[79] S. Jayaraman, A. Jain, M. Ulaganathan, E. Edison, M.P. Srinivasan, R. Balasubramanian, et al., Li-ion vs. Na-ion capacitors: a performance evaluation with coconut shell derived

mesoporous carbon and natural plant based hard carbon, Chem. Eng. J. 316 (2017) 506–513.

[80] J. Ding, Z. Li, K. Cui, S. Boyer, D. Karpuzov, D. Mitlin, Heteroatom enhanced sodium ion capacity and rate capability in a hydrogel derived carbon give record performance in a hybrid ion capacitor, Nano Energy 23 (2016) 129–137.

[81] S. Chen, J. Wang, L. Fan, R. Ma, E. Zhang, Q. Liu, et al., An ultrafast rechargeable hybrid sodium-based dual-ion capacitor based on hard carbon cathodes, Adv. Energy Mater. 8 (2018) 1800140.

[82] R. Wang, S. Wang, X. Peng, Y. Zhang, D. Jin, P.K. Chu, et al., Elucidating the intercalation pseudocapacitance mechanism of MoS_2–carbon monolayer interoverlapped superstructure: toward high-performance sodium-ion-based hybrid supercapacitor, ACS Appl. Mater. Interfaces. 9 (2017) 32745–32755.

[83] J. Gao, Y. Li, Y. Liu, S. Jiao, J. Li, G. Wang, et al., The dual-function sacrificing template directed formation of MoS 2/C hybrid nanotubes enabling highly stable and ultrafast sodium storage, J. Mater. Chem. A 7 (2019) 18828–18834.

[84] J. Cui, S. Yao, Z. Lu, J. Huang, W.G. Chong, F. Ciucci, et al., Revealing pseudocapacitive mechanisms of metal dichalcogenide SnS_2/graphene-CNT aerogels for high-energy Na hybrid capacitors, Adv. Energy Mater. 8 (2018) 1702488.

[85] M. Goswami, C. Nithya, N. Sathish, S. Kumar, N. Singh, A.K. Srivastava, et al., Electrochemical studies on crystalline CuS as an electrode material for non-aqueous Na-ion capacitors, N. J. Chem. 44 (2020) 5278–5284. Available from: https://doi.org/10.1039/c9nj06465f.

[86] L. Que, F. Yu, Z. Wang, D. Gu, Pseudocapacitance of TiO_{2-x}/CNT anodes for high-performance quasi-solid-state Li-ion and Na-ion capacitors, Small. 14 (2018) 1704508.

[87] Z. Le, F. Liu, P. Nie, X. Li, X. Liu, Z. Bian, et al., Pseudocapacitive sodium storage in mesoporous single-crystal-like TiO_2–graphene nanocomposite enables high-performance sodium-ion capacitors, ACS Nano 11 (2017) 2952–2960.

[88] Y. Li, H. Wang, L. Wang, Z. Mao, R. Wang, B. He, et al., Mesopore-induced ultrafast Na + -storage in T-Nb_2O_5/carbon nanofiber films toward flexible high-power Na-ion capacitors, Small. 15 (2019) 1804539.

[89] S. Dong, L. Shen, H. Li, P. Nie, Y. Zhu, Q. Sheng, et al., Pseudocapacitive behaviours of $Na_2Ti_3O_7$@ CNT coaxial nanocables for high-performance sodium-ion capacitors, J. Mater. Chem. A 3 (2015) 21277–21283.

[90] H. Wang, D. Mitlin, J. Ding, Z. Li, K. Cui, Excellent energy–power characteristics from a hybrid sodium ion capacitor based on identical carbon nanosheets in both electrodes, J. Mater. Chem. A 4 (2016) 5149–5158.

[91] X. Wang, S. Kajiyama, H. Iinuma, E. Hosono, S. Oro, I. Moriguchi, et al., Pseudocapacitance of MXene nanosheets for high-power sodium-ion hybrid capacitors, Nat. Commun. 6 (2015) 1–6.

[92] C. Wang, F. Wang, Z. Liu, Y. Zhao, Y. Liu, Q. Yue, et al., N-doped carbon hollow microspheres for metal-free quasi-solid-state full sodium-ion capacitors, Nano Energy 41 (2017) 674–680.

[93] E. Lim, C. Jo, M.S. Kim, M. Kim, J. Chun, H. Kim, et al., High-performance sodium-ion hybrid supercapacitor based on Nb_2O_5@ carbon core–shell nanoparticles and reduced graphene oxide nanocomposites, Adv. Funct. Mater. 26 (2016) 3711–3719.

[94] R. Ding, L. Qi, H. Wang, An investigation of spinel $NiCo_2O_4$ as anode for Na-ion capacitors, Electrochim. Acta. 114 (2013) 726–735.

Electrochemical energy storage systems

9

Marm Dixit[1], Nitin Muralidharan[1], Anand Parejiya[1,2], Rachid Essehli[1], Ilias Belharouak[1,2] and Ruhul Amin[1]
[1]Electrification and Energy Infrastructures Division, Oak Ridge National Laboratory, Oak Ridge, TN, United States, [2]Bredesen Center for Interdisciplinary Research and Graduate Education, University of Tennessee, Knoxville, TN, United States

Chapter Outline

Nomenclature and abbreviation

A	area
Al	aluminum
ASSB	all-solid-state battery
CNT	carbon nano tube
Cu	cupper

Emerging Trends in Energy Storage Systems and Industrial Applications. DOI: https://doi.org/10.1016/B978-0-323-90521-3.00011-9

d	interelectrode distance
DOE	Department of Energy
E	cell potential
E_0	equilibrium potential
ε_0	dielectric constant
EDL	electrochemical double layer
EDLC	electrochemical double layer capacitor
F	Faraday's constant
HOMO	higher occupied molecular orbital
kWh	kilo watt hour
Li-Air	lithium-air
Lib	lithium-ion battery
$Li\text{-}CO_2$	lithium-air
Li_xC	lithium carbon
LUMO	lower unoccupied molecular orbital
mAh/g	milli ampere-hour per gram
μm	micrometer
μ**Wh**	micro watt hour
n	number of electrons transferred in reaction
NKK	Nippon Kodoshi Corporation
OCV	open circuit voltage
ORNL	Oak Ridge National Laboratory
ox	oxidizing
P	power of a device
Pseudo-C	pseudo-capacitor
Q	Q charge stored
R	universal gas constant
R&D	research and development
red	reducing
RFB	redox flow battery
SC	supercapacitor
SEI	solid electrolyte interfaces
SHE	standard hydrogen electrode
SIB	sodium-ion battery
SSB	solid-state battery
T	temperature
t_{dis}	discharge time
V	voltage
Wh	watt hour
Zn	zinc

Highlight

- Role, necessity, and viability of modern energy storage platforms are discussed
- Solid-state and emerging battery technologies are canvassed
- Different energy storage systems and types are displayed

9.1 Introduction to electrochemical energy storage

The ability of a civilization to harness energy effectively and sustainably, its storage, and utilization is the key driving factor that determines continual progress and growth. The modern civilization we know today is largely dependent on a depleting source of carbon-based reserve energy, viz; fossil fuels such as coal, natural gas, and oil, ($\sim$85% of today's energy generation), accumulated over millions of years. Our continuing utilization of this reserve energy has extracted a terrible price in the form of global warming (from greenhouse gas emissions) resulting in unpredictable climate trends that threaten our very existence. Increasing urbanization, rising population, and economic growth around the world in the coming decades are expected to significantly increase our energy consumption, thus, exacerbating this dire threat [1–6]. Almost all the present greenhouse gas emissions contributing to global climate change are directly related to the energy sector [7]. Transportation, electricity generation, and industrial sectors accounted for the majority of greenhouse emissions in 2018 (1869 million metric tons of CO_2 equivalent) in the United States alone [8]. Despite this alarming scenario, continual utilization of energy is essential for civilizational growth and thus requires novel solutions using clean energy sources such as renewables. The pursuit of combined clean energy utilization and decarbonization of crucial energy-consuming sectors can effectively tackle the challenges of both energy consumption and greenhouse gas emissions respectively. Theoretical analyses predict high electrification and decarbonization potential for transportation, industrial, commercial, and residential sectors [9]. This, however, would require significantly modifying the current energy generation and distribution by reducing/eliminating the reliance on fossil fuels and increasing the implementation of renewable resources and sustainable technologies. Implementing renewables poses unique challenges owing to the intermittent nature of these sources and thus requires the use of accompanying energy storage systems to offset such drawbacks.

Energy storage systems can also be deployed to enable the utilization of energy in spatial and temporal regions distinct from energy production [10]. In this regard, energy storage systems are ubiquitous in this era with almost all energy consumption occurring at places and times that are different from their production. Modern energy storage systems are broadly classified based on the mechanism they employ to store energy viz. thermal, mechanical, gravitational, chemical, and hydro, among others. A key parameter for energy storage systems is round-trip efficiency which is defined as the ratio of energy used to the energy stored and reflects the irreversible losses in the system. One of the most successful and widely deployed energy storage technology is the electrochemical energy storage system. Their unique advantages include scalability, affordability, and high coulombic efficiencies thus making them attractive for a wide range of applications [10]. These systems rely on storing electrical energy in the form of chemical energy through certain reactions involving the transfer of electrons. Electrochemical energy systems have applications from micro-electronics [11] all the way up to grid storage [12]. In this chapter, we discuss the key fundamentals of electrochemical energy storage systems and describe the key energy storage technologies from a materials and systems perspective.

9.2 Electrochemical energy storage technologies

The primary classification of electrochemical energy storage devices is based on the charge storage mechanism which can be Faradaic or non-Faradaic (Fig. 9.1)

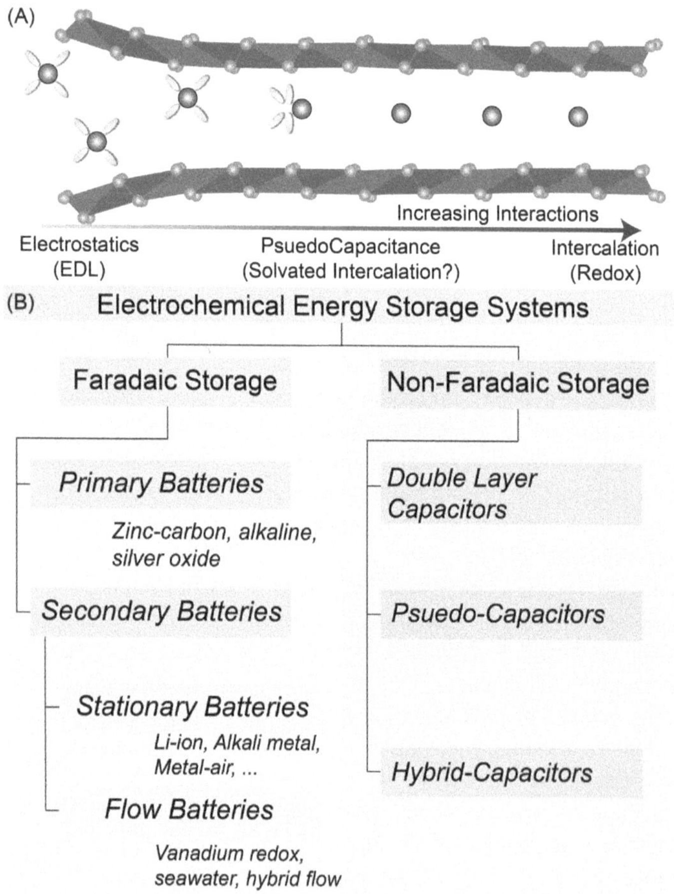

Figure 9.1 (A) Schematic diagram showing the fundamental mechanisms of charge storage in electrochemical energy storage systems. (B) Classification of key energy storage systems by the mechanism of charge storage: faradaic which involves chemical storage of charge and non-Faradaic which involves a physical storage of charge.

[13]. Faradaic charge storage typically involves a redox reaction that involves a chemical transformation of the species involved, while non-Faradaic charge storage involves only physical storage of charge through electrostatics. The key technologies for faradaic and non-faradaic charge storage are outlined in Fig. 9.1.

9.2.1 Supercapacitors

Supercapacitors store electrochemical energy by non-Faradaic charge storage mechanisms on the principles of electrostatics and capacitance [14]. On application of voltage, charge is accumulated on the electrode surface and diffused into the pores of the active material. Physical charge storage on high specific area materials, where charges are separated by small distances results in high specific capacitance with very high cyclability. The charge storage mechanism for two electrodes E_{S1} and E_{S2}, with A^- and C^+ anion and cation, and the interface are represented by // is given by the following equations:

$$E_{S1} + A^- \rightarrow E_{S1}^+//A^- + e^-$$

$$E_{S2} + C^+ + e^- \rightarrow E_{S2}^-//C^+$$

$$E_{S1} + A^- + E_{S2} + C^+ \rightarrow E_{S1}^+//A^- + E_{S2}^-//C^+$$

The charge stored in a non-Faradaic system is given by [15],

$$Q = CV = \frac{\varepsilon_0 A}{d} V$$

where Q is the charge stored, C is capacitance, V is the voltage applied, ε_0 is the dielectric constant, A is the area, and d is the interelectrode distance.

9.2.2 Batteries

Batteries are the most fundamental electrochemical energy storage systems wherein electrochemical energy is stored by a Faradaic charge storage mechanism [16]. Faradaic energy storage systems are developed based on these underlying fundamental redox mechanisms wherein a chemical species in reduced form is able to provide electrons and transform into an oxidized species. A redox couple is characterized through an equilibrium potential which is governed by the Nernst Law [13], given as,

$$E = E_0 + \left(\frac{RT}{nF}\right) \ln\left(\frac{ox}{\text{red}}\right)$$

where E is the cell potential, E_0 is the equilibrium potential, R is the universal gas constant, T is temperature, n is the number of electrons transferred in the reaction, F is the Faradays constant, and ox and red represent the concentration of the oxidizing and reducing species present in the system respectively. The electrochemical energy storage system stores and provides energy equivalent to the difference in free energies of the two species under consideration. In an ideal cell, the negative terminal is connected to a material that can undergo reduction and provide electrons to the circuit,

$$\mathrm{red_{anode}} \rightarrow \mathrm{ox_{anode}} + ne^-$$

This electrode is referred to as the anode of the system. Similarly, the positive terminal is connected to a material that can undergo oxidation and consume electrons during operation,

$$\mathrm{ox_{cathode}} + ne^- \rightarrow \mathrm{red_{cathode}}$$

This electrode is referred to as the cathode of the system. The equilibrium potential of each electrode is given by the Nernst equation as described above related to the redox couple red/ox. The global reaction for this cell can be given as,

$$\mathrm{ox_{cathode}} + \mathrm{red_{anode}} \rightarrow \mathrm{red_{cathode}} + \mathrm{ox_{anode}}$$

Cell potential for this elementary reaction is given as the difference between the cathode and anode potential. On consumption of one mole of reactants at standard conditions, maximum electrical energy of $n \times F \times (E2 - E1)$ J is obtained, where $E2-E1$ is the cell potential which is equal to the variation of the standard free enthalpy of the global cell reaction. Depending on the reversibility of this global equation, we obtain a primary energy storage system (single use) or a secondary energy storage system (rechargeable) [17].

Faradaic and non-Faradaic energy storage systems have their unique applications that pertain to the mechanisms of charge storage. Non-faradaic technologies like electric double-layer capacitors, pseudocapacitors, and hybrid capacitors are used for high power, low energy applications where high-rate discharges are of importance. In contrast, Faradaic energy storage systems are used for applications that require higher energy density and relatively lower power densities [18,19]. A comparison of key metrics for major electrochemical energy storage technologies is provided in Fig. 9.2.

9.3 Primary batteries

Primary batteries are single-use, non-rechargeable battery cells. The first prototype battery cells were demonstrated in the 1800s by Volta which had layers of silver

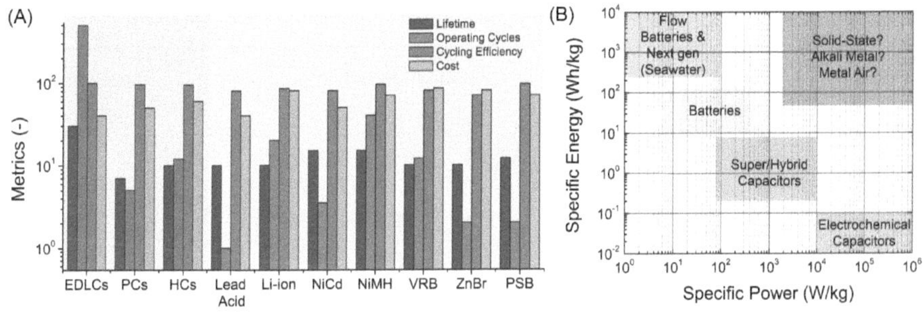

Figure 9.2 (A) Comparison of key electrochemical energy storage technologies in terms of lifetime, operating cycle, efficiencies, cost. (B) Ragone plot for typical energy storage technologies.

and zinc stacked up with a brine-soaked paper separator well known as Volta pile [20]. Further development of these cells leads to a primary battery consisting of zinc and carbon as electrode materials and a paste of ammonium chloride (NH_4Cl) as electrolyte and a fibrous separator [21]. In this cell, the carbon rod is encapsulated by a mixture of carbon and manganese dioxide (MnO_2). This cell was called a Leclanche cell. Chemical reactions at anode, cathode, and overall reactions in this cell are given below,

Anode:

$$Zn(s) \rightarrow Zn^{2+} + 2e^-$$

Cathode:

$$2NH_4(aq) + 2MnO_2(s) + 2e^- \rightarrow Mn_2O_3(s) + 2NH_3(g) + H_2O(l)$$

Total:

$$Zn(s) + 2MnO_2(s) + 2NH_4(aq) \rightarrow Mn_2O_3(s) + H_2O(l) + Zn^{2+} + 2NH_3(g)$$

Since some metals lose electrons more readily than others, a metal that oxidizes easily is utilized for the anode in a primary cell. Here, chemical potentials cause the Zn electrode to oxidize in the presence of electrolyte paste. Zn^{2+} ions from this reaction flow freely in the electrolyte until the electrode terminals are connected to a circuit. Some common primary batteries with their chemical reactions and open-circuit voltage are given in Table 9.1.

Primary batteries can potentially provide very high energy density due to the simplicity of cell design. Inactive elements required for accommodating recharging can be eliminated which reduces battery cost and provides ease of assembly. These batteries are mainly utilized in applications where low energy drainage is required such as toys, flashlights, hearing aids, calculators, remote controls, car keys, etc.

Table 9.1 Common primary battery system and chemistry.

Name	System (anode/electrolyte/cathode)	Chemical reaction	Open circuit voltage (V)
Dry cell (Leclanche)	$Zn/ZnCl_2$, NH_4Cl/MnO_2, C	$Zn + 2MnO_2 + 2NH_4 \rightarrow Mn_2O_3 + H_2O + Zn^{2+} + 2NH_3$	1.55
Alkaline-MnO_2	$Zn/KOH/MnO_2$, C	$Zn + 2MnO_2 + 2H_2O \rightarrow ZnO + Mn_2O_3$	1.5
Ruben-Mallory	Zn/KOH/HgO, Hg	$Zn + HgO \rightarrow ZnO + Hg$	1.35
Silver-Zinc	$Zn/KOH/Ag_2O$	$Zn + Ag_2O + H_2O \rightarrow Zn(OH)_2 + 2Ag$	1.6
Zinc-Air	$Zn/NaOH/O_2$(air), C	$2Zn + O_2 \rightarrow 2ZnO$	1.45

Table 9.2 Typical examples of primary batteries.

Name	Advantage	Application
Leclanche cell	• Inexpensive material • Suitable for wide range of consumer applications • Interchangeable with alkaline batteries • Long traditional reliability	Telephones, toys, remote controls, flashlight, clocks, etc.
Alkaline batteries	• Constant capacity over wide range of current range • Functions at sub-zero temperatures • Good Shelf life • Suitable for high drain rate applications	Toys, remote controls, MP3 player, CD player, Digital camera, etc.
Silver-Zinc batteries	• Low self-discharge • Longer operational life • Flat discharge curve • Higher voltage than Zinc-mercury cell	Smaller batteries: watches, hearing aids, etc. Larger batteries: missiles, submarines and aerospace application
Zinc-air batteries	• Excellent shelf life when sealed properly • Higher energy density • Low cost	Hearing aids, watches, marker lights, remote signaling devices, railway signals, etc.

Also, the devices which are designed for single-use can be powered by primary batteries such as missiles and weapon systems. The most common primary cells with major advantages and applications are listed in Table 9.2.

9.4 Supercapacitor

Supercapacitors (SCs) are one of the most promising energy storage systems [22]. They are also called ultracapacitors or electrochemical capacitors and they exhibit high power density (> 10 kW/kg), long cycle life, and relatively less energy density compared to conventional batteries. Supercapacitors do not store energy through chemical reactions and therefore, offer several advantages over the long-established secondary batteries. Supercapacitors bridge the gap in the special applications of batteries and fuel cells. Based on the energy storage mechanism SCs can be classified into (1) an electrochemical double-layer capacitor (EDLC) [23] and (2) a pseudocapacitor [15,24,25]. In EDLCs, the electric charge store at the electrode/electrolyte interface [26,27] through electrostatic interactions. The SC consists of two electrodes separated by an electron impermeable separator in an electrolytic solution. When a potential difference is applied in the SC cell, the electrons move from the negative electrode to the positive electrode through the external circuit, and anions in the electrolytic solution move towards the positive electrode surface, while the cations accumulate at the negative electrode surface. The reverse action takes place during the discharging process as shown in Fig. 9.3. Noting that no redox reactions are happening at the electrode/electrolyte interface and no charge transfer occurs for EDLC. Therefore, no volume or morphology change is expected to happen at the electrode and consequently results in long cycle life of EDLC. Unlike EDLCs, faradaic redox electrochemical reactions take place in the pseudocapacitors during the charge-discharge process, more like batteries. This process helps to store more energy density of the pseudocapacitor than the EDLC. On contrary, the power density of the pseudocapacitor is less than the EDLC which is due to the slow charge transfer process at the electrode/electrolyte interface and redox reactions. The conducting polymers or transition metal oxides are used as electrodes for pseudocapacitance. An electrochemical process takes at electrode very much like batteries. It should be noted that the supercapacitors are not commonly used as primary electrical energy storage devices and rather use as backup power sources to supplement fuel cells and secondary batteries for many applications.

Governing equations that correlate the energy and power density of SC with other parameters are given in Table 9.3. It is obvious from the equations that the

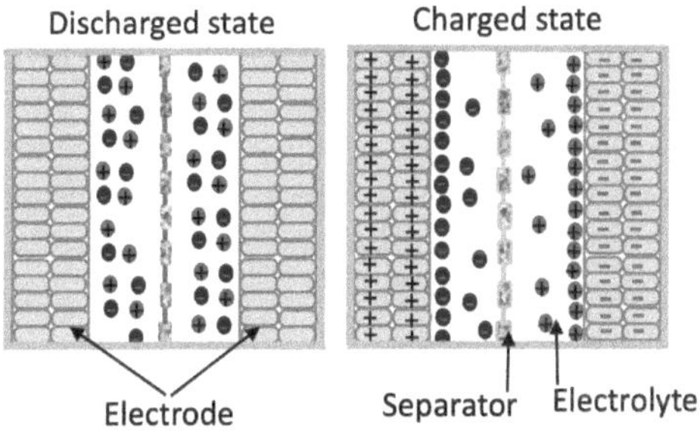

Figure 9.3 Schematic of a supercapacitor at the charged and discharged state.

Table 9.3 Main equations and components used to describe the capacitance, energy, power and physical significant for the performances of the supercapacitor.

Governing equations.	Description of parameters
$C = \frac{\epsilon \epsilon_0 A}{d}$ (i)	Where ε_0 is the permittivity of vacuum, ε is the permittivity of the electrolyte, A is the area of the electrode, and d is the effective thickness of the EDL
$E = \frac{1}{2}CV^2$ (ii)	Where E is the energy, and V is the applied voltage to the device
$P = \frac{E}{t_{dis}}$ (iii)	Where t_{dis} is the discharge time and P is the power of the device

electrode surface area and applied potential are the keys to increasing the energy of a supercapacitor.

Some components of the parameters given in Table 9.1 affect the performance of the supercapacitor, such as the morphology, electric conductivity, specific surface area, and pore size distribution of electrochemically active materials. The intrinsic properties of the electrolyte and the interface between the electrode and electrolyte have great influences on the performance of SC [28]. A good supercapacitor should exhibit a high specific capacitance and high-power density, excellent cyclability and rate performances (within seconds), low cost, and no self-discharging and safe operation [29]. It is discernible from equation (i) that a large specific surface area is a key point for electrode materials and can facilitate the contact between the electrode and electrolyte. Therefore, nanostructure materials are highly recommended as an electrode for supercapacitors which can provide large surface area, more active sites, and lead to an enhanced capacitance [26]. The energy of a supercapacitor also depends on the applied potential of the cell as seen in equation (ii) in Table 9.1. However, the operational safe voltage of the supercapacitor lies between the range of 1 and 3 V for both aqueous and organic electrolytes. It should be mentioned that the electrolyte stability window plays an important role to increase the energy density of the supercapacitor. The higher the applied voltage higher will be the energy density of the SC with the same electrode surface area. Aqueous electrolytes frequently used in supercapacitors include sulfuric acid and potassium hydroxide which have low resistance and offer a large power density. Organic electrolytes have higher resistance and deliver lower power density compared to aqueous electrolytes. The acetonitrile or propylene carbonate-based organic solvents are the most used organic electrolytes. The power density also depends on the properties of the separator which plays an important role in supercapacitor. The separators promote the movement of ions through the pores, and it needs to be wetted with electrolyte, therefore, the selection of separator is important based on the electrolyte employed. The most used separators are Polypropylene, Cellulose-like sheets, NKK, and Nomex. In addition, electrolyte development is a challenge to get high energy and power density. The different electrode materials used for flexible supercapacitors and their properties and performances are compared in Table 9.4.

Table 9.4 Comparison of specific capacitance, power density, energy density of different electrode materials for flexible supercapacitors.

Type of SC	Electrode material	Specific capacitance ($F\ g^{-1}$)	Power density	Energy density	References
EDLC	Activated carbon	39.8	0.0142 ($mW\ cm^{-2}$)	205 ($\mu Wh\ cm^{-2}$)	[30]
EDLC	Activated carbon	32.1	5 ($mW\ cm^{-3}$)	2.5 ($mWh\ cm^{-3}$)	[31]
EDLC	Activated carbon	90.0	—	—	[32]
EDLC	Graphene	81.7	1.5 ($W\ g^{-1}$)	7.13 ($mWh\ g^{-1}$)	[33]
EDLC	Graphene	40.0	0.1 ($mW\ cm^{-2}$)	0.2 ($\mu Wh\ cm^{-2}$)	[34]
EDLC	CNT	22.0	1.5 ($W\ g^{-1}$)	13 ($mWh\ g^{-1}$)	[35]
EDLC	Carbon aerogel	200.0	4 ($W\ g^{-1}$)	20 ($mWh\ g^{-1}$)	[36]
Pseudo-C	MnO_2	141.8	0.23 ($W\ cm^{-3}$)	0.3 ($mWh\ cm^{-3}$)	[37]
Pseudo-C	MnO_2	29.8	62 ($mW\ g^{-1}$)	6.8 ($mWh\ g^{-1}$)	[38]
Pseudo-C	Poyanaline	771.0	0.33 ($mW\ cm^{-2}$)	15 ($mWh\ m^{-2}$)	[39]
Pseudo-C	Polyprrole	114.0	—	2.5 ($mWh\ g^{-1}$)	[40]
EDLC	Graphene (water)	273.1	776.8 ($kW\ kg^{-1}$)	150.9 ($Wh\ kg^{-1}$)	[41]
EDLC	Nitrogen-enriched porous C/graphene composite	381.6	25.0 ($W\ kg^{-1}$)	13.2 ($Wh\ kg^{-1}$)	[42]

This section is envisioned to give a basic understanding of supercapacitors with a description of the key components. The large surface area, good mechanical properties, excellent conductivity, and affordability of carbon materials make them a promising candidate for supercapacitors. Further research activities are expected to be centralized on the engineering of carbon electrodes either by pore distributions, morphology, or cost-effective production methods.

9.5 Lithium-ion batteries

Lithium-ion batteries (LIBs) have been the key driving force behind the standard of living experienced by our modern civilization [43]. From portable electronics, renewables, grid storage, and electric vehicles to space exploration, LIBs have left an unparallel mark in history. The key factors that make this technology attractive include its high energy densities (> 250 Wh/kg), high power densities, high coulombic efficiencies, and high degrees of versatility and customizability. The concept of a LIB was first devised in the 1970s, following which five decades of continuous improvements and R&D efforts lead us to the modern LIB which typically consists of Li host cathodes (e.g., compositional variants of $LiCoO_2$, $LiNiO_2$ $LiFePO_4$, and $LiMn2O_4$) [44–48], and anodes (e.g., graphite, silicon, etc.) [49,50]. A typical LIB comprises an anode, cathode, separator, electrolyte, and the respective current collectors (positive—usually Al, negative—usually Cu). The anode and cathode can store and release lithium-ion depending on whether the battery is charged or discharged. The electrolyte transports the lithium-ion from one electrode to the other (cathode to anode during charge and vice versa during discharge) through the separator [51]. In liquid electrolyte-based LIBs the separator is usually made up of a porous polymeric membrane made up of polyethylene, polypropylene, etc. [52]. The separator ensures that the two electrodes are held apart from each other thus providing a potential difference while avoiding a short. Electrical current flows through an external circuit that can power the desired application when there is ion movement from one electrode to another in response to the potential change between the two electrodes. The specific reactions accompanying energy storage in a LIB are as follows:

Overall reaction:

$$\mathrm{LiMO_2} + C \underset{\text{Discharge Charge}}{\leftrightarrow} \mathrm{Li}_x\mathrm{C} + \mathrm{Li}_{1-x}\mathrm{MO_2}$$

(M in this case is a transition metal oxide)

Cathode reaction:

$$\mathrm{LiMO_2} \underset{\text{Discharge Charge}}{\leftrightarrow} \mathrm{Li}_{1-x}\mathrm{MO_2} + \mathrm{xLi}^+ + xe^-$$

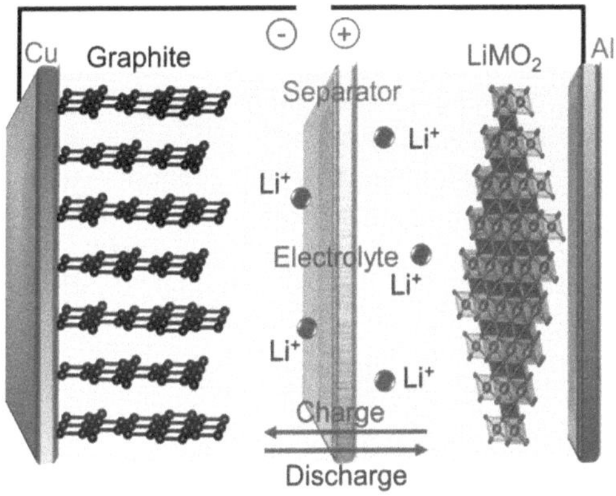

Figure 9.4 Schematic representation of the working principle of a lithium-ion battery.

Anode reaction:

$$C + x\mathrm{Li}^{+} + xe^{-} \underset{\text{Discharge Charge}}{\leftrightarrow} \mathrm{Li}_{x}C$$

The most important metrics relevant to batteries are energy and power delivered by the cell per unit mass, measured in watt-hours per kilogram, Wh/kg, and watts per gram, W/g respectively. Currently, the state-of-the-art lithium-ion batteries (LIBs) offer gravimetric and volumetric energy densities of up to 260 Wh/kg and 770 Wh/L, respectively [53]. The key parameters of interest that govern battery R&D efforts include (1) increasing energy and power densities, (2) reducing critical material dependency thus reducing cost, and (3) environmentally friendly manufacturing. A schematic representation of the working principle of a standard LIB is shown in Fig. 9.4. The following sections would briefly discuss the history of material development pertaining to the major components of a LIB battery.

9.5.1 Lithium-ion battery anode

The most obvious choice of anode for a LIB cell would be lithium metal as it has a high specific capacity (3860 mAh/g) that can be accessed at a very low electrode potential of −3.04 V versus SHE, thus, making it an attractive candidate for high energy LIB cells [54]. However, almost all known nonaqueous electrolyte formulations undergo severe reduction at the Li surface because of the electrochemical potential of Li^{+}/Li situated above the electrolyte LUMO orbital. This results in the formation of unstable solid electrolyte interface layers that would continuously undergo damage during cycling, thus, exposing fresh Li metal to further attack [55,56]. If this is left unchecked in a functioning cell, catastrophic failure could occur due to lithium dendrite growth and short-circuiting resulting in fires. Subsequently, materials with low electrochemical redox potentials that can alloy with lithium were also investigated. Despite their impressive capacities, large volume changes of these alloying anodes during operation limit their applicability in practical cells [57]. The quest for materials that can host lithium-ion through intercalation phenomena led scientists in the early years of LIB research to the implementation of graphite anodes (capacity 372 mAh/g) which can host lithium metal cations with minimal structural issues resulting in a stable and cyclable anode material [58]. Graphite proved to be a versatile anode material to such an extent that most modern LIBs to this day, employ graphite anodes as Li intercalation hosts.

9.5.2 Lithium-ion battery cathode

Cathodes that accommodate lithium ions through a conversion reaction (metal fluorides/sulfides/oxides) were initially considered to support the high lithium metal anode capacities. However, these cathodes suffer from undesired structural transitions which proved detrimental during long-term cycling. Cathodes of the general formula, MX_2 type chalcogenides (e.g., TiS_2), were also briefly considered during the early days of LIB research [59]. However, around the 1980s, $LiCoO_2$ (theoretical capacity $\sim$ 274 mAh/g) [60] gained enormous traction due to the reversible intercalation reaction the material underwent to accommodate lithium ions with minimal structural degradation with long-term cycling. But in modern times, due to the ever-growing supply chain constrain around cobalt, the scientific community gradually moved to substitute cobalt with other transition metals. Present-day commercially available LIBs employ high-capacity cathodes (150–200 mAh/g) derived from the $LiNiO_2$ [61] parent structure such as $LiNi_xCo_yMn_zO_2$ and $LiNi_xCo_yAl_zO_2$, where $x + y + z = 1$ [62]. Higher energy densities are a highly sought-after goal in battery research and to this extent, several new cathode formulations that employ lithium-rich ($Li_{1.2}Mn_{0.54}Ni_{0.13}Co_{0.13}O_2$) and nickel-rich ($Ni > 0.8$) formulations are also being increasingly pursued [63,64]. Additionally, to meet the rapidly evolving demands with the advent of electric vehicles and renewable technologies, several types of LIB chemistries such as Li-sulfur (capacity $\sim$1600 mAh/g) [65–68], Li-air [69], organic electrode [70], Li-CO_2 [71] and solid-state batteries [72,73] are also being extensively investigated. These chemistries, however, require systematic investigations and R&D efforts directed towards solving the crucial issues which hinder commercial deployment -viz; unwarranted phase transitions, unstable SEI formation, undesirable volumetric changes, electrode surface reconstruction, and degradation, cycling and temperature stabilities, etc.

9.5.3 Lithium-ion battery electrolyte

The third and one of the most critical components of a LIB is the electrolyte which is usually chosen by the following criteria—the LUMO and HOMO orbitals of the electrolyte should lie between the electrochemical potentials of the anode and cathode respectively. Moreover, the electrolyte should also form stable passivating layers of solid electrolyte interfaces (SEI) on the surface of the electrodes that prevent detrimental surface reactions. Present-day lithium-ion batteries contain nonaqueous organic electrolytes comprising either standalone or combinations of propylene carbonate, ethylene carbonate, ethyl methyl carbonate, dimethyl carbonate, etc. [74].

9.6 Redox flow batteries

Conventional battery technologies such as LIBs offer a direct and efficient pathway for energy storage. Thus, they are ideally suited for portable electronics and automotive applications. While LIB systems are increasingly penetrating the long-duration energy storage arena, the high capital costs are prohibitive when considering large volume installations despite their energy density advantages. Battery systems that can effectively handle grid load requirements have to be designed to different sets of metrics when compared to the ones designed for portable systems. In this context, redox flow batteries (RFBs) are considered a promising prospect to meet the requirements of such long-term energy storage applications [75–77]. RFBs are excellent candidates when there is a need to separate power and energy requirements. For example, power in an RFB is controlled by the load whereas energy is stored using reactants. This offers great flexibility and a range of customizable variables that can be selectively optimized to cater to specific end-use requirements. Broadly, RFBs can be classified based on the reacting species or the solvent used (aqueous or nonaqueous) and in a typical cell, there are two major components—the anolyte and the catholyte (schematic picture of an RFB cell is displayed in Fig. 9.5). These reactants are stored in separate tanks which are then pumped to the anode and cathode surface using mechanical pumps. The potential of each electrode is maintained close to that of the potentials of the half-reactions occurring in an RFB cell. The general functioning of an RFB can be described using the following equations (A is an anode and C is a cathode):

Reaction at the anode:

$$A^{n+} + xe^- \overleftrightarrow{\text{Discharge Charge}} A^{(n-x)+} (n > x)$$

Reaction at the cathode:

$$C^{m+} - ye^- \overleftrightarrow{\text{Discharge Charge}} C^{(m+y)+}$$

Typical losses in an RFB system can occur from the kinetics of the charge transfer process, charge and mass transport through the separator, and in some cases active species crossing over from the catholyte to the anolyte and vice versa. RFB electrodes do not typically undergo physical changes that are a key feature of alkali ion batteries instead the catholyte and anolytes react adjacent to the electrode surfaces thus enabling the simplification of electrode surfaces in an RFB cell. Thus, cycle life in an RFB cell is not a consequence of the number of operational cycles

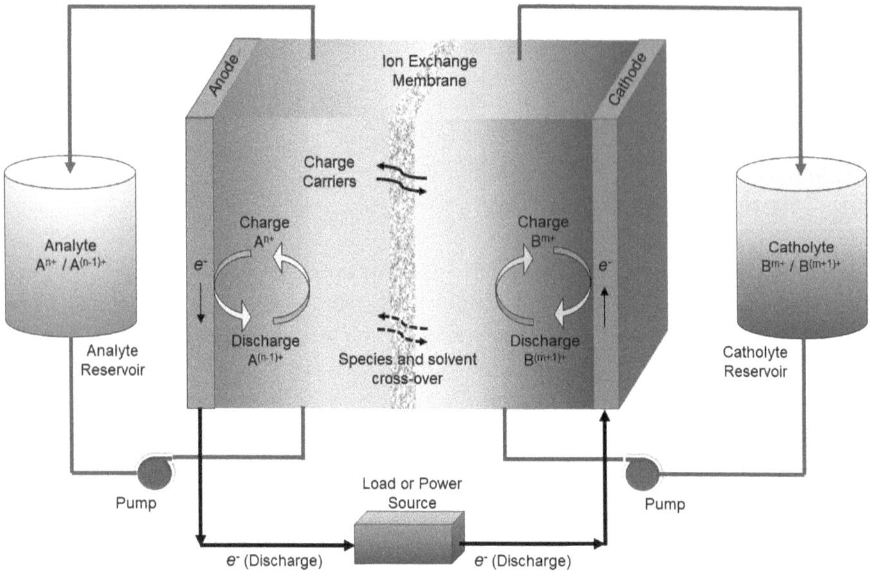

Figure 9.5 Schematic representation of the working mechanism of a typical redox flow battery [78].
Source: Reproduced with permission from A.Z. Weber, M.M. Mench, J.P. Meyers, P.N. Ross, J.T. Gostick, Q. Liu, J. Appl. Electrochem. 41 (2011) 10–41.

like an alkali ion battery. Despite these benefits, RFB systems require complicated designs and optimized engineering with mechanical balance-of-plant components that pump the often-corrosive electrolyte formulations. Consequently, the gravimetric and volumetric energy densities of RFB systems are significantly lower than LIB counterparts.

9.6.1 Redox flow battery cell chemistries

Several cell chemistries are widely deployed in RFB systems and some of which are as follows: (1) iron-chromium, (2) bromine-polysulfide, (3) all vanadium, (4) vanadium-bromine, (5) hydrogen-based, (6) hybrid redox flow, (7) zinc-bromine, (8) soluble lead-acid, (9) all iron and (10) nonaqueous type RFBs. The iron-chromium system (OCV 0.9V–1.2 V) employs the ferric/ferrous (Fe^{2+}/Fe^{3+}) and the chromic/chromous (Cr^{2+}/Cr^{3+}) redox couples at the positive and negative electrodes with a hydrochloric acid supporting electrolyte [78,79]. For the bromine-polysulfide RFB, the electrolytes are generally based on sodium bromide and sodium polysulfide and typically have operating voltages in the range of 1.7 to 2.1 V [80]. The all-vanadium system employs V(II)/V(III) at the negative electrode and V(IV)/V(V) redox couple at the positive electrode to mitigate the ion crossover issues that affect most RFBs using dissimilar anolyte and catholyte [81,82]. Like the all-vanadium system, the all-iron system also employs the redox couples of one element—iron [83]. On one electrode iron is converted from $a+2$ to $a+3$ state while iron undergoes plating at the other electrode. Such a system would be comparatively low in cost due to the use of inexpensive iron, however, low operational voltages of such RFBs have proven to be detrimental for wide-scale deployment. Higher operational voltages can be achieved by employing nonaqueous systems—for example, researchers have explored a zinc/cerium system with an operational voltage between 2 and 2.5 V enabled using methane sulfonic acid [78,84]. Despite the voltage benefits, poor electrolyte conductivities and higher costs of nonaqueous electrolytes limit the practical viability of such RFB cells.

Overall, despite the limitations, RFB systems are attractive candidates for long-duration energy storage installations where conventional LIB systems become cost-prohibitive for effective deployment.

9.7 Emerging technologies

These energy storage technologies were envisioned as improvements to the modern state-of-art LIB systems. These technologies are in various states of technological maturity where some are only a few iterations away from a commercial deployment while some are in their proof-of-concept infancy. The following sections would discuss briefly, the prospects of next-generation battery technologies which if practically feasible would be a dominating game-changer for decades to come.

9.7.1 Sodium-ion batteries

Sodium-ion batteries (SIB) have been largely sought after due to the similarities in electrochemical properties between sodium (Li) and lithium (Na) metals [85]. The predominant driving factors for SIB R&D stem from the fact that sodium is abundantly present on the earth's crust when compared to lithium. Li and Na are group 1 elements in the periodic table, and each contains one electron in their respective valance shells thus making them ideal for use in electrochemical operations. The reduction potential of sodium is around -2.7 V versus SHE which is around 330 mV higher than lithium. Thus, electrolytes that form detrimental SEI in LIBs may not do the same in the case of SIBs (e.g., Glyme ethers [86–88]). SIBs operate like that LIBs but differ in key aspects. For example, graphite which is a widely used LIB anode may not be applied to a SIB due to the poor capacities ($\sim$35 mAh/g) offered by Na^+ ion during intercalation into graphite reaction when compared to Li^+ ion intercalation into graphite (372 mAh/g). This anomalous behavior was extensively studied by Liu et al., who concluded that Na^+ suffered from weak binding with graphite when compared to Li^+ resulting in poor capacities [89]. Thus, the quest for the selection of a suitable SIB anode led researchers to investigate sodium alloying compounds (Sn, Si, Sb, Pb, etc.) as well as sodium intercalating hard carbons as viable candidates. Despite the plethora of research initiatives in this area, hard carbon anodes have emerged as a comparatively better alternative for SIB systems owing to their low first cycle loss (10%–20%) when compared to the alloying anodes. Cathodes for SIBs are also challenging to develop owing to the larger size of Na atom when compared to Li. Promising candidates belonging to the P2 (e.g., $Na_{2/3}(Mx_{1/2}My_{1/2})O_2$ where Mx and My are transition metals) and O_3 (e.g., $Na[Ni_xCo_yMn_z]O_2$) type cathodes have been widely studied for potential use in SIBs, however, these cathodes suffer from low sodium incorporation during the synthesis process [90,91]. Sodium deficiency in these synthesized cathodes unlike similar lithium cathodes causes processing challenges during SIB fabrication. Presodiation of hard carbon anodes and addition of sodium enrichers to the cathodes are some of the promising directions presently investigated in literature to achieve fully-functional commercially viable SIBs. Despite these positives, attractive SIB metrics such as lower cost when compared to LIBs may not be a compelling argument when considering the energy output of the two systems. LIBs have a better $/kWh value when compared to SIBs and therefore further R&D efforts into materials and process development need to be systematically performed before SIBs get the chance to replace LIBs anytime soon.

9.7.2 Solid-state batteries

There is no question that, within the next few years, conventional LIBs, based on liquid electrolytes, will have reached their performance limits. With the ever-increasing demand for higher energy and power densities, solid-state batteries (SSBs) are projected to be one of the solutions [92]. By replacing conventional organic flammable liquid electrolytes with a completely solid electrolyte capable of high Li-ion conduction, lithium metal could be directly enabled in an SSB translating to huge energy gains coupled with increased battery safety. A schematic representation of a typical SSB is shown in Fig. 9.6. Enabling Li metal anodes in a solid-state battery promises to attain specific cell energy densities above 500 Wh/kg and 1500 Wh/L [93]. When combined with long cycle life, improved safety, and low cost (<$100/kWh) the value proposition of an all-solid-state lithium battery becomes more and more relevant [94]. There are, however, significant materials and processing challenges that disrupt the materialization of working SSBs at present. Present-day energy and power densities of most reported SSBs fall drastically short in comparison with the state-of-the-art Li-ion batteries. SSBs are complex systems due to the interplay between kinetics, charge transfer, mechanics as well as Li-ion transport in the vicinity of the multiple inter- and intra- material junctions present in SSBs [95]. The key scientific gaps that need to be systematically and synergistically addressed before SSBs can be commercialized include [96]: (1) *solid-electrolyte*—high ionic conductivities (> 1 mS/cm) and high transference numbers (close to 1) [97], endure high critical current densities (> 1 mA/cm^2) with low thicknesses (<25 μm) [98], and have stable electro-chemo-mechanical response during operation [99–101] (2) *Lithium metal anode*—better morphology control, mechanistic understanding of mass transport and flow of Li-ions, and processing of thin Li metal films (5–25 μm) in a scalable fashion [96,102] (3) *Cathode composite* - high active material loadings (4–10 mAh/cm^2), low tortuosity,

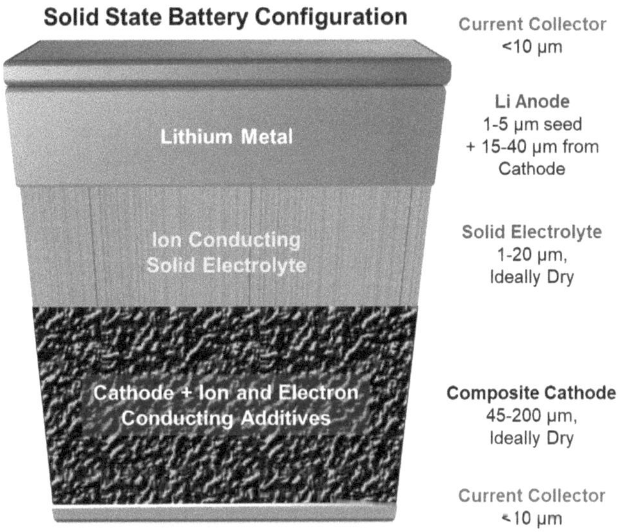

Figure 9.6 Schematic representation of a typical SSB [94].
Source: Reproduced with permission from P. Albertus, V. Anandan, C. Ban, N. Balsara, I. Belharouak, J. Buettner-Garrett, et al., ACS Energy Lett. 6 (2021) 1399−1404.

and enhanced Li-ion transport [103]; (4) *Interfaces and interphases*—stability and resilience of the numerous interfaces in SSBs and their electro-chemo-mechanical endurance during repeated charge and discharge of SSBs.

9.7.3 Multivalent cation systems

Cations that can be accommodated in battery electrode hosts in 2 or 3 electron reactions are increasingly sought after in energy storage research as this prospect offers the final energy densities achieved in such batteries to theoretically be double or triple that of conventional LIBs [104]. Fig. 9.7 shows some of the performance metrics and advantages offered by alternative and multivalent ion chemistries when compared to LIBs. Thus, multivalent ion-based rechargeable batteries are a promising direction for battery R&D to advance towards. Presently, there are several candidates' elements considered such as Mg, Ca, Al, etc. all of which are in various stages of technological maturities. With regards to Mg-based systems, there have been several studies that explored cathodes, electrolytes, and cell architectures. Mg-based organic cathodes, as well as sulfur cathode studies, have also yielded promising results towards the practical feasibility of such systems. Despite the promise, Mg systems suffer from interface stabilities and poor electrolyte conductivities that need systematic R&D efforts to enable the prospects of commercial viability [105]. Ca-based anodes have been demonstrated to show reversible plating and stripping processes but with kinetic and coulombic efficiency limitations. However, the

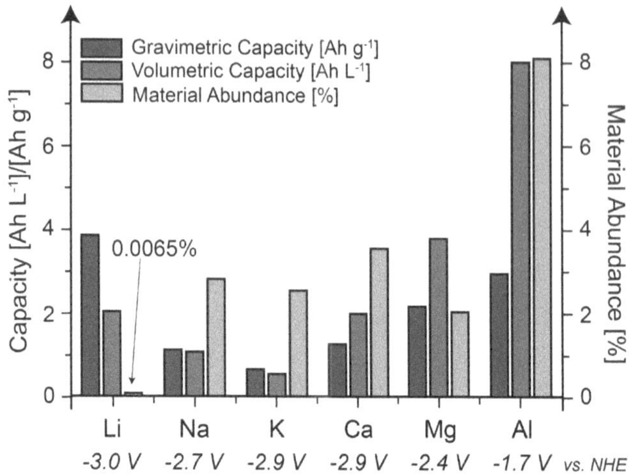

Figure 9.7 Electrochemical redox potential, volumetric energy densities and gravimetric capacities of the alternative ion battery chemistries in comparison with lithium-ion battery.

pathway toward a viable cathode for Ca batteries is still being pursued with only limited success [106]. Al batteries on the other hand presently require functional high Al^{3+} conducting electrolytes that are not Cl based to avoid cell/current collector corrosion issues [107]. Additionally, the successful development of a cathode material capable of accommodating the +3 cation in its host framework would certainly provide the necessary impetus toward Al-battery development. In all cases, systematic R&D efforts that target electrolyte optimization, cathode development through a comprehensive understanding of the cation complexation and transport properties in electrolytes, and interfacial challenges would be essential for future development of multivalent cation-based batteries and energy storage devices.

9.8 Outlook and conclusions

In this chapter, we discussed the current major electrochemical energy storage systems from a material and system perspective. As we transition towards a decentralized, renewable-based, sustainable energy sector, electrochemical energy storage systems are going to find increased importance. Development of next-generation systems with high energy and power densities like Li-air and Li-CO_2 systems as well as solid-state batteries are expected to revolutionize the secondary energy storage systems. In addition, the development of affordable large-scale electrochemical energy storage systems like seawater batteries is required to effectively couple with renewable energy sources to achieve the deep decarbonizations required to mitigate climate change. Further work on the development of optimized materials and systems and scalable manufacturing of this next-generation energy storage system is anticipated in the coming decade to promote sustainable development.

Acknowledgments

This work at the Oak Ridge National Laboratory (ORNL), managed by UT Battelle, LLC, for the US Department of Energy (DOE) under Contract No. DE-AC05−00−23 OR22725, was sponsored by the Laboratory Directed Research and Development Program at Oak Ridge National Laboratory.

Reference

[1] T.T. Mai, P. Jadun, J.S. Logan, C.A. McMillan, M. Muratori, D.C. Steinberg, et al., Natl. Renew. Energy Laboratory (2018) 151.
[2] R.B. Jackson, C. Le Quéré, R.M. Andrew, J.G. Canadell, J.I. Korsbakken, Z. Liu, et al., Environ. Res. Lett. 13 (2018).
[3] M. Åhman, L.J. Nilsson, B. Johansson, Clim. Policy 17 (2017) 634−649.
[4] Z.A. Needell, J. McNerney, M.T. Chang, J.E. Trancik, Nat. Energy 1 (2016).
[5] Z.J. Schiffer, K. Manthiram, Joule 1 (2017) 10−14.
[6] S. Billimoria, L. Guccione, M. Henchen, L. Louis-Prescott, Rocky Mountain Institute, 2018.
[7] V. Gowrishankar, A. Levin, Natural Resources Defense Council, 2017.
[8] IEA, Global EV Outlook 2020, 2020.
[9] Electric Power Research Institute, U.S. National Electrification Assessment, 2018.
[10] Q. Abbas, M. Mirzaeian, M.R.C. Hunt, P. Hall, R. Raza, Energies 13 (2020) 5847.
[11] J.F.M. Oudenhoven, L. Baggetto, P.H.L. Notten, Adv. Energy Mater. 1 (2011) 10−33.
[12] N.C. Hoyt, R.F. Savinell, J.S. Wainright, Chem. Eng. Sci. 144 (2016) 288−297.
[13] A.J. Bard, L.R. Faulkner, Electrochemical Methods Fundamentals and Applications, New York, 2001.
[14] P. Simon, Y. Gogotsi, B. Dunn, Science 343 (2014) 1210−1211.
[15] K.B. Hatzell, M. Beidaghi, J.W. Campos, C.R. Dennison, E.C. Kumbur, Y. Gogotsi, Electrochim. Acta 111 (2013) 888−897.
[16] M. Winter, R.J. Brodd, Chem. Rev. 104 (2004) 4245−4269.
[17] H.A. Kiehne, Battery Technology Handbook, CRC Press, 2003.
[18] J. Liu, J. Wang, C. Xu, H. Jiang, C. Li, L. Zhang, et al., Adv. Sci. 5 (2018).
[19] S. Balasubramaniam, A. Mohanty, S.K. Balasingam, S.J. Kim, A. Ramadoss, Comprehensive Insight into the Mechanism, Material Selection and Performance Evaluation of Supercapatteries, Springer, Singapore, 2020.
[20] W.M. Sudduth, Ambix 27 (1980) 26−35.
[21] S.A.G.R. Karunathilaka, N.A. Hampson, R. Leek, T.J. Sinclair, J. Appl. Electrochem. 10 (1980) 357−363.
[22] S. Dsoke, X. Tian, C. Täubert, S. Schlüter, M. Wohlfahrt-Mehrens, J. Power Sources 238 (2013) 422−429.
[23] N. Muralidharan, E. Teblum, A.S. Westover, D. Schauben, A. Itzhak, M. Muallem, et al., Sci. Rep. 8 (2018) 1−8. 1−9.
[24] A. Douglas, N. Muralidharan, R. Carter, K. Share, C.L. Pint, Nanoscale 8 (2016) 7384−7390.
[25] C. Meng, N. Muralidharan, E. Teblum, K.E. Moyer, G.D. Nessim, C.L. Pint, Nano Lett. 18 (2018) 7761−7768.

[26] G. Wang, L. Zhang, J. Zhang, Chem. Soc. Rev. 41 (2012) 797–828.
[27] B.E. Conway, Electrochemical Supercapacitors, 1999.
[28] L.L. Zhang, X.S. Zhao, Chem. Soc. Rev. 38 (2009) 2520–2531.
[29] A.G. Pandolfo, A.F. Hollenkamp, J. Power Sources 157 (2006) 11–27.
[30] S. Yong, J. Owen, S. Beeby, Adv. Eng. Mater. 20 (2018) 1700860.
[31] W. Ma, S. Chen, S. Yang, W. Chen, W. Weng, M. Zhu, ACS Appl. Mater. Interfaces 8 (2016) 14622–14627.
[32] K. Jost, C.R. Perez, J.K. McDonough, V. Presser, M. Heon, G. Dion, et al., Energy Environ. Sci. 4 (2011) 5060–5067.
[33] W. Liu, X. Yan, J. Lang, C. Peng, Q. Xue, J. Mater. Chem. 22 (2012) 17245–17253.
[34] Y. Meng, Y. Zhao, C. Hu, H. Cheng, Y. Hu, Z. Zhang, et al., Adv. Mater. 25 (2013) 2326–2331.
[35] V.L. Pushparaj, M.M. Shaijumon, A. Kumar, S. Murugesan, L. Ci, R. Vajtai, et al., Proc. Natl. Acad. Sci. 104 (2007) 13574–13577.
[36] Y.M. Fan, W.L. Song, X. Li, L.Z. Fan, Carbon 111 (2017) 658–666.
[37] X. Lu, M. Yu, G. Wang, T. Zhai, S. Xie, Y. Ling, et al., Adv. Mater. 25 (2013) 267–272.
[38] Y. He, W. Chen, X. Li, Z. Zhang, J. Fu, C. Zhao, et al., ACS Nano 7 (2012) 174–182.
[39] X. Zang, X. Li, M. Zhu, X. Li, Z. Zhen, Y. He, et al., Nanoscale 7 (2015) 7318–7322.
[40] C. Zhao, K. Shu, C. Wang, S. Gambhir, G.G. Wallace, Electrochim. Acta 172 (2015) 12–19.
[41] X. Yang, J. Zhu, L. Qiu, D. Li, Adv. Mater. 23 (2011) 2833–2838.
[42] Q. Xie, S. Zhou, A. Zheng, C. Xie, C. Yin, S. Wu, et al., Electrochim. Acta 189 (2016) 22–31.
[43] Bloomberg, Lithium-Ion Battery Costs and Market, 2017.
[44] N. Muralidharan, R. Carter, L. Oakes, A.P. Cohn, C.L. Pint, Sci. Rep. 6 (2016) 1–6. 1–9.
[45] N. Muralidharan, C.N. Brock, A.P. Cohn, D. Schauben, R.E. Carter, L. Oakes, et al., ACS Nano 11 (2017) 6243–6251.
[46] U. Nisar, N. Muralidharan, R. Essehli, R. Amin, I. Belharouak, Energy Storage Mater. 38 (2021) 309–328.
[47] N. Muralidharan, R. Essehli, R.P. Hermann, A. Parejiya, R. Amin, Y. Bai, et al., J. Power Sources 471 (2020) 228389.
[48] N. Muralidharan, R. Essehli, R.P. Hermann, R. Amin, C. Jafta, J. Zhang, et al., Adv. Mater. 32 (2020) 2002960.
[49] D. Parikh, T. Christensen, C.-T. Hsieh, J. Li, J. Electrochem. Soc. 166 (2019) A3377–A3383.
[50] D.L. Wood, M. Wood, J. Li, Z. Du, R.E. Ruther, K.A. Hays, et al., Energy Storage Mater. 29 (2020) 254–265.
[51] L. Hu, Z. Zhang, K. Amine, J. Power Sources 236 (2013) 175–180.
[52] A. Sheidaei, X. Xiao, X. Huang, J. Hitt, J. Power Sources 196 (2011) 8728–8734.
[53] B.D. McCloskey, J. Phys. Chem. Lett. 6 (2015) 4581–4588.
[54] S. Jiao, J. Zheng, Q. Li, X. Li, M.H. Engelhard, R. Cao, et al., Joule 2 (2018) 110–124.
[55] A.U. Chowdhury, N. Muralidharan, C. Daniel, R. Amin, I. Belharouak, J. Power Sources 506 (2021) 230173.
[56] M.A. McArthur, S. Trussler, J.R. Dahn, J. Electrochem. Soc. 159 (2011) A198.
[57] A.F. Gonzalez, N.-H. Yang, R.-S. Liu, J. Phys. Chem. C. 121 (2017) 27775–27787.
[58] K. Zaghib, G. Nadeau, K. Kinoshita, J. Electrochem. Soc. 147 (2000) 2110.

[59] M.S. Whittingham, Lithium Ion Batteries, John Wiley & Sons, Ltd, 2007, pp. 49–66.
[60] Z. Chen, J.R. Dahn, Electrochim. Acta 49 (2004) 1079–1090.
[61] H. Arai, S. Okada, Y. Sakurai, J.I. Yamaki, Solid State Ion 95 (1997) 275–282.
[62] B. Wang, F.-l Zhang, X.-a Zhou, P. Wang, J. Wang, H. Ding, et al., J. Mater. Chem. A 9 (2021) 13540–13551.
[63] J. Zheng, S. Myeong, W. Cho, P. Yan, J. Xiao, C. Wang, et al., Adv. Energy Mater. 7 (2017) 1601284.
[64] D. Darbar, E.C. Self, L. Li, C. Wang, H.M. Meyer, C. Lee, et al., J. Power Sources 479 (2020) 228591.
[65] R. Carter, L. Oakes, A. Douglas, N. Muralidharan, A.P. Cohn, C.L. Pint, Nano Lett. 17 (2017) 1863–1869.
[66] R. Carter, L. Oakes, N. Muralidharan, A.P. Cohn, A. Douglas, C.L. Pint, ACS Appl. Mater. Interfaces 9 (2017) 7185–7192.
[67] R. Carter, L. Oakes, N. Muralidharan, C.L. Pint, J. Phys. Chem. C 121 (2017) 7718–7727.
[68] S.S. Zhang, J. Power Sources 231 (2013) 153–162.
[69] L. Oakes, N. Muralidharan, A.P. Cohn, C.L. Pint, Nanotechnology 27 (2016) 495404.
[70] H. Wu, S.A. Shevlin, Q. Meng, W. Guo, Y. Meng, K. Lu, et al., Adv. Mater. 26 (2014) 3338–3343.
[71] A. Ahmadiparidari, R.E. Warburton, L. Majidi, M. Asadi, A. Chamaani, J.R. Jokisaari, et al., Adv. Mater. 31 (2019) 1902518.
[72] M.B. Dixit, A. Parejiya, N. Muralidharan, R. Essehli, R. Amin, I. Belharouak, Energy Storage Mater. 40 (2021) 239–249.
[73] A. Parejiya, R. Essehli, R. Amin, J. Liu, N. Muralidharan, I. Harry, et al., ACS Energy Lett. 6 (2021) 429–436.
[74] Y. Wang, W.-H. Zhong, ChemElectroChem 2 (2015) 22–36.
[75] M. Youssry, L. Madec, P. Soudan, M. Cerbelaud, D. Guyomard, B. Lestriez, J. Power Sources 274 (2015) 424–431.
[76] M. Youssry, L. Madec, P. Soudan, M. Cerbelaud, D. Guyomard, B. Lestriez, Phys. Chem. Chem. Phys. 15 (2013) 14476.
[77] X. Chen, B.J. Hopkins, A. Helal, F.Y. Fan, K.C. Smith, Z. Li, et al., Energy Environ. Sci. 9 (2016) 1760–1770.
[78] A.Z. Weber, M.M. Mench, J.P. Meyers, P.N. Ross, J.T. Gostick, Q. Liu, J. Appl. Electrochem. 41 (2011) 10–41. 1137–1164.
[79] M. Bartolozzi, J. Power Sources 27 (1989) 219–234.
[80] A. Price, S. Bartley, S. Male, G. Cooley, Power Eng. J. 13 (1999) 122–129.
[81] M. Skyllas-Kazacos, G. Kazacos, G. Poon, H. Verseema, Int. J. Energy Res. 34 (2010) 182–189.
[82] M. Skyllas-Kazacos, M. Rychcik, R.G. Robins, A.G. Fane, M.A. Green, J. Electrochem. Soc. 133 (1986) 1057.
[83] L.W. Hruska, R.F. Savinell, J. Electrochem. Soc. 128 (1981) 18.
[84] T. Raju, C.A. Basha, Ind. Eng. Chem. Res. 47 (2008) 8947–8952.
[85] X. Ma, K. An, J. Bai, H. Chen, Sci. Rep. 7 (2017) 162.
[86] A.P. Cohn, N. Muralidharan, R. Carter, K. Share, L. Oakes, C.L. Pint, J. Mater. Chem. A 4 (2016) 14954–14959.
[87] A.P. Cohn, T. Metke, J. Donohue, N. Muralidharan, K. Share, C.L. Pint, J. Mater. Chem. A 6 (2018) 23875–23884.
[88] A.P. Cohn, N. Muralidharan, R. Carter, K. Share, C.L. Pint, Nano Lett. 17 (2017) 1296–1301.

[89] Y. Liu, B.V. Merinov, W.A. Goddard, Proc. Natl. Acad. Sci. 113 (2016) 3735–3739.
[90] M. Li, Z. Du, M.A. Khaleel, I. Belharouak, Energy Storage Mater. 25 (2020) 520–536.
[91] M. Li, D.L. Wood, Y. Bai, R. Essehli, M.R. Amin, C. Jafta, et al., ACS Appl. Mater. Interfaces 12 (2020) 23951–23958.
[92] M.B. Dixit, J.-S. Park, P. Kenesei, J. Almer, K.B. Hatzell, Energy Environ. Sci. (2021). Available from: https://doi.org/10.1039/D1EE00638J.
[93] F. Shen, M. Dixit, X. Xiao, K. Hatzell, ACS Energy Lett. (2018). acsenergylett.8b00249.
[94] P. Albertus, V. Anandan, C. Ban, N. Balsara, I. Belharouak, J. Buettner-Garrett, et al., ACS Energy Lett. 6 (2021) 1399–1404.
[95] M. Dixit, N. Muralidharan, A. Parejiya, R. Amin, R. Essehli, I. Belharouak, Energy Storage Devices, IntechOpen, 2021.
[96] K.B. Hatzell, X.C. Chen, C. Cobb, N.P. Dasgupta, M.B. Dixit, L.E. Marbella, et al., ACS Energy Lett. 5 (2020) 922–934.
[97] M.B. Dixit, M. Regala, F. Shen, X. Xiao, K.B. Hatzell, ACS Appl. Mater. Interfaces 11 (2018) 2022–2030.
[98] M.B. Dixit, W. Zaman, Y. Bootwala, Y. Zheng, C. Marta, (2019) 1–32.
[99] M.B. Dixit, W. Zaman, N. Hortance, S. Vujic, B. Harkey, F. Shen, et al., Joule 4 (2020) 207–221.
[100] W. Zaman, N. Hortance, M.B. Dixit, V. de Andrade, K.B. Hatzell, J. Mater. Chem. A 7 (2019) 23914–23921.
[101] M.J. Palmer, S. Kalnaus, M.B. Dixit, A.S. Westover, K.B. Hatzell, N.J. Dudney, et al., Energy Storage Mater. 26 (2020) 242–249.
[102] M.B. Dixit, A. Verma, W. Zaman, X. Zhong, P. Kenesei, J.S. Park, et al., ACS Appl. Energy Mater. 3 (2020) 9534–9542.
[103] F. Shen, M.B. Dixit, W. Zaman, N. Hortance, B. Rogers, K.B. Hatzell, J. Electrochem. Soc. 166 (2019) A3182–A3188.
[104] Y. Liang, H. Dong, D. Aurbach, Y. Yao, Nat. Energy 5 (2020) 646–656.
[105] J. Eaves-Rathert, K. Moyer, M. Zohair, C.L. Pint, Joule 4 (2020) 1324–1336.
[106] M. Wang, C. Jiang, S. Zhang, X. Song, Y. Tang, H.-M. Cheng, Nat. Chem. 10 (2018) 667–672.
[107] N. Jayaprakash, S.K. Das, L.A. Archer, Chem. Commun. 47 (2011) 12610–12612.

Energy harvesting and storage for stand-alone microsystems 10

R.H. Gounella[1], *Y.A.O. Assagra*[1], *L.M. Gonçalves*[2] *and J.P. Carmo*[1]
[1]Department of Electrical Engineering (SEL), University of São Paulo (USP), São Carlos, Brazil, [2]Department of Industrial Electronics (DEI), University of Minho (UMinho), Guimarães, Portugal

Chapter Outline

10.1 Introduction

It is not new to anyone that since the last decade, wireless communication microsystems (MSTs) with a high density of nodes and simple protocols have emerged for low data rate distributed sensor network applications such as those in home automation [1,2] and industrial control [3,4]. In this context, the grown interest to implement the paradigm of using independent operation, management, and maintenance in wireless sensors networks, pushed the requirements in WSN nodes to present even more low-power consumption [5,6]. Additionally, there is also an increasing interest in ubiquitous electronic devices in the everyday life [7,8]. Also, the complexity and the requirements of these devices do not know limits. The use of batteries cannot be enough to ensure an uninterruptible working cycle due to its limited capacity. For example, the capacity of commercial class AA batteries (the most suitable for powering stand-alone microsystems for ready deployment) is limited to 3000 mAh [9]. This is of major importance, especially in the operation of stand-alone microsystems to be supplied by the energy harvested from the surrounding environment. Thus, the association of such devices with the use of some kind of energy recovering system can reveal an interesting approach [10]. Energy scavengers are currently emerging for a number of applications from biomedical to automotive [11,12]. The three most important energy sources of interest for use in remote micro/nanosystems are thermoelectric (from thermal gradients), piezoelectric (from mechanical vibrations), and photovoltaic (from solar light). The piezoelectric harvesting systems are widely

Emerging Trends in Energy Storage Systems and Industrial Applications. DOI: https://doi.org/10.1016/B978-0-323-90521-3.00002-8

settled and validated, however, direct thermal-to-electric energy conversion without moving mechanical parts is attractive for a wide range of applications because it provides compact and distributed power, quiet operation, and is usually environmentally friendly. Thus, it is not strange the fact worldwide efforts are being undertaken to expand the technology of thermoelectric devices into the field of microsystems technologies and more covered in this chapter than other harvesting techniques. Today, almost all of these platforms are designed to run on batteries that not only have a very limited lifetime but are also in many areas a cost-prohibitive solution. An attractive alternative is powering the sensors with energy harvested from the environment. Thus, a solution for energy generation through energy harvesting by taking advantage of temperature differences must be found. Since many wireless sensors are powered on a peak basis (e.g., the transmission of data needs much more current than standby or receiving mode) and the environmental condition could not always be present, the energy can be stored in a rechargeable thin-film battery of Li-ion type (integrated into the system). Ultra-low-power electronics perform DC–DC rectification with a variable conversion factor and recharge the battery under optimal conditions. All of these issues are discussed further in this review.

10.2 Energy harvesting systems

Energy harvesters for stand-alone microsystems are currently emerging for a number of applications from medicine to automotive. Typically, one can distinguish between two types of energy scavengers, for example, macro-energy scavengers that are typically in the cm^3 range, whereas the micro-energy scavengers that are typically in the mm^3 range and manufactured using micromachining techniques. Micro-energy harvesters are small electromechanical devices that harvest ambient energy and convert it into electricity [12]. Energy harvesters can collect different types of energies. The solar energy can be collected and stored by means of photovoltaic solar cells with a charge-integrating capacitor for periods of darkness [13], mechanical energy can be collected with piezoelectric or electrostatic converters [14], electromagnetic energy can be collected through radio frequency resonators [15], and finally, thermal energy can be collected with thermoelectric generators [16]. The majority of the micro-energy harvesters are still in the research and development phase. However, thermoelectrics was the first one to appear in the market [12]. This was due to the easiness to fabricate these devices with solid-state technology and because their working principle is based on a well-established physical theory. Table 10.1 resumes the majority of features that characterizes each type of energy harvesting presented in this review. Next, in the further sections, a few energy harvesting types are presented.

10.2.1 Thermoelectric

In 1821, Seebeck observed that the needle of a magnet is deflected in the presence of dissimilar metals that are connected (electrically in series and thermally in

Table 10.1 The list with the majority of features that characterizes each type of energy harvesting is presented in this review.

Energy sources in the environment	Main challenge(s)	Magnitude for the typical impedance	Magnitude for the typical voltage	Magnitude for the typical output power
Light (solar type)	Conformal with small surface areas, wide range for the input voltage	Varying with the light intensity, from a few kΩ to dozens of kΩ	DC: from 0.5 V up to 5 V, depending on the number of PC cells	Outdoors: from a few μW up to a few mW Indoors: from a few hundredths of mW up to a few mW
Mechanical vibration (piezoelectric type)	Narrow-band and high selective: pose problems of frequency variability in mechanical vibrations	Almost constant: from a few dozens Ω up to a few hundreds of Ω	AC: few dozens of V	Ranging from a few μW up to a few dozens of mW
Thermoelectric	Small thermal gradients, efficient heat skinking	Almost constant: from few Ω up to a few hundreds of Ω	DC: from both few dozens of mW up to V	Gradient of 20°C: from a few hundreds of μW up to a few dozens of mW

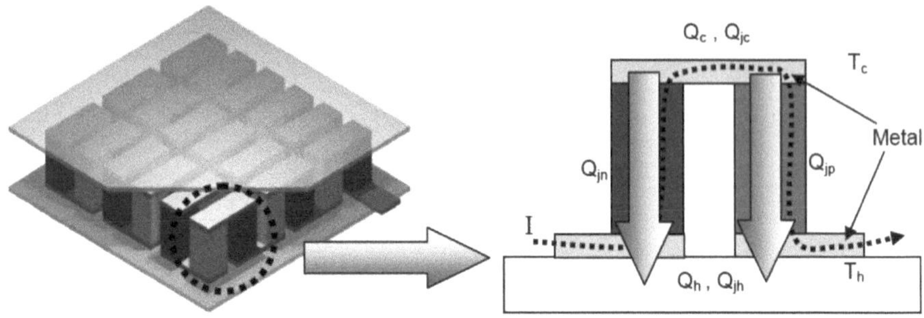

Figure 10.1 An artist's impression of a typical thermoelectric generator and a zoomed view of a thermoelectric pair (e.g., a thermoleg).

parallel) and exposed to a temperature gradient [16]. The effect he observed is the basis for thermoelectric power generation. Fig. 10.1 shows the basic cell of a typical thermoelectric microgenerator, which is composed of two thermoelectric columns (n-type and p-type). As shown in Fig. 10.1, if the junctions at the bottom are heated and those at the top are cooled (producing a temperature differential), electron/hole pairs will be created at the hot end and absorb heat in the process. The pairs recombine and reject heat at the cold ends. A voltage potential, the Seebeck voltage, which drives the hole/electron flow, is created by the temperature difference between the hot and cold ends of the thermoelectric elements. The net voltage appears across the bottom of the thermoelectric element legs.

The physics of these energy converters are widely known and dominated, imposing the research effort in the discovery and research of new materials and structures to make the process of current extraction from thermal gradients the most efficient as possible. This implies obtaining materials (both in the bulk [17], nanostructured [18], and/or superlattice [19] forms) that simultaneously are good electric conductors to minimize heat losses by the Joule effect, poor thermal conductors to retain the most heat possible at the site of the junction and at the same time presenting a maximum Seebeck effect, to produce the required voltage [20]. The performance of thermoelectric devices depends on the figure of merit ZT of the material [19,21] and is a dimensionless quantity, which is given by:

$$ZT = \frac{\alpha^2}{\rho\lambda} T \tag{10.1}$$

where α [V K^{-1}] is the Seebeck coefficient, ρ [Ω m] the electrical resistivity (e.g., the inverse of the electrical conductivity S [S m^{-1}]), λ [W m^{-1} K^{-1}] the thermal conductivity and T [K] the temperature. Another performance factor is more appropriate on thermoelectric devices, for example, the power-factor PF [W K^{-2} m^{-1}], expressed in terms of:

$$PF = \frac{\alpha^2}{\rho} \left[\mathrm{W\ K^{-2} m^{-1}}\right] \tag{10.2}$$

Analyzing the two equations above, it is desirable to select a material or group of materials that maximize both the *ZT* and *PF*, Unfortunately, the three physical quantitiesα, ρ and λ trade between them, meaning that their mutual interdependencies must be taken when a simultaneous optimization of these properties is desired. The simple fact the electrons carry unwanted heat and electric charge (traduced as electrical current) will decrease the Seebeck effect while the electrical conductivity increases. For bulk materials at both macro- and micro-scales, the highest performance is obtained in the presence of heavily doped semiconductors, such as bismuth telluride [22] or silicon-germanium [23]. For the case of semiconductors, the most desirable situation is when the base materials are both *n*- and *p*-doped to apply the same material system on both sides of the junctions [16]. Also, to be used in microsystems, a thermoelectric generator must be small in size, light in weight, and possibly have fabrication compatibility with the microsystem elements. Thin-film generators are the most suited for microsystems applications since they give the advantage of obtaining modules with minimum size and weight [20].

A lot of new developments about thermoelectric converters for energy harvesting can be found in the literature. This review tries to focus on the most updated and the ones with the most potential for a technological breakthrough, as well as in the case of authors' work. The work developed by Yang et al. [23] can be considered unique in the fact it presents a thermoelectric converter directly fabricated on top of a CMOS device. Thanks to the used material based on polysilicon (in fact, as reproduced in Fig. 10.2A, using the polysilicon-1 layer serves as p-type thermoleg, and the polysilicon-2 layer serves as n-type thermoleg, both represented respectively by red and green strips) [23]. In order to force the heat flow to take in the lateral region, a cavity was etched below the thermoelectric pairs, in a way to form a

Figure 10.2 For the work developed by Yang et al.: (A) an artist impression showing the schematic diagram of a micro thermoelectric converter (working as thermogenerator) with thermal isolation cavity on the bottom of thermoelectric pairs in polysilicon; and (B) a photo of a microdevice prototype (with a square area of 4 mm^2) containing 10 thermoelectric converters (red square on bottom) and test structures (the remaining area delimited by red lines distributed around the perimeter) [23].
Source: Reproduced with the previous authorization of the publisher (*Elsevier*).

membrane with minimum thermal conductivity. The hot side of the device is the upper plane in the region above the thermoelectric pairs in silicon, while the cold side is the plane located below the cavity. The underneath cavity ensures thermal isolation to prevent heat loss, maintain the temperature gradient and increase the efficiency by improving the output power [23]. Their simulations were applied on a thermoelectric junction (e.g., thermocouple) with optimal dimensions of $71 \times 4 \times 0.275/0.18\ \mu m$ (defined as length × width × thickness for p-/n-thermolegs) showed a maximum power factor of $0.047\ \mu W\ cm^{-2}\ K^{-2}$ and a voltage factor of $2.788\ V\ cm^{-2}\ K^{-1}$. They also observed a power factor of $0.042\ \mu W\ cm^{-2}\ K^{-2}$ from a thermocouple (measuring $60 \times 4\ \mu m$) within a functional prototype of a microdevice fabricated in the TSMC 0.35 μm CMOS process with 2 layers of polysilicon and 4 layers of metal [23].

The field of human body applications experiments a grown in the demand in last years, ranging from the acquisition of physiological information of several kinds [24–30]. This field is also known as Body-Area Networks or BAN [31–33]. Thus, it is also not strange the growth of these applications using energy harvesting [31,34]. The authors of the work found in [34] made an optimized design of a full-fledged wearable miniaturized thermoelectric specifically for use in human body applications. As illustrated in Fig. 10.3A, such a converter has a shape with a rim structure. This structure was obtained by surface micromachining applied to the polycrystalline silicon–germanium (poly-SiGe). The photos of these converters and a few zoomed views can be observed in Fig. 10.3B. In terms of fabrication details, each thermoelectric converter can contain 2350 or 4700 thermocouples, thermally connected in parallel and electrically in series. In their experimentations, the authors made a flip-chip bond of the converter into an electronic microdevice coated with a thin-layer of BCB. Such a configuration demonstrated the ability to deliver an open-circuit voltage of $12.5\ V\ K^{-1}\ cm^{-2}$ and output power of $0.026\ \mu W\ cm^{-2}\ K^{-2}$ on a matched external load [34]. The experimental tests also showed that under real field conditions, for example, being worn on the human body, this thermoelectric generator can deliver an open-circuit output voltage of ≈ 0.15 V and output power of ≈ 0.3 nW when supplying a matched load.

An outstanding application of thermoelectric converters on BAN can be found in [31], where the authors designed and fabricated readout circuits for the acquisition of biopotential signals. These circuits were optimized for low-power and high-performance, to allow stand-alone applications outside clinical environments and without the need for any kind of intervention from health professionals. Fig. 10.4A shows the photo of a functional prototype of a wireless EEG acquisition system with eight channels, occupying only $1\ cm^3$ [31]. In Fig. 10.4B, it is possible to observe a thermoelectric generator (composed of 10 sections, each measuring 1.6×4 cm) to supply the previous EEG acquisition system. It is interesting to observe the conformity of the generator with the shape of the subject's forehead [31]. According to the Ref. [18] of their work, the thermoelectric generator was designed to work at room temperatures up to 26°C. Additionally, it is claimed the ability to deliver a power of about 2 mW (e.g., a power density of $30\ \mu W\ cm^2$) for

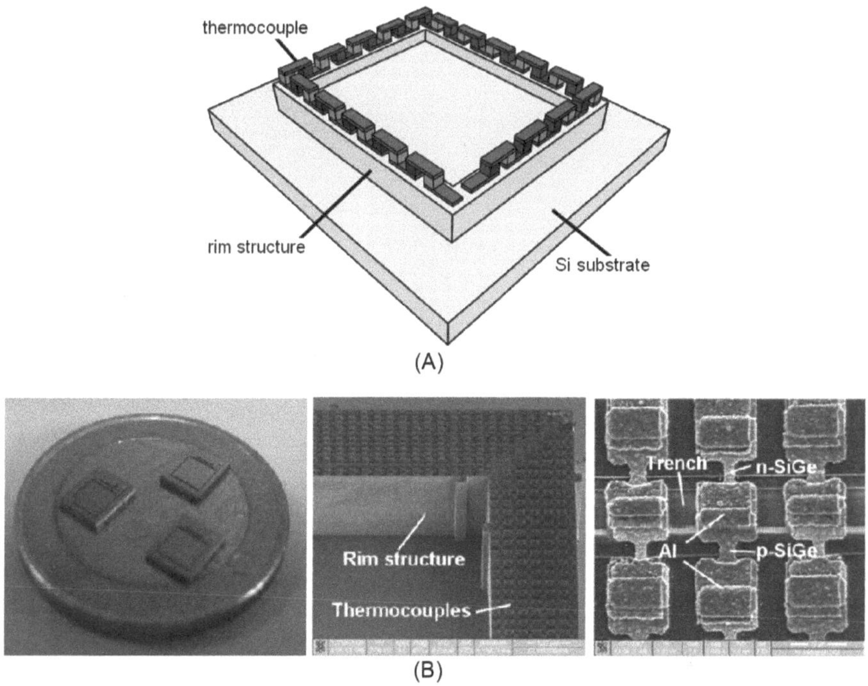

Figure 10.3 For the work developed by Wang et al.: (A) an artist impression showing the schematic of their thermoelectric converter, where it can be observed the several micromachined thermocouples after being fabricated on top of a rim structure with a height of 250 μm. (B) Progressively zoomed photos (from left to right) of a few functional prototypes of micromachined thermoelectric pairs; the middle and right photos only can be observed through SEM microscopy, where the middle one shows the rim structure and thermocouples; the photo on right shows a more deep close-up view of few micromachined thermocouples [34].
Source: Reproduced with the previous authorization of the publisher (Elsevier).

a room temperature of 23°C. This power is more than enough to supply the EEG acquisition system, whose power consumption is only 0.8 mW.

Thermoelectric converters based on bismuth tellurides [35] and both bismuth [36]/antimony [37] tellurides allow the fabrication of converters with high figures-of-merit *ZT* at room temperatures (e.g., 300K). In the context of this review, a high value of *ZT* is defined as one to be close to unity [19]. Unfortunately, this class of materials is not compatible with silicon, thus, requiring back-end techniques for their integration with microdevices based on silicon (e.g., CMOS technology). Despite this drawback, there are still excellent works of thermoelectric converters for energy harvesting using tellurides. The energy converter developed by Takashiri et al. [35] is a classical work focused on this class of materials and it was fabricated using the flash evaporation method through thermoelectric generators fabricated by a flash evaporation method. The n- and p-type powders of $Bi_{2.0}Te_{2.7}Se_{0.3}$ and

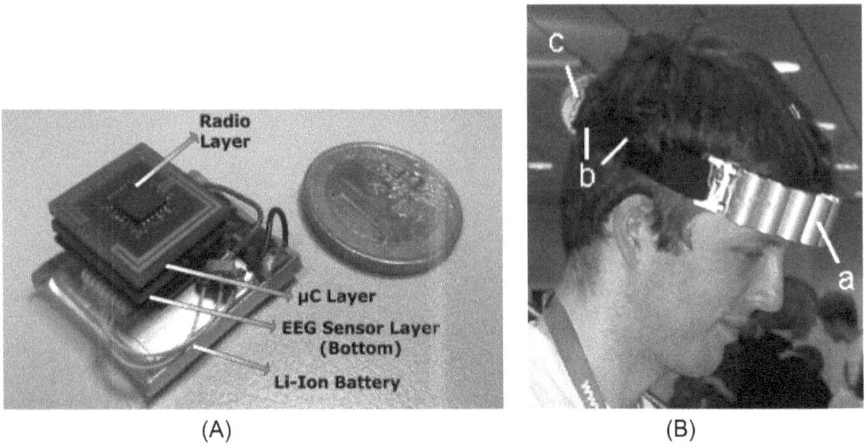

Figure 10.4 For the work developed by Yazicioglu et al.: (A) a photo of a data acquisition system with a capacity to acquire 8 EEG channels and (B) a photo of a thermoelectric converter placed in the forehead of a test subject to power the data acquisition system on left [31].
Source: Reproduced with the previous authorization of the publisher (Elsevier).

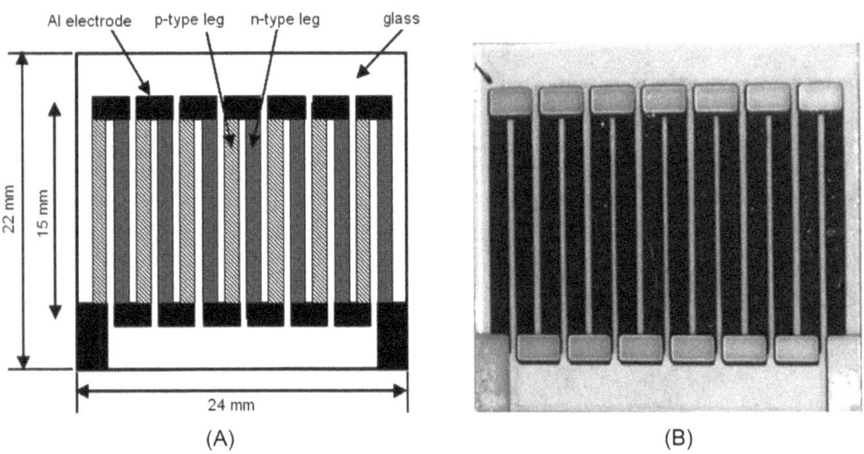

Figure 10.5 For the work developed by Takashiri et al.: (A) schematic and (B) photo of their functional prototype of thermoelectric converter [35].
Source: Reproduced with the previous authorization of the publisher (*Elsevier*).

$Bi_{0.4}Te_{3.0}Sb_{1.6}$ were respectively prepared for flash evaporation of thin films to form the thermoelectric legs (to make the thermoelectric pairs). As shown in Fig. 10.5 (schematic and photo), their thermoelectric converter is composed of seven pairs of thermolegs, electrically connected by electrodes made of aluminum [35]. Each leach measures $15 \times 1 \times 1$ mm (length $\times$ width $\times$ thickness) and their physical arrangement and electrical connection result in a thermoelectric converter measuring 20 mm by 15 mm. After the fabrication of thermolegs, the thermoelectric characteristics of several functional prototypes were improved by submitting them to a hydrogen annealing at temperatures between 25°C and 250°C. For example, the functional prototype whose annealing temperature was of $T_a = 250°C$ provided a maximum output voltage of 83.4 mV and an estimated power of 21 μV, when subjected to a temperature gradient of 30°C.

More works based on tellurides, more precisely planar thermoelectric generators based on bismuth and antimony tellurides can be found in [21] and [38]. Both converters were obtained using the deposition of thin-films. The difference relies on the fact that in the first case microsystems techniques were used (lithography and etching [21]), while the shadow mask technique was used in the second case [38]. Fig. 10.6A shows a magnified photo of a thermoelectric generator composed of eight pairs of n- and p-type elements [22]. A set of measurements were made for the Bi_2Te_3 and Sb_2Te_3 thin-films, resulting respectively in values of *ZT* at room temperatures of 0.97 and 0.56, and in power factors of 0.49 and 0.28 $\mu W\ cm^{-2}\ K^{-2}$. As it can be observed, the thermolegs were deposited on top of a polyimide foil with a thickness of 25 μm. The technique of shadow masks is simple and well used in thin-film depositions, where the distance precision is not critical. The n- and p-type deposited thin films are spaced apart between them by 500 μm. The contact area of the thermoelectric structures measures 1 mm^2, while the metal contacts made of aluminum occupy an area of 7 mm^2. A prototype proposed by Silva et al. [38] makes reminds the coiled-up thermoelectric generator fabricated by Weber et al. [39]. The prototype in Fig. 10.6B was obtained by sputtering thermolegs of bismuth and antimony on polyimide foils, whose geometry was based on the principle of the coiling-up to provide high values of voltages with smaller generator areas. This kind of geometries applied on a large strip of

(A) (B)

Figure 10.6 (A) Photo of a thermogenerator with eight pairs of n- and p-type elements, fabricated with bottom contacts [22]. (B) Photo of a coiled-up thermoelectric generator prototype [39].
Source: The photos on the left and right were reproduced with the previous authorization of the publisher (*Elsevier*).

polymer foil allows the fabrication of many thousands of thermocouples and the corresponding surround around the target structure where the temperature gradient is present [39].

A fascinating application using thermoelectric generators was validated by Wojtas et al. [40]. In strict terms, this application was already known on its fundaments, for example, by harvesting energy from the waste heat present in industrial processes [38] and vehicles [39]. The amazing difference relies on the involved scale, that is while the known approaches were implemented at the large size scale, the work of these authors [40] was implemented at the microscale. As illustrated in the photos Fig. 10.7, this application combines a microfluidic heat transfer system with a thermoelectric generator and is the first one of its kind reported in the literature. Their functional prototype was tested and characterized, considering two physical quantities: the measured thermal and hydrodynamic performance and the correspondent generated power [40].

The field of energy harvesting doesn't know limits when looking at the application of these systems on planes. In this context, a stand-alone sensorial node was designed and fabricated by Samson et al. [41] to operate on short-range flights. The work was validated on laboratory environment to deal with plane fuselage's temperatures in the range of −21.8°C to +20.4°C. The details about the choice of this temperature range as well as the specificities to deal with the several physical

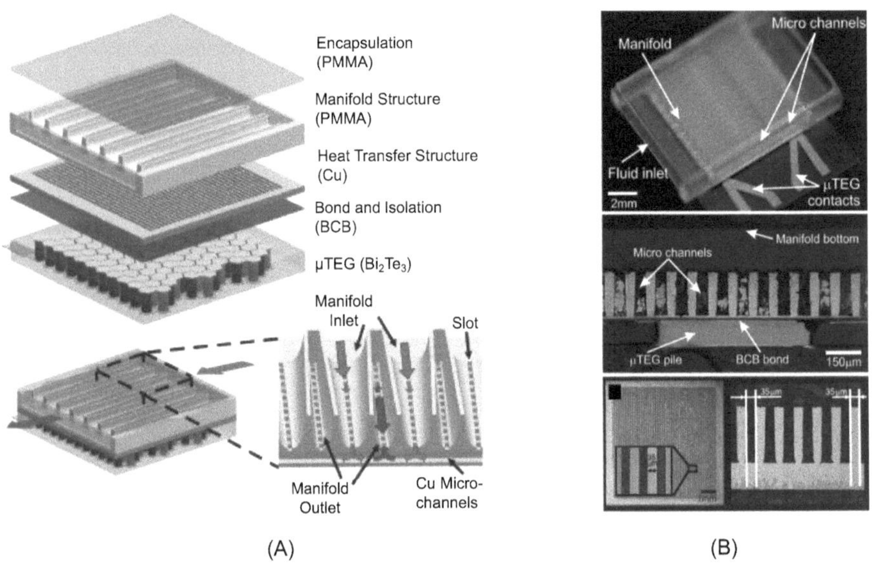

Figure 10.7 For the work developed by Wojtas et al.: (A) Exploded view illustrating the combination of a microfluidic heat transfer system with a thermoelectric generator, and the cross-section of the path where the hot fluid passes through; (B) photo of the functional prototype after being assembled (top), a cross-section of the path where the hot fluid passes through (middle) and top view (bottom) [40].
Source: Reproduced with the previous authorization of the publisher (*Elsevier*).

phenomena during [short-range] flights are fully detailed in their publication [41]. In this sequence of ideas and taking into account the temperature profile, the thermogenerator selection fell on a customized-made device from a well-known manufacturer for matching both the temperature range and profiles.

It must be noted that the voltage at the output of the thermogenerator is not constant and can drop below 1.5 V, thus, a power management chip is required to store the harvested energy and at the same time to make upconversion of the voltage to operate the sensorial node. In this work, a regulated voltage of 3.3 V is required. The tests in the laboratory environments have proven to be promised, thus, in the future is expected real-time tests in real-field operations [41]. Fig. 10.8 shows the block diagram and the respective photo of a stand-alone prototype mounted on an aircraft hull for future tests. The sensorial node is composed of a power management chip (implemented by a Texas Instruments MSP430 series microcontroller for intelligent management) and the sensorial node itself (composed of a crack wire sensor and by the Texas Instruments' CC2420 RF transceiver for the 2.4 GHz ISM band able to implement the standard IEEE 802.15.4 relative to the two lowest levels of the ZigBee's protocol). The experimental tests showed that the whole function prototype has a total power consumption of about 189 μW for an output voltage of 3.3 V [41].

New developments in materials with high *ZT*s (close to 1) are the only future way to follow to improve the characteristics of thermoelectric converters (both for energy harvesting and cold generation) and pave the way to increase the market and at the same time to decrease the costs (by providing high-volume production). It is undeniable the importance that was achieved up to now with telluride-based materials, both on bulk and microscale, but unfortunately, these materials are not compatible with the available technologies of microelectronics based on silicon [42]. Therefore, it is of major interest to develop materials compatible with silicon.

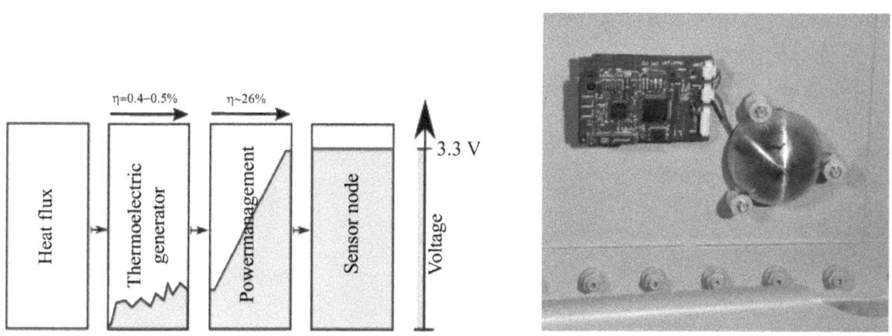

(A) (B)

Figure 10.8 For the stand-alone prototype developed by Samson et al. and composed of the thermogenerator and the energy management chip (e.g., the MSP420 microcontroller) and the sensorial node (e.g., the crack sensor and the RF transceiver) [41]: (A) block diagram and (B) respective photo of a stand-alone prototype mounted on an aircraft hull for future tests. *Source*: Reproduced with the previous authorization of the publisher (*Elsevier*).

New innovative research was proposed by Ashby et al. [42], which proposes and claims (based on outstanding argumentation and supported by experimentation) a promising alternative to telluride-based materials (more specifically, with relation to the Bi_2Te_3), for example, materials based on silicon nanoparticles. In these materials, the thermal conductivity is low thanks to the high number of interfaces created [42]. Randomly distributed nanostructures in nanocomposite materials (see the concept in Fig. 10.9) can lead to a reduction in the thermal conductivity below that of an alloy of the same overall chemical stoichiometry [38,43]. This happens thanks to the high number of interfaces that are created, therefore increasing phonon scattering [42]. A thermoelectric figure-of-merit *ZT* of 0.6 at 300K was measured after subjecting silicon nanoparticles to a process of surface functionalization with phenylacetylene [42], whose value can be classified as significant (if not outstanding) for a bulk silicon.

10.2.2 Solar

One of the most important renewable energy sources is those based on photovoltaic (PV) cells [44–46] because their technology is well established in terms of maturity, applications, and markets. More recently, after the publication of the smart dust work [47], applications of solar photovoltaic to power stand-alone Microsystems started to know the interest for application on distributed sensors networks. In this classical work done by Warneke et al. [47] at the University of Berkley, a 16 mm^2-microsystem was fabricated for stand-alone operation through an integrated solar-cell. This microsystem was designed and fabricated to prove the concept in which it is possible to integrate three microdevices: (1) an SOI (Silicon-On-Insulator) solar cell to provide the power supply; (2) an ASIC fabricated on a 0.25 μm CMOS process, comprising a photosensor (a n-well/p-substrate CMOS photodiode), signal processing (amplification, analog-to-digital conversion) and control electronics (a state-machine with 13 states); and (3) by an optical transmitter [47]. The optical receiver was implemented within the CMOS ASIC through a second n-well/p-substrate photodiode because this is the CMOS photodetector configuration with the highest responsivity (e.g., highest sensitivity), while the optical

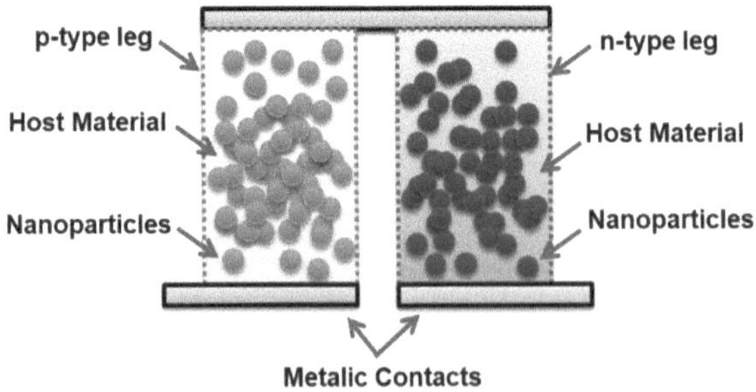

Figure 10.9 A thermocouple composed of a pair of n- and p-type legs, both made of nanocomposite materials.

transmitter was specifically implemented on a third microdevice composed of a micromachined four-quadrant corner-cube retroreflector (named as CCR chip) [47]. The optical transmission was made passive with respect to the target base-station to minimize the power consumption of the whole microsystem, for example, an interrogator in the base-station sends a light beam towards the microsystem in the direction of a set of 3 orthogonal mirrors. The bottom mirror is electrostatically actuated, thus the microsystem does nothing (saves power) for a "1" transmission. For a "0" transmission the bottom mirror is slightly actuated and displaced, breaking the orthogonality and thus spreading the light beam across the space [47]. This spreading is detected in the base-station because the received intensity is smaller for a "0" than for a "1." This ingenious solution results in a very low-power consumption microsystem.

Fig. 10.10 shows (A) the concept of an autonomous flexible thin-film (FTF) and (B)-(D) photos of an FTF functional prototype: (B) just after the application of polydimethylsiloxane (better known as PDMS), (C) after the PV cell placement, and (D) showing the maximum thickness of the FTF prototype [47]. Fig. 10.10A highlights the set of flexible substrates that compose the FTF, for example, a PV cell, a flexible PCB (fPCB), the PDMS material (a key-element between the three

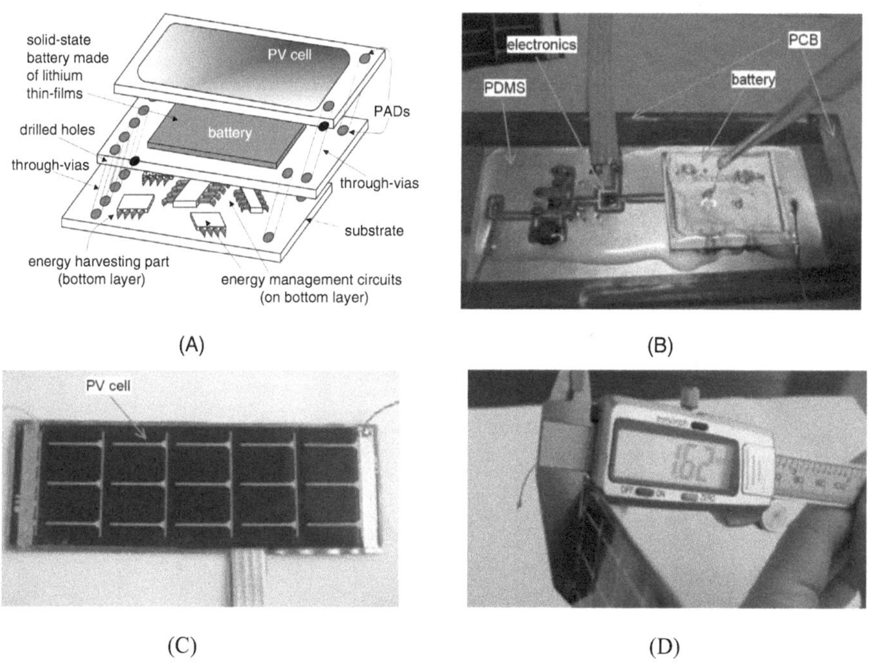

Figure 10.10 (A) Autonomous flexible thin-film (FTF) concept. Photos of the FTF prototype: (B) just after the PDMS application, (C) after the photovoltaic cell placement, and (D) showing the maximum thickness of the FTF [46].
Source: (B) and (C) were reproduced with the previous authorization of the publisher (*Elsevier*).

layers to serve as the adhesion layer) and the respective constructive concept. This FTF was designed and fabricated taking into account the growing need for deformable devices and transducers (either stretchable, foldable, or even compressible) [47–50]. The fPCB provides physical support to the electronics and battery, while the fPCB-PV cell adhesion is provided by a layer of PDMS. As shown in Fig. 10.10D, this TFT has a thickness of 1.62 mm, measures $37 \times 114\ mm^2$, and weighs 10 g. This TFT was able to provide a voltage of 3.8 V under solar irradiations higher than $6\ W\ m^{-2}$, while values below the power were provided for the battery [47]. A certain degree of "intelligence" was thus required to make this energy management: a DC–DC step-up converter model LTC3105 incorporates an adjusted control for maximum power point tracking (MPPT) and was used to provide a regulated voltage for battery charging and power supply to the exterior. Additionally, it was necessary to either protect the battery from overvoltage peaks (during the charge) or undervoltage valleys (during the discharge). The LTC4071 was engaged to detect zones of danger and thus disconnect the battery from the charge/discharge circuit. The complete FTF schematic can be observed in Fig. 10.11.

This kind of energy harvester was designed and fabricated for applications where the need for maintenance should be reduced and solar energy is available and abundant. High-precision agriculture is a target field of applications matching these requirements [47–49].

A cutting-edge idea related to energy harvesting combines the physical characteristic to measure with the harvesting mechanism itself. This philosophy was followed by Emadi et al. [50] to design a self-powered Sun sensor to detect the satellite's position with respect to the Sun. The conventional approach shares with this one the use of 4 photodetectors on a matrix 2×2. Contrary to the former, the latter approach remains simple but ingenious because instead of a pin-hole aperture to allow only one photodetector to be illuminated when the array surface is aligned with the Sun, this last one takes into account the complementary situation [50]. In other words, only one photodetector is blocked and dark when the array is aligned

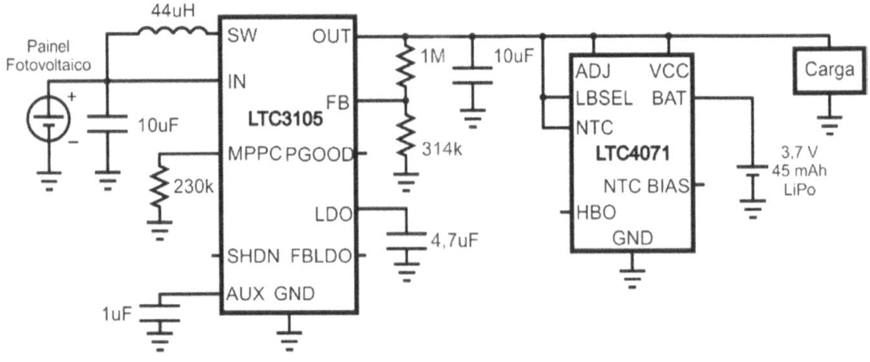

Figure 10.11 Electronic circuit composed of an maximum power point tracking step-up converter, a battery power management, and an under/over-voltage protection [46].

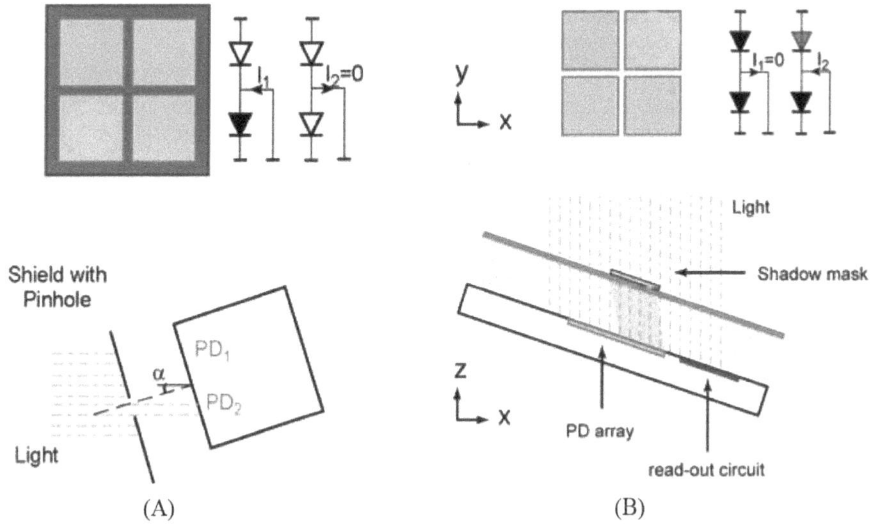

Figure 10.12 For the transduction on Sun sensors to align satellite panels: (A) conventional and (B) novel concepts [50].
Source: Reproduced with the previous authorization of the publisher (*Elsevier*).

with the Sun, allowing the collected photocurrents of the remaining 3 photodetectors to be used for both sensing and powering of the sensor. Fig. 10.12 illustrates the concept behind the (A) conventional and the (B) new approach [50].

Despite the abundant availability of solar light worldwide, this harvesting technique cannot be used in dark environments (oil wells, buried in the soil below the surface, embedded inside mechanical/physical structures, and so on). Another drawback related to this environmental power source is the lack of low-power/low-voltage solutions solar harvesting circuits observed at the industrial level, as concluded in a study conducted by Fröhlich et al. [51].

10.2.3 Piezoelectric

Piezoelectricity is the ability of a material to convert mechanical strain into electric current or voltage [52–55]. This effect is also reversible, by allowing the conversion of electricity into motion. While the first phenomenon is used on mechanical sensors and energy harvesters [56], the second phenomenon is used on mechanical actuators [57]. The most know class of applications is in footstep and footwear power generation [58,59]. Mechanical vibrations were proven to be one of the environmental energy sources with one of the highest delivered power and efficiency [60].

The most common materials used on both energy harvesting and mechanical actuation include: the lead-zirconate-titanate (PZT) [61], lead-titanate ($PbTiO_2$) [62], lead-zirconate ($PbZrO_3$) [63], barium-titanate ($BaTiO_3$) [64] and PVDF [65].

To date, the most used piezoelectric material is the PZT due to its inherent huge electromechanical coupling ability [66].

Despite the larger power handling, piezoelectric harvesting suffers from a huge drawback. As shown in Table 10.1, this kind of energy harvesting is strongly dependent on the natural frequency of oscillating, which is mainly imposed by the geometry of the movable part [67,68]. Behind the narrow-band feature, this poses a lot of problems concerned with miniaturization and integration in microsystems, for example, normally the resonance frequency increases with the object's size reduction [68]. Further improvements can be achieved by optimizing the shape of vibrational parts (for example, the shape of a cantilever and or a mass) [69]. However, nothing is more cutting-edge than the bimorph cantilever made of thick-films of PZT to explore the multimodality, allowing the piezoelectric converters to harvest energy from broadband vibrations [70].

The piezoelectricity is always surprising in the field of electronics. This can be confirmed through an emergent field in aeronautics, consisting of the electricity conversion from the aeroelastic vibration of the plane's wings [71–74]. Aeroelasticity is divided into static and dynamic aeroelastic phenomena. In the context of dynamic phenomena, it exists the flutter, which can present a high destructive potential. The flutter consists of self-sustained oscillations, happening on sustentation surfaces (e.g., on wings) when the flight speed is higher than a threshold (named critical speed of flutter). Below this threshold, the oscillations will be damped. A disturbance of any kind happening with flight speeds larger than the critical speed of flutter will result in auto-excited oscillations (e.g., like positive feedback in a physical system) with a fatal result for the sustaining structure. More recently and despite the potentially destructive nature of flutter, the recent research in aeronautics has confirmed its high potential to extract energy from the aeroelastic vibrations (resultant from the airflow) [72,75] to convert them into electrical energy for providing useful self-powered aeroelastic control methods [76]. In general terms, the idea is to harvest the energy on aeroelastic vibrations and use it to produce the same vibrations in the opposite direction. This implementation is no more than negative feedback towards the control of aeroelastic control.

10.2.4 Electronics and storage

A harvesting system composed only of the energy converter it is not enough to allow a microsystem to be fully stand-alone. Two additional energy processing stages comprising energy storage and power management are needed for a successful energy harvesting implementation [77,78]. In this context, the direct deposition of microbatteries into a microdevice (responsible for the energy management) was proposed by Lhermet et al. [79], making this work the first of this kind to be widely known in the literature. In Fig. 10.13 is possible to observe the block diagram of their microdevice/microbattery prototype [79], which forms a microsystem able to harvest (by transforming it into the electrical form) two power sources (e.g., inductive [80] and/or thermal [81]) and to manage the harvested energy through the microdevice. The microsystem is also able to store energy through microbattery.

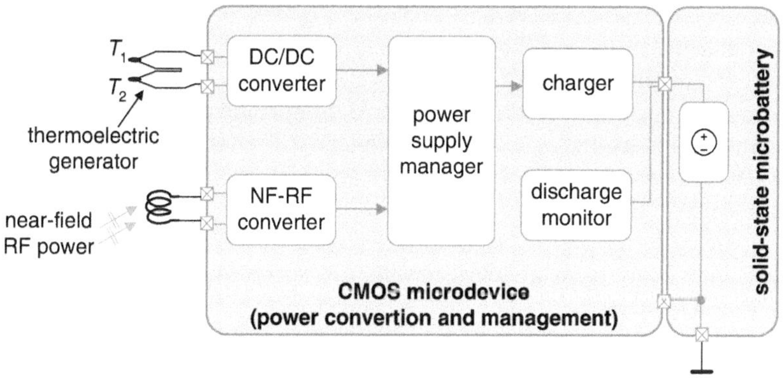

Figure 10.13 Block diagram of the microdevice/microbattery prototype fabricated by Lhermet et al. [79].

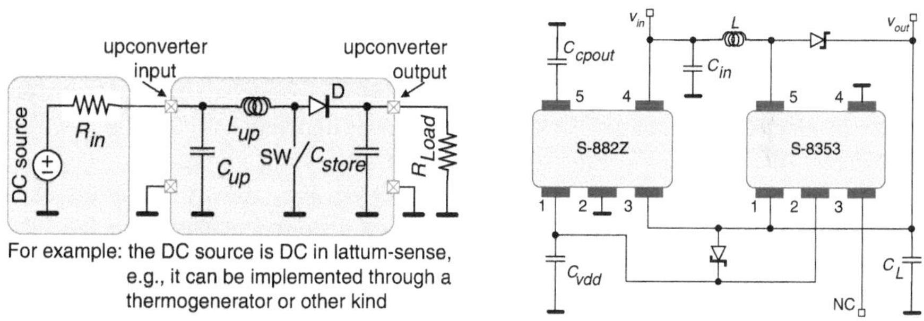

(A) (B)

Figure 10.14 (A) Schematic of a simple voltage upconverter circuit. For the functional prototype of a voltage upconverter, using the Seikos' microdevices S-882Z and S-8353: (B) schematic.

In terms of power management, the microdevice can (1) schedule which external source will be used to charge the battery, even when both are available. The microsystem can also (2) provide a compatible voltage to the battery; make decisions about (3) stopping and (4) starting the charging process, as well as, (5) switch the provenience of power between the battery and the harvested source towards the load to supply it. This last control action allows the energy stored in the microbattery to be spared for situations when external power sources are not available.

After considering the specificities related to the energy type to be harvested, the block diagram of Fig. 10.13 can be considered a base for any power management circuit design. In this context and independently of the application, the DC/DC converter must be able to upconvert (or downconvert when necessary) the voltage from the external source. The voltage downconversion is straightforward to understand, thus, the focus will be placed on the upconversion. Fig. 10.14A illustrates a schematic of a simple upconverter to help the physical principle behind this block [20]. An inductance L_{up} will be required to provide voltage peaking to achieve the upconversion, while a capacitor C_{up} [idealy] prevents the voltage to drop and thus, completing the upconversion process. In terms of operation, the current supplied by a DC source (for example, a thermogenerator) will charge C_{up}. Meanwhile, the switch SW closes and opens synchronously at a high frequency. When closed, the switch remains in this state for a short amount of time to reduce losses. This implies that, when closed, the signal responsible for commanding the switch must have a very low duty-cycle to avoid the capacitor C_{up} from over-discharge. The stored energy in the inductor L_{up} forces the capacitor C_{up} to discharge through the diode D, during the time the switch is opened (e.g., the diode makes a DC rectification). Finally, the current will charge the capacitor, C_{store}, which connects to the load (in the simplest configuration) or a DC regulator (on a more advanced configuration). The charge-capacity of C_{store} must be high to sustain [above a certain level] the voltage when deliver current to the load. This last configuration is very simple and easy to understand the principle behind the upconversion, using the step-up technique. On practical circuits, this configuration is very unsuitable due to the huge number of parameters to trade with the generator itself (switch on/off frequency, inductor L_{up}, capacitors C_{up}, and C_{store}, the forward voltage of diode D and the quality of electronic switch—for example, the open- and short-circuit losses within the switch). As illustrated in Fig. 10.14B, a more interesting solution can be implemented, using components off-the-shelf (COTS) [20]. Fig. 10.14B shows the electronic schematic of a circuit prototype for voltage upconversion through DC–DC step-upping, using the S-882Z (start-up microdevice for connecting into a step-up converter) and S-8353 (DC–DC step-up converter with PWM control and built-in switch) circuits from the Seiko manufacturer [82]. Table 10.2 shows few results for several sets of input voltages and currents, and as it can be observed the output voltage is very close to those required by a lithium battery ($\approx$3.7 V). A significant feature of the prototype based on these microdevices is their efficiency is always above 85%: Such efficiencies mean internal power losses below 15%, resulting in acceptable losses for these kinds of applications.

The combination of energy harvesting systems (comprising energy conversion and energy management systems) with batteries will expand and capability and autonomous features of stand-alone microsystems. However, this poses problems in terms of fully battery/microsystem integration because the battery technology didn't

Table 10.2 Few tests were made to the circuit of Fig. 10.14B–D for a few combinations of voltages and currents.

V_{in} [V]	I_{in} [ma]	V_{out} [V]	I_{out} [ma]	P_{in} [mW]	P_{out} [mW]	η [%]
0.31	40.11	3.20	3.32	12.43	10.62	85
0.39	45.30	3.40	5.05	17.57	17.17	98
0.45	50.50	3.60	5.91	22.70	20.68	91
0.50	70.10	4.50	7.55	35.05	33.97	97

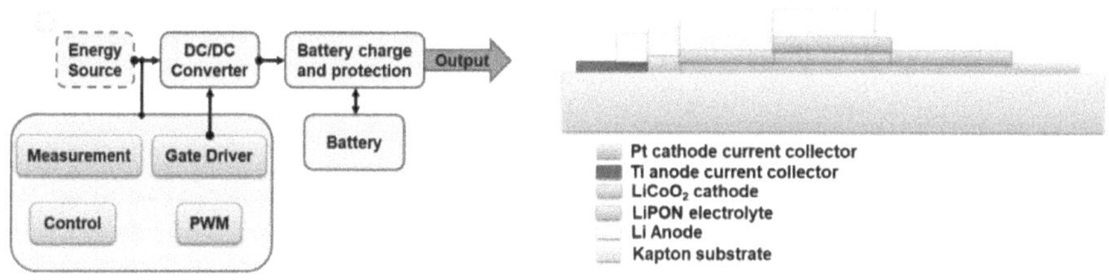

(A) (B)

Figure 10.15 For solid-state microbattery made of thin-films fabricated by Ribeiro et al. [85]: (A) main blocks of a target application for this microbattery. (B) an artist's impression of the selected materials (not in the real scale for a better illustration).

follow the spectacular development of the microelectronics industry [83]. Nonetheless, this did not prevent the research and development of microbatteries for integration with microdevices as can be confirmed in Fig. 10.13 [79], illustrating the block diagram of the microdevice/microbattery prototype fabricated by Lhermet et al. [79]. Their microbattery was fabricated using microsystems techniques, more specifically the deposition of thin-films. The microbattery is of lithium type and was obtained through a stack of lithium/LiPON/TiSO, forming the cathode/electrolyte/anode respectively. More specific details about the fabrication of the microbattery can be found in their publications [78] and [84]. Ribeiro et al. [85] also have fabricated with success a solid-state microbattery on a flexible substrate made of polyimide (Kapton from DuPont). Fig. 10.15A illustrates the main blocks of a target application for this microbattery. The blocks of the energy harvesting system are comprised of a PV cell, which is intended to have a MPPT controller, a DC–DC converter, a battery charger and protection circuit and a microbattery itself. In this target application, the MPPT controller measures the voltage and current in the terminals of an energy source and calculates the necessary duty-cycle of the DC–DC converter. The battery charger and protection circuit control the flux of energy between the energy source, the battery, and the output of the energy harvesting system. The materials that compose the microbattery are a succession of layers made of $LiCoO_2$/LiPON/lithium that act as cathode/electrolyte/anode. In the future, a flexible solar cell is intended to be used as a substrate, where on its back side the microbattery was previously deposited using thin-film techniques by Ribeiro et al. elsewhere [85]. Further details about the deposition of the several materials that compose this battery can be found on [9,86].

10.3 Conclusions

This chapter has presented an overview of the state of the art related to energy harvesting systems for stand-alone microsystems (MSTs). It was intended to present hot-topics concerning this theme, especially those related to thermal, piezoelectric, and solar photovoltaic. In the end, we presented a few integration topics concerning microelectronic systems and microbatteries for adding energy conversion/management and storage functions, respectively.

References

[1] B.R. Stojkoska, A.P. Avramova, P. Chatzimisios, Application of wireless sensor networks for indoor temperature regulation, Int. J. Distrib. Sens. Netw. 2014, 1–10. Paper #502419.
[2] Y. Cui, M. Kim, Y. Gu, J.-J. Jung, H. Lee, Home appliance management system for monitoring digitized devices using cloud computing technology in ubiquitous sensor network environment, Int. J. Distrib. Sens. Netw. 2014, 1–10, Paper #174097.
[3] T. Lu, X. Guo, Y. Li, Y. Peng, X. Zhang, F. Xie, et al., Cyberphysical security for industrial control systems based on wireless sensor networks, Int. J. Distrib. Sens. Netw. 2014 (2014) 1–17. Paper #438350.
[4] Z. Luo, P.S. Min, S.J. Liu, Target localization in wireless sensor networks for industrial control with selected sensors, Int. J. Distrib. Sens. Netw. 2013 (2013) 1–9. Paper #304631.
[5] K.-T. Yang, W.K. Lai, S.-M. Li, Y.-C. Lin, Event-based clustering architecture for power efficiency in wireless sensor networks, Int. J. Distrib. Sens. Netw. 2014, 1–12, Paper #612590.
[6] G. Yi, D. Yu, N. Kim, Adjusting control packet transmission intervals in low power sensor systems, Int. J. Distrib. Sens. Netw. 2014, 1–8, Paper #139682.
[7] Y. Cui, M. Kim, Y. Gu, J.-J. Jung, H. Lee, Home appliance management system for monitoring digitized devices using cloud computing technology in ubiquitous sensor network environment, Int. J. Distrib. Sens. Netw. 2014, 1–10, Paper #174097.
[8] K.H. Park, A ubiquitous motion tracking system using sensors in a personal health device, Int. J. Distrib. Sens. Netw. 2013 (2013) 1–6. Paper #298209.
[9] S. Akbari, Energy harvesting for wireless sensor networks review, in: Proceeding of the 2014 Federated Conference on Computer Science and Information Systems, vol. 2, Warsaw, 2014, pp. 987–992.
[10] L. Mateu, C. Codrea, N. Lucas, M. Pollak, P. Spies, Human body energy harvesting thermogenerator for sensing applications, in: Proceeding of the 2007 International Conference na Sensor Technologies and Applications, Valencia, Spain, October 2007, pp. 366 272.
[11] L. Collins, Harvest for the world, IEEE Power Eng. J. 20 (1) (2006) 34–37.
[12] J. Bouchaud, R. Dixon, Micro-energy harvesters: overview, applications and markets, in: MSTnews: International Newsletter on Micro Nano Integration, No. 1|08, February 2008, pp. 18 19.
[13] D. Dondi, A. Bertacchini, D. Brunelli, L. Larcher, L. Benini, Modeling and optimization of a solar energy harvester system for self-powered wireless sensor networks, IEEE Trans. Ind. Electron. 55 (7) (2008) 2759–2766.
[14] L. Mateu, F. Moll, Appropriate charge control of the storage capacitor in a piezoelectric energy harvesting device for discontinuous load operation, J. Sens. Actuators A: Phys. Sens. 132 (2006) 302–310.
[15] J. Liu, J. Yao, Wireless RF identification system based on SAW, IEEE Trans. Ind. Electron. 55 (2) (2008) 958–961.

[16] L. Bell, Cooling, heating, generating power, and recovering waste heat with thermoelectric systems, Science 321 (2008) 1457–1461.

[17] B. Poudel, Q. Hao, Y. Ma, Y. Lan, A. Minnich, B. Yu, et al., High-thermoelectric performance of nanostructured bismith antomony telluride bulk alloys, Science 320 (2008) 634–637.

[18] M.S. Dresselhaus, G. Chen, M.Y. Tang, R. Yang, H. Lee, D. Wang, et al., New directions for low-dimensional thermoelectric materials, Adv. Mater. 19 (2007) 1–12.

[19] A. Böttner, G. Chen, R. Venkatasubramanian, Aspects of thin-film superlattice thermoelectric materials, devices, and applications, MRS Bull. 31 (2006) 211–217.

[20] J.P. Carmo, L.M. Goncalves, J.H. Correia, Thermoelectric microconverter for energy harvesting systems, IEEE Trans. Ind. Electron. 57 (2010) 861–867.

[21] J.P. Carmo, J. Antunes, M.F. Silva, J.F. Ribeiro, L.M. Goncalves, J.H. Correia, Characterization of thermoelectric generators by measuring the load-dependence behavior, Measurement 44 (2011) 2194–2199.

[22] J.P. Carmo, L.M. Goncalves, R.F. Wolffenbuttel, J.H. Correia, A planar thermoelectric power generator for integration in wearable microsystems, Sens. Actuators A 161 (2010) 199–204.

[23] S.M. Yang, T. Lee, C.A. Jeng, Development of a thermoelectric energy harvester with thermal isolation cavity by standard CMOS process, Sens. Actuators A 153 (2009) 244–250.

[24] R. Carta, P. Jourand, B. Hermans, J. Thoné, D. Brosteaux, T. Vervust, et al., Design and implementation of advanced systems in a flexible-stretchable technology for biomedical applications, Sens. Actuators A 156 (2009) 79–87.

[25] N.S. Dias, J.P. Carmo, P.M. Mendes, J.H. Correia, A low power/low voltage CMOS wireless interface at 5.7 GHz with dry electrodes for cognitive networks, IEEE Sens. J. 11 (2011) 755–762.

[26] N.S. Dias, J.P. Carmo, P.M. Mendes, J.H. Correia, Wireless instrumentation system based on dry electrodes for acquiring EEG signals, Med. Eng. Phys. 43 (2012) 972–978.

[27] M.S. Fernandes, J.H. Correia, P.M. Mendes, Electro-optic acquisition system for ECG wearable sensor applications, Sens. Actuators A 203 (2013) 316–323.

[28] P. Zhao, N. Deng, X. Li, C. Ren, Z. Wang, Development of highly-sensitive and ultra-thin silicon stress sensor chips for wearable biomedical applications, Sens. Actuators A 216 (2014) 158–166.

[29] Y.D. Lee, W.Y. Chung, Wireless sensor network based wearable smart shirt for ubiquitous health and activity monitoring, Sens. Actuators B 140 (2009) 390–395.

[30] M. Ozawaa, K. Sampeia, C. Cortesa, M. Ogawaa, A. Oikawaa, N. Miki, Wearable line-of-sight detection system using micro-fabricated transparent optical sensors on eyeglasse, Sens. Actuators A 205, 208 214, 1014.

[31] R.F. Yazicioglu, T. Torfs, P. Merken, J. Penders, V. Leonov, R. Puers, et al., Ultra-low-power biopotential interfaces and their applications in wearable and implantable systems, Microelectron. J. 40 (2009) 1313–1321.

[32] Body Area Networks, IMEC's 2012 activity report. [online] http://annualreport.imec.be/Domains/page.aspx/1203.

[33] R. Carta, M. Sfakiotakis, N. Pateromichelakis, J. Thoné, D.P. Tsakiris, R. Puers, A multi-coil inductive powering system for an endoscopic capsule with vibratory actuation, Sens. Actuators A 172 (2011) 253–258.

[34] Z. Wang, V. Leonov, P. Fiorini, C.V. Hoof, Realization of a wearable miniaturized thermoelectric generator for human body applications, Sens. Actuators A 156 (2009) 95–102.

[35] M. Takashiri, T. Shirakawa, K. Miyazaki, H. Tsukamoto, Fabrication and characterization of bismuth-telluride-based alloy thin film thermoelectric generators by flash evaporation method, Sens. Actuators A 138 (2007) 329–334.
[36] L.M. Gonçalves, C. Couto, P. Alpuim, A.G. Rolo, F. Völklein, J.H. Correia, Optimization of thermoelectric properties on Bi_2Te_3 thin films deposited by thermal co-evaporation, Thin Solid. Films 518 (2010) 2816–2821.
[37] L.M. Goncalves, P. Alpuim, A.G. Rolo, J.H. Correia, Thermal co-evaporation of Sb_2Te_3 thin-films optimized for thermoelectric applications, Thin Solid. Films 519 (2011) 4152–4157.
[38] M.F. Silva, J.F. Ribeiro, J.P. Carmo, L.M. Gonçalves, J.H. Correia, Thin-films for thermoelectric applicationsISBN: 978-3-642-25413-0 in: B. Bhushan (Ed.), Invited Book Chapter in the book Scanning Probe Microscopy in Nanoscience and Nanotechnology 3, Springer, 2012.
[39] J. Weber, K. Potje-Kamloth, F. Haase, P. Detemple, F. Völklein, T. Doll, Coin-size coiled-up polymer foil thermoelectric power generator for wearable electronics, Sens. Actuators A 132 (2006) 325–330.
[40] N. Wojtas, E. Schwytera, W. Glatz, S. Kühne, W. Escher, C. Hierold, Power enhancement of micro thermoelectric generators by microfluidic heat transfer packaging, Sens. Actuators A 188 (2012) 389–395.
[41] D. Samson, M. Kluge, T. Becker, U. Schmid, Wireless sensor node powered by aircraft specific thermoelectric energy harvesting, Sens. Actuators A 172 (2011) 240–244.
[42] S. Ashby, J.A. Thomas, J. García Cañadas, G. Min, J. Corps, A.V. Powell, et al., FD 176: Bridging silicon nanoparticles and thermoelectrics: phenylacetylene functionalization, in: Faraday Discussions, Accepted for publication on 26th June 2014. Available from: https://doi.org/10.1039/C4FD00109E.
[43] L.M. Gonçalves, Thermoelectric Microsystem Based on Bismuth and Antimony Tellurides (PhD thesis), University of Minho, 2008.
[44] R. Eke, A.S. Kavasoglu, N. Kavasoglu, Design and implementation of a low-cost multi-channel temperature measurement system for photovoltaic modules, Measurement 45 (2012) 1499–1509.
[45] M. Gagliarducci, D.A. Lampasi, L. Podestà, GSM-based monitoring and control of photovoltaic power generation, Measurement 40 (2007) 314–321.
[46] J.P. Carmo, J.M. Gomes, L.M. Gonçalves, J.H. Correia, A flexible thin-film for powering stand-alone electronic devices, Measurement 46 (10) (2013) 4145–4151.
[47] B. Warneke, M. Scott, B. Leibowitz, Z. Lixia, C. Bellew, J. Chediakl, et al., An autonomous 16 mm^3 solar-powered node for distributed wireless sensor networks, in: Proceedings of the IEEE Sensors, vol. 2, June 2002, pp. 1510 1515.
[48] R. Morais, M.A. Fernandes, S.G. Matos, C. Serôdio, P.J.S.G. Ferreira, M.J.C.S. Reis, A ZigBee multi-powered wireless acquisition device for remote sensing applications in precision viticulture, Comput. Electron. Agric. 62 (2008) 94–106.
[49] R. Morais, S.G. Matos, M.A. Fernandes, A.L.G. Valente, S.F.S.P. Soares, P.J.S.G. Ferreira, et al., Sun, wind and water flow as energy supply for small stationary data acquisition platforms, Comput. Electron. Agric. 64 (2008) 120–132.
[50] H. Wu, A. Emadi, G. de Graaf, J. Leijtens, R.F. Wolffenbuttel, Self-powered sun sensor microsystems, Procedia Chem. 1 (2009) 1363–1366.
[51] A.A. Frolich, E.A. Bezerra, L.K. Slongo, Experimental analysis of solar energy harvesting circuits efficiency for low power applications, in: Computers and Electrical Engineering, Accepted for publication on 5th October 2014. Available from: https://doi.org/10.1016/j.compeleceng.2014.09.004.

[52] R. Caliò, U.B. Rongala, D. Camboni, M. Milazzo, C. Stefanini, G. de Petris, et al., Piezoelectric energy harvesting solutions, Sensors 14 (2014) 4755–4790.

[53] A. Venugopal, A. Agrawal, S.V. Prabhu, Performance evaluation of piezoelectric and differential pressure sensor for vortex flowmeters, Measurement 50 (2014) 10–18.

[54] I. Korhonen, R. Lankinen, Energy harvester for a wireless sensor in a boiler environment, Measurement 58 (2014) 241–248.

[55] B.V.M.P.S. Kumar, K. Suresh, U.V. Kumar, G. Uma, M. Umapathy, Resonance based DC current sensor, Measurement 45 (2012) 369–374.

[56] A. Bounouh, D. Bélières, Resonant frequency characterization of MEMS based energy harvesters by harmonic sampling analysis method, Measurement 52 (2014) 71–76.

[57] M.L. Jansen, G.R. Willmott, I. Hoek, W.M. Arnold, Fast piezoelectric actuation of an elastomeric micropore, Measurement 46 (2013) 3560–3567.

[58] J.G. Rocha, L.M. Gonçalves, P.F. Rocha, M.P. Silva, S. Lanceros-Méndez, Energy harvesting from piezoelectric materials fully integrated in footwear, IEEE Trans. Ind. Electron. 57 (2010).

[59] L. Xie, M. Cai, Increased piezoelectric energy harvesting from human footstep motion by using an amplification mechanism, Appl. Phys. Lett. 105 (2014) 1–4. Paper #143901.

[60] P.C.-P. Chao, Energy harvesting electronics for vibratory devices in self-powered sensors, IEEE Sens. J. 11 (2011) 3106–3121.

[61] R. Medeiros, M.L. Ribeiro, M. Sartorato, G. Marinucci, V. Tita, Computational simulation using PZT as sensor elements for damage detection on impacted composite cylinders, in: Proceedong of the 1st International Symposium on Solid Mechanics (MecSol2013), Porto Alegre, Rio Grande do Sul RS, Brazil, 2013, pp. 1–11.

[62] G.R. Blake, A.R. Armstrong, E. Sastre, W. Zhou, P.A. Wright, The preparation of a novel layered lead titanate and its conversion to the perovskite lead titanate $PbTiO_3$, Mater. Res. Bull. 36 (2001) 1837–1845.

[63] A. Kimoto, J.D. Maddumapatabendi, A proposal of multi-imaging system of electrical and ultrasonic properties, Measurement 44 (2011) 2000–2007.

[64] A. Stashans, H. Pinto, Analysis of radiation-induced hole localisation in titanates, Radiat. Meas. 33 (2001) 553–555.

[65] D. Chen, T. Sharma, J.X.J. Zhang, Mesoporous surface control of PVDF thin films for enhanced piezoelectric energy generation, Sens. Actuators A 216 (2014) 196–201.

[66] A. Nechibvute, A. Chawanda, P. Luhanga, Piezoelectric energy harvesting devices: an alternative energy source for wireless sensor, Smart Mater. Res. 2012, 1–13, Paper #853481.

[67] A. Tabesh, L.G. Frechette, A low-power stand-alone adaptive circuit for harvesting energy from a piezoelectric micropower generator, IEEE Trans. Ind. Electron. 57 (2010) 840–849.

[68] A. Khaligh, P. Zeng, C. Zheng, Kinetic energy harvesting using piezoelectric and electromagnetic technologies state of the art, IEEE Trans. Ind. Electron. 57 (2010) 850–860.

[69] W.G. Li, S. He, S. Yu, Improving power density of a cantilever piezoelectric power harvester through a curved L-shaped proof mass, IEEE Trans. Ind. Electron. 57 (2011) 868–876.

[70] M. Baù, M. Ferrari, E. Tonoli, V. Ferrari, Sensors and energy harvesters based on piezoelectric thick films, Procedia Eng. 25 (2011) 737–744.

[71] M. Rosa, C. De Marqui Junior, Modeling and analysis of a piezoelectric energy harvester with varying cross-sectional area, Shock. Vib. 2014 1–9. Paper #930503.

[72] J.A.C. Dias, C. De Marqui Junior, A. Erturk, Hybrid piezoelectric-inductive flow energy harvesting and dimensionless electroaeroelastic analysis for scaling, Appl. Phys. Lett. 102 (2013) 1–5. paper #044101.
[73] C. De Marqui Junior, A. Erturk, Electroaeroelastic analysis of airfoilbased wind energy harvesting using piezoelectric transduction and electromagnetic induction, J. Intell. Mater. Syst. Struct. 24 (2013) 846–854.
[74] A.J. Dias, C. De Marqui Junior, A. Erturk, Three-degree-of-freedom hybrid piezoelectric-inductive aeroelastic energy harvester exploiting a control surface, Am. Inst. Aeronautics Astronautics J. (2014) 1–11.
[75] M. Bryant, E. Wolff, E. Garcia, Aeroelastic flutter energy harvester design: the sensitivity of the driving instability to system parameters, Smart Mater. Struct. 20 (2011) 1–12.
[76] D. D'Assunção, C. De Marqui Junior, Applied self-powered semi-passive control for a 2-degree-of-freedom aeroelastic typical section using shunted piezoelectric materials, J. Intell. Mater. Syst. Struct. (2014) 1–13. Available from: https://doi.org/10.1177/1045389X14526797.
[77] J.P. Carmo, J.F. Ribeiro, M.F. Silva, L.M. Goncalves, J.H. Correia, Thermoelectric generator and solid-state battery for stand-alone microsystems, J. Micromech. Microeng. 20 (2010) 1–8.
[78] R. Salot, S. Bancel, S. Martin, G. Savelli, C. Salvi, M. Plissonnier, Thermoelectric and microbattery hybrid system with its power management, in: Proceeding of the Symposium Design Test Integration & Packaging of MEMS/MOEMS 349 353, 2006.
[79] H. Lhermet, C. Condemine, M. Plissonnier, R. Salot, P. Audebert, M. Rosset, Efficient power management circuit: from thermal energy harvesting to above IC microbattery energy storage, IEEE J. Solid. State Circuits 43 (2008) 246–255.
[80] H. Lhermet, C. Condemine, R. Salot, M. Plissonnier, S. Bacquet, P. Audebert, On chip post processed microbattery powered with RF and thermal energy through a power management circuit, in: Proceeding of the IEEE International Conference on Integrated Circuit Design an Technology, 2007, pp. 184–186.
[81] G. Savelli, M. Plissonnier, J. Bablet, C. Salvi, J.M. Fournier, Energy conversion using new thermoelectric generator, in: Proceeding of the Symposium Design Test Integration & Packaging of MEMS/MOEMS, 2006, pp. 369–374.
[82] Seiko Instruments Inc. [online]. http://www.sii-ic.com/en/semicon/.
[83] M. Armand, J.M. Tarascon, Building better batteries, Nature 451 (2008) 652–657.
[84] S. Oukassi, X. Gagnard, R. Salot, S. Bancel, J.P. Pereira Ramos, Above IC micro power generators for RF MEMS, in: Proceeding of the Symposium Design Test Integration & Packaging of MEMS/MOEMS, 2006, pp. 1–4.
[85] J.F. Ribeiro, R. Sousa, J.A. Sousa, L.M. Goncalves, M.M. Silva, L. Dupont, et al., Flexible thin-film rechargeable lithium batteries, in: Proceeding of the Transducers 2013, Barcelona, Spain, 2013.
[86] J.F. Ribeiro, R. Sousa, J.P. Carmo, L.M. Gonçalves, M.F. Silva, M.M. Silva, et al., Enhanced solid-state electrolytes made of lithium phosphorous oxynitride films, Thin Solid. Films 552 (2012) 85–89.

Techno-economic appraisal for large-scale energy storage systems

11

Chun Sing Lai[1] *and Giorgio Locatelli*[2]
[1]Brunel Interdisciplinary Power Systems Research Center, Department of Electronic and Electrical Engineering, Brunel University London, London, United Kingdom, [2]School of Management, Politecnico di Milano, Italy

Chapter Outline

Highlights

- Reviewed the state-of-the-art grid energy storage technologies
- Presented a discounted cash flow model for energy storage techno-economic study
- Discussed energy storage economic appraisal of transport and energy sectors

11.1 Introduction

To reach a low-carbon economy, the intermittent renewables penetration and low-carbon energy sources, including wind [1], solar [2], and nuclear [3], are anticipated to continue to grow in electrical power systems over the coming decades. Simultaneously, the uncertainty for electricity demand is growing due to the increasing reliance on

Emerging Trends in Energy Storage Systems and Industrial Applications. DOI: https://doi.org/10.1016/B978-0-323-90521-3.00020-X

electric vehicles (EV) [4,5], electrical appliances, and services [6]. The power grid is increasingly facing greater planning and operational challenges.

To address the aforementioned challenges, large-scale (e.g., megawatts) energy storage is crucial in maintaining the grid's power and energy balance. However, the amount of installed energy storage capacity is still very little (but growing quickly) in comparison to the power generation capacity [7]. For example, in 2020, the UK has planned to install 164 MW of new energy storage capacity consisting mainly of batteries [7].

Techno-economic appraisals are important for investors to identify if, when, and how an energy project can be economically viable. Techno-economic results can also aid engineers and policymakers in decision-making. Many techno-economic appraisals have been performed for energy storage systems [8–10].

Precise techno-economic appraisal of energy storage must address the uncertainties associated with technical performance (e.g., efficiency and lifetime) and economic performance (e.g., construction and operating cost). Therefore, several models were created to study energy storage economics, with several studies focusing on the Levelized Cost of Electricity (LCOE) [8,9,11–14]. However, these models made simplified assumptions regarding the economic aspects of energy storage, including the financing strategy or the cash-flow analysis. A crucial aspect of the successful deployment of energy storage is a detailed consideration of the financial characteristics, i.e., the account of assets and liabilities and their distribution across the project lifetime and stakeholders [15–18].

The financing of an electrical energy storage system is examined in [16,17], and a literature review is given in [19]. However, these works excluded some of the key metrics, e.g., Net Present Value (NPV), Internal Rate of Return (IRR), and LCOE. Also, the modeling of energy storage in many techno-economic appraisal works (e.g., [16,20–22]) excludes energy storage degradation.

With regard to the application of large-scale energy storage in power systems, most of the techno-economic appraisals mainly focus on limited aspects. Kapila et al. [23] developed a data-intensive techno-economic model to study the economic feasibility of pumped hydro energy storage and compressed air energy storage. The costs for the system were constructed by a bottom-up approach, with detailed economic and technical data for the energy system components. The economy of scale factor was applied to each component. The authors estimated the LCOE of the electricity produced by energy storage to repay the capital investment.

Zurita et al. [24] investigated the techno-economic merit of a megawatt utility-scale hybrid power plant consisting of concentrated solar power and photovoltaic power plant. The power plant is integrated with a large-scale battery located in northern Chile. The focus is on the integration of solar with energy storage. The investment costs of the battery make this hybrid system unprofitable. To have this system economically viable, the cost of storage needs to be reduced by 60%–90% from 300 USD/kWh.

Olateju et al. [25] conducted a techno-economic appraisal for a 563 MW integrated wind-hydrogen power plant with energy storage. The model employs real-time wind energy data to determine the number of electrolyzer units, the optimal size of the electrolyzer, and the battery capacity that minimizes the hydrogen production cost, whilst working in a liberalized electricity market with dynamic prices.

Using the Energy PLAN tool, Pupo-Roncallo et al. [26] performed a techno-economic appraisal on grid-scale energy storage in a power system with renewable energy sources and high penetration of hydropower in Colombia. Energy storage is needed to accommodate higher penetrations of wind and solar into the power system, by increasing flexibility and reducing carbon intensity. In the Colombia case study, pumped hydro energy storage is determined to be the most cost-effective option.

With a novel techno-economic appraisal methodology, Locatelli et al. [10] examined and compared the economics of compressed air energy storage and pumped hydro energy storage systems with the UK electricity price data. The focus is on electricity arbitrage. The work examined the trade-off between "reserve capacity and profitability" and "optimum size capacity." The authors found that none of the existing energy storage technologies is economically viable without subsidies.

Focusing on coupling large-scale wind power generation and energy storage, Lai and Locatelli [15] examined two types of wind power and energy storage technologies with a state-of-the-art cash flow model. The model considers three categories of uncertainties, i.e., technical, economic, and financing. Based on a study in the UK, wind farms without energy storage have better economic and financial performance.

Generation integrated energy storage (GIES) system is a new and specific category of integrated energy system consisting of a generator and an energy storage system. GIES aims to be an economic solution for storing MWh energy. This is done by "storing energy at some point along with the transformation between the primary energy form and electricity" [27]. The storage medium is non-electrochemical such as compressed air energy storage (kinetic) and thermal storage (heat). The key idea is to transform the primary energy into another form that is more convenient to store than electricity, e.g., wind with thermal pumped energy storage (Wind-TP) system [2,3]. The second and mature class of energy storage for low carbon energy is the non-GIES system. Non-GIES system is the popular grid-scale energy storage solution consisting of electrochemical energy storage e.g., batteries. Batteries including Lithium-ion (Li-ion) have several benefits including high power and energy density, and round-trip efficiency. Nevertheless, Li-ion battery usage may contribute to adverse environmental impact with the consumption of natural resources for assembly and the pollution it creates during disposal. Li-ion batteries contain several toxic chemicals such as Cobalt [4] and have a short life caused by cell degradation [5]. Assessment of energy storage options for low carbon power generation is a highly relevant topic.

From the aforementioned literature, it is clear that current techno-economic appraisal for large-scale energy storage focuses on electricity arbitrage and low-carbon energy integration, e.g., accommodating high penetration of renewables. The deployment of megawatt-scale energy storage is becoming prominent not only for accommodating large-scale power generation but also installed in distribution networks or low voltage networks for other emerging system applications including electrification of transport and the built environment.

The rest of the chapter will be as follows: Section 11.2 presents an overview of available energy storage options and emerging technologies. The traditional applications of energy storage are reviewed. The models for techno-economic appraisal of large-scale energy storage and power system are presented in Section 11.3.

Section 11.4 presents the future research agenda for techno-economic appraisals for large-scale energy storage. Section 11.5 concludes the chapter.

11.2 Energy storage technologies for smart grids

Firstly, this section briefly introduces the benefits and challenges associated with smart grids. Secondly, the benefits of smart grids integrating with large-scale energy storage are described. Finally, this section outlines the available energy storage options for smart grids.

11.2.1 Benefits and costs associated with smart grids

The smart grid can be considered as the evolution of the power system with the presence of smart meters, renewable energy resources, smart appliances, and energy-efficient resources [28]. The resources, devices, and appliances are interconnected with the means of information and communication technology infrastructure [29].

The benefits associated with smart grids include [30]:

- Higher transmission efficiencies of electricity;
- Rapid electricity restoration after power disturbances;
- Minimized operations and management costs for utilities, and eventually reduced electricity costs for end-users;
- Minimized peak demand and electricity price;
- Maximized integration of large-scale renewable energy systems;
- Enhanced integration of distributed generation systems (e.g., renewable energy systems);
- Enhanced system security of power supply.

The costs and challenges associated with smart grids are as follows:

- The capital cost incurred for the replacement of analog meters with sophisticated electronic meters (e.g., smart meters and advanced metering infrastructure);
- Lack of regulatory norms for standards for technologies smart grid;[29]
- Data could be breached and difficult to manage for a complex cyber-physical-social system;[31]
- Energy consumers may need additional training and knowledge to interreact with smart grid devices (e.g., smart appliances and home monitoring systems) and services (e.g., dynamic energy tariffs).

11.2.2 Benefits of smart grids integrating large-scale energy storage

The benefits of using energy storage for smart grids have been widely discussed [32,33]. Defined by the US Department of Energy, the smart grid "represents an unprecedented opportunity to move the energy industry into a new era of reliability, availability, and efficiency that will contribute to our economic and environmental health" [30].

The benefits of smart grids integrating large-scale energy storage can be described as follows: [33]

- Time and peak shifting: to reduce power utilities' cost when energy is stored at off-peak hours and discharged during peak hours; [34]
- Power quality: To maintain power voltage and frequency within specific limits. Energy storage can provide frequency regulation by varying supply to the fluctuating demand;
- Off-grid systems: Energy storage supports the power utility to maintain stable power to the energy end-users;
- Industrial consumer's peak shaving: Energy storage can be operated to minimize the energy bill of the factory that is related to their highest power demand; [35]
- Continuity of energy supply and limitation of disturbances: A charged energy storage can supply electricity to the power network when there is a fault or outage in the grid. Energy consumers will experience higher satisfaction with minimized disturbances;
- Reactive power compensation: With power electronics converter, energy storage can provide reactive power to enhance the voltage levels within the distribution network.

11.2.3 Energy storage options for smart grids

There are several types of energy storage that exist for smart grids. Luo et al. [36]. developed a comprehensive review of the technical and economic features of energy storage systems. Although many demonstration projects on energy storage systems were completed, there is still a lack of corresponding detailed techno-economic analysis to build useful databases to aid decision-making. Amirante et al. [37]. contributed an overview of the following energy storage technologies: mechanical systems (pumped hydro, flywheel, and compressed air); electrochemical systems (Lithium-ion batteries, lead-acid batteries, and flow batteries); electrical systems (supercapacitors and superconducting magnetic); and hydrogen energy storage.

Governments should consider the long-term planning of future energy systems with the range of energy storage options available. Energy storage should be included in the energy policy landscape and policy roadmaps of every country. Lai et al. [32]. provided a comprehensive review of long-term electrical power system modeling with energy storage. These models allow analyses to be performed for decarbonization studies and energy technology appraisals. Current models are bottlenecked in assessing the power system with energy storage, in many cases, on consider one or two types. Lai et al. [32]. also identified that there are many choices of energy storage for a particular type of primary energy. For example, Fig. 11.1 shows that the kinetic energy from wind power can be stored in a mechanical form (e.g., flywheel) or be converted into electricity by a direct current (DC) generator and stored in electrochemical cells (batteries).

Although pumped hydro energy storage and compressed air energy storage systems were built before the prominence of electrochemical energy storage (e.g., lithium-ion), the most popular energy storage option for smart grids is electrochemical energy storage [33,38,39]. This is due to the maturity of battery technologies and plummeting costs. The most popular batteries are lithium, lead-acid nickel-cadmium, and sodium-sulfur [2]. The major issues for the popularity of these batteries

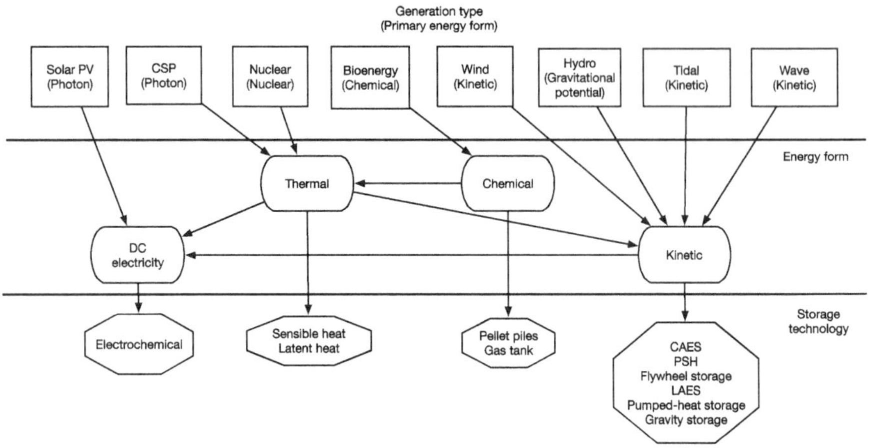

Figure 11.1 Energy transformation processes for energy storage.
Source: Reproduced from Lai CS, Locatelli G, Pimm A, Wu X, Lai LL. A review on long-term electrical power system modeling with energy storage. Journal of Cleaner Production. 2020:124298.

are the high cost as compared to alternative energy storage options presented in Table 11.1, the restricted number of charging cycles or lifetime, the low energy density, the depth of discharge, and the environmental impact of materials used.

Alternative energy storage technologies are developed to tackle the shortcomings and environmental impact of electrochemical energy storage [32]. Several non-electrochemical energy storage technologies are currently in the research and development stage, or prototype/pilot stage, with some, described as follows [32]:

- Flywheel energy storage uses angular kinetic energy. Present research develops flywheel systems that increase efficiency in electrical-mechanical energy conversion, e.g., continuously variable transmission systems [40,41];
- Underwater compressed air energy storage is based on conventional compressed air energy storage but uses hydrostatic pressure in deep seawater. Pimm et al. [42]. developed a methodology to identify the optimal shape and cost of cable-reinforced fabric vessels;
- Gravity energy storage involves moving fluids or alternative great masses upwards in opposition to gravity. There are many types of gravity energy storage in development, including a system using a water-filled container, usually in a dug hole in the ground, and a heavy piston [16,43,44];
- Pumped thermal energy storage was presented at the end of the 1970s [44,45]. However, the technology is still in the demonstration stage. The system consists of two thermal energy storage vessels (one high temperature and one low temperature) and a reversible heat pump/heat engine Smallbone et al. [46] trialed a pilot-scale system and described that the techno-economic appraisal needs to be investigated.

In summary, this chapter shows that many forms of energy storage are available. Some of these technologies are commercially available (batteries and pumped hydro), while several others are at different stages of research and development.

Table 11.1 Overnight and operating costs for grid energy storage [15].

Type	Energy storage	Overnight cost (£/kWh)	Fixed O&M cost (£/kW-yr)
Mechanical	Pumped-storage hydroelectricity	3.8−38.2	2.3−7.8
	Compressed air energy storage	5.5−31.8	1.6−10.6
	Flywheel	339.5−733	3.7−5.1
Heat	Pumped-thermal energy storage	11−17.8	0.0022
Electrochemical	Lithium-ion	130−300	1.7−7.8
	Vanadium redox flow battery	113−650	2.6−12.6

The need for techno-economic appraisals is growing with the technologies available and the applications of smart grids.

The next section introduces several challenges associated with techno-economic models for energy storage and power systems and presents a novel model to address such challenges.

11.3 Techno-economic models for energy storage and power systems

Techno-economic appraisals for energy systems and energy storage are commonly conducted by commercial, open-source, or bespoke models [32]. Ahmad et al. [47]. used HOMER, a commercial model to perform techno-economic analysis for a wind/photovoltaic/biogas generator hybrid energy system for rural electrification. The studies include the cost of energy, net present cost, and LCOE. Based on a case study for a 50 MW system, the hybrid energy system has an LCOE of 0.058 $/kWh. A grid-connected photovoltaic/biomass/wind system can provide a smaller LCOE than a grid-connected photovoltaic/biomass system. This signifies that wind energy can reduce the cost of hybrid energy systems in Pakistan. However, this work has not examined energy storage. In another study, Shahzad et al. [48]. employed HOMER to conduct a techno-economic appraisal for a photovoltaic/biomass off-grid system situated in Pakistan. The cost of energy and payback period are examined. The capital cost constitutes the largest proportion of the net present cost, subsequently the replacement, and at last the operating cost. Batteries are considered in this study, but there is a lack of detail on the LCOE for energy storage or the attention to battery degradation. Das et al. [49]. provided a techno-economic appraisal for an off-grid biogas generator/photovoltaic/diesel/battery hybrid system for remote areas in Bangladesh. Cow dung is used for biogas production and is highly accessible [49]. The study considers the cost of energy, payback period, net present cost, and emissions. The optimal system has photovoltaic, diesel generation, and biogas generation. The capacity shares of 49%, 36%, and 15%, respectively.

From the studies described above, the HOMER model shows to be a common tool for smart grid planning and techno-economic appraisal of energy systems. However, commercial models are generally black-box models, and the internal operation mechanisms are difficult to modify. This includes the optimization algorithm and cost calculation methodologies [9].

To address the aforementioned challenges, Lai and Locatelli [15] developed a state-of-the-art discounted cash flow model to examine the economics and financial performance of generation-integrated energy storage systems. Generation integrated energy storage system *"is a class of energy storage that stores energy at some point along with the transformation between the primary energy form and electricity"* [27]. Lai and Locatelli [15] applied this model in the case of integrating wind power and energy storage with a case study in the UK.

A discounted cash flow analysis is a standard method to examine the economic and financial performance of an investment in energy infrastructure [17]. A discounted cash flow analysis can determine the present value of expected future cash flows based on a discount rate. The total present value of all the discount cash flows (cost and revenues) is known as net present value. The investment requires a positive NPV to allow debt and equity holders remuneration. The discounted cash flow model developed by Lai and Locatelli [15] calculates three key indicators:

1. LCOE: The electricity price needed to meet all the life cycle costs.
2. NPV to the firm: The total of the unlevered cash flows discounted with the weighted average capital cost (WACC) or the "free cash flow to the firm."
3. NPV to the equity: The sum of the levered cash flows. Discounting the "free cash flow to the equity" at the cost of equity gives the NPV to the equity.

The schematic of the discounted cash flow model is provided in Fig. 11.2 [15]. The inputs to the model include economic, financial, and technical parameters. In the enhanced techno-economic analysis, the realistic cost of the system can be captured by considering the financing options. Project financing is commonly employed for financing large-scale energy projects [50].

Cost analysis is relatively common, and revenues analysis is less frequent. The revenue sources for a wind power and energy storage system in the UK are given in: [15]

Wholesale market/ spot price: the "traditional" price of electricity paid to the generator at each point in time.

Contract for difference: A low-carbon electricity generation investment incentive giving certainty and stability to revenue streams [51]. This revenue cannot be received by energy storage projects.

Short Term Operating Reserve: A contracted balancing service where the provider gives a contracted level of power as initiated by the National Grid Electricity System Operator to fulfill energy reserve needs [52].

Fast Reserve: Delivers rapid active power by raising the power generation or lowering the power demand as initiated by an electronic dispatch instruction from the National Grid Electricity System Operator [53].

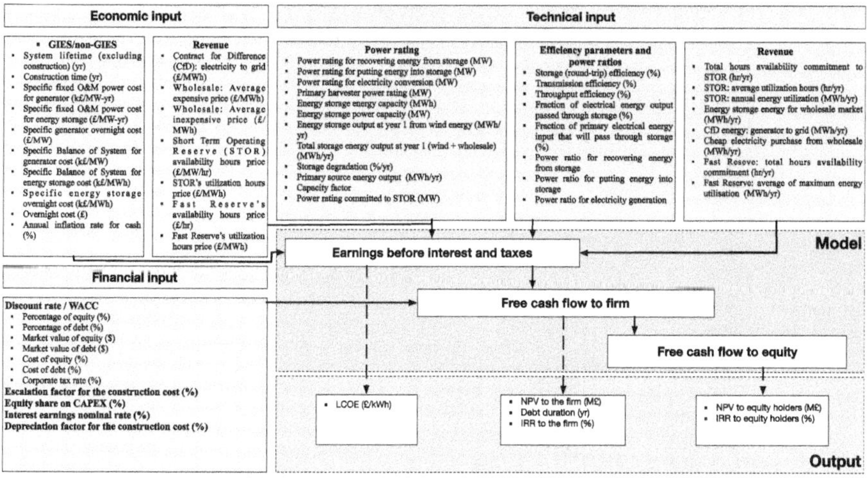

Figure 11.2 Discounted cash flow model for techno-economic appraisal and financial assessments of generation integrated energy storage.
Reproduced from Lai CS, Locatelli G. Economic and financial appraisal of novel large-scale energy storage technologies. Energy. 2020;214:118954.

Considering the above revenue sources, Lai and Locatelli [15] determined that wind farms excluding energy storage lead to better economic and financial performance. This is largely due to the generous incentives available to low-carbon power generators (i.e., contract for difference) and the general lack of financial support for energy storage systems as compared to electricity generators.

The risk analysis provides probability distribution functions (PDFs) for several financial and economic indicators such as NPV, LCOE, and IRR. Risk analysis is effective as part of techno-economic appraisal when there are input uncertainties. The risk analysis is conducted with Monte Carlo analysis with random sampling according to the probability distribution set for individual input. The Monte Carlo stopping criteria is according to the convergence requirement, with a confidence level of 95% and a convergence tolerance of 1%. More than 50,000 iterations are conducted. As GIES is a novel technology, the technical, economic, and financing uncertainties are higher as compared to conventional non-GIES technology. As such, Fig. 11.3 shows that there is a larger LCOE uncertainty for GIES as it is presented in the study and feasibility phase. Conversely, non-GIES is in the bid/tender stage which has more cost certainty. The LCOE and its uncertainty are minimized as the technology becomes mature.

In summary, the typical challenges associated with techno-economic appraisal techniques for energy storage and power systems were discussed, and a novel model addressing such challenges was discussed. The next section examines the research agenda for large-scale energy storage systems techno-economic appraisal.

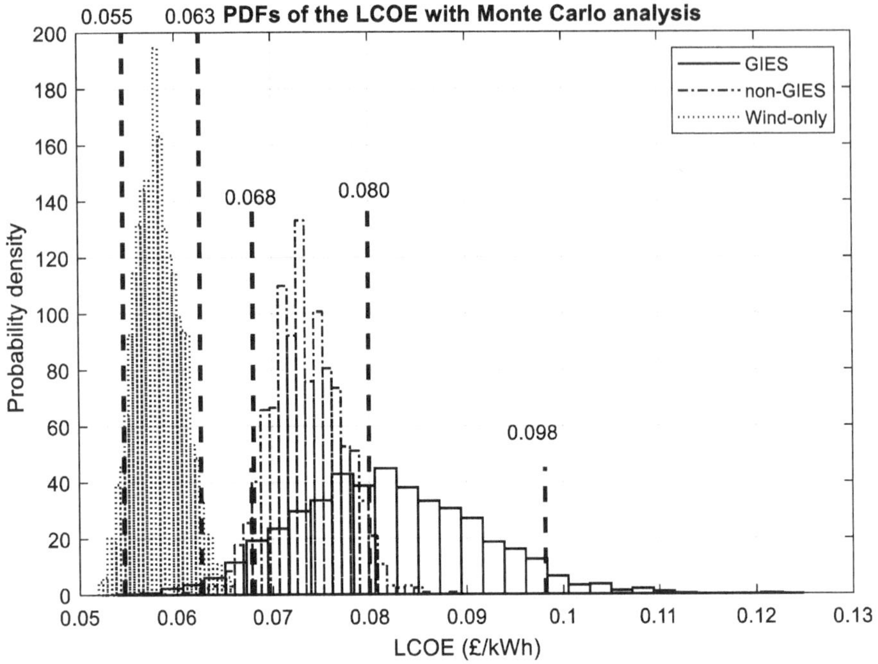

Figure 11.3 PDFs of the LCOE with Monte Carlo analysis.
Reproduced from Lai CS, Locatelli G. Economic and financial appraisal of novel large-scale energy storage technologies. Energy. 2020;214:118954.

11.4 Future techno-economic appraisals of energy storage for smart grids

The advent of energy storage technologies and smart grid applications increases the need for techno-economic appraisals for large-scale energy storage. Present large-scale energy storage techno-economic appraisals generally focus on a few main smart grid applications, including the integration of low carbon generation, energy arbitrage, and off-grid energy systems. The following presents some discussions on the future research agenda:

11.4.1 Electrification of transport and electric vehicles

Current power networks for several countries cannot fulfill the growing energy demand for EVs. *"By 2040, it is estimated that there will be 500 million EVs globally and there will be 36 million"* electric vehicles in the UK [54,55]. Electric vehicles are changing the way citizens commute, consume and interact with energy and affect local and national energy networks [55]. With the changes in demand and supply occurring, it is challenging for congested zones of the grid to balance the increasing demand for prolific electric vehicle charging [55].

The advancement of Internet of Things platforms and vehicle-to-grid charging technologies are expected to facilitate the smart management of assets. The UK Power Networks, a distribution network operator in the UK, is taking part in the world's largest trial of commercial EVs, named Optimize Prime [56]. The project seeks to comprehend and reduce the effect that the electrification of commercial vehicles could bring to distribution networks.

Present research and trials focus on using electric vehicles as decentralized energy storage systems by acting as a load or as an energy provider. Electric vehicles can function in two different modes when connected to the distribution network: [33]

- "Grid-to-Vehicle" operation: EVs are charged by consuming electricity from the grid.
- Vehicle-to-Grid" operation: vehicles are discharged to an electric power grid when parked.

As drivers may respond well to these operations in response to grid prices and signals, the network operation is, however largely uncertain due to customer behavior. A solution is for the distribution network operator to invest and build large-scale energy storage infrastructures to deliver energy for the quick charging demand [55]. By delivering a buffer to relieve the electricity transmission along overloaded lines, large-scale battery storage can facilitate the active management of distribution networks to evade grid overloads [55].

The costs of distributed energy storage options (e.g., with electric vehicles) would be seen as lower than investing in centralized large-scale energy storage, as individual vehicle owners would purchase the cars and batteries. Also, large-scale energy storage requires civil engineering works and a balance of systems for proper operations. However, several benefits can be derived from large-scale energy storage when coupled with electric vehicle charging stations, including maximizing local renewable energy generation, energy arbitrage with the grid, and participating in grid services. Large-scale energy storage will play a key role in enabling the active management of distribution networks to maximize the distribution network performance. Techno-economic appraisal of large-scale energy storage when paired with charging stations in the distribution network can be useful for network operators.

11.4.2 Heating and cooling of the built environment

As data, the internet, and cloud computing play an increasingly important role in technological advancement and social development; there is a need for energy to maintain the growing amount of global data centers. Although data centers are generally energy-efficient, they still require a great amount of electricity from the grid to operate [57,58].

Traditionally, energy storage has played an important role in maintaining the reliability of data centers as an uninterruptable power supply. Oró et al. [59]. examined the potential of integrating direct air-free cooling strategy and thermal energy storage systems into data centers located in different European locations. The cooling electricity cost is being reduced up to 51%, depending on the location. However, the techno-economic appraisal for thermal energy storage is missing.

Also using thermal energy storage, Zheng et al. [60]. developed a peak power shaving framework that uses both uninterruptable power supply (batteries) and thermal energy storage for several data centers. Cooling of data centers is achieved with stored cold water or ice, which generally constitutes up to 40% of the total power cost of a data center. Using thermal energy storage can reduce the capital and operating expenditures of data centers. However, a proper techno-economic study considering the financial dimension is missing.

At present, there are very few projects on large-scale energy storage systems for data centers. However, at the end of 2020 [61,62], it is reported that Google will be investing in large-scale lithium-ion batteries to perform backup power solutions at data centers in St. Ghislain, Belgium. The 3 MW batteries will replace the functionality of a diesel generator for backup power. The project aims to demonstrate that (1) batteries are a feasible substitute for conventional data center backup generators; (2) large-scale energy storage can aid balance the local electricity network. Large-scale energy storage can also integrate greater penetration of renewables when working with data centers. Data centers need to store a great amount of energy on-site for backup. There is a limited option for a diesel generator which only can work for backup power purposes. For the case of large-scale energy storage, and electricity network supplied with renewables can have data centers with energy storage systems connected to it; to store excess renewable energy and discharge it when there is a lack of renewable energy being generated. There is a lack of techno-economic studies for large-scale energy storage systems for data centers, in particular examining various data center operation scenarios

11.4.3 Energy storage for nuclear power

Nuclear power produces low-carbon energy for many applications [63,64]. In a recent study, it was determined that nuclear reactors could help to reduce carbon emissions by using surplus energy to produce hydrogen, heat homes, and decarbonize several industries [63].

In addition, nuclear power provides electricity systems with flexible operation and in particular useful to accommodate high penetrations of intermittent renewables. Cogeneration permits a nuclear plant to change from electricity generation to cogeneration applications, including hydrogen production. Small Modular Nuclear Reactors (SMRs) (i.e. modular reactors generating up to 300 MW) are more suitable for cogeneration than traditional nuclear reactors. Considering their smaller size (therefore, more SMRs are required to reach the same power output), it is feasible to use some of the SMRs for cogeneration, allowing SMRs to operate at full nominal power and highest conversion efficiency [65,66]. SMRs are expected to be economically competitive as compared to traditional nuclear reactors, compensating for the lack of the economy of scale with SMR's peculiar characteristics (e.g., modularisation and modularity) [67]. Small modular reactors also provide greater flexibility in locating stations and operated with greater flexibility to address regional or industrial energy needs.

Presently, techno-economic studies for various types of technically viable and cost-reduced large-scale energy storage (e.g., batteries) still need to be conducted and justified. For example, the Smarter Network Storage project developed by UK Power Networks consists of 50,000 lithium-ion batteries providing 6 MW/10MWh storage capacity [68]. The large battery allows UK Power Networks to manage peak electricity demand without building excess capacity. It would be relevant to see how small modular reactors operate in distribution networks, in particular pairing with large batteries.

11.5 Conclusions

Several energy storage technologies have experienced great technological breakthroughs and cost reductions. Energy storage continues to play a critical role in providing flexibility for smart grid power management. Large-scale energy storage systems can be built to address several key challenges for the deeper decarbonization of society.

Techno-economic appraisal for large-scale energy storage systems is challenging when the appraiser must consider the task from multiple dimensions (e.g., technical, economic, and financing aspects).

Research and development of energy storage have been invested heavily in recent decades with novel technologies, including gravity energy storage and pumped heat energy storage being developed. Batteries technology is also reducing in cost and is being considered for installation by distribution network operators, data center owners, etc. It is worth noting that present techno-economic appraisals focus on a limited number of scenarios, with a strong focus on accommodating high penetration of renewables and electricity arbitrage. Several emerging key functionalities of smart grids have not been addressed in techno-economic appraisals, such as using large-scale energy storage for transport electrification, data center energy management, and pairing with nuclear power plants (e.g., SMRs). A cash flow model is presented for an enhanced techno-economic appraisal, by reflecting the cost of energy storage in a more realistic manner (i.e., financing terms). In addition to the LCOE, the cash flow can also give key indicators, including IRR and NPV. Future work will utilize the cash flow model to appraise large-scale energy storage systems for various smart grid applications.

Acknowledgments

This work is sponsored by the EPSRC grant "GIES: Generation Integrated Energy Storage: A Paradigm Shift" (EP/P022049/1). The authors are grateful to Mr. Benito Mignacca for proofreading the chapter and providing constructive comments. The following two open access articles are reproduced in this chapter: 1. Lai CS, Locatelli G. Economic and financial appraisal of novel large-scale energy storage technologies. Energy. 2020;214:118954. 2. Lai CS, Locatelli G, Pimm A, Wu X, Lai LL. A review on long-term electrical power system modeling with energy storage. Journal of Cleaner Production. 2020:124298.

References

[1] D. Wang, X. Gao, K. Meng, J. Qiu, L.L. Lai, S. Gao, Utilisation of kinetic energy from wind turbine for grid connections: a review paper, IET Renewable Power Generation 12 (6) (2018) 615–624.

[2] C.S. Lai, Y. Jia, L.L. Lai, Z. Xu, M.D. McCulloch, K.P. Wong, A comprehensive review on large-scale photovoltaic system with applications of electrical energy storage, Renewable and Sustainable Energy Reviews 78 (2017) 439–451.

[3] G. Locatelli, T. Sainati, B. Mignacca, Developing UK strategy for nuclear SMRs, Policy Leeds (2021). Brief No. 6.

[4] Y. Jia, Y. Gao, Z. Xu, K.P. Wong, L.L. Lai, Y. Xue, et al., Powering China's sustainable development with renewable energies: Current status and future trend, Electric Power Components and Systems 43 (8–10) (2015) 1193–1204.

[5] D. Kucevic, S. Englberger, A. Sharma, A. Trivedi, B. Tepe, B. Schachler, et al., Reducing grid peak load through the coordinated control of battery energy storage systems located at electric vehicle charging parks, Applied Energy 295 (2021) 116936.

[6] J. Zheng, C.S. Lai, H. Yuan, Z.Y. Dong, K. Meng, L.L. Lai, Electricity plan recommender system with electrical instruction-based recovery, Energy. 203 (2020) 117775.

[7] C.S. Lai, G. Locatelli, Are energy policies for supporting low-carbon power generation killing energy storage? Journal of Cleaner Production (2021).

[8] V. Jülch, Comparison of electricity storage options using levelized cost of storage (LCOS) method, Applied Energy 183 (2016) 1594–1606.

[9] C.S. Lai, M.D. McCulloch, Levelized cost of electricity for solar photovoltaic and electrical energy storage, Applied Energy 190 (2017) 191–203.

[10] G. Locatelli, E. Palerma, M. Mancini, Assessing the economics of large Energy Storage Plants with an optimisation methodology, Energy. 83 (2015) 15–28.

[11] A.S. Mundada, K.K. Shah, J. Pearce, Levelized cost of electricity for solar photovoltaic, battery and cogen hybrid systems, Renewable and Sustainable Energy Reviews 57 (2016) 692–703.

[12] N. Singh, E.W. McFarland, Levelized cost of energy and sensitivity analysis for the hydrogen–bromine flow battery, Journal of Power Sources 288 (2015) 187–198.

[13] M. Obi, S. Jensen, J.B. Ferris, R.B. Bass, Calculation of levelized costs of electricity for various electrical energy storage systems, Renewable and Sustainable Energy Reviews 67 (2017) 908–920.

[14] C.S. Lai, Y. Jia, Z. Xu, L.L. Lai, X. Li, J. Cao, et al., Levelized cost of electricity for photovoltaic/biogas power plant hybrid system with electrical energy storage degradation costs, Energy Conversion and Management 153 (2017) 34–47.

[15] C.S. Lai, G. Locatelli, Economic and financial appraisal of novel large-scale energy storage technologies, Energy. 214 (2020) 118954.

[16] A. Berrada, K. Loudiyi, I. Zorkani, Profitability, risk, and financial modeling of energy storage in residential and large scale applications, Energy. 119 (2017) 94–109.

[17] G. Locatelli, D.C. Invernizzi, M. Mancini, Investment and risk appraisal in energy storage systems: A real options approach, Energy. 104 (2016) 114–131.

[18] S.M. Alirahmi, S.B. Mousavi, A.R. Razmi, P. Ahmadi, A comprehensive techno-economic analysis and multi-criteria optimization of a compressed air energy storage (CAES) hybridized with solar and desalination units, Energy Conversion and Management 236 (2021) 114053.

[19] L. Miller, R. Carriveau, A review of energy storage financing—Learning from and partnering with the renewable energy industry, Journal of Energy Storage 19 (2018) 311–319.

[20] R. Dufo-López, J.L. Bernal-Agustín, Techno-economic analysis of grid-connected battery storage, Energy Conversion and Management 91 (2015) 394–404.

[21] Avendano-Mora M, Camm EH, editors. Financial assessment of battery energy storage systems for frequency regulation service. Power & Energy Society General Meeting, 2015 IEEE; 2015: IEEE.

[22] S. Xia, K. Chan, X. Luo, S. Bu, Z. Ding, B. Zhou, Optimal sizing of energy storage system and its cost-benefit analysis for power grid planning with intermittent wind generation, Renewable Energy 122 (2018) 472–486.

[23] S. Kapila, A.O. Oni, A. Kumar, The development of techno-economic models for large-scale energy storage systems, Energy. 140 (2017) 656–672.

[24] A. Zurita, C. Mata-Torres, C. Valenzuela, C. Felbol, J.M. Cardemil, A.M. Guzmán, et al., Techno-economic evaluation of a hybrid CSP + PV plant integrated with thermal energy storage and a large-scale battery energy storage system for base generation, Solar Energy 173 (2018) 1262–1277.

[25] B. Olateju, A. Kumar, M. Secanell, A techno-economic assessment of large scale wind-hydrogen production with energy storage in Western Canada, Int. J. Hydrogen Energy 41 (21) (2016) 8755–8776.

[26] O. Pupo-Roncallo, D. Ingham, M. Pourkashanian, Techno-economic benefits of grid-scale energy storage in future energy systems, Energy Reports 6 (2020) 242–248.

[27] S.D. Garvey, P. Eames, J. Wang, A. Pimm, M. Waterson, R. MacKay, et al., On generation-integrated energy storage, Energy Policy 86 (2015) 544–551.

[28] M.L. Tuballa, M.L. Abundo, A review of the development of Smart Grid technologies, Renewable and Sustainable Energy Reviews 59 (2016) 710–725.

[29] C.S. Lai, Y. Jia, Z. Dong, D. Wang, Y. Tao, Q.H. Lai, et al., A review of technical standards for smart cities, Clean Technologies 2 (3) (2020) 290–310.

[30] The Smart Grid. SmartGrid.gov, U.S. Department of Energy, [Online]. Available: https://www.smartgrid.gov/the_smart_grid/smart_grid.html.

[31] R. He, H. Xie, J. Deng, T. Feng, L.L. Lai, M. Shahidehpour, Reliability Modeling and Assessment of Cyber Space in Cyber-Physical Power Systems, IEEE Transactions on Smart Grid 11 (5) (2020) 3763–3773.

[32] C.S. Lai, G. Locatelli, A. Pimm, X. Wu, L.L. Lai, A review on long-term electrical power system modeling with energy storage, Journal of Cleaner Production (2020) 124298.

[33] D. Kolokotsa, N. Kampelis, A. Mavrigiannaki, M. Gentilozzi, F. Paredes, F. Montagnino, et al., On the integration of the energy storage in smart grids: Technologies and applications, Energy Storage 1 (1) (2019) e50.

[34] D. Yang, M. Wang, R. Yang, Y. Zheng, H. Pandzic, Optimal dispatching of an energy system with integrated compressed air energy storage and demand response, Energy. 234 (2021) 121232.

[35] F. Xu, W. Zhu, Y.F. Wang, C.S. Lai, H. Yuan, Y. Zhao, et al., A new deregulated demand response scheme for load over-shifting city in regulated power market, Applied Energy 311 (2022) 118337.

[36] X. Luo, J. Wang, M. Dooner, J. Clarke, Overview of current development in electrical energy storage technologies and the application potential in power system operation, Applied Energy 137 (2015) 511–536.
[37] R. Amirante, E. Cassone, E. Distaso, P. Tamburrano, Overview on recent developments in energy storage: Mechanical, electrochemical and hydrogen technologies, Energy Conversion and Management 132 (2017) 372–387.
[38] C.S. Lai, G. Locatelli, A. Pimm, Y. Tao, X. Li, L.L. Lai, A financial model for lithium-ion storage in a photovoltaic and biogas energy system, Applied Energy 251 (2019) 113179.
[39] A.G. Olabi, M.A. Abdelkareem, Energy Storage Systems Towards 2050, Elsevier, 2020.
[40] A. Dhand, K. Pullen, Analysis of continuously variable transmission for flywheel energy storage systems in vehicular application, Proceedings of the Institution of Mechanical Engineers, Part C: Journal of Mechanical Engineering Science 229 (2) (2015) 273–290.
[41] L. Mangialardi, G. Mantriota, The advantages of using continuously variable transmissions in wind power systems, Renewable Energy 2 (3) (1992) 201–209.
[42] A.J. Pimm, S.D. Garvey, R. Drew, Shape and cost analysis of pressurized fabric structures for subsea compressed air energy storage, Proceedings of the Institution of Mechanical Engineers, Part C: Journal of Mechanical Engineering Science 225 (5) (2011) 1027–1043.
[43] A. Berrada, K. Loudiyi, I. Zorkani, System design and economic performance of gravity energy storage, Journal of Cleaner Production 156 (2017) 317–326.
[44] A. Benato, A. Stoppato, Pumped Thermal Electricity Storage: a technology overview, Thermal Science and Engineering Progress (2018).
[45] A. Benato, Performance and cost evaluation of an innovative Pumped Thermal Electricity Storage power system, Energy. 138 (2017) 419–436.
[46] A. Smallbone, V. Jülch, R. Wardle, A.P. Roskilly, Levelised Cost of Storage for Pumped Heat Energy Storage in comparison with other energy storage technologies, Energy Conversion and Management 152 (2017) 221–228.
[47] Ahmad J, Imran M, Khalid A, Iqbal W, Ashraf SR, Adnan M, et al. Techno economic analysis of a wind-photovoltaic-biomass hybrid renewable energy system for rural electrification: A case study of Kallar Kahar. 2018;148:208-34.
[48] Shahzad MK, Zahid A, ur Rashid T, Rehan MA, Ali M, Ahmad MJRE. Techno-economic feasibility analysis of a solar-biomass off grid system for the electrification of remote rural areas in Pakistan using HOMER software. 2017;106:264-73.
[49] Das BK, Hoque N, Mandal S, Pal TK, Raihan MAJE. A techno-economic feasibility of a stand-alone hybrid power generation for remote area application in Bangladesh. 2017;134:775-88.
[50] T. Sainati, G. Locatelli, N. Smith, Project financing in nuclear new build, why not? The legal and regulatory barriers, Energy Policy 129 (2019) 111–119.
[51] Policy paper contracts for difference. Department for Business, Energy & Industrial Strategy, Jan. 2019. [Online]. Available: https://www.gov.uk/government/publications/contracts-for-difference/contract-for-difference.
[52] Short term operating reserve (STOR). nationalgridESO, [Online]. Available: https://www.nationalgrideso.com/balancing-services/reserve-services/short-term-operating-reserve-stor.
[53] Fast reserve. nationalgridESO, [Online]. Available: https://www.nationalgrideso.com/balancing-services/reserve-services/fast-reserve.
[54] Briefing paper number CBP07480, Electric vehicles and infrastructure. David Hirst, House of Commons Library, 4th December 2020, Available: https://commonslibrary.parliament.uk/research-briefings/cbp-7480/.

[55] Andersson J. How large-scale storage will facilitate the EV revolution. Energy Storage News, [Online]. Available: https://www.energy-storage.news/blogs/how-large-scale-storage-will-facilitate-the-ev-revolution; 2020.

[56] Optimise Prime. UK Power Networks, Available: https://innovation.ukpowernetworks.co.uk/projects/optimise-prime/.

[57] C. Koronen, M. Åhman, L.J. Nilsson, Data centers in future European energy systems —energy efficiency, integration and policy, Energy Efficiency 13 (1) (2020) 129−144.

[58] N. Jones, How to stop data centers from gobbling up the world's electricity, Nature 561 (7722) (2018) 163−167.

[59] E. Oró, V. Depoorter, N. Pflugradt, J. Salom, Overview of direct air free cooling and thermal energy storage potential energy savings in data centers, Applied thermal engineering 85 (2015) 100−110.

[60] Zheng W, Ma K, Wang X, editors. Exploiting thermal energy storage to reduce data center capital and operating expenses. 2014 IEEE 20th International Symposium on High Performance Computer Architecture (HPCA); 2014 2014: IEEE.

[61] Google says batteries are a 'multi-talented' option for data centers. Andy Colthorpe, Energy Storage News, 18 Dec 2020. [Online]. Available: https://www.energy-storage.news/news/google-says-batteries-are-a-multi-talented-option-for-data-centers.

[62] Google looks to batteries as replacement for diesel generators. Rich Miller, Data Center Frontier, 16 Dec 2020. [Online]. Available: https://datacenterfrontier.com/google-looks-to-batteries-as-replacement-for-diesel-generators/.

[63] Nuclear cogeneration: Civil nuclear energy in a low-carbon future policy briefing. The Royal Society, Oct. 2020. Available: https://royalsociety.org/-/media/policy/projects/nuclear-cogeneration/2020-10-7-nuclear-cogeneration-policy-briefing.pdf.

[64] G. Locatelli, S. Boarin, A. Fiordaliso, M.E. Ricotti, Load following of Small Modular Reactors (SMR) by cogeneration of hydrogen: A techno-economic analysis, Energy. 148 (2018) 494−505.

[65] G. Locatelli, A. Fiordaliso, S. Boarin, M.E. Ricotti, Cogeneration: An option to facilitate load following in Small Modular Reactors, Progress in Nuclear Energy 97 (2017) 153−161.

[66] B. Mignacca, G. Locatelli, T. Sainati, Deeds not words: Barriers and remedies for Small Modular nuclear Reactors, Energy. 206 (2020) 118137.

[67] B. Mignacca, G. Locatelli, Economics and finance of Small Modular Reactors: A systematic review and research agenda, Renewable and Sustainable Energy Reviews 118 (2020) 109519.

[68] Smarter Network Storage (SNS). UK Power Networks. Available: https://innovation.ukpowernetworks.co.uk/projects/smarter-network-storage-sns/.

Battery energy storage systems in microgrids

12

E.S.N. Raju P[1] *and Trapti Jain*[2]
[1]School of Energy Science & Engineering, IIT Guwahati, Guwahati, Assam, India,
[2]Discipline of Electrical Engineering, IIT Indore, Indore, Madhya Pradesh, India

Chapter Outline

Highlights

- Utilization of a battery energy storage system in islanded AC microgrids.
- Dynamic performance enhancement of islanded AC microgrids against large load disturbances.
- Mode control algorithm for the load-leveling application.
- A dynamic model of islanded AC microgrids with battery energy storage system and static active as well as passive and dynamic induction motor loads.
- Time-domain simulations under MATLAB®/Simulink® environment.
- State-of-charge in the range of a safe charging status, that is, $20\% < \text{SoC} \leq 80\%$.
- Harnessing maximum power from inverter-interfaced distributed generation units and avoiding their overloading.

Nomenclature

ACMGs	AC microgrids
CPL	Constant power load
ESSs	Energy storage systems
IIDG	Inverter-interfaced distributed generation
MMS	Microgrid management system
R	Resistive

Emerging Trends in Energy Storage Systems and Industrial Applications. DOI: https://doi.org/10.1016/B978-0-323-90521-3.00009-0

RIAL	Rectifier interfaced active load
RMS	Root mean square
SoC	State-of-charge
BESS	Battery energy storage system
DQ	Synchronous reference frame
IACMGs	Islanded AC microgrids
IM	Induction motor
PLL	Phase-locked loop
RESs	Renewable energy sources
RL	Inductive
SMES	Superconducting magnetic energy storage
VSI	Voltage source inverter

12.1 Introduction

Energy storage systems (ESSs) are commonly implemented as the energy buffers in AC microgrids (ACMGs) due to the uncertain behavior of renewable energy sources (RESs) based on inverter-interfaced distributed generation (IIDG) units [1]. Furthermore, ESS is one of the most desirable solutions to maintain the power balance, improve stability, and tackle both the frequency as well as voltage variations in ACMGs [2]. Depending on the forms of energy used, ESS technologies can be classified as chemical energy storage, mechanical energy storage, electrical energy storage, and thermal and electrochemical energy storage [3]. ESSs can be utilized in ACMGs for various applications such as energy arbitrage, peak shaving, load leveling, spinning reserve, voltage support, black start, frequency regulation, power quality, power reliability, time shifting, and capacity firming [4]. However, this chapter has considered the application of ESSs for load leveling, wherein, electrochemical energy storage has been preferred due to their high power and energy densities. The load leveling application involves storing the surplus power during non-disturbance periods and delivering it during periods of disturbance.

ESSs can be configured in islanded AC microgrids (IACMGs) as aggregated, distributed, and a hybrid of the two. In an aggregated configuration, all ESSs are combined as a single ESS. This configuration helps to smooth out the power fluctuations at the bus to which ESS is connected [5]. Due to this, in this chapter, the aggregated ESS has been considered to have a better response to compensate for deviations in voltage and frequency than the distributed ESS. The main drawback of this configuration is that the cost of manufacturing rises tremendously with the increasing size of the ESS. In a distributed configuration, ESSs are dispatched in a distributed way, that is, ESSs are connected directly to the individual IIDG units through a power electronic interface. The main advantages of distributed ESSs are low cost, higher efficiency, and simplicity. However, this configuration still suffers from power losses in both the power electronic interface as well as line impedance between ESS and the connected IIDG unit [5]. To optimize the total cost and efficiency of the ACMG system, a hybrid configuration of ESSs, which is a combination of both aggregated as well as distributed configurations, could be a possible solution [6]. Control of ESSs can be classified based on the configuration of ESSs and their applications. Based on their applications, control of ESSs can be categorized as PQ control, conventional droop control and state-of-charge (SoC) weighted droop control [7,8]. Whereas, control strategies for IACMGs with distributed ESSs can be broadly classified, based on their architecture, into three categories: decentralized, centralized, and distributed multi-agent [9].

The dynamic performance of IACMGs under the small-signal disturbances has been enhanced with the help of decentralized, centralized, distributed, and hierarchical controllers in the current literature [10–15]. However, apart from the small-signal disturbances, the IACMG system may be subjected to large load disturbances as well. When the IACMG system is subjected to large load disturbances, the controllers based on linearized models are generally not fully reliable. The reason for this lies in the fact that the large load disturbance reduces the accuracy of the linearized models based on the conventional Taylor series expansion method [16]. Therefore the utilization of energy storage viz. battery, superconducting magnetic energy storage (SMES) [17], hydrogen [3,4], and supercapacitor [18], etc., are usually required in the IACMG system for its performance enhancement against large load disturbances [19]. In this chapter, the battery energy storage system (BESS) has been used due to its advantages in terms of low capital cost, high energy density, and mature technology [3,4].

A cooperative control strategy of IIDG units and the ESS was proposed for stabilizing IACMGs [7]. The proposed control strategy has a two-layered cooperative control structure: the primary control action in the ESS and the secondary control action in the microgrid management system (MMS). The constant frequency and constant voltage control method were applied in the primary control of the ESS, while, the secondary control of MMS regulates the power output of the ESS to be zero. A novel control strategy was proposed for improving the stability of IACMGs in different modes of operation [8]. Aggregated BESS was used to balance load demands and power generation in IACMGs and to address the slow response problem of IIDG units to load changes. Furthermore, droop control was adopted in the primary level control to enhance cooperation between power electronic inverters and to improve load dispatch and voltage regulation. However, they remain limited since the simultaneous connection of multiple types of loads combing various static active as well as passive and dynamic induction motor (IM) loads has not been considered.

Given the above observations, this chapter proposes the utilization of an aggregated BESS for the dynamic performance enhancement of IACMGs, feeding both the static active as well as passive and dynamic IM loads, against large load disturbances. A mode control algorithm has been developed to control the charging and discharging modes of the aggregated BESS for maintaining the SoC of the battery in the range of safe charging status, that is, $20\% < \mathrm{SoC} \leq 80\%$. The aggregated BESS model includes a battery, bi-directional DC/DC converter and its local control, DC-link interface, and bi-directional AC/DC converter and its local control To evaluate the utilization of BESS and its mode control algorithm under different SoCs of the battery as well as different disturbance locations, time-domain simulations have been performed for the studied IACMG system, shown in Fig. 12.1, with the aggregated BESS and multiple types of static and dynamic loads.

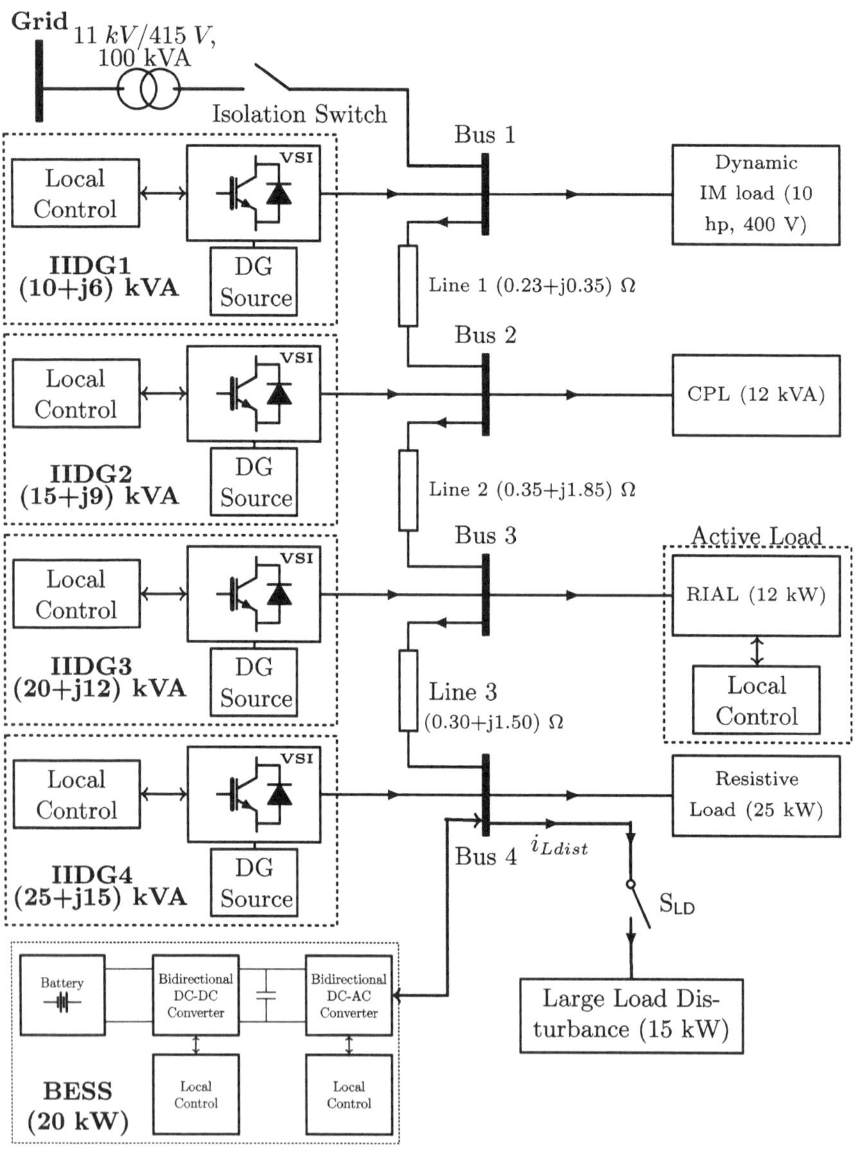

Figure 12.1 Schematic diagram of the studied microgrid system.

12.2 Dynamic model of an IACMG system with BESS and static and dynamic loads

Fig. 12.1 shows a schematic diagram of a studied IACMG system operating at a frequency of 50 Hz and a voltage of 230 V (per phase RMS). The IACMG system includes four IIDG units, three lines, and locally connected loads viz. resistive (R)/inductive (RL), constant power load (CPL), rectifier interfaced active load (RIAL), and dynamic induction motor (IM) load. Each IIDG unit is represented by a DG source, which may be a photo-voltaic, micro-turbine, or a fuel cell generator, etc., a voltage source inverter (VSI), and a local controller, which is composed of power-sharing, voltage, and current controllers. The microgrid is interfaced to the main utility grid bus through an isolation switch and a 415 V/11 kV step-up transformer at its point of common coupling, which is bus-1.

The IACMG can be operated either in an islanded mode or a grid-connected mode based on the status of the isolation switch. The islanded mode of operation is realized by opening the isolating switch, which disconnects the microgrid from the main grid, as shown in Fig. 12.1. In the island mode of operation, IIDG units are responsible for maintaining the system frequency and voltage along with meeting the total power demand. The parameters of the studied IACMG system are given in the Appendix. The inverter parameters are referred from Refs. [20–23], while, the RIAL parameters are referred from Refs. [24,25]. The parameters of 400 V, 10 hp, and 7.5 kW dynamic induction motor load have been referred from Ref. [16].

The complete dynamical non-linear model of the studied IACMG system with the BESS has been developed in a synchronously rotating frame (DQ) from the generalized modeling given in [23]. This modeling divides the whole IACMG system into four sub-modules, that is, generation sub-module, network sub-module, load sub-modules (passive, active, and dynamic), and BESS sub-module, virtually. The transformation used to convert from the ABC reference frame to the DQ reference frame is a variant of the Park–Clark transformation [26], as given in Eq. (12.1). It is the power invariant and the *D*-axis is plotted on the horizontal axis of a two-axis plot.

$$[T_{dq}] = \sqrt{\frac{2}{3}} \begin{bmatrix} \cos(\omega t + \theta) & \cos\left(\omega t + \theta - \frac{2\pi}{3}\right) & \cos\left(\omega t + \theta + \frac{2\pi}{3}\right) \\ -\sin(\omega t + \theta) & -\sin\left(\omega t + \theta - \frac{2\pi}{3}\right) & -\sin\left(\omega t + \theta + \frac{2\pi}{3}\right) \\ \frac{1}{\sqrt{2}} & \frac{1}{\sqrt{2}} & \frac{1}{\sqrt{2}} \end{bmatrix} \tag{12.1}$$

The generation sub-module involves the modeling of each IIDG unit on its individual local DQ reference frame. The IIDG unit model consists of power processing and local control sections, as shown in Fig. 12.2 [22,23]. The power processing section includes a three-leg VSI, an output LC filter, and a coupling inductor (L_c), while, the control section can be divided into the outer power-sharing controller and inner voltage and current controllers. The dynamics of these power processing and control sections of the IIDG unit lead to 13th-an older model, as given in Eqs. (12.2) and (12.3) [20–25]. It is worth mentioning that the DC bus dynamics of IIDG units are neglected with the assumption that DC bus voltage fluctuations are suppressed by the generator side converter which feeds the VSI DC bus [20]. Furthermore, the switching process of the VSI has also been neglected with the realization of high switching frequencies (4–10 kHz) [20].

$$\frac{d([\mathbf{x_{IIDGi}}]_{13\times1})}{dt} = [\mathbf{A_{IIDGi}}]_{13\times13}[\mathbf{x_{IIDGi}}]_{13\times1} + [\mathbf{B_{IIDGi}}]_{13\times2}[\mathbf{v_{bDQi}}]_{2\times1} + [\mathbf{B_{iwcom}}]_{13\times1}[w_{com}]_{1\times1} + [\mathbf{B_{consti}}]_{13\times1} \quad \forall i \tag{12.2}$$

$$\begin{bmatrix} w_i \\ \mathbf{i_{oDQi}} \end{bmatrix}_{3\times1} = \begin{bmatrix} [\mathbf{C_{IIDGwi}}]_{1\times13} \\ [\mathbf{C_{IIDGi}}]_{2\times13} \end{bmatrix}_{3\times13} [\mathbf{x_{IIDGi}}]_{13\times1} \quad \forall i \tag{12.3}$$

where $\mathbf{A_{IIDGi}}$ represents the state matrix of the ith IIDG unit, $\mathbf{B_{IIDG_i}}$, $\mathbf{B_{iwcom}}$, and $\mathbf{B_{consti}}$ are input matrices of the ith IIDG unit corresponding to the two voltage inputs, the frequency input, and constant, respectively. $\mathbf{C_{IIDGi}}$ and $\mathbf{C_{IIDGwi}}$ are output matrices of the ith IIDG unit corresponding to the two output currents $\mathbf{i_{oDQ}}$ and frequency ω_i, respectively. Details of these matrices can be found in Refs. [20–25]. In Eqs. (12.2) and (12.3), the state vector, $\mathbf{x_{IIDGi}}$, which includes the state-space models of the outer power-sharing controller, inner voltage and current controllers, and output LCL filter, is given by a 13-element vector.

$$\mathbf{x_{IIDGi}} = \begin{bmatrix} \delta_i & P_i & Q_i & \phi_{\mathbf{dqi}} \\ \gamma_{\mathbf{dqi}} & \mathbf{i_{ldqi}} & \mathbf{v_{odqi}} & \mathbf{i_{odqi}} \end{bmatrix}^T \quad \forall i \tag{12.4}$$

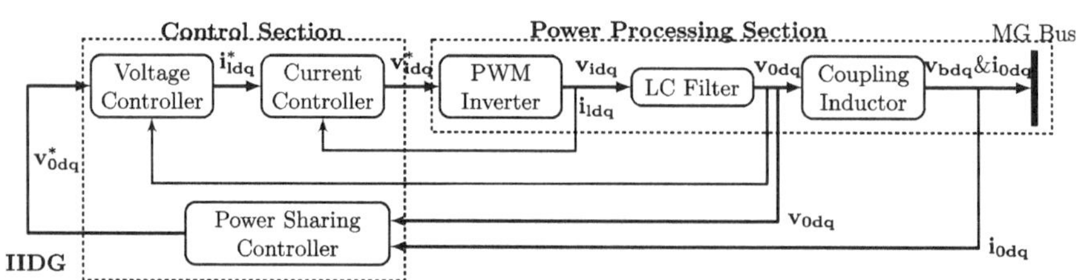

Figure 12.2 Power processing and control sections of IIDG unit. *IIDG*, Inverter-interfaced distributed generation.

The network sub-module includes the modeling of each line, while the passive load sub-module involves the modeling of passive loads on their common DQ reference frame. The interconnected lines and passive loads (R/RL and CPL) have been represented, individually, by a second-order model [20–25]. The dynamical equations of ith the line connected between jth kth nodes are given in Eqs. (12.5) and (12.6).

$$\frac{\mathrm{d}i_{lineDi}}{\mathrm{d}t} = \frac{-R_{linei}}{L_{linei}} i_{lineDi} + \frac{1}{L_{linei}}(v_{bDj} - v_{bDk}) + \omega i_{lineQi} \quad \forall\, i \tag{12.5}$$

$$\frac{\mathrm{d}i_{lineQi}}{\mathrm{d}t} = \frac{-R_{linei}}{L_{linei}} i_{lineQi} + \frac{1}{L_{linei}}(v_{bQj} - v_{bQk}) - \omega i_{lineDi} \quad \forall\, i \tag{12.6}$$

The dynamical equations of ith passive load connected at n_N^{th} bus or node are given in Eqs. (12.7) and (12.8).

$$\frac{\mathrm{d}i_{PLDi}}{\mathrm{d}t} = \frac{-R_{PLi}}{L_{PLi}} i_{PLDi} + \frac{1}{L_{PLi}} v_{bDn_N} + \omega i_{PLQi} \quad \forall\, i \tag{12.7}$$

$$\frac{\mathrm{d}i_{PLQi}}{dt} = \frac{-R_{PLi}}{L_{PLi}} i_{PLQi} + \frac{1}{L_{PLi}} v_{bQn_N} - \omega i_{PLDi} \quad \forall\, i \tag{12.8}$$

where subscript PL will be replaced by $Rload$, $RLload$, and CPL for passive R, RL, and CPL loads, respectively. Here, $R_{CPL} = -r_{CPL}cos\alpha$ and $X_{CPL} = -r_{CPL}\sin\alpha$, where α is the delay angle of a controlled rectifier.

The active load sub-module involves the modeling of each RIAL on its individual local dq reference frame. The RIAL model includes dynamics of DC voltage as well as AC controllers, output LC filter, and coupling inductor, as shown in Fig. 12.3, leading to 10th-order model [24,25]. The dynamical model of a ith RIAL connected at the n_N^{th} node, in a common reference frame, is given in Eqs. (12.9) and (12.10)

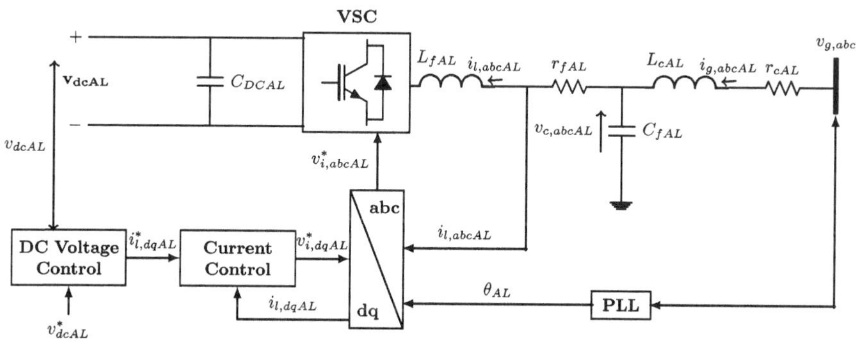

Figure 12.3 Rectifier interfaced active load circuit and its control.

$$\frac{\mathrm{d}\left([\mathbf{x_{ALi}}]_{10\times 1}\right)}{\mathrm{d}t} = [\mathbf{A_{ALi}}]_{10\times 10}[\mathbf{x_{ALi}}]_{10\times 1} + [\mathbf{B_{ALvi}}]_{10\times 2}\left[\mathbf{v_{bDQn_N}}\right]_{2\times 1} + [\mathbf{B_{ALui}}]_{10\times 2}[\mathbf{u_{ALi}}]_{2\times 1} + [\mathbf{B_{ALwi}}]_{10\times 1}[w_{com}]_{1\times 1} \qquad \forall i \tag{12.9}$$

$$\begin{bmatrix} \mathbf{i_{DQALi}} \\ v_{DCALi} \end{bmatrix}_{3\times 1} = \left[\mathbf{C_{ALi}}\right]_{3\times 10}[\mathbf{x_{ALi}}]_{10\times 1} \qquad \forall i \tag{12.10}$$

where $\mathbf{A_{ALi}}$, the state matrix of ith RIAL, $\mathbf{B_{ALvi}}$, $\mathbf{B_{ALui}}$ and $\mathbf{B_{ALwi}}$ are input matrices of ith RIAL corresponding to the voltage, control, and frequency inputs, respectively [24,25]. $\mathbf{C_{ALi}}$ is the output matrix of ith RIAL corresponding to the currents and DC voltage. In Eqs. (12.9) and (12.10),

$$\mathbf{x_{ALi}} = \begin{bmatrix} \phi_{ALi} & \gamma_{\mathbf{dqALi}} & \mathbf{i_{ldqALi}} \\ \mathbf{v_{cdqALi}} & \mathbf{i_{gdqALi}} & v_{dcALi} \end{bmatrix}^T \qquad \forall i \tag{12.11}$$

$$\mathbf{u_{ALi}} = \left[v^*_{dcALi} \quad i^*_{lqALi} \right]^T \qquad \forall i \tag{12.12}$$

The dynamic load sub-module involves the modeling of each dynamic IM load on the common DQ reference frame. The stator, rotor, and electromagnetic as well as load torque dynamics of a dynamic IM load, given in Eqs. (12.13)–(12.18), lead to fifth-order model [23,26].

$$v_{Qsi} = r_{si}i_{Qsi} + L_{ssi}\frac{\mathrm{d}i_{Qsi}}{\mathrm{d}t} + L_{mi}\frac{\mathrm{d}i_{Qri}}{\mathrm{d}t} + \omega L_{ssi}i_{Dsi} + \omega L_{mi}i_{Dri} \qquad \forall i \tag{12.13}$$

$$v_{Dsi} = r_{si}i_{Dsi} + L_{ssi}\frac{\mathrm{d}i_{Dsi}}{\mathrm{d}t} + L_{mi}\frac{\mathrm{d}i_{Dri}}{\mathrm{d}t} - \omega L_{ssi}i_{Qsi} - \omega L_{mi}i_{Qri} \qquad \forall i \tag{12.14}$$

$$v_{Qri} = r_{si}i_{Qri} + L_{rri}\frac{\mathrm{d}i_{Qri}}{\mathrm{d}t} + L_{mi}\frac{\mathrm{d}i_{Qsi}}{\mathrm{d}t} + s_i\omega L_{rri}i_{Dri} + s_i\omega L_{mi}i_{Dsi} \qquad \forall i \tag{12.15}$$

$$v_{Dri} = r_{ri}i_{Dri} + L_{rri}\frac{\mathrm{d}i_{Dri}}{\mathrm{d}t} + L_{mi}\frac{\mathrm{d}i_{Dsi}}{\mathrm{d}t} - s_i\omega L_{rri}i_{Qri} - s_i\omega L_{mi}i_{Qsi} \qquad \forall i \tag{12.16}$$

$$T_{ei} = \frac{3}{2}\frac{P_i}{2}L_{mi}(i_{Qsi}i_{Dri} - i_{Dsi}i_{Qri}) \qquad \forall i \tag{12.17}$$

$$T_{ei} - T_{Li} = J_i\frac{\mathrm{d}((1 - s_i)\omega_i)}{\mathrm{d}t} \qquad \forall i \tag{12.18}$$

where r_{si} & L_{ssi} and r_{ri} & L_{rri} are the resistance & inductance of stator and rotor of the ith IM load, respectively, whereas, L_{mi}, s_i, and ω_i are the linkage inductance, rotor slip, and the stator supply angular frequency, respectively. In Eqs. (12.17) and

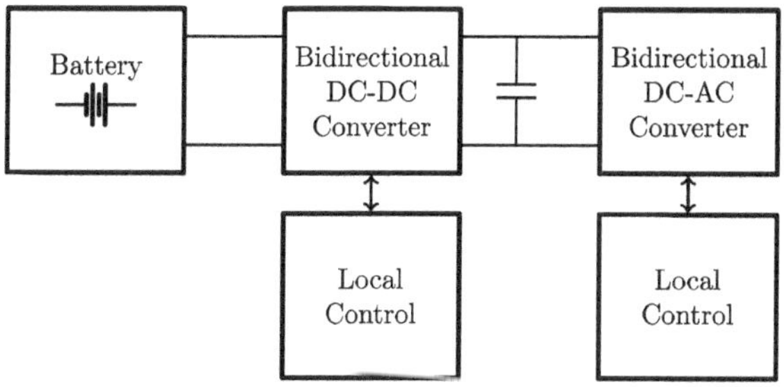

Figure 12.4 BESS model. *BESS*, Battery energy storage system.

(12.18), P_i, T_{Li} and J_i are the number of poles, load torque, and combined motor and load inertia, respectively.

The BESS sub-module involves the modeling of the BESS model. Fig. 12.4 shows the configuration of the BESS model including battery, bidirectional DC/DC converter along with its local control, DC-link interface, and bidirectional AC/DC converter with its local control. The most popular types of rechargeable battery models are lead-acid, lithium-ion (Li-ion), nickel-cadmium (NiCd), and nickel-metal hydride (NiMH) [27]. In this chapter, the Li-ion model, shown below, has been used due to the reason that it offers the best solution for high-power and high-energy applications. However, the choice of battery technology does not affect other aspects of the mode control algorithm.

$$\text{Discharge: } V_{bat} = E_0 - R\cdot i_{bat} - K\frac{Q}{Q-it}\cdot(it + i*) + A\exp(-B\cdot it)$$

$$\text{Charge: } V_{bat} = E_0 - R\cdot i_{bat} - K\frac{Q}{it-0.1Q}\cdot i* - K\frac{Q}{Q-it}\cdot it + A\exp(-B\cdot it)$$

where V_{bat} is the battery voltage (V); E_0 is battery constant voltage (V); K is polarization constant (Ah^{-1}) or polarization resistance (Ω); Q is maximum battery capacity (Ah); $it(=\int idt)$ is actual battery charge (Ah); i^* is filtered current (A); $V_{exp}(t)$ is exponential zone voltage (V); A is exponential zone amplitude (V); B is exponential zone time constant inverse (Ah^{-1}). Here, constants, A and B, can be calculated from the typical discharge characteristic of a Li-ion battery [27].

Fig. 12.5A shows the schematic diagram of the bidirectional DC/DC converter [28–30], utilized for interfacing the battery to the DC link. The low-voltage side of the bidirectional DC/DC converter is connected to the battery, while, its high-voltage side is connected to the DC link. The converter consists of an energy-storage inductor, L_S, energy-storage capacitors, C_{DC1} and C_{DC2}, and two switches, S_1 and S_2 [28–30]. These switches are operated in a complementary manner to enable the feature of the bidirectional power flow of the converter. It can be operated in both the boost as well as buck mode. In boost mode of operation, the switch S_1 acts as a diode, while, S_2 acts as a switch and the power flows from the battery

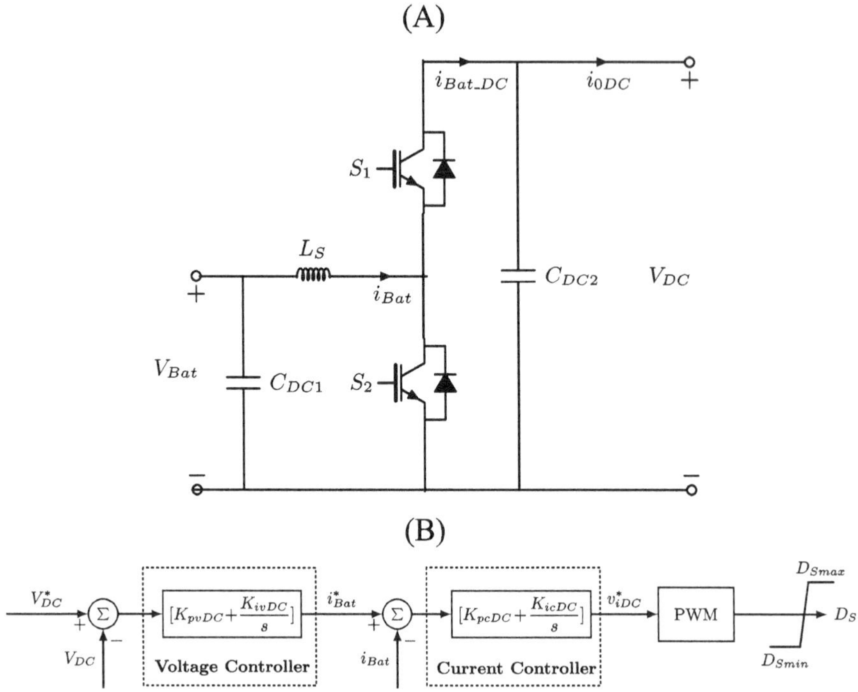

Figure 12.5 Bidirectional DC/DC converter of BESS. (A) Schematic diagram. (B) Local control block diagram. *BESS*, Battery energy storage system.

to the DC link i.e., discharging mode. Fig. 12.6 shows the dynamic average-value model of the bidirectional DC/DC converter under boost mode of operation. The dynamic equations for the boost mode of operation are given in Eqs. (12.19)–(12.21).

$$\frac{di_{Bat}}{dt} = -\frac{R_S}{L_S} i_{Bat} + \frac{V_{Bat}}{L_S} - (1 - D_S)\frac{V_{DC}}{L_S} \tag{12.19}$$

$$\frac{di_{Bat\_DC}}{dt} = (1 - D_S) i_{Bat} \tag{12.20}$$

$$\frac{dV_{DC}}{dt} = \frac{(1 - D_S) i_{Bat}}{C_{DC2}} - \frac{i_{oDC}}{C_{DC2}} \tag{12.21}$$

Whereas, in buck mode of operation, the switch S_1 acts as a switch, while, S_2 acts as a diode and the power flows from the DC link to the battery, that is, charging mode. Fig. 12.7 depicts the dynamic average-value model of the bidirectional DC/DC converter under buck mode. The dynamic equations for the buck mode of operation are given in Eqs. (12.22)–(12.24).

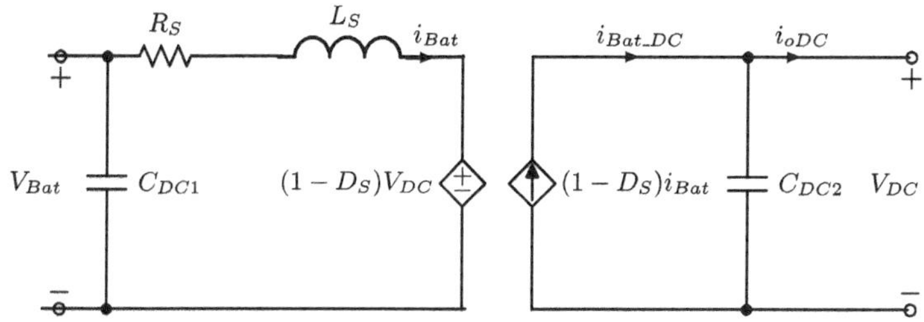

Figure 12.6 Dynamic average-value model of the bidirectional DC/DC converter in boost mode.

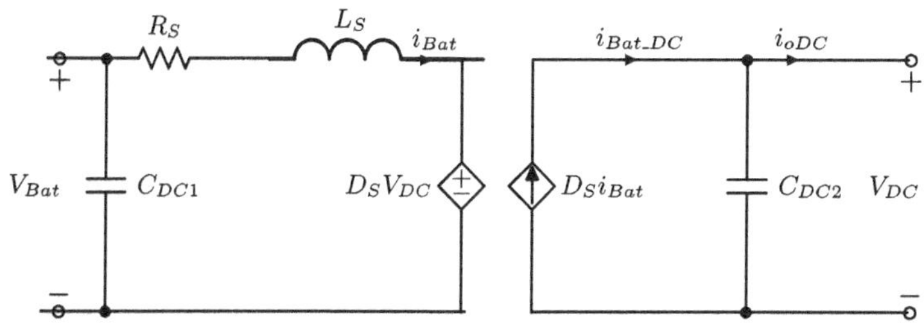

Figure 12.7 Dynamic average-value model of the bidirectional DC/DC converter in buck mode.

$$\frac{di_{Bat}}{dt} = -\frac{R_S}{L_S} i_{Bat} + \frac{V_{Bat}}{L_S} - D_S \frac{V_{DC}}{L_S} \tag{12.22}$$

$$\frac{di_{Bat\_DC}}{dt} = D_S i_{Bat} \tag{12.23}$$

$$\frac{dV_{Bat}}{dt} = \frac{i_{Bat}}{C_{DC1}} - \frac{i_{oDC}}{C_{DC1}} \tag{12.24}$$

In Eqs. (12.19)–(12.23), R_S and L_S are the parasite resistance and inductance of the energy-storage inductor, respectively; i_{Bat_DC} is the current at the DC-link side; D_S is the duty ratio of the bidirectional DC/DC converter. It should be noted that in the above modeling of the bidirectional DC/DC converter, the detailed switching manners of the switches are neglected with the realization of high switching frequencies [30].

The local control block diagram of the bidirectional DC/DC converter includes an outer voltage controller and an inner current controller, as shown in Fig. 12.5B. The outer voltage controller maintains the DC-link voltage, V_{DC}, at its reference value, V_{DC}^*, whereas, the inner current controller regulates the battery current, i_{Bat}. When the DC-link voltage, V_{DC}, deviates from its reference value, V_{DC}^*, the control of the DC-link voltage can be acquired by regulating the deviation of battery current, i_{Bat}, from its reference value, i_{Bat}^*, which is generated by the outer voltage controller [28–30]. The dynamical equations of outer voltage controller and inner current controllers are given in Eqs. (12.25) and (12.26).

$$i_{Bat}^{*} = K_{pvDC}(V_{DC}^{*} - V_{DC}) + K_{ivDC}\int (V_{DC}^{*} - V_{DC})dt \tag{12.25}$$

$$v_{iDC}^{*} = K_{pcDC}(i_{Bat}^{*} - i_{Bat}) + K_{icDC}\int (i_{Bat}^{*} - i_{Bat})dt \tag{12.26}$$

Fig. 12.8 shows the schematic diagram of the bidirectional AC/DC converter of BESS employed for interfacing the DC link to the IACMG system. The AC/DC converter consists of a DC link, a VSI, a series LC filter, and a coupling inductor. It can be operated in both the inverter as well as rectifier modes. The local control section of the bidirectional AC/DC converter is composed of power, DC voltage, and AC current controllers. In discharging mode, the power controller will be activated, while in charging mode, the DC voltage controller will be activated in place of the power controller, as shown in Fig. 12.3. The schematic diagram of the bidirectional AC/DC converter in the inverter mode (discharging mode) is shown in Fig. 12.8 [31,32]. It includes power and current controllers, a phase-locked loop (PLL), and a combined LC filter and coupling inductor (LCL filter). The PLL is employed to provide the reference angle for the rotating frame, while, the power controller is used to generate the inductor reference currents, $\mathbf{i}_{\mathbf{Ldq}}^{*}$. Whereas, the current controller is used for power quality improvement, while, the LCL filter is essential to attenuate switching frequency ripple. The complete 14th-order model of the inverter mode AC/DC converter has been obtained by combining the state-space models of the PLL, power controller, current controller, inverter switching bridge, and LCL filter [31,32]. It is presented in Eq. (12.27).

$$\frac{d\mathbf{x_{INV}}}{dt} = \mathbf{A_{INV}}\mathbf{x_{INV}} + \mathbf{R_{INV}}(\mathbf{x_{INV}}, \mathbf{u_{INV}}) \tag{12.27}$$

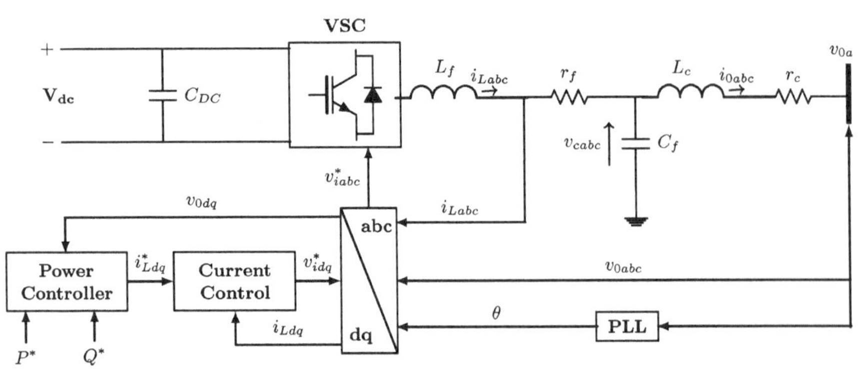

Figure 12.8 Schematic diagram of the bidirectional AC/DC converter of BESS. and its local control block diagram in inverter mode (discharging mode). *BESS*, Battery energy storage system.

$$\mathbf{y}_{\mathbf{INV}} = \mathbf{S}_{\mathbf{INV}}(\mathbf{x}_{\mathbf{INV}}, \mathbf{u}_{\mathbf{INV}}) \tag{12.28}$$

where the state vector, $\mathbf{x}_{\mathbf{INV}}$, the input vector, $\mathbf{u}_{\mathbf{INV}}$, and the output vector, $\mathbf{y}_{\mathbf{INV}}$, can be defined as in Eqs. (12.29)–(12.31). Details of the above matrices, $\mathbf{A}_{\mathbf{INV}}$, $\mathbf{R}_{\mathbf{INV}}(\mathbf{x}_{\mathbf{INV}}, \mathbf{u}_{\mathbf{INV}})$ and $\mathbf{S}_{\mathbf{INV}}(\mathbf{x}_{\mathbf{INV}}, \mathbf{u}_{\mathbf{INV}})$ can be found in Refs. [31,32].

$$\mathbf{x}_{\mathbf{INV}} = \begin{bmatrix} \theta & \Phi_{PLL} & \mathbf{i}^{*}_{\mathbf{Ldq}} & \mathbf{q}_{\mathbf{3Ldq}} \\ \mathbf{q}^{\mathbf{err}}_{\mathbf{Idq}} & \mathbf{i}_{\mathbf{Ldq}} & \mathbf{v}_{\mathbf{cdq}} & \mathbf{i}_{\mathbf{odq}} \end{bmatrix}^{T} \tag{12.29}$$

$$\mathbf{u}_{\mathbf{INV}} = \begin{bmatrix} V_{dc} & \mathbf{v}_{\mathbf{oabc}} & P* & Q* \end{bmatrix}^{T} \tag{12.30}$$

$$\mathbf{y}_{\mathbf{INV}} = \begin{bmatrix} \mathbf{i}_{\mathbf{Labc}} & \mathbf{i}_{\mathbf{oabc}} \end{bmatrix}^{T} \tag{12.31}$$

In rectifier mode (charging mode), the bidirectional AC/DC converter is the same as that in inverter mode except the power controller is replaced by the DC voltage controller. The complete state-space model of the AC/DC converter in rectifier mode is the same as the model of the RIAL, shown in Fig. 12.3, and its dynamical model, given in Eqs. (12.9) and (12.10), has been used.

Therefore, by integrating dynamical equations of IIDG units, lines, loads, and BESS units and applying Kirchhoff's current law at each node, the complete dynamical model of the IACMG system can be described by a non-linear set of differential-algebraic equations (DAEs), as given in Eq. (12.32).

$$\frac{d\mathbf{x}_{\mathbf{MG}}}{dt} = \mathbf{f}(\mathbf{x}_{\mathbf{MG}}, \mathbf{y}_{\mathbf{MG}}, \mathbf{u}_{\mathbf{MG}}) \tag{12.32a}$$

$$\mathbf{g}(\mathbf{x}_{\mathbf{MG}}, \mathbf{y}_{\mathbf{MG}}, \mathbf{u}_{\mathbf{MG}}) = 0 \tag{12.32b}$$

where $\mathbf{x}_{\mathbf{MG}}$ is a vector of state variables representing the dynamic states of four IIDG units, three lines, two passive loads, one RIAL, one dynamic IM load, BESS unit, and other system components; $\mathbf{y}_{\mathbf{MG}}$ is a vector of algebraic variables, such as bus voltage magnitudes and angles; and $\mathbf{u}_{\mathbf{MG}}$ is a set of other parameters of the IACMG system that represents load disturbance value, controller set-points, and so on. The vector, $\mathbf{f}$, represents the system dynamic equations, while, the vector, $\mathbf{g}$, stands for the system algebraic equations.

12.3 Mode control of the BESS for load leveling application

The primary goal of the BESS is to improve the dynamic performance of the IACMG system against large load disturbances. It is intended to deliver power during the large load disturbance periods and store the surplus power during periods of non-disturbance. Furthermore, it is also required to maintain the SoC in the range of safe charging status, that is, 20% < SoC ≤ 80%. Accordingly, the BESS mode control algorithm has been built, as shown in Fig. 12.9. The activated local controllers and switches corresponding to the charging and discharging modes are shown in Table 12.1. When the IACMG system is subjected to a large load disturbance and SoC > 20%, the local controllers and switches corresponding to the discharging mode will be activated. In the discharging mode, the bidirectional DC/DC converter, shown in Fig. 12.5A, will be operated in the boost mode in which the switch S_1 acts as a diode, while, S_2 acts as a switch and the power flows from the battery to the DC link. Whereas, the bidirectional AC/DC converter, shown in Fig. 12.8A, will be operated in the inverter mode and its model consists of a local control section that includes power and current controllers and a power processing section that contains PLL and combined LC filter and coupling inductor (LCL filter), as shown in Fig. 12.8B.

On the other hand, if the SoC goes below the lower limit, that is, 20% under the discharging mode, then the additional large load disturbance will be curtailed by opening the static switch, S_{LD}, to improve the battery's life and the local controllers and

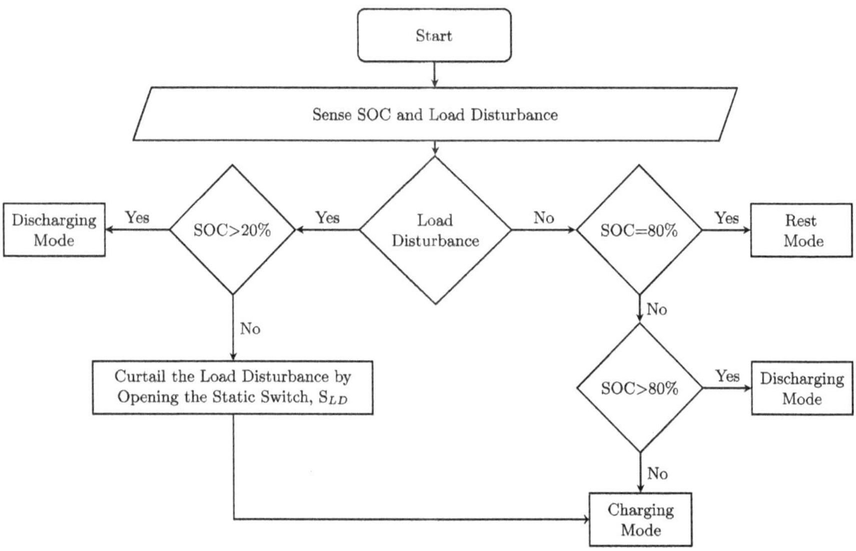

Figure 12.9 Flowchart of the BESS mode control for load leveling application. *BESS*, Battery energy storage system.

Table 12.1 Activated switches and controllers in charging and discharging modes.

BESS mode	Bidirectional DC/DC converter	Bidirectional AC/DC converter
Charging mode	Buck mode: switch S_1 acts as a switch and S_2 acts as a diode	Rectifier mode: DC voltage and AC current controllers
Discharging mode	Boost mode: switch S_1 acts as a diode and S_2 acts as a switch	Inverter mode: power and current controllers

switches corresponding to the charging mode will be activated. In the charging mode, the bidirectional DC/DC converter, shown in Fig. 12.5A, will be operated in the buck mode in which the switch S_2 acts as a diode, while, S_1 acts as a switch and the power flows from the DC link to battery. While, the bidirectional AC/DC converter, shown in Fig. 12.8A, will be operated in the rectifier mode and its model comprises a local control section that includes DC voltage and AC current controllers and a power processing section that consists of PLL and combined LC filter and coupling inductor (LCL filter), as shown in Fig. 12.3. In contrast to the above two modes, when the IACMG system is not subjected to the large load disturbance and if the SoC = 80%, then the BESS remained in the rest mode, while, the BESS will be in charging and discharging modes for the SoC < 80% and SoC > 80%, respectively, and the aforementioned local controllers and switches corresponding to these modes will be activated.

12.4 Results and discussions

A studied IACMG system including BESS connected at bus 4, shown in Fig. 12.1, has been built for the time-domain simulations under MATLAB/Simulink environment. In MATLAB/Simulink, ode23tb, which is an implementation of Trapezoidal Rule with the second order Backward Difference Formula' (TR-BDF2) [33], has been used to solve DAEs, given in Eq. (12.32). The TR-BDF2 is an implicit Runge–Kutta formula with a trapezoidal rule step as its first stage and a backward differentiation formula of order two as its second stage [33]. The parameters of the BESS are tabulated in Table 12.2 [28–32]. To evaluate the utilization of the BESS

Table 12.2 Test parameters of the BESS.

Battery model Battery-side voltage range: 260–300 V (DC); battery type: lithium-ion (Li-ion) model; battery rated capacity: 6.5–7 Ah.
Bidirectional DC/DC converter and its local control Switching device and frequency: IGBT; 10 kHz, DC-bus-side voltage range: 650–750 V (DC); $R_S = 0.001\Omega$, $L_S = 5mH$, $C_{DC} = 2040\mu F$, $K_{pvDC} = 2$ and $K_{ivDC} = 120$.
Bidirectional AC/DC converter and its local control Converter type: three-phase PWM converter; switching device and frequency: IGBT, 10 kHz; rated-output AC voltage: 400 V_{LL} (ac); Rated-output DC voltage range: 650–750 V (DC); PLL parameters: $K_p^{PLL} = 0.25$ and $K_i^{PLL} = 2.0$; power controller parameters: $w_c = 31.41\text{rad/s}$; current controller parameters: $K_P^d = K_P^q = 7$ and $K_I^d = K_I^q = 25e3$; DC voltage controller parameters: $K_{pv} = 0.5$ and $K_{iv} = 150$. $L_f = 1.35$ mH, $C_f = 50$ μF, $R_f = 0.1\ \Omega$, $R_c = 0.03\ \Omega$, $L_c = 0.35$ mH.

and its mode control for load leveling application under different SoCs of the battery, the studied IACMG system including the BESS has been implemented for the three cases: BESS having SoC of 70%, 80%, and 90%. Furthermore, to show the impact of the large load disturbance on the dynamic performance of the IACMG system, simulations have also been performed for the studied IACMG system without BESS. These simulation results have been used to compare the dynamic performances of the IACMG system with and without BESS.

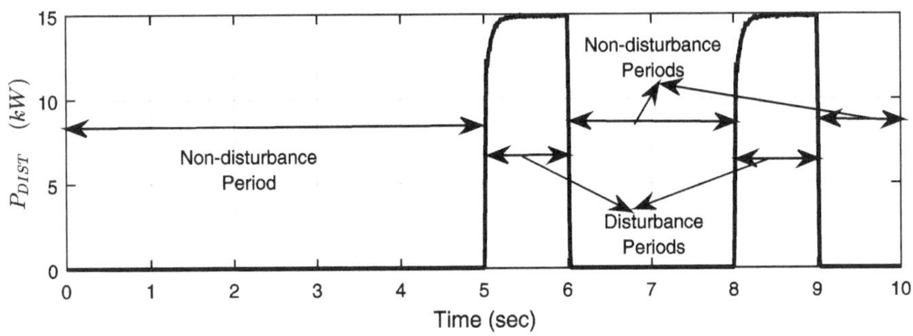

Figure 12.10 Disturbance and non-disturbance periods.

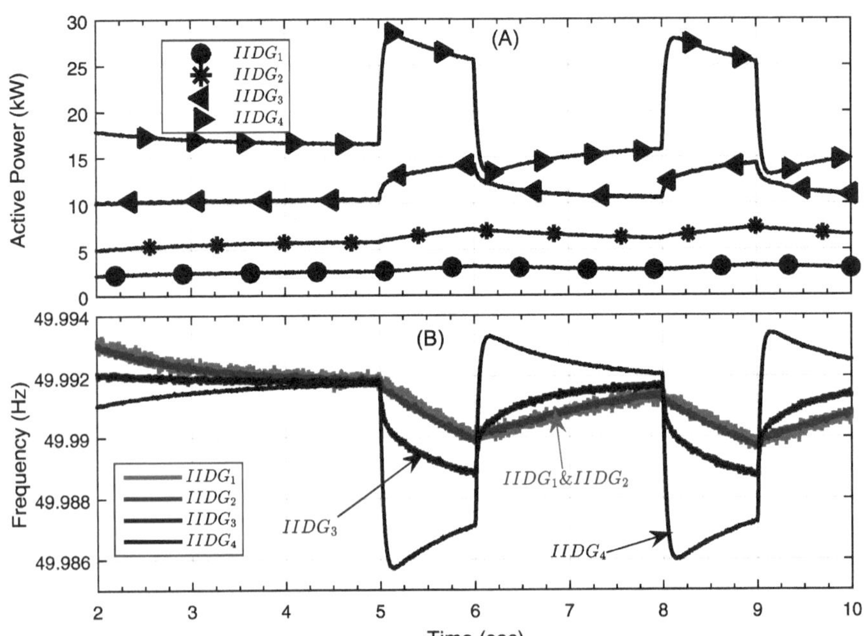

Figure 12.11 Dynamic response of the output of IIDG units for the IACMG system without BESS (A) active power. (B) Frequency. *IACMG*, islanded AC microgrids; *IIDG*, Inverter-interfaced distributed generation; *BESS*, battery energy storage system.

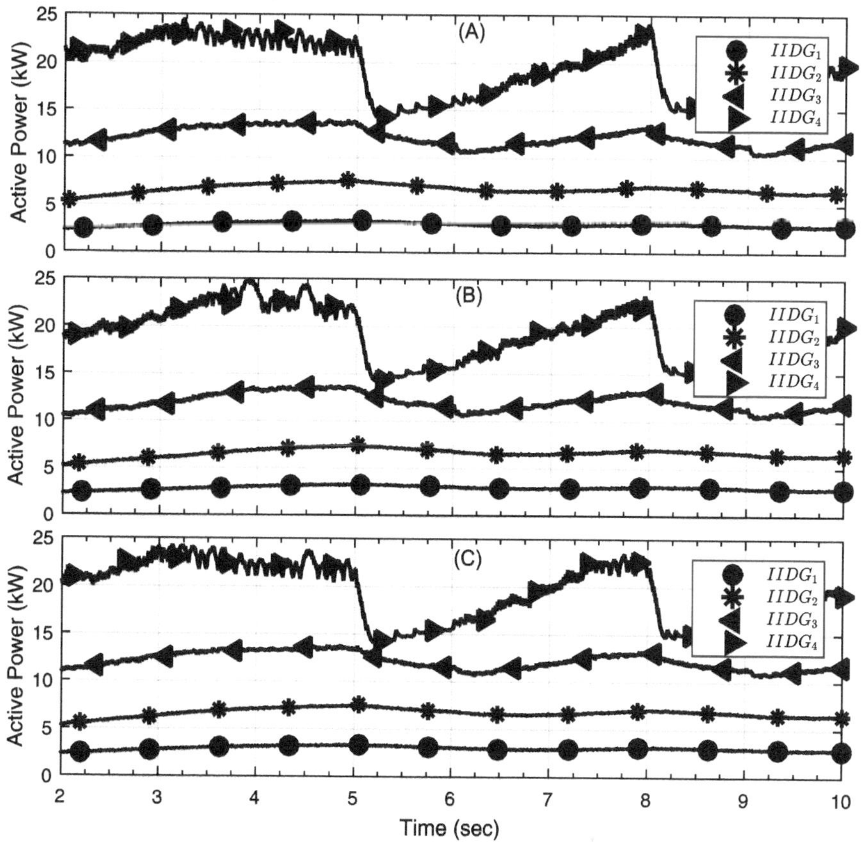

Figure 12.12 Dynamic response of the active power output of IIDG units for the IACMG system with the BESS having SoC of (A) 70%. (B) 80%. (C) 90%. *IACMG*, Islanded AC microgrids; *IIDG*, Inverter-interfaced distributed generation; *BESS*, battery energy storage system; *SoC*, state-of-charges.

The time-domain simulations have been performed by considering that the IACMG system is operating under nominal loading in the periods, 0-5.0, 6.0-8.0, and 9.0-10.0 seconds. These intervals are considered non-disturbance periods. The large load disturbance of 15 kW has been applied at bus 4 by closing the static switch, S_{LD}, in the intervals, 5.0-6.0 and 8.0-9.0 seconds and are referred to as disturbance periods. These non-disturbance and disturbance periods are shown in Fig. 12.10. It should be noted that time-domain simulations have been performed for different disturbance locations. However, the main conclusions drawn from the time-domain simulations are the same for different disturbance locations. Therefore the simulation results have been presented for one disturbance location, that is, at bus 4, as shown in Figs. 12.11–12.16.

Fig. 12.11 shows the dynamic response of active power and frequency outputs of IIDG units for the IACMG system without BESS. It can be seen from Fig. 12.11A

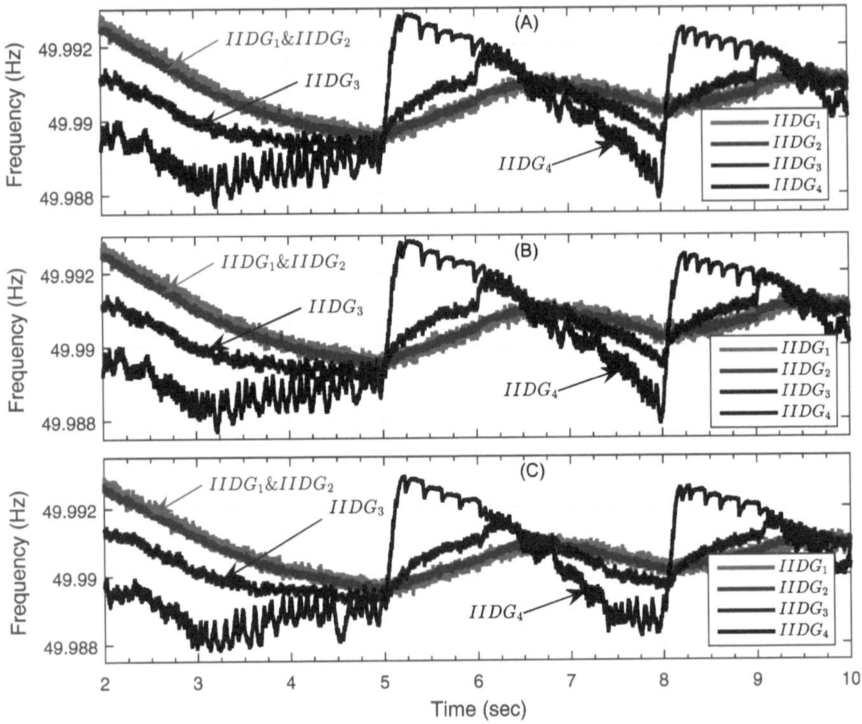

Figure 12.13 Dynamic response of the frequency output of IIDG units for the IACMG system with the BESS having SoC of (A) 70%. (B) 80%. (C) 90%. *IACMG*, Islanded AC microgrids; *IIDG*, Inverter-interfaced distributed generation; *BESS*, battery energy storage system; *SoC*, state-of-charges.

that the active power delivered by the $IIDG_4$ unit exceeds its maximum limit (25 kW) during disturbance periods. Furthermore, $IIDG_4$ the unit shares 66.66% of the load disturbance (10 out of 15 kW), while, 33.33% of the load disturbance (5 out of 15 kW) has been shared by the $IIDG_3$ unit. This leads to more drop in the frequency response of $IIDG_4$ and $IIDG_3$ units as compared with that of $IIDG_1$ and $IIDG_2$ units, as shown in Fig. 12.11B. The reason for this lies in the fact that $IIDG_3$ and $IIDG_4$ are connected to bus 4 and bus 3 corresponding to the connections of large load disturbance and nearest to it, respectively.

The dynamic response of active power and frequency outputs of IIDG units for the IACMG system with the BESS having different SOCs are shown in Figs. 12.12 and 12.13, respectively. It can be observed from Fig. 12.12 that the maximum power has been harvested from the $IIDG_4$ unit without overloading irrespective of the SoC of the battery. Furthermore, the frequency response of $IIDG_4$ and $IIDG_3$ units, shown in Fig. 12.13, does not lead to more drop as compared with that of $IIDG_1$ and $IIDG_2$ units, unlike for the case of the IACMG system without BESS, as shown in Fig. 12.11B. The reason for this lies in the fact that the BESS releases

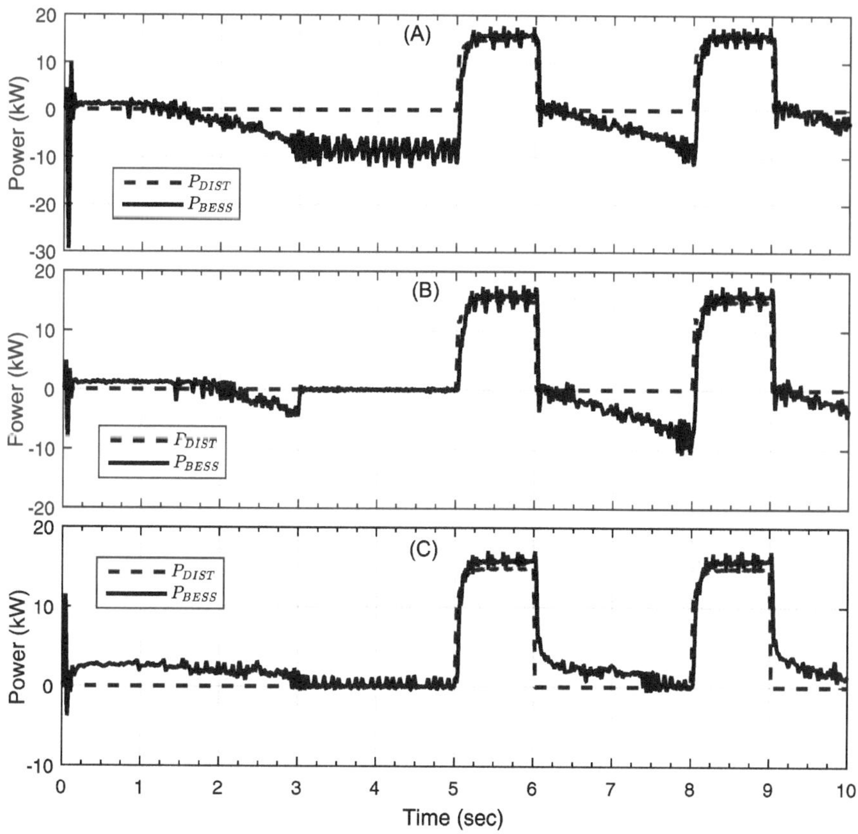

Figure 12.14 Dynamic response of disturbance (P_{DIST}) and BESS (P_{BESS}) active power outputs for the IACMG system with the BESS having SoC of (A) 70%. (B) 80%. (C) 90%. *IACMG*, Islanded AC microgrids; *BESS*, battery energy storage system; *SoC*, state-of-charges.

power (discharging mode) to the IACMG system in the disturbance periods and stores the excess power (charging mode) from the IACMG in the non-disturbance periods, as shown in Fig. 12.14. Thus the utilization of BESS for the load leveling application during large load disturbances has been verified.

Fig. 12.15 shows the dynamic response of the battery current for different SoCs of the battery. It can be seen from Fig. 12.15 that the current in the positive region indicates the discharging of the battery during the disturbance periods, while, in the negative region represents the charging of the battery during the non-disturbance periods for the SoCs less than 80%. Whereas, the current is always in the positive region for the SoC greater than 80%, as shown in Fig. 12.15C. Furthermore, this can be verified from the response of the SoC of the battery, shown in Fig. 12.16. It can be observed from Fig. 12.16A and B that when the SoC is less than 80% then it is increasing during the non-disturbance periods, while, it is decreasing during the

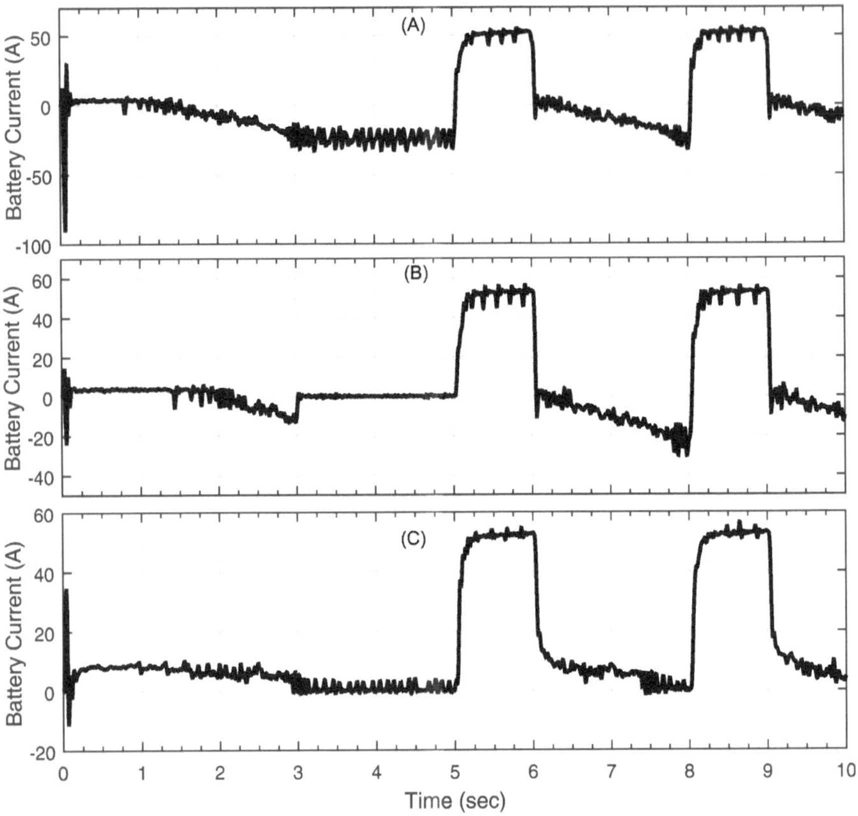

Figure 12.15 Dynamic response of the battery current for the IACMG system with the BESS having SoC of (A) 70%. (B) 80%. (C) 90%. *IACMG*, Islanded AC microgrids; *BESS*, battery energy storage system; *SoC*, state-of-charges.

disturbance periods. Whereas, it can be seen from Fig. 12.16C that since the SoC is greater than 80%, it is decreasing continuously irrespective of the load disturbance. Therefore SoC has been maintained in the range of the safe charging status, that is, 20% < SoC ≤ 80%, thus, validating the mode control algorithm of the BESS. It is worth mentioning that the additional large load disturbance should be curtailed by opening the static switch, S_{LD}, for the SoC by less than 20%, to improve the battery's life. However, the additional large load disturbance will be reconnected by closing the static switch, S_{LD}, once the SoC of the battery reaches greater than 20%.

Thus it can be concluded from the above observations that the utilization of the BESS led to enhancement in the dynamic performance of the IACMG system in terms of harnessing maximum powers from IIDG units and avoiding overloading of IIDG units.

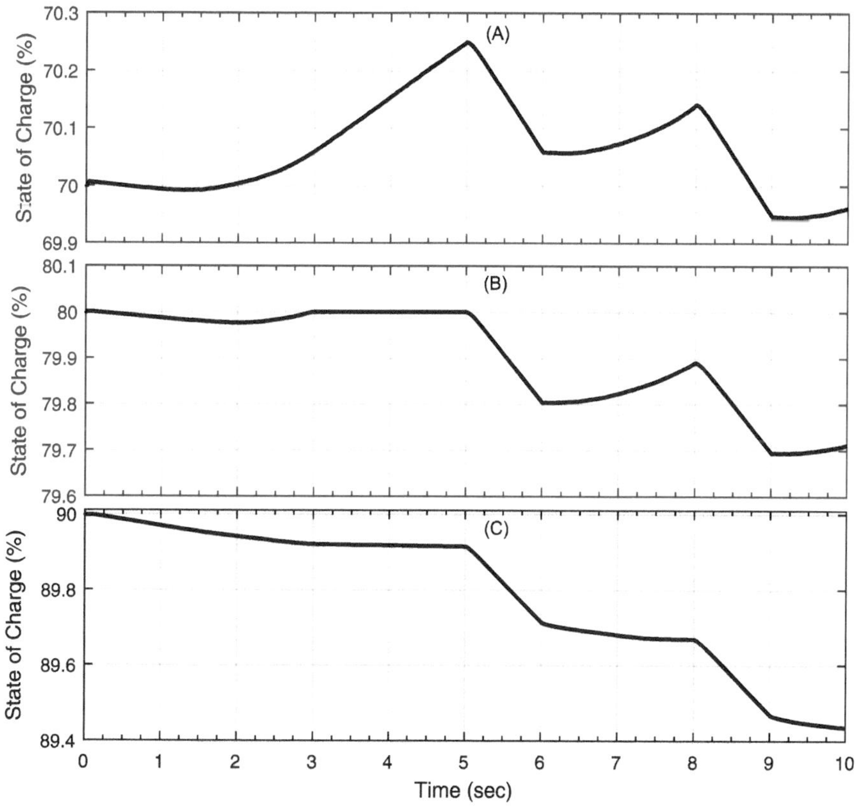

Figure 12.16 Response of the SoC for the IACMG system with the BESS having SoC of (A) 70%. (B) 80%. (C) 90%. *IACMG*, Islanded AC microgrids; *BESS*, battery energy storage system; *SoC*, state-of-charges.

12.5 Conclusion

This chapter proposes the utilization of a BESS to enhance the dynamic performance of IACMGs against large load disturbances. To explore this, a complete dynamic model of the BESS including battery, bidirectional DC/DC converter and its local control, DC-link interface, and bidirectional AC/DC converter and its local control has been developed. The developed model of the BESS has been incorporated into the studied IACMG system. The BESS is intended to deliver power during the large load disturbance periods and store the surplus power during periods of non-disturbance. Furthermore, it is also required to maintain the SoC in the range of safe charging status, that is, $20\% < \text{SoC} \leq 80\%$. Accordingly, the BESS mode control algorithm has been built. The proposed utilization of BESS and its mode control algorithm have been evaluated through the time domain simulations under different SoCs of the battery as well as different disturbance locations. Following are the main conclusions drawn from these time-domain simulations.

- During the disturbance periods, active power delivered by the IIDG unit, connected to the bus at which large disturbance was applied, exceeds its maximum limit when the BESS was not present in IACMGs. Whereas, in the case of IACMGs equipped with the BESS, the maximum power has been harvested from that IIDG unit without overloading irrespective of the SoC of the battery.
- The BESS releases power to the IACMG system during the disturbance periods and imports the excess power from the IACMG and stores it during the non-disturbance periods by maintaining the SoC of the battery in the range of safe charging status.
- Finally, the dynamic performance of IACMGs against large load disturbances has been enhanced in terms of harnessing maximum powers from IIDG units and avoiding the overloading of IIDG units.

Appendix

IIDG units ratings: $IIDG_1$ -(10 + j6) kVA; $IIDG_2$ -(15 + j9) kVA; $IIDG_3$ -(20 + j12) kVA; $IIDG_4$ -(25 + j15) kVA;

Static active and reactive power droop gains: $m_{P1} = 6.28e\text{-}4$ rad/s/W, $m_{P2} = 4.18e\text{-}4$ rad/s/W, $m_{P3} = 3.14e\text{-}4$ rad/s/W, $m_{P4} = 2.52e\text{-}4$, $n_{Q1} = 1.66e\text{-}3$ V/VAR, $n_{Q2} = 1.11e\text{-}3$ V/VAR, $n_{Q3} = 8.33e\text{-}4$ V/VAR and $n_{Q4} = 6.66e\text{-}4$ V/VAR;

IIDG unit parameters: $L_f = 1.35$ mH, $C_f = 50$ μF, $R_f = 0.1$ Ω, $f_{sw} = 8$ kHz, $w_c = 31.41$ rad/s, $K_{pv} = 0.05$, $K_{iv} = 390$, $K_{pi} = 10.5$, $K_{ii} = 16e3$, $F = 0.75$, $f_{nl} = 50.5$ Hz, $R_c = 0.03$ Ω, $L_c = 0.35$ mH;

RIAL parameters: $L_f = 2.3$ mH, $C_f = 8.8$ μF, $R_f = 0.1$ Ω, $f_{sw} = 10$ kHz, $w_c = 31.41$ rad/s, $K_{pv} = 0.5$, $K_{iv} = 150$, $K_{pi} = 7$, $K_{ii} = 25e3$, $R_c = 0.03$ Ω, $L_c = 0.93$ mH;

Line parameters: Line 1: (0.23 + j0.11) Ω, Line 2: (0.35 + j0.58) Ω, Line 3: (0.30 + j0.47) Ω;

Load parameters: induction motor load: 10 hp, 400 V, 50 Hz, $r_s = 0.7834\Omega$, $L_{ss} = 127.1mH$, $r_r = 0.7402\Omega$, $L_{rr} = 127.1mH$, $L_m = 124.1mH$, $P = 4$, $T_L = 47.75$ N.m; *CPL*: 12 kVA, $r_{CPL} = 13.224$ Ω /phase and $\cos\alpha = 0.85$; *RIAL*: 12 kW and $R_{RIAL} = 40.833$ Ω; *R* Load: 25 kW, $R_{RLoad} = 6.347$ Ω/phase and $V_{DC} = 700V$.

References

[1] Gao Jun-Ting, et al., An active and reactive power controller for battery energy storage system in microgrids, IEEE Access. J. 10 (2022) 10490–10499.

[2] I. Serban, C. Marinescu, Control strategy of three-phase battery energy storage systems for frequency support in microgrids and with uninterrupted supply of local loads, IEEE Trans. Power Electron. 29 (9) (2014) 5010–5020.

[3] X. Lin, R. Zamora, Controls of hybrid energy storage systems in microgrids: critical review, case study and future trends, J. Energy Storage 47 (103884) (2022) 1–27.

[4] X. Luo, J. Wang, M. Dooner, J. Clarke, Overview of current development in electrical energy storage technologies and the application potential in power system operation, Appl. Energy 137 (2015) 511–536.

[5] R. Georgious, et al., Review on energy storage systems in microgrids, Electronics 10 (17) (2021) 1–20.
[6] X. Tan, Q. Li, H. Wang, Advances and trends of energy storage technology in microgrid, Int. J. Electr. Power Energy Syst. 44 (1) (2013) 179–191.
[7] J.Y. Kim, et al., Cooperative control strategy of energy storage system and microsources for stabilizing the microgrid during islanded operation, IEEE Trans. Power Electron. 25 (12) (2010) 3037–3048.
[8] M.H. Moradi, M. Eskandari, S.M. Hosseinian, Cooperative control strategy of energy storage systems and micro sources for stabilizing microgrids in different operation modes, Int. J. Electr. Power 78 (2016) 390–400.
[9] T. Morstyn, B. Hredzak, V.G. Agelidis, Control strategies for microgrids with distributed energy storage systems: an overview, IEEE Trans. Smart Grid 9 (4) (2018) 3652–3666.
[10] D.E. Olivares, et al., Trends in microgrid control, IEEE Trans. Smart Grid 5 (4) (2014) 1905–1919.
[11] P.E.S.N. Raju, T. Jain, Robust optimal centralized controller to mitigate the small-signal instability in an islanded inverter based microgrid with active and passive loads, Int. J. Electr. Power Energy Syst. 90 (2017) 225–236.
[12] P.E.S.N. Raju, T. Jain, Optimal decentralized supplementary inverter control loop to mitigate instability in an islanded microgrid with active and passive loads, Int. J. Emerg. Electr. Power Syst. 18 (01) (2017).
[13] P.E.S.N. Raju, T. Jain, A two-level hierarchical controller to enhance stability and dynamic performance of islanded inverter-based microgrids with static and dynamic loads, IEEE Trans. Ind. Inform. Early Access. (2018). Available from: https://doi.org/10.1109/TII.2018.2869983.
[14] M.A. Sofla, G.B. Gharehpetian, Dynamic performance enhancement of microgrids by advanced sliding mode controller, Int. J. Electr. Power Energy Syst. 33 (1) (2011) 1–7.
[15] Y. Wang, X. Wang, Z. Chen, F. Blaabjerg, Distributed optimal control of reactive power and voltage in islanded microgrids, IEEE Trans. Ind. Electron. 53 (01) (2017) 340–349.
[16] P. Kundur, Power System Stability and Control, McGrawHill, New York, 1994.
[17] Z. Nie, et al., SMES battery energy storage system for conditioning outputs from direct drive linear wave energy converters, IEEE Trans. Appl. Supercond. 23 (3) (2013). Art. 5000705.
[18] G. Wang, M. Ciobotaru, V.G. Agelidis, Power smoothing of large solar PV plant using hybrid energy storage, IEEE Trans. Sustain. Energy 5 (3) (2014) 834–842.
[19] R. Majumder, Some aspects of stability in microgrids, IEEE Trans. Power Syst. 28 (3) (2010) 3242–3253.
[20] N. Pogaku, M. Prodanovic, T.C. Green, Modeling, analysis and testing of autonomous operation of an inverter-based MG, IEEE Trans. Power Electron. 22 (2) (2007) 613–625.
[21] N. Pogaku, Analysis, Control and Testing of Inverter-Based Distributed Generation in Standalone and Grid-Connected Applications, Ph.D. dissertation, University of London, Imperial College London, London, United Kingdom, 2006.
[22] P.E.S.N. Raju, T. Jain, Impact of load dynamics and load sharing among distributed generations on stability and dynamic performance of islanded AC microgrids, Electr. Power Syst. Res. 157 (2018) 200–210.
[23] P.E.S.N. Raju, T. Jain, Development and validation of a generalized modeling approach for islanded inverter-based microgrids with static and dynamic loads, Int. J. Electr. Power Energy Syst. 108 (2019) 177–190.

[24] N. Bottrell, M. Prodanovic, T. Green, Dynamic stability of a microgrid with an active load, IEEE Trans. Power Electron. 28 (11) (2013) 5107–5119.
[25] N. Bottrell, Small-signal Analysis of Active Loads and Large-Signal Analysis of Faults in Inverter Interfaced Microgrid Applications, Ph.D. dissertation, University of London, Imperial College London, London, United Kingdom, 2014.
[26] P.C. Krause, O. Wasynczuk, S.D. Sudhoff, S. Pekarek, Analysis of Electric Machinery and Drive Systems, 75, John Wiley & Sons, 2013.
[27] O. Tremblay, L.A. Dessaint, Experimental validation of a battery dynamic model for EV applications, World Electr. Veh. J. 03 (2009) 289–298.
[28] S.K. Kim, et al., Dynamic modeling and control of a grid-connected hybrid generation system with versatile power transfer, IEEE Trans. Ind. Electron. 55 (4) (2008) 1677–1688.
[29] H.L. Do, Non-isolated bidirectional zero-voltage-switching DC-DC converter, IEEE Trans. Power Electron. 26 (9) (2011) 2563–2569.
[30] L. Wang, Q.S. Vo, A.V. Prokhorov, Dynamic stability analysis of a hybrid wave and photovoltaic power generation system integrated into a distribution power grid, IEEE Trans. Sustain. Energy 8 (1) (2017) 404–413.
[31] N. Kroutikova, C.A. Hernandez-Aramburo, T.C. Green, State-space model of grid-connected inverters under current control mode, IET Electr. Power App 1 (3) (2007) 329–338.
[32] M.A. Hassan, M.A. Abido, Optimal design of microgrids in autonomous and grid-connected modes using particle swarm optimization, IEEE Trans. Power Electron. 26 (3) (2011) 755–769.
[33] M.E. Hosea, L.F. Shampine, Analysis and implementation of TR-BDF2, Appl. Numer. Math. 20 (1996) 21–37.

Battery energy storage in micro-grids

13

Yongxi Zhang[1], *Y. Xu*[1], *Hongming Yang*[1] *and Z.Y. Dong*[2]
[1]School of Electrical and Information Engineering, Changsha University of Science and Technology, International Joint Laboratory of Ministry of Education for Operation and Planning of Energy Internet Based on Distributed Photovoltaic-Storage Energy, Hunan Provincial Engineering Research Center for Electric Transportation and Smart Distribution Network, Changsha, P. R. China, [2]School of Electrical & Electronic Engineering, Nanyang Technological University, Singapore

Chapter Outline

Emerging Trends in Energy Storage Systems and Industrial Applications. DOI: https://doi.org/10.1016/B978-0-323-90521-3.00005-3

Highlight

Basic concepts for Micro-grids and the recent developing trend of key energy storage technologies are introduced.

An expansion planning model for BESS in micro-grids considering battery degradation effects is established.

A two-stage operation optimization framework of BESS to cope with high-level renewable power generation is presented.

Nomenclature

Indices

$t,\ T$	Index and the total number of time intervals
Δt	Time duration of each time interval
$\xi,\ \Omega$	Index and total numbers for scenarios
$i,\ j$	Index of system load points

Sets

Φ	Set of system buses

Parameters

Γ_R	Rated charge life of the battery
C_R	Amp-hour capacity of a battery at rated discharge current (kWh)
C_A	Amp-hour capacity of a battery at a given discharge current (kWh)
D_R	Percent depth of discharge at which rated cycle life was determined (%)
D_A	Actual discharge as a percentage of rated power capacity $D_A = 1 - \text{SOC}$ (%)
d_{eff}	Effective amp-hour discharge as adjusted for depth and rate of discharge (Ah)
d_{act}	Actual amp-hour discharge (Ah)
L_R	Cycle life at a rated depth of discharge and rated discharge current (times)
$L_B(d_B)$	Battery lifetime when the depth of discharge is d_B
Γ_{eff}	Effective amp-hour life of a cell in the time period of system operation T
L_{time}	lifetime (years) of the cell under actual operating conditions
$\boldsymbol{d}$	Real interest rate
$\boldsymbol{N_{BE}}$	Time span of planning project
$P^{\xi}_{buy,t}$	Power purchased from the external grid at time t in scenario ξ (MW)
$P^{\xi}_{load,t}$	Load demand at node i at time t in scenario $\mathcal{E}$ (MW);
$P^{\xi}_{DG,i,t}$	Output power of ith DG unit(MW)
I_{ij}	Line current ij (A)
v_i	Voltage of bus i of the system (V)
P_{li}	Active power at bus number i (MW/MVA)
f_l	Power flow on line l (MW)
SOC^{ξ}_t	State-of-charge of battery at time t in scenario $\mathcal{E}$
$I_{ij,max}$	Line ij maximum current (A)
$v_{\min},\ v_{\max}$	Minimum and maximum voltage limitation(v)
$f^{\max}_l,\ f^{\min}_l$	Minimum and maximum Power flow limitations
P^{ξ}_{loss}	Power loss at time t for scenario ξ (MW)

$r_{ij},\ x_{ij}$ Line resistance and reactance between line i and j
$G_{ij},\ B_{ij}$ Conductance and susceptance matrix between bus i and j
MP_t Time-varying electricity price ($/MW)
C_{dep} Per unit BESS depreciation cost ($/MW)
C_E Capital cost of BESS per unit ($MWh)
$C^{\xi}_{bat,t}$ Energy stored at time t in scenario $\mathcal{E}$
C^{ξ}_{gen} Cost of local micro-turbine units
C^{ξ}_{pur} Costs of purchasing power from the external grid
C^{ξ}_{dep} BESS degradation cost
$C_{O,BESS}$ Operation & maintenance cost per day for BESS ($/MWh)
η_{loss} Power loss cost ($MW)
$\eta_{ch},\ \eta_{dis}$ BESS Storage charging/discharging efficiency
CAP Capital investment budget ($)

First Layer Decision Variables
P^{i}_{batmax} Rated power capacity of ith BESS
C^{i}_{batmax} Rated energy capacity of BESS(MW)
L_{BES} Candidate system buses for BESS installation

Second Layer Decision Variables
$P^{\xi}_{ch,t},\ P^{\xi}_{dis,t}$ Charging/discharging power of BESS at time t for scenario ξ (MW)

Acronyms
CVaR Conditional value at risk
ESS Energy storage systems
BESS Battery energy storage systems
LV Low voltage
MV Medium voltage
SO System operator

Indices and Sets
i, I Index and numbers for micro-turbines
k, K Index and numbers for batteries
$t,\ T$ Index and the total number of time intervals in the day-ahead market
$\omega,\ S$ Index and numbers for total scenarios

Parameters
Γ_R Rated charge life of the cell (times)
$\lambda_{g,t}$ Purchasing price of external gird ($/kWh)
$\eta_{ch},\ \eta_{dis}$ Battery charging and discharging efficiency (%)
$P_{grid,\ max}$ Maximum allowable energy from upstream
c_{age} Per unit capital cost of BESS ($/MWh)
$C^{BESS}_{k,t,w}$ Energy stored in kth battery at time t in scenario ω (MWh)
$c^{BESS}_{k,age}$ Per unit capital cost of kth BESS ($/kWh)
$C_{inv,BESS}$ Capital cost of BESS
d_{eff} Effective amp-hour discharge as adjusted for depth and rate of discharge (Ah)
$P^{pre}_{PV,t}$ Predicted photovoltaic (PV) power output (kW)
$P^{(\omega)}_{PV,t}$ PV generation at time t in scenario $C^{BESS}_{k,t,w}$ (kW)

$P^{DA}_{Load,t}$ Load demand at time t in day-ahead market (kW)
$P^{(\omega)}_{Load,t}$ Load demand at time t in intra-day market in scenario ω (kW)
$P_{shortage,t}$ Power shortage in intra-day energy market (kW)
$P^{BESS}_{k,max}$ Rated power capacity of kth BESS
$p(\omega)$ Probability of ωth scenario (%)
$SOC^{BESS}_{t,\omega}$ State-of-charge of battery at time t in scenario ω

Decision Variables
$P^{DA}_{grid,t}$ Power bought from the external grid at time t in the day-ahead market (kW)
$P^{DA}_{i,t}$ Power output of ith micro-turbine at time t in the day-ahead market (kW)
$P^{BESS}_{k,ch/dis,t,\omega}$ Charging/discharging power of kth BESS in the intra-day market at time t in the scenario ω.
$P^{ID}_{grid,t,\omega}$ Power bought from the external grid in the intra-day market at time t in the scenario ω.

13.1 Introduction

13.1.1 Definitions of micro-grids

The traditional power system was comprised of generation, transmission, and distribution systems. With the growing penetration of distributed generation (DG) at medium voltage (MV) and low voltage (LV) in the last few decades, the definition of a modern power system has changed. The DGs, such as photovoltaic (PV) modules, wind turbines (WT), other small renewables (such as biomass, tide), heat and electricity storage, and controllable loads are expected to be on the rise in the future distribution system. The cluster of multiple DGs integrated or islanded with the grid is called Micro-grids. The emerging technologies can significantly reduce carbon emissions, thereby contributing to the commitments of most developed countries to satisfy their carbon emission reduction targets. Also, the presence of generation close to demand can increase the power quality and system reliability, and reduce power losses. These benefits are the key motivations and drivers for DG deployment.

While the application of DGs can potentially provide major opportunities, controlling a potentially huge number of DGs brings new challenges to planning and operating the system. This problem can be mitigated by Micro-grids, which are entitles that coordinate DGs more smartly and flexibly, thus alleviating the control burden on the grid and allowing them to provide their potential benefits. Generally, a micro-grid contains an LV($\approx$1 kV) or MV($\approx$1–69 kV) locally controlled cluster of DGs and operates safely and efficiently within its local distribution of customers (load), but it is also capable of islanding.

Fig. 13.1 illustrates a typical micro-grid schematic. A Micro-grid consists of sources and loads. Micro-grid is often locally controlled. A micro-grid mainly uses local energy generation to satisfy load demand. It also can reduce power losses due to the reduction of long-distance transmission lines. When a blackout occurs in the main grid, a microgrid can disconnect immediately. Thus, higher reliability is achieved.

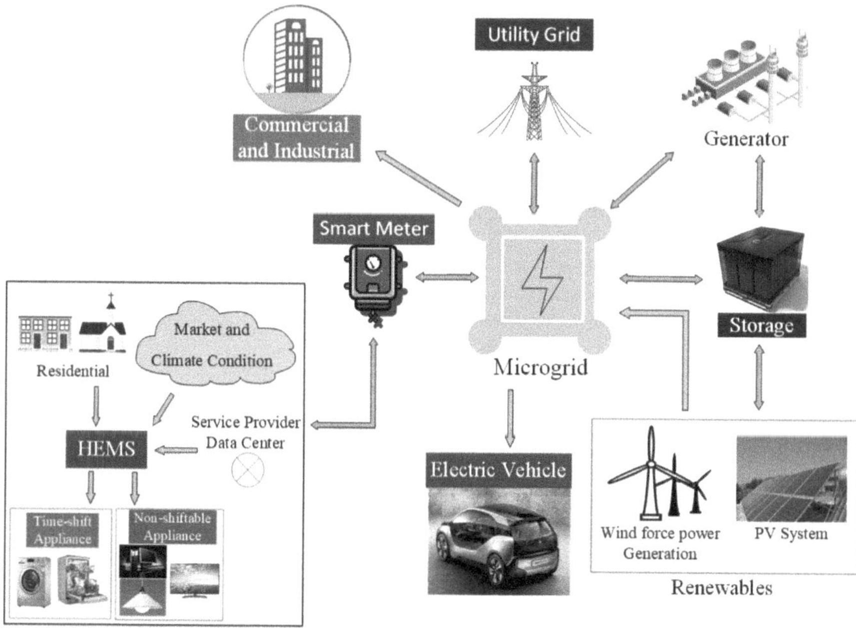

Figure 13.1 Framework of a typical micro-grid.

The concepts of micro-grid are often defined differently in different countries/organizations. According to the Consortium for Electrical Reliability Solutions (CERTS) concept, a micro-grid can separate and isolate itself from the system seamlessly with little or no disturbance when a system failure occurs (e.g., in the CERTS Micro-grid concept, no impacts on power quality) [1]. Then, when the system recovers, the micro-grid automatically resynchronizes and reconnects itself to the grid, in an equal way. In the view of the US Department of Energy Microgrid Exchange Group, a micro-grid is a group of interconnected distributed energy units and loads that acts as a single control unit connected to the main grid. According to CIGRÉ C6.22 Working Group definition, micro-grids are electricity distribution systems with loads and small DGs, (such as distributed generators storage devices, or controllable loads) that can be operated in a controlled, coordinated manner either connected to the main power network or is islanded.

Based on [2], micro-grids are classified based on their area of application, such as (Table 13.1):

Micro-grid operation and planning demand innovative skills and technology, for the planning of Micro-grids, power engineering technologies are needed to meet the demand and realize the control strategy. The engineering technologies involve distribution generation, electric vehicles, energy storage systems, demand response instruments, and related power electronics. The operation of Micro-grids is mainly governed by IEEE 1547 standard [3], which provides essential points relevant to the performance, operation, control, testing, safety, and maintenance. This standard

Table 13.1 Micro-grid classification.

Type	Purpose
Commercial/industrial	Reducing load demand and operation cost and helps to provide uninterrupted power supply for critical loads
Community/utility	Reliability improvement of critical infrastructure
University campus	Every campus has its DG system. The laboratories require an energy system to increase the reliability
Public institutional	Public places like police stations, hospitals, fire stations, etc. require uninterrupted power. These types of micro-grids are efficient for such areas
Military	The communication system, cyber security, and lighting system in army camps require an uninterrupted power supply.
Rural	Remote micro-grids were independent of the main grid. Diesel generators or renewable energy are mainly used for supplying power.

is applicable for all types of distributed energy sources, with an aggregate capacity less than/equal to 10 MVA.

13.1.2 Battery energy storage systems technology

Energy storage systems (ESS) have been utilized in power systems for decades for providing backup power or contingency services. In recent years, the constant growth of load demand and increasing penetration of renewable energy generation have propelled the further development of ESS technology. Its process involves converting the energy in the electrical form to a storable form, while the stored energy can be converted back into electrical energy when necessary.

ESSs are widely applied particularly in transmission and distribution systems. The applications include but are not limited to, energy management, load management, frequency regulation, voltage regulation, and grid stabilization. Based on the different purposes of different power system stakeholders, ESS can participate in different systems [4]. For example, a storage device can be accommodated in the distribution system to provide power to a feeder or customers; or by the end-user to enhance power quality and reliability, or collaborate with other renewable energy resources to mitigate the fluctuation of renewable energy resources in a micro-grid [5].

Energy storage technology may be applied to many fields according to different power and energy requirements. Various ESS technologies include: Lead-Acid Batteries, Nickel-Cadmium Batteries, Sodium-Sulfur Batteries, flywheel energy storage, superconducting magnetic energy storage, ultra battery, compressed air energy storage (CAES), pumped-hydro storage, and electrochemical capacitors (which belong to solid-state batteries). ESS combined with renewable energy technologies is typically best suited for long-term storage applications, such as the battery technologies listed above [6]. The flywheel and capacitor storage systems are typically

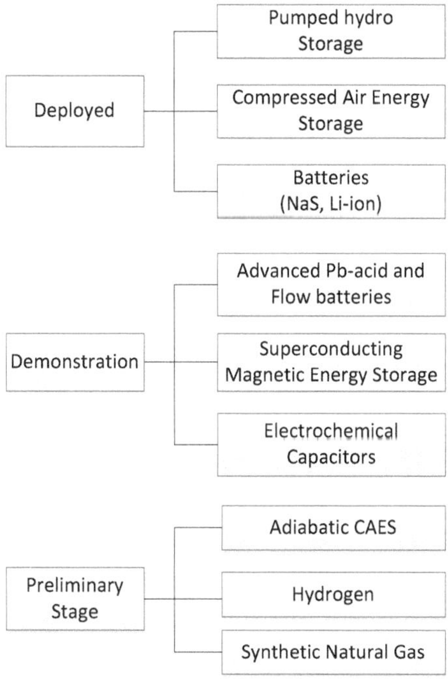

Figure 13.2 Technical readiness of energy storage system.

used for short-duration power requirements due to their very high initial energy rating cost, where battery technologies are not well suited. Meanwhile, the most efficient application of CAES is associated with a thermal combustion power plant. Each energy storage technology has its unique characteristics and advantages and thus can be applied in different fields. Fig. 13.2 illustrates the levels of maturity and development of different energy storage technologies.

Among various types of ESS technologies, BESS is the most cost-effective one and is well established for operation in a micro-grid environment [7]. Batteries are made up of a series of stacked cells where the chemical energy is transferred into electrical energy, and vice-versa. According to the most recent report from research and consulting from DOE Global Energy Storage Database [8], the global battery market is expected to reach $135.43 billion by the year 2027.

Generally, the batteries are rated in terms of their power and energy ratings. In most cases, the power and energy ratings are not independent and the ratio is fixed. Other important features of batteries include: energy efficiency, life-span (or expressed as the numbers of total charge/discharge cycles), operating temperatures, self-discharge rate, and energy density. The cost of the ESS depends on many factors and is primarily categorized into capital cost and operation and maintenance costs [9], as shown in Table 13.2. The capital costs include the power rating initial cost and the energy rating cost. The variable/ fixed O&M costs are also normalized on a per-year basis.

Table 13.2 Parameters of some types of battery energy storage systems.

Battery technology	Power rating initial cost ($kW)	Power rating initial cost ($kWh)	Fixed O&M fixed cost ($kW/year)	Variable O&M cost ($kW/year)	Life span (years)
Lead-acid batteries	315	325	17.6	6.5	15–30
Lithium-ion batteries	1500	600	10	30	20
Nickel-cadmium batteries	640	1197	26.5	1.0	10–15
Zinc-bromine batteries	366	467	30	8.8	10
Sodium-sulfur batteries	305	550000	13.8	6.7	15
Vanadium redox batteries	2125	213	54.8	7	10
Polysulfide-bromide batteries	1312	131	11.6	1.9	10

Of all the battery technologies, the lead-acid battery is the most mature with the longest history and has been widespread both in industry and academia society [10]. The Li-ion, NaS, and VRB batteries are newly developed technologies. The Li-ion battery seems to be perfectly suitable for portable devices, while the NaS is applied mostly for high-power-density applications. VRB is suitable for bulky sizes and is ready to be used for grid energy storage. To be more specific, the battery energy storage application in a distribution network should be classified based on different purposes according to different time frames:

- Peak shaving in LV grid: Typical duration 2 hours
- Load leveling in LV grid: Typical duration up to 8 hours
- Integration of renewable energy into the grid (hours)
- Frequency regulation: quick response (no more than 20 minutes)
- Voltage control: quick response (minutes)

BESS is emerging as an indispensable part of a flexible and efficient micro-grid through a diverse range of potential applications. The current transformation underway for the smart grid will lead to a huge demand for possible BESS applications. Due to its flexible size and high efficiency, the BESS can be utilized in various aspects [11], such as line upgrade deferral, load management, ancillary service, peak shaving, frequency control, voltage regulation, and maximizing renewable power utilization levels. However, a majority of battery storage units still suffer from the high initial investment cost. Large-scale BESS installation requires large sums of money which cannot be covered by income even in the long run. Thus, how to reasonably integrate BESS into a micro-grid is worth studying.

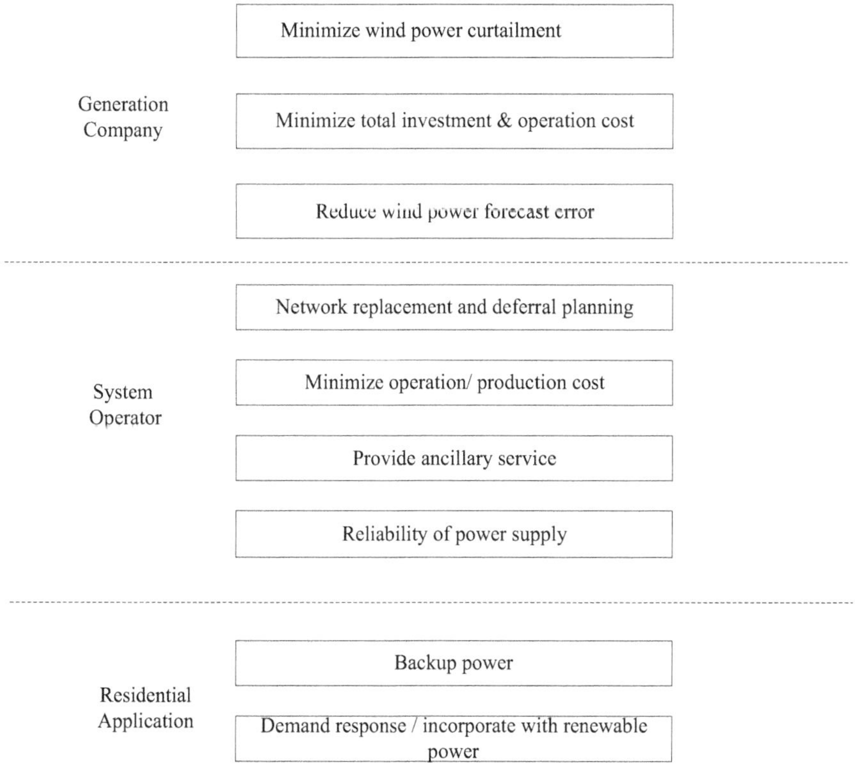

Figure 13.3 Battery energy storage systems application in power systems.

Different investors, such as generation companies or system operators, invest in BESS based on different requirements, different system levels, as well as different technologies of the BESS type. There is no sole target for allocation of BESS or general standard. Based on different perspectives of different investors or stakeholders, the purpose of BESS integration can be divided into three categories, which is shown as following (Fig. 13.3):

The location and sizing of BESS involve the planning phase as well as the operation phase, which is a mixed-integer, non-deterministic polynomial-time hard problem, and cannot be solved directly [12]. A number of different methods have been developed in the literature, which can be mainly classified into the following categories:

- Analytic method: denoting the models that do not use specific mathematical optimization tools, whereas based on the historical data set
- Mathematical programming: denoting numerical methods using the approximated model to find the optimal solution, such as:
- Linear programming
- Dynamic programming
- Stochastic programming

- Exhaustive search: usually combined with mathematical programming or simulation method for finding the optimal solution in a limited discrete search.
- Evolution algorithm: denoting the heuristic algorithms:
- Genetic algorithm
- Particle swarm optimization
- Differential evolution algorithm (DE)

13.2 Optimal planning of battery energy storage systems considering battery degradation effects

13.2.1 Related works

In general, the planning of a power system in tradition is essential to fulfilling this requirement based on the forecast load and system constraints.

Expansion planning is an important topic that has been largely discussed in the literature [13]. Long-term expansion planning is aiming to appropriately add new generation units and other facilities to meet the growing demand continuously with high economic efficiency, reliability, and stability. Expansion planning is based on the existing system, power production, and the appropriate estimation of future demand and renewable energy generation. The key factor in expansion planning is to make investment decisions on when to install the components, what type to select, and where to allocate new electric elements/equipment to supply the load demand adequately and securely at minimum cost. Planning solutions are obtained based on the planner's knowledge, experience, and optimization techniques. Consideration is given to different fields, methods to be used, the elements to be affected and the time horizon to be observed [14].

Battery energy storage systems planning to facilitate renewable energy penetrations has been extensively studied in the literature. According to different objectives, the existing models can be categorized into three types: economic, operational, and technical. The economic type is usually based on the cost-benefit analysis, aiming to minimize the lifecycle planning cost or maximize the total revenue [15]. The operational objected planning considers different indices, such as loss of load expectation [16], expected power not supplied [17], etc. The technical-based planning usually considers the technical improvements after BESS integration, such as voltage deviation mitigation and frequency adjustment [18]. Various optimization algorithms have been employed for solving optimization problems, including the decomposition method, mixed-integer nonlinear programming, and heuristic algorithms.

The battery's remaining cycling number is limited. BESS lifetime duration depends on Frequent usage and depths of discharge (DOD) [19]. Different use modes affect battery life duration. Ref. Zhang et al. [20] find that frequent usage & in-depth charging/ discharging will greatly shorten the battery service life.

In this section, we propose a framework for the BESS investment plan in microgrid with abundant renewable energy resources considering battery degradation effects. The objective of the proposed method is to minimize the total planning cost. The uncertainties of solar power and load demand variations are modeled through the scenario analysis.

13.2.2 Battery lifetime modeling

The main methods for modeling battery lifetime degradation include: a data drive, semi-empirical model, and electrochemical model [21]. The factors influencing the battery life cycle can be categorized into two types: non-operational factors (ambient temperature, ambient humidity, calendar aging) and operational factors (cycle depth, over charge/discharge, and average state of charge). In planning problems, we only consider the operating factor since the battery working condition is assumed to be ideal. A widely used approximately lifetime estimation model is adopted in this chapter. Fig. 13.4 shows the relationship between the numbers of life cycle versus DOD of Nickel-cadmium batteries [22].

Research results suggest that the life cycle numbers of each battery are finite, and can be expressed by the sum of the effective ampere-hours throughput at the rated DOD and rated discharge rate during its useful life. Any discharge event can be converted to equivalently effective ampere-hour discharge, depending on the actual DOD relative to the rated depth of discharge. Once the cumulative ampere-hour equals the rated charge life of BESS, the BESS needs to be scrapped. The rated charge life is shown as:

$$\Gamma_R = L_R D_R C_R \tag{13.1}$$

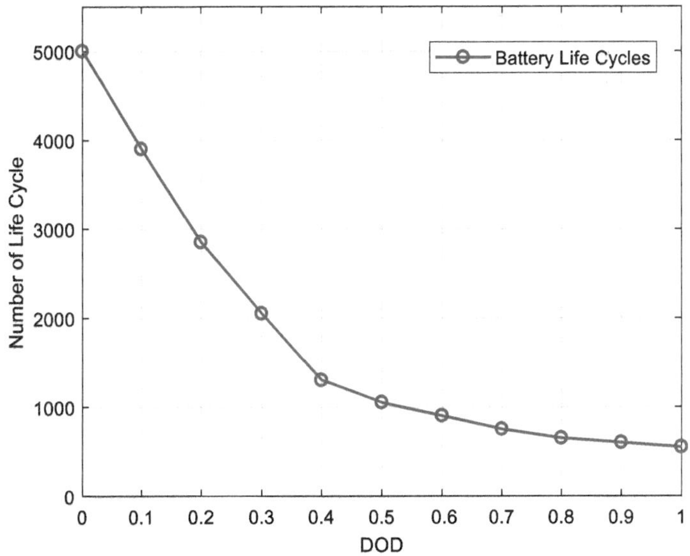

Figure 13.4 Fitting curve for lifecycles and depths of discharge of a Ni-Cd battery.

The functional relationship between the numbers of BESS life cycle and battery DOD can be expressed as:

$$L_B(d_B) = L_R \times \left(\frac{D_R}{D_A}\right)^{u_0} \times e^{u_1\left(1-\frac{D_A}{D_R}\right)} \tag{13.2}$$

where: u_0, u_1, and u_2 are the life cycle curve fitting parameters of batteries.

The actual life cycle can be calculated as a function of the depth of discharge at which it is cycled, shown as:

$$d_{\text{eff}} = \left(\frac{D_A}{D_R}\right)^{u_0} \times e^{u_1\left(\frac{D_A}{D_R}-1\right)} \frac{C_R}{C_A} d_{\text{act}} \tag{13.3}$$

Combining Eqs. (13.2) and (13.3), the BESS actual lifetime under a specific usage pattern containing n discharge events in system operation time T, can be calculated as follows:

$$L_{\text{time}} = \frac{\Gamma_R}{\Gamma_{\text{eff}}/T} = \frac{L_R D_R C_R}{\sum_{i=1}^{n} d_{\text{eff}}(i)} T \tag{13.4}$$

13.2.3 Taguchi's orthogonal array testing-based uncertainty modeling

Taguchi's orthogonal array testing (TOAT) is a statistical method [23] for providing representative uniformly distributed scenarios of combinations for all uncertain variables. In this chapter, this approach is used to represent the uncertain power output of renewable power generation. The computation is simpler, convenient than stochastic optimization/robust optimization.

Assuming an uncertainty set includes F stochastic variables, with B different output levels, therefore, the uncertainty set has B^F combinations in total. When the number of variable F is large, the number of scenarios increases rapidly. Therefore, TOAT is utilized to generate selected scenarios according to orthogonal arrays (OAs) as L_H (B^F). H represents the number of optimized scenarios, B and F denotes the level and number of variables, respectively. The detailed OA formulation rules can be found in the following URL [23].

To depict the uncertainty of DG and load demand, this work employs TOAT to generate an optimal set of scenarios. The uncertainty is modeled as three levels indicating maximum, average, and minimum values of generation output and load demand, respectively. It should be noticed that in this work, each variable modeled through TOAT presenting one period (e.g., 1 hour), and independently with each other. The combination of all the variables forms the scenario set, each scenario includes 24 time periods.

13.2.4 Problem formulation

In this chapter, the objective is to find the optimal planning scheme of BESS to minimize total investment & operation cost considering battery degradation effects. Uncertainties such as random PV power and load variation are generated using the TOAT method. In particular, a scenario is defined as a possible realization of all the random variables which represent a 24-h interval. At first, a set of planning solutions representing the allocation scheme of BESS is randomly generated; subsequently, the operation costs can be calculated under all the different scenarios. The generalized representation of the proposed model is shown in Fig. 13.5.

13.2.4.1 Mathematic formulation

This section proposes a two-layer planning model of BESSs in micro-grid, aiming to minimize the total life cycle considering battery degradation effects during operation. Specifically, the objective function is classified into three parts as follows:

(1) BESS unit investment cost.
(2) The operation & maintenance cost is determined based on the capacities of BESS.
(3) The operation cost under selected representing scenarios.

It should be noticed that: (1) we assume BESS needs to be decommissioned when its capacity degrades to 20% of its initial nominal capacity; (2) The renewable generation serves the local load demand firstly, the excess power is used to charge the battery, and battery release power to the grid during peak-load periods.

The proposed planning model is a co-optimization problem that contains planning and operation problems, respectively. In the upper layer, the decision variables

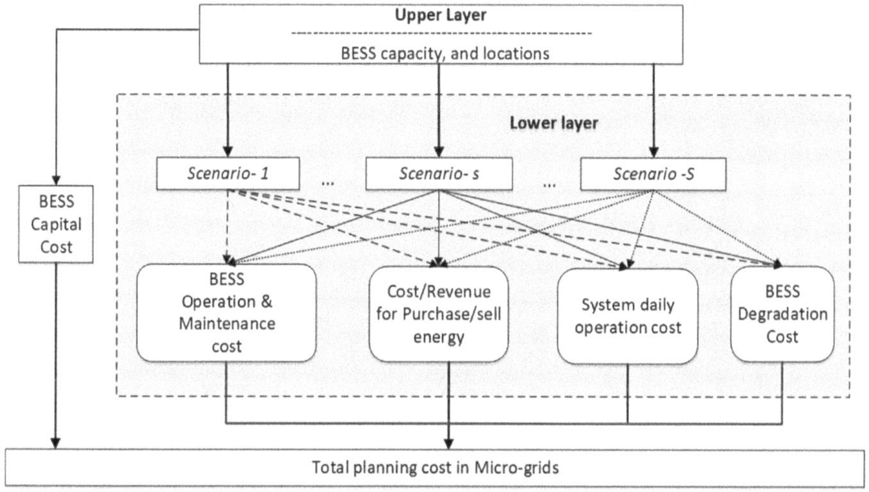

Figure 13.5 Proposed two-layer planning framework.

include the location and capacity of BESS units. In the lower layer, the decision variables are charging/discharging powers of the BESS, the purchasing power from the external grid, and the MT generation power on an hourly basis. The operation costs are calculated based on a group of planning solutions through the lower operation layer. Meanwhile, the detailed degradation model is embedded in the lower layer.

13.2.4.2 Upper layer model

In the upper layer, the total planning cost including the investment and maintenance costs of BESSs:

$$\min\{(C_{\text{inv}} + C_{\text{opm}}) + \text{Exp}_{\xi}[Q(x, \xi)]\} \tag{13.5}$$

The total investment costs include the capital cost of BESSs shown as Eq. (13.6). The daily maintenance is proportional to the power capacity of BESS. For consistency, the annual capital cost is normalized on a daily basis.

$$C_{\text{inv}} = \frac{1}{365}\frac{d(1+d)^{N_{\text{BES}}}}{(1+d)^{N_{\text{BES}}+1} - 1}\sum C_E C^i_{\text{batmax}} \tag{13.6}$$

$$C_{\text{opm}} = \sum \left(P^i_{\text{bat,max}} C_{O,\text{BESS}}\right) \tag{13.7}$$

where $\text{Exp}_{\xi}[Q(x, \xi)]$ is the expected operating cost under the operational scenario ξ, which will be calculated in the lower layer.

The objective function (13.5) is subjected to the overall investment budget constraint and the location constraint,

$$\sum C_E C^i_{\text{batmax}} \leq \text{CAP} \tag{13.8}$$

$$L_{\text{BES}} \in \Phi \tag{13.9}$$

13.2.4.3 Lower layer model

The objective of the lower layer is to minimize the total operation cost considering the battery degradation cost. In the lower layer model, the optimal solutions are tested for each scenario ξ; the expected value of system operational costs is calculated, shown in Eq. (13.10). Eq. (13.10) consists of four terms: the cost of local microturbine units C^{ξ}_{gen}[Eq. (13.11)], costs of purchasing power from the external grid C^{ξ}_{pur} [Eq. (13.12)], network power loss cost C^{ξ}_{loss} [Eq. (13.13)] and BESS degradation cost C^{ξ}_{dep} [Eq. (13.14)]. The objective function of the Lower layer is as follows:

$$\min\left(\text{Exp}_{\xi}[Q(x, \xi)]\right) = \min\sum_{\xi \in \Omega}\left(C^{\xi}_{\text{gen}} + C^{\xi}_{\text{pur}} + C^{\xi}_{\text{loss}} + C^{\xi}_{\text{deg}}\right) \tag{13.10}$$

$$C_{\text{gen}}^{\xi} = \sum_{i \in I} \left(a_i + b_i P_{i,t}^{\text{DA}} \right) \tag{13.11}$$

where a_i,b_i refers to the cost coefficients of ith micro-turbine.

$$C_{\text{buy}} = \sum\nolimits_{t=1}^{T} MP_t P_{\text{buy},t}^{\xi} \tag{13.12}$$

$$C_{\text{loss}} = \sum\nolimits_{t=1}^{T} \eta_{\text{loss}} P_{\text{loss},t}^{\xi} \tag{13.13}$$

Here:

$$P_{\text{loss},t}^{\xi} = \sum_{i \in \Phi} \sum_{j \in \Phi} \Delta v_{ij}^2 \frac{r_{ij}}{x_{ij}^2} \tag{13.14}$$

$$C'_{\text{deg}} = \left(C_{\text{inv}} / \Gamma_R \right) \times \sum\nolimits_{\xi} \sum\nolimits_{t} \Gamma_{\text{eff}}(t) \tag{13.15}$$

Eq. (13.10) is subjected to the following constraints:

(1) Active Power balance equations

$$P_{i,t}^{\xi} + P_{\text{DG},t}^{\xi} + P_{\text{ch/dis},i,t}^{\xi} + P_{\text{buy},t}^{\xi} - P_{\text{load},t}^{\xi} = U_{i,t} \sum U_{j,t} \left(G_{ij,t} \cos\theta_{ij,t} + B_{ij,t} \sin\theta_{ij,t} \right) \tag{13.16}$$

where $P_{\text{DG},t}^{\xi}$ refers to the active power output at time t of ith DG in scenario $\mathcal{E}$. G_{ij} and B_{ij} refers to conductance and susceptance matrix between bus i and j

(2) System operation constraints:

- Line current constraint

$$I_{ij,t}^{\xi} \leq I_{max} \tag{13.17}$$

- Node voltage constraint

$$v_{min} \leq v_{i,t}^{\xi} \leq v_{max} \tag{13.18}$$

- Line transmission power constraint

$$-f_l^{\min} \leq f_{l,t}^{\xi} \leq f_l^{\max} \tag{13.19}$$

- Load balance equation:

$$P_{i,t}^{\xi} + P_{\text{DG},t}^{\xi} + P_{\text{buy},t}^{\xi} + P_{\text{ch/dis},i,t}^{\xi} = P_{\text{load},t}^{\xi} + P_{\text{loss},t}^{\xi} \tag{13.20}$$

- Reverse power flow constraint

$$0 \leq f_t^{\xi} \leq f_s^{\max} \tag{13.21}$$

(3) BESS operation constraints:

- BESs power constraint

$$0 \leq P^{\xi}_{\text{ch},i,t}, \left|P^{\xi}_{\text{dis},i,t}\right| \leq P^{i}_{\text{batmax}} \tag{13.22}$$

- BESS SOC constraint

$$\text{SOC}^{\xi}_{i,t} = C^{\xi}_{\text{bat},i,t} / C^{i}_{\text{bat},max} \tag{13.23}$$

- BESS energy transition constraint

$$C^{\xi}_{\text{bat},t} = C^{\xi}_{\text{bat},t-1} + \left(P^{\xi}_{\text{ch},i,t}\eta_{\text{ch}} - P^{\xi}_{\text{dis},i,t}/\eta_{\text{dis}}\right)\Delta t \tag{13.24}$$

(4) DG power output constraints:

- DG active power output constraint

$$0 \leq P^{\xi}_{\text{DG},i,t} \leq P_{\text{DG},i,rated} \tag{13.25}$$

where $P^{i}_{\text{DG,rated}}$ denotes the rated capacity of ith DG units.

13.2.5 Solution approach

The formulation of the proposed objective function is essentially an optimization problem contains two layers. The upper layer determines the candidate solution of BESS units while the lower layer calculates the output power of BESS units, MT generations, and power purchased from the external grid on an hourly basis. The proposed model is a two-layer mixed-integer nonlinear programming problem; such a problem cannot be solved by conventional optimization methods. In this paper, a two-layer DE is selected to find the optimal solution for the objective function. The flowchart is shown in Fig. 13.6 summarized the proposed algorithm.

13.2.6 Simulation results

Numerical experiments are carried out to validate the effectiveness of the proposed two-layer planning model.

13.2.6.1 Parameter setting

The modified IEEE 11 kV, the 15-bus radial system is used to validate the proposed model. The topology of the system is shown in Fig. 13.7. The parameters of the system are obtained from [24]. The main substation at bus 1 is used to feed a rural area. The distribution system consists of two micro-turbine units located at nodes 4 and 6,

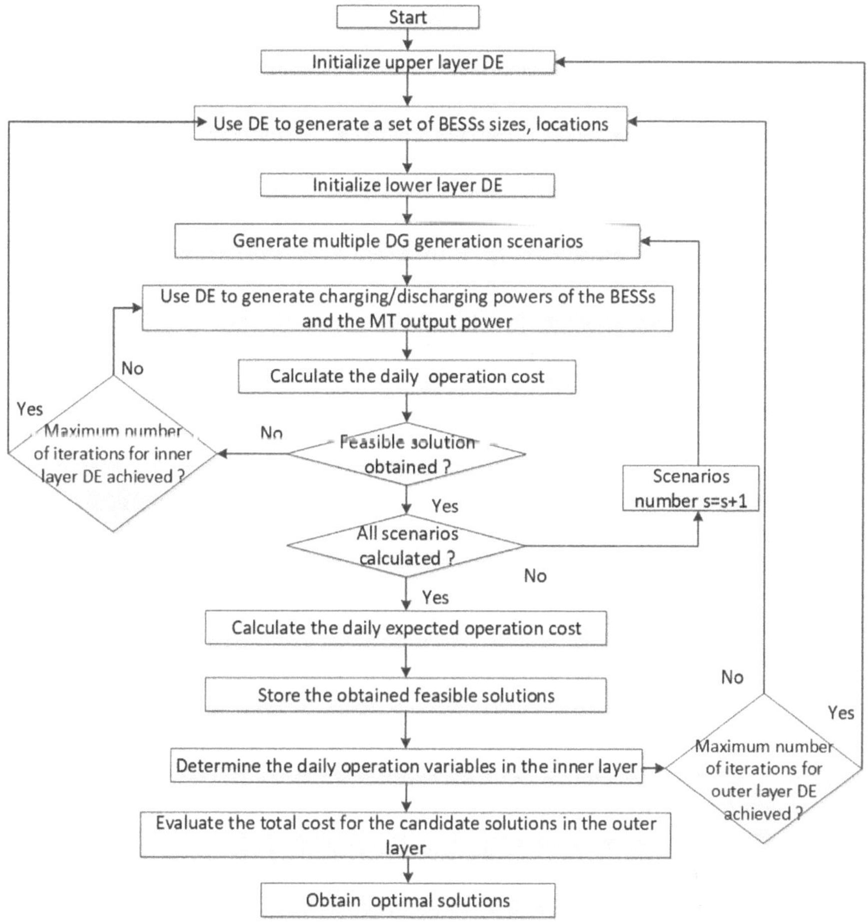

Figure 13.6 Flowchart of the proposed solution approach.

respectively; and one PV module connected to bus 11. Parameters of the micro-turbine units are given in Table 13.3.

In this study, two types of batteries are considered. Type I is the lead-acid battery and type II is the zinc-bromine battery. The parameters of batteries in the simulation are given in Table 13.4. Generally, the power capacity is proportional to the energy capacity of the Li-ion battery. In this work, we assume P^i_{batmax} equals to a sixth of the total amount of $C^i_{\text{bat max}}$. The solar panels considered for this study are the BP3200SX panels.

The population of the bi-level optimization problem is set as 100 for each layer; the condition for the termination of the calculations is attaining accuracy of 10^{-4} within 100 iterations. In this work, the total number of scenarios S are assumed as 100.

A 10-year hourly load data are collected from [25] to represent the load variation and different demand patterns. Hourly solar radiation data of 10-year time-series in

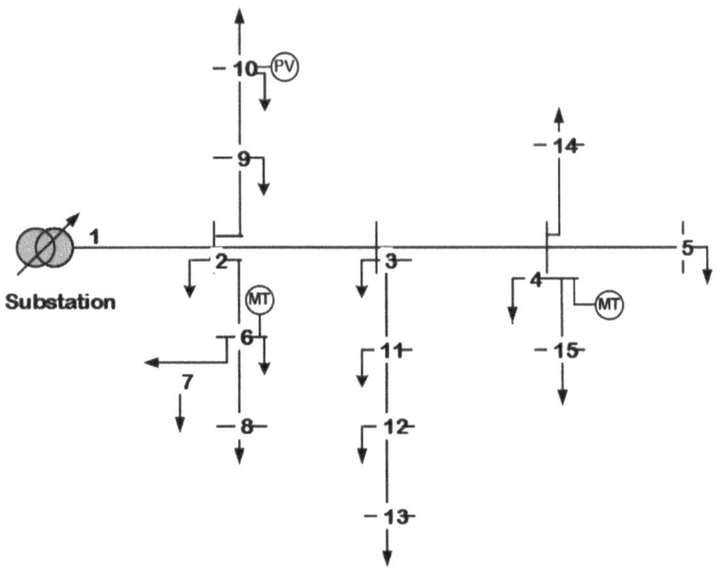

Figure 13.7 Modified IEEE 15-bus distribution radial system.

Table 13.3 Micro-turbine generator data.

Gen #	a_i ($)	b_i ($/MW)	$P_{i,\min}$ (MW)	$P_{i,\max}$ (MW)
1	30	130	0.1	0.6
2	50	350	0.1	0.9

Table 13.4 Simulation parameter setting of battery energy storage systems.

Battery type	L/A	Zn/Br
BESS Operational Parameters		
SOC upper limit (%)	80	
SOC lower limit (%)	20	
Initial SOC of BESS (%)	50	
Round trip efficiency (%)	75	77%
Life cycle (times)	2000	1500
Self-Discharge rate [%Energy/day]	0.24%	0.18%
BES Capital Parameters		
C_E ($/kWh)	225	325
C_{O_BESS} ($ per day)	20	40

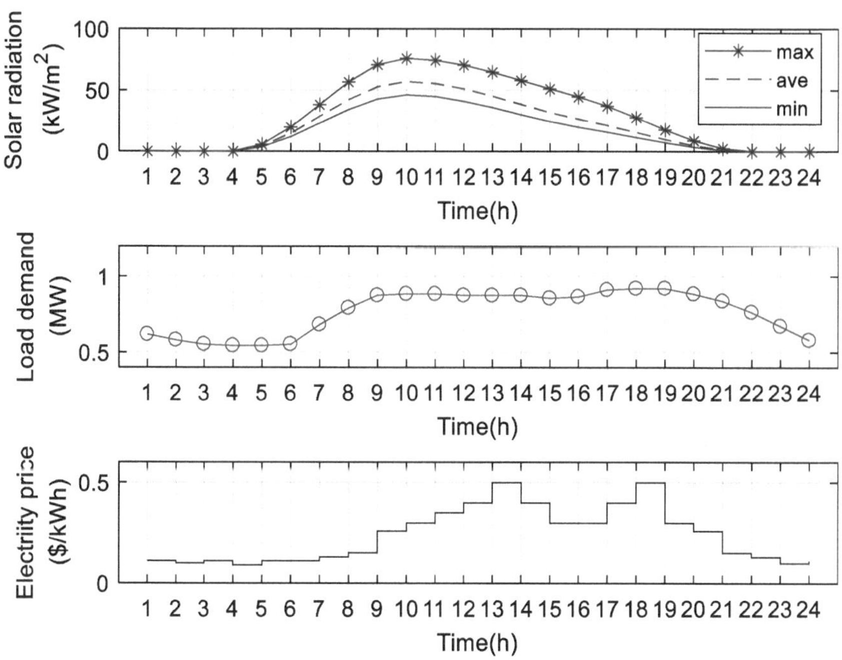

Figure 13.8 Typical daily solar radiation, load demand, and electricity price curves.

eastern Australia are collected from the RDA website to simulate the variation of PV output. Note that there is no PV output almost half of the time during the day, the valid PV generation periods start from 6:00 am and end at 19:00 pm. Therefore, the number of variables F equals 13 and each variable has two levels as mentioned before. The PV power generation during the rest time of the day is set as zero. Consequently, OAL_{27} (3^{13}) is selected and there are a total of 27 testing scenarios. In this paper, the dynamic electricity pricing is considered, and the time-varying market prices for purchasing energy from the external grid are obtained from [26]. Load profiles, three levels of solar radiation, and time-varying electricity prices on an hourly basis are shown in Fig. 13.8.

13.2.6.2 Results analysis

In the case study section, two cases are conducted: (I) Base case, planning neglecting BESS degradation cost; (II) Case I, BESS optimal planning considering BESS degradation;

Firstly, we conduct the optimization for each type of battery separately. The BESS allocation results of the two types of batteries in case I are shown in Table 13.5. The numerical results show that the Lead/Acid batteries would be a better choice due to the relatively low investment cost. The MT generator outputs and the power purchased from a substation in one typical day with/without L/A BESS installation are shown in Fig. 13.9. The results clearly show that the power bought from the

Table 13.5 Battery energy storage systems allocation optimization results under the base case.

BESS type	**LA**		**Zn/Br**	
Battery location	10	14	10	27
Energy capacity (MWh)	1.31	1.09	1.25	1.07
Power loss (MW)	2.34		2.31	
Investment cost ($/day)	244.9		236.8	
Maintenance cost ($/day)	90.1		116.3	
Operation cost($/day)	14504		15349	
Battery life cycle degradation cost($/day)	171.9		196.7	
Total cost($/day)	16866		17862	

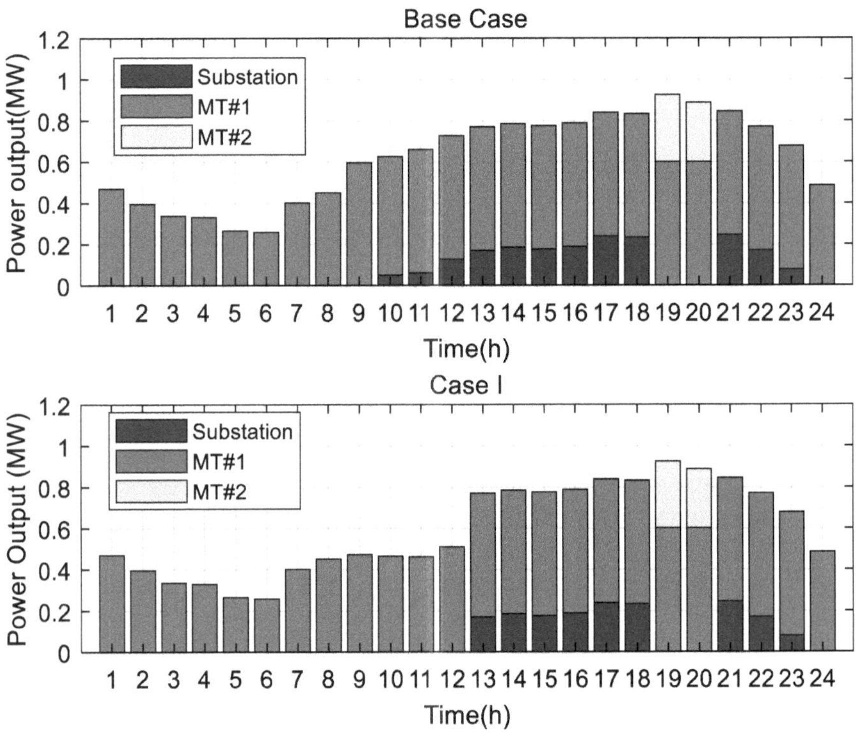

Figure 13.9 Hourly profiles for system variables.

substation largely reduced with the installation of BESS (hour 10–12), which therefore reduces the system operation cost. To illustrate the efficiency of the L/A BESS, the SOC profiles of each BESS unit in a specific scenario are shown in Fig. 13.10.

Then, we conduct the optimization for L/A batteries considering battery degradation effects. The BESS allocation results in case 6 are shown in Table 13.6. It can be

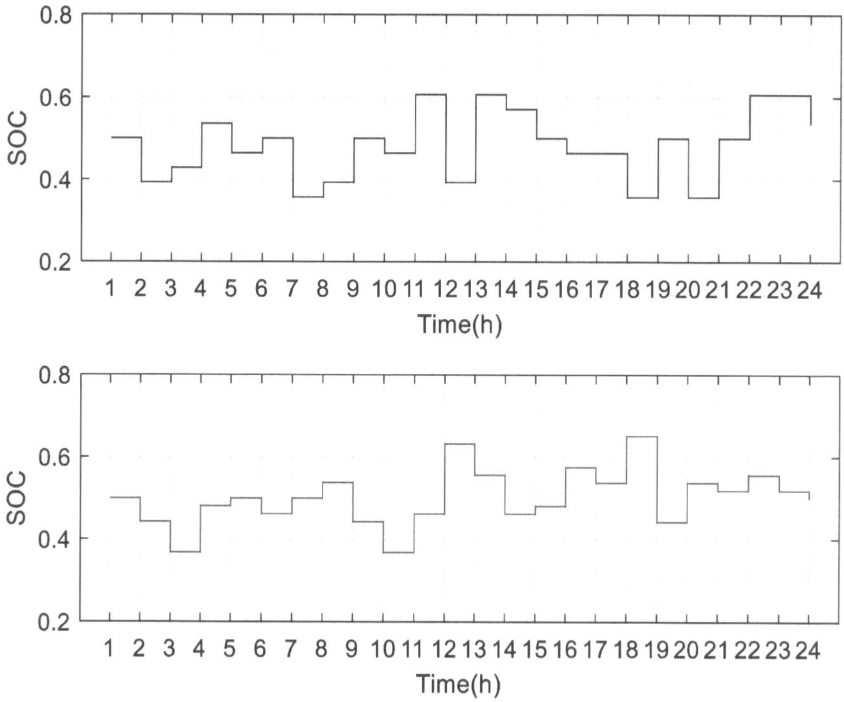

Figure 13.10 Hourly SOC profiles for battery energy storage systems units without degradation consideration.

Table 13.6 Battery energy storage systems allocation optimization results under case I.

BESS type	LA	
Battery location Energy capacity (MWh)	10 1.21	12 1.15
Power loss (MW) Investment cost ($/day) Maintenance cost ($/day) Operation cost ($/day) Battery life cycle degradation cost ($/day) Total cost ($/day)	2.31 239.65 87.39 14504 121.7 15,921	

seen from the results that battery life cycle cost was reduced by 29.20% compared to the case I. Meanwhile, the total operation cost was reduced by 5.6%, as well.

To illustrate the efficiency of the L/A batteries in case II, the SOC profiles of each BESS unit under a specific day are shown in Fig. 13.11. From Fig. 13.11 we can find that BESS charges in off-peak-price hours and discharges in peak-price hours and all BESS units go through a complete charge/discharge cycle under case

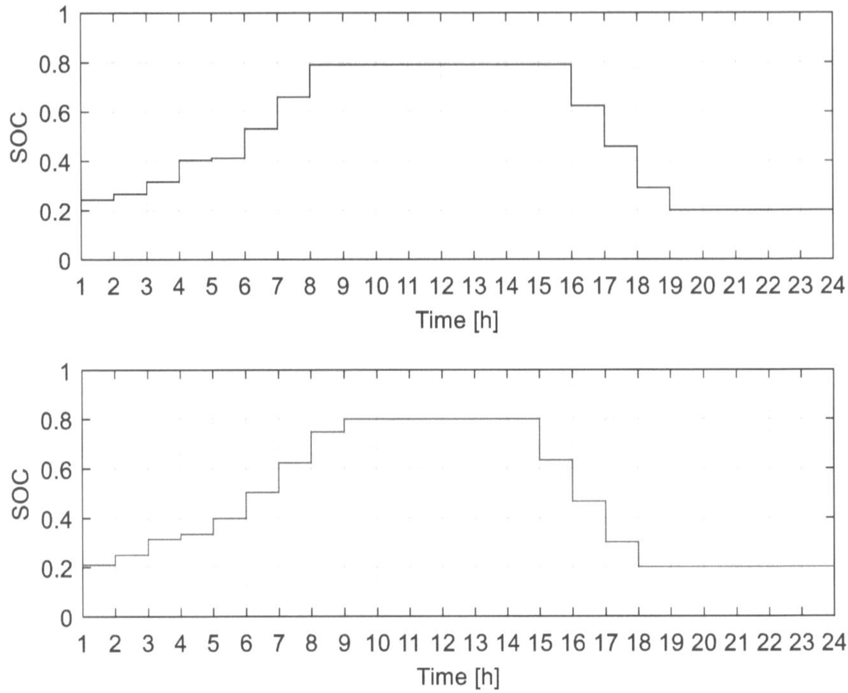

Figure 13.11 Hourly SOC profiles for battery energy storage systems units with degradation consideration.

II. Compared to Case I, a lower charging/discharging rate declines the BESS degradation rate, thus reducing battery degradation effects.

13.2.7 Discussion

In this section, a BESS planning model considering the battery degradation effect is proposed. Uncertainties are modeled using scenario analysis. A Monte–Carlo embedded two-layer DE approach is applied to solve the proposed problem.

A series of simulation studies are conducted on a modified IEEE 15-bus benchmark system. The simulation results demonstrate the performance of the proposed model and the method.

13.3 Risk-constrained two-stage coordinated operation of battery energy storage systems

13.3.1 Related works

Recently, there are a lot of work on the BESS optimal scheduling in micro-grid with renewable energy resources. BESS is commonly used for voltage regulation

[27], power loss minimization [28], renewable energy integration facilitation [29], and other purposes to improve micro-grid operation. While other works focus on economic dispatch problems, such as network reconfiguration and optimal placement. The integration of BESS has complicated the short-time operation since more variables and uncertainties must be taken into account deals while making decisions. Such kinds of problems are often formulated as a two-stage stochastic programming model. There are many examples of two-stage stochastic programming methods applied to power systems using ESS in the literature. Ref. Lopez et al. [30] establish a joint optimization of wind generation and pumped storage units in electricity markets. Ref. Najafi et al. [31] propose risk-based scheduling of an ESS and a fleet of EVs in day-ahead energy and regulation markets of a smart energy hub. Ref. Zhang et al. [32] propose a two-stage coordinated operation method to coordinate combined cooling, heat, and power (CCHP) plants, flexible electric and thermal loads, and thermal storage considering multiple uncertainties. Ref. Huiling et al. [33] propose a control optimization method in distribution systems with PVs/DGs considering power loss and voltage offset risks. In the past few years, significant attention has been paid to the theoretical and practical development for solving two-stage optimization problems, such as Benders decomposition, column-and constraints algorithm, and Evolution algorithm.

On the other hand, since the capital cost of BESS is still expensive (around 150−200$/kWh), it would be reasonable for investors to consider the operational risk in advance (load variation, DG output uncertainty). Using the expected values as an objective function cannot guarantee an optimal solution, which will inevitably lead to small probability but high-risk scenarios with overall high cost.

In this section, we propose a framework of two-stage operation optimization in micro-grid to optimally coordinate the power output of thermal generators, BESS, while hedging the risk induced by renewable power generation and load demand variation. Specifically, in the first stage, the optimal scheduling strategy for day-ahead dispatch is formulated and the decision variables include thermal generator output and power purchased from the external grid; in the second stage, the intraday dispatch scheme is modeled according to the updated renewable power generation information and load demand variation, the decision variables include batteries charging/discharging power and real-time power purchased from the external grid, respectively. Risk analysis is realized through the CVaR theory, which is incorporated in the objective model to control the operation cost such as un-served load. CVaR is an improved version of Value at Risk, which has superior mathematical properties and has been used in power system scheduling and electricity market problems [34]. In this study, the CVaR-based risk-aversion model is established and compared to the risk-neutral model.

13.3.2 Battery energy operational cost modeling

The costs of battery are distributed over the number of deep discharge cycles. The cost per discharging energy unit of BESS depends on battery capital cost, degradation model, and end-of-life (EOL) condition.

As mentioned in Section 13.2.2, The BESS life cycle model is expressed as Eqs. (13.1−13.4), therefore, combining Eqs. (13.1−13.4), the cost per kWh of batteries can be obtained as:

$$c_{\text{age}} = \frac{d_{\text{eff}}}{\Gamma_R} C_{\text{inv,BESS}} \tag{13.26}$$

where $c_{\text{age},k}$ denotes per unit capital cost of BESS

13.3.3 Problem formulation

This paper proposes a risked constrained two-stage optimal scheduling model with energy storage systems in micro-grids, as shown in Fig. 13.12.

In the first stage, SO takes charge of the whole system and schedules the micro-turbine, power purchased from the external grid based on the forecasted data of renewable energy output and load demand in the day-ahead market. The decision variables include the output power of micro-turbine units, which are made before the uncertainties are revealed. Due to the variation between forecasted data and actual value, the system generation may meet shortfalls to balance. In the second stage, the uncertain factors are modeled using scenario analysis, and the output power of BESSs and power purchased from the external grid are recalculated in a different scenario. The risk aversion of SO is captured using the CVaR approach.

It should be noticed that: (1) The scheduling decision of when and which batteries will function depends on the electricity market price. If the per-unit energy price

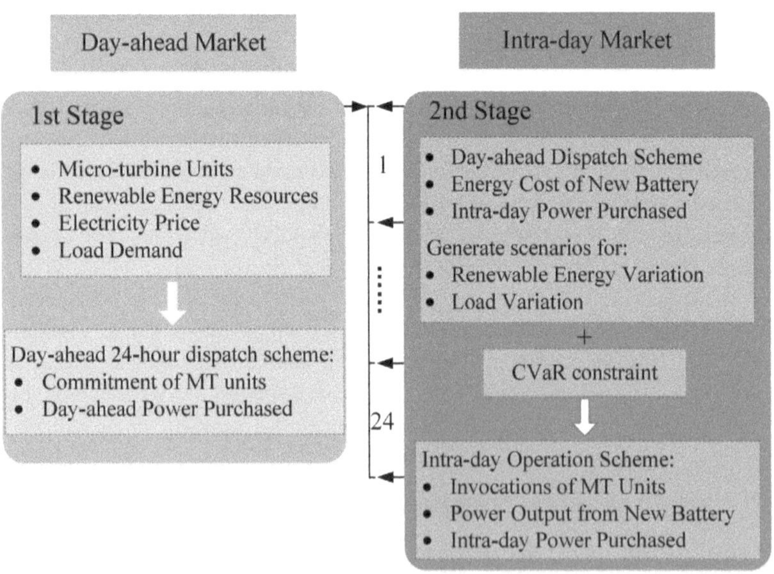

Figure 13.12 Two-stage optimal scheduling framework.

is higher than the per-unit discharging cost in batteries, the SO may decide to release the energy from batteries to support load demand, and vice versa. (2) The degradation cost of batteries are depending on the current cycling number and the EOL criterion; (3) The excess renewable energy output can be utilized to charge the batteries.

A. Uncertainty modeling

(1) Solar power prediction model:

Solar generation: Solar power can be calculated as [35]:

$$P_{PV,t} = \eta_{PV} S_{PV} I_t \tag{13.27}$$

where $P_{PV,t}$ denotes PV generation at time t, $\overline{I}_t$ denotes the estimated value of solar irradiance (kW/m^2), η_{PV} is the solar energy conversion efficiency (%), and S_{PV} is the solar panel size (m^2). In addition, we model the actual solar radiation I_t in time t as:

$$I_t = \overline{I}_t + \mu(I_t) \tag{13.28}$$

where μI_t is assumed to obey zero-mean Gaussian distribution which represents the estimation error.

(2) Power Demand Prediction Model:

Historical data can be used to fit the distribution function to represent the load forecasting error. In literature, the normal distribution is usually applied to model the load forecast uncertainty. The load demand $P_{load,t}$ at time t (kW) is calculated:

$$P_{Load,t} = \overline{P}_{load,t} + \delta_{load,t} \tag{13.29}$$

where $\overline{P}_{load,t}$ and $\delta_{load,t}$ represent the predicted load value and the prediction error, respectively. In this work, the load prediction error is assumed to follow a truncated normal distribution with zero mean and a standard deviation of 5% [36].

B. Objective function

13.3.3.1 First stage- day ahead dispatch model

The first stage problem is modeled as follows:

$$F^{DA} = \sum_{t \in T} \left(\sum_{i \in I} \left(a_i + b_i P_{i,t}^{DA} \right) + \lambda_{g,t} P_{grid,t}^{DA} \right) \tag{13.30}$$

where a_i, b_i refers to the cost coefficients of ith micro-turbine. $P_{grid,t}^{DA}$ and $P_{i,t}^{DA}$ refers to Power bought from the external grid at time t and power output of ith micro-turbine at time t in the day-ahead market. $\lambda_{g,t}$ refers to the time-varying price of purchasing power from external gird ($MW/h).

The dispatch cost in the day-ahead market is shown as Eq. (13.30). The objective function contains two parts: the micro-turbine generation costs and the cost of purchasing energy from the external grid.

Model (17) is subjected to the following constraints:

(1) Power balance constraints:

$$\sum_{i \in I} P_{i,t}^{\mathrm{DA}} + P_{\mathrm{PV},t}^{\mathrm{pre}} + P_{\mathrm{grid},t}^{\mathrm{DA}} = P_{\mathrm{Load},t}^{\mathrm{DA}} \tag{13.31}$$

where $P_{\mathrm{PV},t}^{\mathrm{pre}}$ refers to Predicted PV power output (MW). $P_{\mathrm{Load},t}^{\mathrm{DA}}$ refers to Load demand at time t in the day-ahead market(MW).

The load demand in the day-ahead market must be equal to the sum of micro-turbine units, solar power generation, and purchased energy from the external grid.

(2) Micro-turbine constraint in the day-ahead market:

The output power of ith micro-turbine is constrained by the generation capacity, expressed as:

$$P_{i,min} \le P_{i,t}^{DA} \le P_{i,max} \tag{13.32}$$

where $P_{i,min}$ $P_{i,max}$ refers to the minimum and maximum output power of ith micro-turbine units.

(3) PV power generation constraints in the day-ahead market:

The PV power generation should not exceed the rated PV panel capacity, expressed as:

$$0 \le P_{\mathrm{PV},t}^{\mathrm{pre}} \le P_{\mathrm{PV,\ rated}} \tag{13.33}$$

(4) Limitation of purchasing power from upstream in Day-ahead market

$$0 \le P_{\mathrm{grid},t}^{\mathrm{DA}} \le P_{\mathrm{grid},max} \tag{13.34}$$

where $P_{\mathrm{grid},max}$ refers to the maximum allowable energy from upstream.

13.3.3.2 Second stage-intra-day dispatch model

The expected value of operation cost in the second stage includes the degradation costs of batteries, and the cost of purchasing electricity with the main grid. The power shortage is the sum of solar power prediction error, which is calculated as:

$$P_{\mathrm{shortage}} = \sum_{\omega \in \Omega} p(\omega)\left(\left(P_{\mathrm{PV},t}^{(\omega)} - P_{\mathrm{PV},t}^{\mathrm{pre}}\right) + \left(P_{\mathrm{load},t}^{\mathrm{DA}} - P_{\mathrm{load},t}^{(\omega)}\right)\right) \tag{13.35}$$

where $P_{\mathrm{PV},t}^{(\omega)}$ and $P_{\mathrm{Load},t}^{(\omega)}$ refers to real-time solar panel output power and Load demand at time t in the intra-day market in the scenario ω, respectively. $p(\omega)$ denotes the probability of ω the scenario.

In this chapter, the second stage objective function can be presented as:

$$F^{\mathrm{ID}} = \sum_{\omega \in \Omega} \left(p(\omega) \sum_{t \in T} \left(\sum\nolimits_{k \in K} c_{\mathrm{age},k}^{\mathrm{BESS}} \bullet P_{k,\mathrm{ch/dis},t,\omega}^{\mathrm{BESS}} + \lambda_{g,\ t} \cdot P_{\mathrm{grid},t,\omega}^{\mathrm{ID}} \right) \right) \tag{13.36}$$

$$\text{where} \begin{cases} P_{\mathrm{grid},t,\omega}^{\mathrm{ID}} = 0, \ P_{\mathrm{shortage}} = P_{k,\mathrm{dis},t,\omega}^{\mathrm{BESS}} & c_{\mathrm{age},k}^{\mathrm{BESS}} < \lambda_{g,\ t} \\ P_{k,\mathrm{dis},t,\omega}^{\mathrm{BESS}} = 0, \ P_{\mathrm{shortage}} = P_{\mathrm{grid},t,\omega}^{\mathrm{ID}} & c_{\mathrm{age},k}^{\mathrm{BESS}} > \lambda_{g,\ t} \end{cases}$$

where $P_{\mathrm{grid},t,\omega}^{\mathrm{ID}}$ and $P_{k,\mathrm{ch/dis},t,\omega}^{\mathrm{BESS}}$ refers to power bought from external grid and charging/discharging power of kth BESS in the intra-day market at time t in the scenario ω. $c_{k,\mathrm{age}}^{\mathrm{BESS}}$ refers to the per-unit capital cost of kth BESS ($/MWh).

(1) Power balance constraints:

$$\sum_{i \in I} P_{i,t}^{\mathrm{DA}} + P_{\mathrm{PV},t}^{(\omega)} + \sum_{k \in K} P_{k,\mathrm{ch/dis},t,\omega}^{\mathrm{BESS}} + P_{\mathrm{grid},t}^{\mathrm{DA}} + P_{\mathrm{grid},t}^{(\omega)} = P_{\mathrm{load},t}^{(\omega)} \tag{13.37}$$

where $P_{\mathrm{load},t}^{(\omega)}$ denotes load demand at time t in the intra-day market in scenario ω (MW)

(2) BESS operation constraint

BESS power constraint

$$0 \le \left| P_{k,\mathrm{ch/dis},t,\omega}^{\mathrm{BESS}} \right| \le P_{k,\max}^{\mathrm{BESS}} \tag{13.38}$$

where $P_{k,\max}^{\mathrm{BESS}}$ denotes the rated power capacity of kth BESS

BESS SOC constraint

$$\mathrm{SOC}_{\min}^{\mathrm{BESS}} \le \mathrm{SOC}_{k,t,\omega}^{\mathrm{BESS}} \le \mathrm{SOC}_{\max}^{\mathrm{BESS}} \tag{13.39}$$

where $\mathrm{SOC}_{t,\omega}^{\mathrm{BESS}}$ refers to the State-of-charge of kth battery at time t in scenario ω

BESS energy transition constraint

$$C_{k,\omega,t}^{\mathrm{BESS}} = C_{k,\omega,t-1}^{\mathrm{BESS}} + \left(P_{k,\omega,ch,t}^{\mathrm{BESS}} \eta_{\mathrm{ch}} - P_{k,\omega,\mathrm{dis},t}^{\mathrm{BESS}} / \eta_{\mathrm{dis}} \right) \Delta t \tag{13.40}$$

where $C_{k,t,w}^{\mathrm{BESS}}$ denotes energy stored in the kth battery at time t in scenario ω (MWh). η_{ch} and η_{dis} refers to BESS charging and discharging efficiency.

(3) Limitation of purchasing power from upstream in real-time market

$$0 \le P_{\mathrm{grid},t}^{\mathrm{DA}} + P_{\mathrm{grid},t}^{(\omega)} \le P_{\mathrm{grid},\ \max} \tag{13.41}$$

13.3.3.3 Risk management

Due to the imperfect prediction of renewable energies and load demand, it is required to address the associated uncertainty in the problem to evaluate the risk for SO. Therefore, typical scenarios are generated using the Monte Carlo simulation method. Solar radiation and load demand are incorporated in each scenario for the period (24 hour) of operation. Each scenario and its probabilities are represented by a discrete distribution function. To model the operation problem of micro-grids using BESS, a risk-based two-stage optimization approach is used. The CVaR theory is used to model the SO's risk aversion as follows:

When the confidence level is β, CVaR can be shown as:

$$\mathrm{CVaR} = \mu + \frac{1}{S(1-\alpha)} \sum_{\omega \in \Omega} \left[F^{\mathrm{RT}} - \mu\right]^{+} \tag{13.42}$$

where μ refers to the risk value(VaR), which is the smallest cost value. The term $f(x,y)$, $[f(x,y)-\mu]^{+}$ are auxiliary variables refers to the cost function, excess cost over μ in the scenario w, respectively. S represents the number of total scenarios and α refers to the confidence level of the CVaR constraint.

The final objective function is described as follows:

$$F = \mathrm{Min}\ \left\{(1-\beta) * \left(F^{\mathrm{DA}} + F^{\mathrm{ID}}\right) + \beta * \mathrm{CVaR}\right\} \tag{13.43}$$

The risk of SO in decision-making is controlled by the risk-aversion parameter β. The value of β varies from 0 to 1. For $\beta = 0$, the decision-maker is neutral. When β increases the DISCO becomes more risk-averse. The constraints (13.31–13.34) and (13.37–13.41) are considered for each scenario.

13.3.4 Numerical results

13.3.4.1 Parameter setting

In this section, we implement the proposed risk-constrained two-stage optimization model introduced in Section 13.3.3. A 10-year hourly load data are collected from [25] to represent the load variation. Hourly solar radiation data of 10-year time-series in eastern Australia are collected to simulate the variation of PV output. The solar panel considered for this study is the BP3200SX panel, whose technical parameters are obtained from [26]. Table. 13.7 lists the hourly electricity pricing for purchasing power from the external grid.

Table 13.7 Electricity price of upstream grid.

Hour	Price ($/kWh)	Hour	Price ($/kWh)	Hour	Price ($/kWh)	Hour	Price ($/kWh)
1	0.14	7	0.17	13	0.65	19	0.39
2	0.13	8	0.19	14	0.52	20	0.34
3	0.14	9	0.34	15	0.39	21	0.19
4	0.12	10	0.39	16	0.39	22	0.17
5	0.14	11	0.45	17	0.52	23	0.13
6	0.14	12	0.52	18	0.65	24	0.14

Table 13.8 Parameters for battery energy storage systems.

Technical & economical parameters	
Rated power/rated energy (kW/kWh)	200/1000
SOC limit (%)	80–40
End of life criterion (%)	80
Initial SOC of BESS (%)	50
Round trip efficiency (%)	75
End-of-life criterion (%)	80
Life cycle (times)	4580
BESS life cycle curve fitting parameters u_0, u_1	1.98,0.016
$C_{\text{inv,BESS}}$ ($/kWh)	175

Table 13.9 Micro-turbine generator data.

Gen #	a_i **($)**	b_i **($/MW)**	$P_{i,\min}$ **(MW)**	$P_{i,\max}$ **(MW)**
1	30	130	0.2	0.6
2	50	350	0.2	0.9

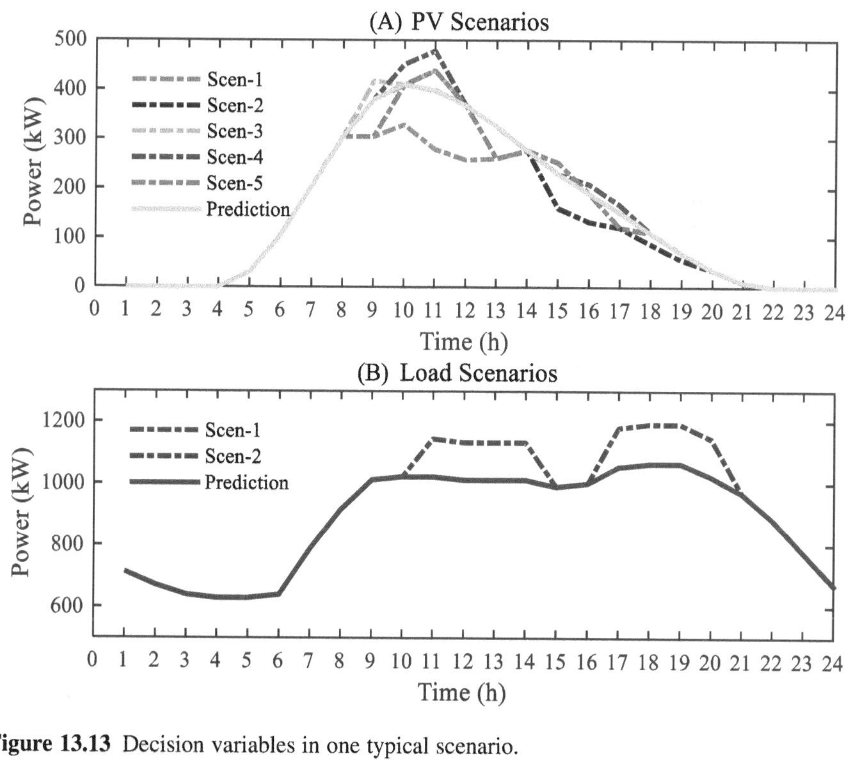

Figure 13.13 Decision variables in one typical scenario.

In this work, the Li-ion battery is selected due to its better performance compared with other types of batteries [7]. The technical & economical parameters for BESS are given in Table 13.8, respectively. The parameters of micro-turbine units are given in Table 13.8.

(Table 13.9)

In the second stage, the Monte-Carlo simulation method is used to generate scenarios representing the uncertainties: five scenarios represent PV variation, and two scenarios indicate load variation, which is 2•5 = 10 scenarios in total. And each scenario is assigned with equal probabilities. The different scenarios are given in Fig. 13.13.

Numerical simulations are implemented in the environment of Matlab on a personal computer (Intel Core i7 3.2 GHz CPU, 16GB RAM), where YALMIP is combined with CPLEX to solve optimization problems.

13.3.4.2 Results analysis

Fig. 13.14 illustrates the detailed decision variables in one typical scenario in the first and second stages, respectively. From the results, we can find that the PV generation consumption reduced from 9:00 am-13:00 pm, and the power shortage is

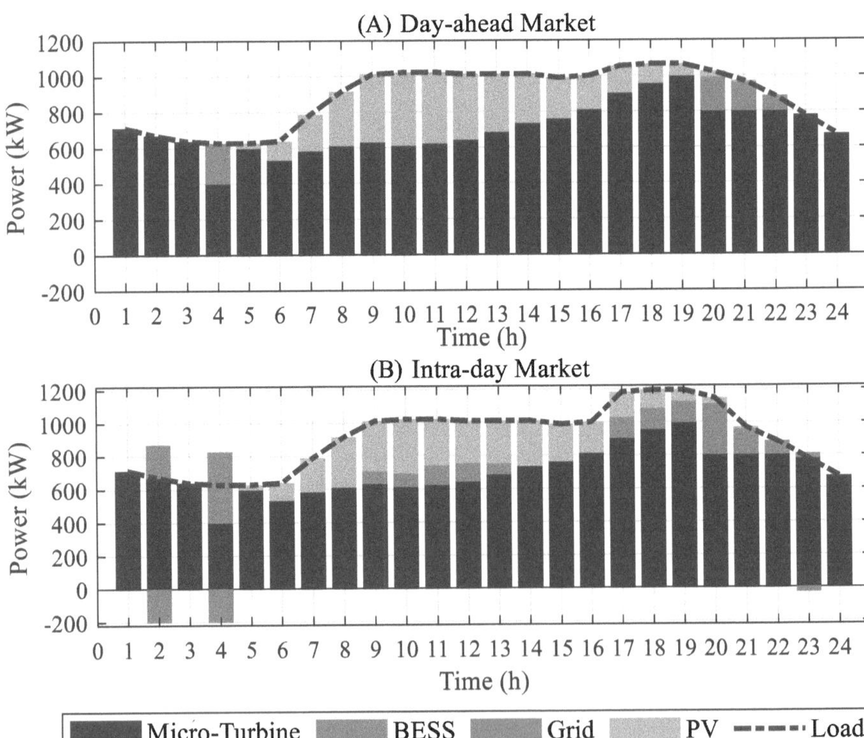

Figure 13.14 Decision variables in one typical scenario.

supplied by the external grid and BESS. Specifically, from 9:00 am-11:00 am the load demand is provided by power purchased from the external grid; during 12:00–13:00 the shortage is supplied by BESS due to the electricity price. Noted that load demand increased during 17:00–20:00 pm; therefore, BESS and external grid all participated in daily operation to provide energy.

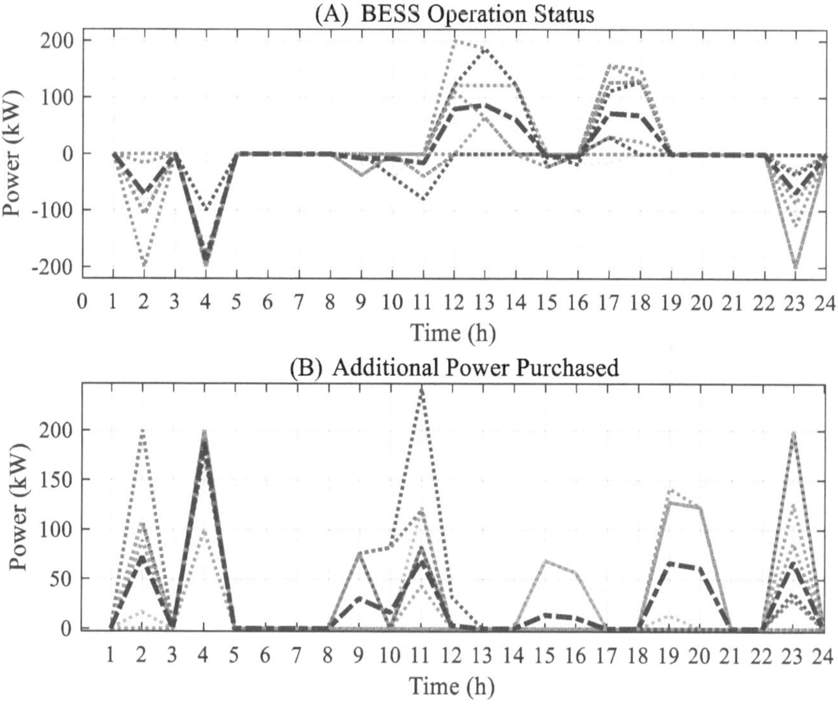

Figure 13.15 Two-stage decision variables in each scenario.

Table 13.10 Objective values under different value of β.

α	β	F_{da}($)	F_{id}($)	F($)	CVaR($)
0.8	0	5418.4	272.1	5690.5	5839.1
0.8	0.1	5433.2	266.5	5712.2	5825.2
0.8	0.2	5447.9	260.8	5731.0	5820.2
0.8	0.3	5462.7	255.2	5747.1	5815.3
0.8	0.4	5477.4	249.6	5760.4	5810.5
0.8	0.5	5492.1	244.1	5771.0	5805.8
0.8	0.6	5506.9	238.5	5778.8	5801.1
0.8	0.7	5521.9	232.9	5784.1	5796.7
0.8	0.8	5537.9	227.4	5787.6	5793.2
0.8	0.9	5553.9	221.9	5790.4	5792.0

To demonstrate the effectiveness of the proposed method, detailed operation solutions during the second stage in each scenario are given in Fig. 13.15, where the black dotted line represents the expected value of BESS output power and power purchased from the external grid in each scenario.

13.3.4.3 Risk analysis

In this section, the effect of the risk aversion parameter β on objective value is analyzed. For this purpose, the value of confidence level α remains unchanged, and β varies from 0 to 1. The variations in the total cost in terms of different values of β are also calculated, listed in Table 13.10 and Fig. 13.16.

It can be seen from the results, that the objective values increase from \$5690.5 to \$5790.4 and the value of CVaR correspondingly decreases as β increases. This is because the larger the risk aversion factor β, the more conservative the SO, meanwhile the greater the overall cost. When β equals 0, the results represent the smallest profit with the greatest risk; when $\beta = 0.9$, which represents minimizing the value of CVaR, resulting in a more conservative solution, with the highest overall cost.

To analyze the impact of the risk aversion factor, detailed operation solutions in one typical scenario under the different values of β are listed in Table 13.11. From the results, we can find that when the value of β increases, SO inclines to dispatch stable power to avoid the risk, for example, #1 MT increases power output from 5:00 pm and is running at full capacity at 15:00 pm. Therefore, #2 MT starts to increase power output from 15:00 pm-19:00 pm. Since the SO tends to use less PV generation to avoid the risk, the power variation in the second stage is reduced. As a result, the BESS output power and power purchased from the external grid decrease significantly.

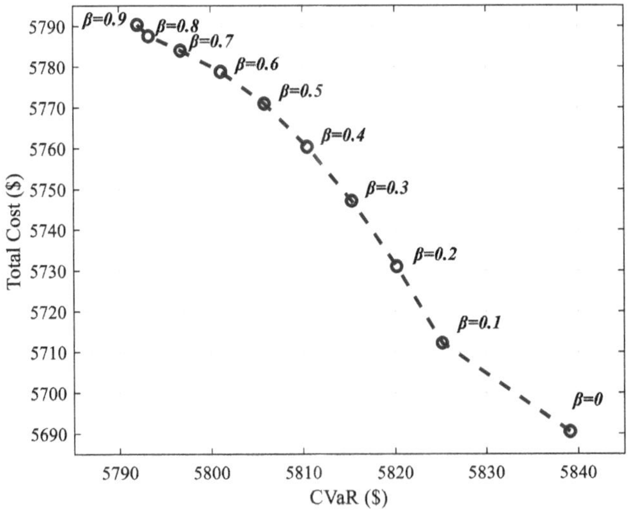

Figure 13.16 Total costs versus conditional value at risk under different values of β.

Table 13.11 Optimal scheduling decisions under different risk levels (kW).

T	$P_{1,t}^{DA}$		$P_{2,t}^{DA}$		$P_{grid,t}^{DA}$		$P_{grid,t,\omega}^{ID}$		$P_{ch/dis,\omega,t}^{BESS}$	
	$\beta=0$	$\beta=0.9$	$\beta=0$	$\beta=0.9$	$\beta=0$	$\beta=0.9$	$\beta=0$	$\beta=0.9$	$\beta=0$	$\beta=0.9$
1	513.6	513.6	200.0	200.0	0.0	0.0	0.0	0.0	0.0	0.0
2	471.0	471.0	200.0	200.0	0.0	0.0	200.0	71.9	−200.0	−71.9
3	439.0	439.0	200.0	200.0	0.0	0.0	0.0	0.0	0.0	0.0
4	200.0	200.0	200.0	200.0	228.4	228.4	200.0	200.0	−200.0	−200.0
5	397.9	404.8	200.0	200.0	0.0	0.0	0.0	0.0	0.0	0.0
6	332.6	356.5	200.0	200.0	0.0	0.0	0.0	0.0	0.0	0.0
7	382.8	429.0	200.0	200.0	0.0	0.0	0.0	0.0	0.0	0.0
8	411.8	480.2	200.0	200.0	0.0	0.0	0.0	0.0	0.0	0.0
9	431.6	517.2	200.0	200.0	0.0	0.0	76.0	7.6	0.0	0.0
10	411.9	504.2	200.0	200.0	0.0	0.0	82.1	8.2	0.0	0.0
11	423.5	513.3	200.0	200.0	0.0	0.0	119.7	12.0	0.0	0.0
12	443.5	526.3	200.0	200.0	0.0	0.0	0.0	0.0	110.5	11.0
13	486.4	559.6	200.0	200.0	0.0	0.0	0.0	0.0	65.1	6.5
14	534.9	597.2	200.0	200.0	0.0	0.0	0.0	0.0	0.0	0.0
15	561.2	600.0	200.0	212.8	0.0	0.0	0.0	0.0	0.0	0.0
16	600.0	600.0	211.8	254.4	0.0	0.0	0.0	0.0	0.0	0.0
17	600.0	600.0	302.7	336.8	0.0	0.0	0.0	0.0	126.5	126.5
18	600.0	600.0	355.0	379.8	0.0	0.0	0.0	0.0	127.8	127.8
19	600.0	600.0	395.9	411.4	0.0	0.0	127.8	127.8	0.0	0.0
20	600.0	600.0	200.0	200.0	188.4	196.1	122.7	122.7	0.0	0.0
21	600.0	600.0	200.0	200.0	159.9	161.9	0.0	0.0	0.0	0.0
22	600.0	600.0	200.0	200.0	84.0	84.0	0.0	0.0	0.0	0.0
23	577.5	577.5	200.0	200.0	0.0	0.0	29.9	0.0	−29.9	0.0
24	471.1	471.1	200.0	200.0	0.0	0.0	0.0	0.0	0.0	0.0

13.3.5 Discussion

The increasing penetration of renewable energy resources opens up the possibility for employ BESS in the future. This section presents a risk-constrained two-stage optimal scheduling model with BESS in the day-ahead and intra-day dispatch scheme, respectively. Firstly, the detailed cost of batteries per unit is calculated based on the battery degradation model. Uncertainties in renewable energy generation and load variation are modeled using the scenario analysis method. The CVaR theory is employed in the second stage to help SO to make decisions considering the tradeoff between risk mitigation and overall operation cost minimization. To test the validity and efficiency of the proposed model, extensive numerical results and discussions have been presented in the MATLAB® environment.

13.4 Conclusions

Micro-grid operation and planning with renewable energy resources integration are the fundamental steps for modern power system research. This chapter is devoted to applying the advanced optimization techniques for BESS planning and operation, the contents mainly comprise the following aspects: (1) BESS expansion planning in micro-grid considering battery degradation effects. (2) a two short-stage operation model mainly concentrating on a risk-constrained operation using BESS with high renewable energy power penetrations. The proposed models and solution methodologies have all been successfully tested on a number of benchmark systems. The practicability and reliability of the proposed models have been proven by simulation. Overall, this chapter lays a solid foundation for the future BESS operational planning and expansion planning in micro-grid.

Acknowledgement

This work was supported in part by the National Natural Science Foundation of China (No. 51807011, No. 71931003, No. 72061147004, and No. 72171026), the Science and Technology Projects of Hunan Province and Changsha City (No. 2021WK2002, No. 2019GK5015, and No. 2020GK1014), the Hunan International Scientific and Technological Cooperation Base of Energy Storage Interconnection (No. 2018WK4010), the Big Data-driven Energy Internet International Science and Technology Cooperation base of the Ministry of Science and Technology (No. 2017D01011).

References

[1] http://certs.lbl.gov/certs-dermicro.html, C. M. C.

[2] Rogers H. L., Maitra A. Electric & power research institute california, F. R. Program on Technology Innovation Microgrid Implementations. Literature Review. 2016 January, pp. 1–172.

[3] T. Basson, "IEEE 1547 and 2030 Standards for Distributed Energy Resources Interconnection and Interoperability with the Electricity Grid" National Renewable Energy Laboratory U.S, December 2014. Pp. 1–14.
[4] M. Kermani, E. Shirdare, G. Parise, M. Bongiorno, L. Martirano, A comprehensive techno-economic solution for demand control in ports: energy storage systems integration, IEEE Trans. Indust. Appl. (2022) 1. -1.
[5] S.M. Lukic, J. Cao, R.C. Bansal, F. Rodriguez, A. Emadi, Energy storage systems for automotive applications, IEEE Trans. Indust. Electr. 55 (2008) 2258–2267.
[6] S. C. Smith, P. Sen, B. Kroposki, Advancement of energy storage devices and applications in electrical power system, in: 2008 IEEE Power and Energy Society General Meeting-Conversion and Delivery of Electrical Energy in the 21st Century, IEEE, 2008, pp. 1–8.
[7] B. Dunn, H. Kamath, J.-M. Tarascon, Electrical energy storage for the grid: a battery of choices, Science 334 (2011) 928–935.
[8] J. Hernández, I. Gyuk and C. Christensen, "DOE global energy storage database — A platform for large scale data analytics and system performance metrics," 2016 IEEE International Conference on Power System Technology (POWERCON), 2016, pp. 1–6, Available from: https://doi.org/10.1109/POWERCON.2016.7754009.
[9] IEC White Papers & Technology Reports, 2022, [Online]. Available from: https://www.iec.ch/basecamp/iec-white-papers-technology-reports.
[10] A. Saez-De-Ibarra, A. Milo, H. Gaztañaga, I. Etxeberria-Otadui, P. Rodr Guez, S. Bacha, et al. Analysis and comparison of battery energy storage technologies for grid applications. PowerTech (POWERTECH), 2013 IEEE Grenoble. IEEE, 2013, pp. 1–6.
[11] M. Yilmaz, P.T. Krein, Review of battery charger topologies, charging power levels, and infrastructure for plug-in electric and hybrid vehicles, IEEE Trans. Power Electr. 28 (2013) 2151–2169.
[12] M. Zidar, P.S. Georgilakis, N.D. Hatziargyriou, T. Capuder, D. Škrlec, Review of energy storage allocation in power distribution networks: applications, methods and future research, IET Gener. Trans. Distrib. 10 (2016) 645–652.
[13] P.S. Georgilakis, N.D. Hatziargyriou, A review of power distribution planning in the modern power systems era: models, methods and future research, Electr. Power Syst. Res. 121 (2015) 89–100.
[14] P. Paliwal, N.P. Patidar, R.K. Nema, Planning of grid integrated distributed generators: a review of technology, objectives and techniques, Renew. Sust. Energy Rev. 40 (2014) 557–570.
[15] R. Mena, M. Hennebel, Y.-F. Li, C. Ruiz, E. Zio, A risk-based simulation and multi-objective optimization framework for the integration of distributed renewable generation and storage, Renew. Sust. Energy Rev. 37 (2014) 778–793.
[16] S. Bahramirad, W. Reder, A. Khodaei, Reliability-constrained optimal sizing of energy storage system in a microgrid. *Smart Grid*, IEEE Trans. 3 (2012) 2056–2062.
[17] L.C.D. Costa, F.S. Thom, J.D. Garcia, M.V.F. Pereira, Reliability-constrained power system expansion planning: a stochastic risk-averse optimization approach, IEEE Trans. Power Syst. 36 (2021) 97–106.
[18] H. Zhao, M. Hong, W. Lin, K.A. Loparo, Voltage and frequency regulation of microgrid with battery energy storage systems, IEEE Trans. Smart Grid 10 (2019) 414–424.
[19] R. S. Sullivan, Power System Planning. McGraw Hill.
[20] Y. Zhang, Y. Xu, H. Yang, Z.Y. Dong, R. Zhang, Optimal whole-life-cycle planning of battery energy storage for multi-functional services in power systems, IEEE Trans. Sust. Energy 11 (2020) 2077–2086.

[21] M.S. Hosen, J. Jaguemont, J. Van Mierlo, M. Berecibar, Battery lifetime prediction and performance assessment of different modeling approaches, iScience 24 (2021) 102060.
[22] P. Jayathissa, M. Luzzatto, J. Schmidli, J. Hofer, Z. Nagy, A. Schlueter, Optimising building net energy demand with dynamic BIPV shading, Appl. Energy 202 (2017) 726–735.
[23] Orthogonal Arrays (Taguchi Designs). Available at: http://www.york.ac.uk/depts/maths/tables/orthogonal.htm.
[24] Y. Zhang, Z.Y. Dong, F. Luo, Y. Zheng, K. Meng, K.P. Wong, Optimal allocation of battery energy storage systems in distribution networks with high wind power penetration, IET Renew. Power Gener (2016).
[25] RDA CISL Research Data Archive [Online]. Available from: http://rda.ecar.edu/.
[26] M. Korpa, A.T. Holen, Operation planning of hydrogen storage connected to wind power operating in a power market, Energy Convers. IEEE Trans. 21 (2006) 742–749.
[27] Y. Zhang, Y. Xu, H. Yang, Z.Y. Dong, Voltage regulation-oriented co-planning of distributed generation and battery storage in active distribution networks, Int. J.Electr. Power Energy Syst. 105 (2019) 79–88.
[28] M. Nick, R. Cherkaoui, M. Paolone, Optimal allocation of dispersed energy storage systems in active distribution networks for energy balance and grid support, IEEE Trans. Power Syst. 29 (2014) 2300–2310.
[29] Y. Zhang, K. Meng, F. Luo, H. Yang, J. Zhu, Z.Y. Dong, Multi-agent-based voltage regulation scheme for high photovoltaic penetrated active distribution networks using battery energy storage systems, IEEE Access 8 (2020) 7323–7333.
[30] J. Lopez, J. Contreras, J.R. Mantovani, Reactive power planning under conditional-value-at-risk assessment using chance-constrained optimisation, IET Gener. Trans. Distrib. 9 (2014) 231–240.
[31] A. Najafi, A. Tavakoli, M. Pourakbari-Kasmaei, M. Lehtonen, A risk-based optimal self-scheduling of smart energy hub in the day-ahead and regulation markets, J.Clean. Prod. 279 (2021) 123631.
[32] C. Zhang, Y. Xu, Z. Li, Z.Y. Dong, Robustly coordinated operation of a multi-energy microgrid with flexible electric and thermal loads, IEEE Trans. Smart Grid 10 (2019) 2765–2775.
[33] T. Huiling, W. Jiekang, C. Lingmin, L. Zhijun, W. Fan, L. Kangxing, et al., A control optimization model for CVaR risk of distribution systems with PVs/DSs/EVs using Q-learning powered adaptive differential evolution algorithm, Int. J. Electr. Power Energy Syst. 132 (2021) 107209.
[34] R.T. Rockafellar, S. Uryasev, Optimization of conditional value-at-risk, J.Risk 2 (2000) 21–42.
[35] L. Xu, X. Ruan, C. Mao, B. Zhang, Y. Luo, An improved optimal sizing method for wind-solar-battery hybrid power system. *Sust. Energy*, IEEE Trans. 4 (2013) 774–785.
[36] M.E. Khodayar, M. Shahidehpour, L. Wu, Enhancing the dispatchability of variable wind generation by coordination with pumped-storage hydro units in stochastic power systems, IEEE Trans. Power Syst. 28 (2013) 2808–2818.

Harmonic distortion effect of large-scale hydropower storage based on doubly fed induction machine in power system

14

Mohsen Alizadeh Bidgoli[1] *and Amin Hajizadeh*[2]
[1]Department of Electrical Engineering, Yadegar-e-Imam Khomeini (RAH) Share Rey Branch, Islamic Azad University, Tehran, Iran, [2]AAU ENERGY, Aalborg University, Esbjerg, Denmark

Chapter Outline

Nomenclature

Abbreviation

DFIM Doubly fed induction machine
FRT Fault ride-through
VS Variable-speed
FS Fixed-speed
HPT Hydro pump-turbine
RPT Reversible pump turbines
PSPP pumped-storage power plant
WT Wind turbines
MSC Machine side converter
GSC Grid side converter

Emerging Trends in Energy Storage Systems and Industrial Applications. DOI: https://doi.org/10.1016/B978-0-323-90521-3.00017-X

PSS Power system stabilizer
SMIB Single machine infinite bus

Parameters

R	Resistance
L	Inductance
i	Current
V	Voltage
s	Slip
S	Laplase transform
ω	Angular speed
δ	Machine power angle
D	Rotor damping constant
f,m	Constant coefficients of state and the input matrix
P	Active power
Q	Reactive power
η	Efficiency
q	Water flow rate of turbine
T	Electric torque
k	Constant control coefficient
e	Error signal
α	Phase of Thevnan voltage
H	The Inertia constant
β	Bandwidth
g	Percentage of opening the gate
F, M	State and input matrix
x,u	State variable and input signal
h	Head
S	Apparent power

Indices

r	Rotor
s	Stator
b	Base
ds,qs	Stator winding on *dq* axes
Dr,qr	Rotor winding on *dq* axes
m	Magnetizing
G	Generator
M	Motor
e	Electric
g	Grid
p	Active power portion
q	Reactive power portion
0	Equilibrium point
tr	Exciting transformer
N	Number of turns of the coil
T,P	Turbine, Pump mode
GSC	Grid side converter

MSC	Machine side converter
n	Nominal
pp	Proportional part of active power loop
ip	Integral part of active power loop
pv	Proportional part of voltage loop
iv	Integral part of voltage loop
pq	Proportional part of reactive power loop
iq	Integral part of reactive power loop
pg	Proportional part of turbine gate loop
ig	Integral part of turbine gate loop
nl, fl	No load, full load
cl	Closed-loop
ol	Open-loop
cr	Current loop
pcr	Proportional part of current loop
icr	integral part of current loop
pvdc	Proportional part of dc-link voltage loop
ivdc	Integral part of dc-link voltage loop
pω	Proportional part of speed loop
iω	Integral part of speed loop

14.1 Introduction

The pumped storage power plants (PSPPs) are one of the most mature technologies among the various methods for storing energy, such as batteries, flywheels, superconductors, compressed air systems, and flywheels. As early as the 20th century, fixed speed (FS) PSPP employing SM was widely implemented in Europe, Asia, and America. Despite its advantages, the FS-PSPP has significant drawbacks, including low efficiency when partially generating and poor frequency control in pumping operations. A variable speed (VS) unit can help overcome the above-mentioned shortcomings. In recent years, there has been an increase in VS-PSPP use, especially in areas where the water head variation is considerable. Also, places that need frequency control when in pumping mode modify the structure to adjust the plant's speed, such as Grimsel 2, developed by the ABB company in Switzerland [1]. FS-PSPPs, however, is preferable to the other case due to the lower investment cost and simpler implementation of VS technology.

Diverse aspects have been studied. Ref. [2] develops a simple mathematical model and compares its results with on-site measurements for Ohkawachi PSPP. Using the DFIM, a dynamic simulation of a VS-PSPP is performed, and the results of the simulation for grid-connected operation and three-phase short circuit are shown in [3], while simulation of the dynamic behavior of VS-PSPPs is compared to that of FS-PSPPs in [4]. Also, a simplified model of the converters is used in [5] to study control strategies. Ref. [6] suggests that wind power fluctuations within an isolated grid can be accounted for by using VS-PSPPs based on synchronous machines and back-to-back voltage source converters, and case studies are conducted to examine control strategies. VS-PSPPs also offer some other promising areas for research, including operation efficiency [7], advanced power electronics

[8], the transient stability of the large DFIM under weak grid conditions [9], reducing wind power fluctuations based on a comparison with FS units [10], and starting and braking of a 250 MW DFIM-based PSPP in pumping mode [11].

However, in the aforementioned papers, challenges ahead for the power system such as power quality (current and voltage harmonics) in VS-PSPP have not been addressed. Harmonic distortion in voltage or current waveform can occur because of converters, that is machine side converter (MSC) and transformer side converter (TSC), in VS-PSPP. At the same time, the FS solution does not suffer from this issue. On the other side, several risks can arise due to existing harmonics in the power network; parallel and series resonances, overheating in capacitor banks and power transformers, neutral conductor overcurrent, and malfunctioning protective devices. However, the state-of-the-art VS units use the multi-level voltage source converter instead of the cyclo-converter to reduce harmonic distortion of current and voltage. Moreover, the leakage inductance of the excitation transformer and parallel filters can eliminate the high-frequency harmonics of the TSC current. In short, currently, few papers focus on the power quality of the large DFIM-based PSPP from the power grid's point of view, which is meaningful and urgently needed. Among the power quality parameters, the harmonics in voltage and current are assessed by the generator's total harmonic distortion (THD) index. Note that this chapter does not discuss the other issues of power quality, which are not significant in the case of the PSPP.

As a result, the performance of VS units in terms of power quality for VS-PSPP is assessed in this chapter. For the simulation study, two stages are designed in this chapter; The waveforms of the VS-PSPP are examined using FFT tools in MATLAB®/Simpower in the first stage of the simulation. Then, the THD findings and the fundamental value related to current and voltage are listed for generating mode. All simulations for studying the things listed above are carried out using comprehensive models in MATLAB/SimPowerSystemTM, which use the fixed-step discrete solver ode23 with a 50-second step size. Harmonic generating data of DFIM and SM-based PSPP, extracted from simulation results of FFT analysis in MATLAB/Simulink, is utilized for harmonic analysis toolbar DIgSILENT *PowerFactory* in the second stage of simulation.

The content of this chapter is organized as follows. In Section 14.2, the principle and history of PSPP are discussed. In Section 14.3, the modeling and controlling method of VSPSPPs are introduced. In Section 14.4, simulations of single machine PSPP, modeled in MATLAB, and multi-machine in Digsilent are done, and simulation results for harmonic analysis are presented. In Section 14.5, conclusions and suggestions are summarized.

14.2 Principle and history of pumped storage power plant

14.2.1 Historical of variable speed-pumped storage power plant

Table 14.1 shows a list of VS-PSPPs created in previous years. The cycloconverter is used in the world's first large-scale variable speed PSHP based on a doubly fed

Table 14.1 Commissioned and under construction variable speed-pumped storage power plant in all of the world.

Unit	Country	HPT (MW)	MG (MVA)	Power electronic[a]		Speed range (rpm)	Ratedspeed (rpm)	Man.[b]	Year
				Type	Size (MVA)				
Narude	Japan	22	22	1	–	190–210		H	1987
Yagisawa	Japan	2 × 82	2 × 85	1	25.8	130–156		T	1990
Takami	Japan	105	140	2	24.3	209–254		M	1992
Ohkawachi	Japan	392	395	1	72	330–390 (± 8.3%)		H	1993
Shiobara	Japan	330	360	1	51.1	356–394 (± 8%)		T	1995
Okukiyotsu	Japan	340	345	2	31.5	408–450 (± 5%)		T	1996
Yanbaru	Japan	30	31.5	2	3.96	423–477 (± 5%)		T	1999
Goldisthal	Germany	2 × 265	2 × 340	1	2 × 57.6	300–347 (−10/ + 4%)	333	V	2003
Avce	Slovenia	185	195	3	10	576–636 (± 4%)	600	ABB, M	2008
Omarugawa III	Japan	2 × 330 2 × 330	2 × 350 2 × 370	3 1	2 × 27.4	576–624 (± 4%)	600	M H	2008 2010
Grimsel II[c]	Switzerland	80	100	4	2 × 43.3	680–765 (-10/ + 2%)	750	ABB	2010
Okutataragi	Kansai Electric Power Co.	2 × 320	–	–	–	–	300	H	2014
Kyogoku	Japan	228	230	2		–		T	2015
Linthal	Switzerland	4 × 250	4 × 280	3	–	470–530 (± 6%)	500	A	2015
Frades II	Portugal	2 × 390	2 × 440	3	–	350–381 (+2%, − 7%)	364	Vo, S	2015
Vende Nova III	Portugal	2 × 380	2 × 420	–	45	–	–	Vo, S	2015

(Continued)

Table 14.1 (Continued)

Unit	Country	HPT (MW)	MG (MVA)	Power electronic[a]		Speed range (rpm)	Ratedspeed (rpm)	Man.[b]	Year
				Type	Size (MVA)				
Le cheylas	France	1 × 250	–	–	–	(± 7%)	300		
Tehri	India	4 × 250	4 × 278	3		214–256 (± 6%)	230.8	GE, T	2016
Kazunogawa	Japan	2 × 475	2 × 475	2	25	480–520	–	T	2017
Iowa Hill	USA	3 × 133	3 × 133	2	–	–	–	–	2018
Nant de Drance	Switzerland	6 × 157	6 × 175	3	–	399–459 (± 7%)	428.6	A	2019
Kozjak	Slovenia	2 × 220	2 × 230	–	–	–	600	V	–
Grimsel III	Switzerland	–	–	–	100	–	–	ABB	2020
Fengning II	China	2 × 300	2 × 330	–	–	–	–	GE	2021

[a]Power Electronic Type: 1- Cyclo-converter, 2- GTO- back to back (VSI), 3- IGCT- back to back (VSI), 4: full converter (VSI).
[b]Manufacturer: H: HITACHI, T: TOSHIBA, M: MITSUBISHI, V: VATECH, Vo: VOITH, A: ALSTOM, *GE*, general electric; *S*, Siemens.
[c]Only pumping mode operation.

induction machine (DFIM) installed in Japan in the 1990s. The first one in Europe, on the other hand, was commissioned in 2004. (Goldisthal in Germany). This practical method has recently been employed in a number of projects around the world that use a state-of-the-art power electronic converter, that is a back-to-back voltage source converter, such as Avce in Slovenia and Linthal in Switzerland [12]. It should be noted that Grimsel 2 is only being modified to work in pumping mode with a full converter coupled to the SM, and the converter is disregarded in generating mode via a short-circuiting circuit.

14.2.2 Principle of variable speed-pumped storage power plant

The electrical part of VS-PSPPs is composed of three-phase wound rotor induction machines, namely DFIM. The three-phase stator winding is directly connected to the power grid. The three-phase rotor winding is also connected to the grid through a partial-scale back-to-back converter depending on the variation range of the slip. The VS-PSPP can operate in two modes, namely super-synchronous and sub-synchronous. As shown in Table 14.2, the rotor frequency sign depends on the direction of the rotating magnetic field, which is determined by sub- or super-synchronous mode operations in generating and motoring states.

14.2.3 Variable speed-pumped storage power plant types

The continuous operation of a DFIM-based PSPP is crucial in the power grid, in which primarily the converters, that is MSC and TSC, dictate one. Such unscheduled shutdown comes from a power electronic converter fault that threatens the power grid security. On the other hand, the primary source of harmonic distortion in a modern power grid is the on/off switches of power converters. Therefore, briefly, the progress of VS-PSPP is considered in the following.

Table 14.2 operation of doubly fed induction machine in sub- and super synchronous mode.

Mode of operation		Speed	Stator power	Rotor power	Rotor frequency
Generating	Sub-synch	$\omega_s > \omega_r$	Negative	Positive	$f_r = f_s - \omega_r{}^*p/120$
	Super-synch	$\omega_s < \omega_r$	Negative	Negative	$f_r = -f_s + \omega_r{}^*p/120$
Motoring	Sub-synch	$\omega_s > \omega_r$	Positive	Negative	$f_r = f_s - \omega_r{}^*p/120$
	Super-synch	$\omega_s < \omega_r$	Positive	Positive	$f_r = -f_s + \omega_r{}^*p/120$

Note: Positive- power flow from grid to machine, Negative- power flow from machine to the grid.
where p is the number of poles, ω_r and ω_s is the mechanical rotor speed and equivalent stator speed in rpm, and f_r and f_s are rotor and stator frequency in Hz.

14.2.3.1 Conventional variable speed-pumped storage power plant

As mentioned before, a VS-PSPP is equipped with a DFIM and a partial-scale converter connecting the rotor and stator winding. Earlier, VS-PSPP used a cycloconverter, as shown in Fig. 14.1, a DC-link-less conventional type of converter to provide variable frequency and AC voltage. The first-worldwide VS-PSPP, installed in Japan in the 1990s, employed gate turn-off thyristors (GTOs) equipped with the cycloconverter. Also, the first one in Europe was commissioned in 2004 (Goldisthal in Germany) with such type of converter. The advantages of cycloconverter are instantaneous PQ control and low on-state power losses. At the same time, its disadvantages are the inability to generate output voltage with a frequency greater than input frequency, high distorted rotor current in DFIM, thereby introducing the large size of filters compared to others, and reactive power consumption from the grid in rotor side. For example, the number of GTO switches used in the cycloconverter is 24 for each phase, that is in three phases is 3*24 = 72 switches, for the minimum distorted required rotor current waveform in Ohkawachi PSPP.

14.2.3.2 State-of-the-art variable speed-pumped storage power plant

As shown in Fig. 14.2, voltage source inverters (VSIs) can be located back to back and linked with a DC-link. These converters can provide variable voltage and variable frequency supply during starting of the machine resulting in the reduction of start-up transients and energy losses [8]. The main advantage of VSI is its ability to control active and reactive power independently, with

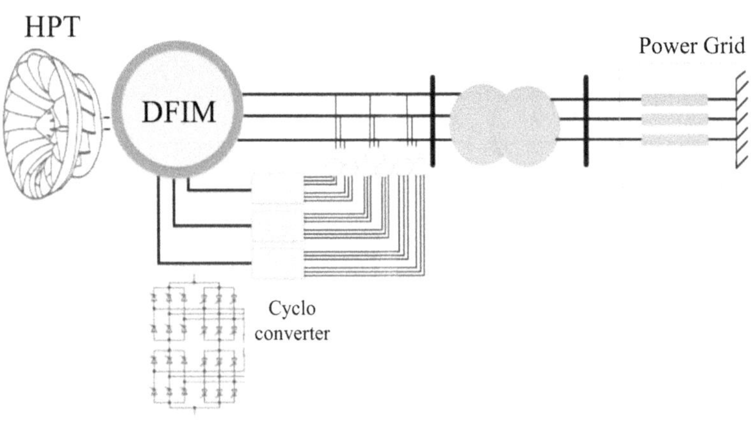

Figure 14.1 Doubly fed induction machine -based pumped storage power plant with cycloconverter.

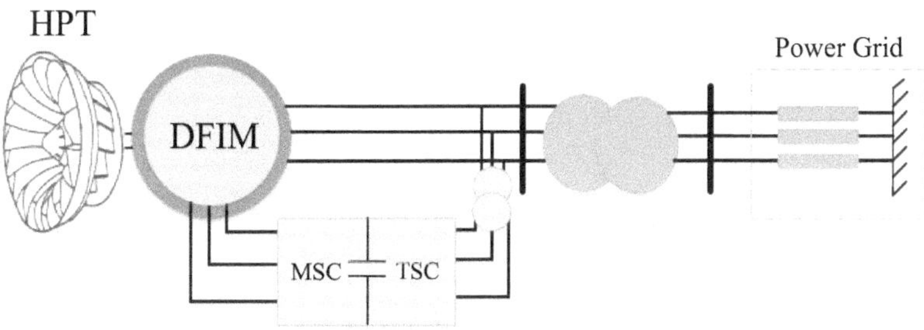

Figure 14.2 Doubly fed induction machine-based pumped storage power plant with back-to-back voltage source inverter.

a significant decrease in current harmonics compared to cycloconverter one. Usually, VSI used in PSPP is two-level (2 L) or three-level (3 L). The former comprises six IGBT or GCT switches with a freewheeling diode across each switch, such as Frades II VS-PSPP in Portugal. The latter belongs to the multi-level converter family. 3L-VSI is maturely applied in large-size drives due to the lesser voltage stress on devices.

Moreover, a reduction in THD of current waveforms compared to cycloconverter and 2L-VSI leads to a decrease in filter size. This topology is employed in Linthal VS-PSPP in Switzerland. Also, the stage III project of Tehri is planned to install 3L-VSI the project.

14.3 Modeling and control of variable speed-pumped storage power plant

In this section, a VS-PSPP equipped with DFIM coupling to hydro pump-turbine (HPT) is modeled and controlled in MATLAB/Simulink/Simpower.

14.3.1 Modeling of doubly fed induction machine-based pumped storage power plant

In this section, the numerical modeling for PSPP is introduced in detail. For this work, the methodology adopted at this stage is the classical modeling approach without considering the influence of the uncertainty caused by the variation of parameter input of the numerical models.

14.3.1.1 Hydraulic System

For the hydraulic system of this paper, including a non-elastic water column with long-length penstock without affecting the surge tank, a nonlinear hydraulic components model is considered [3]. The state-space equations of the hydraulic subsystem in turbine mode are as Eqs. (14.1)–(14.2) [3]:

$$\begin{cases} \dot{q}_t = (h'_s - h_t)/T_w \\ h'_s = h_s - f_{\text{eq}} q_t^2 \\ h_t = \left(q_t/(g/(g_n - g_{\text{nl}}))\right)^2 \\ T_w = L q_n/(g' A h_n) \end{cases} \tag{14.1}$$

$$\dot{g} = -k_a g + k_a v_{\text{sm}} \tag{14.2}$$

where h'_s, h_s, and h_t are the net static head, static head which is equal 1, and net head in the operating point of the turbine; Also, q is the water flow, g is the gate opening, v_{sm} is the servomotor voltage, T_w is water starting time of the pipe, k_a is the constant gain for servomotor modeling, and f_{eq} is the equivalent frictional coefficient. Additionally, L is the length of the pipe, A is the area of the pipe, and g' is gravitational acceleration, which is equal to 9.81 m/s^2; note that subscripts n, nl, and t denote the nominal, no-load, and turbine mode values, respectively. Additionally, the mechanical power of the turbine is defined as follows:

$$\begin{cases} P_m = \eta p_r h_t (q_t - q_{\text{nl}}) \\ p_r = \dfrac{P_n(MW)}{S_n(MVA)} \end{cases} \tag{14.3}$$

where η is the efficiency of the turbine and p_r is a constant value for per unit value conversion from the turbine base to DFIM; Also, P_n and S_n are nominal power of the turbine and apparent nominal power of the machine in generating mode, respectively. The hydro turbine model and its controller are presented in Fig. 14.3 are used in Simulink/Simpower.

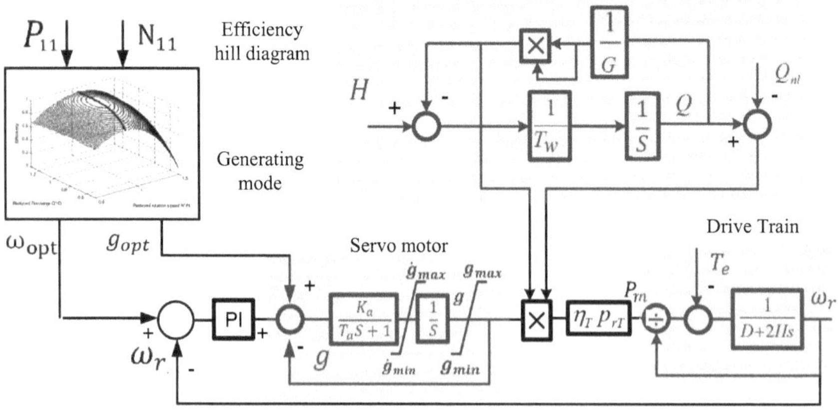

Figure 14.3 Hydro turbine model.

14.3.1.2 Doubly fed induction machine and machine side converter

Assuming the positive direction for the stator and rotor currents to be motor conversion (into the machine), the fourth-order electrical model of the DFIM (shown in Fig. 14.2), namely the full order model, is written in the synchronous reference frame, d-q axis as Eqs. (14.4)–(14.7). Note that all parameters and equations are given in per unit except for time is sec.

$$V_{ds} = R_s i_{ds} - \varphi_{qs} + p\varphi_{ds}/\omega_b \tag{14.4}$$

$$V_{qs} = R_s i_{qs} + \varphi_{ds} + p\varphi_{qs}/\omega_b \tag{14.5}$$

$$V_{dr} = R_r i_{dr} - s\varphi_{qr} + p\varphi_{dr}/\omega_b \tag{14.6}$$

$$V_{qr} = R_r i_{qr} + s\varphi_{dr} + p\varphi_{qr}/\omega_b \tag{14.7}$$

where i_{ds}, i_{qs}, v_{ds}, v_{qs} are the d-q axis stator currents and voltages; i_{dr}, i_{qr}, v_{dr}, v_{qr} are the d-q axis rotor currents and voltages; ρ is derivative operator (*d/dt*); R_s and R_r are the stator and rotor resistances; s is the slip, and $\omega_b = 377$ (rad/s). Additionally, φ_{ds}, φ_{qs}, φ_{dr}, φ_{qr} are the d-q axis stator and rotor flux magnitude, which can be written as follows:

$$\begin{cases} \varphi_{ds} = L_s i_{ds} + L_m i_{dr} \\ \varphi_{qs} = L_s i_{qs} + L_m i_{qr} \\ \varphi_{dr} = L_r i_{dr} + L_m i_{ds} \\ \varphi_{qr} = L_r i_{qr} + L_m i_{qs} \end{cases} \tag{14.8}$$

where $L_s L_r$ are the stator and the rotor inductances; L_m is the magnetizing inductance; Additionally, the swing equation is defined as follows.

$$\dot{\omega}_r = \frac{1}{2H}(T_m - T_e) \tag{14.9}$$

where T_m, T_e, and ω_r are the mechanical torque, electric torque, and rotor speed, respectively. Also, the electrical torque T_e, the stator reactive power Q_s, and the real electrical power P_s are defined as follows:

$$T_e = (L_m/L_s)\left(-\varphi_{ds} i_{qr} + \varphi_{qs} i_{dr}\right) \tag{14.10}$$

$$\begin{aligned} P_s &= -v_{ds} i_{ds} - v_{qs} i_{qs} \\ Q_s &= v_{ds} i_{qs} - v_{qs} i_{ds} \end{aligned} \tag{14.11}$$

14.3.1.3 Transformer side converter

Assuming the positive direction of the TSC is into the dc-link, and all parameters and equations per unit except for time is sec, the dc-link and the excitation transformer dynamics, shown in Fig. 14.2, can be expressed in the synchronous reference frame, *d-q* axis, as follows [3]:

$$v_{ds} = R_{tr} i_{dtr} - L_{tr} i_{qtr} + (L_{tr}/\omega_b)\rho i_{dtr} + v_{dtr} \quad (14.12)$$

$$v_{qs} = R_{tr} i_{qtr} + L_{tr} i_{dtr} + (L_{tr}/\omega_b)\rho i_{qtr} + v_{qtr} \quad (14.13)$$

$$\rho V_{dc} = (P_{tr} - P_r)/(V_{dc} C) \quad (14.14)$$

where L_{tr} and R_{tr} are leakage inductance and resistance of the excitation transformer. Also P_r and P_{tr} in Eq. (14.16) are defined as $P_r = v_{dr} i_{dr} + v_{qr} i_{qr}$, $P_{tr} = v_{ds} i_{dtr} + v_{qs} i_{qtr}$.

14.3.2 Control of doubly fed induction machine-based pumped storage power plant

Doubly fed induction machines, coupled to the hydro turbine, use a back-to-back converter, that is MSC and TSC, whose power rating is the partial scale of the DFIM rated power. This converter is located between the rotor and stator terminal that can transfer the mechanical power through both the stator and rotor, as shown in Fig. 14.2.

14.3.2.1 Turbine

There are two controllable components in DFIM-based PSPPs for the generating mode. One is the wicket gate position of the turbine, which is governed by a servomotor, and the other is the rotor current of the DFIM, which the MSC controls. Unlike FS-PSPPs, with these two control elements in the turbine mode, two quantities, that is the rotor speed and the real output power, can be controlled independently, as shown in Fig. 14.4A and B.

Since the response of the real power should be fast to improve the frequency regulation of the grid and free from the water hammer phenomenon, it is desired to control the real power with the MSC [13]. In other words, as shown in Fig. 14.4A, It is preferred to maintain the rotor speed by the wicket gate position of the turbine, and the real power is governed by the rotor voltage of the MSC [7]. Therefore, the servomotor voltage of the hydro turbine, shown in Fig. 14.3, is defined as follows.

$$v_{sm} = k_{ps}\left(\omega_r^* - \omega_r\right) + k_{is}\int\left(\omega_r^* - \omega_r\right) + G * \quad (14.15)$$

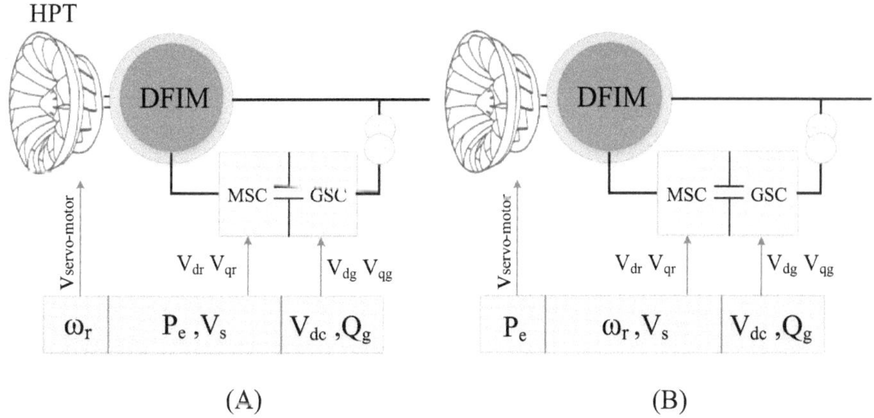

Figure 14.4 Two possible methods for control of active power and rotor speed in doubly fed induction machine-based pumped storage power plant.

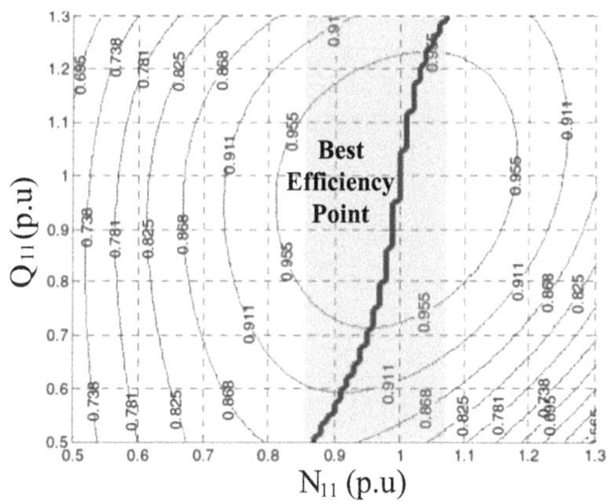

Figure 14.5 Hill diagram of a hydro turbine [7].

where k_{ps} and k_{is} are proportional and integral gains of the controller in the hydro turbine governor, respectively; ω_r^* and G^* are the reference value of the rotor speed and the wicket gate of the hydro turbine, which are obtained from the hill diagram of the hydro turbine shown in Fig. 14.5 [7]. Also, the controller coefficient values of this study are given in Table 14.3.

14.3.2.2 Machine side converter

The controller of the MSC consists of two decoupled controllers, and each controller is comprised of two loops. First, the rotor current control loop is usually called a fast or inner loop. Second, there are two outer or slow loops in the MSC controller. The former is the electrical real power regulator, whereas the latter loop is the stator

Table 14.3 PI coefficient values of controller loops of doubly fed induction machine used in simulations.

Machine side converter		**Transformer side converter**		**Hydro turbine**	
K_{p_p}	2	K_{p_vdc}	8	K_{p_w}	5
K_{i_p}	3	K_{ivdc}	400	K_{i_w}	0.3
K_{p_v}	1	K_{p_q}	0		
K_{i_v}	0	K_{i_q}	0		
K_{p_cr}	60	K_{pcr}	50		
K_{i_cr}	0.6	K_{icr}	1.5		

voltage regulator in the generating mode. It should be noted that the PSPP under study is considered for voltage regulator mode, not in power factor correction mode, that is its reference signal is V_s^* not Q_s^* [3]. As above explanations, the voltage and current of the MSC in dq axis are derived as follows.

$$v_{dr}^* = k_{pcr}\left(i_{dr}^* - i_{dr}\right) + k_{icr}\int\left(i_{dr}^* - i_{dr}\right) - sL_r^{'}i_{qr} - s\frac{L_m}{L_s}\varphi_s \tag{14.16}$$

$$v_{qr}^* = k_{pcr}\left(i_{qr}^* - i_{qr}\right) + k_{icr}\int\left(i_{qr}^* - i_{qr}\right) + sL_r^{'}i_{dr} \tag{14.17}$$

$$i_{dr\_g}^* = \frac{L_s}{\omega_r L_m \varphi_s}\left[k_{pp}\left(P_e^* - P_e\right) + k_{ip}\int\left(P_e^* - P_e\right)\right] \tag{14.18}$$

$$i_{qr\_g}^* = -k_{pv}\left(V_s^* - V_s\right) - k_{iv}\int\left(V_s^* - V_s\right) - Q_s \tag{14.19}$$

where $L_r' = L_r - \frac{L_m^2}{L_s}$ is the transient inductance of the rotor winding; k_{pcr} and k_{icr} are proportional and integral gains of the MSC inner (current) loop, respectively; k_{pp} and k_{ip} are proportional and integral gains of the outer active power loop, respectively; k_{pv} and k_{iv} are proportional and integral gains of the outer voltage loop, respectively. All values of this study are given in Table 14.3.

14.3.2.3 Transformer side converter

The main aim of the TSC is to maintain the voltage of the dc-link constant in the nominal value regardless of the rotor power direction. Moreover, the reactive power of the unit can be shared between the MSC and TSC to reduce the rated rotor current value and, therefore, design a smaller DFIM [14]. However, the extra reactive current for the TSC usually is higher than the MSC reactive current rating reduction due to the winding ratio (for this study 1.6) [15]. In this study, the nominal power factor of the TSC is supposed to be unity. It means that the reactive power of

DFIM-based hydropower is only controlled through the stator winding, and the reference reactive current of the TSC is equal to zero. As above explanations, the voltage and current of the TSC in dq axis are derived as follows.

$$v_{dg}^{*} = -k_{pcg}\left(i_{dg}^{*} - i_{dg}\right) - k_{icg}\int\left(i_{dg}^{*} - i_{dg}\right) + L_{tr}i_{qg} + v_{ds} \tag{14.20}$$

$$v_{qg}^{*} = -k_{pcg}\left(i_{qg}^{*} - i_{qg}\right) - k_{icg}\int\left(i_{qg}^{*} - i_{qg}\right) - L_{tr}i_{dg} \tag{14.21}$$

$$i_{dg}^{*} = k_{pvdc}\left(V_{dc}^{*} - V_{dc}\right) + k_{ivdc}\int\left(V_{dc}^{*} - V_{dc}\right) \tag{14.22}$$

$$i_{qg}^{*} = k_{ppf}\left(Q_{g}^{*} - Q_{g}\right) + k_{ipf}\int\left(Q_{g}^{*} - Q_{g}\right) \tag{14.23}$$

where k_{pcg} and k_{icg} are proportional and integral gains of the TSC inner (current) loops, respectively; k_{pvdc} and k_{ivdc} are proportional and integral gains of the outer (DC-link voltage) loop, respectively; k_{ppf} and k_{ipf} are proportional and integral gains of the outer (output power factor) loop, respectively. All values of this study are given in Table 14.3.

14.4 Simulation results

In this section, a performance assessment of VS-PSPP in terms of power quality is undertaken. As mentioned before, for the simulation study two stages are designed in this chapter; in the first stage, the waveforms of the VS-PSPP are evaluated via FFT tools in MATLAB/Sim-power. Then, the results of THD and the fundamental value corresponding to current and voltage are listed for generating mode. In the second stage of simulation, harmonic generation data of DFIM-based PSPP, extracted from simulation results of FFT analysis in MATLAB/Simulink, is used in the harmonic analysis toolbar of DIgSILENT *PowerFactory*.

14.4.1 MATLAB/Simulink/Simpower

This section performs all simulations through detailed models by applying MATLAB/SimPowerSystem which uses the fixed-step discrete solver ode23 with a step size 50 μs.

14.4.1.1 Case study

This chapter considers a 343 MW HPT coupled to a 381 MVA DFIM. A detailed model using components of Simpower toolbox in MATLAB/Simulink increases the validity of this work for real setup. Moreover, for this aim, signal monitoring, the inverter dead time, and measuring device delays are modeled as a first-order transfer function with a small-time constant added to the simulation. The detailed information and parameter values of the study case are shown in Fig. 14.6 can be found in p.u. in the following.

- DFIM [5]: $R_s = 0.00174$, $R_r = 0.002$, $L_s = 4.46$, $L_r = 4.472$, $L_m = 4.2$, $H = 7.7$ s, $f_n = 60$ Hz, 2P = 16, $S_n = 381.1$ MVA.
- HPT (Francis turbine): $D_I = 4.41$ m, $S_n = 340$ MW, $N_n = 450$ rpm, $Q_n = 54.38$ m^3/s, $T_n = 7.5503 \times 10^6$ N.m, $H_n = 709.2$ m.

14.4.1.2 Simulation results

Among the power quality issues, harmonic distortion is a significant problem that is considered in this paper. Harmonic distortion in voltage or current waveform can occur because of converters, that is MSC and TSC, in VS-PSPP. On the other side, several risks can arise due to existing of harmonics in the power network; parallel and series resonances, overheating in capacitor banks and power transformers, neutral conductor overcurrent, and malfunctioning protective devices [16]. However, the state-of-the-art VS units use the multi-level voltage source converter instead of the cyclo-converter to reduce harmonic distortion of current and voltage [17]. Moreover, the leakage inductance of the excitation transformer and parallel filters can eliminate the high-frequency harmonics of the TSC current. In this study, the resistance and inductance of the excitation transformer, shown in Fig. 14.6, are 0.005 and 0.15 p.u, respectively. Also, a capacitor bank with 50 MVAR is connected to the low voltage side of the excitation transformer. It should be noted that both the MSC and TSC are considered two-level in this study and simulations. As shown in Table 14.4, and Fig. 14.7, waveforms of the TSC and stator current and stator voltage for three phases (A, B, and C) have been evaluated via FFT tools in MATLAB/Sim-power. According to Table 14.4, the THD of the TSC current, as expected, is much more than the stator current.

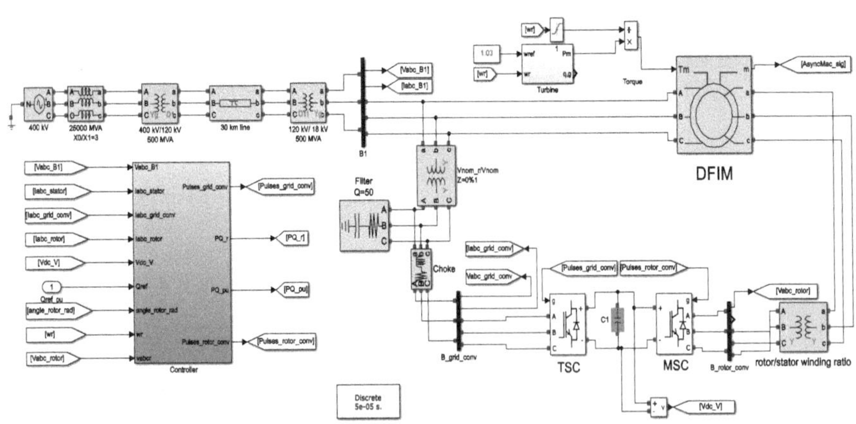

Figure 14.6 MATLAB/Simulink/Simpower simulation diagram.

Table 14.4 Harmonic analysis for under study doubly fed induction machine-based PSHP.

DFIM			**Generator**									
	$v_{s\text{-}a}$	$i_{s\text{-}a}$	$i_{g\text{-}a}$	$i_{\text{out-}a}$	$v_{s\text{-}b}$	$i_{s\text{-}b}$	$i_{g\text{-}b}$	$i_{\text{out-}b}$	$v_{s\text{-}c}$	$i_{s\text{-}c}$	$i_{g\text{-}c}$	$i_{\text{out-}c}$
Base (p.u)	1.01	0.79	0.04	0.81	1.01	0.79	0.04	0.81	1.01	0.79	0.04	0.81
THD (%)	1.47	1.1	28.4	0.98	1.37	0.9	20.7	1.18	1.7	1.1	20.9	1.2

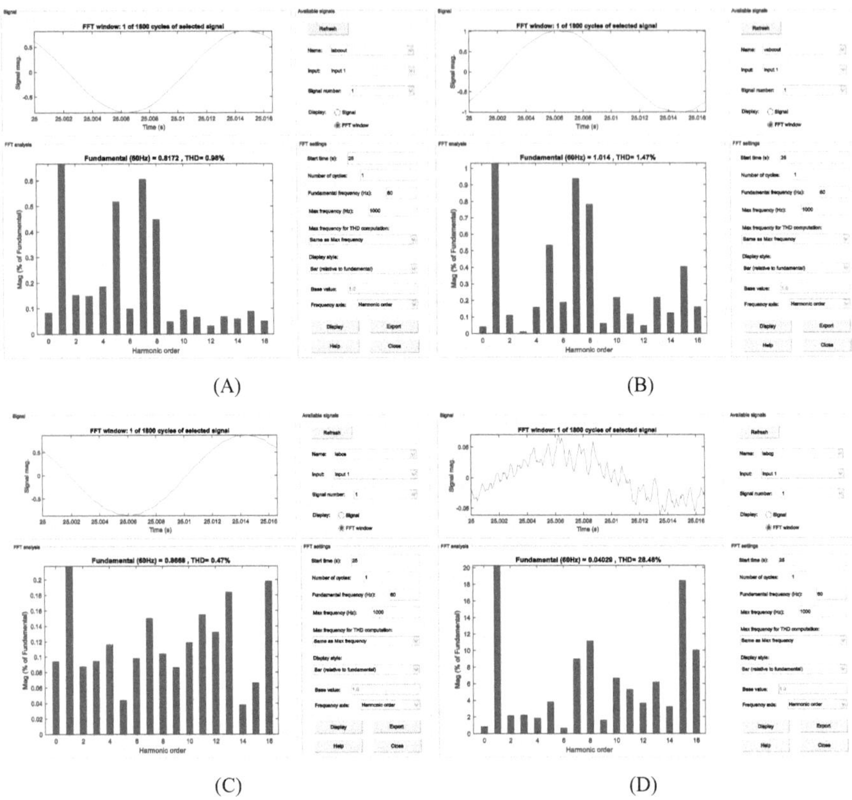

Figure 14.7 FFT analysis of waveform in MATLAB for (phase a), (A) output current, (B) output voltage, (C) stator current, (D) transformer side converter current.

14.4.2 DIgSILENT power factory

To analyze the power quality, all network components must be carefully modeled. In other words, the harmonic model and nonlinear model of elements are identified to inject harmonic current into the network, such as renewable energy and power electronic converters. In DIgSILENT *PowerFactory*, the user can define a current harmonic source. It is also possible to consider the harmonic pollution of a grid power that can be modeled with voltage sources. In this chapter, harmonic generation data of DFIM, extracted from Section 14.4.1, is used to model a current harmonic source for DFIM in the harmonic analysis toolbar of DIgSILENT *PowerFactory*.

The injection of harmonic currents is frequently modeled by a current source and various elements are available in PowerFactory as current sources. Also, the current injections are defined for each harmonic order, based on the corresponding current spectrum. In DIgSILENT *PowerFactory*, the user can specify the harmonic current injection spectrum using three methods; IEC, Balanced, and Unbalanced spectrums. The IEC spectrum is available for many power system studies, and the balanced harmonic load flow calculation based on the IEC spectrum (magnitude only) is sufficient for the assessment. In the load flow calculation based on the IEC spectrum, the harmonics multiple of 3 is shifted into the positive sequence [18].

14.4.2.1 Case study

Modified New England 39-bus transmission network consists of 10 synchronous generators, three photovoltaic (PV), 12 transformers, and 46 transmission lines. The technical parameters for the generation system and the transmission network are obtained from [19]. As shown in Fig. 14.8, three DFIG-based wind turbine (WT) is added to the New England IEEE-39 bus system at buses 5, 15, and 33 as a study system in which the rating power of the WT farms is 50 MVA (10 WTs 2.5 MW). Note that the standard harmonic model represents the WTs as a current source in the library of DigSILENT for harmonic analysis. In contrast, the DFIM-based PSPP is modeled as a current source with harmonic order data manipulated from MATLAB/Simulink. As found in [19], all generators except G_{10}, that is G_1-G_9, are steam plants, while the last generator (G_{10}), located at bus 30, is the hydro plant. In this study, G_{10} is substituted by DFIM-based PSPP in two cases which are explained as follows.

- Case 1: New England system with a DFIM-based PSPP in bus 30
- Case 2: New England system with three DFIG-based WT in buses 5,14 and 33
- Case 3: New England power network as shown in Fig. 14.8.

14.4.2.2 Simulation results

Fig. 14.9A–C show the THD value of all buses for cases 1 through 3. As is evident in these figures, harmonic distortion is much higher for cases 1 and 3 than in case

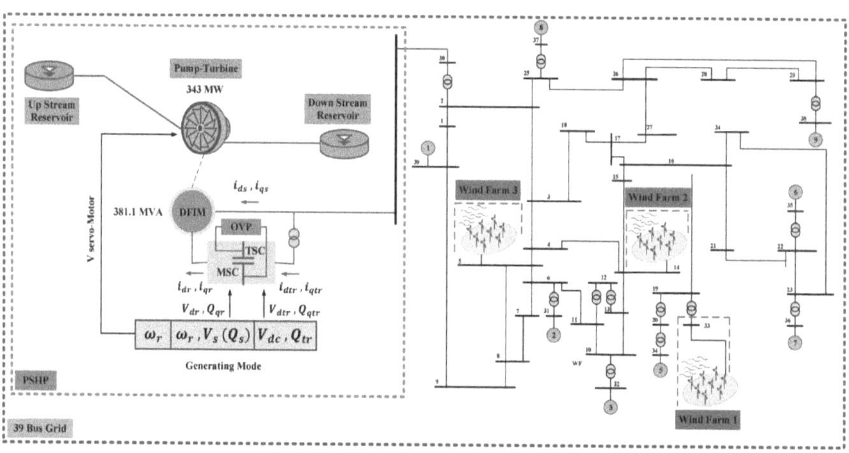

Figure 14.8 Modified New England 39 bus test system for harmonic analysis study.

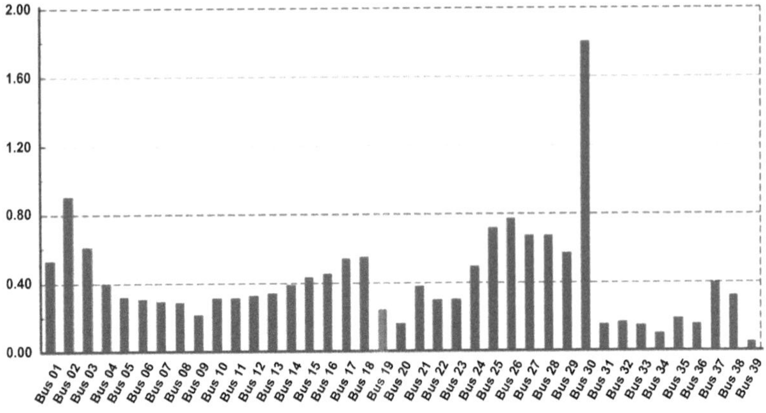

(A)

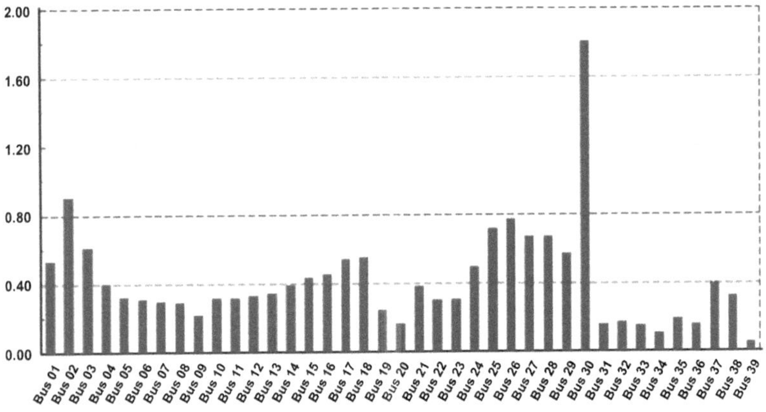

(B)

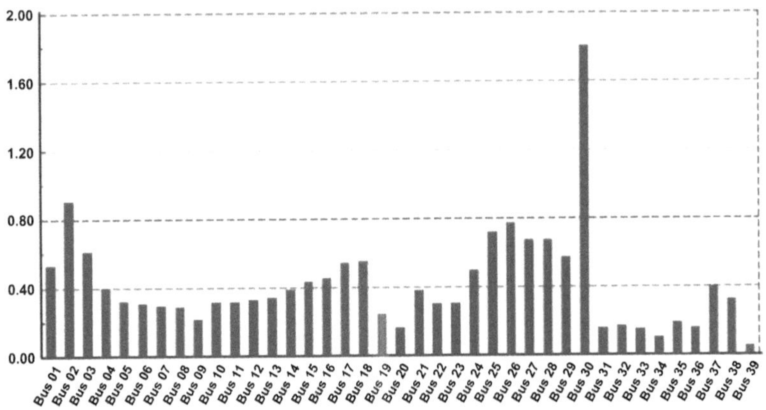

(C)

Figure 14.9 Total harmonic distortion in % for all 39 buses in balanced type for all cases, (A) case 1, (B) case 2, and (C) case 3.

2. Case 3 can be a modern power grid consisting of a VS-PSPP and three WTs based on a power electronic converter. By comparing cases 1 and 3 in Fig. 14.9A and B, it is found that a VS-PSPP has a higher effect on the harmonic distortion in a power grid. The rated power of DFIM (380 MVA) is much larger than WTs (3 × 50 MVA), which are located decentralized. Also, it shows that the designing of the output filter of the VS-PSPP is essential. It should be noted that the filter design of the VS may be an exciting topic for future research in this area [20]. Using the PowerFactory, a researcher is able to design either an active or passive filter for a polluted power grid with harmonic distortion.

Fig. 14.10A–D show harmonic order of buses #30, #25, #2, and #1 of the test system in balanced type for all cases. Bus 30 is a bus that the VS-PSPP is located on one. As shown from Fig. 14.10A, once the VS-PSPP is located in bus 30, the fifth and seventh harmonic orders of this bus are much higher than when three decentralized VS-WTs are applied. Another note of this figure is that DFIM-based PSPP and DFIG-based WT have a similar harmonic order spectrum of the injected current into the power grid. Three other figures (Fig. 14.10B–D) are shown the results of buses that are far from bus 30. The same effects can be observed from these figures for harmonic orders.

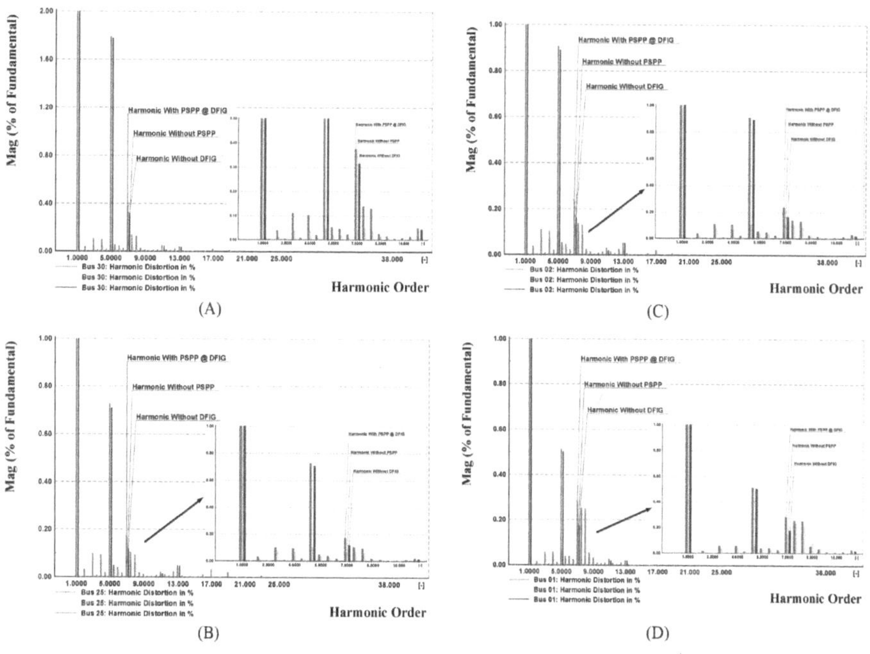

Figure 14.10 Harmonic order of four buses of the test system in balanced type for all cases, (A) bus 30, (B) bus 25, (C) bus 2, and (D) bus 1.

14.5 Conclusion

In this chapter, harmonic analysis of DFIM-based PSPP is conducted using the detailed model in MATLAB/Simulink. The effect of such harmonic on a large power grid and DFIG-based WTs is considered. This chapter thoroughly assesses the various types of VS-PSPP commissioned worldwide. That shows that the state-of-the-art VS units use the multi-level voltage source converter instead of the cyclo-converter and 2L-VSI to reduce harmonic distortion of current and voltage. Also, it is shown that there are not many published kinds of research on power quality study of the large DFIM-based PSHP, which is meaningful and urgently needed, especially for a smart and modern power grid. All simulations for investigating the harmonic order of DFIM-based PSPP are performed through detailed models by applying MATLAB/SimPowerSystem. The results show that the leakage inductance of the excitation transformer and parallel filters can eliminate the high-frequency harmonics of the TSC current. Also, the harmonic analysis toolbar of DIgSILENT *PowerFactory* is used to compare the effect of harmonic generation of VS-PSPP and VS-WT in a modern power grid. The results show that VS-PSPP has a higher impact on the harmonic distortion in a power grid due to the rated power of DFIM being much larger than a WT farm. Moreover, other publishers might extend this study, such as VS-PSPP under pumping mode operation and VS-PSPP with multi-level converters.

References

[1] H. Schlunegger, A. Thöni, 100 MW full-size converter in the Grimsel 2 pumped-storage plant, Hydro (2013).

[2] J.K. Lung, Y. Lu, W.L. Hung, W.S. Kao, Modeling and dynamic simulations of doubly fed adjustable-speed pumped storage units, IEEE Trans. Energy Convers. 22 (2) (2007) 250–258. Available from: https://doi.org/10.1109/tec.2006.875481.

[3] M.A. Bidgoli, H.A. Mohammadpour, S.M.T. Bathaee, Advanced vector control design for DFIM-based hydropower storage for fault ride-through enhancement, IEEE Trans. Energy Convers. 30 (4) (2015) 1449–1459.

[4] A.C. Padoan, et al., Dynamical behavior comparison between variable speed and synchronous machines with PSS, IEEE Trans. Power Syst. 25 (3) (2010) 1555–1565. Available from: https://doi.org/10.1109/TPWRS.2009.2039586.

[5] Y. Pannatier, B. Kawkabani, C. Nicolet, J.-J. Simond, A. Schwery, P. Allenbach, Investigation of control strategies for variable-speed pump-turbine units by using a simplified model of the converters, IEEE Trans. Ind. Electron. 57 (9) (2010) 3039–3049. Available from: https://doi.org/10.1109/TIE.2009.2037101.

[6] L. Belhadji, S. Bacha, I. Munteanu, A. Rumeau, D. Roye, 'Adaptive MPPT applied to variable-speed microhydropower plant', IEEE Trans. Energy Convers. 28 (1) (2013) 34–43. Available from: https://doi.org/10.1109/TEC.2012.2220776.

[7] K. Grotenburg, F. Koch, I. Erlich, U. Bachmann, Modeling and dynamic simulation of variable speed pump storage units incorporated into the german electric power system, Eur. Conf. Electron. Appl. (2001) 1–10.

[8] A. Joseph, T.R. Chelliah, A review of power electronic converters for variable speed pumped storage plants: configurations, operational challenges, and future scopes, IEEE J. Emerg. Sel. Top. Power Electron. 6 (1) (2018) 103–119.
[9] V. Azbe, R. Mihalic, Transient stability of a large doubly-fed induction machine in a pumped-storage plant, Electr. Power Syst. Res. 142 (2017) 29–35.
[10] W. Yang, J. Yang, Advantage of variable-speed pumped storage plants for mitigating wind power variations: integrated modelling and performance assessment, Appl. Energy 237 (2019) 720–732.
[11] A. Joseph, R. Selvaraj, T.R. Chelliah, S.V.A. Sarma, Starting and braking of a large variable speed hydro-generating unit subjected to converter and sensor faults, IEEE Trans. Ind. Appl. 9994 (i) (2018) 1–11.
[12] S. Auber, Power on tap from variable speed pumped water storage scheme, *Energize*, July 2012.
[13] T. Kuwabara, A. Shibuya, H. Furuta, E. Kita, K. Mitsuhashi, Design and dynamic response characteristics of 400 MW adjustable speed pumped storage unit for Ohkawachi Power Station, IEEE Trans. Energy Convers. 11 (2) (1996) 376–382. Available from: https://doi.org/10.1109/60.507649.
[14] D. Plants, M. Kayıkc, Reactive power control strategies for, IEEE Trans. Energy Convers. 22 (2) (2007) 389–396.
[15] S. Engelhardt, I. Erlich, S. Member, C. Feltes, Reactive power capability of wind turbines based on doubly fed induction generators, IEEE Trans. Energy Convers. 26 (1) (2011) 364–372.
[16] H.A. Mohammadpour, E. Santi, Sub-synchronous resonance analysis in DFIG-based wind farms: definitions and problem identification - part I, IEEE Energy Convers. Congr. Expo. (2014) 812–819.
[17] A. Hodder, Double-fed Asynchronous Motor-Generator Equipped with a 3-Level VSI Cascade (Ph.D. thesis), Ecole polytech. f'(ed.) Lausanne, Lausanne, Switzerland, 2004.
[18] DIgSILENT GmbH, User manual powerfactory, V15.0. Germany, 2013.
[19] M.A. Pai, Energy Function Analysis for Power System Stability, Springer Science & Business Media, 2012.
[20] A. Bagheri, M.A. Bidgoli, Designing a passive filter for reducing harmonic distortion in the hybrid micro-grid including wind turbine, solar cell and nonlinear load, Przegląd Elektrotech. 95 (12) (2019) 9–12.

Advanced energy storage system in smart grids: power quality and reliability

15

Mohammadhossein Deihimi, Navid Rezaei, Mehrdad Gholami and Hadi Tarimoradi
Department of Electrical Engineering, University of Kurdistan, Sanandaj, Kurdistan, Iran

Chapter Outline

Highlights

- The chapter provides a comprehensive comparison of the advanced energy storage system (ESS) technologies.
- The effects of ESS technologies on the power quality and reliability indices are analyzed.
- The role of storage devices in reliability and power quality improvement is addressed.

Emerging Trends in Energy Storage Systems and Industrial Applications. DOI: https://doi.org/10.1016/B978-0-323-90521-3.00010-7

- A helpful framework for decision-making on selecting an appropriate ESS is presented.
- The requirements and capabilities of storage devices for proper performance are itemized.

Nomenclature

AENS	Average energy not supplied due to interruption
APF	Active power filter
ASAI	Average service/system availability index
ASIDI	Average system interruption duration index
ASIFI	Average system interruption frequency index
ASUI	Average system unavailability index
BESS	Battery energy storage system
CAESS	Compressed air energy storage system
CAIDI	Customer average interruption duration index
CELID-t	Customers experiencing long total interruption durations
CEMSIn	Customers experiencing multiple sustained interruption and momentary interruption events
CENS	Cost of energy not supplied
CI/I	Customer interrupted per interruption index
DG	Dispersed generator
DSTATCOM	Distribution static compensator
DVR	Dynamic voltage restorer
ENS	Energy not supplied
ESS	Energy storage system
FBESS	Flow battery energy storage system
FDR	Power frequency deviation ratio
FESS	Flywheel energy storage system
FOR	Forced outage rate
HESS	Hydrogen energy storage system
H-ESS	Hybrid energy storage system
LOEE	Loss of energy expectation
LOLE	Loss of load expectation
MAIFI	Momentary average interruption frequency index
PHESS	Pumped-hydroelectric energy storage system
RIB	Reliability improvement benefit
SAIDI	System average interruption duration index
SAIFI	System average interruption frequency index
SCESS	Supercapacitor energy storage system
SDR	Symmetrical components deviation ratio
SMESS	Superconducting magnetic energy storage system
TESS	Thermal energy storage system
UPHESS	Underground pumped-hydroelectric energy storage system
UPQC	Unified power quality conditioner
UPS	Uninterruptible power supply
WDR	Waveform distortion ratio

15.1 Introduction

The reliability and power quality are two main concerns of energy management services in modern power systems [1]. Increasing nonlinear loads on the distribution grids as the source of distortions along with penetration of renewable energy resources into the power systems due to the inherent uncertainty and relatively rapid fluctuations in their output power, may reduce the power quality and have deteriorative impacts on the associated reliability indices. The state-of-the-art technologies of energy storage systems (ESSs) redound the crucial efficacies toward the improvement of reliability and power quality becoming more important in the smart grids. Fig. 15.1 shows some considerable benefits of ESSs.

The ESSs are widely used to cause synergy accompanied by renewable energy resources that can make a balance between the power sources and demands. Thus, the ESSs help to increase the penetration of dispersed generators (DGs) in the modern active distribution networks. The potential ESS implementations in renewable-energy-based system operations have been clarified in [2]. The capability in the operation of the ESSs along with the DGs can enhance the reliability of power systems because of the flexibility in stand-alone versus grid-connected operation of such cooperation. The ESSs provide a prompt response to large power generation of DGs during periods of energy surplus and cope with power consumption during times of energy shortage, especially for the inadequacy of supplies due to the

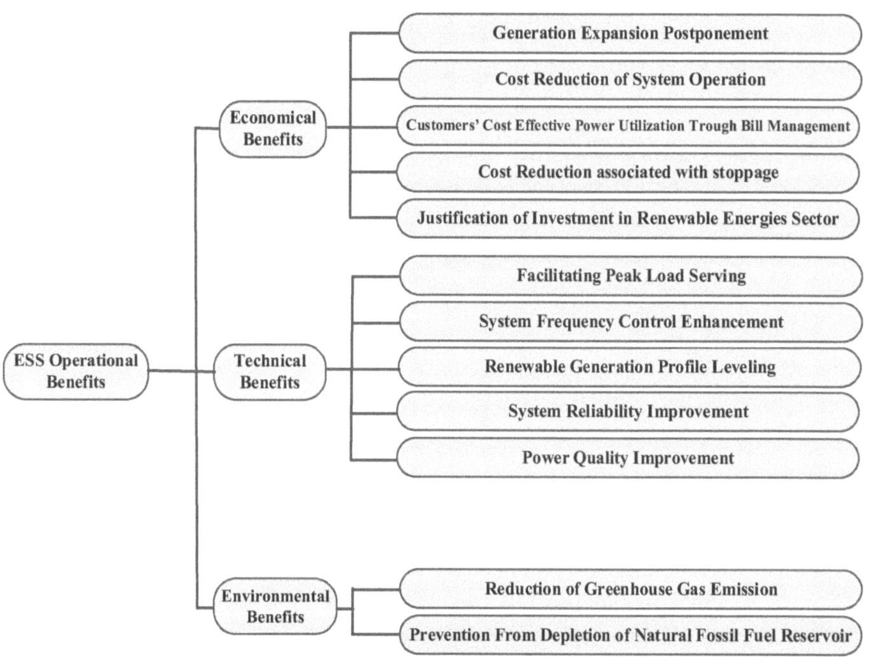

Figure 15.1 Typical energy storage system operational benefits.

intermittency of renewable energy resources. Reference [3] has studied the innovative viewpoint for electricity generation and stationary grid storage in two systems. The selection of the ESS is based on the time horizon of intermittency, the required capacity, and the type of single/hybrid application as well as the availability of technology, infrastructure, and environmental impacts [4]. The smart grids should be planned and contrived so that they can accomplish their aimed functions passably with a rational risk under the conditions encountered. The risk of outages and interruptions made by energy shortage decreases by providing adequate reserve power capacity. From the viewpoint of smart power grid operation, the ESSs can bestow a considerable portion of the required practical capacity which can be considered as a substitution for spinning reserve. Furthermore, since the ESSs are mostly embedded nearby the consumption areas in distribution networks, the occurrence of faults and failure of equipment in the transmission lines have no impact on the electrical energy delivery. Also, the difficulties due to limitations in bulk generation as well as electricity flow in transmission systems are mitigated by their application. Thus, the ESSs enhance the reliability indices that quantitate the risk of interruption in the smart grids. Generally, the concept of reliability pertains to the relative period in which the end-use consumers have a sustained interruption due to a lack of energy supply.

Moreover, power quality is an influential issue for both utilities and customers. The power quality arena involves a wide range of the power grid operational irregular distortions. Generally, any deviations in the voltage, current, and/or frequency of the system that result in the failure or malfunction of the end-user equipment, can be considered power quality irregularities. The smart grid consists of a large number of nonlinear loads that have deteriorative impacts on the power quality and may be harmful to sensitive loads. Substantially, the deficiency of power quality can cause many problems in the distribution grids. The reduction of power quality indices leads network operators toward the utilization of effective improver approaches. One of the applicable solutions for enhancing power quality is the suitable employment of ESSs. Reference [5] has itemized the most common consequences of the power quality issues and the role of ESSs to cope with such problems. The ESSs can contribute to rescuing almost all power quality problems in the smart grids except impulsive transients, imbalance, and noise. Indeed, the ESSs can affect the unfavorable voltage waveform through proper charging/discharging for energy exchange with the power grid. There are various technologies of the ESSs such as battery energy storage system (BESS), superconducting magnetic energy storage system (SMESS), ultra-capacitor, compressed air energy storage system (CAESS), flywheel energy storage system (FESS), pumped-storage hydroelectricity, a thermal energy storage system (TESS), electrolysis of water, and so on. Each technology has individual technical characteristics and capabilities that can be determinant in the selection of one type to attain the aims for embedment of the ESS in a scheme. In other words, the ESSs have distinctive power and energy capacity, energy efficiency, response time, lifetime, maintenance, and other technical features. Therefore each ESS type may be appropriate for improving special indices of reliability and power quality in the smart grids. For example, the pumped-storage hydroelectricity can be suitable for the long-term variations because of its higher energy capacity; whereas, the batteries might be a better solution for the short duration variations due to greater power and energy density as well fast response time. On the other hand, power quality indices may require different capabilities and characteristics of ESSs. For example, for improving voltage sag and voltage interruption, technical features such as response time and power capacity characteristics may be more important than others. Nevertheless, the ESS should inject a large amount of energy into the grid in a short time interval and be capable to charge quickly for re-operation. Therefore the selection of the proper type of ESS for each reliability or power quality index is a very critical task for grid operators.

This chapter aims to provide a comprehensive comparison of the ESS technologies thereby analysis of the ESS effects on the reliability of the smart grids can be clarified and a helpful framework for decision-making on selecting an appropriate ESS can be introduced to cope with power quality problems. First, various types of ESS technologies are explained in detail considering the general principles and the rules governing them. The types of ESSs are compared and categorized in terms of important parameters and characteristics. Next, a review of the emerging advanced structure of ESS technologies in the smart grids is presented. Then, power quality and reliability indices are explained and the impact of ESS technologies on smart grid power quality and reliability indices are analyzed. The role of the storage devices in reliability and power quality improvement is addressed and the requirements and capabilities of a certain storage device for proper performance are itemized. Moreover, the considerations on the storage capacity and the response time according to power quality severity will be given along with related relationships. Lastly, some concluding remarks of the whole chapter are stated along with future horizons of the ESSs applications.

15.2 A brief overview of basic energy storage system technologies

The task of the ESS is actually to exchange energy with the power system through the charging and discharging process at the required times. Depending on the characteristics of the exchanged energy, the ESS technologies can be classified into different categories. The features can be defined as the maximum amount of the energy exchanged per cycle, the maximum power rate, the speed of the response to the command and system feedback, the lifetime of the ESS, investment and operational costs, and the power and energy densities (which determines the volume required by the ESS). From the grid's perspective, the various energy storage technologies can be classified into three categories based on the size of storage capacity (see [6]):

- Large-scale power (> 50 MW) and energy capacities (> 100 MWh)
- Medium-scale power (1-50 MW) and energy capacities (5-100 MWh)
- Medium-scale power and small-scale energy capacities (<5 MWh)

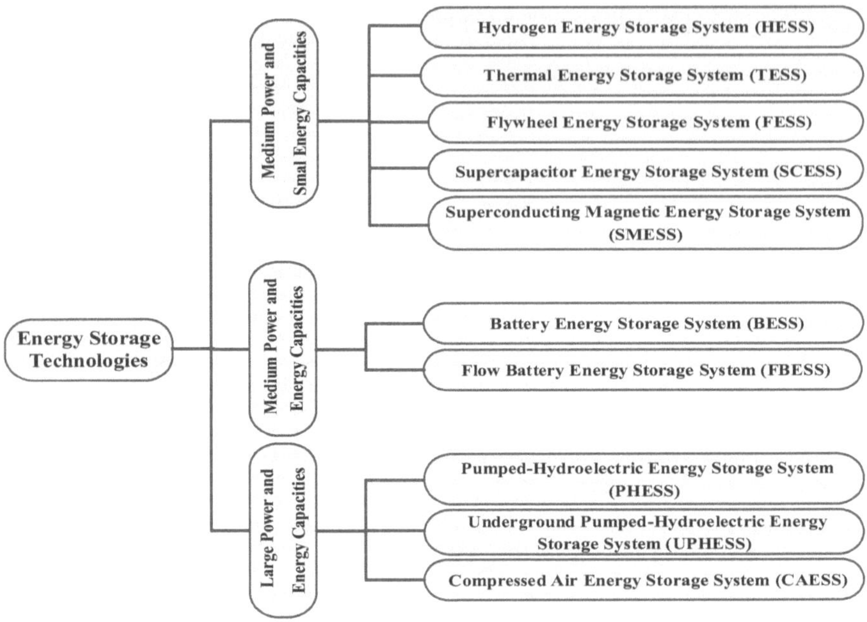

Figure 15.2 Classification of energy storage system technologies based on the size of storage capacity.

Fig. 15.2 shows the energy storage technologies in each category based on [6].

On the other hand, depending on the storing process, the ESSs technologies include electrical, electrochemical, mechanical, thermal, and magnetic structures. Each structure has individual characteristics. The basic structures of ESSs are shown in Fig. 15.3.

Depending on the purpose and type of application in the power grid, the desired attributes of the exchanged power and energy (in terms of rate, volume, and response quickness) may require different ESS technologies that may be suitable for certain applications. Therefore, in this section, different types of ESS technologies are inspected along with their functional features. Then, depending on the different features, we categorize them so that the appropriate technology can be selected for different purposes of reliability and power quality improvement in the smart grids.

15.2.1 Electrochemical energy storage system technologies

This category includes a variety of batteries and hydrogen gas storage systems.

15.2.1.1 Batteries

The batteries are one of the most popular types of ESSs used in a wide range of applications. To achieve the proper electrical characteristics, they may be connected

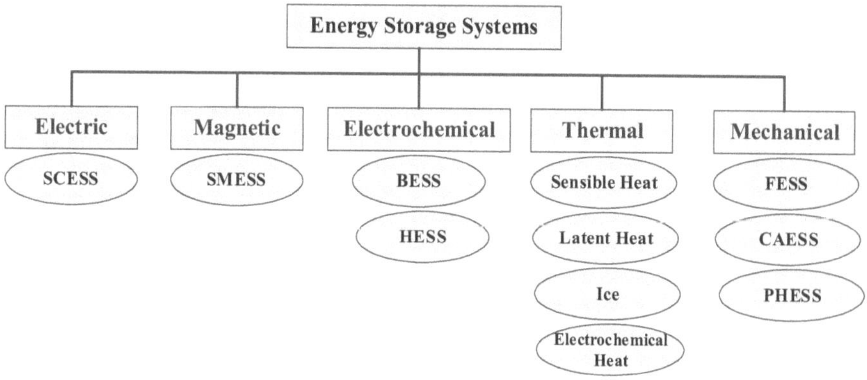

Figure 15.3 Energy storage system technology classification based on storing process.

in series or parallel. In a general classification, the batteries can be divided into two categories. The first group involves a unified electrical and chemical structure in form of "Lead-acid," "Lithium-ion," "Nickel-Cadmium," and "Sodium-Sulfur," and the second group engages an external tank for electrolyte solution called flow batteries that allow their active substances to be stored outside the base and flow through the cells as needed by the pumps [7]. Mostly, the batteries of the first group have medium energy and power density and have relatively good response quickness whereas their lifetime is relatively short. However, the advancement in their technologies has improved their specifications [8]. The lead-acid battery is the cheapest type which has a low energy density and low lifetime limiting its grid-side applications. The Lithium-ion battery has a higher energy density, long lifetime, and high efficiency. The main drawback of this battery is its high cost due to the specific type of packaging and the internal overload protection circuits. The "Nickel-Cadmium" battery has high energy density and low toxins. It has a long lifetime in permanent applications. The "Sodium-Sulfur" battery has a very high energy density and its charge-discharge efficiency is close to 90%. The cost of this battery is high. These types of batteries are usually suitable for large-scale, nonmobile applications such as grid energy storage and wind farm applications (see [9]). The main aspect of flow batteries that distinguishes them from other batteries is the independent capacity of the battery power and energy due to the separation of the battery cells and the electrolyte tank. More involved battery cells increase the power capacity and enlargement of the electrolyte tank, leading to increased battery energy capacity. The benefits of the flow batteries include a long lifetime (more than 10,000 cycles), high efficiency (about 70%), greater discharge depth (20%–100% wide operating range), low ratio of charge to discharge time as well a high power and energy density. They also have less cost ($ kWh^{-1}) than other batteries due to their long lifetime [10]. Moreover, the compatibility of flow batteries with environmental concerns is high due to lack of the cadmium or lead in their solutions. The Zinc Bromine (ZnBr) and Vanadium Redox Batteries (VRB) are two types of flow batteries. A typical flow battery structure has been shown in Fig. 15.4.

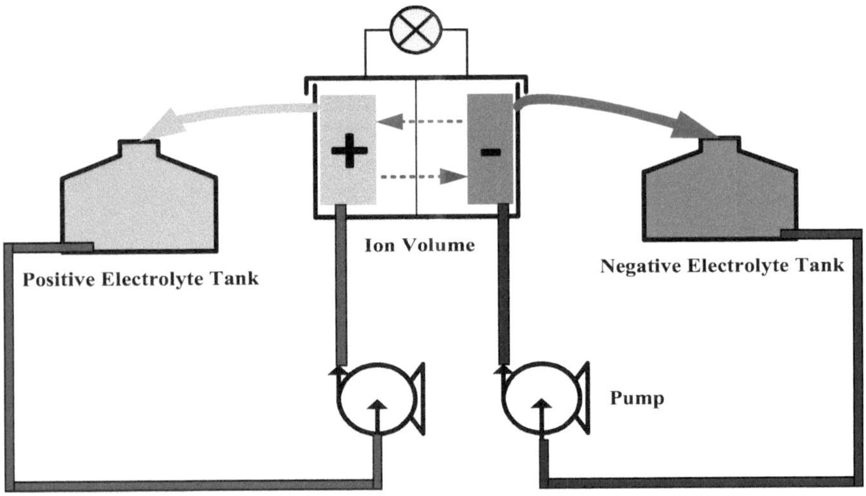

Figure 15.4 A typical flow battery structure.

15.2.1.2 Hydrogen energy storage systems

This type of ESS includes the production, storage, and utilization of hydrogen for the required application. As a well-known scheme, a fuel cell converts the stored chemical energy of hydrogen into electrical energy. The fuel cell system consists of three parts: a water electrolyzer system (for hydrogen production), a fuel cell stack (for electricity generation), and a hydrogen storage tank. A typical application of a fuel cell-hydrogen energy storage system (FC-HESS) has been shown in Fig. 15.5. The fuel cells have medium energy density and low power density. The fuel cell operation does not make any pollutants. Moreover, the failure occurs rarely which results in high reliability in normal conditions.

15.2.2 Magnetic energy storage system technologies

The superconductive magnetic energy storage system (SMESS) stores energy in the magnetic field of a large coil with a DC. This system consists of a large superconducting coil, a cooling system, and a bidirectional AC/DC converter [11]. A typical SMESS has been shown in Fig. 15.6. The SMESS has a very fast response and high efficiency. Of course, their cost is very high right now and they have low energy density. Due to its fast response and high power density, this system can be useful for short-term applications such as power quality improvement.

15.2.3 Thermal energy storage system technologies

There are several types of thermal ESSs such as sensible heat, latent heat, ice-based, and electrochemical-based thermal ESSs. These systems have high power and energy capacity and very high energy density. Therefore they are suitable for

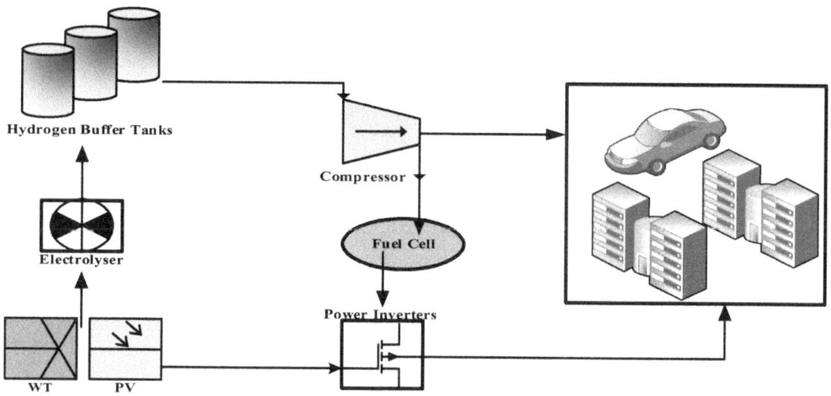

Figure 15.5 A typical application of fuel cell-hydrogen energy storage system

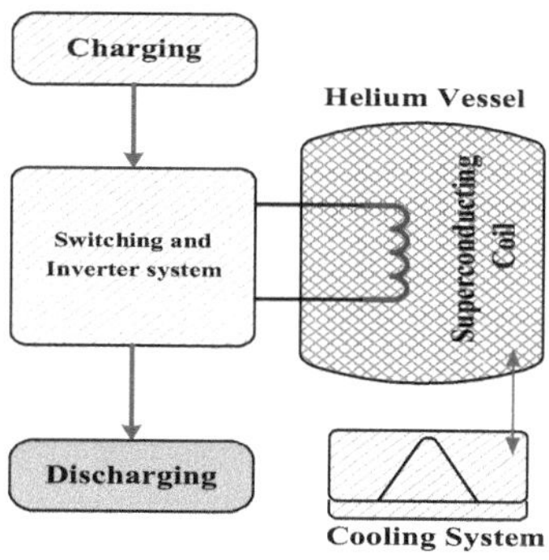

Figure 15.6 Structure of a typical superconducting magnetic energy storage system.

grid-scale applications such as load leveling. Given the high efficiency of these systems, they can be a better option than other grid-scale technologies such as PHESS and CAESS. However, this technology is currently being developed and its capacity is lower than PHESS (see [12,13]).

15.2.4 Mechanical energy storage system technologies

15.2.4.1 Pumped hydroenergy storage system

A typical pumped hydro-ESS (PHESS) consists of two interconnected reservoirs (lakes), tunnels (for directing water from one reservoir to another), valves, pumps, and generators. The stored energy is proportional to the product of the amount of water transferred in the height difference between the two lakes. This system requires special topography, differences in height, and the possibility to create a

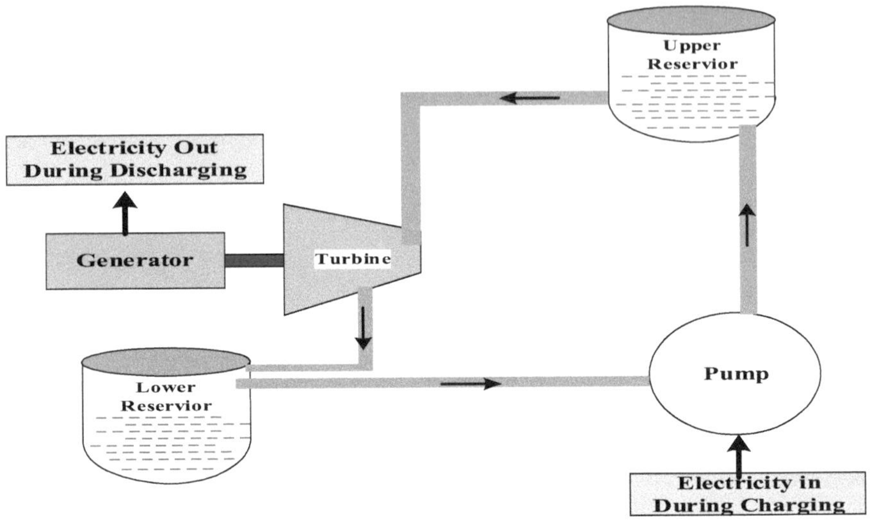

Figure 15.7 Structure of a typical pumped-hydroelectric energy storage system.

reservoir at a high altitude. The structure of PHESS and technological aspects has been elucidated in Ref. [14]. A typical PHESS has been shown in Fig. 15.7.

The energy stored in these plants is very high but their charge and discharge rate is low. Hence, PHESS has high energy capacity and low response quickness. This system can be useful for long-term applications such as load balancing, daily fluctuation compensation of wind farms, and so on.

15.2.4.2 Compressed air energy storage system

Compressed air ESS (CAESS) compresses the air to be used as an energy source. The CAESS technology is a maximum efficiency gas turbine power plant that uses gas to produce a specified power of about 40% less than the conventional gas turbines; because in conventional gas turbines 2/3 inlet fuel is used for compression of air, whereas in CAESS technology the air is already compressed by electricity at a lower cost during off-peak times and when needed (e.g., peak consumption). Similar to PHESS technology, it has high energy capacity and can be used for energy management purposes in the grid, so that it can compress the air during the nonpeak time and convert it to electrical energy through the gas turbine at peak time. A typical CAESS has been shown in Fig. 15.8. The components and operating parameters of CAESSs have been elucidated by [15].

15.2.4.3 Flywheels

Flywheel energy storage systems have a large cylindrical rotor mounted on the stator by magnetic float bearings, which eliminates the problem of bearings erosion and extends the lifetime of the system. To reduce friction and thus maintain

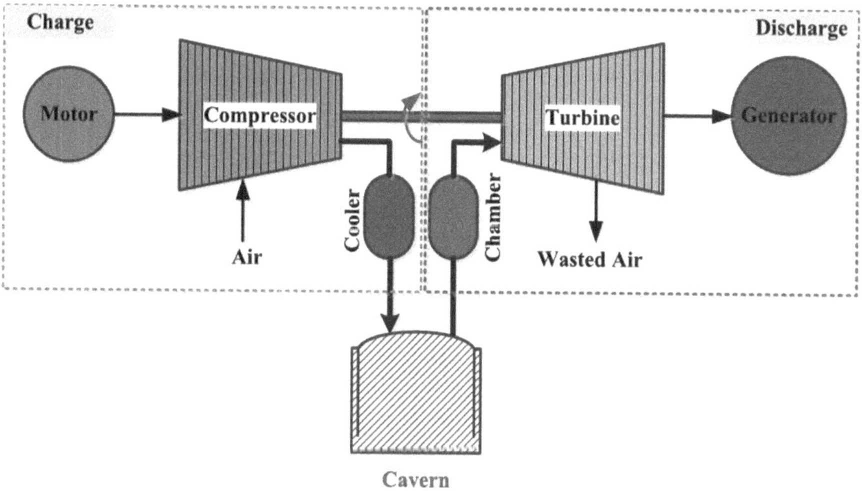

Figure 15.8 Structure of a CAESS. *CAESS*, Compressed air energy storage system.

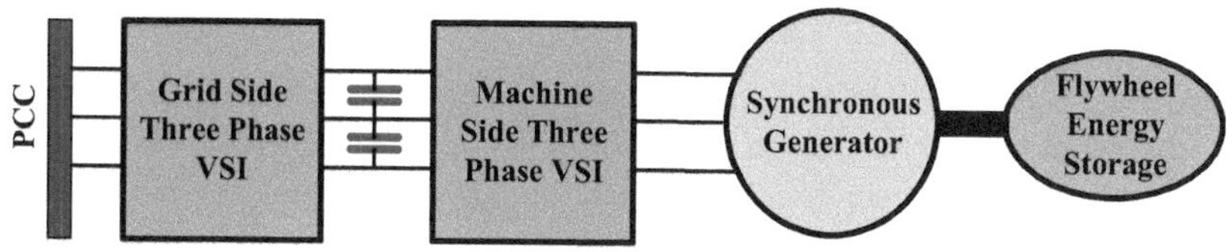

Figure 15.9 Structure of a typical flywheel energy storage system.

efficiency over time, the flywheel system works in a vacuum environment. The flywheel is connected to a generator motor, mounted on the stator so that it can exchange energy with the grid through a converter [12]. Important features of FESS are a long lifetime (20 years or tens of thousands of complete cycles), poor maintenance, and no environmental problems. However, the cost of the system is high and its energy density is low. This system can be used for frequency regulation purposes. Atypical FESS has been shown in Fig. 15.9.

15.2.5 Electrical energy storage system technologies

The electrochemical capacitors have a high capacity and high storage efficiency (higher than 95%) that can be cycled innumerable times without reduction in energy storage capacity [13]. Also, their energy density is tens of times that of ordinary capacitors. The supercapacitors consist of electric double-layer capacitors (ELDC) between the electrodes and the electrolyte ions. In ELDC technology, the charge separation occurs at several angstroms and the surface area is very high, so the capacitance of these capacitors is much greater than that of conventional capacitors [9]. Fig. 15.10 shows a typical ELDC. The merits of these capacitors versus

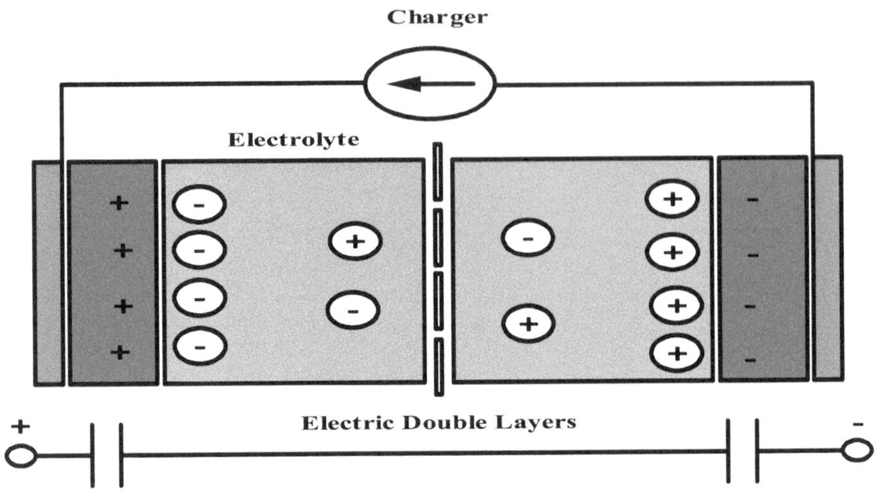

Figure 15.10 Structure of a typical electric double-layer capacitors.

batteries include much higher charge/discharge cycles, high response quickness, and high power capacity.

However, due to their high cost, their current energy capacity is low. Concerning their characteristics, they can be very useful for short-term applications such as short-term high power fluctuation in wind turbine output power, power-line flicker, and so on.

15.2.6 Energy storage technologies comparison

The implementation of an energy storage scheme in a smart grid is influenced by technological specifications such as energy and power rating, energy and power density, response time, energy efficiency, round-trip efficiency, the lifetime, the initial cost, the maintenance cost, the required location, and environmental requirements issues. These features can be described as follows.

15.2.6.1 Energy rating

The maximum energy stored in an ESS is called the energy rating. Larger capacity systems are more suitable for long-term and grid-scale applications such as spinning reserve. Among the available systems, PHESS and CAESS have the highest energy capacity; therefore these systems can be used for the grid applications such as peak shaving.

15.2.6.2 Power rating

The maximum power which an ESS can deliver in the long run is called power rating. Usually, the power ratings of most ESSs are proportional to their energy capacities. For example, a PHES system that has a large power rating also has a large energy capacity. In other words, it has a large discharge duration and can deliver energy in minutes or hours. However, there are several ESSs, such as supercapacitors and flywheels, whose nominal power is much greater than their nominal energy so that they can deliver relatively large power in a short period of time and their discharge time durations are few seconds. For example, supercapacitors may be capable to deliver the same power as some batteries while having much less power than them. Therefore they may play a different role in power grid applications.

15.2.6.3 Energy and power density

Energy density is the ratio of the energy capacity of an ESS to the total required volume. The power density is also defined as the ratio of the power capacity to the required volume. For example, PHESSs has the lowest energy and power density while superconductors have the high power density. Network applications may have limitations for installation space, so, these features should be taken into account when selecting an ESS. Fig. 15.11 shows a comparison between power and energy density for fuel cells, batteries, and supercapacitors (see [9,13]).

15.2.6.4 Response time

Response time is defined as the length of time in which a system can respond to a network request. Some storage devices, such as SCs, SMESSs, and FESSs, have fast responses. These systems can handle fast profiles such as power supply issues and voltage fluctuations, whereas some ESSs, such as PHESS and CAESS, have a low response time and are not suitable for rapid applications.

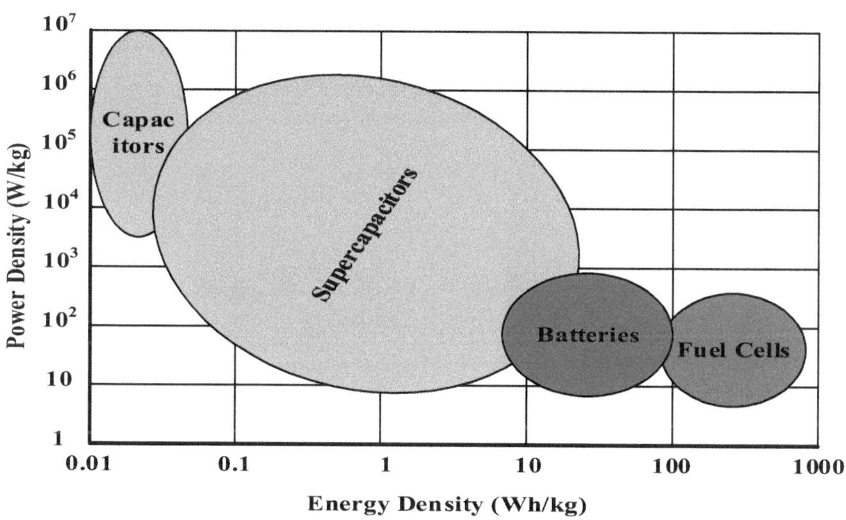

Figure 15.11 Power & energy density comparison; batteries, supercapacitors, fuel cells.

15.2.6.5 Lifetime

The life span of an ESS is defined either by the number of complete charge-discharge cycles that the system can provide rating power and energy or by the number of years. Some ESSs, such as batteries, have shorter lifespans due to chemical deterioration, so they may not be suitable for frequent charge/discharge applications, as they need to be replaced in short periods. In contrast, superconductors and magnetic storage systems have a large number of charge-discharge cycles. After them, mechanical ESSs such as PHESS, CAESS, and FEES and fuel cell systems have a relatively large number of cycles. However, since mechanical systems such as PHEES and CAESS are not used for fast-changing applications, because of low response time, so they have a longer life span in terms of the number of years, despite fewer cycles than supercapacitors.

15.2.6.6 Capital and operating costs

In addition to the initial investment, maintenance costs and replacement costs (according to their lifetime) are also important in selecting ESSs. These costs are known as operating costs and are usually expressed in terms of cost per "kWh" or "kW" or "kWh per cycle." The latter option, cost per kWh per cycle, is more convenient to compare because it also takes into account the number of cycles (i.e., lifetime). For example, Lithium-ion batteries and supercapacitors have approximately the same $\$\,kWh^{-1}$ cost, while $\$\,kWh^{-1}$ per cycle for supercapacitors is much lower as the number of charge cycles in the supercapacitor is far higher than the batteries. So, for short-term applications, in addition to the problem of response time, considering the cost per cycle, supercapacitors can be a good option. Of course, The first two options, $\$\,kWh^{-1}$ and $\$\,kW^{-1}$ are also important to determine the minimum capacity by which an ESS can be cost-effective. For example, the PHESS, despite its lower energy cost ($\$\,kWh^{-1}$), has a high power cost ($\$\,kW^{-1}$), so it is not economical to use it for low energy capacities [11–13]. The comparison of the characteristics of the most popular technologies is presented in Table 15.1 based on [5,9,11–13,16,17].

15.3 Review of emerging advanced structure of energy storage system technologies in a smart grid environment

The advancement of technologies in designing and manufacturing of ESSs as well as investigation and implementation of innovative schemes to achieve the desired characteristics in power grid result in the employment of newcomer ESSs in the smart grids.

Table 15.1 Comparison of energy storage system technologies characteristics [5,9,11–13,16,17].

ESS	Mechanical			Electrochemical						Electrical-magnetic	Thermal
	PHESS	CAESS	FESS	Lead-acid	Lit-ion	NaS	FBs	FC	SCES	SMESS	Latent-molten salt
Power rating (MW)	10–5000	110–300	≤0.25	0.05–10	0.015–50	0.05–8	0.25–12	0.1–50	0.05–0.3	1–10	≤350
Energy rating (MWh)	500–8000	580–2860	0.0052–5	0.01–40	0.0015–50	0.05–34	2–120	≤120	0.01	0.001–0.015	≤50
Energy density (kWh m^{-3})	0.2–2	0.5–20	0.3–400	25–90	90–750	20–1700	10–70	500–3000	2–35	0.2–14	100–370
Power density (kW m^{-3})	0.1–1.5	0.2–10	40–5000	10–700	60–10000	1–180	0.5–25	>500	>100000	2600–4000	-
Efficiency	70–80	70	85–95	75–80	80–85	80–85	65–75	35–40	75–85	95–98	≤90
Response time	Sec-min	≤15 min	≤10 *ms*	≤5*ms*	≤5 *ms*	≤5 *ms*	≤10 min	≤5 *ms*	≤10 *ms*	≤10 *ms*	≤10 min
Lifetime (cycles)	$<5\times10^4$	$<3\times10^4$	$<10^7$	< 3500	< 3500	< 2500	< 13000	$<10^4$	$<10^6$	$<10^5$	-
Lifetime (year)	30–50	30–40	20	5–15	15	12–20	15–20	15	8–10	20	≤30
Capital Cost ($ kW^{-1})	600–2000	400–800	250–350	300–600	1200–4000	1000–3000	600–1500	500–10000	100–300	200–300	400–700
Capital Cost ($ kWh^{-1})	5–100	3–70	1000–5000	200–400	900–1300	200–400	600	-	6800–20000	1000–10000	30–60
Capital Cost ($ kWh^{-1}) per cycle	0.1–1.4	2–4	3–25	-	15–100	8–20	5–80	-	2–20	-	-

15.3.1 Hybrid energy storage systems

A hybrid energy storage system (H-ESS) [5] is constituted by a useful combination of two or more ESSs with supplementary desired characteristics (e.g., energy efficiency, energy, power density, self-discharge rate, lifetime, etc.). Since each storage technology has some merits and demerits, combinations of various storage technologies can improve the performance of ESSs.

In the H-ESS, one ESS unit may be dedicated to coping with high demanded power, fast fluctuations, and transients; thus, should have a fast response time, high efficiency, and long cycle lifetime. Accordantly, the second ESS can have a high energy capacity with a low self-discharge rate. For instance, with a combination of flywheel and battery technologies within the smart grid, the fluctuations at the point of common coupling can be reduced ($>80\%$) with respect to the output profile of renewable power plants [18]. The merits of an H-ESS are [19]:

- Since the energy and power aspects decouple in the H-ESS, total investment costs decrease.
- The resultant energy efficiency of H-ESS is higher compared to every single component.
- The energy storage capacity and system lifetime enlarge in H-ESS.

A typical H-ESS has been introduced in [20] which uses biogas for long-term and batteries for short-term storage. Table 15.2 shows the arrangement of six configurations for H-ESSs successfully implemented in [21] to coordinate several components in a DC-nano grid.

The Liquid air ESS explicated in [22] is considered a versatile ESS that can be substantially integrated with other thermal ESSs.

15.3.2 New emerging energy storage system schemes

Some studies have focused on the materials in ESSs [23–25]. Moreover, there are worthwhile investigations for newfangled schemes in the ESS arena (see [26–28]). For instance, crystalline porous materials have been inspected in the fields of

Table 15.2 The hybrid energy storage systems and their components implemented in Ref. [21].

	Components of hybrid ESS				
Hybrid scheme	**Lithium battery**	**Supercapacitor**	**Flow battery**	**Hydrogen ESS**	**Thermal ESS**
1	✓				
2	✓	✓			
3	✓	✓	✓		
4		✓		✓	
5	✓	✓		✓	
6	✓	✓			✓

electrochemical ESSs. The metal–organic frameworks and covalent-organic frameworks have been studied in [29] for batteries and supercapacitors. Reference [30] has explained the advanced dielectric polymers for utilization in ESS structures. Another emerging ESS scheme is plasma (i.e., ionized gas) technology that can be used in energy storage in cooperation with renewable energies because it can easily be switched on/off [31]. Recent advances in the applications of biochar-based materials in FC-HESS, supercapacitors, and lithium/sodium-ion batteries, have been discussed in [32]. The metal hydride storage and compression in hydrogen ESSs have been studied by [33] to clarify the aspects of integrated ESSs involving the metal hydride-based components. A special evaporator with latent thermal energy storage ability has been introduced by [34].

15.4 Power quality and reliability indices

15.4.1 Power quality-based index

A simple classification of the power quality issues has been presented in Fig. 15.12 based on IEEE std 1159–2009 [35]. For each power factor issue, several indices have been defined in the literature (e.g., in [36–39]). Total harmonic distortion, power factor, K-factor, crest factor, and unbalance factor are samples of common power quality indices [36]. The conventional Power quality indices cannot perfectly fulfill practical aspects. Thus, a set of three novel PQ indices has been introduced

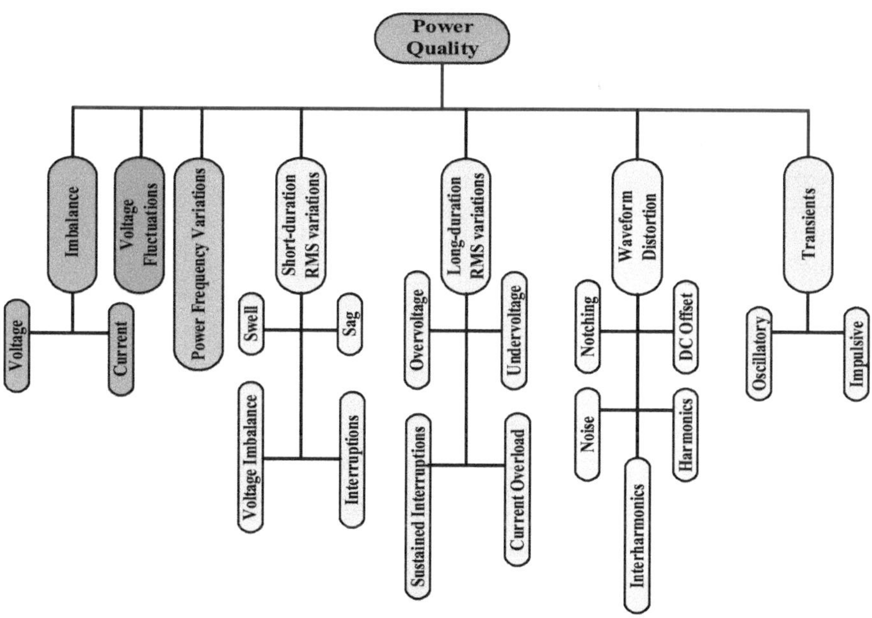

Figure 15.12 Classification of power quality issues.

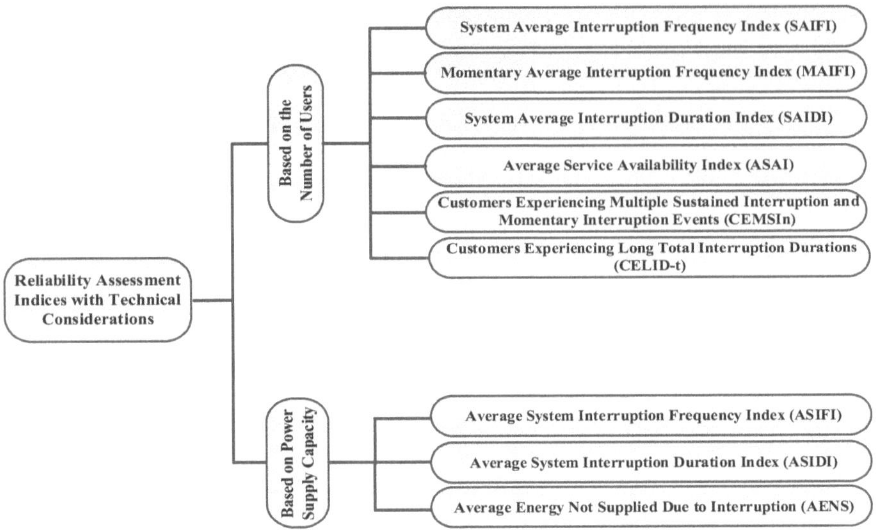

Figure 15.13 Classification of reliability assessment indices with technical considerations [40].

by [37] that can describe impressively all characteristics of power disturbances in smart grids and include:

- Power frequency deviation ratio (FDR)
- Waveform distortion ratio (WDR)
- Symmetrical components deviation ratio (SDR)

15.4.2 Reliability-based index

The reliability can be assessed by various indices. There are two different perspectives for the categorization of reliability indices that emerge grid-side and consumer-side reliability indices. A typical classification of the assessment index system for distribution network reliability can be seen in Fig. 15.13 based on [40] in which the definition of each index has been explicitly presented.

Reference [41] has categorized the reliability indices into "reliability contribution" and "reliability loss" indices. The reliability evaluation index types can be seen in Fig. 15.14. based on [41].

15.5 Impact of energy storage system technologies on smart grid power quality and reliability indices

15.5.1 Technical viewpoint in power quality

Improving power quality indices is usually done by custom power devices such as dynamic voltage restorer (DVR), distribution static compensator (DSTATCOM), active

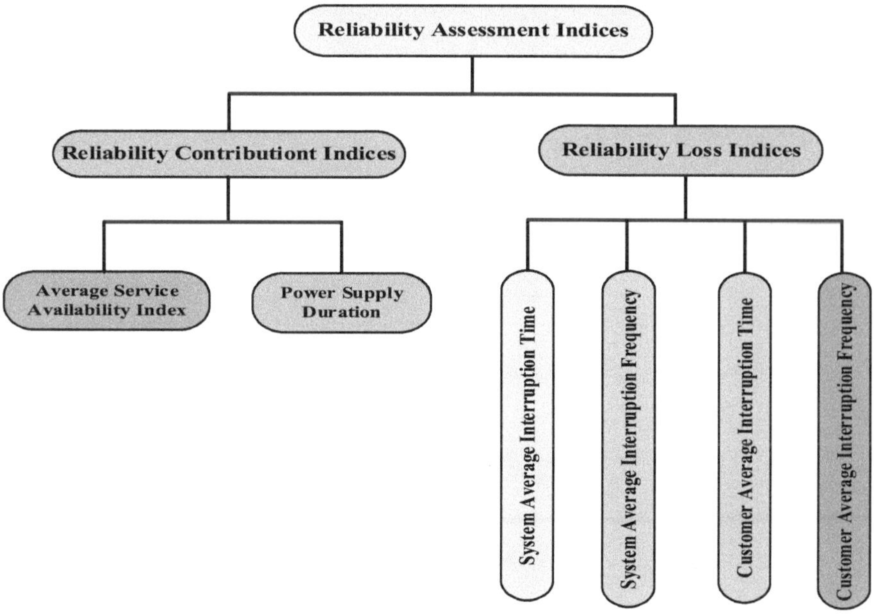

Figure 15.14 Categorization of reliability assessment indices for distribution grid [41].

power filter (APF), and unified power quality conditioner (UPQC), and uninterruptible power supply (UPS). All aforementioned devices include the ESS unit. A simple classification of power quality applications of ESSs is shown in Fig. 15.15.

The DVR, DSTATCOM, and UPQC are connected to the grid via series, shunt, and series/shunt transformers to inject the required voltage/current into the grid to efface the power quality problem and restore the waveform to normal condition. On the other hand, the UPS acts as an interface between the load and the grid. During a power outage, power is supplied through it. The UPSs are three types online, offline, and line-interactive. The online UPS can compensate for all power quality phenomena. The other two types are suitable for mains fails compensation and under/overvoltage conditions. They isolate the load from the network; thus, the load always experiences normal conditions. In these schemes, when a power outage occurs, these UPSs feed the load. Therefore, in these conditions, the load experiences a power outage for several cycles. The schematic structures of DVR, DSTATCOM, UPQC, offline, online, and line-interactive UPS have been shown in the following six figures including from Figs. 15.16–15.21.

Custom power devices have different capabilities for improving power quality problems, for example, DVR can address the sag, swell, and unbalanced voltages. The capabilities of these devices have been shown in Table 15.3 (see [42,43]).

In these devices, the injection of energy is done by charging and discharging ESSs. The amount of required power and its duration varies depending on the type of power quality phenomenon, so different systems may be considered for each phenomenon. The power quality indices have different characteristics in terms of

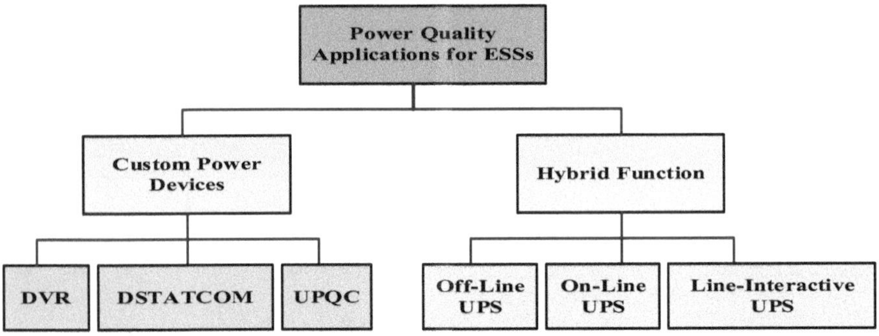

Figure 15.15 Simple classification of power quality applications for energy storage systems.

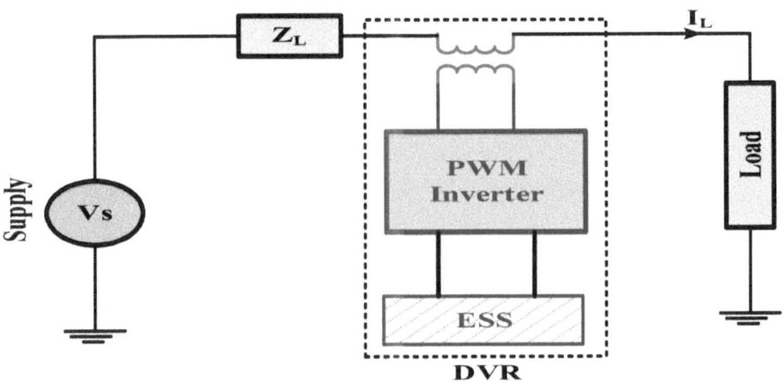

Figure 15.16 Schematic structure of dynamic voltage restorer.

duration of event, intensity, and fluctuations. For example, the voltage sag occurs over a short period of time and may vary greatly in intensity, so a large amount of energy is required to compensate for that in a short period. While over or undervoltage conditions occur over a longer period and are less severe. Thus, compensating does not require a quick response and much power. Notwithstanding different phenomena have different time periods and different intensities, almost all of them occur in less than 1 hour. Therefore, since their compensation is done by charging and discharging the ESSs, the characteristics of the phenomenon should be considered in determining the appropriate type of ESSs. To determine the appropriate types of storage devices, the different characteristics of these systems in terms of energy and power capacity, and response time have been shown in Fig. 15.22 (see [11,12,16]).

The appropriate ESS must be selected for each phenomenon according to the specifications of power quality phenomena. Since most power quality phenomena require less than 1 hour, due to the limited response time, only some ESSs can be proper to improve the specific power quality indices including fast response ESSs

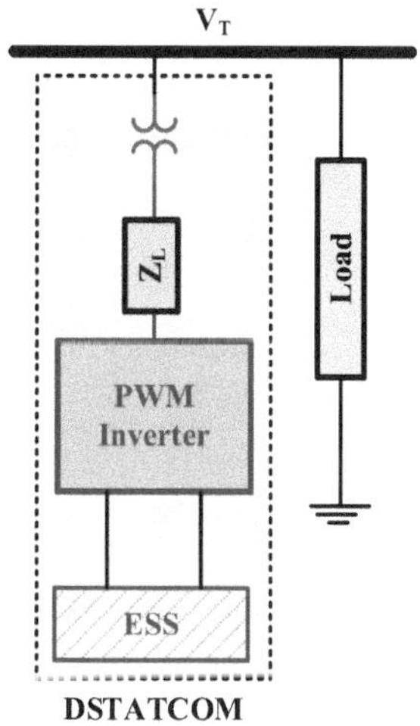

Figure 15.17 Schematic structure of distribution static compensator.

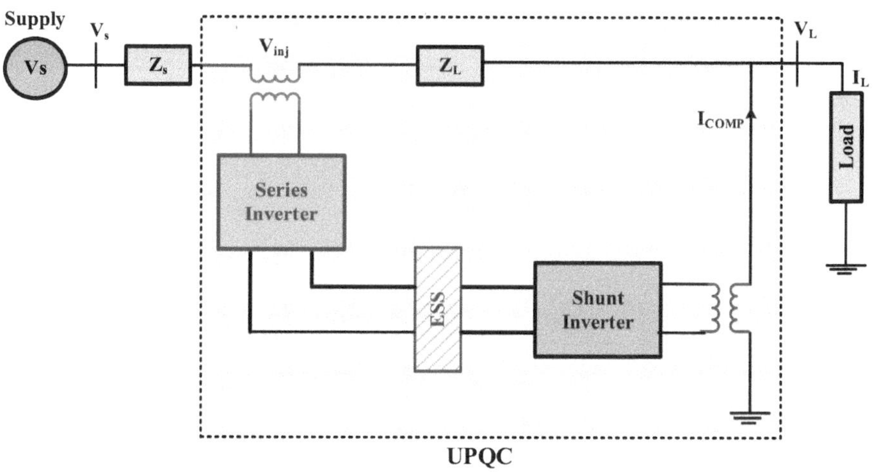

Figure 15.18 Schematic structure of unified power quality conditioner.

such as supercapacitors, most batteries, SMES, flywheels, and small CAESSs. The most suitable options for each phenomenon have been shown in Table 15.4.

For transient phenomena, due to the need for a very fast response, the only suitable option is supercapacitors. While for voltage sag and swell phenomena, three options, SC, SMESS, and FES, can be used. The batteries and CAESS can also be a proper option for flicker, as it occurs over a longer period and requires more energy capacity and discharge time. Finally, for an interruption, since the

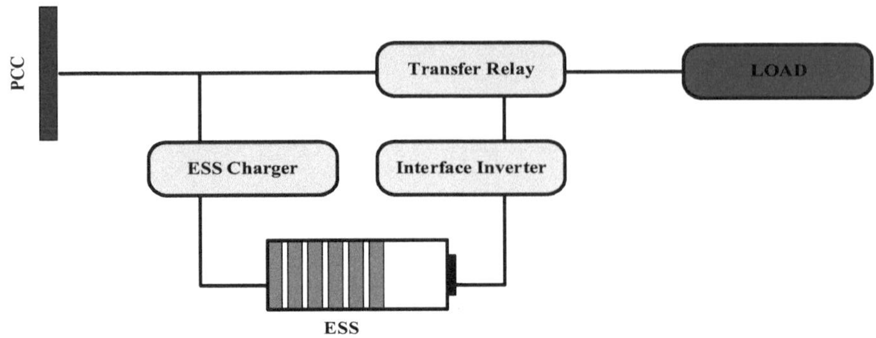

Figure 15.19 Structure of offline uninterruptible power supply.

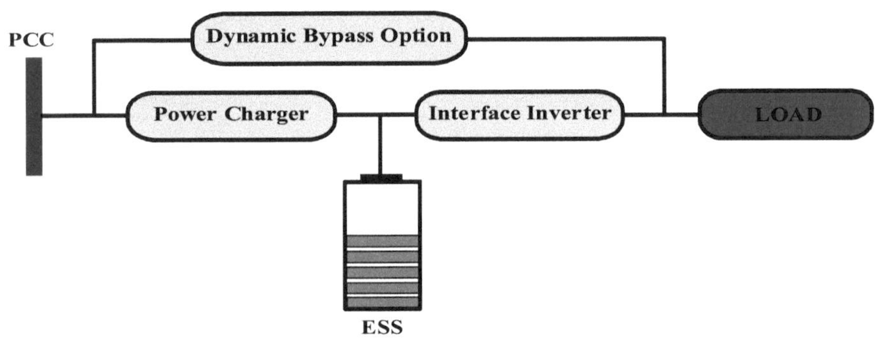

Figure 15.20 Structure of online uninterruptible power supply.

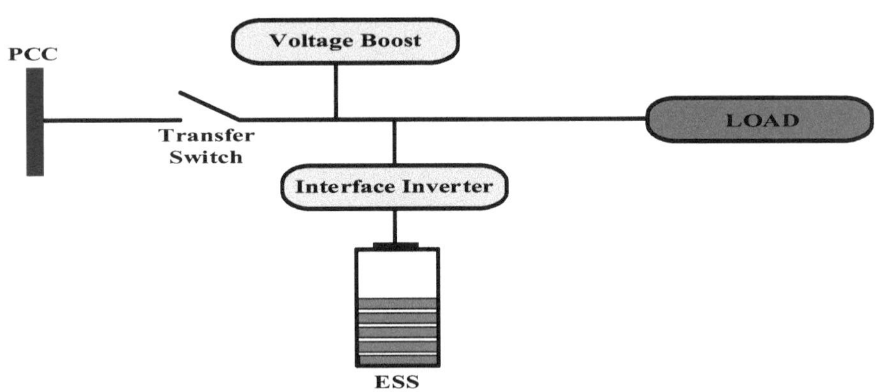

Figure 15.21 Structure of line-interactive uninterruptible power supply.

Table 15.3 The capability of custom power devices to improve power quality indices.

Device	Power quality issues								
	Sag	**Swell**	**Interruption**	**Undervoltage**	**Overvoltage**	**Voltage unbalance**	**Harmonics**	**Fluctuation**	**Flicker**
DVR	✓	✓	✓			✓			
DSTATCOM	✓					✓			✓
UPQC	✓	✓	✓			✓	✓		✓
Online-UPS	✓	✓	✓	✓	✓	✓	✓	✓	✓
Offline-UPSS			✓	✓	✓	✓			
Line-interactive UPS			✓	✓	✓	✓			

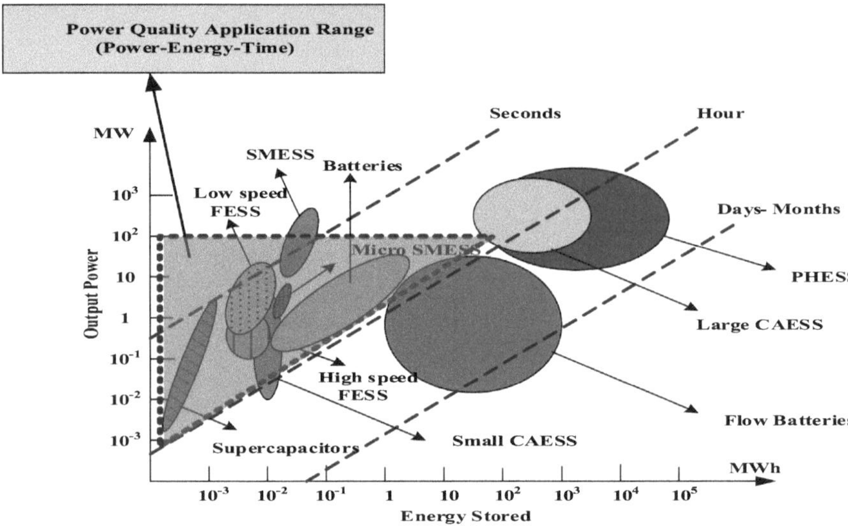

Figure 15.22 Energy storage system technologies versus power quality technical range based on [11,12,16].

Table 15.4 Most suitable energy storage systems for different power quality indices.

	ESS				
Power quality issues	**SC**	**FESS**	**SMESS**	**Batteries (Lead-acid, Li-ion, NaS)**	**Small CAESS**
Transients	✓				
Voltage sag	✓	✓	✓		
Voltage swell	✓	✓	✓		
Under/ overvoltage				✓	✓
Flicker				✓	✓
Interruption				✓	✓

event duration may be long and all the load power must be provided by the ESS, it is necessary to select a system with a larger energy capacity such as batteries and CAESSs.

15.5.2 Technical viewpoint in reliability

To analyze the effect of ESSs on the reliability of the power grid, the related risk can be quantitatively evaluated by applying suitable risk indices such as loss of load expectation (LOLE) and loss of energy expectation (LOEE) that can be determined through direct analytical methods or Monte Carlo simulation [44]. A numerical approach has been suggested in [45] to assess and improve the reliability of the distribution system from a planning point of view. The expected energy supplied by ESS (kWh year^{-1}) is an ascending function of ESS capacity. The LOLE is the anticipated time duration of a certain time horizon in which the load demand exceeds the total generated power [46]. The LOLE (hours per year) is a descending function of ESS capacity (kWh), but the incremental benefit attained by ESS becomes minimal when the energy capacity of ESS exceeds a certain value depending on system characteristics. Table 15.5 shows the reliability indices that have been regarded by literature for evaluation or consideration of ESSs in distribution power grids. Table 15.6 shows the reliability indices that can be affected by The ESSs in the smart grids. The conventional risk indices have been used in [47–49]. The indices for availability and interruption have been used in [50]. References [51–53] have used a single reliability index. The concepts are almost alike in [50] and [54].

15.5.3 Economic viewpoint

In addition to technical issues, the cost is also important in choosing the type of ESSs. For example, when several ESS options may be technically appropriate for a phenomenon, one can be selected based on the minimum cost. From the economical point of view, the assessment of energy storage types can be accomplished on the basis of capital-recovery factors for investment, degradation costs, the day-ahead market-clearing prices, and the levelized costs of energy storage [55].

The ESSs can increase the reliability of smart grids by reducing load shedding and improving the sparing aspects because of storing energy at off-peak periods in which the power prices are low and rendering the stored energy at peak hours periods in which the power prices are high. Despite high capital costs, the ESSs provide economic benefits for smart grids. The ESSs offer low-cost electrical energy to customers that result in effacing the necessity of situ generation or energy import from the mains. Another aspect of the contribution of ESSs in the improvement of reliability is trend optimization schemes for sizing ESSs regarding their capacities as well as siting them in distribution networks based on reliability considerations along with interruption and operational costs objectives (e.g.,; [5,16,47,52,56]). Fundamentally, minimizing the total cost donates straightforwardly optimal reliability of the system [50].

The ESSs may either be used solely to improve power quality or in combination with other schemes, for example, energy management. The ESSs can be applied in a hybrid framework including energy management (in normal status) and emergency power supply schemes which can help to be attained the aims of energy management in normal conditions and provide energy in case of mains failure [57]. Thus, the total power conversion system cost is reduced via a hybrid model. In the restructured energy environment, the aggregators aim to provide their customers reliably with rational costs which fulfill the related objective functions. With the integration of ESSs, an aggregator can be present in the power market with more flexibility in appropriate decision-making for electrical energy transactions [48].

Table 15.5 Reliability indices for reflecting the effect of energy storage systems.

References	Reliability index										
	LOLE	**LOLP**	**FOR**	**RIB**	**ASAI**	**ASUI**	**SAIDI**	**SAIFI**	**CAIDI**	**CENS**	**CIII**	**ENS**
[44]	✓	✓										
[45]				✓								
[46]	✓											
[47]	✓											
[48]		✓										
[49]		✓										
[50]					✓	✓	✓	✓	✓			
[51]							✓					
[52]										✓		
[53]			✓									
[54]							✓	✓	✓		✓	✓

LOLE, loss of load expectation; *LOLP*, loss of load probability; *FOR*, forced outage rate; *RIB*, reliability improvement benefit; *ASAI*, average system availability index; *ASUI*, average system unavailability index; *SAIDI*, system average interruption duration index; *SAIFI*, system average interruption frequency index; *CAIDI*, customer average interruption duration index; *CENS*, cost of energy not supplied; *CIII*, customer interrupted per interruption indices; *ENS*, energy not supplied.

Table 15.6 Reliability indices practically affected by different energy storage systems.

ESS	Reliability index								
	SAIFI	MAIFE	SAIDI	ASAI	CEMSIn	CELID-t	ASIDI	ASIFI	AENS
FC-HESS	✓	✓		✓					✓
TESS	✓	✓		✓					✓
FESS	✓	✓		✓					✓
SCESS	✓	✓		✓					✓
SMESS	✓	✓		✓					✓
BESS	✓	✓	✓	✓	✓			✓	✓
FBESS	✓	✓	✓	✓	✓			✓	✓
PHESS				✓	✓	✓	✓	✓	✓
UPHESS				✓	✓	✓	✓	✓	✓
CAESS				✓	✓	✓	✓	✓	✓

15.6 Conclusions and future trends

In the present chapter an inclusive elucidation of the effect of ESSs on the reliability and power quality of the smart grids has been presented. The concept of reliability in presence of ESS has extensive scope in the smart grids. The ESSs are often used beside the renewable energy resources to mitigate the inherent intermittency of such generators' output power. Thus, the system risk decreases significantly. Furthermore, modern distribution grids must have the capability to work in a standalone situation. The ESSs enhance this capability of the smart grids because of providing energy for local loads. Also, they can function as a substitution for spinning reserve in the power system. Hence, the ESSs improve the reliability indices of smart grids. The reliability is not restricted to the energy provider. The improvement of power quality by ESSs can mitigate the failures and malfunctions of equipment and load devices. This issue reduces the damages due to stoppages which can improve the reliability of electrical energy delivery in turn. The principles of basic and advanced ESS technologies have been explained. To select an appropriate ESS for different power quality problems, the types of ESSs have been compared and categorized in terms of important parameters and characteristics. Moreover, the power quality and reliability indices have been determined and the effect of ESS technologies on smart grid power quality and reliability indices has been clarified. Consequently, the ESSs improve the reliability of the power grid. Also, the ESSs are very helpful to cope with power quality problems. The effect of hybrid ESSs in custom power devices can be investigated in the future to achieve the best response for different situations and distortions. Also, the role of ESSs in market-based demand response can be studied to reduce the risk in deregulated environments.

References

[1] F. Mohamad, J. Teh, Impacts of energy storage system on power system reliability: a systematic review, Energies. 11 (7) (2018) 1749.

[2] A.G. Olabi, C. Onumaegbu, T. Wilberforce, M. Ramadan, M.A. Abdelkareem, A.H. Al – Alami, Critical review of energy storage systems, Energy. 214 (2021) 118987.

[3] S.M. Jordaan, J. Park, S. Rangarajan, Innovation in intermittent electricity and stationary energy storage in the United States and Canada: a review, Renew. Sustain. Energy Rev. 158 (2022) 112149.

[4] A.G. Olabi, M.A. Abdelkareem, Energy storage systems towards 2050, Energy. 219 (2021) 119634.

[5] C.K. Das, O. Bass, G. Kothapalli, T.S. Mahmoud, D. Habibi, Overview of energy storage systems in distribution networks: placement, sizing, operation, and power quality, Renew. Sustain. Energy Rev. 91 (2018) 1205–1230.

[6] D. Connolly, A review of energy storage technologies, University of Limerick, Ireland, 2009.

[7] L. Barote, R. Weissbach, R. Teodorescu, C. Marinescu, M. Cirstea (Eds.), Stand-alone wind system with vanadium redox battery energy storage, in: 2008 11th International Conference on Optimization of Electrical and Electronic Equipment, 22–24 May 2008, 2008.

[8] B. Roberts, Capturing grid power, IEEE Power Energy Mag. 7 (4) (2009) 32–41.
[9] F. Díaz-González, A. Sumper, O. Gomis-Bellmunt, R. Villafáfila-Robles, A review of energy storage technologies for wind power applications, Renew. Sustain. Energy Rev. 16 (4) (2012) 2154–2171.
[10] Y.R. Challapuram, G.M. Quintero, S.B. Bayne, A.S. Subburaj, M.A. Harral (Eds.), Electrical equivalent model of vanadium redox flow battery, in: 2019 IEEE Green Technologies Conference (GreenTech), 3–6 April 2019, 2019.
[11] O. Palizban, K. Kauhaniemi, Energy storage systems in modern grids matrix of technologies and applications, J. Energy Storage 6 (2016) 248–259.
[12] M. Aneke, M. Wang, Energy storage technologies and real life applications—a state of the art review, Appl. Energy 179 (2016) 350–377.
[13] S. Koohi-Fayegh, M.A. Rosen, A review of energy storage types, applications and recent developments, J. Energy Storage 27 (2020) 101047.
[14] S. Rehman, L.M. Al-Hadhrami, M.M. Alam, Pumped hydro energy storage system: a technological review, Renew. Sustain. Energy Rev. 44 (2015) 586–598.
[15] A.G. Olabi, T. Wilberforce, M. Ramadan, M.A. Abdelkareem, A.H. Alami, Compressed air energy storage systems: components and operating parameters—a review, J. Energy Storage 34 (2021) 102000.
[16] L.A. Wong, V.K. Ramachandaramurthy, P. Taylor, J.B. Ekanayake, S.L. Walker, S. Padmanaban, Review on the optimal placement, sizing and control of an energy storage system in the distribution network, J. Energy Storage 21 (2019) 489–504.
[17] Z. Zhang, T. Ding, Q. Zhou, Y. Sun, M. Qu, Z. Zeng, et al., A review of technologies and applications on versatile energy storage systems, Renew. Sustain. Energy Rev. 148 (2021) 111263.
[18] L. Barelli, G. Bidini, D.A. Ciupageanu, D. Pelosi, Integrating hybrid energy storage system on a wind generator to enhance grid safety and stability: a levelized cost of electricity analysis, J. Energy Storage 34 (2021) 102050.
[19] T. Bocklisch, Hybrid energy storage systems for renewable energy applications, Energy Procedia 73 (2015) 103–111.
[20] M. Wegener, J. Villarroel Schneider, A. Malmquist, A. Isalgue, A. Martin, V. Martin, Techno-economic optimization model for polygeneration hybrid energy storage systems using biogas and batteries, Energy. 218 (2021) 119544.
[21] R. Ciavarella, G. Graditi, M. Valenti, A. Pinnarelli, G. Barone, M. Vizza, et al., Modeling of an energy hybrid system integrating several storage technologies: the DBS technique in a nanogrid application, Sustainability. 13 (3) (2021) 1170.
[22] E. Borri, A. Tafone, A. Romagnoli, G. Comodi, A review on liquid air energy storage: history, state of the art and recent developments, Renew. Sustain. Energy Rev. 137 (2021) 110572.
[23] T. Xu, H. Du, H. Liu, W. Liu, X. Zhang, C. Si, et al., Advanced nanocellulose-based composites for flexible functional energy storage devices, Adv. Mater. 33 (48) (2021) 2101368.
[24] R. Kumar, A. Pérez del Pino, S. Sahoo, R.K. Singh, W.K. Tan, K.K. Kar, et al., Laser processing of graphene and related materials for energy storage: state of the art and future prospects, Prog. Energy Combust. Sci. (2022) 100981.
[25] X. Li, Y. Li, S. Xie, Y. Zhou, J. Rong, L. Dong, Zinc-based energy storage with functionalized carbon nanotube/polyaniline nanocomposite cathodes, Chem. Eng. J. 427 (2022) 131799.

[26] L. Trahey, F.R. Brushett, N.P. Balsara, G. Ceder, L. Cheng, Y.-M. Chiang, et al., Energy storage emerging: a perspective from the joint center for energy storage research, Proc. Natl Acad. Sci. 117 (23) (2020) 12550–12557.
[27] B. Zheng, X. Lin, X. Zhang, D. Wu, K. Matyjaszewski, Emerging functional porous polymeric and carbonaceous materials for environmental treatment and energy storage, Adv. Funct. Mater. 30 (41) (2020) 1907006.
[28] J. Peng, W. Zhang, S. Wang, Y. Huang, J.-Z. Wang, H.-K. Liu, et al., The emerging electrochemical activation tactic for aqueous energy storage: fundamentals, applications, and future, Adv. Funct. Mater. (2022) 2111720. n/a(n/a).
[29] J. Zhou, B. Wang, Emerging crystalline porous materials as a multifunctional platform for electrochemical energy storage, Chem. Soc. Rev. 46 (22) (2017) 6927–6945.
[30] X. Wu, X. Chen, Q.M. Zhang, D.Q. Tan, Advanced dielectric polymers for energy storage, Energy Storage Materials. 44 (2022) 29–47.
[31] A. Bogaerts, E.C. Neyts, Plasma technology: an emerging technology for energy storage, ACS Energy Lett. 3 (4) (2018) 1013–1027.
[32] W.-J. Liu, H. Jiang, H.-Q. Yu, Emerging applications of biochar-based materials for energy storage and conversion, Energy Environ. Sci. 12 (6) (2019) 1751–1779.
[33] B.P. Tarasov, P.V. Fursikov, A.A. Volodin, M.S. Bocharnikov, Y.Y. Shimkus, A.M. Kashin, et al., Metal hydride hydrogen storage and compression systems for energy storage technologies, Int. J. Hydrog. Energy 46 (25) (2021) 13647–13657.
[34] B.D. Mselle, G. Zsembinszki, D. Vérez, E. Borri, L.F. Cabeza, A detailed energy analysis of a novel evaporator with latent thermal energy storage ability, Appl. Therm. Eng. 201 (2022) 117844.
[35] IEEE Recommended Practice for Monitoring Electric Power Quality. IEEE Std 1159–2019 (Revision of IEEE Std 1159–2009), 2019, pp. 1–98.
[36] C.T. Heydt, W.T. Jewell, Pitfalls of electric power quality indices, IEEE Trans. Power Deliv. 13 (2) (1998) 570–578.
[37] L. Tao, A. Domijan, On power quality indices and real time measurement, IEEE Trans. Power Deliv. 20 (4) (2005) 2552–2562.
[38] Y. Shin, E.J. Powers, M. Grady, A. Arapostathis, Power quality indices for transient disturbances, IEEE Trans. Power Deliv. 21 (1) (2006) 253–261.
[39] A. Baggini, Handbook of Power Quality, John Wiley & Sons, 2008.
[40] S. Jiang, Y. Wang, D. Wang, J. Yin, H. Yan, J. Wang, Reliability assessment of distribution network considering differentiated end-users demand for reliability, in: IOP Conference Series: Earth and Environmental Science, vol. 645, 2021, pp. 012026.
[41] S. Jiang, W. Yang, J. Deng, Y. Zhao, J. Li, P. Li, et al., Reliability index evaluation system of distribution network based on maximum deviation and application, in: IOP Conference Series: Earth and Environmental Science, vol. 645, 2021, pp. 012034.
[42] Y. Pal, A. Swarup, B. Singh (Eds.), A review of compensating type custom power devices for power quality improvement, in: 2008 Joint International Conference on Power System Technology and IEEE Power India Conference, 12–15 October 2008, 2008.
[43] E. Hossain, M.R. Tür, S. Padmanaban, S. Ay, I. Khan, Analysis and mitigation of power quality issues in distributed generation systems using custom power devices, IEEE Access. 6 (2018) 16816–16833.
[44] B.R. Bagen (Eds.), Impacts of energy storage on power system reliability performance, in: Canadian Conference on Electrical and Computer Engineering, 1–4 May 2005, 2005.

[45] W. Liu, S. Niu, H. Xu, Optimal planning of battery energy storage considering reliability benefit and operation strategy in active distribution system, J. Mod. Power Syst. Clean. Energy 5 (2) (2017) 177–186.
[46] F.A. Bhuiyan, A. Yazdani, Reliability assessment of a wind-power system with integrated energy storage, IET Renew. Power Gener. [Internet]. 4 (3) (2010) 211–220. Available from. Available from: https://digital-library.theiet.org/content/journals/10.1049/iet-rpg.2009.0070.
[47] S. Bahramirad, W. Reder, A. Khodaei, Reliability-constrained optimal sizing of energy storage system in a microgrid, IEEE Trans. Smart Grid 3 (4) (2012) 2056–2062.
[48] Y. Xu, C. Singh, Power system reliability impact of energy storage integration with intelligent operation strategy, IEEE Trans. Smart Grid 5 (2) (2014) 1129–1137.
[49] J. Jurasz, A. Beluco, F.A. Canales, The impact of complementarity on power supply reliability of small scale hybrid energy systems, Energy. 161 (2018) 737–743.
[50] M.A. Abdulgalil, M. Khalid, J. Alshehri (Eds.), Microgrid reliability evaluation using distributed energy storage systems, in: 2019 IEEE Innovative Smart Grid Technologies —Asia (ISGT Asia), 21–24 May 2019, 2019.
[51] H. Abniki, S.M. Taghvaei, S.M. Mohammadi-Hosseininejad, Reliability improvement in smart grid through incorporating energy storage systems in service restoration, Int. Trans. Electr. Energy Syst. 29 (1) (2019) e2661.
[52] P. Lata, S. Vadhera, Reliability improvement of radial distribution system by optimal placement and sizing of energy storage system using TLBO, J. Energy Storage 30 (2020) 101492.
[53] T.T. Pham, T.-C. Kuo, D.M. Bui, Reliability evaluation of an aggregate battery energy storage system in microgrids under dynamic operation, Int. J. Electr. Power Energy Syst. 118 (2020) 105786.
[54] M.R. Tur, Reliability assessment of distribution power system when considering energy storage configuration technique, IEEE Access. 8 (2020) 77962–77971.
[55] Z. Topalović, R. Haas, A. Ajanović, A. Hiesl, Economics of electric energy storage, Case West. Balkans. Energy. 238 (2022) 121669.
[56] P. Lata, S. Vadhera, TLBO-based approach to optimally place and sizing of energy storage system for reliability enhancement of radial distribution system, Int. Trans. Electr. Energy Syst. 30 (5) (2020) e12334.
[57] S. Kim, M. Kwon, S. Choi, Operation and control strategy of a new hybrid ESS-UPS system, IEEE Trans. Power Electron. 33 (6) (2018) 4746–4755.

The battery storage management and its control strategies for power system with photovoltaic generation

16

Hongming Yang[1], Shijie Zhang[1], Bowen Chen[1], Sheng Xiang[1], Yan Xu[2], Bangzhe Yin[1] and Emmanuel Ackom[3]

[1]School of Electrical and Information Engineering, Changsha University of Science and Technology, International Joint Laboratory of Ministry of Education for Operation and Planning of Energy Internet Based on Distributed Photovoltaic-Storage Energy, Hunan Provincial Engineering Research Center for Electric Transportation and Smart Distribution Network, Changsha, P.R. China, [2]Nanyang Technol Univ, Sch Elect & Elect Engn, Singapore, Singapore, [3]UNEP DTU Partnership, Technical University of Denmark, Copenhagen, Denmark

Chapter Outline

Emerging Trends in Energy Storage Systems and Industrial Applications. DOI: https://doi.org/10.1016/B978-0-323-90521-3.00013-2

Highlights

- The grid-connected control strategy of ESS is investigated to improve the frequency stability of the power system with the PV generation.
- The voltage control strategy with distributed ESS is investigated to improve the reactive output capacity of the PV/ESS system.
- The current compensation method for a series array of PV-supercapacitors is proposed so that the output of the PV array works at the maximum power point under shadow occlusion.

Nomenclature

AC	Alternating current
CAES	Compressed air energy storage
DC	Direct current
FES	Flywheel energy storage
EES	Electrochemical energy storage
EMS	Energy management system
ESS	Energy storage system
GMPPT	Global maximum power point tracking
MPP	Maximum power point
MPPT	Maximum power point tracking
PCS	Power conversion system
PI	Proportional integral controller
PN junction	Positive-negative junction
PSPS	pumped storage power station
PV	Photovoltaic
PVA	photovoltaic array
PWM	Pulse width modulation
SCSD	Solar cell–supercapacitor devices
SMES	Superconducting magnetic energy storage
SOC	State of charge
VRB	Vanadium redox flow battery
VSI	Voltage-source inverter

16.1 Introduction

Energy is the cornerstone of social development and an important material base for humankind's existence, which affects and determines the economy, national defense security, and sustainable development of a country. To handle increasingly urgent challenges of global energy security, environmental pollution, and climate change, many actions become more and more important, such as reducing dependence on fossil energy, optimizing the restructuring of energy structure, improving energy efficiency, and enhancing the sustainability of energy. Therefore many countries have proposed carbon-neutral solutions. At the 75th session of the United Nations General Assembly in 2020, Chinese President Xi Jinping took a solemn oath to the world that "With China already committing to peak carbon dioxide emissions before 2030 and achieve carbon neutrality before 2060." Solar and wind energy are representative of renewable energy, which is the core of the transition from energy systems to sustainable and low carbon. According to the International Energy Agency, compared with 2019, global wind and PV installed capacity in 2020 increased by more than 111 GW (16.1%) and 78 GW (11.3%) respectively [1]. Therefore, with the vigorous development of renewable energy, wind power, PV and other renewable energy will become the dominant power supply in worldwide energy structure transformation and safe operation.

PV generation becomes one of the fastest-growing renewable energy sources because of its rich resources, large storage capacity, and mature development technology [2]. However, PV generation is susceptible to weather, temperature, season, day and night, shadow masking, and other factors. Its radiation intensity and direction are unstable, making PV generation with strong randomness and high volatility characteristics. On the one hand, a large-scale PV generation grid will directly affect the stability and reliability of power system operation, and there comes seriously abandoned optical phenomena. On the other hand, PV is susceptible to partial shadows such as leaves and sand, resulting in reduced output power in the actual operation of the photovoltaic array (PVA), multiple peaks in the power−voltage curve, and PVA that cannot fully operate at the MPP. Improving the stability of the output power of PV generation not only benefits the safety and reliability operation of the power grid but increases the utilization of PV power.

In recent years, the rapid development of materials and chemistry has promoted the obvious progress of energy storage technology, more efficient and reliable, reasonable price of energy storage devices are constantly used in power systems. By late 2020, the cumulative installed capacity of global energy storage projects was 191.1 GW, up 3.4%. The cumulative scale of electrochemical energy storage (EES) exceeded 10 GW for the first time (cumulative installed capacity of 14.2 GW). Some energy storage projects have been established in various countries, Such as Zhang Bei Wind/PV/Energy storage/Transmission in China (14 MW iron phosphate lithium battery, 2 MW full-molybdenum liquid flow battery), the United States New York Frequency Modulation (FM) power station (20 MW flywheel energy storage), Hokkaido, Japan PV/energy storage system (ESS) station (19 MW lithium-ion battery), etc., these projects show the safety and stability of power system operation have been effectively improved. Due to the excellent dynamic response performance of the energy storage device, it can be a primary candidate for the voltage and frequency control in the power system. Therefore energy storage devices enhance the absorption of PV generation with maintaining safety and steady operation in the power system. On the other hand, by combining energy storage devices, the power loss under partial shadow conditions can be reduced, which improves the efficiency of PV panels. Therefore the "PV and ESS" mode will become one of the main features in the future power system. In this context, this chapter applies energy storage technology to the stability control of PV generation and studies the related technologies to improve the stability of PV generation with energy storage devices.

16.2 Characteristics analysis of power system with high penetration of photovoltaic generation

Since the production, transmission, distribution, and consumption of electric energy must be carried out at the same time, the real-time load/generation balance is required to be maintained. The load/generation balance is an important prerequisite for ensuring the safe and stable operation of the power system. PV output is highly related to climatic factors such as sunshine conditions and seasons. Uncertain changes in climate cause random fluctuations in PV output. In the short term, the intraday PV generation curve does not match the load curve, and even exhibits anti-peaking characteristics. The comparison of intraday PV generation and load curves are shown in Fig. 16.1. The load curve presents double peaks per day, the morning peak and the evening peak. The PV generation is high during the diurnal day, which decreases since sunsets and until reaches 0. From a long-term perspective, load demand peaks occur in summers and winters, while PV generation peaks occur in springs and winters. The characteristics of PV generation and load demand show obvious seasonal mismatch contradictions. The average daily output curve of PV in different seasons is shown in Fig. 16.2. Therefore, with the increasing proportion of grid-connected PV generation, the stable operation of the power system faces serious challenges.

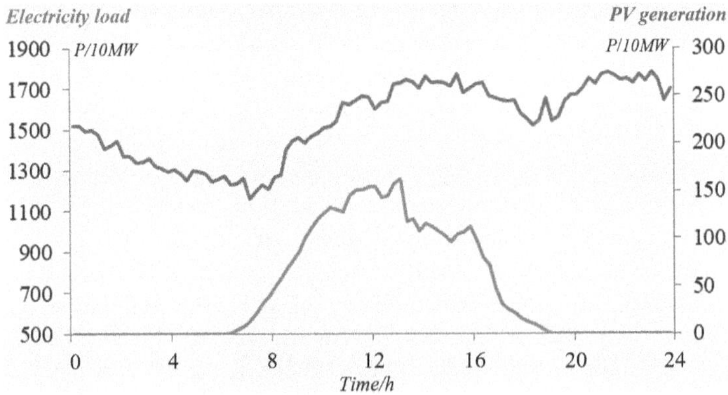

Figure 16.1 Intraday curves of photovoltaic generation and load power.

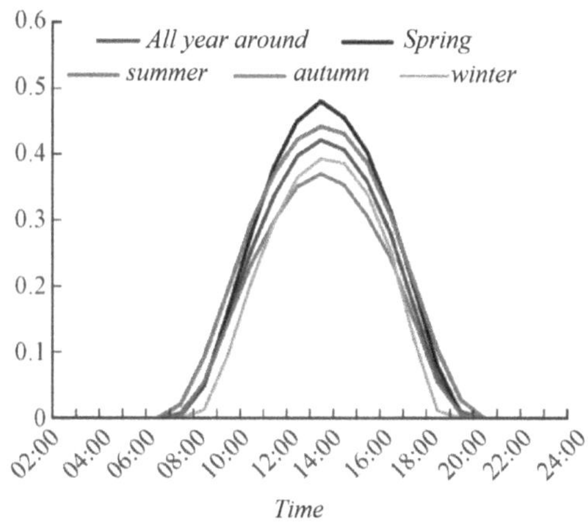

Figure 16.2 Average daily curves of photovoltaic output power in different seasons.

16.3 Classification of energy storage devices and their regulation ability

Energy storage is a physical device with a storage energy function. The energy storage device combines the dual functions of power supply and loads via charge/discharge. When the power supply on the generation side is oversupplied, the energy storage device acts as a load, and the electric energy is absorbed and converted into mechanical energy, electrochemical energy, electromagnetic energy, and other forms of storage. When there is a great shortage of electricity supply on the generation side, the energy storage devices act as a power source, converting the stored energy into electrical energy to be fed back into the power system. According to the form of energy storage, energy storage devices can be divided into physical energy storage, EES, and electromagnetic energy storage [3,4]. The characteristics of different types of energy storage are illustrated in Fig. 16.3.

16.3.1 Physical energy storage

16.3.1.1 Pumped storage

At present, pumped storage technology has become the most widespread energy storage method in the world due to its mature technology, low cost, large scale of storage, high application level, environmental protection, clean, and high energy conversion efficiency.

The pumped storage power station (PSPS) consists of device units such as upper and lower reservoirs, drainage systems, power plants, and turbine units [5–8]. The hydropower potential energy and electrical energy can be easily interconverted through turbine units. The principle of pumped storage technology is shown in Fig. 16.4. If the load is at a low level, the pumping storage unit operates in the

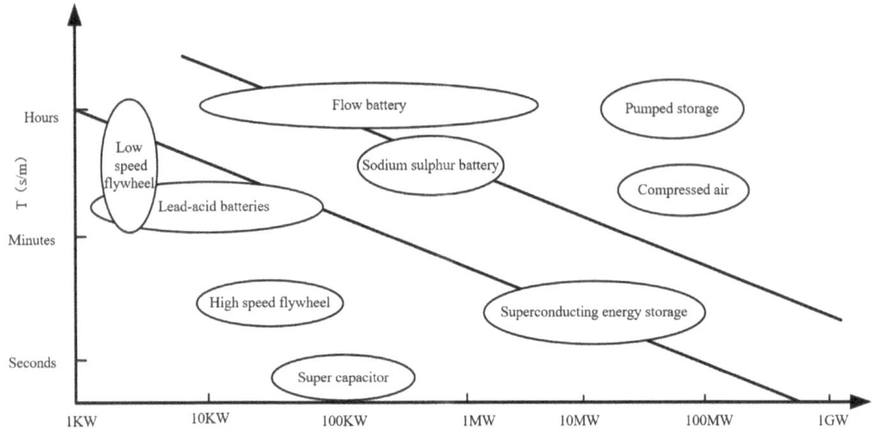

Figure 16.3 Power and energy characteristics of different energy storages.

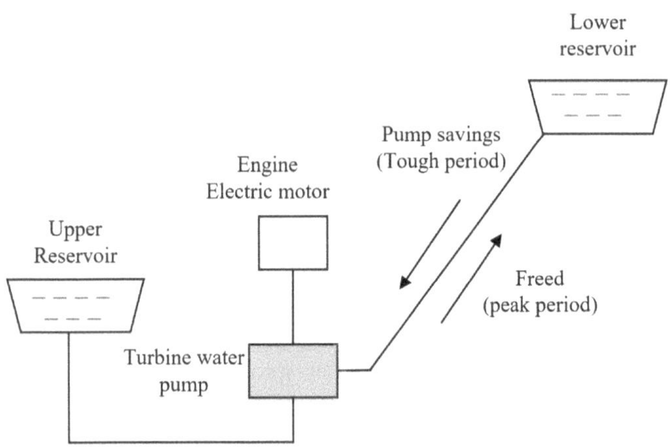

Figure 16.4 Operation principle of pumped storage technology.

pump condition and pumps water from the lower reservoir to the upper reservoir, thus converting the electrical energy into the gravity potential energy of the water for storage. If the load is at its peak, the pumped storage unit operates in the working conditions of the turbine and releases the water stored in the upper reservoir for power generation. Pumped storage has a short start-up time and flexible operation mode, which can be used for peak shaving, valley filling, frequency control, phase adjustment, emergency backup, and so on. The installed capacity of the PSPS is related to the drop height between reservoirs and their capacities [9]. Hence, the area and terrain of the upper reservoir determine the installed capacity of pumped storage stations. Therefore the development and construction of pumped storage technologies are limited to the terrain conditions, the initial investment is large, and the construction period is long, far away from the load center.

16.3.1.2 Compressed air energy storage

Compressed air energy storage (CAES) technology has attracted widespread attention due to its large-scale energy storage, flexible operation mode, fast start-up speed, short construction period, low investment cost, and low environmental pollution. The developing directions of CAES are high parameters and large capacity [10–12].

The operation principle of CAES technology is illustrated in Fig. 16.5. During the valley period, the air is compressed with high-pressure and stored in a large and sealed container, so that the nonstorable electrical energy can be converted into the storable pressure potential energy During the peak period or power supply is tight, the high-pressure air is released from the air storage container and heated by the heat exchanger, then is burned in the combustion chamber mixed with fuel to generate electricity [13,14].

16.3.1.3 Flywheel energy storage

Flywheel energy storage (FES) technology has the advantages of fast start-up capacity, low maintenance cost, high life, no pollution, high energy storage, fast charging, and infinite charge/discharge times. In the microgrid, it has a broad application prospect [15–17].

FES technology through the motor and flywheel drive each other rotation, to achieve the mutual conversion of electrical and mechanical energy. As shown in Fig. 16.6, the structure of FES is composed of a wheel rotor, bearing, motor/generator, power converter, and vacuum chamber. When external electric energy is abundant, the motor is driven by an electric electronic device to rotate the flywheel and

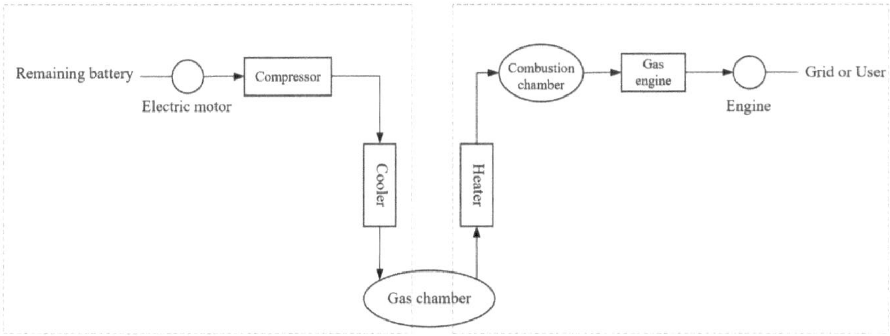

Figure 16.5 Operation principle of compressed air energy storage technology.

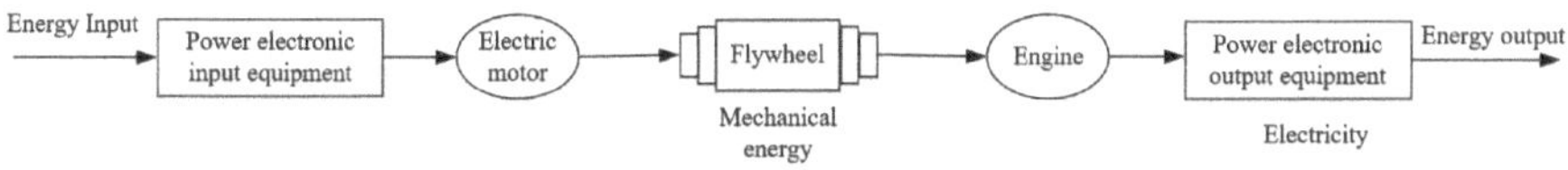

Figure 16.6 Operation principle of flywheel energy storage technology.

convert the electrical energy into storable mechanical energy. When energy needs to be released, the flywheel reverses the generator and transforms the stored mechanical energy into electrical energy. The frequency and voltage of the output electric energy are controlled by an electrical electronic device to meet the requirements [18]. However, the static loss of the flywheel is large, the relative energy density is low and the technology is not mature, which limits the application of the flywheel.

16.3.2 Electrochemical energy storage

Electrochemical energy storage technology is a technology that converts electric energy and chemical energy into energy storage and releases it through chemical reactions [19]. Among them, the battery is the main carrier of energy conversion, which is composed of a positive electrode, an electrolyte, a separator, and a negative electrode. There are many types of batteries. According to the different chemical substances used in the batteries, batteries can be divided into lead-acid batteries [20], nickel-cadmium batteries [21], nickel-metal hydride batteries [22], lithium-ion batteries [23], etc. Taking the Vanadium Redox Flow Battery (VRB) as an example, the principle of the cascaded VRB is shown in Fig. 16.7. It includes storage batteries, AC/DC converters and their control systems, safety and environmental protection, and other auxiliary devices.

Compared with other energy storage technologies, EES has the advantages of flexible energy and power configuration, low environmental impact, easy large-scale utilization, short construction period, and fast response. It is considered to be

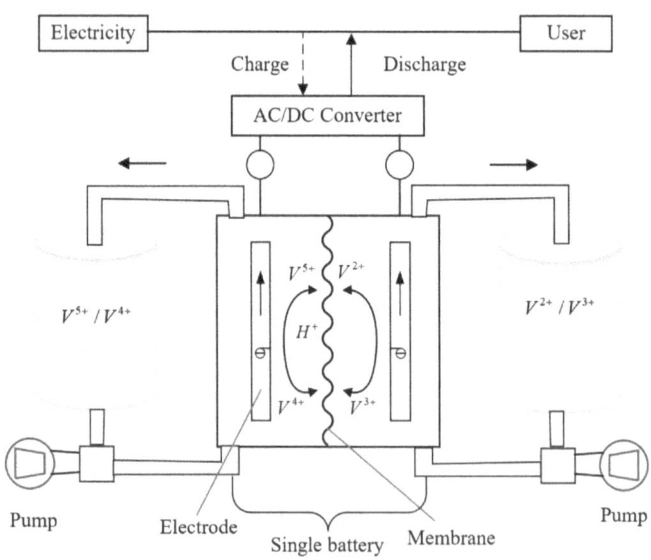

Figure 16.7 Operation principle of cascaded vanadium redox flow battery.

the most promising technology. So it has achieved rapid development in recent years [24]. However, EES technology has disadvantages such as limited battery life and high cost. These disadvantages are the challenges of the application of electrochemical storage devices and are required to be focused on by researchers.

16.3.3 Electromagnetic energy storage

16.3.3.1 Superconducting magnetic energy storage

The operation principle of superconducting magnetic energy storage (SMES) is illustrated in Fig. 16.8 [25]. Among them, superconducting energy storage coils and power conversion systems are the key components of SMES. During normal operation, the superconducting inductor is charged through rectification, and then maintains a constant current. When a transient voltage drop or transient active power imbalance occurs in the power grid, energy can be extracted from superconductivity and converted to the power grid.

The superiority of SMES technology is a high power, small size, lightweight, low loss, and fast response. As the SMES superconductor has zero resistance and extremely low energy storage loss, it shows the best performance of other energy storage devices, and it has received widespread attention. However, due to the high price and complex maintenance, the proportion of applications in the power grid is small, mostly experimental applications.

16.3.3.2 Supercapacitor

With the high power density, long service life, fast charging speed, high charging/discharging efficiency, maintenance-free, safe, and nontoxic, supercapacitor become a potential energy storage device and is widely studied worldwide [26].

A supercapacitor, a high power density, and environment-friendly energy storage device, are between the traditional electrostatic capacitor and secondary battery. Its principle is shown in Fig. 16.9. When a DC power supply is used to charge the supercapacitor monomer, the positive and negative ions in the electrolyte gather on

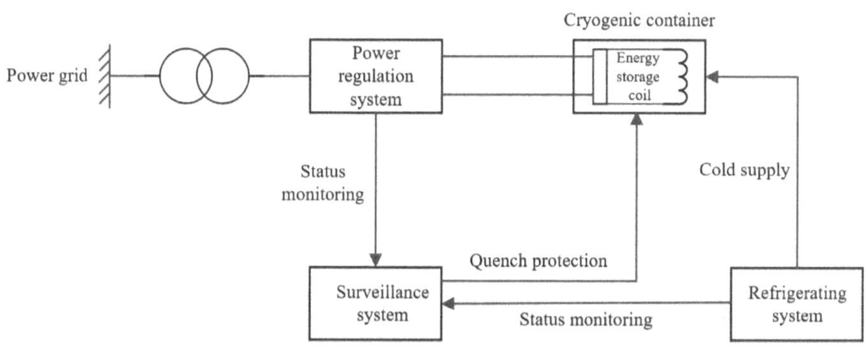

Figure 16.8 Operation principle of superconducting magnetic energy storage.

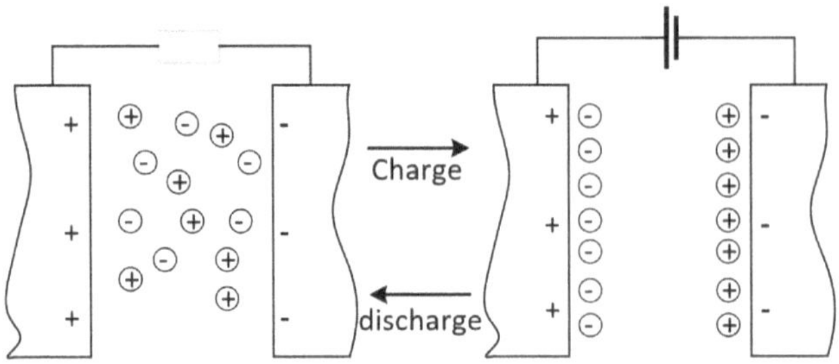

Figure 16.9 Operation principle of Supercapacitor.

the surface of the solid electrode, forming an "electrode/solution" electric double-layer to store the electric charge. When the electric field is removed, the positive and negative charges on the electrodes attract the oppositely charged ions in the solution to stabilize the electric double layer, and a relatively stable potential difference is maintained between the positive and negative electrodes. For a certain electrode, an ion charge equal to the charge on the electrode will be generated within a certain distance to keep it electrically neutral. When the two poles are connected to an external circuit [27], the charge on the electrodes migrates to generate a current in the external circuit. However, supercapacitors have disadvantages such as a high self-discharge rate and immature technology. The energy storage level is limited by the withstand voltage, and the supercapacitor is expensive for applications.

16.4 Battery storage management and its control strategies for power systems with large-scale photovoltaic generation

Photovoltaic generation not only depends on the component performance but also depends on lighting conditions, which makes the PV generation demonstrates the randomness, volatility, and intermittency. With the increase of PV generation installed capacity, the grid-connected PV generation power brings great challenges to the safe and stable operation of the power system.

Energy storage technology enables the time transfer of power and energy, which embodies many advantages, such as flexible control and rapid response. According to the requirements of power systems, the energy storage device can be operated in four quadrants to improve the static power characteristics and dynamic response characteristics of PV, which helps to realize the power and energy balance in power systems under various conditions. The configuration of suitable energy storage devices in the power system improves the system's stability.

Many researchers focus on the application of energy storage devices to support grid-connected PV. Generally, the relevant researches mainly focus on the optimization of PV/ESS system configuration, optimization of operation control, application mode, and other fields. The objectives of these researches include smoothing the short-term fluctuations of PV generation power, compensating for the prediction error of PV generation, assisting the peaking problem of the PV generation system, improving the capacity of low voltage ride through in PV stations, and exploring the application mode of energy storage devices based on the rationality and economy of the ESS. The economic operation of ESS benefits the maximum utilization of large-scale grid-connected PV generation, which is significant to the renewable-dominated power system.

16.4.1 Grid-connected configuration of energy storage in photovoltaic/energy storage system

At present, there are two main ways to improve the dynamic regulation capacity of PV stations by energy storage devices. The first way is distributed compensation, that is each group of PVA with an ESS. The output power of the grid-connected PV station is smoothed by suppressing that of each group of PVA. The second way is centralized compensation, that is, the ESS is installed at the outlet bus of the PV station. The output power of the entire PV station is smoothed through the independent ESS. The ability of the equivalent inertia characteristics, dynamic frequency support, and voltage support of PV stations is enhanced by the rapid response of energy storage devices. Fig. 16.10 shows the configurations of distributed compensation and centralized compensation schemes in PV/ESS systems.

If a large-scale PV station with distributed compensation of ESS, the output performance of individual PVA is improved by the connected ESS. If one or more PV arrays are off-grid, the impacts on the PV station are limited, in other words, the stability is high. And, the distributed compensation can share grid-connected

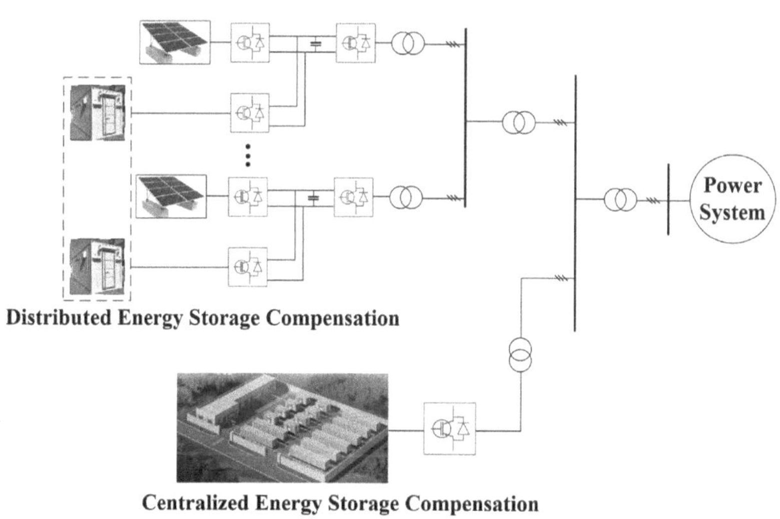

Figure 16.10 A configuration diagram of distributed compensation and centralized compensation schemes.

Table 16.1 Comparison of distributed compensation and centralized compensation for photovoltaic stations.

Configuration	Equipment cost	Installation	Schedulability	Reliability
Distributed compensation	Lower cost	Simple (little space)	Independent scheduling	Higher
Centralized compensation	High cost	Complex (large space)	Centralized scheduling	Lower

devices with photovoltaic systems, which reduces the cost. If the PV station with centralized compensation of ESS, the cost is high. And, there has a risk to deteriorate the static power and dynamic response performances of the PV station with the ESS being off-line. The distributed compensation and centralized compensation for PV stations are compared as shown in Table 16.1.

16.4.2 Capacity configuration of energy storage system

From the perspective of the power system, the larger the installed capacity of ESS is, the better the compensation performance of the power system with PV generation when a disturbance occurs. However, from the perspective of ESS, the difficulty and cost are increasing with the increase of the capacity of ESS. Therefore it is necessary to minimize the capacity of ESS for improving the dynamic performance of power systems with large-scale grid-connected PV generation.

Some research on the capacity configuration of ESS has been investigated. The economic performance of ESS as the target and the minimum cost configuration of ESS is presented with the satisfaction of system operation constraints. Another configuration of ESS is also presented to improve the dynamic stability of system frequency in the PV/ESS system. The economic operation can be realized via reasonable power control of ESS. Optimizing the charging/discharging strategy of the ESS can be an important topic of energy storage technologies, which helps the realization of economic operations. The strategies include: maintaining a reasonable range of the state of charge (SOC), avoiding or reducing excessive charging/discharging, increasing the effective operating time of ESS, and reducing the unit operating time cost of ESS.

Taking the inertia support in a PV station as an example, the capacity configuration process is detailed as follows.

In power systems, the inertia constant H is generally used to characterize the influence of the inertia of a synchronous generator on its dynamic behaviors, which can be expressed as:

$$H = \frac{E_{\mathrm{ks}}}{S_N} = \frac{J\omega_s^2}{2S_N} \tag{16.1}$$

where E_{ks} is the rotor kinetic energy at the rated speed of the generator, J is the moment of inertia, ω_s is the rated speed, and S_N is the rated capacity.

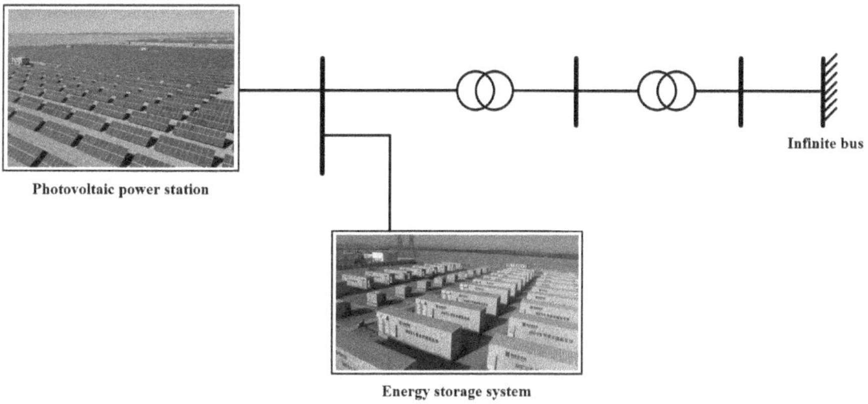

Figure 16.11 Configuration diagram of photovoltaic station with energy storage system

The ESS is connected in parallel to the AC bus of the PV station, as shown in Fig. 16.11. The integration of PV station and ESS act as a supplier in the system. The PV/ESS station shows a similar inertia response effect on the frequency to the traditional synchronous generator, through the reasonable control of the active power of the ESS.

For synchronous generators, H is the ratio of rotor kinetic energy at the rated speed of the generator to the rated capacity S_N. The virtual inertia H_{PV_ESS} of the PV/ESS station can be obtained by analogy to the synchronous generator.

$$H_{PV\_ESS} = \frac{\sum_{i=1}^{n} E_{pvi} + E_{ESS}}{S_{N\_PV}} \tag{16.2}$$

where n is the rotor kinetic energy at the rated speed of the generator, E_{pvi} is the rotational kinetic energy of the ith PVA in the system, E_{ESS} is the equivalent kinetic energy stored by the energy storage device at the rated frequency, S_{N_PV} is the rated capacity of the PV station.

According to Eqs. (16.1) and (16.2), it is obvious that when the PV station employs ESS to replace the conventional generator set, the configured capacity has to provide equivalent rotor kinetic energy to synchronous generators. Otherwise, the ESS cannot produce the same virtual inertia effect on the power system. The speed of the synchronous generator is directly coupled with the system frequency when adjusting the frequency with synchronous generators. The rotation speed varies within a small range, and the convertible kinetic energy is very small. For ESS, the energy interaction occurs in a wide range, and almost all the stored energy can be released (only limited by SOC). Therefore a small configured ESS will provide the same frequency control services as a large synchronous generator.

The principle of energy conservation provides a theoretical foundation for the control of energy exchange between PV/ESS stations and conventional power stations. When the external environmental factors change or the load changes, it can be regarded as the PV/ESS station having an energy collision with the grid. The original equilibrium state of the power grid is broken, and the system frequency changes. Similar to the impulse theorem in physics, if the exchanged energy between the PV/ESS station and the grid (case 1) is the same as the exchanged energy between the equal-capacity synchronous generator set and the grid (case 2) during short duration, it can be assumed that the impulses in two cases are the same. The inertia effect produced by the two cases can be assumed to be the same for the grid. Therefore the ESS will provide the same frequency control performance with synchronous generators with suitable control strategies.

Since the upward and downward fluctuations of the grid frequency are similar, and the range of the upward fluctuation is smaller than the downward fluctuation, the following only analyzes the downward fluctuation of the frequency. In China, the safety operation frequency cannot be lower than 48 Hz, and the rotation speed (per unit value) of the synchronous generator during the frequency modulation process is normally limited to $0.96 \sim 1$. Therefore the maximum releasable rotor kinetic energy of a synchronous generator can be written as:

$$\Delta E_{k\max} = \frac{1}{2}J(1-0.96)\omega_s^2 = 0.039J\omega_s^2 \tag{16.3}$$

When the generator is running at its rated speed, the stored rotor kinetic energy is

$$E_{\mathrm{ks}} = \frac{1}{2}J\omega_s^2 = \frac{1}{2}P_N T_J \tag{16.4}$$

where P_N is the rated speed of the generator, T_J is the generator inertia time constant.

Let the ESS releases the same energy as the synchronous generator at the same time.

$$\Delta E_{\mathrm{ESS}} = \Delta E_{k\max} = P_{\mathrm{ESS}} \cdot \Delta t = 0.0392 P_N T_J \tag{16.5}$$

where P_{ESS} is the power released by the ESS.

The time span of inertia provides frequency control services is about 10 seconds, so the ESS does not need to provide long-term services. Assuming the same duration as the inertia time constant of synchronous power generation, that is $\Delta t = T_J$, then the required capacity of the ESS is

$$P_{\mathrm{ESS}} = 0.0392 P_N \tag{16.6}$$

Therefore if the control strategy is reasonable, the configured capacity of ESS can be limited to 5% of the total capacity of the PV/ESS station [a little larger than the Eq. (16.6)]. Then, the station could provide the same inertia as the traditional synchronous generator plant with the same installed capacity. Furthermore, the operating state of PV modules does not be changed when the station is providing virtual inertia. Hence, it avoids internal coordination and communication problems, which increases the reliability of the control system and reduces the difficulty of implementation.

16.4.3 Control strategies of energy storage to frequency/voltage regulation of power system with photovoltaic generation

16.4.3.1 Grid-connected control strategy of power conversion system

A power conversion system (PCS) is the exchange hinge of the energy reserving element and grid interconnection, which is the physical foundation to support grid frequency/voltage. PCS is normally formed a by three-phase voltage-source inverter (VSI). The topology of three-phase VSI mainly consists of a two-level inverter, Neutral Point Clamped three-level inverter, modular multi-level inverter, etc. Among them, the two-level converter is widely used in the actual industrial environment. The topology and control strategy of PCS are shown in Fig. 16.12.

To distinguish the expression of the control variables of the PCS under different coordinate systems, the foot signs "abc," "$\alpha\beta$," and "dq" are employed to distinguish between the three-phase still frame, two-phase stationary frame, and variables in the rotating frame.

Fig. 16.12, V_{dc} represents the DC bus voltage in the PCS, it is affected by the output power of the power generation unit (i.e., battery pack in the EES power

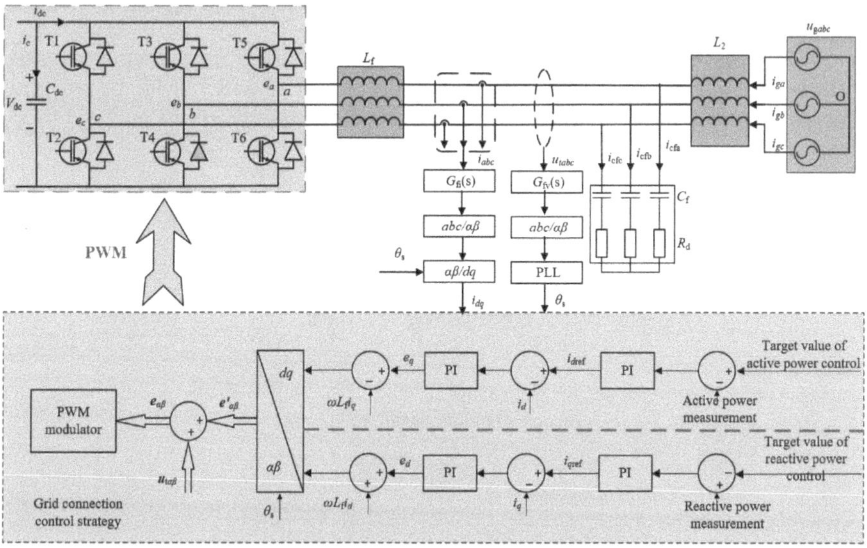

Figure 16.12 Topology and control strategy of power conversion system.

station), T1 ~ T6 are corresponding to the switch tubes of each bridge arm of the PCS, L_f filters inductance at the AC output side of the PCS, C_f corresponding to the filter capacitor at the AC output side of the PCS, R_d is the damping resistor, L_2 is the internal resistance of the grid. u_{gadc} is the voltage vector of the ideal voltage source, u_{tabc} is the AC terminal voltage vector of the PCS, e_a, e_b and e_c are the three-phase voltage output of the PCS, i_{abc} is the AC filter inductor current, i_{cfabc} is the current vector flowing through the AC filter capacitor (each phase is marked in a scalar form in Fig. 16.12), i_{gabc} is the current vector injected into the ideal grid (i.e., the current flowing on L_2, each phase is marked in a scalar form in Fig. 16.12).

For a typical PCS, its control loops typically include an outer loop (e.g., an active power control loop and a reactive power control loop) and an inner loop (e.g., a current control ring). The advantages of PCS to absorb or release active/reactive power with fast speed can be achieved by the control loops.

16.4.3.2 Control strategy of energy storage for system frequency regulation

ESS has a fast power response speed, and be used to generate virtual inertia for primary frequency control, which increases the stability of system frequency with large-scale grid-connected PV generation. Some researchers propose a practical ESS control strategy with an additional frequency response based on the principle of the traditional governor, which does not change the existing flat/smooth control of the ESS. The control strategy is shown in Fig. 16.13, which includes three parts:

1. Existing flat/smooth control strategy. The power of the PV station is taken as the input signal. The output power of the ESS is generated to suppress the fluctuation of the PV/ESS station according to different time scales.
2. An additional control strategy based on frequency response. This strategy takes the system frequency as the input signal and obtains the target power of the ESS with the power/frequency response factor K. Therefore primary frequency control can be provided by the

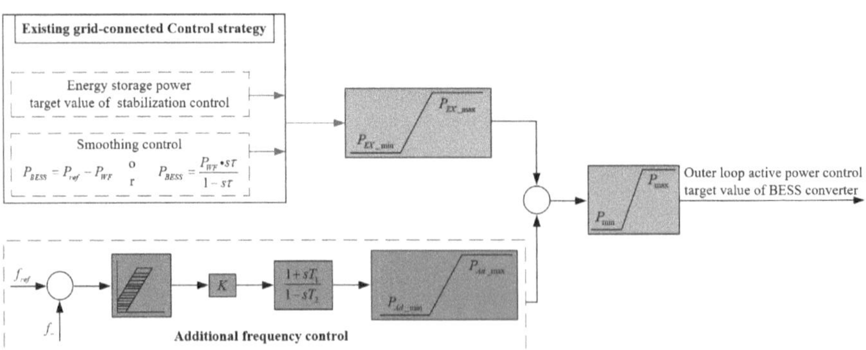

Figure 16.13 Grid-connected control strategy of energy storage system based on additional frequency control.

PV/ESS, which increases the dynamic stability of the system via active control of system disturbances.

The phase compensation can be provided by the lead/lag loop to respond to the requirement of system stability, which increases the system damping.

To achieve a rapid response to system frequency regulation, the ESS is charged/discharged frequently, which will result in a decrease in device lifespan. Therefore the strategy includes a dead band as a traditional governor. The reasonable dead-band ε not only helps the ESS respond to the large disturbance but also avoids the device changing its state frequently. The dead band is set according to the governor, and its default range generally is 0.001 to 0.002 (per unit value).

The strategy also is an amplitude limitation to suppress the power fluctuation of the PV station. The limitation avoids the impacts of additional frequency regulation on the existing smooth control. The limit values of P_{au_max} and P_{au_min} is adjusted dynamically by the energy management system according to the actual operating state (e.g., SOC, dischargeable multiple).

3. The overall amplitude limitation. Due to the introduction of the additional frequency control strategy, the control target of the ESS becomes restraining power fluctuations and improving transient stability. The upper and lower limits of the overall amplitude limitation can be dynamically adjusted according to the actual operating status of the energy storage device. Therefore the ESS capacity can be allocated reasonably to restrain the power fluctuation of the PV station and improve the stability of the power system. Hence, The ESS is optimized used.

The objective of active power control is eventually obtained based on the additional frequency control strategy. Then, the fluctuation is restrained and the stability is increased through the adjustment of ESS with the outer loop control.

In addition, the ESS is a multi-functional system with four-quadrant power adjustment, which can quickly adjust active and reactive power at the same time. The price of the PCS is relatively high and is not suit to increase the reactive capacity of the ESS with a large capacity PCS. But the ESS is normally not working at full power operation, and it still has the ability to reactive power control. The target of the outer reactive power control loop of the PCS can be set as a certain bus voltage, thereby stabilizing the bus voltage fluctuation. The main objective of control strategies is active power control, and reactive power control is a supplementary control. Therefore the coordinate ability of the ESS can be made full use.

16.4.3.3 Control strategy of energy storage for system voltage regulation

As the ESS can be controlled to absorb or release reactive power, it can be employed to provide voltage control services in the power system.

The topology structure of the grid-connected PV generation system with distributed ESS is shown in Fig. 16.14. The hardware part includes PVA, energy storage devices, DC–DC converter (take Buck/boost as an example), grid-side converter, and filter circuit. The PVA is connected to the DC bus of the converter through the Boost converter and is connected in parallel with the energy storage device with Buck/boost converter, and the energy is fed into the grid by the inverter.

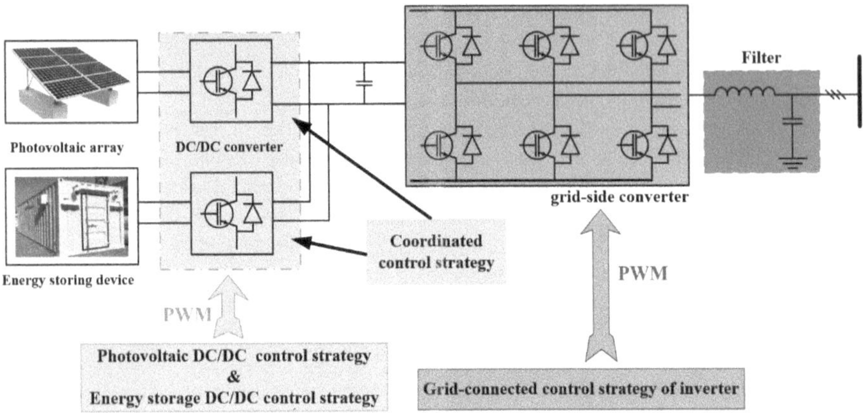

Figure 16.14 Topology of grid-connected photovoltaic generation system with distributed energy storage.

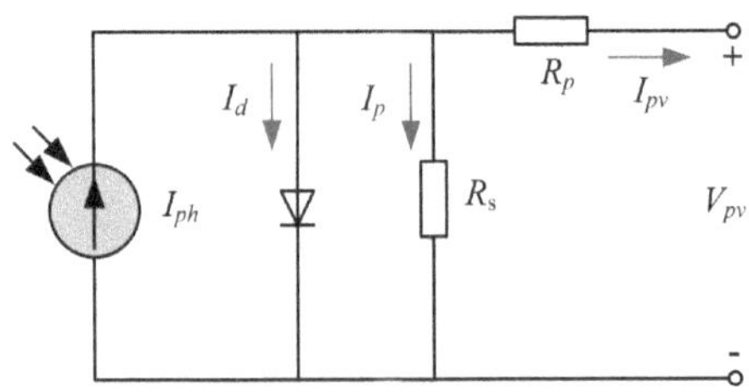

Figure 16.15 Equivalent circuit of a photovoltaic module.

1. PVA structure and its control strategy

The equivalent circuit diagram of the PVA is shown in Fig. 16.15. According to the theory of electronic circuits, the *I-V* equation of the PV module can be obtained:

$$I_{\text{pv}} = I_{\text{ph}} - I_0\left[\exp\left(\frac{V_{\text{pv}} + \text{IR}_s}{\text{nkTN}_s/q}\right) - 1\right] - \frac{V_{\text{pv}} + I_{\text{pv}}R_s}{R_p} \tag{16.7}$$

where V_{pv} and I_{pv} are the output voltage and current of the PVA respectively, I_{ph} is the photocurrent of the PVA, I_0 is the diode reverse saturation current, R_s is equivalent series resistance, R_p is equivalent parallel resistance, n is the ideality factor of the diode ($1 \leq n \leq 2$), q is the amount of charge ($q = 1.602 \times 10^{-19}C$), k is the Boltz constant ($k = 1.381 \times 10^{-23}\text{J/K}$), N_s is the number of PV units connected in series, and T is the battery temperature $(^\circ F)$.

The *I-V* curve of the PVA can be obtained by Eq. (16.7). It is necessary to know the parameters of I_{ph}, I_0, R_s, R_p and N_s, and their relationships with the temperature and light intensity of the PVA.

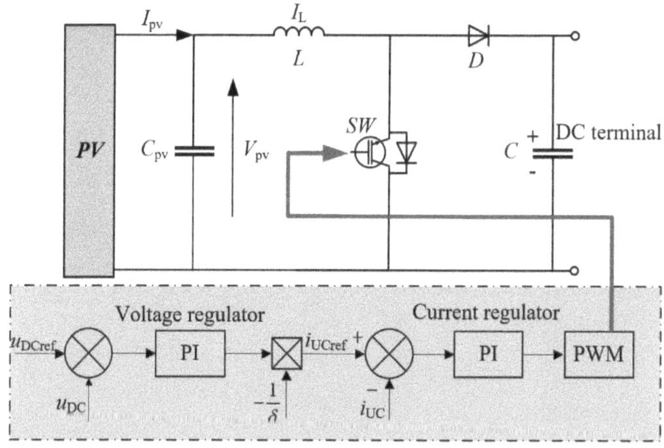

Figure 16.16 Topology of boost circuit and diagram of the control block.

The DC/DC connected to the PVA adopts a Boost circuit to realize the maximum power point tracking (MPPT) control of PV generation. The Boost DC/DC converter topology and control block diagram are shown in Fig. 16.16. C_{pv} is the parallel output capacitor of the PVA, I_L is the inductor current, L and C are the inductance and capacitance of the Boost circuit respectively, D is a diode, and SW is the power tube. The input capacitor C_{pv} mainly absorbs the switching ripple that affects the inductor current. Under steady-state conditions, the inductor current does not change, that is $I_{C_{pv}} = 0$, thus $I_{pv} = I_L$.

2. Energy storage device control strategy

Buck/boost bidirectional DC/DC converter is the core component of the energy storage device. By regulating the on/off state of power electronic switching devices, the DC voltage can be maintained at a constant. The topological structure and control block diagram of the Buck/boost bidirectional DC/DC converter of the ESS are shown in Fig. 16.17.

The control strategy of the ESS adopts the voltage and current double closed-loop control, the voltage outer loop tracks the DC bus voltage, and its deviation from the set value of the DC voltage is passed through the PI regulator to generate the DC bus current control signal. According to the duty cycle of the converter and the principle of power conservation, the charging or discharging current control command for UC is obtained. The current output signal adjusted by the current inner loop PI is PWM modulated to generate complementary control signals for the switching elements SC_1 and SC_2 of the DC converter. The UC system can absorb or release energy by controlling the alternate switching of SC_1 and SC_2. When SC_1 is turned on and SC_2 is turned off, the DC/DC converter is in the buck working state, and the energy storage device absorbs the redundant power in the system to keep the DC voltage constant. When SC_1 is turned off and SC_2 is turned on, the DC/DC converter is in a boost working state, and the energy storage device releases energy to the converter to maintain a constant DC voltage.

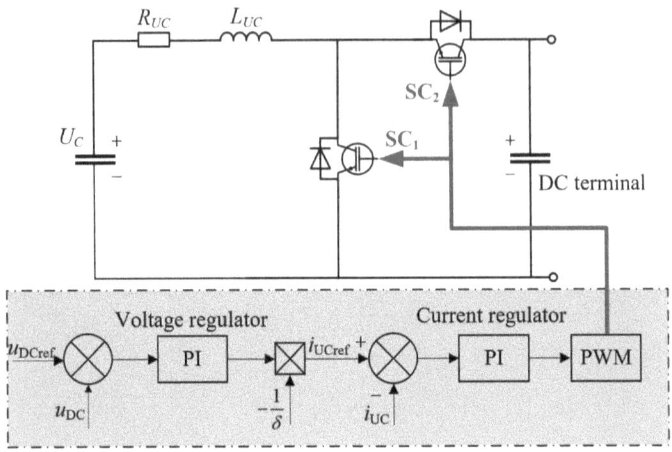

Figure 16.17 Control block diagram of buck/boost bidirectional DC/DC converter.

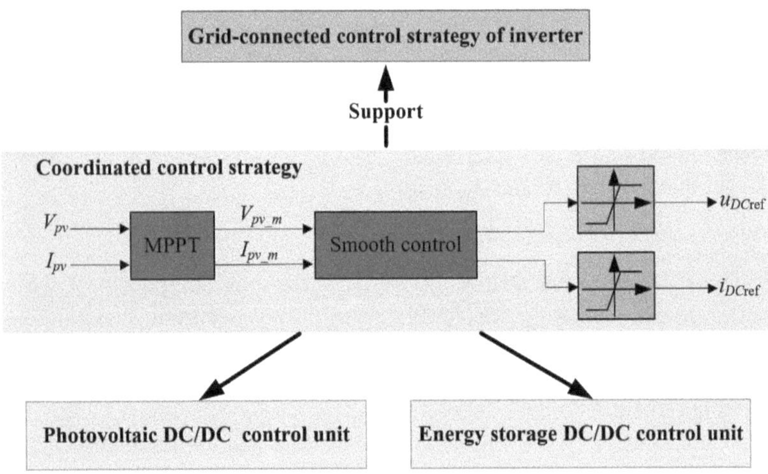

Figure 16.18 Block diagram of cooperative control.

3. Coordinated control strategy

There are three main tasks of coordinated control strategy: (1) Determine the MPPT of the PVA. (2) Smoothing the impact of PVA power fluctuations on system stability in a short time. (3) Control the SOC of the energy storage device to maintain sufficient capacity for the voltage regulation in the power grid. The block diagram of cooperative control is shown in Fig. 16.18.

From Eq. (16.7), the maximum power equation can be obtained:

$$P_{pv\_max} = V_m I_m = V_{pv\_m}\left\{I_{ph} - I_0\left[\exp\left(\frac{V_{pv\_m} + I_{pv\_m}R_s}{nkTN_s/q}\right) - 1\right] - \frac{V_{pv\_m} + I_{pv\_m}R_s}{R_p}\right\} \tag{16.8}$$

where V_{pv_m} and I_{pv_m} are the voltage and current corresponding to the MPP P_{pv_max} respectively.

According to the current/voltage reference value of the current MPPT of the PVA, the DC/DC control signal of the PVA and the energy storage device are obtained through smooth control. When the light intensity increases, the PV power increases. If the output voltage of the grid-connected inverter does not change, the PVA is still working at the MPP, and the energy storage device is charged to maintain a constant DC bus voltage through a coordinated control strategy. The control principle is the same when the light intensity decreases.

4. Voltage support control strategy for grid-side converters

In the grid voltage-oriented vector control mode, the injected power from the station to the grid is:

$$\begin{cases} P_g = u_{dg}i_{dg} \\ Q_g = -u_{dq}i_{dq} \end{cases} \tag{16.9}$$

where Q_g is the output power of the grid-side converter, u_{dg} and u_{dq} are the voltage components of the d-axis and q-axis on the grid side, respectively, i_{dg} and i_{dq} are the current components of the d-axis and q-axis on the grid side, respectively.

According to Eq. (16.9), the grid-side converter controls the d-axis and q-axis current components respectively to realize the decoupling of active and reactive power on the grid-side. The block diagram of reactive power control of the grid-side converter is shown in Fig. 16.19. The grid-side converter adopts double-loop cascade control, including reactive power outer loop and current inner loop. In Fig. 16.19, Q_{gref} the reactive power reference of the outer loop $i_{gq\ ref}$ is the inner loop shaft current reference command of the grid-side converter current. The power outer loop tracks the reactive power changes on the grid-side. After the reactive power deviation is adjusted by PI, the shaft

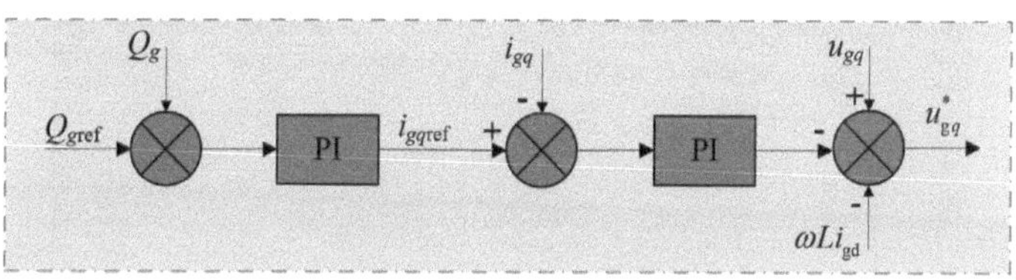

Figure 16.19 Block diagram of reactive power control of grid-side converter.

current reference value is generated. The current inner loop adopts feedforward decoupling control and provides a PWM trigger signal. The reactive power outer loop controls the reactive power exchange between the converter and the grid.

16.4.4 Demonstration projects of energy storage system and photovoltaic generation

Zhang Bei national wind/PV/ESS/transmission demonstration project in China aims to build “grid-friendly” renewable generation. It is currently the largest comprehensive renewable energy demonstration project in the world, which integrates wind generation, PV generation, energy storage, and power transmission projects. The first stage of the project had built 98.5 MW of wind power, 40 MW of PV generation, and 20 MW of energy storage devices (Including 14 MW/63 MWh lithium-ion battery and 2 MW/8 MWh all-vanadium flow battery), a 220 kV smart substation was built too. Through the large-scale energy storage power station monitoring system, the coordinated control and energy management of a variety of energy storage devices are realized. It has various functions such as smoothing the power fluctuation of renewable generation, auxiliary renewable power according to the planned curve power, peak shaving, valley filling, etc. The panorama of the project is shown in Fig. 16.20.

The upper-level energy management strategy realizes the complementary of wind, solar, and energy storage. The operation results show that less than 7% of wind-solar combined power fluctuation and less than 3% error of tracking power generation plan can be achieved. The monitoring interface of the project is shown in Fig. 16.21.

Figure 16.20 Panorama of Zhang Bei national wind/photovoltaic/electrochemical energy storage/transmission demonstration project in China.

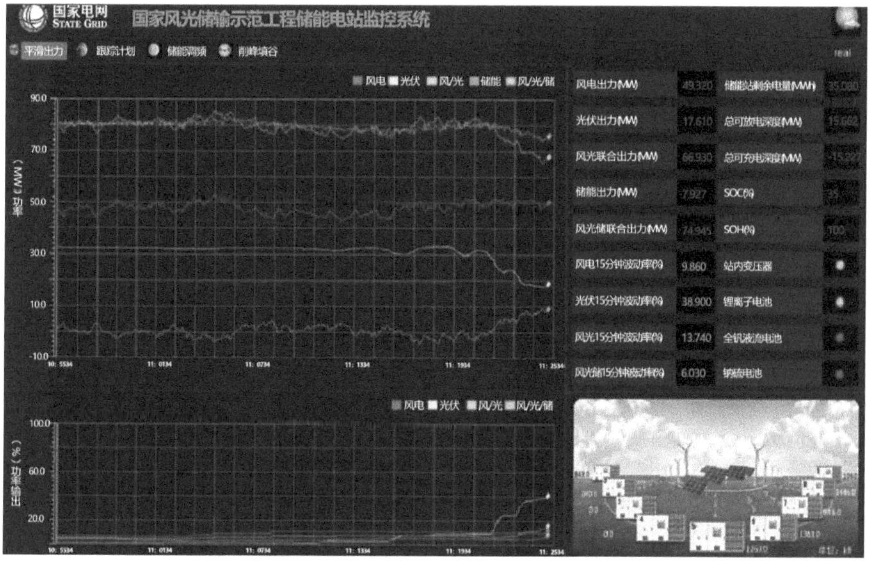

Figure 16.21 Interface of the management system of the demonstration project.

16.5 Current compensation of solar cell–supercapacitor devices series array to improve photovoltaic generation efficiency under partial shading

The efficiency of PV generation being limited by solar cell manufacturing materials and manufacturing processes for a long time. Although polycrystalline silicon, monocrystalline silicon, and thin-film modules are commonly used, their PV conversion efficiencies are low, only about 18%. A lot of research work had been carried out on conversion materials, the internal structure of cells, and manufacturing processes. The highest efficiency is reaching about 30%. It is difficult to make breakthroughs in PV conversion efficiency due to material constraints in the short term. Therefore researchers are not only exploring more efficient PV conversion materials and preparation processes but also trying to update the control method and optimize the module structure to increase the output power of the PVA.

In addition, solar cells are easily blocked by leaves, dust, or clouds in practice [28,29], and the output characteristics of solar cells are not consistent due to different light intensity and ambient temperatures. The cell with a small output power cannot contribute to the output of the PVA, what's more, it becomes a load and consumes electricity, resulting in the hot spot effect [30]. Therefore the efficiency of the array and the life span of solar cells is reduced, which also reduces the stability of PV systems.

In recent years, many scholars had carried out a series of researches on the optimization method of PVA based on parallel power compensation units to solve the problem caused by partial shading. According to the type of compensation unit, these methods can be divided into two kinds: PVA self-compensation and adjunct unit compensation. The self-compensation method compensates for the voltage or output power of the shadow masking battery by using the shadowless masking battery in the PVA [31,32]. The parallel ESS acts as an auxiliary unit to compensate power generation systems. The main principle of the self-compensation method is that when components are mismatched, the control storage unit and mismatched components are in parallel to compensate for the array output power. The method exerts the fast response speed and flexible control of ESS to meet the randomness and volatility of PV generations, which lets the output of PV generation more flexible.

To solve the decrease of output power and multi-peak of PVA under partial shading, this chapter proposed a series array with SCSD as the basic unit with the consideration of the topological reconstruction and parallel compensation. Through the analyses of the equivalent circuit of the SCSD unit and the working mode of the supercapacitor, the current compensation control method of the SCSD under partial shading is proposed.

Compared with the traditional global MPPT (GMPPT) algorithm, the proposed method increases the cost of PVA but avoids the power loss of PV generation under partial shading. Compared to the compensation structure with plus energy storage batteries, although the installation cost of a supercapacitor is 5–10 times of a lead-acid battery, the average cost of the entire life cycle is only 1/10 of the lead-acid battery due to the extremely long charge/discharge life of the supercapacitor, resulting in a decrease in the overall cost of the SCSD system. T Numerical simulations and experiments demonstrate the proposed SCSD array current compensation control method is effective, which lets each cell in the SCSD array work at the MPP, and improves the efficiency of PV generation.

16.5.1 Equivalent circuit of solar cell–supercapacitor devices unit and operating mode of its supercapacitor

16.5.1.1 Physical structure and equivalent circuit of solar cell–supercapacitor devices

The physical structure of SCSD is shown in Fig. 16.22. Port 1 is the common end of the solar cell and supercapacitor, port 2 is the cathode of the solar cell, and port 3 is the cathode of the supercapacitor. The screen-printed aluminum electrode of the solar cell acts as an anode substrate for the supercapacitor, combining the solar cell and the supercapacitor into a three-port device. Compared with the general four-port device architecture (solar cells have 2 ports, supercapacitors have 2 ports), the purpose of simplifying the connection structure between PV modules and energy storage battery packs, easy integration, and reducing control complexity is achieved. The granular aluminum electrodes increase the surface area of the supercapacitor and effectively increase the energy storage density. At the same time, aluminum electrodes, as an isolation layer, can effectively prevent damage to the solar cell unit during the production of supercapacitor units. An integrated structure

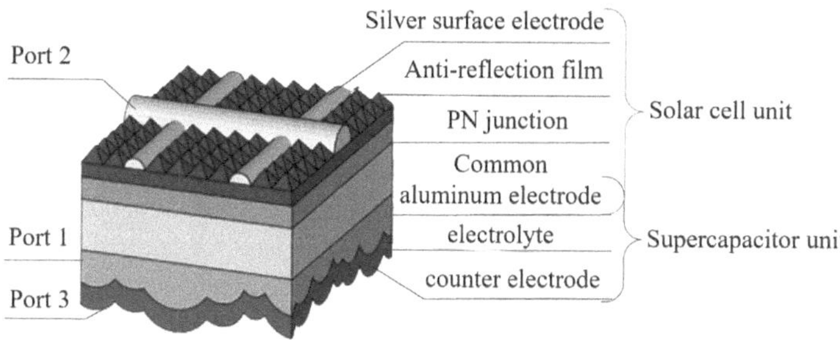

Figure 16.22 Physical structure of solar cell−supercapacitor.

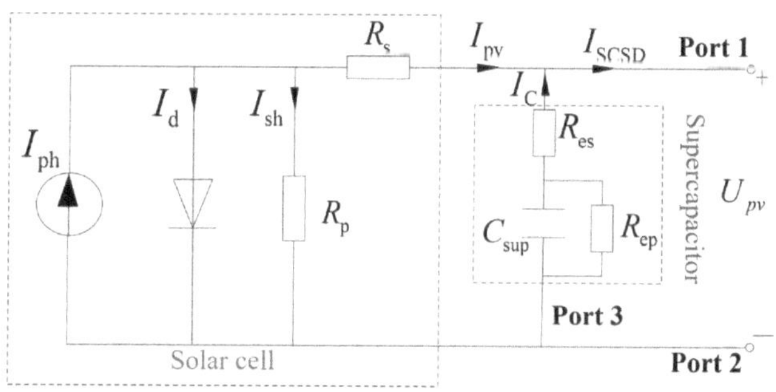

Figure 16.23 Equivalent circuit of solar cell−supercapacitor.

without an isolation layer is reported to have at least 10% of the open-circuit voltage loss, resulting in a loss of about 10% of the photoelectric conversion efficiency.

Solar cells are semiconductor diodes with PN junctions that generate electricity by the PV effect. Light causes the PN junction to form electron holes and electrons, and the directional movement of holes and electrons causes potential differences between uneven semiconductors or between different parts of semiconductor-metal binding. PV modules are powered by these semiconductors and metals. The PV cell of the SCSD can be equivalented as a single diode application model, and the supercapacitor can be equivalented as the classic model, then the equivalent circuit of the SCSD is shown in Fig. 16.23.

$$I_{SCSD} = I_{pv} + I_C = I_{ph} - I_d - I_{sh} + I_C = I_{ph} - I_0\left[\exp\left(\frac{q(U_{pv} + R_s I_{pv})}{kAT}\right) - 1\right] - \frac{U_{pv} + R_s I_{pv}}{R_p} + I_C \tag{16.10}$$

where I_{ph} is the photocurrent, which is positively correlated with the irradiance and temperature of the solar cell. I_d is the diode saturation current, I_{sh} is the bypass current, R_p and R_s are the equivalent parallel resistance and series resistance of the solar cell respectively, I_0 is the diode reverse current, A is the ideality factor of the diode, between 1 and 2, q is the amount of charge(the value is 1.6×10^{-19}C), k is Boltzmann's constant (the value is 1.38×10^{-23}J/K), T is the thermodynamic temperature of the battery, U_{pv} and I_{pv} is the output voltage and output current of the solar cell port. I_C is the supercapacitor compensation current, when the supercapacitor is discharged, its value is positive. When the supercapacitor is charged, the direction of the current is opposite to that in Fig. 16.23, and its value is negative.

16.5.1.2 Equivalent circuit of supercapacitor

The supercapacitor of the SCSD unit in Fig. 16.23 is directly connected in parallel to the solar cell port, that is port 2 and port 3 are directly connected, and the output impedance of the SCSD unit is expressed as:

$$Z_{SCSD} = \frac{Z_{pv}Z_C}{Z_{pv} + Z_C} < Z_{pv} \tag{16.11}$$

The bode diagram of the equivalent output impedance of SCSD is shown in Fig. 16.24, where C1 > C2 > C3. Supercapacitor reduces the output impedance of

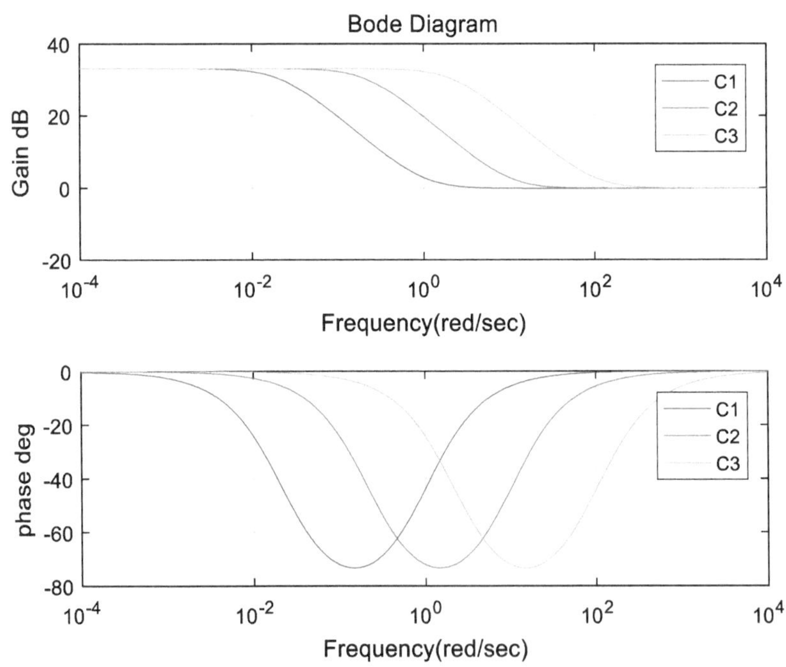

Figure 16.24 Bode diagram of the equivalent output impedance of solar cell–supercapacitor devices.

the SCSD, and the larger the supercapacitor, the lower the response frequency. This feature has the following advantages: (1) the supercapacitor keeps the port voltage of the SCSD unchanged when solar cells are affected by short-term light intensity changes which avoid a sudden reduction in output generation with a compensating current. (2) When a sudden change in load causes the voltage of the SCSD output port to drop sharply, the supercapacitor provides more power than the MPP to enhance e the system overload capacity.

The parallel supercapacitor at the solar cell port reduces the output impedance of the SCSD port, providing more generations than the MPP of solar cells. However, if the voltage of the supercapacitor is not equal to the voltage of the solar cell, it not only causes the circulation inside the SCSD unit, resulting in power losses, but also causes the power–voltage curve to various, and the MPP of the solar cell cannot be tracked anymore.

To solve the problem of imbalance between supercapacitor voltage and solar cell voltage, the three ports of the SCSD unit are connected separately to the three ports of the bidirectional DC/DC converter, as shown in Fig. 16.25.

In supercapacitor discharge mode, the bidirectional DC/DC converter operates in the BUCK state, that is the control switch K_1 receives the PWM control signal and the switch is K_2 closed. The discharge circuit of the supercapacitor is shown in Fig. 16.26 (The red line indicates the discharge circuit of the supercapacitor when S_1 is on, and the blue line represents the discharge circuit of the supercapacitor when S_1 is off).

In the supercapacitor charge mode, the bidirectional DC/DC converter is controlled to work in the BOOST state, the PWM control signal is received, and the switch K_2 is turned off. At this time, the supercapacitor is equivalent to a load and is charged by the solar cell. The charging circuit of the supercapacitor is shown in Fig. 16.27 (The circuit represented by the red line is the charging circuit of the supercapacitor when S_2 is on, and the circuit represented by the blue line is the charging circuit of the supercapacitor when S_2 is off.).

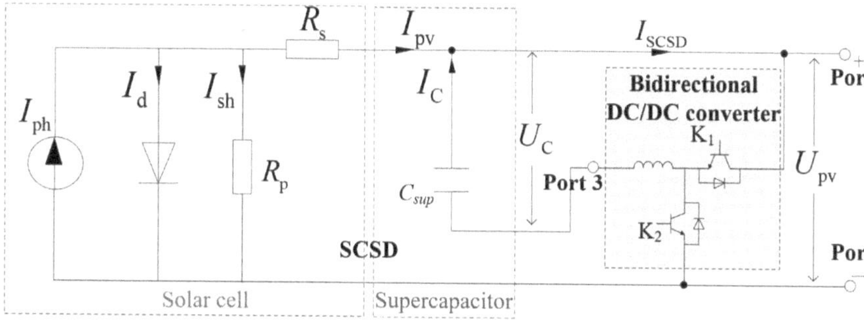

Figure 16.25 Equivalent circuit of solar cell–supercapacitor devices with bidirectional DC/DC Converter.

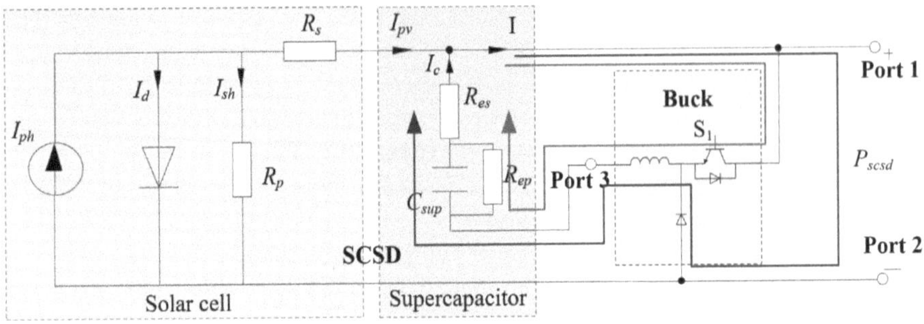

Figure 16.26 Discharge circuit of supercapacitor.

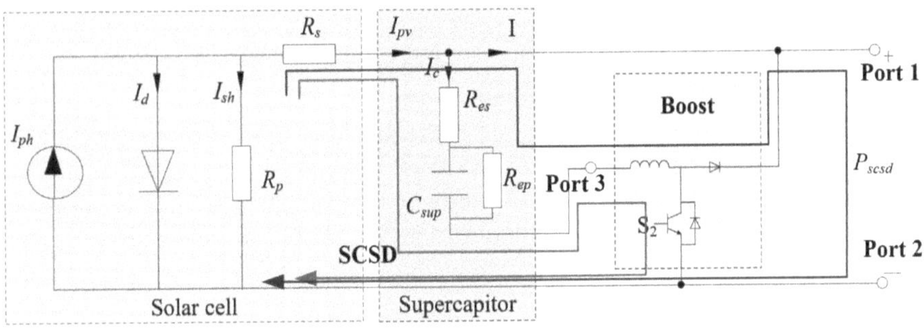

Figure 16.27 Charging circuit of supercapacitor.

16.5.2 Characteristics analysis of output power of solar cell–supercapacitor devices series array under partial shading

The working mode of the supercapacitor in the SCSD depends on the operating conditions of the solar cell. In the SCSD series/parallel array, if the solar cell operating conditions are the same, the supercapacitor and the connected bidirectional DC/DC converter do not work. Whenever there have shaded cells in the array, the supercapacitor is charged or discharged to match the characteristics of the solar cell.

The light intensity is used to characterize the shadow occlusion, and describe the quantitative relationship between the partial shadow occlusion and the output voltage and current of the solar cell. Engineering mathematical models are usually employed to describe the functional relationships between solar cell output characteristics, light intensity, and temperature, that is:

$$\begin{cases} I_{\mathrm{sc}} = I_{\mathrm{sc.ref}}\dfrac{G}{G_{\mathrm{ref}}}, V_{\mathrm{oc}} = V_{\mathrm{oc.ref}}\ln(\mathrm{e} + b\Delta G) \\ I_{\mathrm{mp}} = I_{\mathrm{mp.ref}}\dfrac{G}{G_{\mathrm{ref}}}, V_{\mathrm{mp}} = V_{\mathrm{mp.ref}}\ln(\mathrm{e} + b\Delta G) \end{cases} \tag{16.12}$$

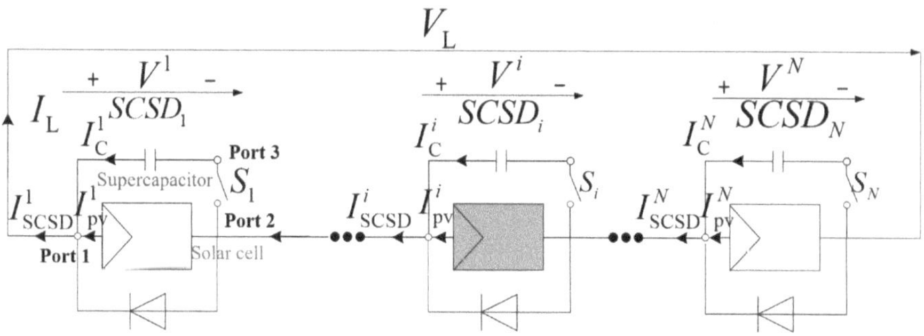

Figure 16.28 Solar cell–supercapacitor devices series array.

where G is the current light intensity (according to the shadow occlusion, the number of areas with different light intensity, set the light intensity sensor), G_{ref} is the standard light intensity (1000 W/m^2),$\Delta G = G\text{-}G_{ref}$. $I_{sc.ref}$, $V_{oc.ref}$, $I_{mp.ref}$, $V_{mp.ref}$ are the short-circuit current, open-circuit voltage, MPP current, and MPP voltage of the solar cell under standard light intensity. I_{sc}, V_{oc}, I_{mp}, V_{mp} are the short-circuit current, open-circuit voltage, MPP current, and MPP voltage of the solar cell under the light intensity G. b is the light intensity correction coefficient (default value is 0.0005 m^2/W).

The SCSD array is an array of several SCSD units composed of solar cells and supercapacitors connected in series or parallel. The SCSD series array is shown in Fig. 16.28. N SCSD units are connected in series to form a series array, $SCSD_i$ representing the ith SCSD unit in the array.

The supercapacitor is connected with the PV cell through a bidirectional DC/DC converter, as shown in Fig. 16.25. In Fig. 16.28, a switch S_i is used to represent a bidirectional DC/DC converter connected to the SCSD unit. The bypass diode to prevent the hot spot effect is connected in parallel between port 1 and port 2 of the SCSD unit. V^i is the terminal voltage of the SCSD unit, I_C^i is the supercapacitor current of the SCSD unit i. The functional relationship between the output current I_{pv}^i and the output voltage V^i of the solar cell in the SCSD unit i is:

$$I_{pv}^i = I_{sc}^i\left\{1 - C_1\left[\exp\left(\frac{V^i}{C_2 V_{oc}^i}\right) - 1\right]\right\} \tag{16.13}$$

where C_1 and C_2 are coefficients, which can be calculated as follows:

$$\begin{cases} C_1 = \left(1 - \dfrac{I_{mp}^i}{I_{sc}^i}\right)\exp\left(-\dfrac{V_{mp}^i}{C_2 V_{sc}^i}\right) \\ C_2 = \left(\dfrac{V_{mp}^i}{V_{oc}^i} - 1\right)\left[\ln\left(1 - \frac{I_{mp}^i}{I_{sc}^i}\right)\right]^{-1} \end{cases} \tag{16.14}$$

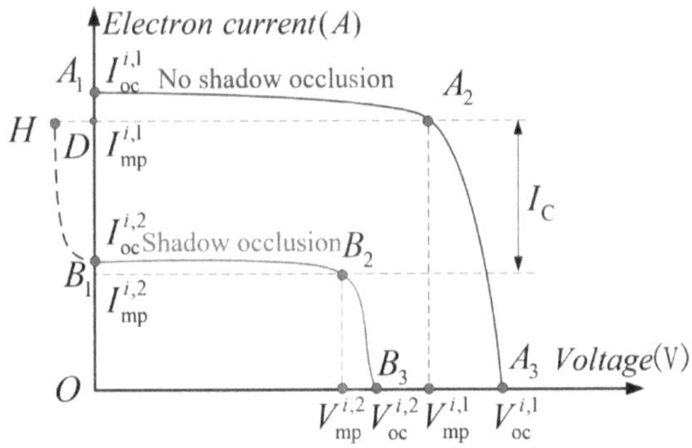

Figure 16.29 *I-V* characteristic curve of a solar cell.

According to Eq. (16.13), the *I-V* characteristic curve of the PV cell is shown in Fig. 16.29. Ω_1 is the set of SCSD units under shadow occlusion. $I_{mp}^{i,1}, V_{mp}^{i,1}, I_{oc}^{i,1}, V_{oc}^{i,1}$, $I_{mp}^{i,2}, V_{mp}^{i,2}, I_{oc}^{i,2}, V_{oc}^{i,2}$ are the MPP current, MPP voltage, open-circuit current, and open-circuit voltage of the solar cell in the *i*th SCSD unit without or with shadow. Curves $A_1A_2A_3$ and $B_1B_2B_3$ represent the output characteristics of the solar cells in the SCSD unit without or with shadow, respectively. As the port voltage gradually increases, the output current slowly decreases at first and then decreases to zero quickly.

The output voltage of the SCSD series array port is:

$$V_L = \sum_{i=1}^{N} V^i \tag{16.15}$$

When the SCSD unit in the series array is not shading, the output characteristics of all solar cells are the same, and the operating characteristics are shown as the curve $A_1A_2A_3$. When $I_{pv}^i = I_L$, $I_C^i = 0$, the output current of the SCSD series array is:

$$I_L = I_{sc}^i \left\{ 1 - C_1 \left[\exp\left(\frac{V_L/m}{C_2 V_{oc}^i} \right) - 1 \right] \right\} \tag{16.16}$$

The product of the abscissa and ordinate of each point on the curve $A_1A_2A_3$ represents the output power of the solar cell. If the solar cell is controlled to work in MPP $A_2\left(V_{mp}^{i,1}, I_{mp}^{i,1}\right)$, the maximum output power $P_{L.max}$ of the SCSD series array under no shadow conditions is:

$$P_{L.max} = N V_{mp}^{i,1} I_{mp}^{i,1} \tag{16.17}$$

If the SCSD series array is shaded, the solar cell $I - V$ characteristic curve becomes the curve $B_1B_2B_3$ by Eqs. (16.15–16.16). The supercapacitor in the SCSD unit does not output power when the supercapacitor is not employed to compensate for output generation, and the solar cell $I-V$ characteristic curve is the SCSD Ω_1 unit $I-V$ characteristic curve. If the solar cell of the SCSD unit in control Ω_1 works at voltage $V_{\mathrm{mp}}^{i,1}$, and the solar cell of the SCSD unit in control Ω_2 works at voltage $V_{\mathrm{mp}}^{i,2}$, then the solar cells of SCSD units in Ω_1 and Ω_2 work at MPP, and the output currents are $I_{\mathrm{mp}}^{i,1}$ and $I_{\mathrm{mp}}^{i,2}$ respectively.

However, since the SCSD units Ω_1 and Ω_2 are in the same series array, the output currents of all SCSD units are the same. The solar cells of the SCSD units the Ω_1 and Ω_2 cannot work at MPP $A_2\left(V_{\mathrm{mp}}^{i,1}, I_{\mathrm{mp}}^{i,1}\right)$ and $B_2\left(V_{\mathrm{mp}}^{i,2}, I_{\mathrm{mp}}^{i,2}\right)$ at the same time when the supercapacitor is not activated for compensation.

16.5.3 Current compensation method of solar cell–supercapacitor devices series array under partial shading

To make all solar cells in the SCSD series array work in MPP and reduce power losses, a current compensation control method based on the SCSD is proposed. According to the output current I_L of the SCSD series array, the compensation can be divided into several cases. The specific strategies are as follows:

Case 1:: When $I_{\mathrm{L}} \in \left(0, I_{\mathrm{mp}}^{i,1}\right)$, the solar cells of the SCSD unit in the set Ω_1 and Ω_2 cannot work at the MPP power, and the supercapacitor of the SCSD units in Ω_1 and Ω_2 discharge at the same time for compensation.

Case 2:: When $I_L = I_{\mathrm{mp}}^{i,1}$, only the solar cells of the SCSD unit in the shadow occlusion set Ω_2 cannot work in the MPP, and only the supercapacitor in Ω_2 is discharged for compensation.

Case 3:: When $I_L \in \left(I_{\mathrm{mp}}^{i,1}, I_{\mathrm{oc}}^{i,1}\right)$, the solar cells of the SCSD units in the shading set Ω_1 and Ω_2 cannot work at MPP power, and the supercapacitor of the SCSD unit in Ω_1 is charged, and the supercapacitor of the SCSD unit in Ω_2 is discharged for coordinated compensation.

16.5.3.1 Discharge compensation of supercapacitor for solar cell–supercapacitor devices unit without or with shading shielding

For case 1, the output current of the SCSD series array is $I_L = I_{\mathrm{mp},1}^{i}, I_{\mathrm{mp},1}^{i} \in \left(0, I_{\mathrm{mp}}^{i,1}\right)$. At this time, the solar cell of the SCSD unit in Ω_1 is working at coordinate $\left(I_{\mathrm{mp},1}^{i}, V_{\mathrm{mp},1}^{i,1}\right)$ ($V_{\mathrm{mp},1}^{i,1}$ is the corresponding working voltage when the solar cell working current of the SCSD unit in Ω_1 is $I_{\mathrm{mp},1}^{i}$.). When $I_{\mathrm{mp},1}^{i} \in \left(0, I_{\mathrm{oc}}^{i,2}\right]$, the solar cell of the SCSD unit in Ω_2 works at coordinate $\left(I^i_{\text{mp},1}, V^{i,2}_{\text{mp},1}\right)$ ($V^{i,2}_{\text{mp},1}$ is the corresponding working voltage when the solar cell working current of the SCSD unit in Ω_2 is $I^i_{\text{mp},1}$). When$I^i_{\text{mp},1} \in \left(I^{i,2}_{\text{oc}}, I^{i,1}_{\text{mp}}\right)$, the SCSD unit in Ω_2 does not output power to the outside. The power loss of the output power of the solar cell of the SCSD unit in Ω_1 relative to the MPP A_2 is:

$$P^1_{\text{loss},1} = V^{i,1}_{\text{mp}} I^{i,1}_{\text{mp}} - V^{i,1}_{\text{mp},1} I^i_{\text{mp},1} \tag{16.18}$$

When $I^i_{\text{mp},1} \in \left(0, I^{i,2}_{\text{oc}}\right]$, the output power of the solar cell of the SCSD unit in Ω_2 relative to the power loss of MPP B_2 is:

$$P^2_{\text{loss},1} = V^{i,2}_{\text{mp}} I^{i,2}_{\text{mp}} - V^{i,2}_{\text{mp},1} I^i_{\text{mp},1} \tag{16.19}$$

When $I^i_{\text{mp},1} \in \left(I^{i,2}_{\text{oc}}, I^{i,1}_{\text{mp}}\right)$, the output power of the solar cell of the SCSD unit in Ω_2 relative to the power loss of MPP B_2 is:

$$P^2_{\text{loss},1} = V^{i,2}_{\text{mp}} I^{i,2}_{\text{mp}} \tag{16.20}$$

At this time, control the solar cells of the SCSD units in Ω_1 and Ω_2 to work in their MPP A_2 and B_2, and the supercapacitor in Ω_1 and Ω_2 are discharging at the same time, and compensate for the current lower than $I^i_{\text{mp},1}$. The discharge current $I^1_{\text{C.ch}}$ of the supercapacitor of the SCSD unit in Ω_1 is:

$$I^1_{\text{C.ch}} = I^{i,1}_{\text{mp}} - I^i_{\text{mp},1} \tag{16.21}$$

The discharge power $P^1_{\text{C.ch}}$ of the supercapacitor in the Ω_1 is:

$$P^1_{\text{C.ch}} = V^{i,1}_{\text{mp}} I^1_{\text{C.ch}} = V^{i,1}_{\text{mp}}\left(I^{i,1}_{\text{mp}} - I^i_{\text{mp},1}\right) \tag{16.22}$$

The discharge current $I^2_{\text{C.ch}}$ of the supercapacitor in Ω_2 is:

$$I^2_{\text{C.ch}} = I^{i,1}_{\text{mp}} - I^i_{\text{mp},1} \tag{16.23}$$

The discharge power $P^2_{\text{C.ch}}$ of the supercapacitor in Ω_2 is:

$$P^2_{\text{C.ch}} = V^{i,2}_{\text{mp}} I^2_{\text{C.ch}} = V^{i,2}_{\text{mp}}\left(I^{i,1}_{\text{mp}} - I^i_{\text{mp},1}\right) \tag{16.24}$$

Assuming there are a SCSD units in Ω_1 and b SCSD units in Ω_2, $a + b = N$. After the supercapacitor is discharged for compensation, the maximum output power of the SCSD series array $P_{\text{L.max,ch}}$ is:

$$P_{\mathrm{L.max,ch}} = aV_{\mathrm{mp}}^{i,1}I_{\mathrm{mp}}^{i,1} + bV_{\mathrm{mp}}^{i,2}I_{\mathrm{mp}}^{i,2} - aP_{\mathrm{C.ch}}^{1} - bP_{\mathrm{C.ch}}^{2} = \left(aV_{\mathrm{mp}}^{i,1} + bV_{\mathrm{mp}}^{i,2}\right)I_{\mathrm{mp}}^{i,1} \tag{16.25}$$

Therefore the difference between the maximum output power of the SCSD array without shading and the maximum output power with supercapacitor in Ω_1 and Ω_2 are charged for compensation under shading conditions is:

$$\Delta P_1 = P_{\mathrm{L.max}} - P_{\mathrm{L.max.ch}} = \left(aV_{\mathrm{mp}}^{i,1} - bV_{\mathrm{mp}}^{i,2}\right)I_{\mathrm{mp}}^{i,1} \tag{16.26}$$

16.5.3.2 Discharge compensation of supercapacitor for solar cell–supercapacitor devices unit with shadow shielding

For case 2, the output current of the SCSD series array is $I_{\mathrm{L}} = I_{\mathrm{mp}}^{i,1}$. At this time, the solar cell of the SCSD unit in Ω_1 works at MPP A_2, and the solar cell of the SCSD unit in Ω_2 does not output power. The generated power of the solar cell in Ω_2 is equal to the power loss of MPP B_2 is:

$$P_{\mathrm{loss},2}^{2} = V_{\mathrm{mp}}^{i,2}I_{\mathrm{mp}}^{i,2} \tag{16.27}$$

At this time, control the solar cells of the SCSD units in Ω_2 to work in their MPP B_2, and control the supercapacitor in Ω_2 discharging at the same time to compensate for the current lower than $I_{\mathrm{mp}}^{i,1}$. The discharge current $I_{\mathrm{C.ch}}^{2}$ of the supercapacitor in Ω_2 is:

$$I_{\mathrm{C.ch}}^{2} = I_{\mathrm{mp}}^{i,1} - I_{\mathrm{mp}}^{i,2} \tag{16.28}$$

The discharge power $P_{\mathrm{C.ch}}^{2}$ of the supercapacitor in Ω_2 is:

$$P_{\mathrm{C.ch}}^{2} = V_{\mathrm{mp}}^{i,2}I_{\mathrm{C.ch}}^{2} = V_{\mathrm{mp}}^{i,2}\left(I_{\mathrm{mp}}^{i,1} - I_{\mathrm{mp}}^{i,2}\right) \tag{16.29}$$

Similarly, if there are a SCSD units in Ω_1 and b SCSD units in Ω_2. After the supercapacitor of the SCSD units in Ω_2 is discharged for compensation the maximum output power of the SCSD series array $P_{\mathrm{L.max,ch}}$ is:

$$P_{\mathrm{L.max,ch}} = aV_{\mathrm{mp}}^{i,1}I_{\mathrm{mp}}^{i,1} + bP_{\mathrm{C.ch}}^{2} = aV_{\mathrm{mp}}^{i,1}I_{\mathrm{mp}}^{i,1} + bV_{\mathrm{mp}}^{i,2}\left(I_{\mathrm{mp}}^{i,1} - I_{\mathrm{mp}}^{i,2}\right) \tag{16.30}$$

Therefore the difference between the maximum output power of the SCSD array without shading and the maximum output power with compensation under shading:

$$\Delta P_{\mathrm{L}} = P_{\mathrm{L.max}} - P_{\mathrm{L.max.ch}} = b\left(V_{\mathrm{mp}}^{i,1}I_{\mathrm{mp}}^{i,1} - V_{\mathrm{mp}}^{i,2}I_{\mathrm{mp}}^{i,1} + V_{\mathrm{mp}}^{i,2}I_{\mathrm{mp}}^{i,2}\right) \tag{16.31}$$

16.5.3.3 Coordinated compensation of supercapacitors for solar cell–supercapacitor devices units

For case 3, the output current of the SCSD series array is $I_L = I^i_{mp,3}, I^i_{mp,3} \in \left(I^{i,1}_{mp}, I^{i,1}_{oc}\right)$. At this time, the solar cell of the SCSD unit in Ω_1 is working at coordinate $\left(I^i_{mp,5}, V^{i,1}_{mp,5}\right)$ ($V^{i,1}_{mp,5}$ is the corresponding working voltage when the solar cell working current of the SCSD unit in Ω_1 is $I^i_{mp,1}$.). The solar cell of the SCSD unit in Ω_2 does not output power to the outside. The power loss of the output power of the solar cell of the SCSD unit in Ω_1 relative to the MPP A_2 is:

$$P^1_{loss,5} = V^{i,1}_{mp} I^{i,1}_{mp} - V^{i,1}_{mp,5} I^i_{mp,5} \tag{16.32}$$

The output power of the solar cell of the SCSD unit in Ω_2 is related to the power loss of MPP B_2 is:

$$P^2_{loss,5} = V^{i,2}_{mp} I^{i,2}_{mp} \tag{16.33}$$

At this time, control the solar cells of the SCSD units in Ω_1 and Ω_2 to work in their MPP A_2 and B_2, and the supercapacitor in Ω_1 is charging at the same time to absorb the current higher than $I^{i,1}_{mp}$. The supercapacitor in Ω_2 is discharging to compensate for the current lower than $I^{i,1}_{mp}$.

The charging current $I^1_{C.ch}$ of the supercapacitor in Ω_1 is:

$$I^1_{C.ch} = I^i_{mp,5} - I^{i,1}_{mp} \tag{16.34}$$

The charging power $P^1_{C.ch}$ of the supercapacitor in Ω_1 is:

$$P^1_{C.ch} = V^{i,1}_{mp} I^1_{C.ch} = V^{i,1}_{mp}\left(I^i_{mp,5} - I^{i,1}_{mp}\right) \tag{16.35}$$

The discharge current $I^2_{C.ch}$ of the supercapacitor in Ω_2 is:

$$I^2_{C.ch} = I^{i,1}_{mp} - I^{i,2}_{mp} \tag{16.36}$$

The discharge power $P^2_{C.ch}$ of the supercapacitor in Ω_2 is:

$$P^2_{C.ch} = V^{i,2}_{mp} I^2_{C.ch} = V^{i,2}_{mp}\left(I^{i,1}_{mp} - I^{i,2}_{mp}\right) \tag{16.37}$$

Similarly, if there are a SCSD units in Ω_1 and b SCSD units in Ω_2. After the supercapacitor in Ω_2 is discharged for compensation, the maximum output power of the SCSD series array $P_{L.max,ch}$ is:

$$P_{L.max,ch} = aV^{i,1}_{mp} I^{i,1}_{mp} + bV^{i,2}_{mp} I^{i,2}_{mp} - aP^1_{C.ch} + bP^2_{C.ch} = 2aV^{i,1}_{mp} I^{i,1}_{mp} - aV^{i,1}_{mp} I^i_{mp,5} + bV^{i,2}_{mp} I^{i,1}_{mp} \tag{16.38}$$

Therefore the difference between the maximum output power of the SCSD array without shading and the maximum output power with supercapacitor in Ω_1 and Ω_2 are discharged for compensation under shading conditions is:

$$\Delta P_1 = P_{\mathrm{L.max}} - P_{\mathrm{L.max.ch}} = (b-a)V_{\mathrm{mp}}^{i,1}I_{\mathrm{mp}}^{i,1} + aV_{\mathrm{mp}}^{i,1}I_{\mathrm{mp},5}^{i} - bV_{\mathrm{mp}}^{i,2}I_{\mathrm{mp}}^{i,1} \tag{16.39}$$

From the Eqs. (16.26), (16.31), (16.39), the PV cell will be maintained at its MPP with any output PVA current via controlling the charge/discharge of supercapacitor in the SCSD series.

16.5.4 Current compensation implementation of solar cell–supercapacitor devices series array

In practical, each SCSD unit has its own light intensity sensor The open-circuit voltage, short-circuit current, MPP voltage and current of each SCSD unit in the array can be calculated by Eq. (16.12). Thereby the reference voltage and reference current for the supercapacitor of the SCSD unit under partial shading conditions are obtained.

The bidirectional DC/DC converter is connected to port 1 and port 3, and port 1 and port 2 of the SCSD unit. The connection method is shown in Fig. 16.25. The supercapacitor Charge absorption control circuit of the SCSD unit without shadows is shown in Fig. 16.30. The bidirectional DC/DC converter works in the boost state, the PI controller tracks the solar cell MPP voltage $V_{\mathrm{mp}}^{i'}$ of the SCSD unit without shadows, and the Charge absorption reference current $I_{\mathrm{C.ch.ref}}$ is obtained by Eq. (16.34). The deviation is obtained by comparing $I_{\mathrm{C.ch.ref}}$ with the actual compensation current $I_{\mathrm{C.ch}}$, than the PI controller and the modulator output the PWM signal, than the switch K_1 receives the PWM signal to control the supercapacitor charging of the SCSD unit without shadow.

The compensation circuit of the discharged supercapacitor in the shadow shielding SCSD unit is shown in Fig. 16.31. The bidirectional DC/DC converter works in

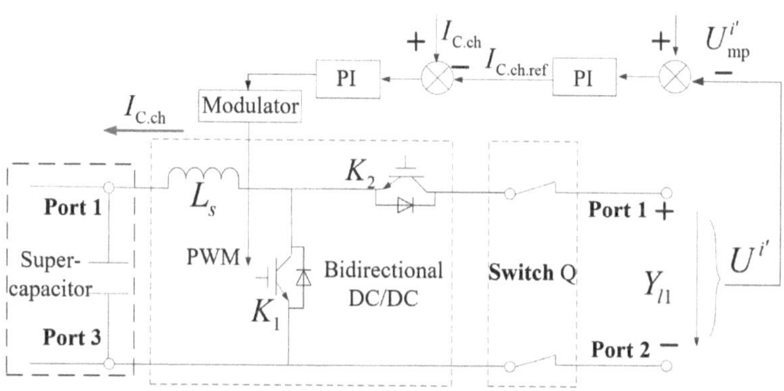

Figure 16.30 Charge compensation of supercapacitor.

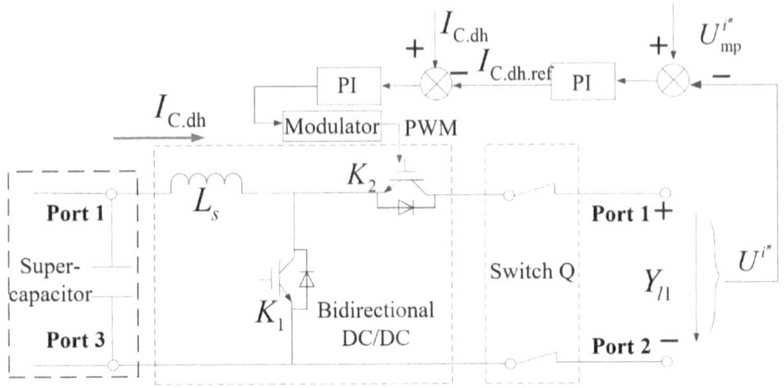

Figure 16.31 Discharge compensation of supercapacitor.

the buck state, the PI controller tracks the solar cell MPP voltage $V^{i,2}_{mp}$ under partial shading, and the reference discharging current $I_{C.dh.ref}$ can be obtained by Eqs. (16.21), (16.23), (16.28), (16.36). The deviation between $I_{C.dh.ref}$ and the actual compensation current $I_{C.dh}$ is sent to the PI controller and the modulator to generate the PWM signal. The switch K_2 receives the PWM signal to control the discharging of the supercapacitor of the SCSD unit in Ω_1.

With the coordination of the charge/discharge process of supercapacitor, the output generation of PV cells can be matched for different shading conditions, then the hot spot effect can be avoided and the efficiency of PV generation is increased.

16.5.5 Case study

16.5.5.1 Parameter setting

10 SCSD units numbered 1, 2, 3, ..., 10 are connected in series to verify the proposed method, and the analysis tool is the MATLAB®/Simulink and the laboratory tests. The simulation model is shown in Fig. 16.32. Assuming the SCSD arrays are all working in uniform conditions, the irradiance under nonshading conditions is 1000 W/m^2, the temperature of all units keep at 25°C, and the theoretical parameter of the supercapacitor is 15 mF/cm^2. The parameters of the SCSD unit under normal light are as follows Table 16.2 shows.

The irradiance of the SCSD array under partial shading condition is 600 W/m^2, and the parameters of the SCSD unit under shadow occlusion are shown in Table 16.3.

16.5.5.2 Analysis of simulation and laboratory test

Set the SCSD array to change from standard light intensity to a partial shading condition with uniform intensity. In 0–0.1 second, the PVA works normally at 1000 W m^{-2} solar irradiance. At 0.1 second, set different numbers of SCSD units to be in the shadow of the same light intensity at 600 W m^{-2} solar irradiance, and

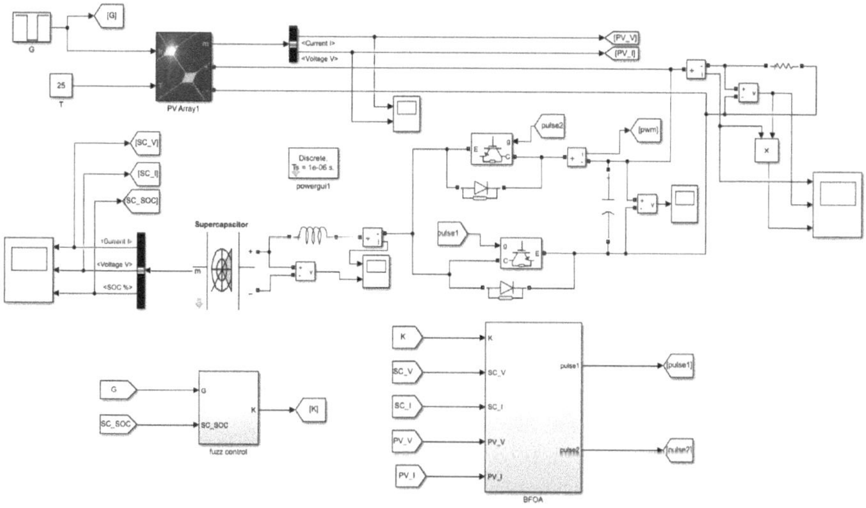

Figure 16.32 The simulation model.

Table 16.2 Solar cell–supercapacitor devices parameters under normal light.

Solar cell parameters	Numerical value	Supercapacitor parameters	Numerical value
Power P/W	270	Capacity C_{sup}/F	1.5
Short-circuit current I_{sc}/A	9.7	Rated voltage U_C/V	48
Open-circuit voltage V_{oc}/V	36	SOC range SOC	10%–90%
The MPP current I/A	9.02	Cycle life/10,000 times	50
The MPP voltage U/V	30		

Note: The SOC is the current state parameter for the calibration of the supercapacitor capacity.

Table 16.3 Solar cell–supercapacitor devices parameters under partial shading.

Solar cell parameters	Numerical value	Supercapacitor parameters	Numerical value
Power P/W	100	Capacity C_{sup}/F	1.5
Short-circuit current I_{sc}/A	5.7	Rated voltage U_C/V	48
Open-circuit voltage V_{oc}/V	24	SOC range SOC	10%–90%
The MPP current I/A	5.02	Cycle life/10,000 times	50
The MPP voltage U/V	20		

Note: The SOC represents the parameter of supercapacitor capacity at current state.

the light intensity of other units remains unchanged. At 0.25 second, according to the output current value of the series array in the partial shading state, the supercapacitor of the SCSD units in Ω_1 and Ω_2 are employed for compensation. The initial SOC values of the supercapacitor of the SCSD units in Ω_1 and Ω_2 are both set to 0.5.

1. Discharging process of supercapacitor in different cases

When the SCSD series array output current $I_L \in \left(0, I_{mp}^{i,1}\right)$ under partial shadow shielding, the supercapacitor of the SCSD units in Ω_1 and Ω_2 are simultaneously discharged for compensation. The output current curves of the solar cell are shown in Fig. 16.33, and the SOC curves of the supercapacitor are shown in Fig. 16.34.

During 0–0.1 second, when the series array is in normal light, the solar cells in the array work at MPP, the output currents are kept at 9.02 and 8.81 A under MATLAB/Simulink and laboratory test. During 0.1–0.25 second, the series array is blocked by partial shading, and the output current of the array is reduced to 4.51 and 4.36 A under MATLAB/Simulink and laboratory test due to the hot spot effect. At this time, the supercapacitor of the SCSD units in Ω_1 and Ω_2 are discharged at the same time to compensate. Starting from 0.25 second, the solar cells of the SCSD units in Ω_1 and Ω_2 work in the corresponding MPP, and the output currents are 9.02, 5.04 and 8.85, 4.93 A under MATLAB/Simulink and laboratory test, respectively.

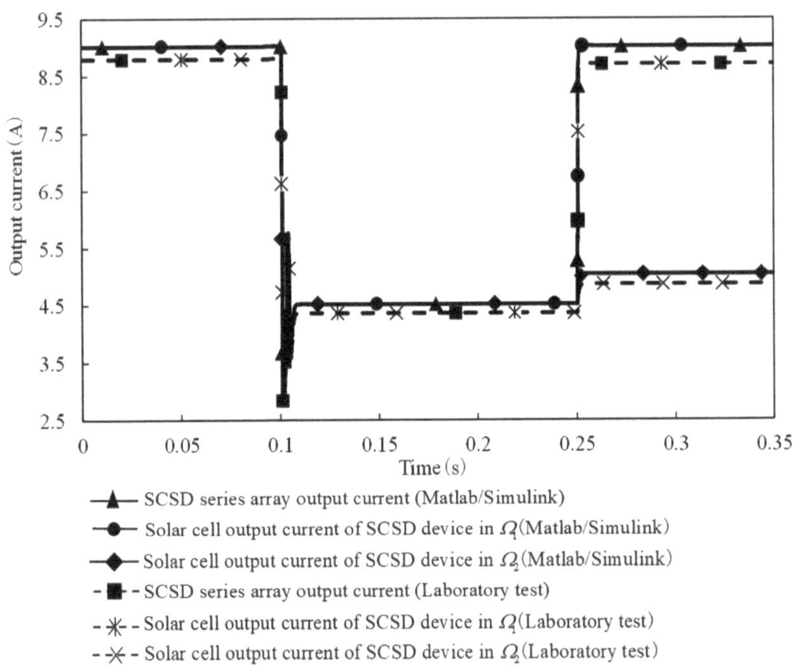

Figure 16.33 Change of output current under the simultaneous discharging compensation of the supercapacitor of solar cell–supercapacitor devices unit in Ω_1 and Ω_2.

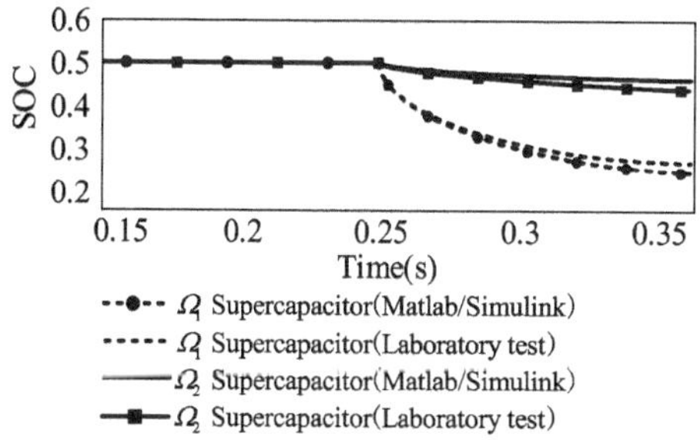

Figure 16.34 Changes of supercapacitor state of charge under the simultaneous discharging compensation of the supercapacitor of solar cell–supercapacitor devices unit in Ω_1 and Ω_2

After the PV output is compensated, the supercapacitor of the SCSD unit in Ω_1 discharges the solar cell, and the SOC of the supercapacitor reduces from 0.5 to 0.27 and 0.29 under MATLAB/Simulink and laboratory test, respectively. The solar cell of the SCSD unit in Ω_2 also discharges the supercapacitor, and the supercapacitor SOC reduced from 0.5 to 0.47 and 0.48 under MATLAB/Simulink and laboratory test, respectively.

2. Coordination compensation for SCSD unit under partial shading

When the SCSD series array output current $I_L = I_{mp}^{i,1}$ under partial shadow shading, it is compensated by the discharge of the supercapacitor in Ω_2. The curves of solar cell output current are shown in Fig. 16.35, and the curves of supercapacitor SOC are shown in Fig. 16.36.

During 0.1–0.25 second, the series array is blocked by partial shadows, but the output current of the array remains unchanged. The output current is compensated by the supercapacitor of the SCSD unit in Ω_2. Since 0.25 second, the solar cells of the SCSD units in Ω_1 and Ω_2 work in the corresponding MPP, and the output currents are 9.02, 5.04 and 8.81, 4.89 A under MATLAB/Simulink and laboratory test, respectively.

After the supercapacitor of the SCSD unit in Ω_2 is discharged for compensation, the supercapacitor of the SCSD unit in Ω_1 remains its SOC. The supercapacitor of the SCSD unit in Ω_2 discharges to compensate the solar cell, and the supercapacitor SOC is reduced from 0.5 to 0.24 and 0.25 under MATLAB/Simulink and laboratory test, respectively.

3. Coordination compensation of supercapacitor charging of SCSD unit without shading and supercapacitor discharging of shadow blocked SCSD unit

When the SCSD series array output current $I_L \in \left(I_{mp}^{i,1}, I_{oc}^{i,1}\right)$ under partial shadow shading, the supercapacitor charging of the SCSD unit in Ω_1 and the supercapacitor discharge of the SCSD unit in Ω_2 are coordinated controlled for compensation. The output current curves are shown in Fig. 16.37, and the SOC curves of supercapacitor are shown in Fig. 16.38.

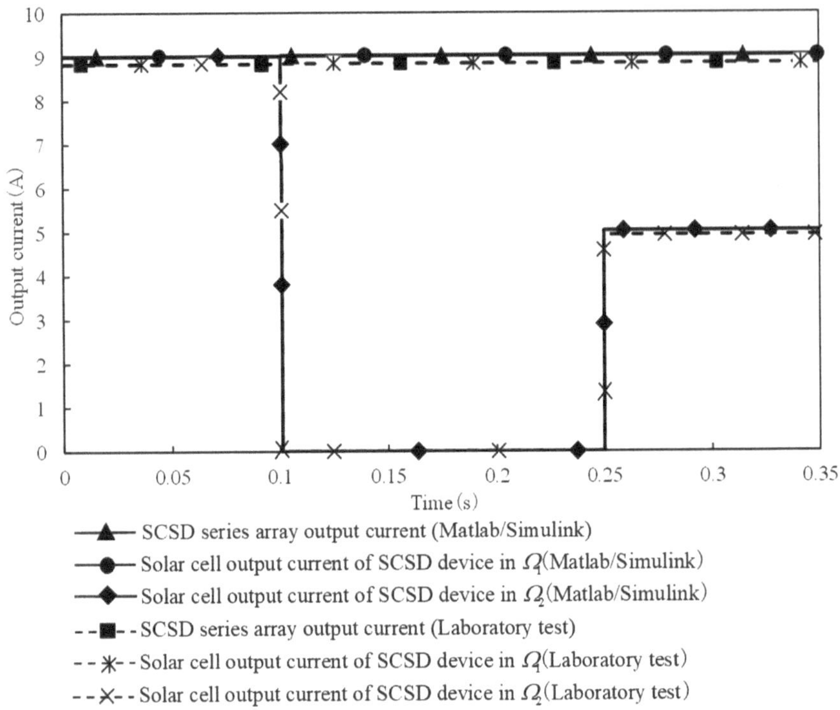

Figure 16.35 Changes of supercapacitor state of charge under the discharging compensation of the supercapacitor of solar cell—supercapacitor devices unit in Ω_2.

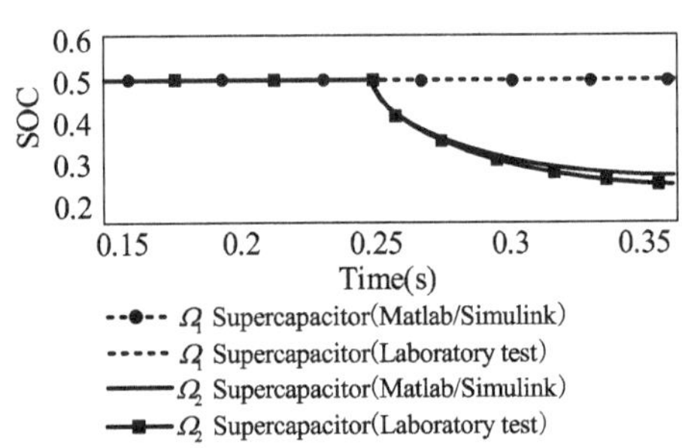

Figure 16.36 Changes of supercapacitor state of charge under the discharging compensation of the supercapacitor of solar cell—supercapacitor devices unit in Ω_2

During 0.1—0.25 second, the series array is blocked by partial shading, and the output current of the array rises to 9.43 A due to the hot spot effect. Since 0.25 second, the solar cells of the SCSD units in Ω_1 and Ω_2 work in the corresponding MPP via the compensation by supercapacitor, and the output currents are 9.02, 5.04 and 8.81, 4.89 A under MATLAB/Simulink and laboratory test, respectively.

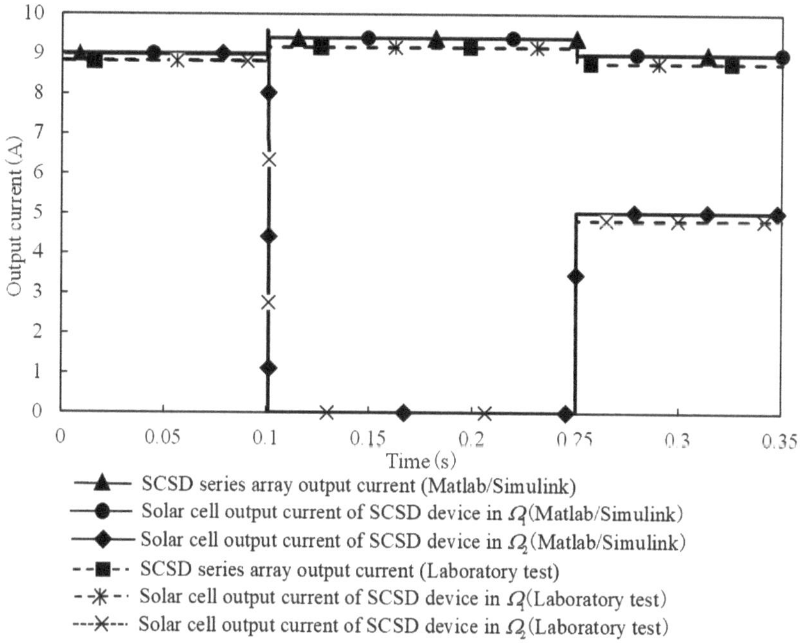

Figure 16.37 Changes of output current under the coordinated compensation of the supercapacitor charge of SCSD unit in Ω_1 and the supercapacitor discharge of SCSD unit in Ω_2. *SCSD*, Solar cell–supercapacitor devices.

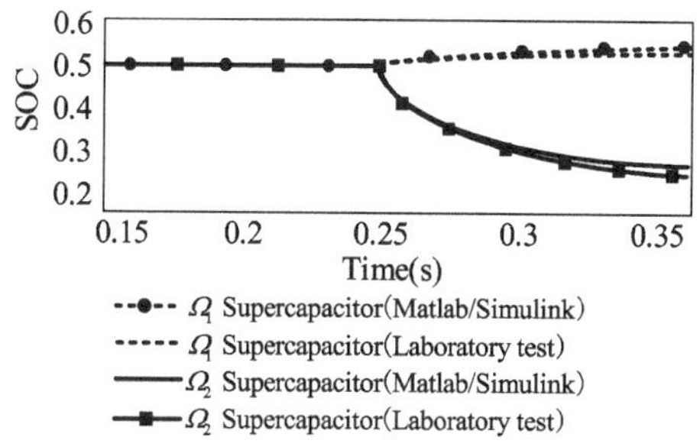

Figure 16.38 Changes of supercapacitor state of charge under the coordinated compensation of the supercapacitor charge of solar cell–supercapacitor devices (SCSD) unit in Ω_1 and the supercapacitor discharge of SCSD unit in Ω_2.

After the output generation is compensated by the supercapacitor of the SCSD unit in Ω_1 and Ω_2, the PV cell of the SCSD unit in Ω_1 charges the supercapacitor, and the SOC of the supercapacitor increases from 0.5 to 0.52 and 0.51 under MATLAB/Simulink and laboratory test, respectively. The supercapacitor of the SCSD unit in Ω_2 discharges the PV cell, and the supercapacitor SOC is reduced from 0.5 to 0.24 and 0.25 under MATLAB/Simulink and laboratory test, respectively.

16.6 Conclusions

Sections 16.1−16.3 summarize the trend of energy development, analyze the characteristics of PV generation and the impact of large-scale grid-connected PV on the power system. The development of energy storage technologies is presented, and the importance of energy storage technology to the stability of the power system is pointed out. Finally, the types of energy storage technologies and their respective characteristics are analyzed in detail.

Section 16.4 mainly studies the energy storage configuration mode and its control strategy under large-scale grid-connected PV generation. First, the access method of energy storage with large-scale grid-connected PV is analyzed from the aspects of hardware cost, the difficulty of implementation, and reliability. Secondly, the capacity configuration method of energy storage in the PV generation system is studied. Finally, the control strategy of energy storage to support the frequency/voltage control with PV generation is developed. The following researches have been carried out:

1. The topology of the power conversion system and the basic control strategy have been carried, which provide a foundation for frequency/voltage control with energy storage devices with PV generation.
2. The grid-connected control strategy of ESS based on additional frequency control is studied. The frequency stability of the PV generation system is improved with the fast response of ESS.
3. Voltage control strategy with distributed ESS is studied. Through the coordinated control of the ESS and the PVA, the reactive power demand of the power system is met and the reactive output capacity of the PV/ESS system is improved.

Section 16.5 proposes a current compensation method for the SCSD series array under partial shadow shading. Through the theoretical analysis of the current compensation, the power loss of the series array under different output currents and the supercapacitor compensation current are determined. The compensation method via the charging/discharging of supercapacitor under partial shadow shading is proposed. Some conclusions can be obtained based on case studies:

1. The supercapacitor compensation lets each PV cell works at the MPP under different shading conditions, which increases the output power of the SCSD series array.
2. The output characteristics of each SCSD tend to be consistent with the current compensation method, which realizes the integrated control of the generation and storage functions of the SCSD.

Acknowledge

This work was supported in part by the National Natural Science Foundation of China (No. 71931003, No. 72061147004, and No.72171026), the Science and Technology Projects of Hunan Province under (No. 2019GK5015, No. 2020GK1014 and No. 2021WK2002), the University-Industry Collaborative Education Program (No. 202102574001, and No. 202102574005), the Hunan International Scientific and Technological Cooperation Base of Energy Storage Interconnection (No. 2018WK4010), the Big Data-driven Energy Internet International Science and technology Cooperation base of the Ministry of Science and Technology (No. 2017D01011).

References

[1] IEA, Renewables 2021: Analysis and Forecast to 2026, OECD Publishing, Paris, 2021. Available from: https://doi.org/10.1787/6dcd2e15-en.
[2] B. Uzum, A. Onen, H.M. Hasanien, et al., Rooftop solar PV penetration impacts on distribution network and further growth factors—a comprehensive review, Electronics. 10 (1) (2021) 55.
[3] K.M. Tan, T.S. Babu, V.K. Ramachandaramurthy, et al., Empowering smart grid: a comprehensive review of energy storage technology and application with renewable energy integration, J. Energy Storage 39 (2021) 102591. Jul.
[4] A.N. Abdalla, M.S. Nazir, H. Tao, et al., Integration of energy storage system and renewable energy sources based on artificial intelligence: an overview, J. Energy Storage 40 (2021) 102811. Aug.
[5] L. Ji, C.B. Li, X.P. Li, et al., Multi-method combination site selection of pumped storage power station considering power structure optimization, Sustain. Energy Technol. Assess. 49 (2022) 101680. Feb.
[6] Z.G. Zhao, J.D. Yang, C.Y. Chung, et al., Performance enhancement of pumped storage units for system frequency support based on a novel small signal model, Energy. 234 (2021) 121207. Jun.
[7] W.C. Guo, D.Y. Zhu, Nonlinear modeling and operation stability of variable speed pumped storage power station, Energy Sci. Eng. 9 (10) (2021) 1703−1718. Oct.
[8] C. Cao, Z.G. Xu, J.R. Cai, et al., Determination method for influence zone of pumped storage underground cavern and drainage system, J. Hydrol. 595 (2021) 126108. Apr.
[9] M. Stocks, B. Lu, A review of pumped hydro energy storage, Prog. Energy 3 (no.2) (2021) 022003. Apr.
[10] A.G. Olabi, T. Wilberforce, M. Ramadan, et al., Compressed air energy storage systems: components and operating parameters—a review, J. Energy Storage 34 (2021) 102000. Feb.
[11] M. Heidari, D. Parra, M.K. Patel, Physical design, techno-economic analysis and optimization of distributed compressed air energy storage for renewable energy integration, J. Energy Storage 35 (2021) 102268. Mar.
[12] S. Mucci, A. Bischi, S. Briola, et al., Small-scale adiabatic compressed air energy storage: control strategy analysis via dynamic modelling, Energy Convers. Manag. 243 (2021) 114358. Sep.
[13] Z.M. Tong, Z.W. Cheng, S.G. Tong, A review on the development of compressed air energy storage in China: technical and economic challenges to commercialization, Renew. Sustain. Energy Rev. 135 (2021) 110178. Jan.
[14] H. Gou, Y.J. Xu, X.J. Zhang, et al., Finite-time thermodynamics modeling and analysis on compressed air energy storage systems with thermal storage, Renew. Sustain. Energy Rev. 138 (2021) 110656. Mar.
[15] S. Fang, Z. Lv, G. Chao, Methods of increasing the energy storage density of superconducting flywheel, IEEE Trans. Appl. Superconduct 31 (8) (2021) 1−5. Art no. 5700705.

[16] X.J. Li, A. Palazzolo, A review of flywheel energy storage systems: state of the art and opportunities, J. Energy Storage 46 (2022) 103576. Feb.
[17] A. Floris, M. Porru, A. Damiano, et al., Energy management and control system design of an integrated flywheel energy storage system for residential users, Appl. Sci. 11 (10) (2021) 4615.
[18] A.G. Olabi, T. Wilberforce, M.A. Abdelkareem, et al., Critical review of flywheel energy storage system, Energies. 14 (8) (2021) 2159.
[19] S.T. Senthilkumar, J.S. Park, R. Marcilla, et al., Using redox electrolytes to extend the charge storage capacity in an aqueous hybrid ion battery, Chem. Eng. J. 411 (2021) 128416. January.
[20] Y. Zhang, C.G. Zhou, J. Yang, Advances and challenges in improvement of the electrochemical performance for lead-acid batteries: a comprehensive review, J. Power Sources 520 (2022) 230800. Feb.
[21] N.E. Galushkin, N.N. Yazvinskaya, D.N. Galushkin, Nickel-cadmium batteries with pocket electrodes as hydrogen energy storage units of high-capacity, J. Energy Storage 39 (2021) 102597. Jul.
[22] N.N. Yazvinskaya, D.N. Galushkin, I.R. Isaev, Neural network-based residual capacity indicator for nickel-metal hydride batteries in electric vehicles, IEEE Trans. Veh. Technol. 1029 (1) (2021) 012091. Jan.
[23] C.M. Costa, J.C. Barbosa, R. Gonçalves, et al., Recycling and environmental issues of lithium-ion batteries: advances, challenges and opportunities, Energy Storage Mater. 37 (2021) 433–465. May.
[24] Y. Liu, J. Li, G. Zhang, et al., State of charge estimation of lithium-ion batteries based on temporal convolutional network and transfer learning, IEEE Access. 9 (8) (2021) 34177–34187.
[25] A. Boudia, S. Messalti, A. Harrag, et al., New hybrid photovoltaic system connected to superconducting magnetic energy storage controlled by PID-fuzzy controller, Energy Convers. Manag. 244 (2021) 114435. Sep.
[26] Y.M. Wang, X.L. Wu, Y.Q. Han, et al., Flexible supercapacitor: overview and outlooks, J. Energy Storage 42 (2021) 103053. Oct.
[27] M.E. Şahİn, F. Blaabjerg, A. Sangwongwanİch, Modelling of supercapacitor based on simplified equivalent circuit, CPSS Trans. Power Electron. Appl. 6 (1) (2021) 31–39.
[28] F. Belhachat, C. Larbes, PV array reconfiguration techniques for maximum power optimization under partial shading conditions: a review, Sol. Energy 230 (2021) 558–582.
[29] Z.R. Tahir, S. Hafeez, M. Asim, et al., Estimation of daily diffuse solar radiation from clearness index, sunshine duration and meteorological parameters for different climatic conditions, Sustain. Energy Technol. Assess. 47 (2021) 101544. Oct.
[30] W.M.O. Narvios, Y.Q. Nguyen, Detection and mitigation system for shading-induced hot spots in household crystalline silicon photovoltaic modules, AIP Conf. Proc. 2406 (1) (2021) 060003.
[31] A. Ai, K. Irshad, M.F. Khan, et al., Artificial intelligence and bio-inspired soft computing-based maximum power plant tracking for a solar photovoltaic system under non-uniform solar irradiance shading conditions—a review, Sustainability. 13 (19) (2021) 10575.
[32] G.S. Krishna, T. Moger, A novel adaptive dynamic photovoltaic reconfiguration system to mitigate mismatch effect, Renew. Sustain. Energy Rev. 141 (2021) 110754. May.

Solar power smoothing using battery energy storage system through fuzzy filtration technique

17

Ammar Atif[1], *Khalid Abdullah Khan*[1] *and Muhammad Khalid*[1,2,3]
[1]Electrical Engineering Department, King Fahd University of Petroleum and Minerals (KFUPM), Dhahran, Saudi Arabia, [2]Center for Renewable Energy and Power Systems, KFUPM, Dhahran, Saudi Arabia, [3]SDAIA-KFUPM Joint Research Center for Artificial Intelligence, Dhahran, Saudi Arabia

Chapter Outline

Nomenclature

a	Proportional gain
BESS	Battery energy storage system
DG	Distributed Generation
EV	Electric vehicle

Emerging Trends in Energy Storage Systems and Industrial Applications. DOI: https://doi.org/10.1016/B978-0-323-90521-3.00004-1

$\mathbf{E_{BESS}}$	Battery energy
FESS	Flywheel energy storage system
FLC	Fuzzy logic controller
LPF	Low pass filter
MA	Moving average
MF	Membership function
MPC	Model predictive control
PCC	Point of common coupling
PSO	Particle swarm optimization
PV	Photovoltaic system
$\mathbf{P^*}$	Reference power
$\mathbf{P_{cap}}$	Name plate capacity of the solar PV
$\mathbf{P_f}$	Targeted filtered output power
$\mathbf{P_G}$	PV power sent to the grid
$\mathbf{P_{shaded}}$	Diffuse output from irradiance
$\mathbf{P_{PV}}$	Rated output power of PV
$\mathbf{\overline{P}_{PV}}$	Nominal solar PV power
RE	Renewable energy
RR	Ramp rate
SMESS	Superconducting magnetic energy storage system
SOC	State-of-charge of battery
SOC_{mh}	Upper SOC limit of battery
SOC_{mL}	Lower SOC limit of battery
T_c	Smoothing time constant
TF	Transfer function of low pass filter
WPGS	Wind power generation system

Highlights

Fuzzy control coupled with a varying filter to reduce BESS operation with enhanced solar power smoothing

17.1 Introduction

With the increased reliance on technology, the energy demand is increased also which arise a challenge of how to meet this colossal energy demand without threatening the globe. Conventional power generation resources like diesel and coal-based power supplies depend mainly on fossil fuel resources which emit greenhouse gases that trap heat in the atmosphere and contribute negatively to the global warming phenomenon. Renewable energy (RE) resources are clean and do not negatively impact the environment, resultantly, there is a great incentive to switch to RE, and hopefully, the manufacturing prices are going down. However, Solar and wind farms are weather-dependent resources that are characterized by high volatility and make the generated power from such resources intermittent, uncontrollable, and unpredictable [1–3]. Dispatching the generated power from such intermittent supplies affects the power grid negatively by deteriorating the stability performance, creating power quality issues, overvoltage, and voltage fluctuations [4]. Also, the intermittency negatively impacts the operation and planning of the power system. In conventional power systems, the load variability and the generation unit's outage are met by the generation reserve as the generated power from such resources is controllable and dispatchable. Integrating RE with the grid imposes additional uncertainties and operational challenges [5]. The variability of RE results in demand-generation mismatch [6]. The uncertainty of the generated power creates scheduling difficulties as they cannot be accurately predicted [7]. Also, high sudden penetration creates grid congestion [8].

Conventional generation power like fired coal generators cannot reduce their generated power when there is excess power generated from RE supplies as they cannot be switched off and start immediately due to efficiency and lifetime constraints. Hydropower plants can accommodate the RE variability as they have a high-power ramp rate, however, irrigational constraints may restrict their smoothing capabilities. A reversal power flow might stem also from the connected distributed generators (DGs) like rooftop solar photovoltaic (PV) panels. Distribution power systems are radial as the power flows in a single direction from the substation to the customer loads. The transformer's voltage settings have been designed to handle the network losses by setting the voltage to be $\pm$ (5%−10%) greater than the customer side voltage. To maintain the voltage within this limit on-load tap changer transformers and automatic voltage relays regulate the voltage of the transformer to keep it within the acceptable limits as violating it will affect the system security [9]. However, a reverse power flow is induced that leads to an increment in the line voltage that violates this limit when a higher level of distributed RE is integrated into the electrical network. The reverse power flow affects the operation of the voltage regulators as they are unidirectional. The reverse power flow may result in a higher short circuit current as well [10]. Moving clouds over solar PV panels can affect the voltage level on the distribution side as it could violate the power quality standards. Solar output power is proportionate to the subjected irradiance into the PV panels which may cause a sudden power drop in a short time and causes voltage fluctuations accordingly because of the voltage change in the point of common coupling (PCC). The voltage fluctuations consist of voltage fall, voltage rise, or unbalanced voltage in distribution feeders and may contribute to appliances' damage [11].

The voltage fluctuation depends on the location of PV, capacity, and configuration. Solar and wind farms are integrated into the grid via power electronic devices. The presence of power electronic switching creates power quality problems like harmonics and voltage rises and dips. Harmonics are complex currents with higher frequency waveforms over the fundamental frequency that overheat cables, transformers, and loads. Also, harmonics are interfaced with the nearby equipment and make electromagnetic interference. Moreover, the voltage fluctuation that is occurred because of the RE intermittency creates lighting flickers [12]. The integration of battery energy storage systems (BESSs) with solar PV systems has been extensively studied to enhance the flexibility of the grid and mitigates the effects of moving clouds. The literature review shows a wide application of the energy storage system (ESS) to enhance system quality, reliability, energy management, and improve system stability, especially from an inertia-free resource. The ESS applications with the RE resources could be summarized in the following points.

17.1.1 Applications of energy storage systems

ESSs are widely adopted to solve the challenges that stem from intermittent power resources. The literature review shows a wide application of the ESS to enhance system quality, reliability, energy management, and improve system stability, especially from an inertia-free resource. The ESS applications with the RE resources are summarized in the following points.

17.1.1.1 Power smoothing

As renewable supplies are weather dependent, the generated power from PV panels for example may drop or rise following the moving clouds over PV panels (i.e., cloud-dependent resources). The generated power from the PV system is dependent on the cloud size and its speed. Similarly, the produced power from wind farms is affected by the wind speed which is also intermittent and discontinuous. The generated power could be smoothed with the aid of ESS before being sent into the grid to alleviate the power fluctuations issues and control the power ramp rate (RR) within acceptable limits [13]. Additionally, it is possible to use the filtering controller to construct a charging and discharging profile that accounts for fluctuating power, which assures compliance with the grid standards and codes.

17.1.1.2 Peak shaving

Energy storage systems store the surplus energy from renewable energy during the daytime and discharge the power during maximum demand hours. The tariff during peak hours is usually higher than the rest of the day. Thus, ESS can reduce the burden of the generators to provide more power during peak demand hours by storing energy from RE sources during off-peak intervals and dispatching the power in peak load intervals [14].

17.1.1.3 Load leveling

ESSs are used to store power in the light loads' interval and supply the grid with power during peak loads which obviates drawing electricity from the grid or increases the grid infrastructure. Load leveling is like peak shaving as the generation cost could be reduced with the aid of the ESS. The load leveling is also referred to as energy arbitrage which helps to make profits by purchasing electricity when the price of energy is low and selling it when the energy price is high [15].

17.1.1.4 Microgrid operation

ESSs help the operation of micro-grids by maintaining the voltage and the frequency at permissible limits. Besides, energy storages act as a frequency regulation source by having a fast time response [16]. Normally, when an island is formed as a subsequent of a fault the dispatchable DG unit is disconnected from the grid as it lost the reference voltage that must follow. The ESS could help to work as a grid firming unit and help the DG to act as a dispatchable unit.

17.1.1.5 Power quality

The use of BESS in power quality improvement is to mitigate the grid disturbances like flickers, harmonics, and voltage dips [17].

17.1.1.6 Black start

ESSs help the power system to restart after a blackout [18].

17.1.1.7 Energy arbitrage

ESSs can be utilized to earn profits during high peak demand which is known as energy arbitrage. During the high demand load, the electricity price is high, so ESS is used to store the energy during low load demand and discharge when the load demand is high. Pumped hydro energy storage has been employed to maintain the property of energy arbitrage as it can be used for high energy capacities of more than 100 MW and discharge over (100–1000) minutes [19].

17.1.1.8 Energy storage-based smoothing architectures

There are extensive studies examining the use of filtering and control techniques to mitigate PV fluctuations and limit the rate of power change below the maximum allowable limits. Filtering methods include first- and second-order low-pass filters and sliding filters (typically moving average filters, moving median filters [20], and Savitzky–Golay filters). Fig. 17.1 depicts a generic smoothing scheme of the intermittent photovoltaic power output. The system involves DC–DC converters to connect the PV system to the DC link, as well as a bidirectional DC–DC converter to absorb/inject energy from/to the battery storage system. Furthermore, the smoothing model comprises a filtering mechanism to determine the charging and discharging power to the battery storage system during ramp-up and ramp-down events, respectively.

Several studies investigate coupling the energy storage system with a filtering or smoothing technique to reduce the variability of PV power fluctuations. The authors of [21] used an MA filter in combination with ESS to smooth and reduce PV variability and reduce the number of taps on the automatic voltage regulator. According to the study findings in [22], the LPF filter overcomes both the simple and exponential moving average filters as a significant delay is observed with the smoothed power output of simple and exponential MA filters. The delay of the smoothed

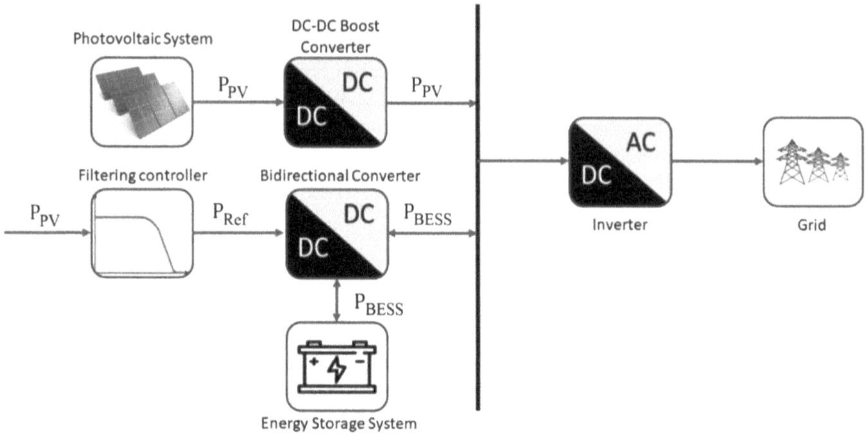

Figure 17.1 General schematic for the smoothing process [13].

power contributes to higher charging/discharging power to the BESS system, implying more battery operation. The study presented in [23] introduces a Savitzky–Golay filter that performs a polynomial fitting to the smoothed PV power and helps to smooth the variation of PV power while preserving the solar power characteristics (i.e., height and width). The results show better performance than MA, LPF, and Gaussian smoothing filters. Although the SG filter gives better smoothing performance, it requires adjusting two parameters, namely window size, and polynomial fitting order. Model predictive control (MPC) is utilized for wind power smoothing as it's employed to predict generated power from wind farms. The predicted power will be fed to the controller, and the maximum power ramp rate limit will be optimized along with the battery level of charge [24]. The given study in [25], evaluates the storage energy requirement when wind power generation systems (WPGSs) are installed adjacent and far from each other. The study demonstrates that a smaller energy capacity is required for power smoothing in the case of distributed wind turbines vs the aggregated arrangement. More investigation has been carried out by authors in [26], where the smoothing of the wind power fluctuations has been performed with two stages MPC controller with hybrid BESS-supercapacitor ESS. The purpose of the first stage is to determine the initial charge and discharge of the battery system whereas the second stage determines the optimum operation of the combined supercapacitor.

The utilization of LPF-based control leads to a consistent charge/discharge of BESS even during the absence of clouds and the fluctuation is minimal or negligible. Therefore, the BESS life is reduced because of the increased BESS consumption and operation. In addition, selecting a nonoptimal filtering time constant leads to poor power tracking or improper smoothing performance, both of which cause the BESS's state of charge (SOC) to fluctuate more. MPC is utilized for wind power smoothing as it's employed to predict generated power from wind farms. The predicted power will be fed to the controller, and the maximum power ramp rate limit will be optimized along with the battery level of charge [24]. The given study in Ref. [25] evaluates the storage energy requirement when wind turbines are installed adjacent and far from each other. The study demonstrates that a smaller energy capacity is required for power smoothing in the case of distributed wind turbines vs the aggregated arrangement. More investigation has been carried out by authors in Ref. [26], where the smoothing of the wind power fluctuations has been performed with two stages MPC controller with a hybrid BESS-supercapacitor ESS.

The purpose of the first stage is to determine the initial charge and discharge of the battery system whereas the second stage determines the optimum operation of the combined supercapacitor. The provided study in [27], shows a coordination approach to optimize the operation of a pumped storage system with a BESS to smooth the variation of wind power. The hybrid energy storage system yields the full advantage of pumped storage and battery storage. Normally, a hybrid energy storage system (HESS) consists of a storage component that is characterized by high specific energy and another which is characterized by a high specific power.

High specific energy components of the storage system can smooth the low-frequency power fluctuations whereas the storage component with high specific power can smooth the high-frequency power fluctuations. Further investigation has been provided in [28], as a two-stage coordinated control method has been implemented to control HESS based on wavelet filtering. The compensating power profile to alleviate the power intermittency is obtained using the first stage whereas the second stage denoises the compensating power. However, the paper ignored the wind power uncertainty which might affect estimating the hybrid system state of charge.

Energy storage based-optimization algorithms have been presented in many studies including employing a holistic optimization methodology to optimize the operation and the economic performance of an offshore wind turbine connected with an energy storage system [29]. The proposed control strategy investigates mitigating rotor unbalanced loads and extends service lifespan which makes it suitable for real application. In Ref. [30] a multiobjective optimization technique is proposed to satisfy the load demand and minimize the cost of microgrid operation that includes a solar PV, WPGS, fuel cell, and diesel generator. Results show that the proposed methodology satisfies the consumer demand whereas meeting the system constraints typically, meets the demand with lower operational cost and emissions.

A trade-off study between three objective functions is conducted to minimize peak loads, minimize annual cost and regulate the voltage for a battery storage system connected with a low voltage network [31]. Results conclude that connecting the battery storage system in low voltage grids reduces voltage deviations and overvoltage events which facilitate the integration of PV in residential networks. A particle swarm optimization (PSO) algorithm is formulated in [32], to optimize the energy storage size for a battery-electric vehicle. The presented results demonstrate an optimal battery size could be obtained while satisfying system constraints. When the PSO is used to predict the SOC of BESS, it aids in the design of an energy management controller for electric-vehicle applications, as demonstrated in [33]. The proposed study in [34] demonstrates that there is a relationship between optimum number and capacity and the size of the batteries. When using a monotonic charging/discharging strategy to maximize the capacity of battery storage while also lowering power changes, the outcomes have been evaluated in this study. In [35], smoothing control of a hybrid generation system typically (wind and PV panels) has been optimized to determine the compensating power and the energy capacity of the storage system connected with an LPF. Linking to that, in [36], A PSO has been integrated with a fuzzy-deduced fitness evaluator technique for optimal battery charging patterns. The objective function is formulated to minimize the charging time and maximize the discharging capacity of the battery. The charging efficiency, charging time and the battery life cycle are improved by 0.4%, 56.8%, and 21%, respectively, using the proposed methodology compared to the conventional constant voltage constant current method.

Improving distributed generation and load hosting capability is investigated with optimal integration of BESS connected in both large-scale and low voltage networks according to cost-based multiobjective function [37]. In [38], an MPC is integrated with a PSO algorithm to optimize and control the battery's SOC of an electric vehicle (EV) and minimize the electricity cost. The study also examined the integration of EVs with photovoltaic panels to produce and sell electricity. A day-ahead dispatch of energy storage is proposed in [39], by formulating a two-stage fuzzy multiobjective problem that minimizes both battery charging/discharging rate and losses while improving the voltage profile and maximizing the peak shaving. The robustness of the proposed topology is investigated with an IEEE 37-node feeder as results are validated with a battery storage system located in Northwest USA. Genetic algorithms are employed also to search for the optimal parameter that estimates the health of a lead-acid battery [40].

17.2 Related work

17.2.1 Filter-based smoothing topologies

Studies in the literature suggest that utilizing ESS to enhance the performance of renewable energy produced by facilitating greater flexibility and a higher degree of controllability over power dispatching is an effective approach to enhance the performance of nondispatchable variable renewable generation systems. The application of the MA filter has been commonly used to smooth the intermittent power variability with the aid of BESS [21,41–45]. In Ref. [41], the MA algorithm uses a 1-hour time interval to average the intermittency of the fluctuated PV power, thus the inverter output was controlled by averaging the solar PV irradiance for the past 1 hour and that means the dispatched power will be delayed, whereas in Ref. [43], the designed algorithm reduces the variability of PV power. However, the study doesn't consider the impact of the algorithm on the charging/discharging rate of the BESS. More investigation has been carried out by authors in Ref. [44], and reducing the PV power fluctuations has been obtained by applying a MA algorithm to smooth a PV power using winter and summer power profiles. The falling irradiance into PV panels reaches 1200 W m^{-2} during summer whereas reaches 550 W m^{-2}

during winter. Consequently, the generated solar power reaches 400 and 150 W in summer and winter respectively. For such reason, the BESS system was subjected to 3.4 cycles in summer and only 1.5 cycles in winter. More investigation has been performed by authors in [45], to mitigate the dispatched power fluctuations, reducing the change in the battery level of charge and reducing the battery capacity by using a simple and half MA filter. An MA filtering has been integrated with a SOC feedback controller to mitigate the changes in the voltage regulator taps and decrease the solar power ramp rate [21]. However, there is a burden in the selection of the window size as it does not guarantee to reduce the power fluctuations within the acceptable ramp rate. Besides, the MA filter with a long window size needs a larger BESS capacity to compensate for the difference between the original solar output power and MA output power, even if there are no fluctuations. Another limitation of using MA is that the filtered power always leads ahead or lags the actual power. That creates more charging and discharging power into and from the BESS system because, MA has a memory effect, where the past values are used in the calculation of the new average data point and the past values may differ from the present value which affects the smoothing performance of the data. In [46], the authors have designed a smoothing controller based on a BESS system and natural gas engine generator where an optimization problem has been formulated to optimize the BESS capacity and the ramp rate reduction level. The Simulation results show that the application of the gas engine is not fast enough to mitigate the solar power variability; however, it helps to reduce the BESS's SOC variability from 25.1% to 12.6%. The reduction of the battery's SOC means it helps to reduce the size of the battery capacity.

Utilizing fuzzy logic control (FLC) with an energy storage system has been used in numerous applications to smooth the PV and wind power fluctuations while taking into account the targeted BESS constraints, SOC, and maintaining an adaptive smoothing [47]. However, the studies did not consider the fluctuation rate of renewable power in adjusting the degree of charging/discharging power. With the investigated approach the smoothing is affected by the level of the BESS energy for exclusively compensating the power fluctuations regardless of the power ramp rate. Whereas in [48], FLC has been used to adjust the battery target power based on the battery's SOC as the reference power is generated with an LPF filtering. The proposed smoothing mechanism suffers from the LPF limitation when a clear irradiance profile is present as the low pass filtering will generate a reference power even if there are no high-power fluctuations, the system schematic is shown in Fig. 17.2.

Kalman filter has been augmented with an FLC to provide an instantaneous intermittent power smoothing based on the battery's SOC level. The methodology helps to adjust the needed reference power for charging and discharging the battery energy storage system (BESS) [49,50]. The smoothing performance is also affected by the BESS's SOC regardless of the present power profile and its ramp rate. The structure of the Kalman filter along with the FLC is illustrated in Fig. 17.3. The fuzzy wavelet transform has been used to smooth the output power of the hybrid solar and wind energy systems using BESS considering a preset BESS size in

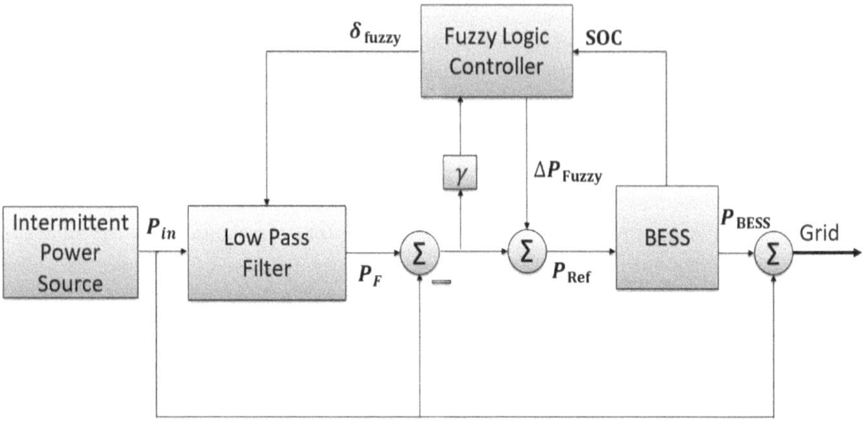

Figure 17.2 Fuzzy logic control and low pass filter to regulate battery level of charge.

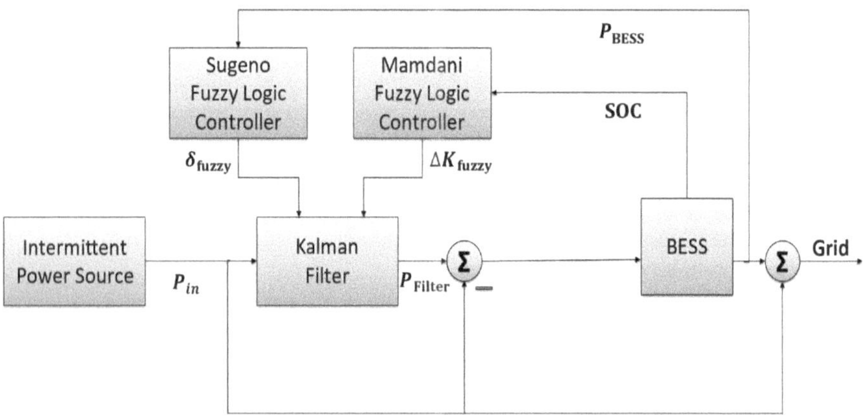

Figure 17.3 Kalman-fuzzy logic control-based intermittent power smoothing.

a subsequent application, which has been carried out in Ref. [51]. The suggested approach necessitates the installation of a substantial BESS capacity for smoothing the output power. The use of a fuzzy neural network technique has enabled the lowering of the battery capacity while optimally smoothing the variability of the RE generation. With a reduced power delay, it is possible to derive an estimate of the output power [52]. The superconducting ESS is utilized for smoothing the variations of the WPGS, and a fuzzy-based control approach is used to regulate the converter associated with the ESS [53], in addition to other applications. Many smoothing based on different energy storage systems is demonstrated in literature like using compressed air energy storage and pumped storage [54,55]. Nonetheless, the above methods require huge capital and have a slow dynamic response which makes them unsuitable for large-scale farms.

Flywheel energy storage systems (FESS) are widely integrated into wind farms for enhancing stability, improving power quality, and mitigating power intermittency [56]. FESS system has a fast response with high efficiency and large capacity. However, there are many limitations when it comes to rotor strength design and reducing the power consumption of the system. Linking to that, FESS has a remarkable self-discharge loss which reaches 20% of the total energy capacity per hour which makes it unsuitable for long-term energy storage [57]. The superconducting magnetic energy storage system (SMESS) has been used also to smooth the wind power variability it has good dynamic characteristics which means a rapid power exchange is maintained with the grid even in event of long-term disturbances [58]. Nevertheless, using it is enormously expensive because of the cooling requirement and the superconducting coil.

Furthermore, there are many technical limitations as it could not be used with WGPS that are already installed. Besides, the high energy of SMESS is required to be used with Current source converters. Application of supercapacitor in smoothing the power variability is presented in many studies as it has many technical capabilities and properties like the long lifetime, high power density, and the fast charging with a long discharge life which makes it a reliable energy storage system [59,60]. But the limitation of supercapacitors stands beyond the fact that the capacity of it could not be compressed as the using supercapacitor is more expensive than compares to BESS systems. Moreover, BESS system possesses many advantages and key properties such as the low investment and the low maintenance requirements, also BESS systems are easy to be installed and integrated with the REs technologies. BESSs have a rapid response rate and low self-discharge rate which makes them suitable for smoothing applications.

17.2.2 Ramp rate control

To alleviate the effect of the solar irradiance intermittency on the dispatched solar power, power ramp rate limits have been incorporated into the power grid code of several utilities around the world. The ramp rate limits are given in a second or a minute interval of the power profile as the maximum permissible limits are defined as the generated solar power to the plant size or defined as a set of the generated power. For instance, the Ireland grid code defines the maximum RR as a positive incremental of the power till 30 MW min^{-1} whereas Hawaii set it as $\pm$ 2 MW min^{-1}. For Germany and Puerto Rico, the maximum RR is given in terms of seconds (100 kW sec^{-1}) [61,62]. Many methods are given to calculate the ramp rate. The first method defines RR as the difference between to endpoints of a one-minute interval. The second method is calculating the difference between the maximum and minimum values of the power for a certain interval. The third method is calculating the difference between the two endpoints of one second. PV plants must reserve additional power to limit the RR to the allowable limit and this is achieved by operating the plant below its rated capacity. In a real application, this is achieved by disconnecting some arrays of the PV panels. Another approach is using the BESS to inject and absorb power to compensate for the solar power variation which allows the plant to be operated at its full capacity without disconnecting a plant. Many methods are presented in the literature to calculate the power fluctuation rate from the PV panels [63,64]. The commonly used methods are the difference between the end values of 60 seconds or the difference between the maximum and minimum values of a certain period. The ramp rate calculation depends on the grid code as it could be represented as a percentage of the full capacity or a per-minute value. The calculation of power RR could be expressed with Eq. (17.1).

$$\mathrm{RR}(t) = \frac{P_{\mathrm{PV}}(t) - \mathrm{P}_{\mathrm{PV}}(t-60)}{t(t) - t(t-60)} \tag{17.1}$$

where $\mathrm{RR}(t) < \mathrm{RR}_{\mathrm{limit}}$, $P_{\mathrm{PV}}(t)$ and $t(t)$ are the simultaneous solar power and time. The index $(t-60)$ represent the power and time values before 60 seconds. The variation rate could also be calculated by considering the maximum power difference in a one-minute interval which is expressed with Eq. (17.2) [65].

$$\in = \frac{\Delta P_{\mathrm{PV,max}}(1\ m)}{P_{\mathrm{PV,max}}} \tag{17.2}$$

where $\in$ defined as the variation rate of solar power and $\Delta P_{\mathrm{PV,max}}$ is the maximum power change in the moment of $(t-60)$ whereas $P_{\mathrm{PV,max}}$ is the full PV capacity. Using ramp rate control has been used in many studies to limit the change of the time derivative of the dispatched power [66–69]. The proposed smoothing mechanism doesn't suffer from the memory effect as there is no delay between the original intermittent power and the power smoothing filter. Ramp rate control can reduce losses and battery cycling compared to MA and LPF. However, ramp rate control has been proved that requires more battery capacity and it does not return to its original state of charge [70].

17.3 Motivation behind using fuzzy logic control combined varying low pass filter for solar power smoothing

Due to the weather-dependent nature of the RE system, the generated power is variable, unpredictable, and discontinuous, resulting in voltage and frequency fluctuations. Voltage fluctuations adversely affect the grid stability, performance, reliability, and power quality. And the impact on the electrical grid increases with the higher penetration of the RS system. Also, when solar energy production exceeds demand, reversal power can occur, increasing the line voltage and affecting the voltage regulation operation. Battery systems present a possible solution to attenuate the stochastic effect of RE systems. Literature studies present a variety of applications for storage technology, including power quality enhancement, ramp rate reduction of dispatchable REs, and reliability improvement. In this study, we

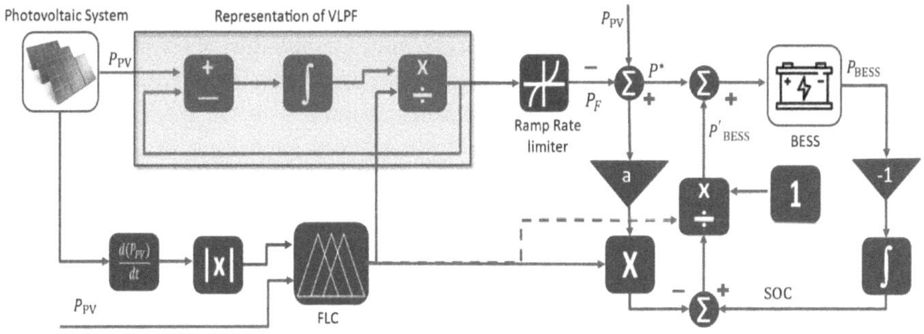

Figure 17.4 Schematic diagram of smoothing photovoltaic power using fuzzy logic controller augmented with varying low pass filter (VLPF).

propose using an innovative smoothing framework that employs FLC and low pass filtering (LPF) to provide a higher degree of smoothing in the case of excessive moving clouds (i.e., highly intermittent PV power) but less smoothing in the case of a sunny day with ramp events. The motivation for combining the connected FLC and LPF is to offer a degree of smoothing proportional to power fluctuations, which increases the degree of smoothing on cloudy days and reduces battery operation on sunny days.

We provide a reconstructed varying low pass filter (VLPF) representation from the standard LPF, as well as the fuzzification and defuzzification processes that regulate rising the time constant value with the current degree of power variations (Fig. 17.4). The superiority of the proposed controller may be described as follows: (1) Reducing the burden on the BESS system while lowering the power variability of a solar PV system by offering a variable degree of smoothing proportionate to the power level and power ramp rate. (2) The controller eliminates the power lag that is often caused by conventional smoothing approaches such as low pass and moving average filters. (3) The controller has a rate limiter block that may be modified to further limit the power ramp rate.

The GECAD Photovoltaic system database was used to import real solar irradiation data for clear, normal, and overcast situations to model the output PV power of a 200-kW system during various months in 2013. A 7-kWh BESS is integrated for smoothing the variability associated with a 200-kW solar panel. The simulated results demonstrate the efficiency of the coupled (FLC + VLPF) toward smoothing the variability of solar output power with variable smoothing time constant based on the measured power fluctuations and compare the smoothing performance, the ramp rate reduction, and the battery's SOC reduction to the normal LPF smoothing.

17.4 Proposed methodology

In response to growing environmental worries related to global warming, the percentage of renewable energy sources as ecologically acceptable resources has increased in recent years. Particularly advantageous is the use of photovoltaic panels since the production prices are decreasing with time. Despite this, PV systems have an intermittent power profile given the fact that they are weather reliant, which causes several difficulties with the grid and has a detrimental impact on the power system. An increase in the penetration of unregulated power sources may result in instabilities in the power system, poor power quality, voltage drops, and frequency oscillations. Smoothing controllers that are based on energy storage may be used to efficiently tackle the difficulties. It is shown in the literature that the use of LPF can be incorporated into BESS to mitigate fluctuations in solar power generation. This turned out to be a simplified approach to real-world applications. Although, the presentation of the filter time constant settings often results in high smoothing performance, setting a higher smoothing time constant on a sunny day can consume more battery and increase SOC variation as the battery will force to charge and discharge power that equals the difference between the original PV power and the filtered power (i.e., the filter output) which reduces the life expectancy of the BESS system.

When the ramp rate of the solar power system exceeds the permitted limit of 10 (percent/min), a smoothing strategy is necessary; otherwise, the power is sent straight into the grid without the need for a smoothing strategy. The system consists of a solar PV panel that is integrated into the system via a boost converter. Similarly, the BESS is incorporated into the system using a DC–DC converter. Average models of the DC–DC and DC–AC converters are utilized to emulate the complete day in terms of hours to reduce the computational complexity and memory size as the converter dynamics are operated in milliseconds. To test the smoothing ability of the controller, we generated an average model combining the PV system, variable low-pass filter, charge controller, SOC feedback controller, and FLC controller as shown in Fig. 17.5. The model also includes a SOC controller linked to BESS as well as the smoothing algorithm. The BESS operation during the ramp-up and ramp-down intervals is dictated based on the difference between the actual PV power and the filtering topology. The power generated because of actual PV and battery production indicates the resultant power required by the grid to maintain a smooth output PV profile. One of the objectives of this research is to smooth the fluctuating PV power while minimizing the

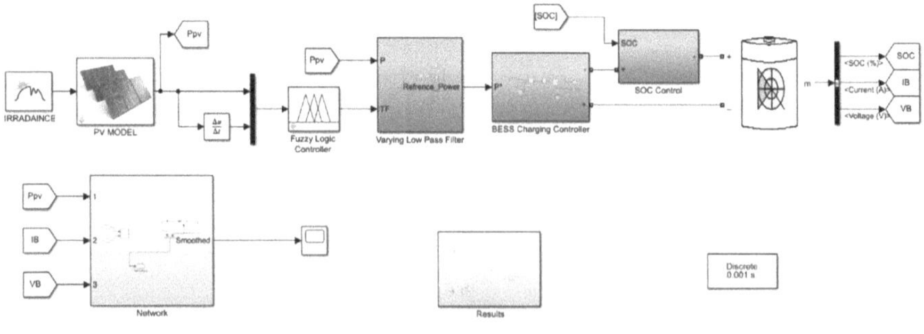

Figure 17.5 Simulink model of the smoothing of photovoltaic variability using battery energy storage system coupled with a charging controller in MATLAB®.

BESS operation through FLC coupled with a VLPF controller that determines the smoothing time constant which helps to smooth out PV fluctuations while keeping the power ramp rate below the maximum allowable limits and hence satisfying the grid code requirement. The simulation of the proposed controller is implemented using MATLAB/Simulink, and the design of the fuzzy logic controller is designed using the Fuzzy Logic Toolbox.

The proposed approach outline the formulation of the FLC utilized for regulating the time constants of the VLPF as well as the operation of the FLC. The controller inputs are the produced PV power and its change with time (i.e., power ramps). The two inputs will then be processed by the fuzzification membership functions to determine the controller output, which is the variable smoothing time constant. The design also demonstrates how to regulate the charging of the battery using a state of charge controller, which aids in avoiding overcharging or depletion of the battery storage system. In accordance with the solar power ramp rate, the smoothing process as shown in Fig. 17.5, indicates how the suggested methodology controls the smoothing of solar power variability by applying the proposed FLC control approach. Changing the smoothing time constant can cause spikes, especially when switching from high to low values, and vice versa, resulting in high-frequency components with battery output power. Therefore, the model is equipped with a ramp rate limiter to eliminate the effects of high frequencies and allows for a soft transition from one time constant to another. The generated reference power is sent to the BESS system to provide charging and discharging power profiles.

17.4.1 Varying low pass filter

Smoothing out the variations that develop in the solar PV generation system by adjusting the BESS power highlights the core function of the proposed controller. The LPF inhibits higher frequency components of solar power from being transmitted while allowing the low-frequency components to flow through it. The transfer function of the low pass filter (TF) could be expressed with Eq. (17.3). The smoothing time constant (T_c), dictates the smoothing level. A low value of the smoothing time constant reduces higher degrees of PV power fluctuations.

$$\mathrm{TF} = \frac{1}{T_c s + 1} \tag{17.3}$$

The LPF has been modified to accommodate a wide range of values for the T_c. By multiplying the numerator and denominator by the constant ($T_c s$). The transfer function is expressed as follows:

$$\mathrm{TF} = \frac{\frac{1}{T_c s}}{1 + \frac{1}{T_c s}} \tag{17.4}$$

The reciprocal of (T_c) is denoted by $G(S)$, which represents an integrator. Accordingly, considering a unity feedback $H(s)$ the transfer function could be expressed as Eq. (17.5):

$$\mathrm{TF} = \frac{\dfrac{1}{T_c s}}{1 + \dfrac{1}{T_c s}} = \frac{G(s)}{G(S).H(s) + 1} \tag{17.5}$$

The filtered output power or the so-called target output is denoted by P_f and expressed in Eq. (17.6). The output of the filter is fed into a rate limiter to soften the transition from a time constant to another time constant value. Based on the FLC controller the value of the filtering time constant is formulated.

$$P_f(s) = \frac{1}{T_c s + 1} * P_{\mathrm{PV}}(s) \tag{17.6}$$

The BESS charging and discharging power will be determined by subtracting the target output or the filtered output from the original PV power. The reference power is given with P^* and expressed in Eq. (17.7)

$$P^*(s) = \frac{-sT_c}{T_c s + 1} * P_{\mathrm{PV}}(\mathrm{s}) \tag{17.7}$$

The energy used by a battery system relevant to its nominal capacity is called the battery state of charge as shown in Eq. (17.8). The consumed battery energy is calculated by integrating the output power of the BESS system P_BESS. during ramp-up periods the battery will be in charging mode and the power will have a negative value, while during the ramp-down interval the power will discharge the stored energy and the power will have a positive value.

$$\mathrm{SOC}(s) = \frac{-P_{\mathrm{BESS}}(s)}{s\,E_{\mathrm{BESS}}} \tag{17.8}$$

In this study, the battery efficiency has been assumed to be 100% and the BESS output power is equivalent to the reference power that has been determined by the smoothing filter. The consumed battery energy (E_{BESS}) could be calculated as in Eq. (17.9).

$$\text{Battery consumed energy} = \frac{-sT_c}{T_c s + 1} * P_{\mathrm{PV}}(\mathrm{s}) * \frac{1}{s} = P_f(s) * T_c \tag{17.9}$$

The charging/discharging power may be equivalent to, less than, or greater than the BESS capacity. if the battery capacity is less than the charging capacity, implies that the BESS system will be overcharged, thus in this controller, the battery capacity was assumed to be greater than the charging power from the solar PV system as shown in Eq. (17.10). Therefore, the product of the smoothing time constant and the nominal solar PV power s less than the BESS capacity. The rated or the nominal solar PV power is given by $\overline{P}_{\mathrm{PV}}$.

$$T_c * \overline{P}_{\mathrm{PV}} < E_{\mathrm{BESS}} \tag{17.10}$$

$$\mathrm{SOC}\ (\%) = \frac{T_c * \overline{P}_{\mathrm{PV}}}{E_{\mathrm{BESS}}} \quad \text{where } (\mathrm{SOC}\ (\%) < 100\%) \tag{17.11}$$

The feedback signal of the battery charge level is used to prevent the battery from being overcharged or depleted. To balance between decreasing the BESS operation and smoothing the solar PV power (i.e., tracking the initial value of the BESS's SOC), a SOC gain is introduced in Eq. (17.12) wherein the initial charge level is taken at 50% of the BESS capacity. The difference between the initial and the measured BESS's SOC represents the tracking error. The SOC tracking signal is calculated by multiplying the proportional gain (a) with the SOC of BESS. The higher and lower SOC boundaries Eq. (17.12), are implemented to deduct the marginal capacity from the utilized capacity of BESS, where $\mathrm{SOC_{mh}}$ and $\mathrm{SOC_{mL}}$ denote the upper and lower limits, respectively.

$$(\mathrm{SOC_{mh}} + \mathrm{SOC_{mL}}) * E_{\mathrm{BESS}} = E_{\mathrm{BESS}} - aT_c * \overline{P}_{\mathrm{PV}}(s) \tag{17.12}$$

The system comprises feedback control associated with SOC in order to govern the BESS power. The discharged BESS power is equivalent to the sum of the storage capacity margin and the smoothed PV power.

$$P'_{\mathrm{BESS}}(s) * T_c = \left(\mathrm{SOC}\,(s) * E_{\mathrm{BESS}} - \frac{aT_c}{T_f S + 1} * P_{\mathrm{PV}}(\mathrm{s})\right) - \left(E_{\mathrm{BESS}} - aT_c * \overline{P}_{\mathrm{PV}}(s)\right) \tag{17.13}$$

In this situation, the coefficient a restricts the BESS capacity, implying that a degree of smoothing is sacrificed to limit the charging/discharging capacity below the BESS capacity.

$$\begin{aligned}(\mathrm{SOC_{mh}} + \mathrm{SOC_{mL}}) * E_{\mathrm{BESS}} = \mathrm{SOC}(s) * E_{\mathrm{BESS}} - (aT_C s + 1) * P_{\mathrm{PV}}(s) \\ - E_{\mathrm{BESS}} - aT_C * \overline{P}_{\mathrm{PV}}(s)\end{aligned} \tag{17.14}$$

The sent solar PV power to the grid (P_G) equals the total of the solar PV power plus the BESS charging/discharging power. The dispatch power is calculated using Eq. (17.13).

$$P_G(s) = P_{\mathrm{BESS}}(s) + P_{\mathrm{PV}}(s) \tag{17.15}$$

The needed BESS requirement to mitigate the power ramp rate under the existence of large moving clouds is calculated as the following approach. The generated output power can drop or rise dramatically in proportion to the subjected fluctuated irradiance. The output PV power from the rated value P_{PV} can be expressed with Eq. (17.16) as the P_{shaded} represent the diffuse output from irradiance.

$$P_{PV}(t) = \max\left(P_{PV}(1 - RR.t), P_{cap}.P_{shaded}\right) \tag{17.16}$$

If the ramp rate of the dispatched PV power with the help of battery output power is limited, the dispatched power will be given in Eq. (17.17).

$$P_{PV+BESS}(t) = \max\left(P_{cap}(1 - RR_{limit}.t), P_{cap}.P_{shaded}\right) \tag{17.17}$$

The needed BESS power can be obtained as shown in Eq. (17.18).

$$P_{BESS}(t) = P_{PV+BESS}(t) - P_{PV}(t) \tag{17.18}$$

The maximum power and the delivered energy for battery sizing can be calculated with Eq. (17.19) and Eq. (17.20).

$$P_{Peak}[MW] = P_{cap}(1 - P_{shaded})\left(1 - \frac{RR_{limit}}{RR}\right) \tag{17.19}$$

$$E_{BESS}[MWh] = P_{cap}\left(\frac{1 - P_{shaded}}{2RR_{limit}}\right)\left(\frac{1\,hr}{60\,min}\right) \tag{17.20}$$

For the study a BESS with 7 kWh and 102 kW peak has been used for a PV plant with a 200-kW capacity. The P_{shaded} has been assumed to be 0.15 p.u. and the $RR_{limit} = 0.1$ p.u. min^{-1} while RR is 0.25 p.u. min^{-1}.

17.4.2 Fuzzy logic controller design

Fuzzy logic has been adopted as an effective tool in solving a complex system with nonlinear behavior, which differs from the crisp logic for having a wide range of values from zero to one instead of having only two values only (zero or one). The proposed fuzzy logic controller is shown in the following phases.

17.4.2.1 Fuzzification

The fuzzification process is the initial stage in creating an FLC, and it involves converting all measured solar power (P_{PV}) and solar power ramp rate

$\frac{d(P_{PV})}{dt}$ to fuzzy premises. Membership functions (MFs) are adjusted to reflect the specific data values using triangle membership functions, as shown in Fig. 17.6 and Fig. 17.7. FLC has two inputs and one output. The inputs include the ramp rate of the solar PV generation system and solar power. The filtering

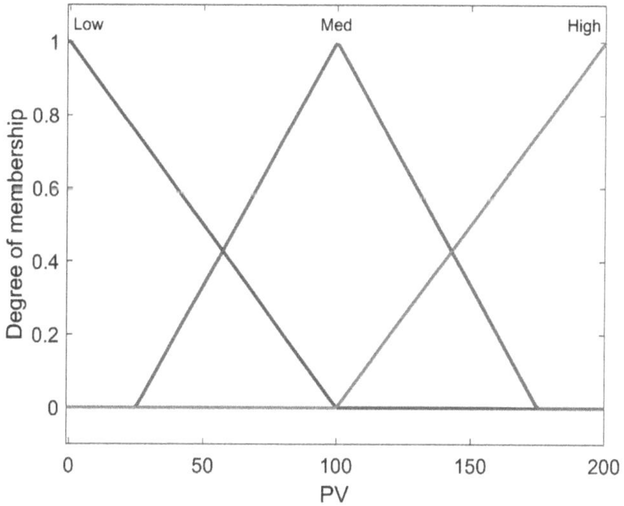

Figure 17.6 Membership function for solar photovoltaic power.

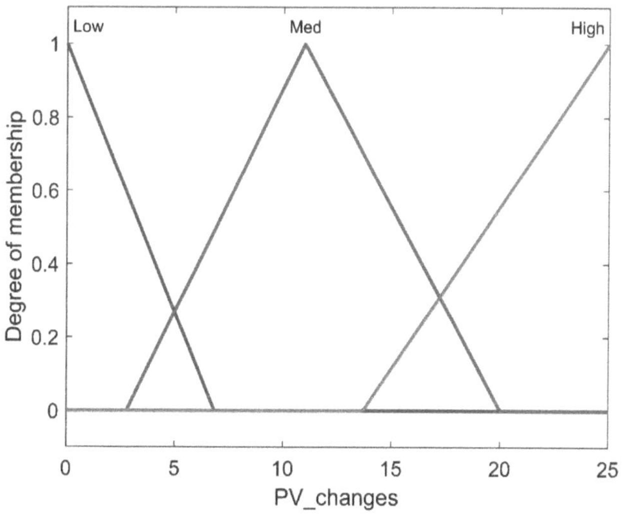

Figure 17.7 Membership function for solar power ramp rate.

time constant is the controller output. In the fuzzification phase of solar power low, medium, and high are the three fuzzy sets with values at 25, 100, and 200 respectively. Accordingly, solar ramp rate utilizes six sets, namely negative low, negative medium, negative high, positive low, positive medium, and positive high with values at −25, −15, −8, 8, 15, and 25 respectively. The level of membership is determined by Eq. (17.21).

$$\mu_{\mathrm{Ai}}(x) = \frac{1}{w}(w - 2|x - h|) \tag{17.21}$$

Where the value of the input variable is represented by x, w denotes the width, and the MF's height is denoted by h, which equals one.

17.4.2.2 Fuzzy rules

Fuzzy rules include the collection of control algorithms that are utilized for executing a closed-loop control of a system. The fuzzy rules are written in the form of IF-THEN syntax as described in Eq. (17.22). Table 17.1 summarizes the fuzzy rules.

$$\text{if}(x_1 \text{ is } A_i) \text{ and } (x_2 \text{ is } B_i) \text{ then } Z = C_i \; i = 1, 2, \ldots, r \tag{17.22}$$

x_1 and x_2 represent the fuzzy inputs are whereas the FLC output is denoted by z. Where A_i and B_i denoted the fuzzy sets in the universe of discourses x_1 and x_2, z respectively. Based on the following criteria, ten consequents are computed using six premises of the ramp rate of the solar PV and three premises of the solar PV power. The resultant surface rule is depicted in Fig. 17.8.

Table 17.1 The developed fuzzy rules.

Condition no.	If P_{PV}	and $\frac{d(P_{PV})}{dt}$	Then Tc
1	High	High	High
2	High	Low	Low
3	High	Med	Med
4	Low	Low	Low
5	Low	Med	Med
6	Low	High	High
7	Med	Med	Med
8	Med	High	High
9	Med	Low	Low

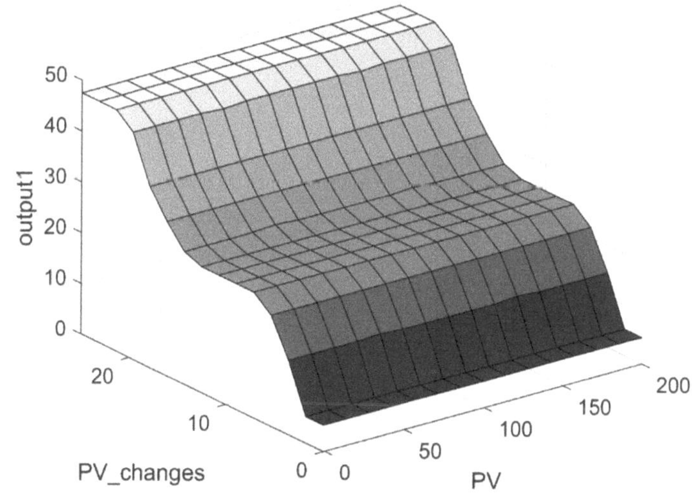

Figure 17.8 Surface rule of the developed T-S type fuzzy system.

17.4.2.3 Fuzzy inference system

The degree of conformity (W_i) is developed using Eq. (17.23) through the Mamdani FLC method.

$$W_i = \mu_{\mathrm{Ai}}(x_1) * \mu_{\mathrm{Bi}}(x_2) \tag{17.23}$$

The MF grade values denoted by with the $\mu_{\mathrm{Ai}}(x_1)$ and $\mu_{\mathrm{Bi}}(x_2)$. The lowest norm is represented by the logic operator AND, whereas the maximal norm is represented by the logic operators OR, which is represented by Eq. (17.24) and Eq. (17.25).

$$\mu_A(x)\mathrm{AND}\mu_B(x) = \mathrm{MIN}\{\mu_A(x), \mu_B(x)\} \tag{17.24}$$

$$\mu_A(x)\mathrm{OR}\mu_B(x) = \mathrm{MAX}\{\mu_A(x), \mu_B(x)\} \tag{17.25}$$

17.4.2.4 Defuzzification

Single numerical values (i.e., crisp outputs) are formulated during the defuzzification of the fuzzy rule sets that represent the control signal which will govern the LPF filtering performance in accordance with the level of PV power fluctuations. The centroid technique is used to create precise control outputs by determining the center of the region as indicated in Eq. (17.26). Fig. 17.9 depicts output membership.

$$Z = \frac{\sum W_i * C_i}{\sum W_i} \tag{17.26}$$

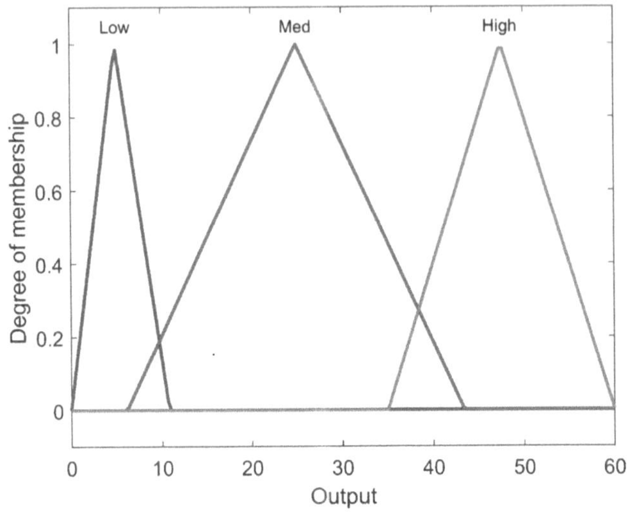

Figure 17.9 Membership function of the proposed fuzzy logic control.

17.5 Simulation and discussion

Three alternative irradiance and cell temperature data are employed to show the viability of the proposed controller, generally, sunny, normal, and cloudy irradiance curves. The data sets are imported from the GECAD Photovoltaic system PV and recorded over 24 hours with 5 minutes resolution. The solar irradiance and temperature are input to the PV system to model the generated PV power. A 7 kAh BESS is been integrated with the solar PV to smooth the output power fluctuation and is evaluated for a 24-hour time interval to present the effectiveness of controller tracking during the first and last hours of the day.

17.5.1 Limitations of using low pass filter and moving average smoothing

The provided results in this section demonstrate the limitations of LPF and MA filters in smoothing the intermittent power profile in the case of a normal day. The smoothed PV power with the help of LPF is shown in Fig. 17.10, the simulated smoothing response highlights the effect of increasing the filtering time constant against the smoothing performance. Three different time constant values have been used typically: 30, 20, and 10 minutes. A smoothed dispatched power is delivered to the utility with the help of BESS by setting a larger time constant of the filter. However, with increasing the time constant the lag of the dispatched power increases, and the difference between the

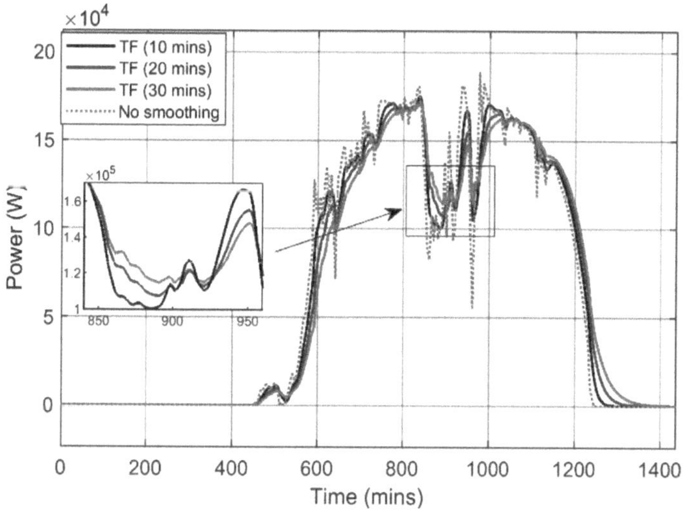

Figure 17.10 Smoothed lagging power using different time constants with a low passing filter.

lagged power and the PV power will increase also which represents the reference power that is responsible to charge and discharge the battery. The required battery power to compensate for the power variability is shown in Fig. 17.11. The produced reference power using 30 minutes results in more charging and discharging power compared to using 20 and minutes. The increment in time constant reflects a proportional increase in the BESS power that projects an increment in the level of charge of BESS which means increased participation and consumption of BESS energy toward the smoothing process. The level of charge is higher in the case of using 30 minutes.

However, it is worth noting that the battery's SOC returns to its initial value (i.e., SOC = 50) at the end of the day which means all the charged energy stated by the controller is utilized for smoothing the power as reflected in Fig. 17.12. Thus, no extra power is charged from the battery nor stored in the battery. Furthermore, smoothing with the help of the MA filter is analogous to smoothing using LPF, the smoothing performance is shown in Fig. 17.13. Also, with increasing the window size the smoothing of the MA is increased. Unlike LPF, smoothing using MA makes the smoothed power precedes the original power and then lag the original power. Preceding and lagging the power at the same time will be reflected in having a more fluctuated SOC of BESS as shown in Fig. 17.14, which indicates that BESS will have a more fluctuating charging/discharging power curve which impacts the performance of BESS. Moving average (MA) like LPF, the battery's SOC returns to its original value at the end of the day, but the battery experience more charging and discharging cycles, and because of the nature of the filter, MA provides the smoothing based on averaging the historical power data.

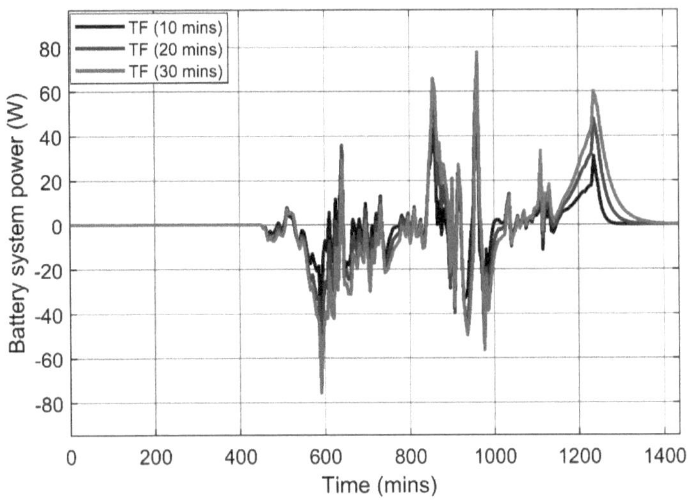

Figure 17.11 Impact of increasing the filtering time constant on the battery energy storage system power.

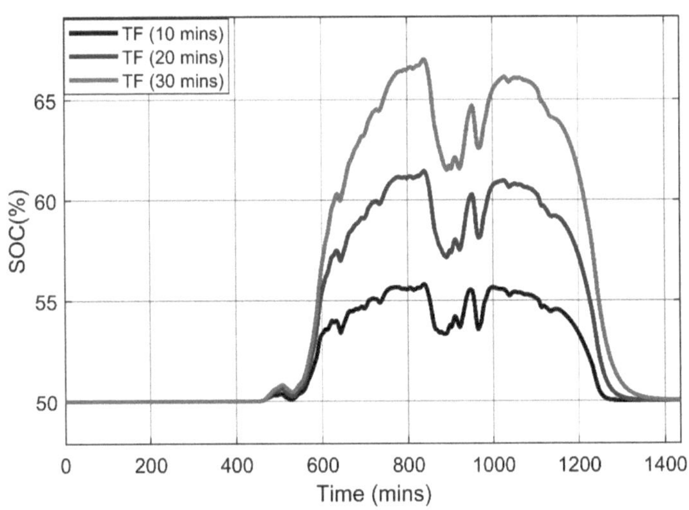

Figure 17.12 Impact of increasing the filtering time constant on the battery level of charge.

17.5.2 Evaluation of the proposed smoothing controller against the conventional low pass filter smoothing considering normal day

The presented results highlight the capability of using the proposed (FLC + VLP) smoothing controller while comparing its performance against the conventional

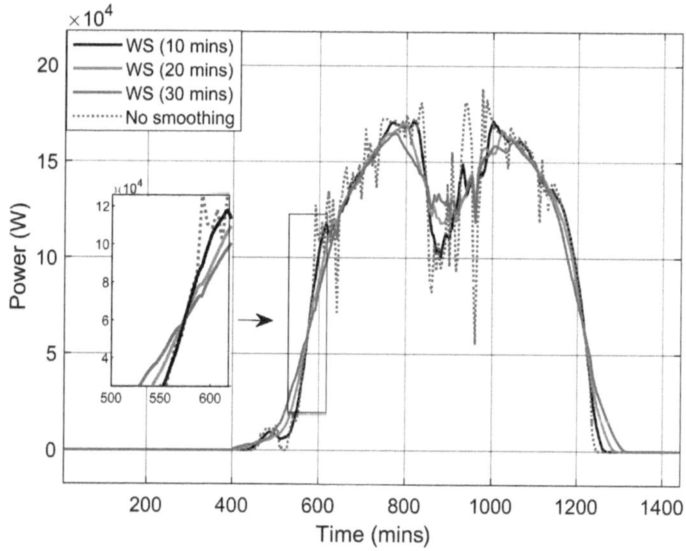

Figure 17.13 Leading and lagging smoothed power using moving average.

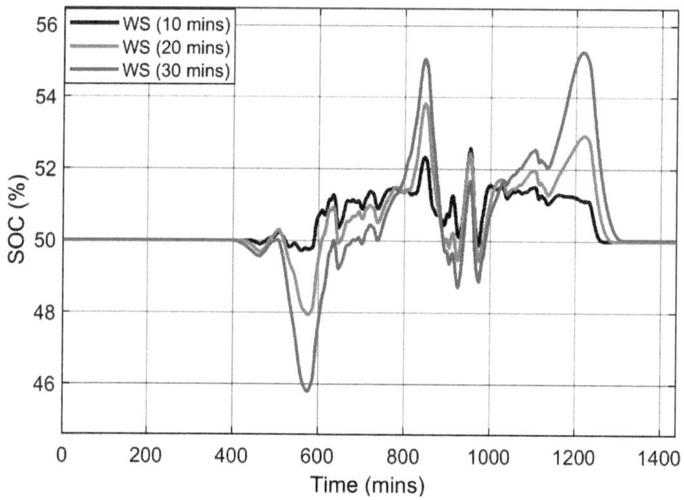

Figure 17.14 The battery level of charge fluctuation stems from moving average.

LPF control strategy considering a moderate irradiance profile. The simulated result in Fig. 17.15 demonstrates that the generated power profile is clear most of the day, however, the PV panels experience a cloudy event from 260 minutes to 480 minutes

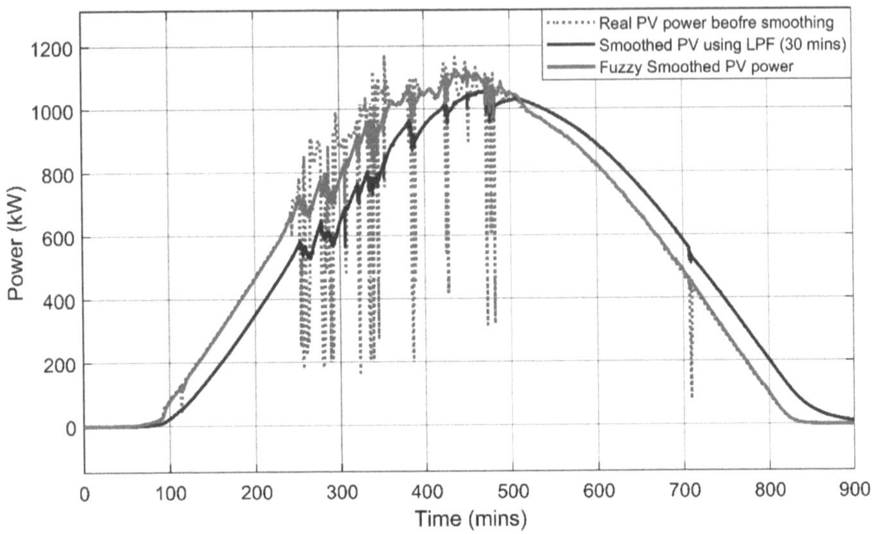

Figure 17.15 Comparison between the conventional low pass filter and proposed fuzzy logic control approach toward the smoothing performance of solar photovoltaic power considering moving clouds on a normal day.

which means the solar power violates the permissible limits during this interval. The (FLC + VLP) framework reduces the PV power variability and is perfectly aligned with the original PV power profile and has no power delay at the beginning and the end of the day. In contrast, the conventional LPF results in a good and smoothed power profile as well.

However, the smoothed power profile lags behind the original power profile. Also, it is worth noting that both techniques result in similar smoothing performance because (FLC + VLP) controller is possessed the LPF characteristic during the cloud events. Apart from that, the smoothed power profile with the proposed smoothing topology has a higher peak than in the case of LPF which helps to maintain the characteristics of the original power profile (i.e., peak) that is usually attenuated because of the smoothing. The effect of power smoothing on the battery performance is shown in Fig. 17.16 the level of charge of (FLC + VLP) is different than the conventional LPF, which discharges the battery from 50% to less than 48%.

The resultant SOC from using LPF is increased to 54% (from 50%) indicating higher BESS participation in the smoothing performance. It is worth noting that, the battery's SOC starts from a reference value equal to 50% of the battery capacity and does not return to its reference value at the end of the day due to the used rate limiter. The existence of the rate limiter helps the smoothed power to be aligned with the original power profile at the end of the day, however, prevents the battery from returning to the initial level of charge. The battery's charging/discharging power of the fuzzy approach is compared to the low pass filter. In the case of LPF

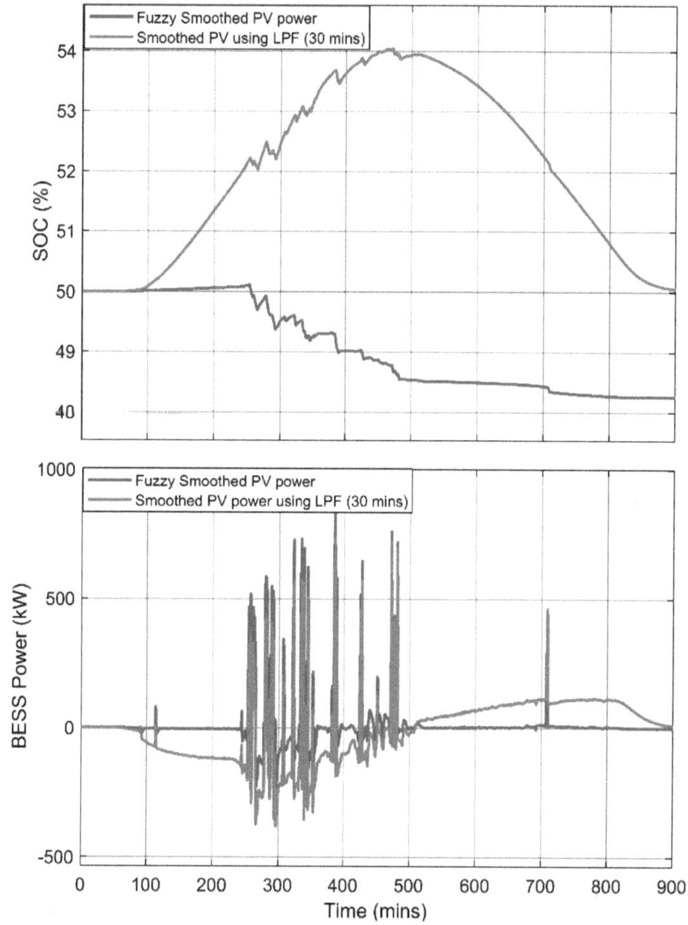

Figure 17.16 Comparison of battery energy storage system performance between the proposed controller and the conventional low pass filter control approach during the smoothing process of the solar photovoltaic power.

strategy, using 30 minutes causes the BESS to start charging from 100 minutes to 300 minutes, after which it will begin to charge and discharge repetitively to adjust for solar variation. Finally, the power was discharged from 500 minutes to 900 minutes, indicating that the battery was operational throughout the interval. In comparison, when employing the fuzzy logic controller, the battery is generally in idle mode for the first 240 minutes before beginning to drain to compensate for solar variations during the next 220 minutes. Furthermore, in the case of the FLC technique, the charging/discharging power is lower, implying that less amount of energy is required by the BESS. The comparison between both strategies is depicted in Fig. 17.17 in terms of ramp rate with the smoothing operation and reduction of the ramp rate. The ramp rate reduction in the case of the proposed FLC-based

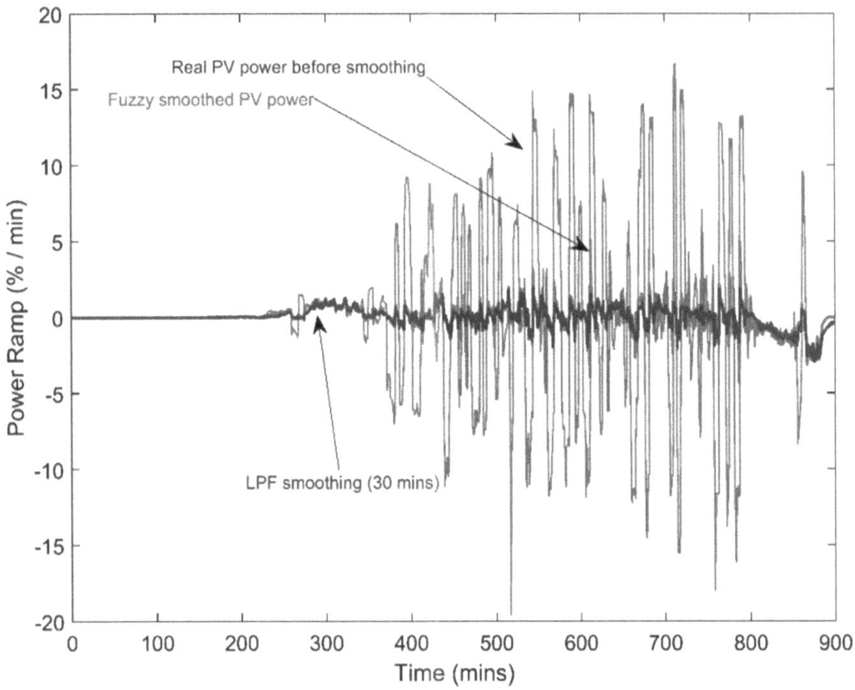

Figure 17.17 Ramp rate reduction achieved with proposed fuzzy logic control and conventional low pass filter smoothing approaches.

controller is similar to the conventional LPF controller. However, the standard LPF is slightly better than the proposed controller.

17.5.3 Evaluation of the proposed smoothing controller against the conventional low pass filter smoothing considering intermittent irradiance day

An additional test is carried out to assess the efficacy of FLC-based controller in the presence of dominant clouds. Comparatively, smoothing the variable output power of solar PVs proves to be more challenging on a cloudy day than on a regular day which may also create excessive pressure on BESS. Power fluctuations last for 450 minutes on a cloudy day and 250 minutes on a sunny day. Fig. 17.18 compares the efficiency of both strategies in minimizing PV variability to the original solar output. Even though both approaches result in the same decrease in PV

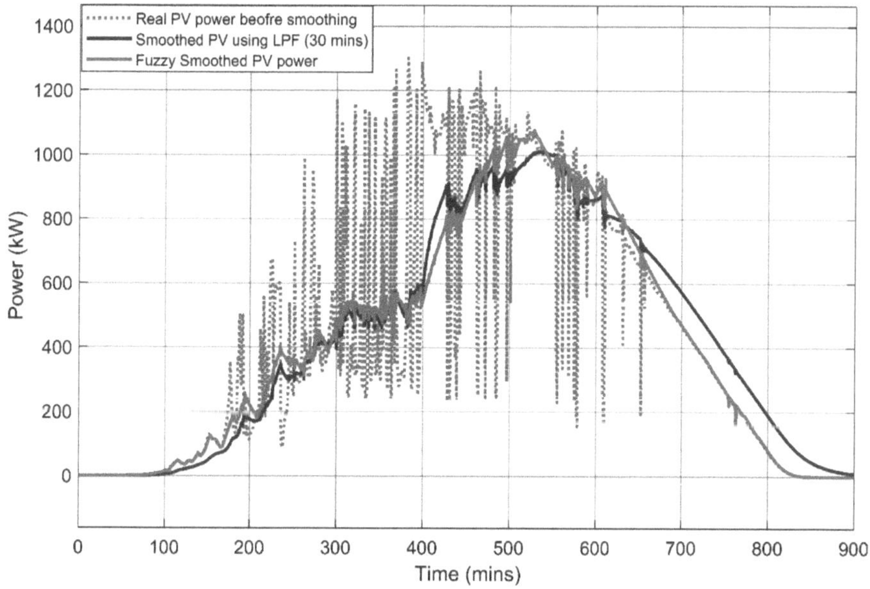

Figure 17.18 A Smoothing comparison between the fuzzy logic control approach to conventional low pass filter smoothing on a cloud day.

fluctuations, the proposed FLC facilitates a higher degree of solar output power smoothing with less delay from the original solar PV power. The dispatched power with the proposed framework helps to have a smoothed power profile with a higher peak than LPF. The effect of smoothing on the BESS performance is illustrated in Fig. 17.19. The BESS charging/discharging profile in the case of the proposed FLC control is similar to the conventional LPF, indicating that the proposed FLC-based smoothing acts like conventional LPF during cloudy days. Nonetheless, BESS changes from 50% to <53.5% in the case of the FLC controller and from 50% to <54% with the conventional LPF which means less BESS operation is realized with the FLC controller.

The presented battery power demonstrates that an FLC absorbs/injects equivalent power to that in LPF during cloud events and provides no power when there are no fluctuations. The BESS is idle from 650 minutes to 900 minutes in the case of the FLC controller whereas the battery is discharging in the case of LPF. The battery participates in the smoothing process and does not return to its reference profile at the end of the day. Compared to the smoothing in the case of a normal day, the battery level of charge is higher than the reference value (50%) and less than the reference value and that happens because of using the rate limiter. The ramp rate reduction of both methods is shown in Fig. 17.20.

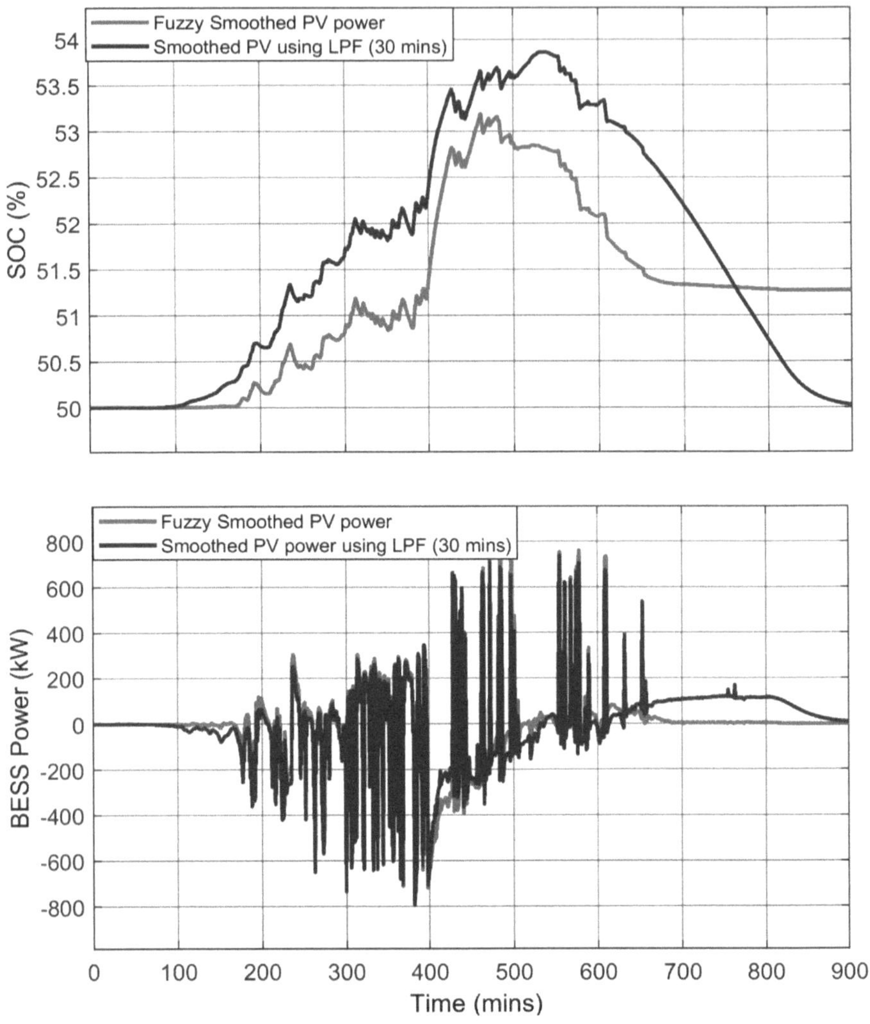

Figure 17.19 Comparing the battery performance of the proposed fuzzy logic approach to the low pass filter smoothing on a cloud day.

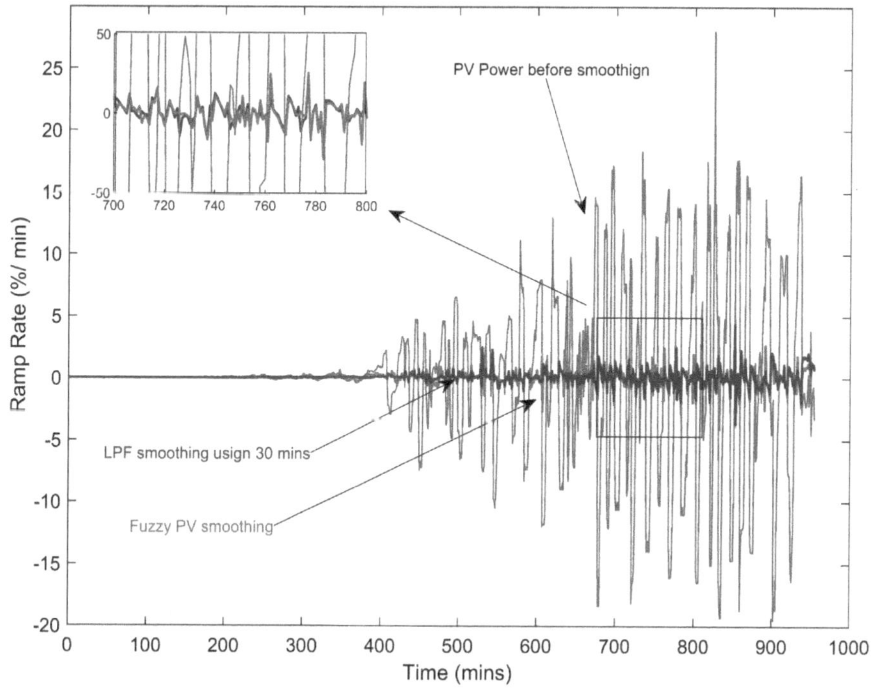

Figure 17.20 Ramp rate reduction using the proposed fuzzy logic control and using low pass filter smoothing approaches.

17.6 Conclusion

The stochastic and intermittent behavior of solar and wind farms imposes many negative effects typically power variations that yield voltage fluctuations. The cloud dependency of solar PV panels introduces a variation of the produced power in proportion to the fallen irradiance into PV panels. When clouds pass over the panels the power drops and when it vanishes the power return to its peak value which creates a fluctuated power profile. Battery storage systems are coupled with smoothing filters to facilitate a firm and smooth the intermittent output solar PV power profile. The used filter will generate the charging and discharging reference signal to the battery system. Conventional LPF smoothing has a preset filter time constant so that the battery system supplies and absorbs power even in the absence of clouds. This implies that the battery system will work longer. The proposed smoothing approach contains a fuzzy logic controller, varying representation of the low pass filter, and rate limiter. To assess the capabilities of the suggested technique, real PV power data were employed. The solar PV power profile as well as its absolute change rate served as the inputs to the fuzzy logic controller that formulated the value of the filtering time constant. The smoothing time constant is varied by the fuzzy logic controller in accordance with the ramp rate of the PV power. The suggested smoothing approach assists the battery in injecting/absorbing power in the presence of moving clouds and extends the battery's idle phase when there are no power fluctuations. The conventional LPF approach utilizes a battery even during the absence of solar power fluctuation resulting in smoothing the power fluctuation on cloudy days as well as during clear days when the ramp rate is low and smoothing operation is not required. The developed fuzzy logic controller with variable low pass filter has a simple approach and is straightforward to enable implementation in a real-world application. The simulated results are provided for different power profiles, normal and cloudy days with the results validating the superiority of the proposed FLC against the smoothing operation of conventional LPF in terms of obviation of delay which means less battery usage. Also, the proposed framework, shows a higher peak could be realized with the proposed FLC control approach.

References

[1] F. Durante, A. Gianfreda, F. Ravazzolo, L. Rossini, A multivariate dependence analysis for electricity prices, demand and renewable energy sources, Inf. Sci. (Ny). 590 (2022) 74–89.

[2] W. Hanif, J. Arreola Hernandez, W. Mensi, S.H. Kang, G.S. Uddin, S.M. Yoon, Nonlinear dependence and connectedness between clean/renewable energy sector equity and European emission allowance prices, Energy Econ. 101 (2021) 105409.

[3] H.A.H. Al-Hilfi, A. Abu-Siada, F. Shahnia, Estimating generated power of photovoltaic systems during cloudy days using gene expression programming, IEEE J. Photovolt. 11 (1) (2021) 185–194.

[4] M. Haghighat, M. Niroomand, H. Dehghani Tafti, An adaptive power ramp rate control method for photovoltaic systems, IEEE J. Photovolt. (2021) 1–8.

[5] A.Q. Al-Shetwi, Sustainable development of renewable energy integrated power sector: trends, environmental impacts, and recent challenges, Sci. Total. Environ. 822 (2022) 153645.

[6] B. Mukhopadhyay, D. Das, Optimal multi-objective long-term sizing of distributed energy resources and hourly power scheduling in a grid-tied microgrid, Sustain. Energy, Grids Netw. (2022) 100632.

[7] A. Alarifi, A. Ali AlZubi, O. Alfarraj, A. Alwadain, Automated control scheduling to improve the operative performance of smart renewable energy systems, Sustain. Energy Technol. Assess. 45 (2021) 101036.

[8] S. Namilakonda, Y. Guduri, Chaotic darwinian particle swarm optimization for real-time hierarchical congestion management of power system integrated with renewable energy sources, Int. J. Electr. Power Energy Syst. 128 (2021) 106632.

[9] T. Xu, Y. Ren, L. Guo, X. Wang, L. Liang, Y. Wu, Multi-objective robust optimization of active distribution networks considering uncertainties of photovoltaic, Int. J. Electr. Power Energy Syst. 133 (2021) 107197.

[10] P. Saini, L. Gidwani, An investigation for battery energy storage system installation with renewable energy resources in distribution system by considering residential, commercial and industrial load models, J. Energy Storage 45 (2022) 103493.

[11] H.I. Alhammad, K.A. Khan, F. Alismail, M. Khalid, Capacity Optimization and Optimal Placement of Battery Energy Storage System for Solar, IEEE, 2021, pp. 847–852.

[12] W. Zhang, S. Liu, O. Gandhi, C.D. Rodriguez-Gallegos, H. Quan, D. Srinivasan, Deep-learning-based probabilistic estimation of solar PV soiling loss, IEEE Trans. Sustain. Energy 12 (4) (2021) 2436–2444.

[13] L. Barelli, G. Bidini, D.A. Ciupageanu, A. Micangeli, P.A. Ottaviano, D. Pelosi, Real time power management strategy for hybrid energy storage systems coupled with variable energy sources in power smoothing applications, Energy Rep. 7 (2021) 2872–2882.

[14] Z. Hong, Z. Wei, J. Li, X. Han, A novel capacity demand analysis method of energy storage system for peak shaving based on data-driven, J. Energy Storage 39 (2021) 102617.

[15] P. Iliadis, S. Ntomalis, K. Atsonios, A. Nesiadis, N. Nikolopoulos, P. Grammelis, Energy management and techno-economic assessment of a predictive battery storage system applying a load levelling operational strategy in island systems, Int. J. Energy Res. 45 (2) (2021) 2709–2727.

[16] S. Som, S. De, S. Chakrabarti, S.R. Sahoo, A. Ghosh, A robust controller for battery energy storage system of an islanded ac microgrid, IEEE Trans. Ind. Inform. 18 (1) (2022) 207–218.

[17] S. Lenka, P. Sinha, C. Jena, A review on power quality improvement of grid connected PV with lithium-ion and super capacitor based hybrid energy storage system using a new control strategy, Renew. Energy Optim. Plan. Control. Springer, 2022, pp. 1–10.

[18] C. Gu, J. Wang, Q. Yang, X. Wang, Assessing operational benefits of large-scale energy storage in power system: comprehensive framework, quantitative analysis, and decoupling method, Int. J. Energy Res. 45 (7) (2021) 10191–10207.

[19] S. Ali, R.A. Stewart, O. Sahin, Drivers and barriers to the deployment of pumped hydro energy storage applications: systematic literature review, Clean. Eng. Technol. 5 (2021) 100281.

[20] A.A. Abdalla, M. Khalid, Smoothing methodologies for photovoltaic power fluctuations, in: 8th International Conference on Renewable Energy Research and Applications ICRERA 2019, 2019, pp. 342–346.

[21] M.J. Reno, M. Lave, J.E. Quiroz, R.J. Broderick, PV ramp rate smoothing using energy storage to mitigate increased voltage regulator tapping, in: 2017 IEEE 44th Photovoltaic Specialist Conference PVSC 2017, 2018, pp. 1–6.

[22] M.E. Haque, M.N.S. Khan, M.R.I. Sheikh, Smoothing control of wind farm output fluctuations by proposed low pass filter, and moving averages, in: ICEEE 2015 – 1st International Conference on Electrical and Electronics Engineering, no. November 2016, pp. 121–124.

[23] A. Atif, M. Khalid, Savitzky–Golay filtering for solar power smoothing and ramp rate reduction based on controlled battery energy storage, IEEE Access. 8 (2020) 33806–33817.

[24] J. Hu, Y. Xu, K.W. Cheng, J.M. Guerrero, A model predictive control strategy of PV-battery microgrid under variable power generations and load conditions, Appl. Energy 221 (2018) 195–203. no. December 2017.

[25] M. Khalid, A.V. Savkin, Minimization and control of battery energy storage for wind power smoothing: aggregated, distributed and semi-distributed storage, Renew. Energy 64 (2014) 105–112.

[26] T. Guo, Y. Zhu, Y. Liu, C. Gu, J. Liu, Two-stage optimal MPC for hybrid energy storage operation to enable smooth wind power integration, IET Renew. Power Gener. 14 (13) (2020) 2477–2486.

[27] J. Li, S. Wang, L. Ye, J. Fang, A coordinated dispatch method with pumped-storage and battery-storage for compensating the variation of wind power, Prot. Control. Mod. Power Syst. 3 (1) (2018).
[28] Q. Jiang, H. Hong, Wavelet-based capacity configuration and coordinated control of hybrid energy storage system for smoothing out wind power fluctuations, IEEE Trans. Power Syst. 28 (2) (2013) 1363–1372.
[29] X. Wang, L. Li, A. Palazoglu, N.H. El-Farra, N. Shah, Optimization and control of offshore wind farms with energy storage systems, IFAC-PapersOnLine 51 (18) (2018) 862–867.
[30] F.A. Mohamed, H.N. Koivo, Online management of MicroGrid with battery storage using multiobjective optimization, in: POWERENG 2007 – International Conference Power Engineering – Energy Electrical Drives Proceeding, 2007, pp. 231–236.
[31] J. Tant, F. Geth, D. Six, P. Tant, J. Driesen, Multiobjective battery storage to improve PV integration in residential distribution grids, IEEE Trans. Sustain. Energy 4 (1) (2013) 182–191.
[32] A. Ostadi, M. Kazerani, S.K. Chen, Optimal sizing of the energy storage system (ESS) in a battery-electric vehicle, in: 2013 IEEE Transportation Electrification Conference and Expo Components, System Power Electron. From Technol. to Bus. Public Policy, ITEC 2013, 2013, pp. 1–6.
[33] N.H.F. Ismail, S.F. Toha, State of charge estimation of a Lithium-ion battery for electric vehicle based on particle swarm optimization, in: 2013 IEEE International Conference Smart Instrumentation, Measure Application ICSIMA 2013, November 2013, pp. 1–4.
[34] A.V. Savkin, M. Khalid, V.G. Agelidis, Optimal size of battery energy storage and monotonic charging/discharging strategies for wind farms, in: 2014 IEEE Conference Control Application CCA 2014, 2014, pp. 1372–1376.
[35] Y. Liao, L. Dong, H. Shen, D.F. Li, An optimization strategy of smoothing control of HPWS active power output, APAP 2011 – Proc. 2011 Int. Conf. Adv. Power Syst. Autom. Prot. 2 (2011) 1182–1187.
[36] S.C. Wang, Y.H. Liu, A PSO-based fuzzy-controlled searching for the optimal charge pattern of Li-Ion batteries, IEEE Trans. Ind. Electron. 62 (5) (2015) 2983–2993.
[37] N. Jayasekara, M.A.S. Masoum, P.J. Wolfs, Optimal operation of distributed energy storage systems to improve distribution network load and generation hosting capability, IEEE Trans. Sustain. Energy 7 (1) (2016) 250–261.
[38] Y. Yoshimura, T. Kondo, M. Kawanishi, T. Narikiyo, A. Sato, Model predictive control of EV storage battery with HEMS based on particle swarm optimization, in: Proc. of 2015 Innovative Smart Grid Technologies – Asia, *ISGT ASIA 2015*, 2016, pp. 1–5.
[39] K.A. Joshi, N.M. Pindoriya, A.K. Srivastava, A two-stage fuzzy multiobjective optimization for phase-sensitive day-ahead dispatch of battery energy storage system, IEEE Syst. J. 12 (4) (2018) 3649–3660.
[40] E. Banguero, A. Correcher, A. Pérez-Navarro, E. Garcia, State of health estimation of lead acid battery bank in a renewable energy system by parameter identification with genetic algorithms, in: 2018 7th International Conference System Control ICSC 2018, no. 3, 2018, pp. 418–423.
[41] T.D. Hund, S. Gonzalez, K. Barrett, Grid-tied PV system energy smoothing, in: Conference Record of the IEEE Photovoltaic Specialists Conference, 2010, pp. 2762–2766.
[42] P. Chanhom, S. Sirisukprasert, N. Hatti, A new mitigation strategy for photovoltaic power fluctuation using the hierarchical simple moving average, in: Proc. of the 2013

IEEE International Workshop on Intelligent Energy Systems IWIES 2013, 2013, pp. 28–33.
[43] A. Ellis, D. Schoenwald, J. Hawkins, S. Willard, PV Power Output Smoothing Using Energy Storage, Sandia National Laboratories (SNL), Albuquerque, NM, and Livermore, CA (United States), 2012.
[44] D.I. Stroe, A. Zaharof, F. Iov, Power and energy management with battery storage for a hybrid residential PV-wind system—a case study for Denmark, Energy Procedia 155 (2018) 464–477.
[45] R.P. Sasmal, S. Sen, A. Chakraborty, Solar photovoltaic output smoothing: using battery energy storage system, in: 2016 National Power System Conference NPSC 2016, 2017.
[46] J. Johnson, et al., PV output smoothing using a battery and natural gas engine-generator, in: Conference Record of the IEEE Photovoltaic Specialists Conference, March 2016, 2013, pp. 1811–1816.
[47] X. Li, N. Li, X. Jia, D. Hui, Fuzzy logic based smoothing control of wind/PV generation output fluctuations with battery energy storage system, in: 2011 International Conference on Electrical Machines and Systems ICEMS 2011, 2011, pp. 1–5.
[48] X. Li, D. Hui, L. Wu, X. Lai, Control strategy of battery state of charge for wind/battery hybrid power system, IEEE Int. Symp. Ind. Electron. (2)(2010) 2723–2726.
[49] D. Lamsal, V. Sreeram, Y. Mishra, D. Kumar, Kalman filter approach for dispatching and attenuating the power fluctuation of wind and photovoltaic power generating systems, IET Gener. Transm. Distrib. 12 (7) (2018) 1501–1508.
[50] H. Xu, C. Wang, C. Lu, Z. Lu, An adaptive wind power smoothing method with energy storage system, IEEE Power Energy Soc. Gen. Meet. 2014 (2014) 1–5. no. October.
[51] X. Li, Y. Li, X. Han, D. Hui, Application of fuzzy wavelet transform to smooth wind/PV hybrid power system output with battery energy storage system, Energy Procedia 12 (2011) 994–1001.
[52] F.J. Lin, H.C. Chiang, J.K. Chang, Y.R. Chang, Intelligent wind power smoothing control with BESS, IET Renew. Power Gener. 11 (2) (2017) 398–407.
[53] P. Mukherjee, V.V. Rao, Fuzzy logic controlled superconducting magnetic energy storage for leveling power fluctuation of grid connected wind generator, in: 2018 International Conference Power Energy, Environmental intelligent control PEEIC 2018, 2019, pp. 665–669.
[54] K.M. Tan, T.S. Babu, V.K. Ramachandaramurthy, P. Kasinathan, S.G. Solanki, S.K. Raveendran, Empowering smart grid: a comprehensive review of energy storage technology and application with renewable energy integration, J. Energy Storage 39 (2021) 102591.
[55] H.M. Ghazal, K.A. Khan, F. Alismail, M. Khalid, Maximizing Capacity Credit in Generation Expansion Planning for Wind Power Generation and Compressed Air Energy Storage System, December 2021, pp. 1–5.
[56] X. Li, A. Palazzolo, A review of flywheel energy storage systems: state of the art and opportunities, J. Energy Storage 46 (2022) 103576.
[57] S. Choudhury, Flywheel energy storage systems: a critical review on technologies, applications, and future prospects, Int. Trans. Electr. Energy Syst. 31 (9) (2021) e13024.
[58] P. Mukherjee, V.V. Rao, Superconducting magnetic energy storage for stabilizing grid integrated with wind power generation systems, J. Mod. Power Syst. Clean. Energy 7 (2) (2019) 400–411.

[59] A. Fathy, D. Yousri, T. Alanazi, H. Rezk, Minimum hydrogen consumption based control strategy of fuel cell/PV/battery/supercapacitor hybrid system using recent approach based parasitism-predation algorithm, Energy 225 (2021) 120316.
[60] K.A. Khan, M. Khalid, Improving the transient response of hybrid energy storage system for voltage stability in DC microgrids using an autonomous control strategy, IEEE Access. 9 (2021) 10460–10472.
[61] S. Sukumar, H. Mokhlis, S. Mekhilef, M. Karimi, S. Raza, Ramp-rate control approach based on dynamic smoothing parameter to mitigate solar PV output fluctuations, Int. J. Electr. Power Energy Syst. 96 (2018) 296–305. no. October 2017.
[62] S. Sukumar, M. Marsadek, K.R. Agileswari, H. Mokhlis, Ramp-rate control smoothing methods to control output power fluctuations from solar photovoltaic (PV) sources—a review, J. Energy Storage 20 (2018) 218–229. no. April.
[63] M.A. Syed, M. Khalid, Moving regression filtering with battery state of charge feedback control for solar PV firming and ramp rate curtailment, IEEE Access. 9 (2021) 13198–13211.
[64] G.C. Mahato, T.R. Choudhury, B. Nayak, D. Debnath, S.B. Santra, B. Misra, A Review on high PV penetration on smart grid: challenges and its mitigation using FPPT, in: ICPEE 2021 – 2021 1st International Conference Power Electronic Energy, January, 2021.
[65] X. Han, F. Chen, X. Cui, Y. Li, X. Li, A power smoothing control strategy and optimized allocation of battery capacity based on hybrid storage energy technology, Energies 5 (5) (2012) 1593–1612.
[66] M.J.E. Alam, K.M. Muttaqi, D. Sutanto, A novel approach for ramp-rate control of solar PV using energy storage to mitigate output fluctuations caused by cloud passing, IEEE Trans. Energy Convers. 29 (2) (2014) 507–518.
[67] R. van Haaren, M. Morjaria, V. Fthenakis, An energy storage algorithm for ramp rate control of utility scale PV (photovoltaics) plants, Energy 91 (2015) 894–902.
[68] I. de la Parra, J. Marcos, M. García, L. Marroyo, Control strategies to use the minimum energy storage requirement for PV power ramp-rate control, Sol. Energy 111 (2015) 332–343.
[69] H.J. Lee, J.Y. Choi, G.S. Park, K.S. Oh, D.J. Won, Renewable integration algorithm to compensate PV power using battery energy storage system, in: 2017 6th International Youth Conference Energy, IYCE 2017, 2017, pp. 1–6.
[70] J. Marcos, I. de La Parra, M. García, L. Marroyo, Control strategies to smooth short-term power fluctuations in large photovoltaic plants using battery storage systems, Energies 7 (10) (2014) 6593–6619.

Multilevel converter-based STATCOM with hybrid storage system

18

A Bharadwaj and S Maiti
Department of Electrical Engineering, Indian Institute of Technology Kharagpur, West Bengal, India

Chapter Outline

Nomenclature

Acronyms

HSS	Hybrid storage system
MMC	Modular Multilevel Converter
WEGS	Wind energy generating system

Symbols

N	Number of submodules per arm
V_c	Nominal voltage of submodule capacitor
V_{dc}	Dc link voltage of MMC
P_g	Active power injected into the grid
P_{es}	Real power support from SESS

Emerging Trends in Energy Storage Systems and Industrial Applications. DOI: https://doi.org/10.1016/B978-0-323-90521-3.00012-0

P_{wp} Real power generated by wind-farm
P_{bat} Active power support from battery modules
P_{uc} Active power support from UC modules
V_{leg}^{dc} Dc component of a leg in MMC
v_{SM_o} Output voltage of a submodule
V_{uc} Terminal voltage of a UC module

18.1 Introduction

The penetration of renewable energy-based power generation like solar farms and wind farms into the grid has seen enormous growth during the last decade. Among these systems, the integration of large-scale wind power generating systems (WPGS) has witnessed a high growth rate due to the development of advanced power converter topologies, higher conversion efficiency, and encouragement from government policies. The WPGS consists of an asynchronous electrical machine and it is coupled to the grid with the ac network through an AC/DC/AC conversion system.

The amount of the power generated by WPGS is dependent on the wind velocity and maximum power point tracking algorithm that is being employed which will result in the generation of variable power. Injection of this unsteady power into the Ac grid may bring some challenges like, scheduling of conventional power generating units, operation of existing ac network, and stable margins of voltage and frequency of the Ac grid [1]. To overcome these issues, certain grid codes have been designed by transmission system operators, such as reactive power support, frequency support, active power support, unbalanced current compensation, harmonic filtering, and ride-through capability during faults [2–4]. Thus, to integrate a large-scale WPGS into a transmission network the respective grid codes have to be satisfied at the common coupling point (CCP).

To provide the required gridcodes, STATCOM and a storage system are required at the CCP of WPGS. Instead of using two different systems, a single system that can provide the features of both STATCOM and the storage system may be employed. This kind of system is termed "E- STATCOM" (ESt). The standard 2-level voltage source converter with battery and supercapacitor modules has been proposed in [5]. The ESt has been employed to provide real and reactive power support for a wind energy conversion system in an electrical distribution network. The respective topology is applicable for medium power at low voltage applications. But, for high power/voltage applications the respective topology has the limitations like low efficiency, the need for passive filters, and a large footprint [6].

The battery storage system for utility applications has been presented in [7] to integrate renewable energy-based power generating units. But for high-energy applications, the battery storage system is not suitable because of its low charging/discharging rate. Thus, a hybrid storage system (HSS) has been proposed in [8] to meet both high power and energy density applications. In [9], the advantages of modular multilevel converter (MMC) for high power and high voltage applications has discussed in detail. In comparison to the two-level voltage source converter, the

MMC has better efficiency and less footprint. The MMC-based topology for ESt applications using a HSS has been proposed in [10]. In the proposed topology, battery modules are integrated at the DC-link of MMC to provide the required energy support. But the storage system will be over-rated and underutilized. Also, the battery modules have to provide the high-frequency component which results in degradation in the lifespan of the battery. Thus, a topology using MMC with both battery and supercapacitor modules has been proposed in [11]. In this topology, the battery and supercapacitor modules are distributed at the DC-link and submodules of MMC using a DC converter. This converter helps in two-way power transfer between the storage modules and the MMC. The limitations of the topologies presented in [5–11] are listed below.

1. The footprint of the configurations is high to support both real and reactive power simultaneously.
2. An interfacing transformer is required to integrate the converter with the high voltage ac network.
3. A fault in any of the storage modules may result in a complete disconnection of the storage system from the main converter.
4. Passive filters are needed to reduce the harmonic components in the real and reactive powers which are being injected into the grid by the ESt.

Hence, this chapter presents a topology to integrate the HSS into MMC with less component count. Apart from this, the required control methodology to operate the MMC as ESt has been discussed in detail. The main contributions of this chapter are as follows.

1. A topology for ESt application using both battery and supercapacitor with less component count.
2. The control principle to operate the presented topology to provide the needed real and reactive power.
3. The methodology to operate the configuration as ESt during a fault in any of the distributed storage modules.
4. Results to present the effectiveness of the configuration presented for ESt application.

18.2 Proposed configuration

This section of the chapter presents the proposed configuration for ESt with a HSS. Fig. 18.1 shows the topology of ESt. The hybrid storage modules are disbursed at the DC-link of MMC. The DC–DC converters are connected between the storage modules and the DC-link to regulate the flow of power between the storage modules and DC-link. The complete control principle is discussed in the next section of this chapter. The main advantages of this configuration are presented below.

- The configuration offers better efficiency in comparison to the topologies presented in [10].
- With the division of storage modules, the redundancy of this configuration is more than the available topologies.

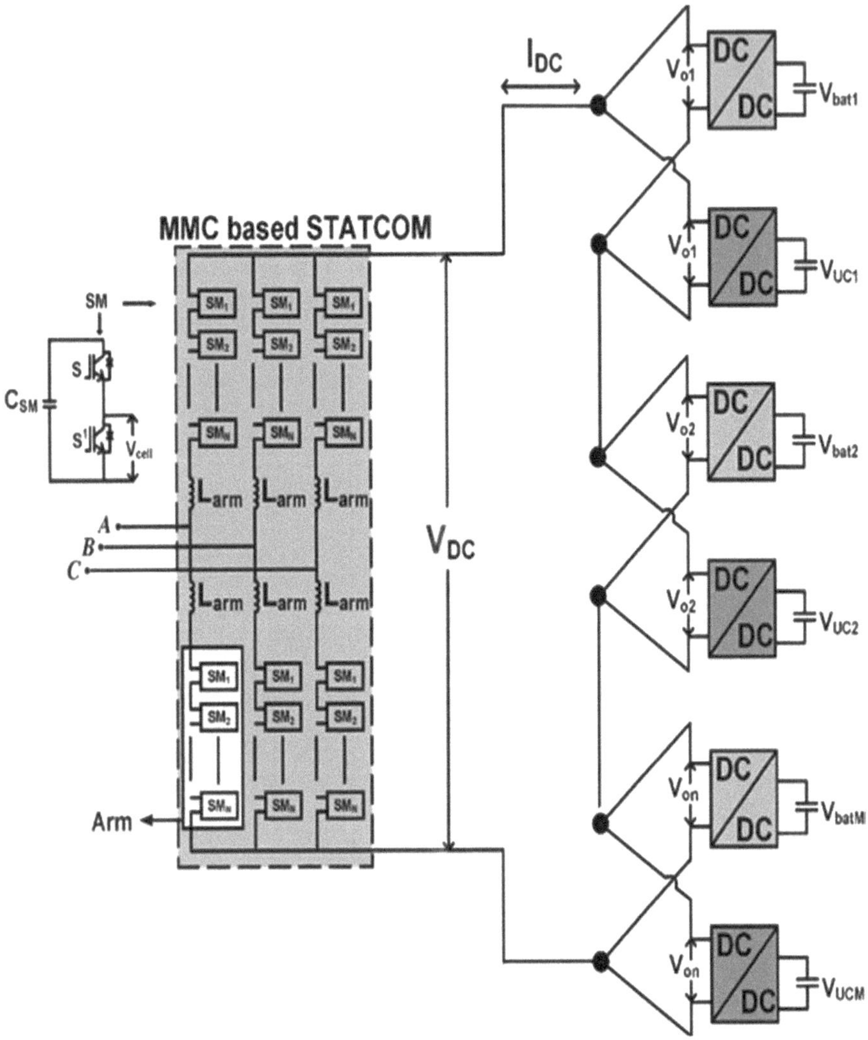

Figure 18.1 Distribution of battery and UC modules at DC-kink of modular multilevel converter.

- The amount of needed real power support from the presented configuration can be raised by the addition of DC converters at the DC-link of MMC.
- The integration of HSS with MMC uses fewer components in comparison to the topologies proposed [10].

To integrate the low voltage storage modules with the MMC, the configuration of the DC–DC converter is given in Fig. 18.2. The converter has the two-way power transfer ability from the storage module to MMC and vice-versa.

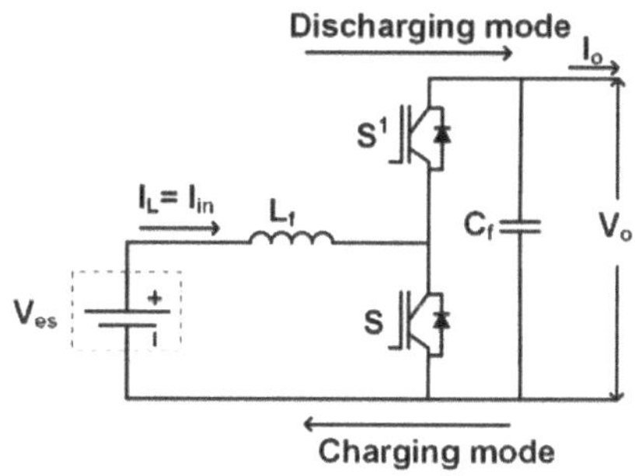

Figure 18.2 Converter to integrate storage modules.

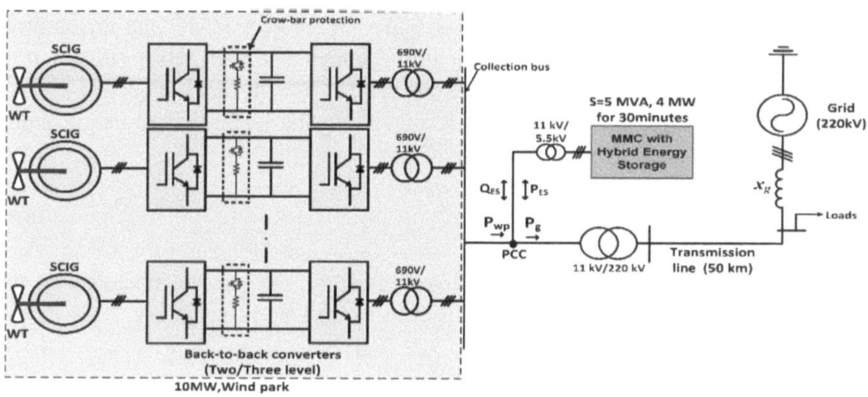

Figure 18.3 Grid-connected wind-park with ESt.

18.3 System for study

The system to evaluate the working of the proposed configuration is presented in Fig. 18.3. The squirrel cage induction machine has been employed as an asynchronous generator(SCIG) and connected to ac system through an ac-to-ac conversion system through an intermittent DC link. The wind park is formed by connecting all the SCIGs to a common ac bus to collect the power generated. The wind park is connected with the existing ac system through a network of transmission lines. At the PCC, the ESt is connected to provide the gridcodes to inject constant active power into the grid. The operation of the wind park is presented in [5] so that the maximum power is extracted.

18.4 Control of the topology

This section presents the control philosophy to operate the proposed configuration in Fig. 18.1 as ESt. The complete control structure for the configuration is presented in Fig. 18.3. The controller regulates the modulation index to support the following requirements at the point of common coupling (PCC) of a wind park (Fig. 18.4).

1. Active power support
2. Voltage regulation
3. Compensation for negative sequence currents
4. Capability to ride during the low voltages

Each of the aforementioned requirements is discussed here.

18.4.1 Active power support

The main aim of ESt is to introduce constant real power into the grid by nullifying the deviations in the power generated by the wind park. This can be achieved by charging/discharging the storage modules at the DC-link of MMC and this control will be explained in the next section of this chapter. The real power support provided by the ESt can be determined using (18.1). The positive sequence current controller is presented in Fig. 18.3 while regulating the positive modulation index to supply the real power needed at the PCC.

$$P_{\mathrm{ES}} = P_g - P_{\mathrm{wp}} \tag{18.1}$$

where P_g, P_{wp}, and P_{ES} are power injected into the grid, power generated by the wind farm, and active power support provided by the ESt, respectively.

18.4.2 Voltage regulation

The reactive power is needed to operate the asynchronous generators in the wind park and also to maintain the voltage stability of the remotely located PCC. Thus, the

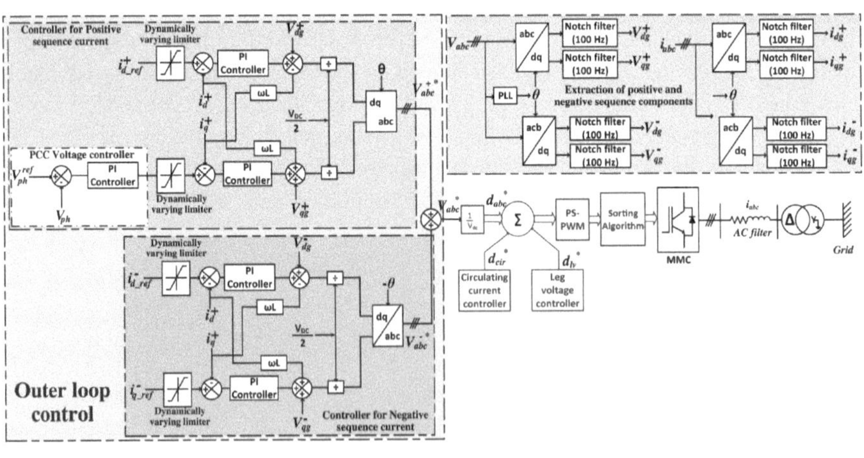

Figure 18.4 Controller to obtain ESt features from modular multilevel converter.

ESt has to regulate the voltage at the common coupling point by providing the required reactive power. This can be achieved by the PCC voltage controller by regulating the q-axis component of the current.

18.4.3 Negative sequence current compensation

Unequal distribution of loads in the electrical distribution network may lead to the flow of negative sequence currents into the transmission network and this may flow into the asynchronous generators of the wind park. This results in the deteriorating performance of the generators. Thus, a controller is needed to limit the flow of negative sequence currents. With the help of ESt provides a compensating current to neutralize the negative sequence currents at the common coupling point. The required control loop is presented in Fig. 18.3.

18.4.4 Ride through capability

One of the main grid codes for a wind park is to ride through capability during the low voltages at the common coupling point. During this period, the wind park is to be connected with the grid and the complete capacity of ESt is used to injected reactive power to support the voltage. This can be achieved by varying the limits of the controller and it is done with the help of dynamically varying limits. The limits of the positive and negative sequence control loops are varying using the algorithm shown in Fig. 18.5. From the figure, it can be observed that when the voltage at the common coupling point is less than 0.8pu then the complete capacity of the ESt will be used to support the reactive power. While for other operating states, the capacity of ESt is decided to support the active power and other requirements.

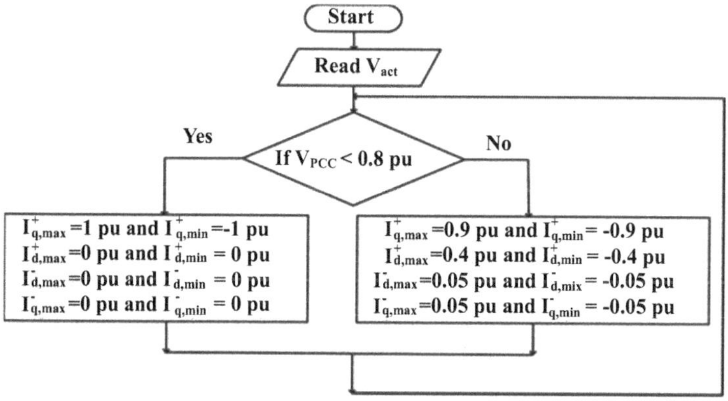

Figure 18.5 Limit changer procedure for current controller.

18.5 Operation of hybrid storage system

This section of the chapter presents the control methodology to operate the distributed HSS in the proposed configuration.

18.5.1 Extraction of frequency components

The charging/discharging characteristics are different from each other. Thus, an energy management mechanism has to be employed to improve the operating efficacy of the storage modules. Hence, the Fig. 18.6 shows the energy management system (EMS). Here the required active power is segregated into low and high-frequency components using a low-pass filter(LPF). The cut-off frequency is kept at 1.25 Hz the dominant frequency component observed in the power generated by the wind park. The high-frequency component of P_{ES} is supplied by the ultracapacitor while the other is supported by the battery storage system. Here P_{bss} and P_{uss} are the amount of active power delivered by the battery and ultracapacitor storage modules.

18.5.2 Control of supercapacitor fed DC converter

The ultracapacitor modules are distributed at the DC-link of MMC with the help of a DC−DC converter as shown in Fig. 18.3. To obtain the required DC-link voltage, these converters are cascaded and the main objective of these converters is to maintain the DC-link voltage at its reference by providing the required active power support. So, the converters are operated in voltage mode of operation (VMO) as presented in Fig. 18.7. Since the DC−DC converters are identical, the output voltage at each converter can be expressed as given below.

$$V_{oi} = V_{dc}/M \tag{18.2}$$

The magnitude of the reference value for the controller can be determined using (18.2). Note that the VMO is configured with an outer voltage loop and inner current loop so that the charging/discharging current of a battery module can be limited. The complete procedure to implement the super-twisting sliding mode controller in the current loop is discussed in [5].

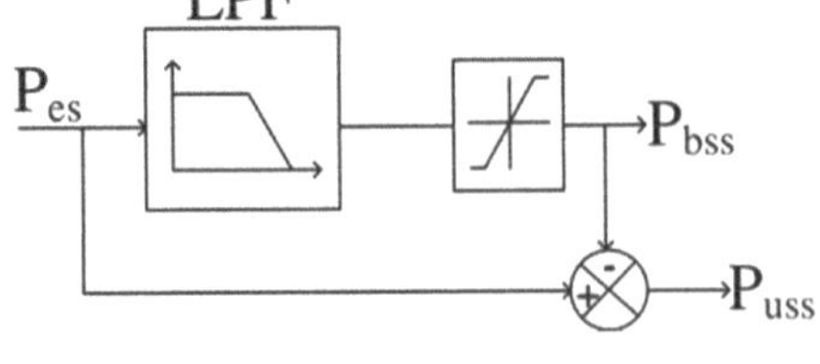

Figure 18.6 Energy management scheme for battery and ultracapacitor modules.

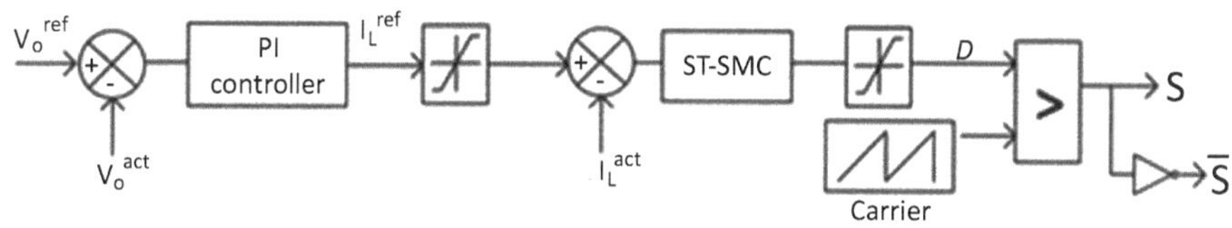

Figure 18.7 Control block diagram of DC–DC converter connected to supercapacitor module.

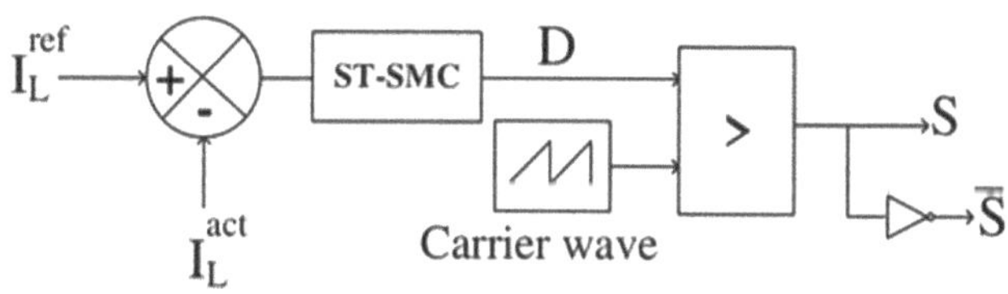

Figure 18.8 Controller for the battery-fed DC–DC converter.

18.5.3 Control of battery module fed DC converter

The battery modules are also distributed at the DC-link of MMC using the DC–DC converter and they are operated in the current mode of operation. The respective control structure is given in Fig. 18.8. The reference value for the controller can be determined by the following expression. The complete derivation of the expression is discussed in [11].

$$I_{ref} = P_{uss}/M \tag{18.3}$$

18.6 Discussion on simulation results

The system presented in Fig. 18.3 is modeled in PSCAD/EMTDC using the parameters given in [10]. The amount of active power support from ESt and the real power injected into the grid are shown in Fig. 18.9. The windpark injects a constant active power (P_G) of 6.75 MW and regulates the PCC voltage with appropriate reactive power support (Q_G). Fig. 18.10 shows the PCC voltage without and without ESt and it can be observed that the PCC voltage is maintained at the reference with the help of ESt.

The DC-link voltage of the MMC is regulated with the help of the DC–DC converter fed from the battery storage system and it is given in Fig. 18.6. The voltage at the terminals of each DC–DC converter is shown in Fig. 18.10. From the figure, it can be noticed that each DC–DC converter maintains 3 kV at its output terminals. Here the total DC-link voltage of 12 kV is shared among the DC–DC

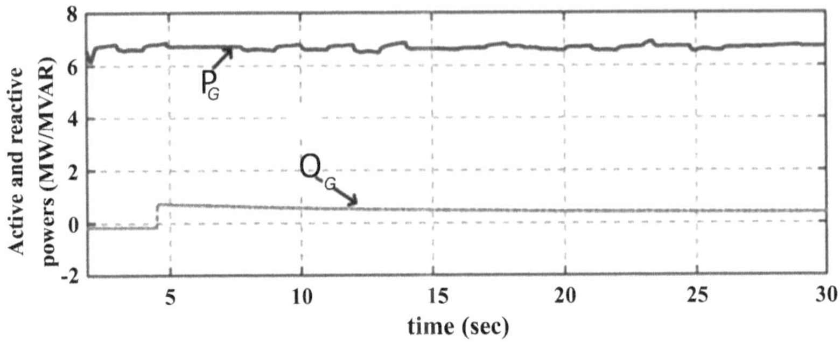

Figure 18.9 Power injected into the grid.

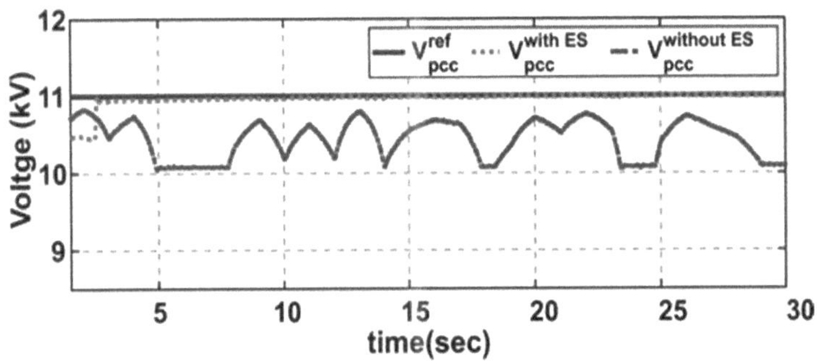

Figure 18.10 Point of common coupling voltage of windpark.

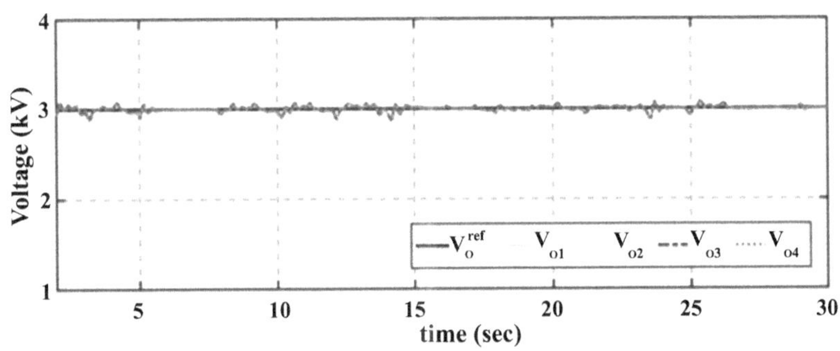

Figure 18.11 Output voltages of battery-fed DC–DC converters at the DC-link of modular multilevel converter.

converter equally and it is given in Fig. 18.11. While the DC-link current drawn by the MMC is given in Fig. 18.12.

The presence of an unbalanced local load at the PCC results in a flow of negative sequence currents into the asynchronous machines of WEGS. This leads to the

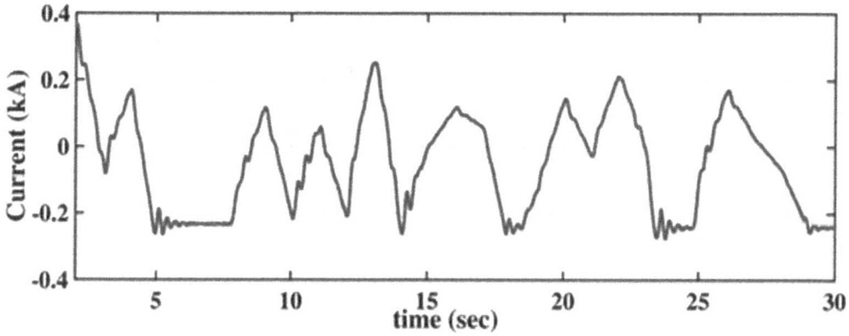

Figure 18.12 DC-link current of modular multilevel converter.

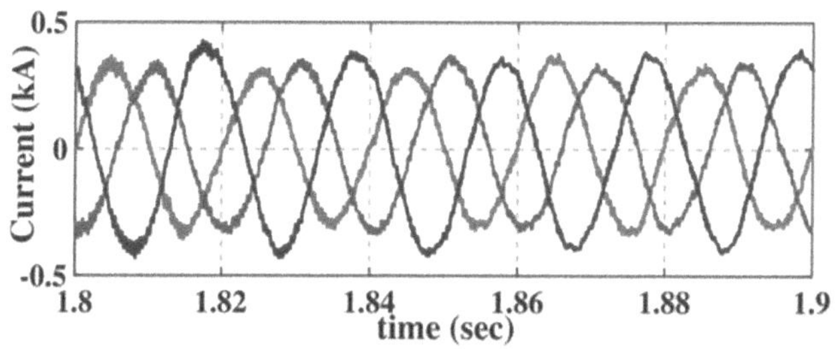

Figure 18.13 Grid currents at the point of common coupling of wind farm without negative sequence current compensation.

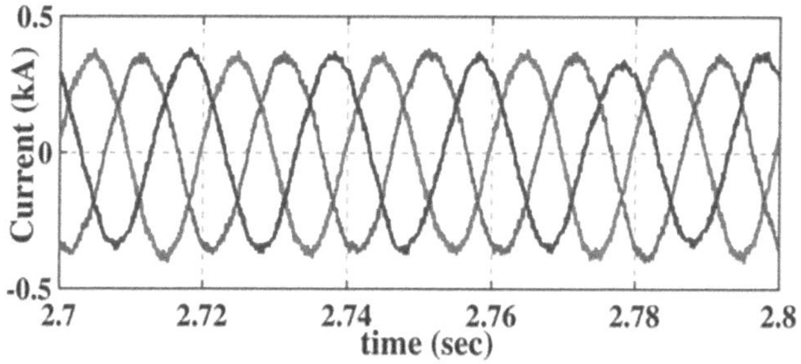

Figure 18.14 Grid currents at the point of common coupling of wind farm with negative sequence current compensation.

flow of negative sequence currents into the stator of asynchronous machines which are connected to the same bus. Fig. 18.13 shows the grid currents at PCC. From the figure, it can be understood that the currents are unbalanced due to the presence of negative sequence components. With the initiation of the negative sequence controller of MMC, the ESt provides the compensating currents to balance the grid currents. The grid currents after the initiation of the negative sequence controller are

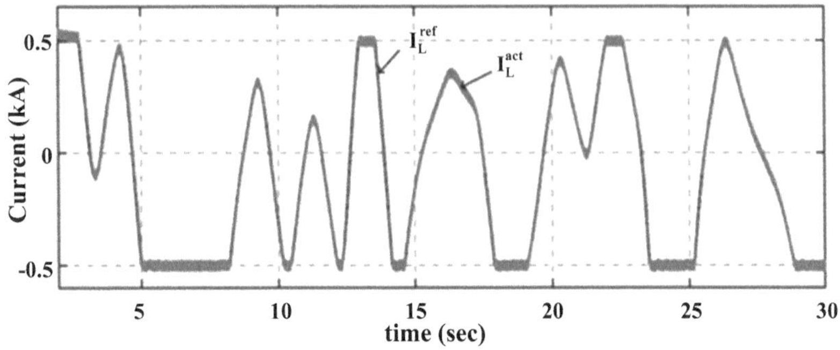

Figure 18.15 Actual and reference inductor currents of a battery module fed DC–DC converter.

presented in Fig. 18.13. Thus, the negative sequence currents are drawn neither from the grid nor from the wind park but it is supplied by ESt and it is given in Fig. 18.14.

The performance of the controller for the DC–DC converter with battery modules is presented in Fig. 18.15. Using the controller presented in Fig. 18.7. Here the ST-SMC is employed to track the reference value of inductor current with less steady-state error.

18.7 Conclusions

This chapter has presented a configuration of ESt applications using MMC with HSS. The configuration is competent enough to contribute the needed active and reactive powers at the same instance. The distribution of HSS at the DC-link will ensure that all the legs of MMC will share equal real power. Also, it can support high power/energy demand. The main benefits like higher efficiency, less component, and better transient performance from HSS. The MMC is operated with P-U_{ac} control and CMO/VMO is employed to operate the DC–DC converters. To distribute the active power demand among the USS and BSS, an EMS is used and it is formed by combining peak-saving and the LPF methods. This assures that the low-frequency component of P_{ES} is drawn from BSS while the USS supplies the rest. A control technique for MMC has been proposed which helps in providing the presented grid codes. A negative sequence current controller is also used in the MMC controller to mitigate the unbalance in the grid currents. To provide the required grid code at the PCC, the dynamically varying limiters are used to regulate the amount of d and q axis currents of MMC. The PSCAD/EMTDC platform has been used to verify the performance of the proposed configuration and the controller presented. The results obtained from the simulation have shown that the configuration is capable to maintain grid codes at the common coupling of a wind park.

Appendix

Table A1 Rating of wind park.

Parameter	Value
Capacity of the plant	10 MW
Collection bus line voltage	11 kV
Rating of each SCIG	1.5 MW
Output voltage of the generator	400 V
Rating of each back-to-back converter	450 kW

Table A2 Parameters of modular multilevel converter.

Parameter	Value
Converter capacity (S)	10 MVA
Active power support for 30 min	4 MW
DC link voltage	12 kV
Line voltage	5.5 kV
Number of submodules	4
Arm inductance (0.1 pu)	2.5 mH
Nominal voltage of a submodule	3 kV
Capacitance of a submodule capacitor	5 mF
Switching frequency of a submodule	500 Hz
Voltage rating of the transformer	5.5 kV/11 kV

Table A3 Rating of battery storage system.

Parameter	Value
No of battery modules	4
Active power support from the UC modules	1 MW
Nominal voltage of each UC module	1.75 kV
Capacitance of each UC module	
Output voltage of DC–DC converter	3 kV
Filter inductance of a DC–DC converter	6 mH
Switching frequency	2.5 kHz

References

[1] V. Yaramasu, B. Wu, P.C. Sen, S. Kouro, M. Narimani, High-power wind energy conversion systems: state-of-the-art and emerging technologies, Proc. IEEE 103 (5) (2015) 740–788.

[2] M. Mohseni, S.M. Islam, Review of international grid codes for wind power integration: diversity, technology and a case for global standard, Renew. Sust. Ener. Rev. 16 (06) (2012) 3876–3890.

[3] W.U. Tareen, S. Mekhilef, M. Seyedmahmoudian, B. Horan, Active power filter (APF) for mitigation of power quality issues in grid integration of wind and photovoltaic energy conversion system, Renew. Sust. Energy Rev 70 (01) (2017) 635–655.
[4] M.A. Saqib, A.Z. Saleem, Power-quality issues and the need for reactive-power compensation in the grid integration of wind power, Renew. Sust. Ener. Rev. 43 (01) (2015) 51–64.
[5] S. Bharadwaj, N. Maiti, Dhal, S. Chakraborty, Control of two level converter based STATCOM with battery and ultracapacitor, in: Proceeding of National Power Electronics Conference (NPEC), 2019, pp. 1–6.
[6] Solar inverters by ABB [online]. Available from: http://new.abb.com/powerconverters-inverters/solar/central/pvs980.
[7] P. Pourbeik, et al., Modeling and dynamic behavior of battery energy storage: a simple model for large-scale time-domain stability studies, IEEE Electrif. Mag. 3 (3) (2015) 47–51.
[8] J.I. LEON, E. Dominguez, L. Wu, A. Marquez Alcaide, M. Reyes, J. Liu, Hybrid energy storage systems: concepts, advantages, and applications, IEEE Ind. Electron. Mag. 15 (1) (2021) 74–88.
[9] A. Dekka, B. Wu, R.L. Fuentes, M. Perez, N.R. Zargari, Evolution of topologies, modeling, control schemes, and applications of modular multilevel converters, IEEE J. Emerg. Sel. Top Power Electron. 5 (4) (2017) 1631–1656.
[10] A.B. Chivukula, S. Maiti, Modular multilevel E-STATCOM considering distributed energy storage at the dc link, in: 2016 IEEE 7th Power India International Conference (PIICON), 2016, pp. 1–6.
[11] A.B. Chivukula, S. Maiti, Analysis and control of modular multilevel converter-based E-STATCOM to integrate large wind farms with the grid. IET Gener Trans. Distr. 13 (20) 4604–4616.

Hybrid battery-supercapacitor energy storage for enhanced voltage stability in DC microgrids using autonomous control strategy

19

Khalid Abdullah Khan[1], *Ammar Atif*[1] *and Muhammad Khalid*[1,2,3]
[1]Electrical Engineering Department, King Fahd University of Petroleum and Minerals (KFUPM), Dhahran, Saudi Arabia, [2]Center for Renewable Energy and Power Systems, KFUPM, Dhahran, Saudi Arabia, [3]SDAIA-KFUPM Joint Research Center for Artificial Intelligence, Dhahran, Saudi Arabia

Chapter Outline

Emerging Trends in Energy Storage Systems and Industrial Applications. DOI: https://doi.org/10.1016/B978-0-323-90521-3.00007-7

Highlights

- Enhanced transient response of hybrid battery-supercapacitor for voltage regulation

Nomenclature

BESS	Battery energy storage system
DG	Distributed generation
ESS	Energy storage system
HESS	Hybrid energy storage system
HPF	High pass filters
LPF	Low pass filter
LUT	Look-up table
MG	Microgrid
NPIC	Nonlinear PI controller
PCC	Point of common coupling
PV	Photovoltaics
P&O	Perturb and observe
RES	Renewable energy sources
SCSS	Supercapacitor energy storage system
A	Interfacial area
B_M	BESS charge/discharge mode
$B_{\max}$	Maximum capacity of battery
C_E	Exponential capacity of battery
C_G	Thermal motion capacitance
C_M	Model capacitance of SCSS
C_p	Polarization constant of battery
C_Z	Exponential zone voltage of battery
ε_p	Material permeability
f_p	Pole frequency
f_{sf}	Switching frequency
f_z	Zero frequency
I_b	Battery current
I_{br}	Battery reference signal
I_{err}	DC-bus error signal
i_f	Battery filter current
I_{hf}	high frequency components
I_{lf}	Low frequency components
$V_{r-\mathrm{dc}}$	Reference voltage
I_{ph}	Photocurrent
$I_{\mathrm{ph},s}$	Standard photocurrent
I_{sat}	Saturation current of PV
I_{SC}	SCSS current
i_{sc}	SCSS series current
i_{sD}	Self-discharge current of SCSS

I_{sD}	Leakage current waveform of the SCSS
I_{ts}	Current-temperature coefficient
k_B	Boltzmann constant
k_i	Integral constant
k_p	Gain constant
λ	Irradiation on the PV device
λ_n	Nominal irradiation on the PV device
N_p, N_s	Number of parallel and series SCSS cells
N_L	Number of electrolyte layers of SCSS
n_s	Series connected PV cells
q	Electron charge
R_p	Battery polarization resistance
R_s	Series resistance of PV
R_{sh}	Shunt resistance of PV
r_m	Molecular radius
S_{sc}	SCSS electric charge
S_{si}	Initial SCSS amount of charge
T	Temperature of the p-n junction
T_A	Transfer function of current controller
T_B	Transfer function of voltage controller
T_{lpf}	Transfer function of Low pass filter
T_s	Operating temperature
μ	Diode's ideality factor
V_{bat}	Battery voltage
V_{BVS}	Controlled voltage source of battery
V_{DV}	Dynamic zone voltage of battery
V_{dc}	DC-bus voltage
V_s	Overpotential

19.1 Introduction

The global transition toward integrating renewable energy sources (RESs) is pertinent to securing a sustainable electric power network to satisfy the ever-increasing need for electricity while preserving the environment. RESs generate electricity from natural or recurring resources that replenish over a finite amount on the human time scale. The RESs also facilitate remote area electrification that in combination with RES's dependency on environmental factors influences their allocation which mainly projects a transition from a centralized power grid to deregulated distributed generation (DG) network. Consequently, the concept of DG superimposes the conventional centralized grid network. Hence, garnering grid flexibility is important to establishing large-scale RES installation [1]. The worldwide population growth is anticipated to raise electric power demand by 30% by 2030 compared to 2012. Thus, the need for incorporating RES intensified over the past few years leading the researchers and legislates to contribute and innovate in this field, to supply and sustain the increasing electric power needs while preserving global climatic conditions and a greener environment for the future generations.

The global electric energy production has therefore experienced an increased integration of renewable sources, ranging from solar, wind, hydro, tidal, geothermal, and biomass with an 8.3% increase in its installed capacity between 2016 and 2017 [2]. In recent decades, photovoltaics (PV) has been prominently used, due to their low cost, ease of installation, maturity, and incentives. The applicability of PV became a viable option for small scale installations and is considered the third most implemented renewable source for power generation, after hydro and wind energy, respectively.

Nevertheless, the unpredictable and uncontrollable intermittency of these RES introduces contributes to various power quality degradation in a distribution network. Primarily the challenges associated are demand-generation mismatch, reliability degradation [3], degradation of voltage and frequency [4], and inadvertently impeding the viability of RES penetration [5]. Voltage regulation, among other technical challenges associated with RES integration, is an important technical aspect that tends to define the allowable limit of RES integration in a power network.

Presently, the concept of microgrid (MG) is extensively adopted for RES integration [6], propagating the flexibility and controllability to establish DG that is necessary to advocate the techno-economic installation of RES. Typically, MG is designed to locally establish exclusive RES control to facilitate a load-following connection between the RES and the grid through energy storage systems (ESSs), power electronic interfaces, and some local loads; and has proved to be highly efficient in economically overcoming the transient power generation of RES.

19.1.1 Energy storage systems

The prerogative advantage of energy buffering achieved through an ESS significantly mitigates the effects of RES transience. In addition, ESS also intensively maintains the power quality of MG, especially in the case of standalone and/or islanded conditions. This characteristic facilitates numerous applications of ESS, both at generation and distribution-based deployment of RES in a power network as shown in Fig. 19.1. However, the scope of a techno-economic installation of ESS, apart from capital cost, operational cost, optimal sizing/siting, reliability, and availability, are largely dependent on their respective technical characteristics against a defined challenge associated with RESs integration.

Moreover, a combination of systematic, technical, and economic barriers has stymied the significantly higher penetration of ESS over the globe [7]. Despite ongoing developments, there is a lack of cost-effective, mass-produced, and, most importantly, fully integrated storage solutions. Consequently, this resource and knowledge gap leads to increased energy costs, either seen as initial costs of installation, or as the levelized cost of energy over the estimated lifetime of storage systems [8].

19.1.2 Hybrid energy storage systems

Pragmatically, choosing the right storage technology for RES integration, its design, and optimal operation is similarly important for the success or failure of any storage

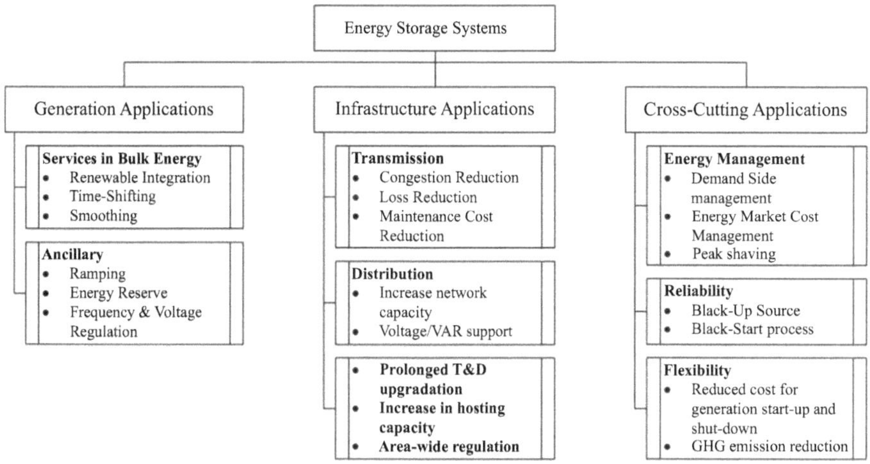

Figure 19.1 Application of energy storage system in various fields of renewable integrated power grids [5].

project across all steps from its initial design to deployment, operation, and end-of-life of a system. Most of the ESS technologies lack the technical properties to solely overcome the vast band of challenges associated to overcome the challenges perceived in MGs, for instance, the commonly used battery energy storage technologies even though, highly feasible for the techno-economic deployment of RES, are detrimentally impacted and oversized in case of peak abrupt transient load/generation disturbances and variations. Among other solutions, hybridization of storage technologies has been proposed to exploit the dominant characteristics of energy storage to overcome various issues faced in renewable integration [9–11].

Among various recommendations for ESS hybridization such as lead-acid and li-ion battery ESSs (BESSs), BESS-superconducting magnetic energy storage, flywheel-BESS, and BESS-supercapacitor ESS (SCSS); the high degree of characteristic complementarity between BESS and SCSS facilitates numerous feasible solutions in RES integrated MGs (Fig. 19.2) [12]. The complementary characteristics of a battery-supercapacitor hybrid ESS (HESS) expedite its application in various challenges and optimization in the field of electrical and electronics. The intrinsic characteristics achieved with this HESS are highly versatile specifically in the field of electric power generation through transient RES in terms of output power smoothing, fault support, voltage regulation, and various other improvements in power quality.

In a DC MG, the point of common coupling (PCC) uses power electronic devices to facilitate the interfacing of numerous distinctive components such as the numerous RESs, loads, and energy storage devices. In accordance, voltage regulation is pertinent for optimized interconnection as well as, maintaining the stability of grid-interconnection through inverters. Therefore, abrupt variations in the DC MG pertaining to the load requirement and RES generation will inherently affect the

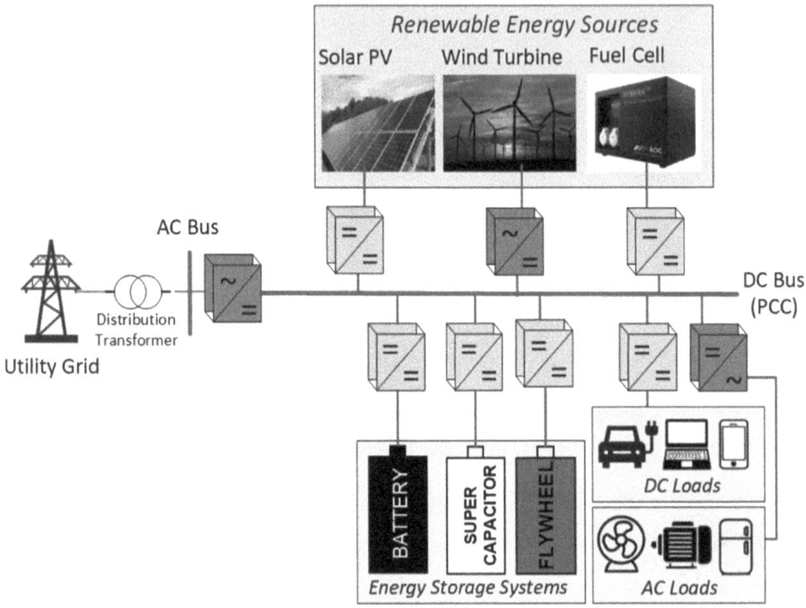

Figure 19.2 Typical DC/AC architecture of renewable integration with numerous types of energy storage systems [11].

efficacy of the integration [13]. The commonly used DC capacitors at the PCC are unable to maintain the degree of demand-generation-based voltage variation [14]. In this case, hybrid BESS-SCSS is considered to be among the best solution [15]. The associated concept regarding this HESS in DC voltage regulation lies in the power allocation, that is, compensation of the long-term average power demand by BESS and allocating the short-term abrupt peak power variation to the SCSS. Nevertheless, the control strategy implemented for power allocation governs the degree of utilization of these HESS features. Firstly, the control technique should maintain the current level of BESS, to increase its operational viability and lifetime by mitigating the BESS current stress. At the same time, the switching of SCSS should correspond to the abrupt load-generation variation to regulate both voltage deviation and magnitude.

19.2 Literature review

19.2.1 Types of integration topology for hybrid energy storage system

The topology of integration of battery supercapacitor HESS substantially impacts their potential performance. The selection of the topology governs the controllability over the HESS which enables the operators to improve the HESS efficiency and exploit the strength of each HESS component while avoiding their weaknesses. Typically, DC bus integration of the HESS is usually preferred as the HESS as well as the most of the RES technologies operate in DC voltage which facilitates minimal requirement of power electronic converter interface [16]. In addition, operation in DC voltage reduces the complexity associated with synchronization [17]. Usually, battery-supercapacitor HESS can be classified with respect to their interconnection to the DC-bus as passive, semi-active, and fully active topologies.

19.2.1.1 Passive topology of hybrid energy storage system

The passive topology of BESS-SCSS consists of direct integration to the DC bus link without the incorporation of DC–DC converters Fig. 19.3. This topology is the simplest and considered the cheapest solution. A number of studies have shown that it may efficiently mitigate the transient current under abrupt load-generation variation while simultaneously increasing peak power and reducing internal losses [18]. The BESS and SCSS share a common terminal voltage that is determined by the SoC and charge/discharge characteristics of BESS. In the case of rural applications, the BESS capacity can be designed to facilitate power for up to a few days as a reserve in the absence of an external power supply [19]. Therefore, the BESS will be operated with a lower depth of discharge and low C-rate for most of the time it is used. Consequently, the variation in DC bus voltage will be limited, resulting in an overall system voltage that is generally steady.

Nevertheless, the capability of utilizing the suitable high energy and high-power density of BESS and SCSS is not optimal due to the lack of controllability over them. Especially, the high-power density of SCSS will be under-utilized as the power to/from the SCSS will be governed by the internal resistance of BESS as well as the SCSS [20]. Therefore, the BESS terminal voltage is low and the full SoC of SCSS will not be appropriately utilized impeding their efficacy [21].

19.2.1.2 Semi-active topology of hybrid energy storage system

In the case of semi-active topology, a bi-directional DC–DC converter is incorporated interconnecting either the BESS or SCSS to the DC-bus. This allows the controllability of power flow between the grid and the HESS [22]. Fig. 19.4 depicts the semi-active BESS topology wherein only the BESS is interconnected to the DC-bus via a bi-directional DC–DC converter whereas the SCSS interconnection is passive

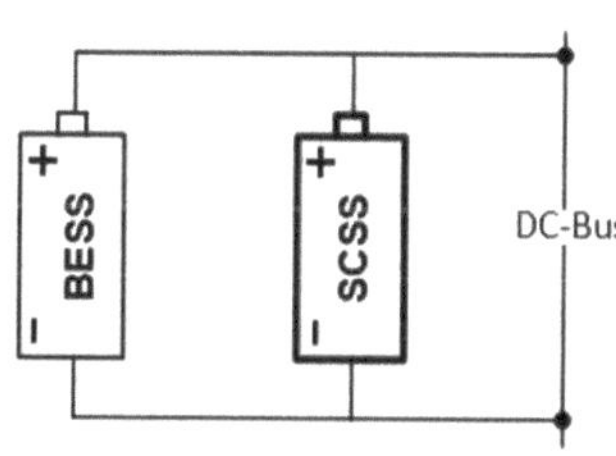

Figure 19.3 Passive battery-supercapacitor topology [20].

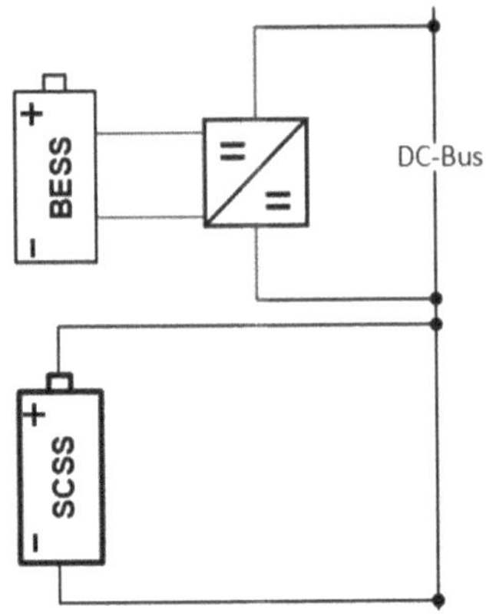

Figure 19.4 Semi-active battery energy storage system topology [25].

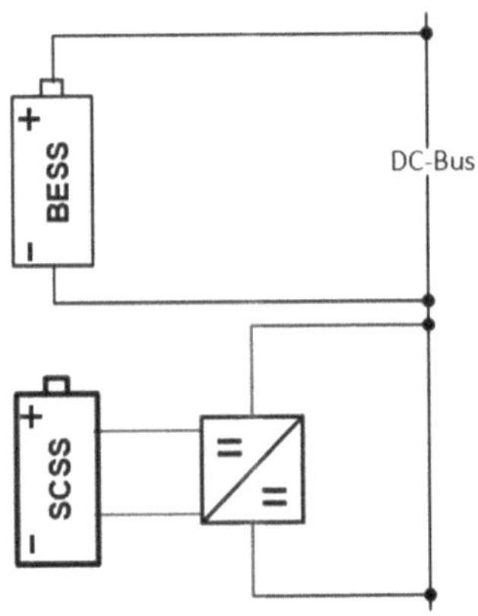

Figure 19.5 Semi-active supercapacitor energy storage system topology [21].

[23]. Therefore, irrespective of the system demand-generation variation the BESS current can be effectively controlled. In addition, flexibility is obtained in the capacity sizing of the BESS as its terminal voltage can be suitably selected and is not needed to match the DC-bus voltage [24]. Nevertheless, the SCSS is not appropriately utilized in accordance with its high-power density to compensate for peak short-term power transience in the grid which will cause large DC-bus voltage variations.

Similarly, semi-active SCSS comprises the incorporation of a bi-directional DC–DC converter for interconnecting the SCSS to the DC-bus whereas the BESS is passively connected as shown in Fig. 19.5 [25]. This topology enables better utilization of SCSS efficacy utilizing its wider range of SoC towards peak short-term power compensation required in the system. The BESS also contributes toward voltage regulation of the DC-bus; however, it incurs high current stress during system power transience due to a lack of controllability over the terminal voltage of the BESS which negatively impacts its operational lifetime [26].

19.2.1.3 Fully active topology of hybrid energy storage system

In this topology, both BESS and SCSS are interconnected to the DC-bus via a bi-directional DC–DC converter. The topology can be further classified based on parallel or series/cascaded configuration of BESS and SCSS. The series active topology of HESS is presented in Fig. 19.6. The DC–DC converter is current-controlled, and voltage-controlled for BESS and SCSS, respectively, in the series active topology. The DC–DC converter allows a limitation of current stress on BESS protecting the BESS from the intermittency of demand-generation variation. However, a large voltage swing is observed between the converter of the SCSS and the DC-bus as a wide range of SCSS is utilized to regulate the abrupt peak intermittency of the system which results in higher losses in the DC–DC converter [27].

The parallel configuration of BESS and SCSS in the active HESS topology is the most reliable and widely implemented, especially for grid-scale applications.

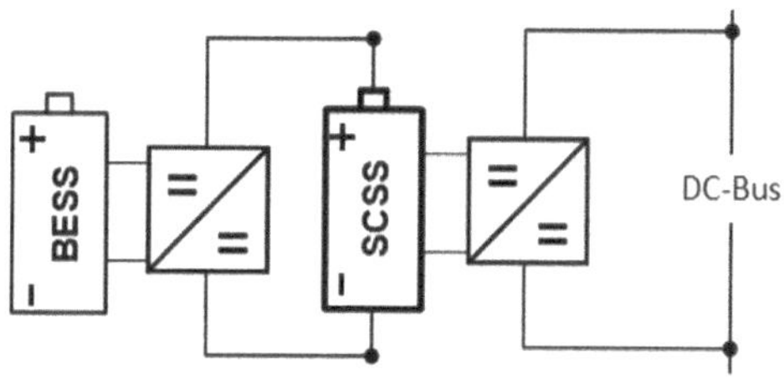

Figure 19.6 Fully active topology of hybrid energy storage system with series battery energy storage system and supercapacitor energy storage system configuration [27].

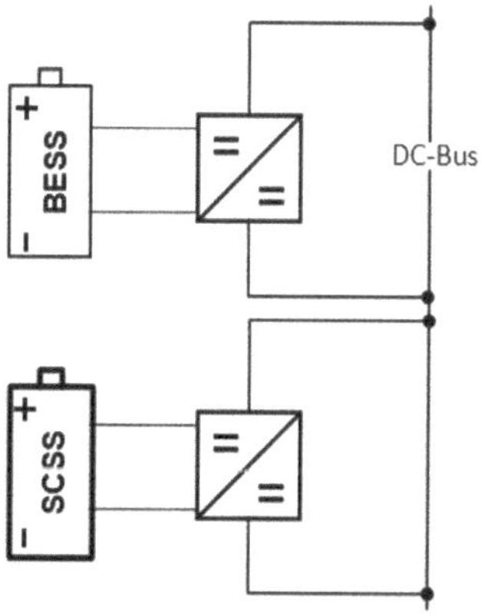

Figure 19.7 Fully active topology of hybrid energy storage system with parallel battery energy storage system and supercapacitor energy storage system configuration [28].

The topology is depicted in Fig. 19.7. A higher degree of controllability is facilitated over both BESS and SCSS. An enhanced performance, BESS life, and DC bus voltage regulation can be achieved through user-designed with preference to the system requirement as well as the storage units. For example, this topology facilitates systematic isolation of high- and low-power frequencies that can be suitably allocated to controlled SCSS and BESS, respectively. The parallel configuration of BESS and SCSS enables a wide range of control logic that can be applied to meticulously utilize both high energy and high-power densities of the HESS with improved efficacy of the overall HESS [28].

Nevertheless, the overall efficiency of the HESS decreases with an increase in the number of converters which is attributable to the losses incurred in the power converters.

19.2.2 Voltage regulation

Several control approaches and algorithms are postulated in the literature for effective utilization of BESS and SCSS towards voltage regulation [29]. The authors in [30], propose the implementation of high pass filters (HPF) to systematically formulate the current reference signal for the SCSS and BESS converters. Using generated error signals between the voltage values of the DC-bus and the desired reference, the high-frequency components are allocated for compensation by the supercapacitor, whereas the resultant low-frequency power is allocated to the battery. The control loops are formulated based on closed PI control. Accordingly, a low pass filter (LPF) is posited for power frequency isolation by primarily allocating the low-frequency power component to BESS and then the remaining high-frequency components to the SCSS. Some studies have provided comparative analysis between the HPF and LPF proving that a better response can be generated by the HESS using the LPF method [31]. The slow dynamics of the battery combined with the delay incurred due to the current and voltage sensing devices hinder the overall response in the LPF and HPF filter-based control approach. Therefore, additional intervention is required to circumvent these delays. On this front, the study in [32] proposes a quantification of the issue faced in the filter-based approach by incorporating an additional function that is the ratio between the BESS and SCSS voltages that is included in the SCSS feedback control loop. This allows the SCSS to compensate for the peak short-term power variations as well as the uncompensated power that results due to the delay observed with the filter-based control approach.

The study in [33], proposes a polynomial control approach for allocating the high-power frequency components of wind generators and the demand to SCSS. Specifically, a method based on LPF is used to generate the reference signal based on the load current and output current of the wind energy system for a particular time period. Finally, based on the inductor voltage and implementing a polynomial estimate function, the current reference for DC voltage regulation is formulated. An averaging approach that distinguishes between the BESS and SCSS current component components is formulated in [34]. A supplemental feedforward compensator for the load current is included to accelerate the reaction time of the load current. Consequently, the signal created by the voltage control loop and the load current feedforward compensator are used to assign the current reference signals to the battery and supercapacitor.

Nevertheless, the above-mentioned strategies are based on filter-based achieve the desired voltage regulation but have certain disadvantages. For example, an optimal cut-off frequency is required to allocate the power between BESS and SCSS. Most filter-based control approaches have an integrated control loop that generates the reference current required for appropriate HESS response. Therefore, the optimal cut-off frequency needs to be meticulously derived for distinctive load-generation and system configurations which proves to be tedious. Furthermore, additional sensors are required in the filter-based approach which in combination with the slow battery response decreases the overall efficacy of the HESS towards voltage regulation. The study in [35], proposes a polynomial-based technique for regulating the DC-bus voltage that is suitable for a wide range of loading scenarios as well as the dynamic intermittency of the system. Control laws are used to calculate duty ratios for BESS and SCSS converters, which are generated by generating reference current for the converters in question. However, sensing devices are necessary for battery and supercapacitor operation.

19.3 Proposed control of hybrid energy storage system

19.3.1 Conventional low pass filter controller

In the conventional strategy, based on the error signal (I_{err}) generated due to deviation from the reference voltage regulation value ($V_{r-\mathrm{dc}}$) at the actual DC-bus voltage V_{dc} caused by the demand-generation mismatch the low (I_{lf}) and high (I_{hf}) frequency components are isolated using the LPF through cut-off frequency (f_l). This is formulated using the filter equation (T_{lpf}), as:

$$T_{\mathrm{lpf}}(s) = \frac{2\pi f_l}{s + 2\pi f_l} \tag{19.1}$$

$$I_{\mathrm{lf}} = T_{\mathrm{lpf}} \cdot \quad (I_{\mathrm{err}}) \tag{19.2}$$

$$I_{\mathrm{hf}} = (I_{\mathrm{err}}) \quad - \quad I_{\mathrm{lf}} \tag{19.3}$$

The cut-off frequency can be derived using numerous approaches that include, through sampling frequency of the renewable energy generator, Ragone plot, or iterative methods [36–38]. The formulation of the control strategy is shown in Fig. 19.8. Apart from deriving distinctive optimal cut-off frequencies for different demand and generation profiles, the major drawback of this method is that due to the slow dynamic rate of response of BESS, the sensed battery current (I_b) cannot instantaneously coordinate with the battery reference signal (I_{br}) which results in a part of current being uncompensated by the BESS. This exacerbates and impedes voltage regulation and additional sensing devices [39].

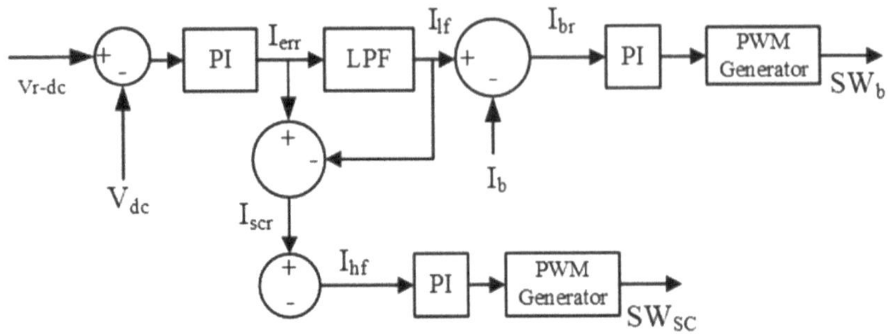

Figure 19.8 Conventional low pass filter controller [32].

19.3.2 Proposed controller strategy

A decoupled control framework is developed for BESS and SCSS to overcome the drawbacks associated with each storage component. Considering the technical constraints of BESS, the proposed controller consists of an inner and an outer control loop for controlling the current and voltage, respectively. The control loop for the current and voltage loops is designed using a type II k-type compensator.

The type II compensator is a robust lead compensator with faster response and nominal overshoot. It provides compensation ranging from 0 degree to 90 degree with a zero steady-state error. The shaping of the control loop is done through the sets of poles and zeroes by modifying the gain and phase characteristics of the open-loop frequency. The related transfer functions and argument (19.4–19.6) are as follows:

$$k_{II}(s) = v\left(\frac{1+\frac{w_z}{s}}{1+\frac{s}{w_p}}\right) \tag{19.4}$$

$$k_{II}(jw) = \frac{\left|1+\frac{jw}{w_z}\right|}{\left|\frac{jw}{w_{po}}\right|\left|1+\frac{jw}{w_p}\right|} \tag{19.5}$$

$$\arg k_{II}(jw) = \tan^{-1}\left(\frac{w}{w_z}\right) - \tan^{-1}\left(\frac{w}{w_p}\right) - \frac{\pi}{2} \tag{19.6}$$

where the poles and zeroes of the compensators are denoted by w_p and w_z in the frequency domain, respectively. The gain value is denoted by v. Using the phase gain and the loop crossover frequency (19.7–19.9), the shape of the controller is designed as per the desired requirement of the controller. The k-factor approach is derived using the pole frequency (f_p) and zero frequency (f_z) as:

$$k = \tan\left(\frac{\text{phase boost}}{2} + \frac{\pi}{4}\right) \tag{19.7}$$

$$f_p = k.\quad f_c = \tan\left(\frac{\text{phase boost}}{2} + \frac{\pi}{4}\right) f_c \tag{19.8}$$

$$f_z = \frac{f_c}{k} = \frac{f_c}{\tan\left(\frac{\text{phase boost}}{2} + \frac{\pi}{4}\right)} \tag{19.9}$$

Therefore, the stability, as well as the response of BESS, is acquired through utilizing the k-factor strategy to design the associated control loop for the battery's current and voltage regulation, considering a low bandwidth for these control loops for comprehending the response of the battery towards system fluctuation. In this respect, the battery facilitates long-term average energy demand resulting in the prolongation of the BESS life cycle via the decrease in current stress level. The functioning of SCSS therefore can be tailored to take advantage of its dominating operational features by controlling BESS such that it provides steady-state voltage regulation to the system at the PCC (as depicted in Fig. 19.9 by the blue line). Firstly, the optimal operation of a supercapacitor is impeded while facilitating steady-state voltage regulation due to low specific energy hindering their technical significance towards the system dynamics. The SCSS response should rapidly achieve a zero value when the DC-bus voltage transitions from short-term peak transience to average long-term energy requirement, as indicated in Fig. 19.9. Secondly, quick transient response is required to cope with sudden power fluctuations. A 1-D look-up table (LUT) is designed to formulate the proposed nonlinear PI controller framework which is used for establishing the nonlinear function for restricting the operation of SCSS during steady-state voltage variation. In accordance with Fig. 19.10, the primary goal of this technique is to identify the tolerable bounds, the bottom, and higher borders of DC buses at the PCC. Because of the observations made by $V_{r-\text{dc}}$ and V_{dc} at the DC link regarding error signals

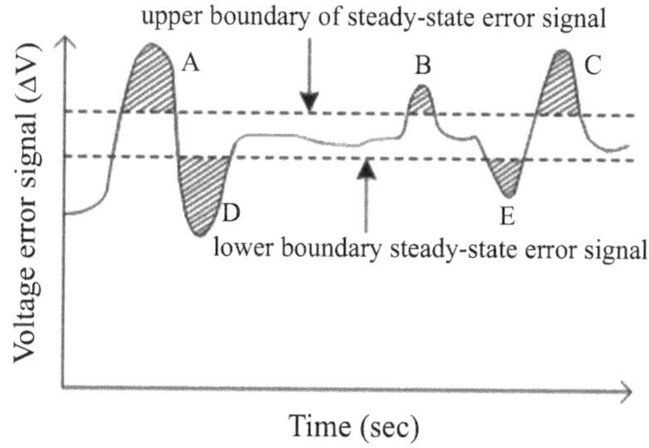

Figure 19.9 Power allocation between battery energy storage system and supercapacitor energy storage system considering DC bus voltage deviation.

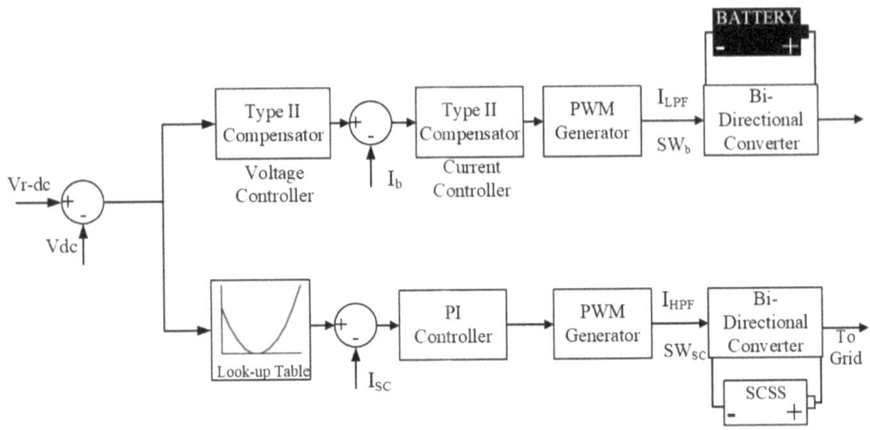

Figure 19.10 Proposed decoupled controller design.

generated, the controller is designed to operationally preclude SCSS for compensating the undervoltage and overvoltage conditions.

This is accomplished by controlling its power injection/absorption, as well as by specifying the controller's lower and upper boundary limits, as appropriate. As a result, the controller associated with the supercapacitor is restricted and does not operate inside the BESS compensation zone (i.e., steady-state). During dynamic load charging, SCSS charges at A, B, and C regions and regulates the voltage at the upper boundary area (Fig. 19.9). A similar process occurs in regions D and E, where voltage is maintained above the lower boundary by the immediate discharge of SCSS as well as the provision of insufficient power. Additionally, this method provides for the limiting of power transition during steady-state faults using appropriate model-based regression techniques, which are essential for this strategy. Section 19.4.4 provides further details into the operational dynamics of the proposed control approach.

19.4 Modeling of microgrid components

The MG consists of loads, solar PV as the generation system, and the HESS that consists of lead-acid batteries and supercapacitors. The schematic diagram of the test system is represented in Fig. 19.11.

19.4.1 Battery energy storage system

Hysteresis phenomena of BESS are used to develop the dynamic model of BESS using a controlled voltage source (V_{BVS}) and internal resistance (R_{bat}). Fig. 19.12 depicts the model of BESS that accurately represents the operation of the battery voltage (V_{bat}). The V_{bat} is calculated (19.10) using the exponential capacity (C_E),

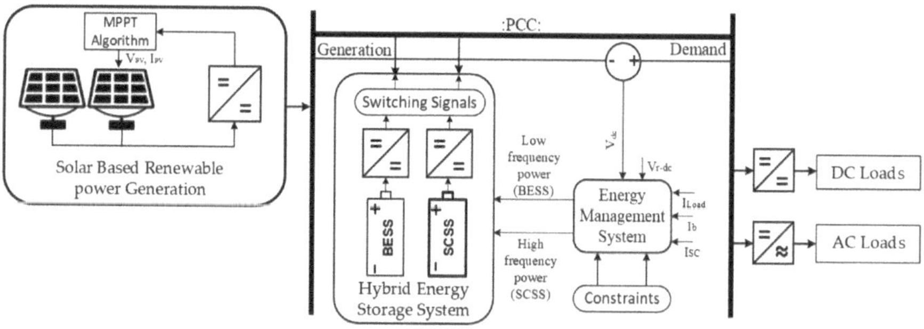

Figure 19.11 Schematic configuration of the microgrid test system integrated with battery-supercapacitor.

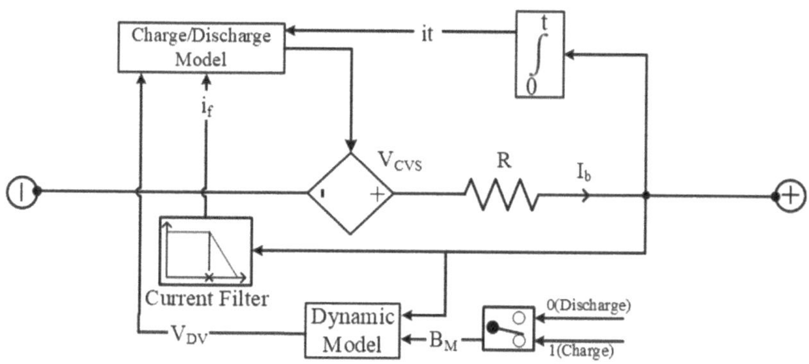

Figure 19.12 Dynamic model of battery energy storage system [40].

exponential zone voltage (C_Z), and the polarization constant (C_p) of the BESS considering the maximum capacity of the battery ($B_{\max}$).

$$V_{\mathrm{bat}} = E_b \frac{C_p B_{\max}}{B_{\max} - it} - R_{\mathrm{bat}} i_b + C_z e^{(-C_E.\mathrm{it})} - \frac{C_p B_{\max}}{B_{\max} - \mathrm{it}} . i_f \qquad (19.10)$$

where the actual charge extracted from the BESS is represented by it. The particularity of the BESS model is exhibited by its voltage dynamics that are emulated through the filter current (i_f) that corresponds to the current step response. The polarization resistance (R_p) is used to design the fluctuation associated with the open-circuit voltage of the battery with respect to its state-of-charge that is nonlinear. In most cases, R_p is calculated using Eq. (19.11).

However, the value of this resistance will be impractical in the situation of a completely charged BESS (i.e., when *it* equals 0). Thus, when compared to the experimental data, the assessed R_p contribution to BESS capacity is around 10% of the total capacity of the system. Additionally, the quick voltage rise that occurs when the BESS approaches full charge is modeled by this technique and is described by Eq. (19.12).

$$R_p = C_p \frac{B_{\max}}{\text{it}} \tag{19.11}$$

$$R_p = C_p \frac{B_{\max}}{\text{it} - 0.1. \quad B_{\max}} \tag{19.12}$$

Regardless of the SoC of the BESS, the exponential region observes a hysteresis phenomenon between the state-of-charge and the discharge/charge process [40]. The dynamic model of BESS is used to quantify this phenomenon (19.13). The constructed charge/discharge characteristics are achieved by considering the developed dynamic zone voltage (V_{DV}), and discharge/charge mode of battery (B_M) (19.14–19.15).

$$\frac{V_{\mathrm{DV}}(s)}{B_M(s)} = \frac{C_Z}{1/(C_E. \quad i_b(t).s + 1)} \tag{19.13}$$

$$V_{\mathrm{ch}} = E_b - C_p.\frac{B_{\max}}{\text{it} + 0.1 + B_{\max}}.i_f - C_p.\frac{B_{\max}}{B_{\max} - \text{it}}.\text{it} + \mathcal{L}^{-1}\left[\frac{V_{\mathrm{DV}}(s)}{B_M(s)}.\frac{1}{s}\right] \tag{19.14}$$

$$V_{\mathrm{dis}} = E_b - C_p.\frac{B_{\max}}{\text{it} - B_{\max}}. \quad i_f - C_p. \quad \frac{B_{\max}}{B_{\max} - \text{it}}.\text{it} + \mathcal{L}^{-1}\left[\frac{V_{\mathrm{DV}}(s)}{B_M(s)}.0\right] \tag{19.15}$$

19.4.2 Supercapacitor energy storage system

In this study the nonlinear dynamic model for developing and modeling SCSS and establishing the relationship between the voltage, current, and available power during the charge/discharge process [41] as shown in Fig. 19.13. The input to the

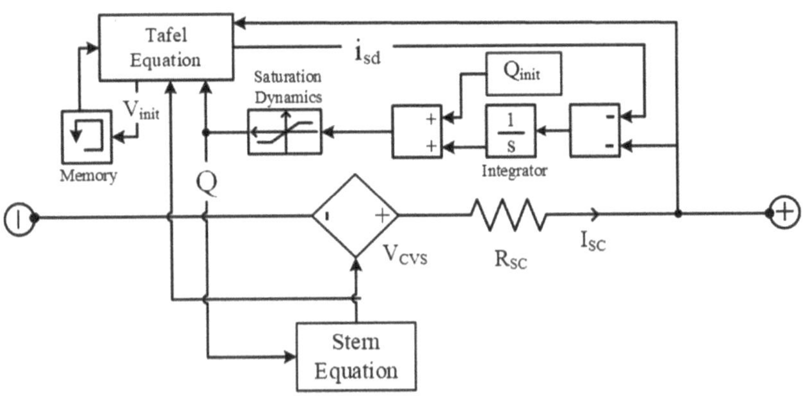

Figure 19.13 Dynamic model of supercapacitor storage system [41].

dynamic model consists the of the SCSS current (I_{SC}), voltage (V_{SC}), and the SCSS SoC. Using the SCSS series current (i_{sc}), self-discharge current (i_{sD}), initial SCSS amount of charge (S_{si}), and the quantification for the SCSS electric charge (S_{sc}) is derived (19.16). The over-charging/over-discharging are obviated using a predetermined range for the amount of SCSS charge through charge limit function.

$$S_{sc} = S_{si} + \int_0^t -(i(t) + i_{sD}(t))dt \tag{19.16}$$

$$\frac{1}{C} = \frac{1}{C_M} + \frac{1}{C_G}\left(1 + \frac{\delta\sigma_A}{\delta\sigma}\right) \tag{19.17}$$

$$i_{SD}(t) = -A.\ i_d.\ N_L.\ \exp\left\{\frac{C_T.\ F}{G.\ T_S}.\left(\frac{V_{sc}}{N_s} - \frac{V_{smax}}{N_s} - \Delta V_s\right)\right\} \tag{19.18}$$

$$V_{scv} = \frac{N_L.\ N_s\ S_{SC}.\ r_m}{N_p.\ N_L^2.\ \varepsilon_p.\ A} + \frac{2.\ N_L.\ N_s.\ G.\ T_s}{F}.\ C_T.\ r_m.$$

$$\sinh\left(\frac{S_{SC}}{N_p.\ N_L^2.\ A\ \sqrt{8.\ G.\ T_s.\ \varepsilon_p\ .\ m_c}}\right) \tag{19.19}$$

The Tafel equation [42], known as a generalized plot has a variety of applications; this study utilizes it to construct the waveform associated with the supercapacitor's leakage current (I_{SD}). The Stern equation is utilized to depict an exact ion alignment by using the modified value of capacitance, which includes the inclusion of the surface charge of ions and the charge density of the ions to formulate the enhanced capacitive summation (19.17). Accordingly, the voltage form of the combined Stern-Tafel model is obtained using Eq. (19.18). The voltage form stated is inserted in the Tafel Eq. (19.19), where F and G are the Faraday and ideal gas constants, respectively, to simulate the SCSS current. The curve obtained is used to model the SCSS current, with the exponent of the curve variation based on overpotential, ideal gas constant, and the temperature ratios (19.16). Furthermore, with the inclusion of the effect associated with the motion related to the thermal ion mobility on the capacitance the Gouy-Chapman model is improved [43]. To construct the integrated Stern-Tafel model of the supercapacitor, the remaining parameters, such as the ion alignment factor, model capacitance (C_M), thermal motion capacitance (C_G), molecular radius (r_m), material permeability (ε_p), operating temperature (T_s), overpotential (V_s), number of parallel and series cells of supercapacitor (N_p, N_s), number of electrolyte layers of supercapacitor (N_L), and interfacial area (A).

19.4.3 Solar photovoltaics generation system

A single-diode five-parameter model of PV cell module is utilized to develop the equivalent electrical model of the PV generation system. In comparison to the other models, such as double-diode seven parameter models [44] that are complex formulations with no contribution towards the model accuracy, the single-diode five parameter model proves to be simplified with resiliency towards better accuracy to experimental data sets [45]. The single-diode five-parameter model comprises a diode, series resistance (R_s), and shunt resistances (R_{sh}). The five parameters consist of the, R_s, R_{sh}, diode's ideality factor (μ), saturation current (I_{sat}), and the photocurrent (I_{ph}). However, R_{sh} or both R_s and R_{sh} are neglected in some studies for simplicity [46]. The mathematical formulation of the model is as the followings:

$$I_{pv} = I_{ph} - I_{sat}\left[\exp\left(\frac{V_{pv} + R_s I_{pv}}{\mu V_{th}}\right) - 1\right] - \frac{V_{pv} + R_s I_{pv}}{R_{sh}} \tag{19.20}$$

$$V_{th} = \frac{n_s k_B T}{q} \tag{19.21}$$

where the thermal voltage of the PV module is denoted by V_{th}. The relationship of V_{th} is established with the electron charge (q), Boltzmann constant (k_B), the temperature of the p-n junction (T), series-connected cells (n_s), and the ideality factor. The range of the ideality factor is taken between 1 and 1.5. Therefore, the current and the solar irradiance are derived using the generated standard photocurrent ($I_{ph,s}$) at 25°C and 1000 W/m^2, current-temperature coefficient (I_{ts}), the difference between the actual and nominal temperature in kelvin (ΔT), W/m^2 irradiation on the device (λ), and nominal irradiation (λ_n) in W/m^2. The standard parameters of the structured Kyocera KC200GT PV module are given in Table 19.1. Accordingly, Perturb and Observe (P&O) MPPT algorithm is used to obtain the MPPT of the PV modules that are incorporated through a boost converter connecting the solar PV at the PCC of the MG.

$$I_{pv,s} = \left(I_{ph,s} + I_{ts}\Delta T\right)\frac{\lambda}{\lambda_n} \tag{19.22}$$

$$I_{ph,n} = \frac{R_{sh} + R_s}{R_{sh}} I_{sat} \tag{19.23}$$

Table 19.1 Standard parameters of the photovoltaics module.

Parameter	Value
Power at MPP	200 W
Current at MPP	26.3 V
Short circuit current	7.6 A
Voltage at MPP	32.9 V
Open circuit voltage	8.2 A

19.4.4 Bi-directional converter

The bidirectional DC–DC converter allows a systematic control over the HESS enabling their respective charging/discharging in accordance with the energy and power requirement of the system. The converter's open-loop transfer function is developed using small signal stability analysis [47]. The following ratios are employed in the modeling process: the ratio of inductor current to duty cycle (19.24) and the ratio of voltage output to duty ratio (19.25).

$$G_{\text{id}} = \frac{\frac{V_{\text{in}}}{1-D}\left(2 + sR_L\right)}{(1-D)^2R_L + sL + s^2R_LC} \tag{19.24}$$

$$G_{\text{vd}} = \frac{\frac{V_{\text{in}}}{R_L(1-D)^2}\left(R_L^2(1-D)^2 - sR_L\right)}{(1-D)^2R_L + sL + s^2R_LC} \tag{19.25}$$

where, V_{in} denotes the input voltage, R_L represents the load resistance, D denotes the duty ratio of the converter, the inductor is represented by L, and C represents the converter's output capacitance. The derived transfer functions for BESS current and voltage loops are expressed in (19.26–19.27), respectively. The control loop of SCSS is given in (19.28).

$$G_{\text{id}} = \frac{190230.4(s + 130)}{(s^2 + 432s + 1412760.4)} \tag{19.26}$$

$$G_{\text{vd}} = \frac{-0.369(s^2 - 11068s - 212703645)}{(s^2 + 432s + 1412760.4)} \tag{19.27}$$

$$G_{\text{id}} = \frac{73166.1(s + 235.9)}{(s^2 + 259.5s + 9897944.2)} \tag{19.28}$$

The k-type compensator design is derived using low bandwidth to comprehend the slow transient response of the BESS. In accordance with the switching frequency ($f_{\text{sf}} = 30\text{kHz}$), $f_{\text{sf}}/67$ and $f_{\text{sf}}/1380$ are selected as the current-voltage bandwidths, respectively. The desired phase margin is kept at 52 degrees. Therefore, the current and voltage controller's transfer is denoted by T_A (19.28) and T_B (19.29), respectively. A PI controller is designed for the SCSS control loop (19.31) with gain constant (k_p) and integral gain (k_i) set to 0.05 and 10, respectively. The bandwidth of the controller is at 3914 rad/sec with a phase margin of 87.7 degrees.

$$T_A = \frac{78.7(s + 1179.8)}{s(s + 6511)} \tag{19.29}$$

$$T_B = \frac{180.9(s + 83.8)}{s(s + 220)} \tag{19.30}$$

$$T_{PI} = \frac{k_p s + k_i}{s} \tag{19.31}$$

To operate the LUT controller, the voltage error on the DC bus is used as an input signal. The range of the input signal is kept within -1 and 1. The resulting output voltage is confined to the range of -3V to 3 V (in this case), and this voltage is used as the LUT controller's input signal after being reduced and normalized to one-third of its original value via reduction and normalization. The signal is amplified by multiplying it by the base value of the current for increasing its transient response, resulting in the generation of the nominal value of the supercapacitor's reference current for the system.

The controller of the supercapacitor is designed to meet the subjective model-based requirements of SCSS with respect to the power requirement of the system. The connection between the input of the LUT controller, which is represented by the voltage error of the DC link, and the produced SCSS current reference output to the LUT controller is shown in the graph shown in Fig. 19.14. During steady-state error intervals, between -0.2 and 0.2 p.u., this output is set to stay constant, resulting in a zero-reference current for SCSS when the DC voltage error is minimal. As a result, the output of the controller rises above these levels, which is set non-linearly in accordance

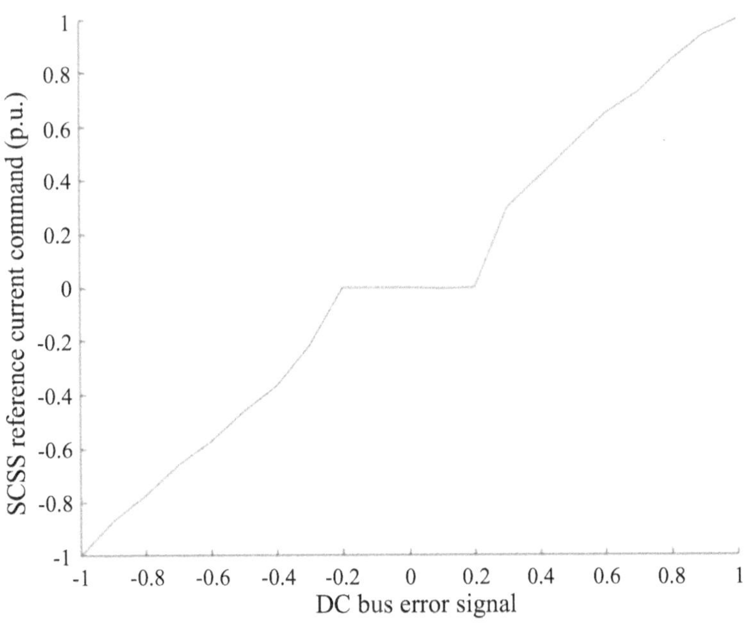

Figure 19.14 Graphical representation of supercapacitor energy storage system control loop: Relationship between the input and the output signal.

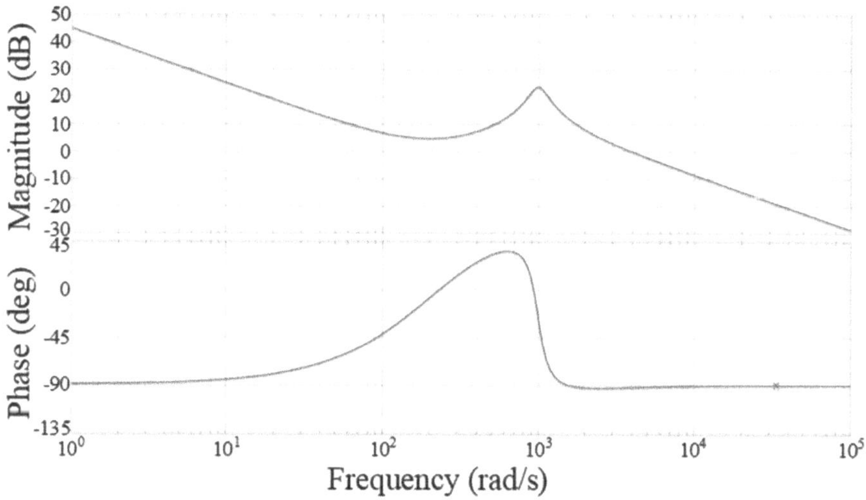

Figure 19.15 Stability analysis of supercapacitor control loop using bode plot.

with the operational requirement considered to be desired for an effectively faster response towards voltage regulation of the DC link bus. The associated stability of the SCSS and BESS controllers are depicted in Figs. 19.15 and 19.16, respectively.

19.5 Results and discussion

The formulated control technique is validated, and a comparative analysis with LPF strategy is a renewable-based MG is performed. The test system comprises of load, a parallel-active integrated battery-supercapacitor HESS, and a solar PV generation source integrated through a DC–DC boost converter as depicted in Fig. 19.11. The desired reference voltage of the DC-bus link ($V_{r-\text{dc}}$) is taken as 48 V. The specifications of BESS and SCSS with their corresponding converter rating is given is Table 19.2. Numerous cases of load and renewable generation are performed. The comparative analysis is performed based on the voltage variation incurred during demand-generation fluctuation considering the DC-link voltage, BESS current, SCSS current, and BESS SoC highlighting the efficacy of DC-bus voltage regulation and BESS stress current reduction.

19.5.1 Renewable power smoothing

19.5.1.1 Step decrease in generation

The load demand is set to 480 W by setting the load resistance to 4.8 Ω during the entire operation of this test case that projects a load demand of 10 A (i.e., 480 W). The solar irradiance is initially set to 509 W/m^2 till time interval, $t = 1$ seconds after which it is increased to 300 W/m^2 that projects a decrement of solar power generation from

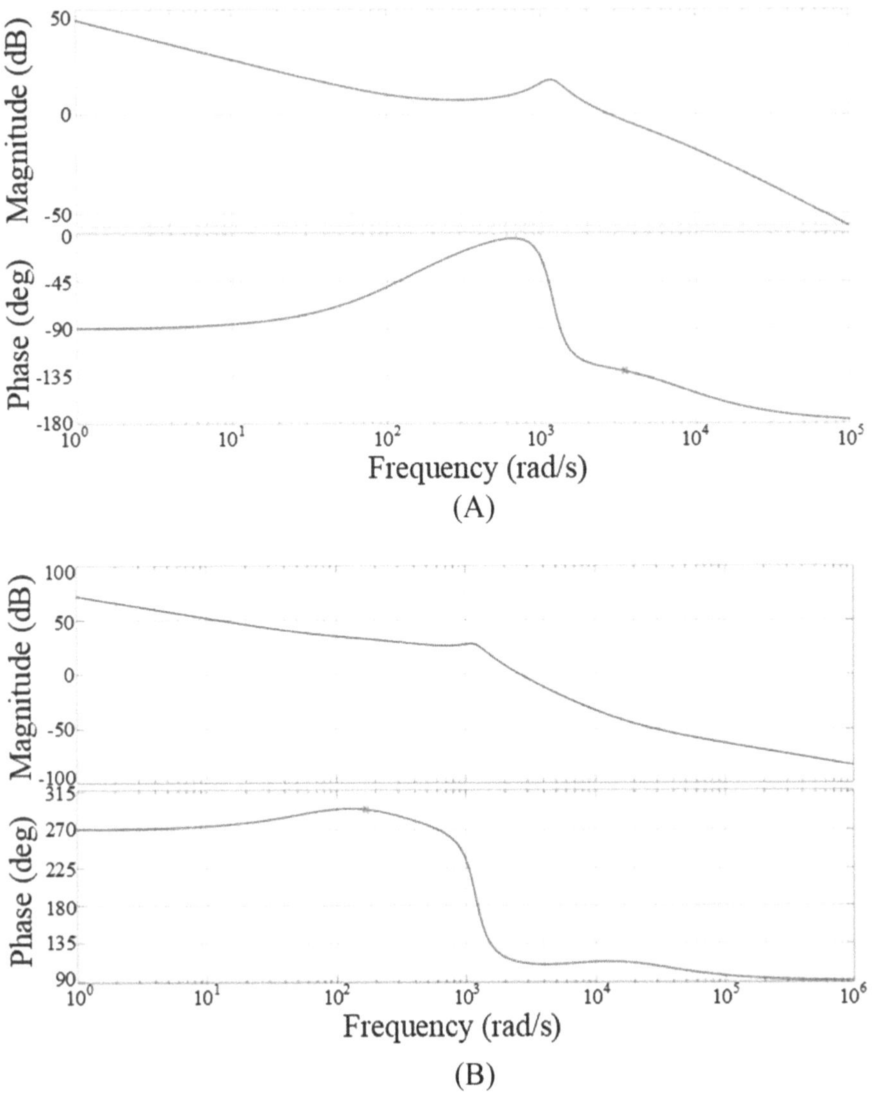

Figure 19.16 Stability analysis of the battery control loop: (A) Current (B) Voltage.

initially from 480 W is reduced to 280 W after the MPPT operation. Therefore, a voltage drop from 48 V is experienced at the DC-link bus at $t = 1$ seconds that needs to be regulated through the HESS discharging operation to satisfy the load demand due to the deficient power from the variable generation source.

With the solar PV generation supplying 5.83 A to the load demand, the BESS supplies the deficient supply of 4.17 A with an initial overshoot (as shown in Fig. 19.17) reaching a steady state at $t = 1.3$ seconds with the LPF control

Table 19.2 Parameters of the bi-directional converters and rating of battery-supercapacitor hybrid energy storage system.

Components	Rating
The equivalent series resistance of a capacitor	0.1 Ω
Parasitic resistance of the inductor	0.1 Ω
Battery energy storage system (BESS)	14 Ah, 24 V
BESS converter output capacitance	2000 F
BESS converter inductance	230 H
Supercapacitor storage system (SCSS)	29 F, 32 V
SCSS converter output capacitance	1000 F
Switching frequency	30 kHz
SCSS converter inductance	600 H

technique. On the other hand, the initial BESS current overshoot is negligible with the BESS reaching a steady state at $t = 1.05$ seconds with the proposed control technique as shown in Fig. 19.18.

The LPF-based control technique is unable to optimally regulate the voltage during the step decrement of the solar PV generation experiencing a voltage drop of 46 V whereas a comparatively less voltage deviation magnitude of 47.5 V is experienced with the proposed control technique. This can be attributable to the slow dynamics of BESS, in the LPF control technique, which inherently hinders the required rapid response resulting in an uncompensated current that contributes to this voltage deviation and in addition has a detrimental effect on the BESS life cycle. Furthermore, the decrease in the BESS SoC is comparatively lower for the proposed control technique.

19.5.1.2 Step increase in generation

The solar irradiance is initially kept at 509 W/m^2 generating a power of 480 W after the MPPT operation. The load demand is set at 480 W by setting the load resistance to 4.8 Ω that projecting the load current to be 10 A. At time interval, $t = 1$ seconds the solar irradiance is increased to 990 W/m^2 increasing the solar energy production to 960 W/m^2. This results in excess power generation that results in voltage increases at the DC-bus link that requires charging of the HESS to regulate the voltage deviation by absorbing the excess generated power through charging.

The BESS compensates for the excess power of 20 A generated by the solar PV generation and reaches a steady current state at $t = 1.2$ seconds with minor current overshoot experienced at the initial stage of compensation as shown in Fig. 19.19. In comparison, the BESS reaches a steady-state comparative faster and instantly (at $t = 1$ seconds), with no current overshoot at the initial stages of charging (Fig. 19.20).

Accordingly, considering the degree of the voltage deviation and its magnitude at the DC-bus link. The voltage reaches 56 V in the case of the LPF control technique whereas in the case of the proposed controller voltage is regulated well below 52 V. The large overshoot incurred in the LPF control technique is mainly due to the large amount of abrupt current generated during the rapid increment in the solar PV

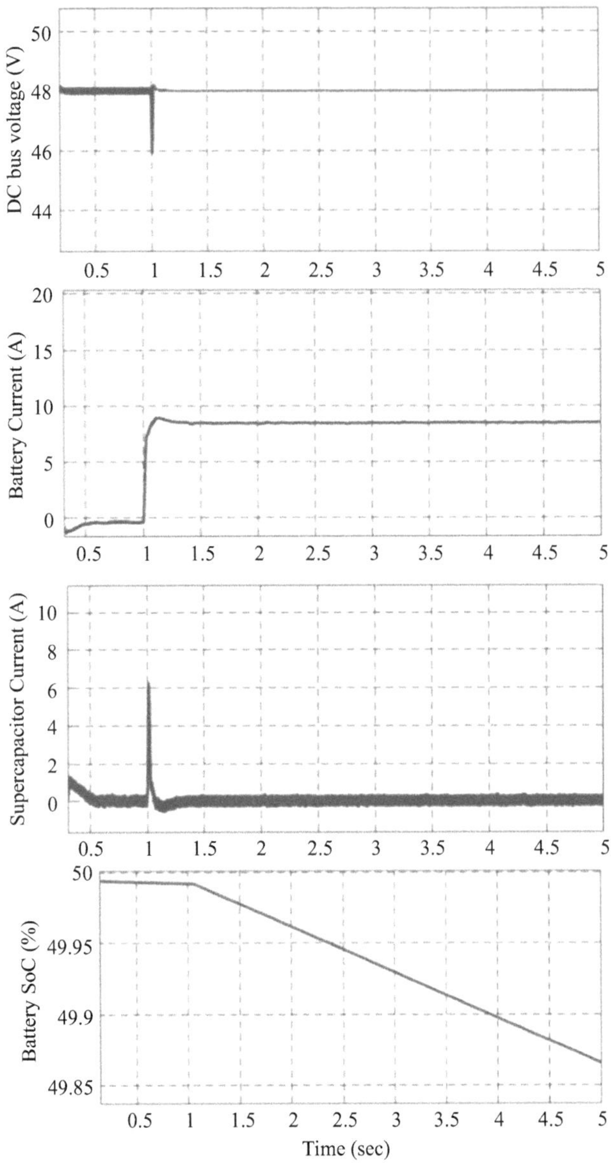

Figure 19.17 Results for step decrease in solar photovoltaics using low pass filter control technique.

generation that remains uncompensated due to the slow dynamics of the BESS and subsequently a delayed response to the interlinked response generated from the SCSS that contributes to this huge increase in the voltage level. The proposed control technique additionally facilitates a comparatively lower BESS SoC increase highlighting the obviation of excess current stress and enhanced life cycle utilization.

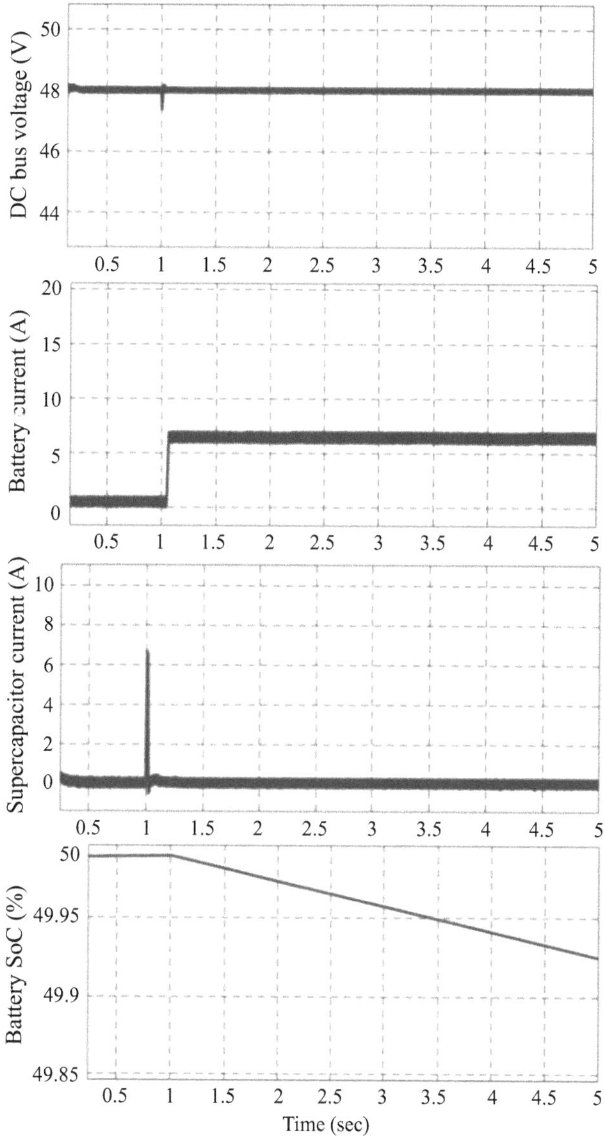

Figure 19.18 Results for the step decrease in solar photovoltaics using the proposed control technique.

19.5.2 Load smoothing

19.5.2.1 Step decrease in load demand

The solar irradiance is kept constant at 509 W/m^2 generating a constant power of 480 W from the solar PV generation system in the entire operation of the test case.

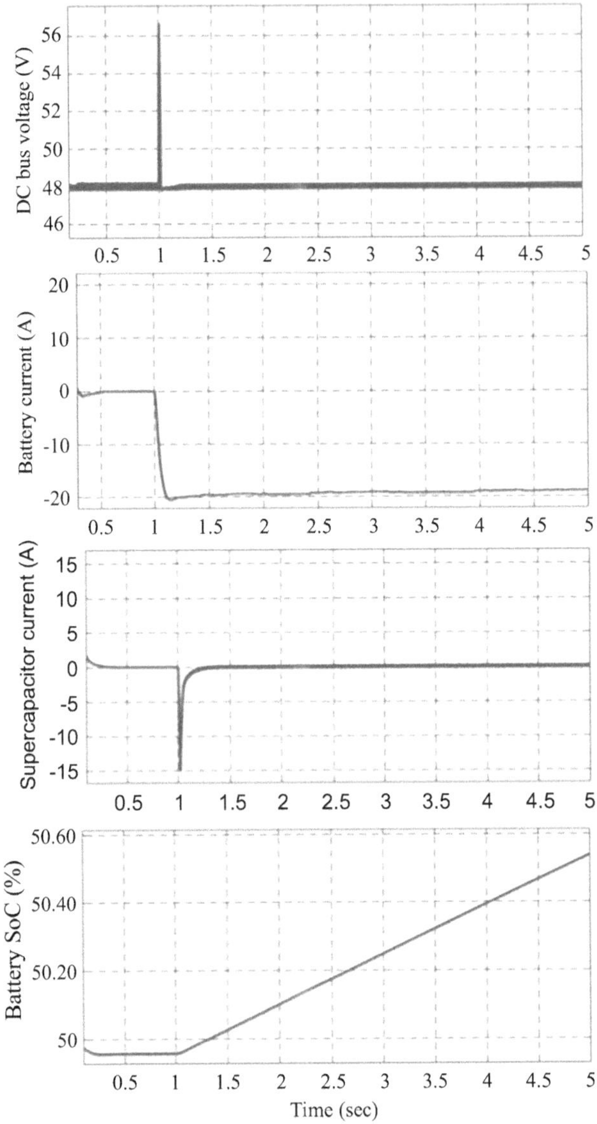

Figure 19.19 Results for step increase in solar photovoltaics using low pass filter control technique.

The load is kept at 480 W during the initial operational interval through a load resistance of 4.8Ω that projects a load current demand of 10 A. At the simulation time interval ($t = 1$ seconds), the load resistance is increased to 9.6 Ω resulting in the load demand decrease to 240 W is 5 A. The initial DC-link voltage at 48 V experiences a voltage rise due to the rapid load decrement that results in excess generated power by the solar PV generation system.

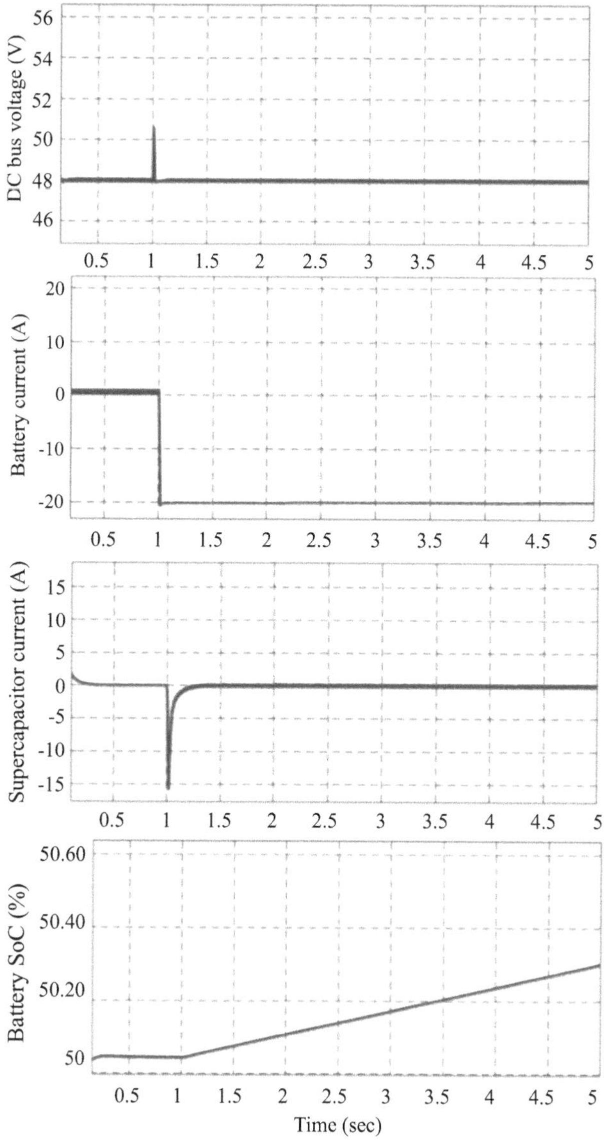

Figure 19.20 Results for step increase in solar photovoltaics using the proposed control technique.

This requires a rapid charging of the HESS components. In accordance with the results obtained using the LPF control technique (Fig. 19.21), the BESS absorbs the generated excess power reaching a steady current state at $t = 2.0$ seconds. On the other hand, the BESS rapidly reaches a steady operating state (within $t = 1$ seconds) as shown in Fig. 19.22. Considering the LPF control technique, it is observed that the SCSS for a longer duration until the BESS reaches a steady operating state. This is attributed to the delayed BESS current reference generation that hinders the operational response of the SCSS. In contrast, the proposed controller is based on a decoupled control for both BESS and SCSS enabling the SCSS to circumvent the delayed response by the BESS sensors and controllers that generates a rapid response of SCSS enabling the optimal utilization of SCSS for short-term power compensation. Accordingly, the voltage deviation magnitude reaches around 50 V in the case of the LPF control technique whereas the proposed controller experiences 47 V fluctuation at the DC-link bus during the abrupt load decrement. Finally, the BESS SoC increase experienced with the proposed controller is lower.

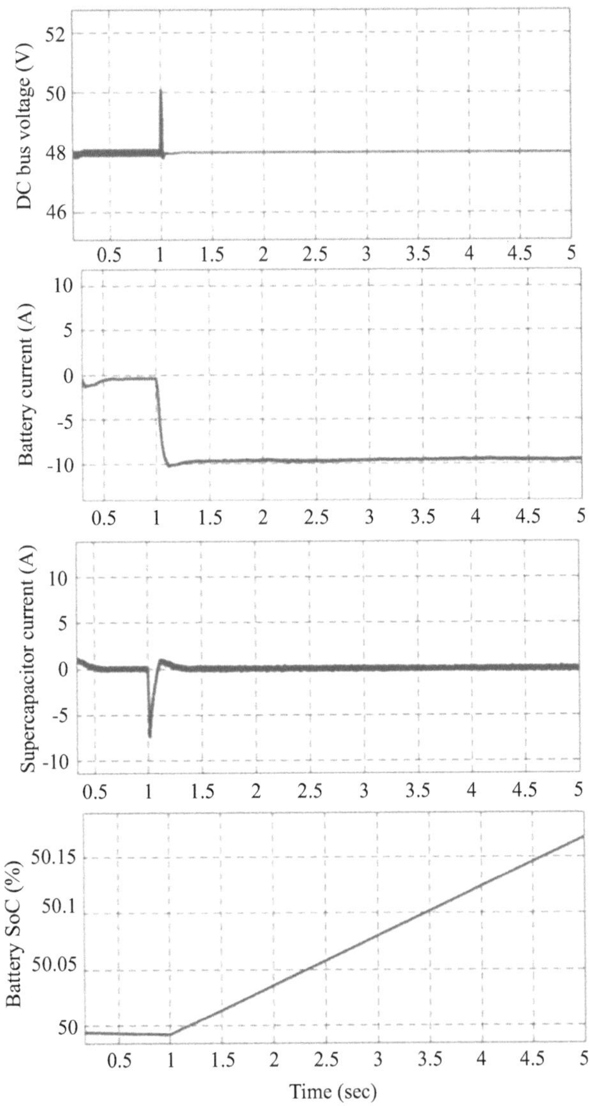

Figure 19.21 Results for step decrease in load demand using low pass filter control technique.

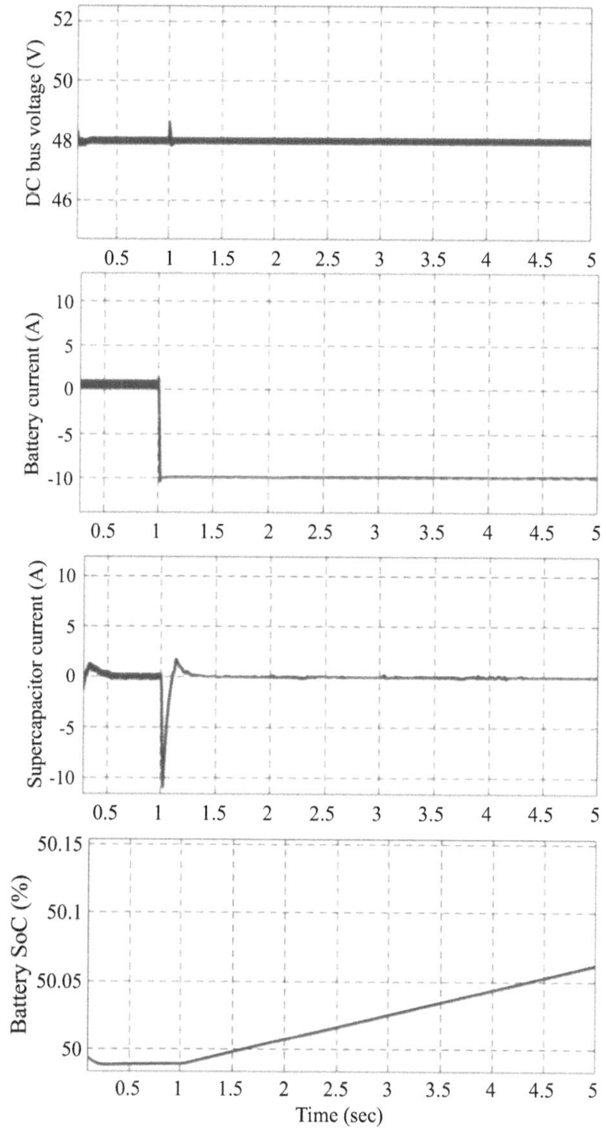

Figure 19.22 Results for step decrease in load demand using the proposed control technique.

19.5.2.2 Step increase in load demand

In this test, the load resistance is initially set to 4.8 Ω, that is, the load demand is 480 W. At time interval $t = 1$ seconds, the load is increased to 780 W by decreasing the load resistance to 2.95 Ω that increases the load current requirement to 16.25 A. The solar irradiance is kept constant at 509 W/m^2 generating a constant power of 480 W during the whole operation of the test. Due to the load increment, the

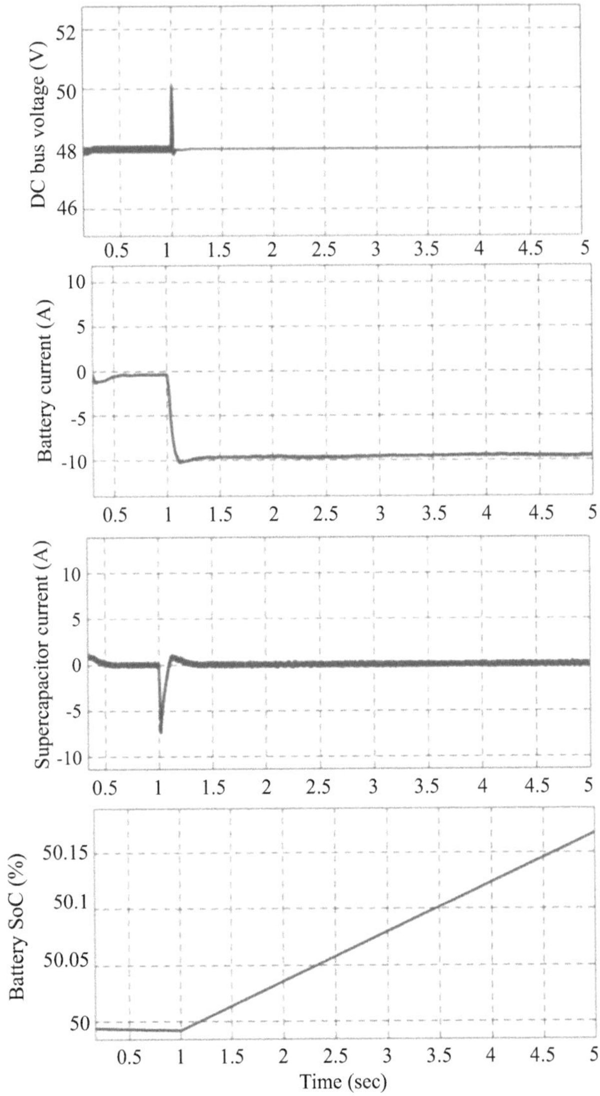

Figure 19.23 Results for step increase in load demand using low pass filter control approach.

DC-link voltage experiences a voltage rise from its initial value of 48 V requiring the deficient power being supplied through the discharging of the HESS.

The BESS provides the average energy demand of the load and reaches a steady operating state at $t = 1.25$ seconds with an initial minor current overshoot in the case of the LPF control approach (Fig. 19.23) whereas with the proposed control approach the BESS steady operating state is achieved rapidly at $t = 1$ seconds with negligible initial current overshoot. Accordingly, the degree of voltage magnitude

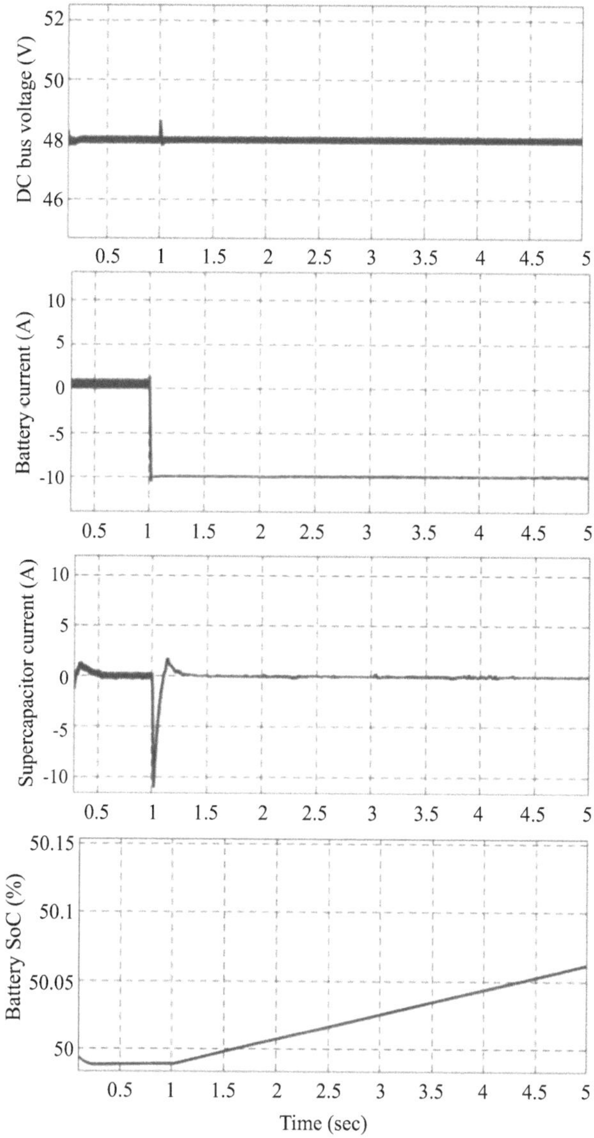

Figure 19.24 Results for step increase in load demand using the proposed control technique.

incurred with the LPF control approach (around 45.9 V) is comparatively greater than the voltage level observed with the proposed control approach (47 V) as shown in Fig. 19.24. The proposed control approach therefore observably proves to facilitate an improved HESS response towards DC-bus voltage regulation while utilizing appropriate power allocation between the SCSS, for short-term peak power compensation, and BESS for average long-term average power requirements. Furthermore, the percentage of BESS SoC decrease experienced with the proposed control approach is lower is comparatively lower highlighting a decreased current stress during the abrupt discharging process due to rapid load increment.

19.6 Conclusion

In this chapter hybridization of BESS with SCSS is presented to highlight the potential enhancement obtained with the HESS. A control approach is proposed to regulate the DC-bus voltage during rapid and transient renewable and load variation cases. The control approach is based on establishing a separate control strategy for BESS and SCSS based on k-type compensators and nonlinear PI control, respectively. The controller aims to utilize the advantages of each energy storage device based on their dominant technical characteristics, that is, the high energy density of BESS and high-power density of SCSS, and enhanced voltage regulation response for transient disturbances observed at the PCC.

The proposed control approach is designed and validated in various cases of solar PV and load demand variation. Additionally, a comparative analysis is presented with the benchmark LPF control approach to highlight the superiority and the enhanced efficacy of the proposed controller in terms of reducing the voltage magnitude and deviation during rapid system variations. The BESS life cycle is observable enhanced due to reduced current stress during the transient variation intervals of the system with an improved state-of-charge profile. The proposed control technique hence proves to provide comparatively enhanced voltage regulation with an overall rapid transient response obviating the need for additional sensing devices with a relatively simple design approach.

References

[1] F. Verástegui, Á. Lorca, D. Olivares, M. Negrete-Pincetic, Optimization-based analysis of decarbonization pathways and flexibility requirements in highly renewable power systems, Energy 234 (2021) 121242.

[2] A. Jawad, N. Al Masood, A systematic approach to estimate the frequency support from large-scale PV plants in a renewable integrated grid, Energy Rep. 8 (2022) 940–954.

[3] A.J. Harker Steele, J.W. Burnett, J.C. Bergstrom, The impact of variable renewable energy resources on power system reliability, Energy Policy 151 (2021) 111947.

[4] H.S. Salama, A. Bakeer, G. Magdy, I. Vokony, Virtual inertia emulation through virtual synchronous generator based superconducting magnetic energy storage in modern power system, J. Energy Storage 44 (2021) 103466.

[5] M. Jafari, A. Botterud, A. Sakti, Decarbonizing power systems: a critical review of the role of energy storage, Renew. Sustain. Energy Rev. 158 (2022) 112077.

[6] J.A. Villanueva-Rosario, F. Santos-García, M.E. Aybar-Mejía, P. Mendoza-Araya, A. Molina-García, Coordinated ancillary services, market participation and communication of multi-microgrids: a review, Appl. Energy 308 (2022) 118332.

[7] H. Kang, S. Jung, M. Lee, T. Hong, How to better share energy towards a carbon-neutral city? A review on application strategies of battery energy storage system in city, Renew. Sustain. Energy Rev. 157 (2022) 112113.
[8] Y.M. Al-Humaid, K.A. Khan, M.A. Abdulgalil, M. Khalid, Two-stage stochastic optimization of sodium-sulfur energy storage technology in hybrid renewable power systems, IEEE Access. 9 (2021) 162962–162972.
[9] M.Z. Iqbal, U. Aziz, Supercapattery: merging of battery-supercapacitor electrodes for hybrid energy storage devices, J. Energy Storage 46 (2022) 103823.
[10] S.C.L. Koh, et al., Higher 2nd life Lithium Titanate battery content in hybrid energy storage systems lowers environmental-economic impact and balances eco-efficiency, Renew. Sustain. Energy Rev. 152 (2021) 111704.
[11] X. Lin, R. Zamora, Controls of hybrid energy storage systems in microgrids: critical review, case study and future trends, J. Energy Storage 47 (2022) 103884.
[12] M. Bagheri-Sanjareh, M.H. Nazari, S.H. Hosseinian, Coordination of hybrid energy storage system, photovoltaic systems, smart lighting loads, and thermostatically controlled loads for microgrid frequency control, Int. Trans. Electr. Energy Syst. 31 (8) (2021) e12976.
[13] J. Fang, Q. Xu, R. Tang, Y. Xia, Y. Ding, L. Fang, Research on demand management of hybrid energy storage system in industrial park based on variational mode decomposition and Wigner–Ville distribution, J. Energy Storage 42 (2021) 103073.
[14] M. Badar, I. Ahmad, A.A. Mir, S. Ahmed, A. Waqas, An autonomous hybrid DC microgrid with ANN-fuzzy and adaptive terminal sliding mode multi-level control structure, Control. Eng. Pract. 121 (2022) 105036.
[15] M.K. Baek, J.H. Roh, J.B. Park, W.C. Jeong, Optimal sizing of battery/supercapacitor hybrid energy storage systems for frequency regulation, J. Electr. Eng. Technol. 17 (1) (2022) 111–120.
[16] D.P. Kothari, A. Pathak, U. Pandey, Design of microgrids for rural electrification, Resid. Microgrids Rural. Electrif (2022) 87–108.
[17] R. Madhana, M. Geetha, Power enhancement methods of renewable energy resources using multiport DC-DC converter: a technical review, Sustain. Comput. Inf. Syst. (2022) 100689.
[18] A. Agrawal, R. Gupta, Coordinated control of hybrid DERs enabled grid-interactive residential PCM with hybrid bus layout, IEEE Syst. J. (2022).
[19] K.U. Jan, A.M. Dubois, D. Diallo, Hybrid battery-SC and battery-battery multistage design and energy management for power sharing, in: IECON Proceeding Industrial Electronics Conference, vol. 2021, October 2021.
[20] S. Ferahtia, et al., Optimal adaptive gain LQR-based energy management strategy for battery–supercapacitor hybrid power system, Energies 14 (6) (2021) 1660.
[21] N. Qi, Y. Yin, K. Dai, C. Wu, X. Wang, Z. You, Comprehensive optimized hybrid energy storage system for long-life solar-powered wireless sensor network nodes, Appl. Energy 290 (2021) 116780.
[22] F. Liu, C. Wang, Y. Luo, Parameter matching method of a battery-supercapacitor hybrid energy storage system for electric vehicles, World Electr. Veh. J. 12 (4) (2021) 253.
[23] J. Li, F. Yao, Q. Yang, Z. Wei, H. He, Variable voltage control of a hybrid energy storage system for firm frequency response in the UK, IEEE Trans. Ind. Electron. (2022).
[24] H. Xu, M. Shen, The control of lithium-ion batteries and supercapacitors in hybrid energy storage systems for electric vehicles: a review, Int. J. Energy Res. 45 (15) (2021) 20524–20544.

[25] M.J. Lencwe, S.P. Daniel Chowdhury, T.O. Olwal, An effective control for lead-acid performance enhancement in a hybrid battery-supercapacitor system used in transport vehicles, Sustainability 13 (24) (2021).
[26] Z. Cabrane, J. Kim, K. Yoo, M. Ouassaid, HESS-based photovoltaic/batteries/supercapacitors: energy management strategy and DC bus voltage stabilization, Sol. Energy 216 (2021) 551–563.
[27] Q. Zhong, C. Xie, S. Jin, B. Li, K. Shi, New optimal control algorithms for battery-supercapacitor HESS based on Wirtinger-based integral inequality, IEEE Access. 9 (2021) 17707–17716.
[28] K. Jayasawal, A.K. Karna, K.B. Thapa, Topologies for interfacing supercapacitor and battery in hybrid electric vehicle applications: an overview, in: 2021 International Conference on Sustainable Energy and Future Electric transportation SeFet 2021, January 2021.
[29] W. Jing, C.H. Lai, S.H.W. Wong, M.L.D. Wong, Battery-supercapacitor hybrid energy storage system in standalone DC microgrids: a review, IET Renew. Power Gener. 11 (4) (2017) 461–469.
[30] W. Li, G. Joós, A power electronic interface for a battery supercapacitor hybrid energy storage system for wind applications, PESC. Rec. - IEEE Annu. Power Electron. Spec. Conf. (2008) 1762–1768.
[31] L. Wang, Y. Li, M. Sheng, Y. Xu, Z. Yang, B. Zhang, Control strategies of HESS for the application of directly driven WTG, Adv. Mater. Res. 512–515 (2012) 694–699.
[32] S.K. Kollimalla, M.K. Mishra, N.L. Narasamma, Design and analysis of novel control strategy for battery and supercapacitor storage system, IEEE Trans. Sustain. Energy 5 (4) (2014) 1137–1144.
[33] A. Tani, M.B. Camara, B. Dakyo, Energy management in the decentralized generation systems based on renewable energy - ultracapacitors and battery to compensate the wind/load power fluctuations, IEEE Trans. Ind. Appl. (2015).
[34] D. Pavković, M. Lobrović, M. Hrgetić, A. Komljenović, A design of DC bus control system for EVs based on battery/ultracapacitor hybrid energy storage, in: 2014 IEEE International Electric Vehicle Conference IEVC 2014, March 2015.
[35] M.B. Camara, B. Dakyo, H. Gualous, Polynomial control method of DC/DC converters for DC-bus voltage and currents management-battery and supercapacitors, IEEE Trans. Power Electron. (2012).
[36] Y. Sun, et al., Application of integrated energy storage system in wind power fluctuation mitigation, J. Energy Storage 32 (2020) 101835.
[37] A.L. Allègre, A. Bouscayrol, R. Trigui, Flexible real-time control of a hybrid energy storage system for electric vehicles, IET Electr. Syst. Transp. 3 (3) (2013) 79–85.
[38] A. Florescu, S. Bacha, I. Munteanu, A.I. Bratcu, A. Rumeau, Adaptive frequency-separation-based energy management system for electric vehicles, J. Power Sources 280 (2015) 410–421.
[39] S.K. Kollimalla, M.K. Mishra, A. Ukil, H.B. Gooi, DC grid voltage regulation using new HESS control strategy, IEEE Trans. Sustain. Energy 8 (2) (2017) 772–781.
[40] T. Çarkıt, M. Alçı, Investigation of V oc and SoH on Li-ion batteries with an electrical equivalent circuit model using optimization algorithms, Electr. Eng. (2022) 1–12.
[41] H. Çınar, I. Kandemir, Active energy management based on *meta*-heuristic algorithms of fuel cell/battery/supercapacitor energy storage system for aircraft, Aerospace 8 (3) (2021).
[42] S. Hamedi, T. Ghanbari, E. Moshksar, Z. Hosseini, Time-varying model of self-discharge in a double layer supercapacitor with blocking layer, J. Energy Storage 40 (2021) 102730.

[43] A. Allagui, H. Benaoum, O. Olendski, On the Gouy–Chapman–Stern model of the electrical double-layer structure with a generalized Boltzmann factor, Phys. A Stat. Mech. Appl. 582 (2021) 126252.
[44] T. Ma, et al., Performance modelling of photovoltaic modules under actual operating conditions considering loss mechanism and energy distribution, Appl. Energy 298 (2021) 117205.
[45] Z. Chen, et al., Rapid and accurate modeling of PV modules based on extreme learning machine and large datasets of I-V curves, Appl. Energy 292 (2021) 116929.
[46] V.K. Yadav, et al., A novel reconfiguration technique for improvement of PV reliability, Renew. Energy 182 (2022) 508–520.
[47] S. Saurav, A. Ghosh, Switched Mode Fourth-Order Buck–Boost Converter Using Type II and Type III Controllers in DC Grid Applications, in: DC—DC Converters for Future Renewable Energy Systems. Energy Systems in Electrical Engineering, Springer, Singapore, 2022, pp. 383–403.

Supervisory control strategy for a customized solar photovoltaic-based microgrid with a battery storage system: an experimental approach

20

Rahul Singh[1], Syed Muhammad Amrr[2], Abdelaziz Salah Saidi[3,4] and M.S. Jamil Asghar[5]

[1]Department of Electrical Engineering, Indian Institute of Technology Jodhpur, Jodhpur, Rajasthan, India, [2]Department of Electrical Engineering, Indian Institute of Technology Delhi, New Delhi, India, [3]Department of Electrical Engineering, King Khalid University, Abha City, Saudi Arabia, [4]University of Tunis El Manar, National School of Engineers of Tunis, Laboratory of Electrical Systems, Tunis, Tunisia, [5]Department of Electrical Engineering, Aligarh Muslim University, Aligarh, Uttar Pradesh, India

Chapter Outline

Emerging Trends in Energy Storage Systems and Industrial Applications. DOI: https://doi.org/10.1016/B978-0-323-90521-3.00018-1

Highlights

1. Enhances SPV energy utilization
2. Works in both islanded and grid-tied/grid-connected modes
3. Reserves power for emergency loads under worst-scenario
4. The control algorithm is realized and validated on hardware setup

Nomenclature

SPV	solar photovoltaic
T&D	transmission and distribution
MPPT	maximum power point tracking
GT	grid-tied
GC	grid connected
GLS	general lighting service
LDR	light dependent resistors
LED	light emitting diode
DC	direct current
BLDC	brushless DC
MG	microgrid
PCC	point of common coupling
BMS	battery management system
EMS	energy management system
PMS	power management system
UPS	uninterrupted power supply
SOC	state-of-charge
V_B	battery voltage
S_1, S_2, **and** S_3	different static relay switches

20.1 Introduction

The energy demand is growing rapidly and tremendous utilization of conventional energy resources (fossil fuels) are drawing wide attention to serious environmental issue [1–4]. According to the BP Statistical Review of World Energy, the existing resources of coal, oil and flammable gases are only accessible for around the next 54 years with respect to the current mining rate [5]. The rate at which the world's oil, gas, coal, and other energy reserves are depleted raises widespread concern on various levels. Furthermore, the use of fossil fuels (produced from petroleum) contributes to pollution and global warming. In the present circumstances, sustainable and renewable power-based generation can satisfy the interest for a longer period. Renewable energy is a promising energy source that can provide power for a very long time, but it has some drawbacks which should also be taken into consideration. Renewable energy-based generation is muchly unpredicted because it depends on the natural environment condition (solar insolation, wind speed, etc.). To resolve this issue and for better utilization of renewable energy systems, better control techniques are needed which can ensure better utilization of existing resources and provide an uninterrupted power supply. A small and self-sustainable system has existed for decades in distant regions where grid power was not available. The developing countries like India where power outage is very frequent and even after higher generation capacity, uninterrupted power is not available to every consumer due to poor distribution structures [6].

A microgrid (MG) is characterized collectively by interconnected loads and distributed energy resources inside defined electrical limits that go about as a solitary controllable entity in the context of the grid. An MG can interface and disengage from the main grid to work in both network-associated (grid-connected) and island modes [7]. The MGs can work separately from the main grid when it is needed or during any faulty condition and in that case, they can fulfill the demand of loads. The motivation behind MG is to coordinate with a sustainable power-based generation system to the main grid, reduce transmission and distribution (T&D) losses, control power theft or burglary, and provide power at a minimal rate [8]. Attractive features of the MGs are their flexibility and saving of a substantial amount of T&D losses when power is supplied from the grid. The MG can act as an independent source of electricity in distant regions, and it can connect/disconnect with the main grid for selling and purchasing power depending on the situation. The association between the MG and the main grid will be through a distinct and controlled standard interface. The MG is answerable for overhauling and addressing the necessities of its purchaser's interest in distant regions or when the grid power is not available it can provide power to the consumers.

India is the world's third-largest consumer of electricity [9] and also the world's third-largest electricity producer [10]. The electricity consumption is 1208 kWh per capita in India [9]. The installed power generation capacity of India has around 393.38 GW by December 2021 [11] and installed renewable power plants are established at 40.1% of total capacity [9]. The installed solar power till January 2021 is 48.55 GW capacity [12]. The target of the Indian government is to increase solar generation and make it up to 100 GW by 2022 [13]. Due to high transmission & distribution (T&D) losses and lack of quality infrastructure, an uninterrupted power supply is not available to consumers. To address the country's lack of adequate electricity supply by March 2019, the Indian government has already begun taking steps to generate more power through solar photovoltaic (SPV) and also initiated a program called "Power for All" [14]. This program will ensure that all families, factories, and commercial organizations have constant and uninterrupted access to power by constructing and strengthening the essential infrastructure. The ministry of new and renewable energy is providing solar panels at a 30% discount to encourage consumers to use them to meet their energy needs [15]. Furthermore, in recent years, the cost of a PV module has been steadily declining. Solar PV generating system-based microgrids are very favorable and commonly employed in both urban and rural areas. In India, there are approximately 300 clear sunny days per year [16]. Therefore, PV units in microgrids are more appealing for large and small-scale applications. The flexibility of microgrids allows SPV units to be located near consumers, reducing T&D losses and making them a cost-effective alternative [17].

For PV-based microgrid applications, there are different types of solar inverters available in the market which are grid-connected types and grid-tied types. In a grid-tied system with an energy storage unit (battery), the batteries can be charged with help of solar energy, grid, and both. The grid-tied solar inverter controls the flow of power in both directions. It means the load can receive power from the grid (mains) and also the surplus solar energy can be supplied to the mains or grid. In a grid-tied solar inverter system, when the batteries are fully charged and the loads are getting sufficient power, surplus energy can be supplied to the grid. However, in a commercially available grid-connected solar inverter system, when the grid power and solar power are available, batteries get charged by grid power and partially by the solar energy and the load gets power from the grid only. In this situation, the available solar power gets underutilized. Thus, there are limitations with these commercially available grid-connected and grid-tied inverters. Moreover, the grid-tied inverters are costlier while the grid-connected inverters offer poor utilization of solar energy. Solar power-based energy generation is growing exponentially, and its adaptability is being increased significantly. Thus, more attention is required in designing control techniques for grid-connected PV systems to get high efficiency and better performance of the inverters. The inclusion of maximum power point tracking (MPPT) in the PV system allows for maximum power harvesting, while the energy storage unit provides better and more stable electricity, improving the microgrid's overall consistency and dynamic nature [8,18]. The control techniques are different for stand-alone and for grid-tied (GT) systems. The primary objective of the GT inverter is to control the injected current to the grid while in a stand-alone system, the principal objective is to control voltage at the point of common coupling (PCC). Because of the twofold mode MG, the controller can control the different parameters in both stand-alone and GT modes and should be able to maximize the harnessing of solar power. For the productive operation of these interconnected MGs and energy storage systems, the control design of microgrids needs considerable attention [19]. The proper control of the smart grid can be achieved by proper prediction, planning, and real-time control [20].

Many control topologies in respect of bus voltage, phase locking, frequency, constant bus voltage, battery management system (BMS), and MPPT are proposed to improve the reliability of the MG system. In [21], solar PV-based low voltage DC-bus MG is developed with the control technique to operate the MG system in island mode as well as in grid-tied mode. In [22], a controller is developed for automatic scheduling and online control of dispersed energy sources but the limitation of this system is that it can only work in stand-alone mode. The primary goal of the power management system (PMS) is to control the state of charge (SOC) of the battery. The system is also verified on different types of active and reactive loads. The independent MG has its constraints with respect to voltage, frequency, and BMS [23]. In the MG system, the energy storage unit plays a vital role, and control of SOC in the batteries is important which increases the battery's life. In [24], the SOC of the battery in a stand-alone MG system is controlled by regulating the battery's terminal voltage to receive power from different distributed renewable energy sources. In [25], real-time management of a commercial grid-tied MG system is proposed with a wind turbine, solar PV modules, and Li-ion batteries. The objective of the proposed system is to improve battery life and decrease the overall cost. The model is developed on MATLAB®/Simulink and the size of a system is estimated with help of HOMER software. In the paper [26], different types of sustainable power sources based on GT microgrids with battery energy storage systems are discussed. The energy storage system helps to optimize the cost of the system and makes the system efficient.

It is important to estimate the load which helps to configure the different power-related applications of MG. In [27], the authors have proposed a bi-level (supervisory-local) MG configuration for low-power residential applications. A control technique is developed to define the set point in such a manner that helps to schedule the solar power generation forecasting and dynamic price. The authors of [28] have proposed a nonlinear controller, based on the adaptive back which helps to control the DC bus voltage of the DC MG system. The proposed system consists of SPV, battery, and diesel generators for backup with a different types of critical and non-critical loads. To mitigate peak-hour loads [29], a microcontroller-based controller has been proposed for peak shaving applications using the SPV system. The proposed cost-effective controller supervises the charging/discharging of the energy storage unit. Moreover, the reduction of peak load demand is a significant aspect of recent and upcoming smart grid technologies. An EMS is proposed in [30] which has less complexity in an application for a residential MG system. The MG is comprised of a grid-tied solar PV inverter system and battery. Instead of overall optimization, the proposed system optimizes the parameters by a two-layer controller. In [31], a supervising controller for commercially available conventional inverter/uninterrupted power supply (UPS) is proposed to increase solar energy utilization and improve backup power reliability. The developed controller works in two modes, that is reliable mode and economy mode. In reliable mode, the battery energy storage is charged from solar energy and the grid. In economy mode, solar energy is used to charge the battery and to fulfill the load demand in the daytime specifically during sunshine hours. During load shedding, extra battery backup is utilized for key loads, which raises the system's cost. In residential loads, there are always certain loads that need a continuous uninterruptable power supply. Therefore, to resolve the issue of uninterrupted power to crucial loads and to increase the utilization of solar power, a microcontroller-based control technique is proposed. The control topology is based on the battery voltage control in which different switches are operating to maximize the harnessing and utilization of solar power generated. Moreover, the surplus generated solar power is also fed back to the main grid power supply.

The main contributions of this chapter are summarized below:

1. A customized SPV MG system is developed to resolve the problem of inefficient solar power utilization in a commercially available solar inverter using a proposed static relay-based switching control strategy.
2. The proposed controller enhances the harnessing of solar energy by employing the SPV power with no restriction on the use of battery power during load shedding hours only. Further, the excess solar energy is also fed back to the main grid power.

3. The advantage of the proposed methodology is that it can be incorporated into an existing backup/UPS system to make it a solar grid-connected inverter.
4. The use of microcontrollers gives the freedom to operate the system under both islanding mode and grid-connected mode with varying operating conditions as per the consumer's need. Therefore, the proposed strategy helps in total harnessing and consumption of solar energy.
5. The designed switching strategy is also successfully tested through both simulation and hardware analysis to verify its effectiveness.

The organization of the remaining chapter is carried out as follows:

1. Section 2.2 discusses the shortcomings and motivation behind the development of the proposed method.
2. Section 2.3 explains the architecture of the proposed system design and its working under different input conditions.
3. Section 2.4 illustrates the working principle of the proposed control algorithm with simulation results.
4. Section 2.5 presents the hardware setup of the proposed methodology with household load applications. The experimental results are also illustrated at different battery voltage conditions to illustrate the working of the proposed control strategy in different input scenarios.
5. Lastly, the chapter is completed with concluding remarks and possible future extensions of this work.

20.2 The motivation behind the design of the proposed strategy

Electric power has become an essential need of humanity. However, power outages are common in developing countries like India. Besides, peak load management and power distribution system is also inadequate. There are almost 4–6 hours of load shedding in urban and smart cities on daily basis. The vast majority of the working-class families use commercially available lead-acid battery-based dc-ac inverter/UPS systems to get power during load shedding. The villages have more than 12 hours of scheduled and unscheduled load shedding and even their battery storage system is not able to supply uninterrupted power. The charging of the battery in a short period, while grid supply is available (before the next load shedding or blackout), is also an issue. The accessibility of a dependable, cost-effective, and continuous power supply is an important issue in the current scenario. The power demand is very high which is why there is a need for eco-accommodating, dependable resources. Therefore, the integration of renewable sources-based power generation can help the reliability of the system and decrease the related life cycle cost [32].

For residential purposes, inverters/UPS which are commercially available are mainly grid-connected inverters. These inverters have combined features of the dc-dc buck-boost converter associated with MPPT to charge the battery from the SPV system. When the battery voltage is below a particular pre-defined voltage level, it is charged by grid power and partially by SPV. If the output voltage of SPV is higher than the rated voltage of the battery, only then battery gets charged from SPV (otherwise from the grid). The solar inverters which are available in markets are a combination of conventional inverters and a solar charge controller (dc-dc) as shown in Fig. 20.1. It does not have the facility to feed surplus power to the grid and also does not have a net metering system. There is unidirectional power flow and loads take power from the grid when batteries are charging, or solar power is not sufficient. The battery gets charged by SPV only when the battery voltage level is higher than a specified value, otherwise, the battery gets charged from the grid, like a conventional inverter.

For the MG, PMS can manage battery storage, distributed generation scheduling, dispatch, and supervise the import-export of electric power between the MG and the main grid, among other things [8]. The configuration of another GT solar inverter is shown in Fig. 20.2 [33], where DC voltage is converted into AC voltage by a GT inverter system which supplies AC output to load and grid to sell the surplus power.

For this system, let us consider the battery gets discharged in the evening and it gets charged by grid power at night such that in next the morning the battery is fully charged. As the SPV power is available the next morning and the whole day, the SPV power is underutilized. In another case, especially in the context of rural areas, the power is provided on a rotation basis, that is one week for the daytime and one week for the night. Therefore, in that situation, the battery does not get fully charged by grid power. Further, there are always some important loads that need an uninterrupted power supply and these loads do not get power in the absence of solar power and grid power with low battery voltage levels. Hence, all these issues need to be resolved.

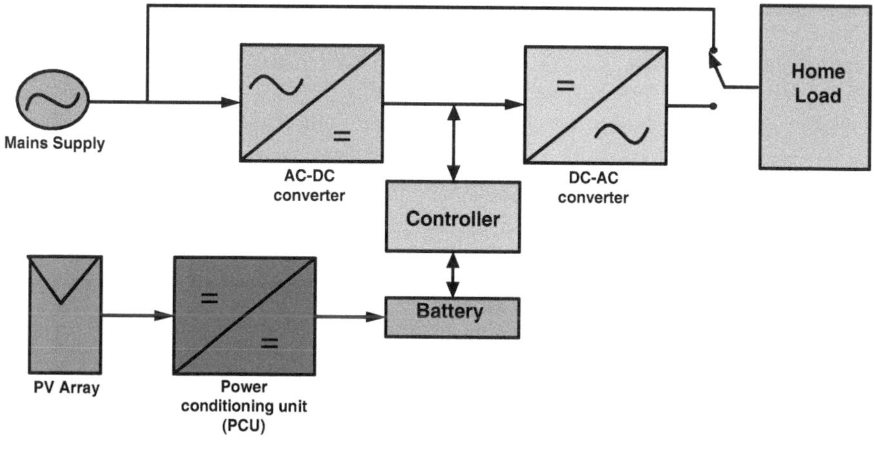

Figure 20.1 Block diagram of the solar grid-connected inverter.

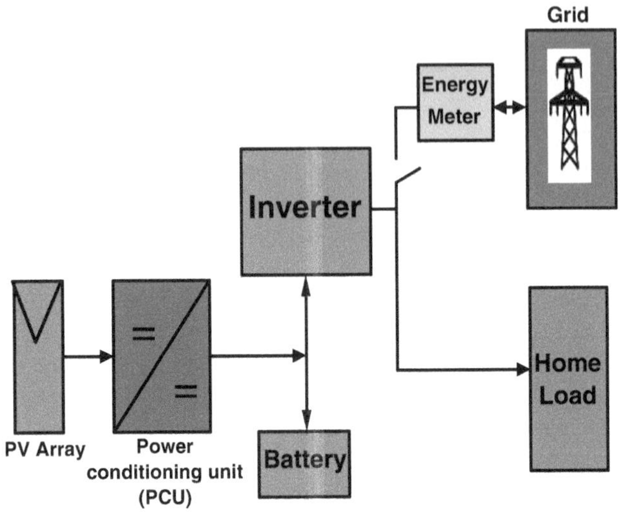

Figure 20.2 Schematic diagrams of the solar grid-tied inverter [33].

20.3 Architecture and working of the proposed supervisory control technique

In this proposed topology, the battery voltage is used as the main reference voltage in addition to the availability (inputs) of solar power and grid power for the controller.

20.3.1 The architecture of the system and control technique

The block diagram of the proposed control system is shown in Fig. 20.3. The proposed controller is integrated with solar PV modules, LDR sensor, AC–DC converter-based (solar) charger, grid supply, battery, and bidirectional AC–DC converter/inverter. The objective of the proposed system is to provide an uninterrupted supply to load with maximum power utilization of solar energy by controlling different static relay switches, that is S_1, S_2, and S_3. In this scheme, the load has been divided into two categories: normal loads (room cooler, T.V, high power consuming fans) and Emergency loads (LED lamps, low power consuming fan, Laptop, Mobiles). In Fig. 20.3, the solid connection lines represent electric power connection, and the dotted connection line indicates the signal of input variables to the controller block. The dash-dot line is the control output information signal, which is fed to the static relay switches. Based on the input variables, that is battery voltage, availability of grid power supply, and solar energy, switching commands are generated using the proposed control scheme. The controller even manages the worst-case scenario, when solar and grid are not available and the battery is also not fully charged, emergency loads can receive power from the battery.

In the proposed setup, there are three static relay switches S_1, S_2, and S_3. Switch S_1 is used to receive grid power when needed and switch S_3 decides the feeding of surplus

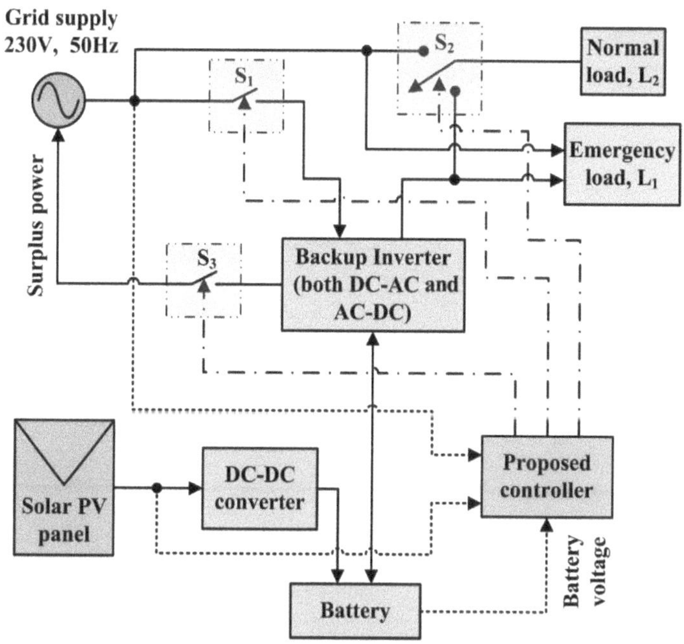

Figure 20.3 Block diagram of the proposed control system.

solar power/energy to the main grid. Switch S_2 determines the power supply to normal loads, that is it gets turned off when both grid and SPV powers are not available and the battery is also not fully charged (battery voltage in the lowest allowable range).

20.3.2 Working on the proposed controller

In this section, firstly the complete functioning of the proposed controller is discussed with the help of a flow chart (Fig. 20.4). The first line action under the proposed scheme is to sense the battery voltage (V_B). According to the value of V_B and the condition of the grid and solar supplies, the next line of the action takes place, as depicted in the flow chart in Fig. 20.4.

The circuit diagram for realizing the proposed control algorithm is illustrated in Fig. 20.5. In this diagram, the battery voltage is attenuated with the help of a potential divider supplied to the Arduino board-based controller. The signal for the availability of grid power is taken from the grid-based AC–DC converter/adapter output. Similarly, the signal for the availability of solar power/insolation level is taken from an LDR sensor. The LDR sensor sends the signal corresponding to the insolation level to the Arduino-based controller which is compared to a pre-set value that helps to ascertain a high or low level of insolation. The output (digital output pins) goes high or low according to different input combinations of the proposed control algorithm. The switching condition algorithm is implemented through the Arduino. The proposed controller operates in accordance with different combinations of battery voltage levels, the

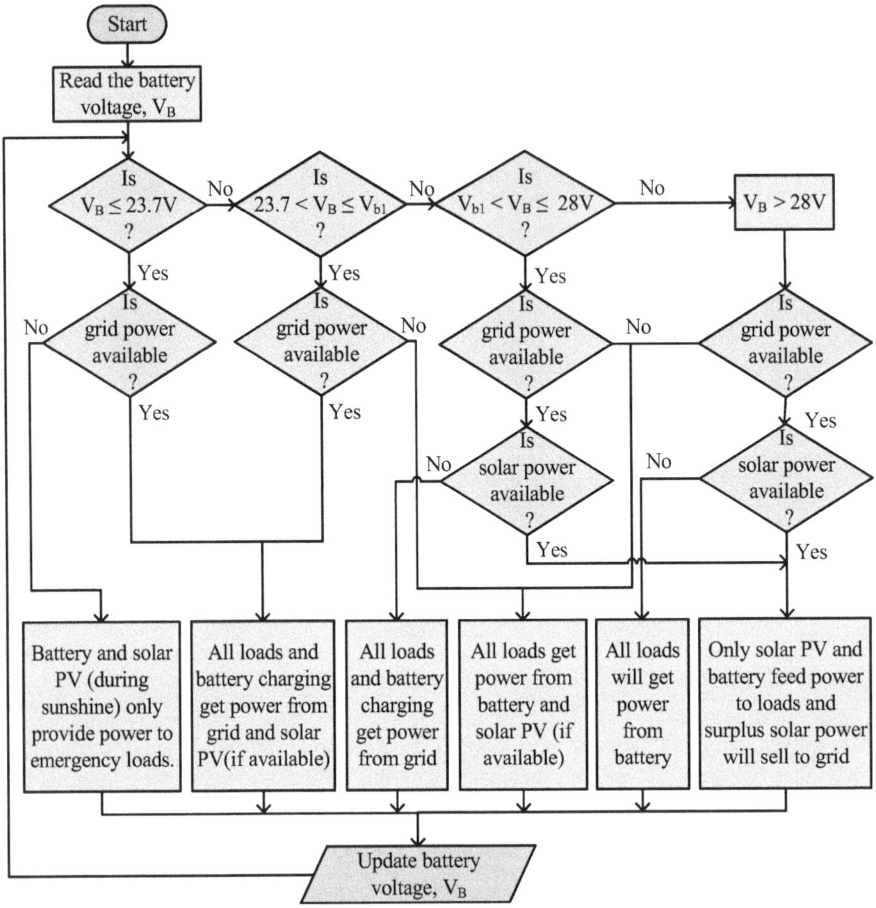

Figure 20.4 Flow chart of the proposed controller.

availability of grid and solar powers and accordingly the static relay switches operate. The battery voltage is categorized into four categories and different switches S1, S2, and S3 will operate at different battery voltages to harness and utilize the maximum of solar power generated from the microgrid and to provide continuous power to loads. Here switch S3 is used to trigger the grid-tied inverter to supply surplus power to the main grid when both solar and grid powers are available and load demand is fulfilled.

20.4 Numerical analysis of the proposed rule-based control algorithm

Fig. 20.6 shows the MATLAB/Simulink model of the system in which the inputs of the controller are battery voltage, grid power, and solar power of the PV-based

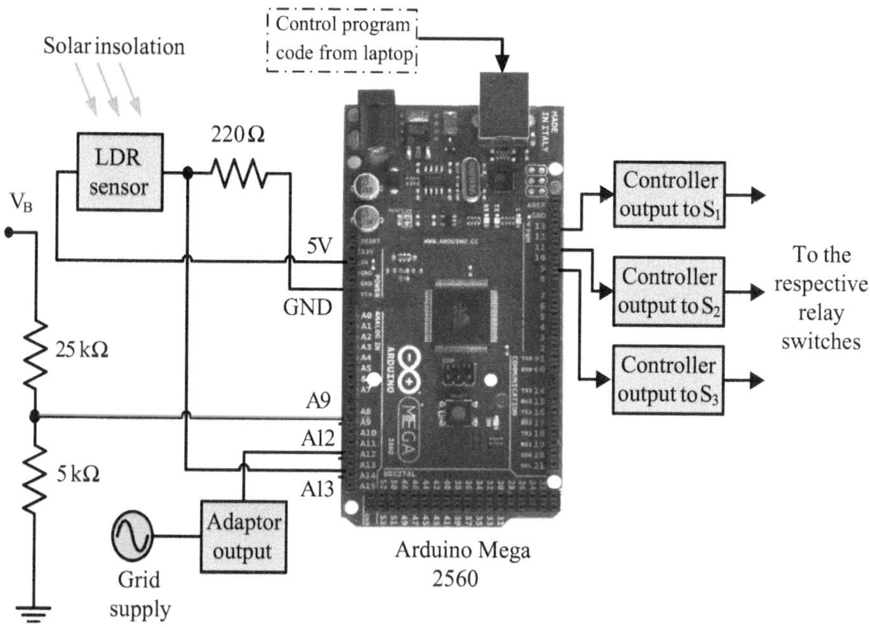

Figure 20.5 Circuit diagram of the proposed controller.

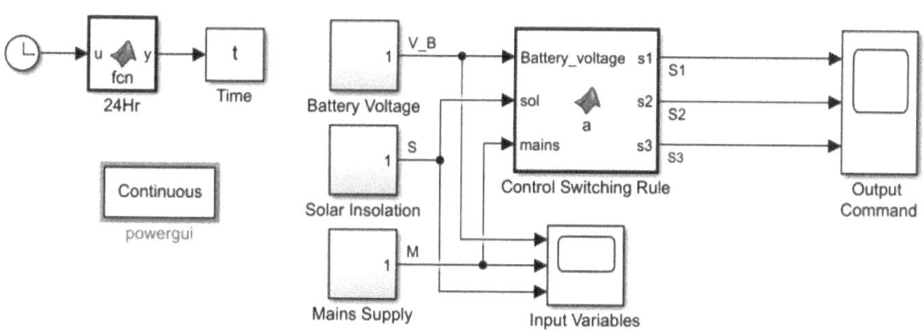

Figure 20.6 Simulink model of controller.

microgrid. In the proposed system, the battery voltage is categorized into four sections and further each section is categorized into four sections which include the combination of grid power and solar power. The Simulink model is simulated for 24 hour (86,400 second) for a complete analysis of the control algorithm. Here, in the Simulink model, there is the third output of the controller (S3) which is added to show the flow of surplus solar power supplied to the grid.

20.4.1 Switching techniques

Here, battery voltage, mains grid supply, and solar insolation are represented by V_B, M, and S, respectively. The base voltage of the battery is 23.7 V. The switching algorithm of the proposed control scheme is given in Table 20.1 A and B. The switching of static switches is based on the battery voltage levels along with the availability of solar and grid powers (i.e., V_B, M, and S). The status of switch S3 indicates the flow of surplus solar power from the consumer side to the mains or grid.

The surplus solar power is supplied to the grid only when the battery is fully charged and grid power is available.

20.5 Simulation results

The Simulink model is simulated for 24 hours, and switches operate according to the input variables of the proposed strategy. To verify the performance of the proposed controller, arbitrary battery voltage, grid power, and solar power are taken as inputs to the controller, as shown in Fig. 20.7. The battery is protected from overcharging, and deep discharge situations, and when the voltage level is below a certain pre-set value, it does not supply power to normal loads.

Figs. 20.7 and 20.8 shows the controller input and controller output, respectively. In the simulation model, 24 V (two series-connected 12 V batteries) is taken as a reference for the SOC measurement. The availability of grid and solar power is identified by the binary output. The simulation analysis is discussed below for five different cases.

Table 20.1 Switching algorithm of static switches.

Static relay switches	$V_B < 23.7$ V				$V_{b1} \leq V_B < V_{b2}$			
	$M = 0$		$M = 1$		$M = 0$		$M = 1$	
	$S = 0$	$S = 1$	$S = 0$	$S = 1$	$S = 0$	$S = 1$	$S = 0$	$S = 1$
S_1	OFF	OFF	ON	ON	OFF	OFF	ON	ON
S_2	OFF	OFF	ON	ON	ON	ON	ON	ON
S_3	OFF	OFF	OFF	OFF	OFF	OFF	OFF	OFF
Static relay switches	$V_{b2} \leq V_B < 28$ V				$V_B \geq 28$ V			
	$M = 0$		$M = 1$		$M = 0$		$M = 1$	
	$S = 0$	$S = 1$	$S = 0$	$S = 1$	$S = 0$	$S = 1$	$S = 0$	$S = 1$
S_1	OFF	OFF	OFF	OFF	OFF	OFF	OFF	OFF
S_2	ON	ON	ON	ON	ON	ON	ON	ON
S_3	OFF	OFF	OFF	OFF	OFF	OFF	OFF	ON

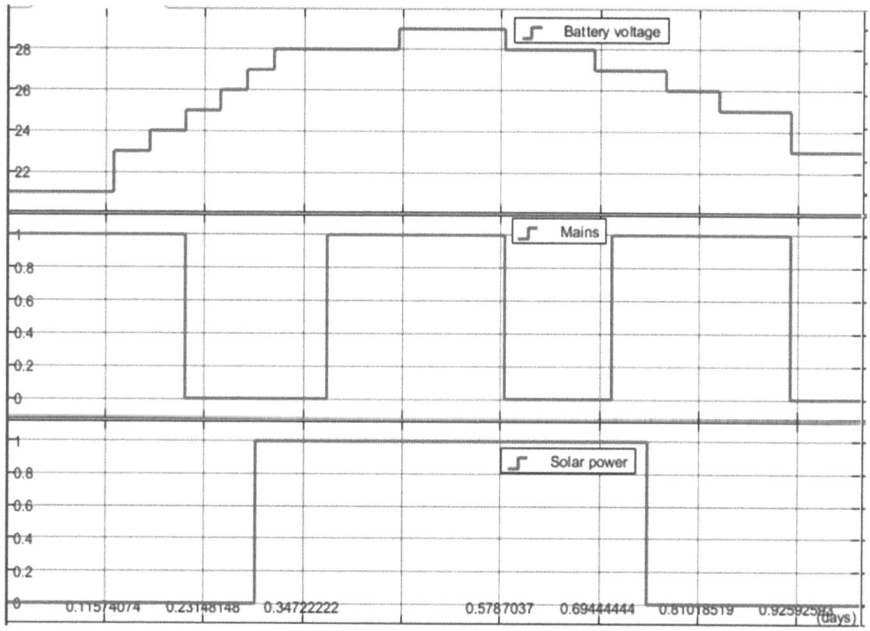

Figure 20.7 Inputs to the proposed controller.

Case 1: When the battery voltage (V_B) is between 22 to 24 V and the grid power is available but solar power is not available, in that condition switch S_1 is turned ON and the system works in grid-connected mode. It receives power from the grid to provide power to both types of loads. Therefore, switch S_2 is also ON and all loads receive power including battery charging, which also takes place through an AC–DC converter.

Case 2: When the battery voltage (V_B) is between 24 to 26 V and grid power as well as solar power both are not available. In this condition switch, S1 is turned OFF. The system works in stand-alone mode and the load receives power from the battery only. The switch S_2 is ON such that power is supplied to both normal loads and emergency loads.

Case 3: When the battery voltage (V_B) is between 26 to 28 V and solar power is available but grid power is not available. In this condition switch, S1 is turned OFF and the system works in stand-alone mode. The switch S_2 is ON and the load receives power from the battery storage as well as from the SPV system.

Case 4: When the battery voltage (V_B) is more than 28 V and grid power, and solar power both are available. In this condition switch, S_1 is turned OFF. If the power generated by solar PV is surplus (more than the total load demand) and a grid-tied inverter system is available, the surplus power is supplied to the grid. As a result, the switch S_3 gets high to enable the flow of surplus power from the MG to the main grid.

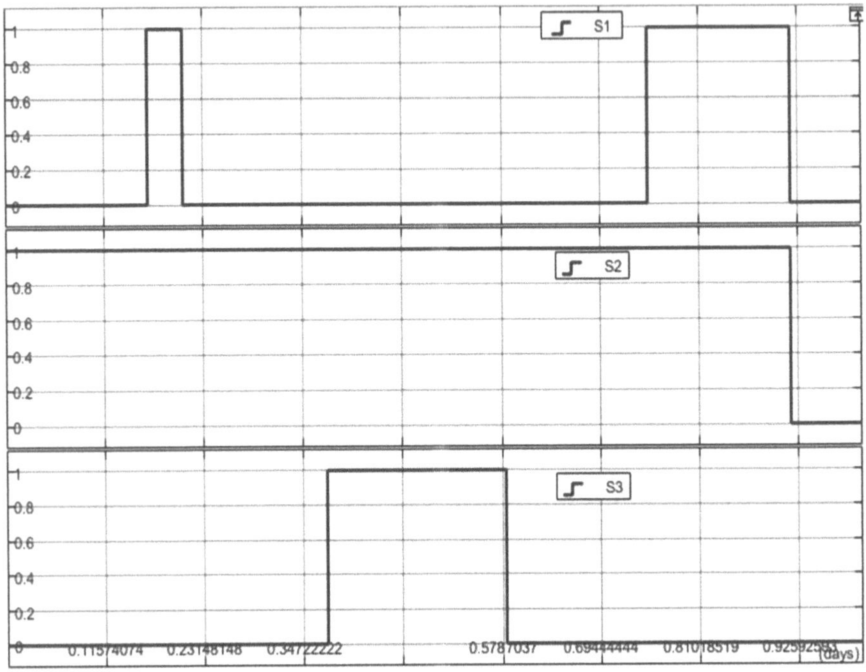

Figure 20.8 Output state of controller and static relay switches.

Case 5: When the battery voltage (V_B) is below 23.7 V and both grid power and solar power are not available, in that condition switches S_1 and S_2 are turned OFF, and the system works in stand-alone mode. In this case, normal loads do not receive power and only emergency load receives power from the battery storage system and the SPV system, if available.

20.6 Experimental validation of the closed-loop microgrid system

For the verification of the proposed system, the hardware is realized with the Arduino Mega (2560) based microcontroller. The system consists of a commercially available residential inverter system, one 250Wp solar PV panel, one solar dc-dc charger, two 12-V lead-acid batteries, and different types of residential loads. All the important hardware components are discussed below.

20.6.1 Proposed controller

Fig. 20.9 shows the Arduino-based proposed controller which works in conjunction with the commercially available inverter system or grid-tied inverter. In addition to

Figure 20.9 Proposed controller.

grid power and solar power as inputs, the proposed controller receives battery voltage input. Based on the proposed control algorithm fed in the Arduino microcontroller, different digital output pins give the output command as high or low to operate static relay switches. The battery voltage is attenuated with a potential divider to the level of 5 V which is given at the analog input pins of Arduino. Solar power availability input is taken from the LDR sensor and grid power availability signal is received input from the AC–DC converter/adapter. The main work of the proposed controller is to detect the battery voltage level during the charging and discharging and to compare the logic conditions to ultimately control the static relay switches.

20.6.2 Battery storage

The Hardware system consists of two 12-V lead-acid batteries of 150AH capacity connected in series. Each battery has 6 cells and the voltage of each cell varies from 2.25 to 2.45 V [34]. Each battery can be charged up to 14.4 V [29].

20.6.3 Residential loads

Fig. 20.10 shows the different types of residential loads, and it is divided into two categories: normal loads and emergency loads. Normal loads consist of room coolers, TV, fans and emergency loads, low power consuming (BLDC motor-based) fans, and LED lamps. For testing purposes, a GLS lamp-based load of 1 kW is connected as a normal load and a 200 W GLS lamp is used as an emergency load. The loads are categorized to provide uninterrupted power to emergency loads and to increase the reliability of the system. The proposed system provides power to all types of loads during the blackout/load shedding and when solar and grid power are

Figure 20.10 Different types of testing load.

not available. Further, when a battery is not fully/sufficiently charged (below a certain level), at least emergency loads receive power in the crucial situation (in the absence of both grid and solar power).

20.6.4 Solar photovoltaic-based system

For a solar PV-based system, a 250W solar PV module is taken as an SPV power source. The maximum current output of the PV module is 8.95 A, the maximum open-circuit voltage is 37.25 V and at maximum power, the voltage and current are 29.95 V and 8.35 A, respectively.

20.6.5 Experimental setup

The experimental setup has been shown in Fig. 20.11 which includes one 24 V commercially available inverter system with two 12-V lead-acid batteries, one solar dc-dc converter/charger, different types of residential loads, AC–DC adapter, and an Arduino microcontroller.

The Arduino (Mega 2560) based microcontroller operates at 5 V DC input. The controller takes input from the battery in addition to solar power and grid power. The main aim of the proposed control system is to increase the utilization (harnessing) of solar power and to provide uninterrupted power to loads even in the worst conditions. First, the controller was tested successfully at the ideal condition with the help of a constant source. Therefore, the controller successfully tested the different levels of battery voltage with the combination of grid and SPV powers.

20.6.6 Experimental results of controller

The switching operation of the proposed system works according to the control algorithm for different input combinations. The proposed controller assesses the

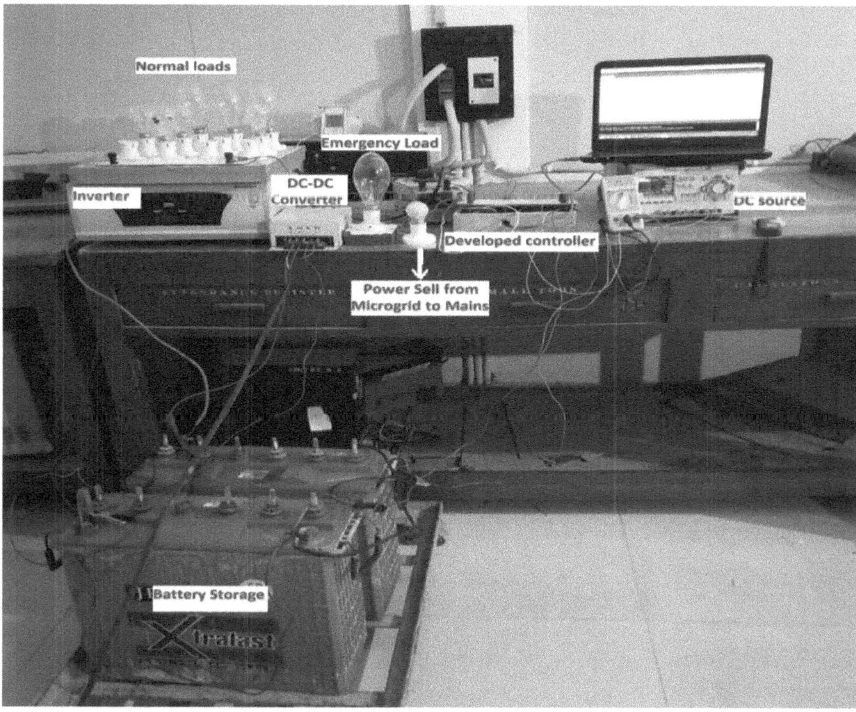

Figure 20.11 Experimental setup with the proposed controller.

battery voltage level and availability of grid power and solar energy. According to the proposed algorithm, the microcontroller gives output signals (high or low) at different output pins. The experimental results are discussed below for four important cases which are based on different battery voltage levels.

Case 1: When the battery voltage is less than 23.7 V as shown in Fig. 20.12 and grid power is not available, the switch S_1, S_2, and S_3 are turned OFF. The battery does not supply power to normal loads, but it supplies power to emergency loads only, and in the meanwhile, if solar power is available, it charges the battery. When the grid power is available, the switches S_1 and S_2 are turned ON and the battery gets charged by both grid power as well as solar power (Fig. 20.12). Both types of loads receive power. Further, since the surplus power is not available, the switch S_3 is remained OFF.

Case 2: When the battery voltage is greater than 23.7 V and less than 26 V as shown in Fig. 20.13. Here, grid power and solar power both are available. The first priority is to charge the battery and to provide continuous power to both types of loads. The controller turns on switches S_1 and S_2, and switch S_3 is kept low to restrict the flow of power to the main grid. On the other hand, if solar power is available and grid power is not available, both loads receive power from solar PV. If the load is more than the generated solar power, it receives power from the

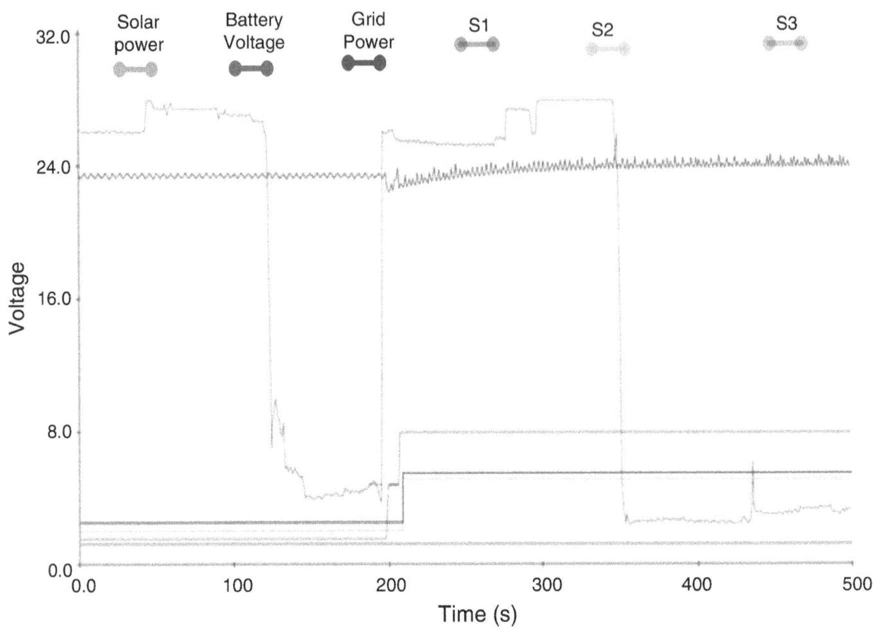

Figure 20.12 Switching operation at battery voltage less than 23.7 V.

battery storage. Fig. 20.13 shows how battery voltage starts decreasing in the absence of a grid as the loads fulfill the power requirement from it. When both grid and solar powers are not available the switches S_1 get OFF and the system works in stand-alone mode. Moreover, switch S_2 is ON to supply power to both loads.

Case 3: When the battery voltage is greater than 26.1 V and less than 28 V as shown in Fig. 20.14 and grid power, and solar power both are not available. At this condition, the battery is sufficiently charged which supplies power to both types of loads for a long time. Thus, S_1 is OFF and S_2 is ON. On the other hand, if solar power is available but grid power is not available, both the loads receive power from solar PV and battery. The SPV power generated is fully utilized. At the night, when solar power is not available, but the grid power is available, switch S_1 turns ON to charge the battery. When solar and grid power both are available, the system becomes isolated from the grid and works as a standalone system. Thus, providing power to both loads with efficient utilization of solar power. Moreover, owing to the fact that the battery is not fully charged, the surplus power is not fed to the main grid. Hence, the switch S_3 is remain low.

Case 4: When the battery voltage is greater than 28.1 V as shown in Fig. 20.15. Here, grid power, as well as solar power both, is not available, and both the loads receive power from the battery storage system. On the other hand, if solar power is available and grid power is not available, the solar power gets utilized to charge the battery and to supply power to both loads. When grid power is available at the night, the battery has enough stored energy to supply power to loads that is why the

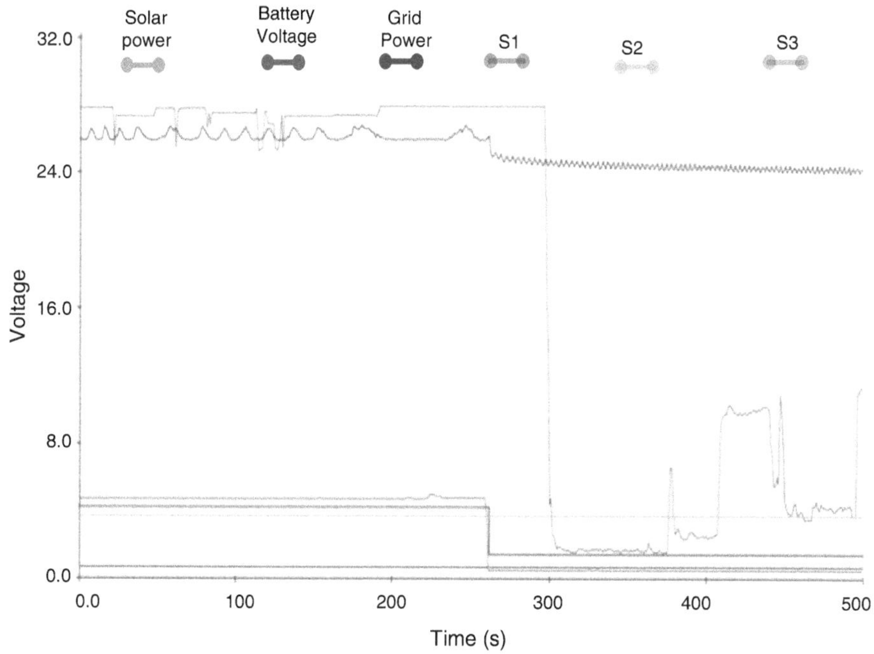

Figure 20.13 Switching operation at battery voltage from 23.7 to 26.2 V.

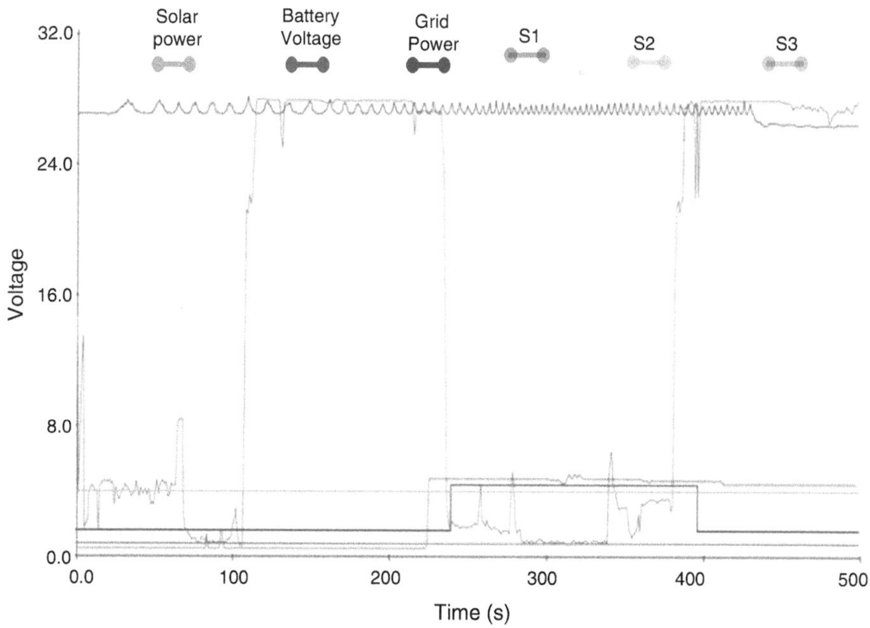

Figure 20.14 Switching operation at battery voltage from 26.2 to 28.1 V.

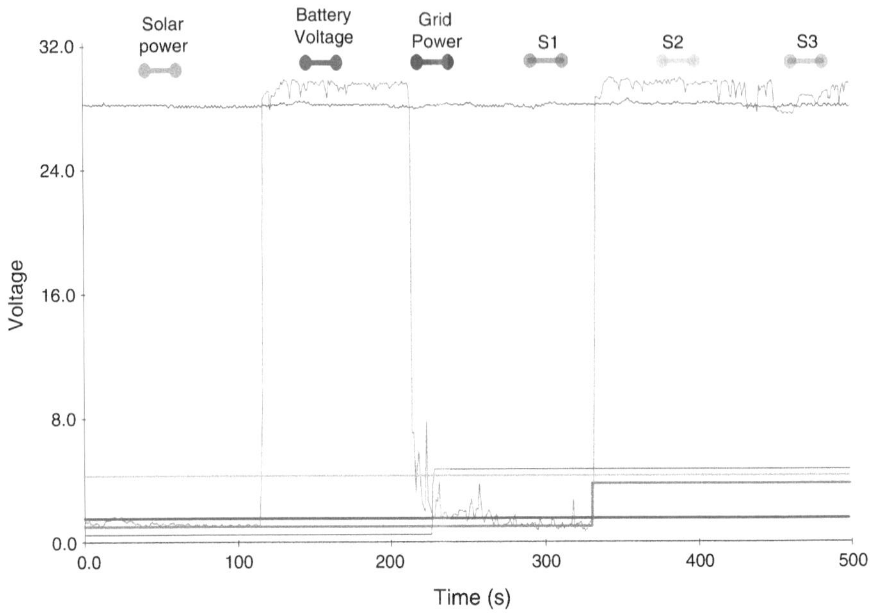

Figure 20.15 Switching operation at battery from greater than 28.1.

system does not require power from the main grid. In a special condition, when both solar and grid powers are available and after fulfilling the load demand, the surplus solar power will be supplied to the power grid by enabling the switching command S_3 is high, as shown in Fig. 20.15 after 300 seconds. Thus, the surplus power is fed to the main grid from the grid-tied inverter system. At this condition, S_1 is OFF and S_2 is ON, and power is supplied to both types of loads.

The proposed system is successfully tested at a different voltage level of the battery in the absence as well as in the presence of solar and grid power. The experimental results match the simulation results, and the controller operates according to the developed control algorithm. The proposed control system makes the consumer less dependent on the grid power, increases the utilization of solar power, and provides uninterrupted power to the loads. Hence, the desired objective of the proposed controller is successfully achieved.

20.7 Conclusion

In a developing country, there is ample scope for harnessing solar energy and SPV-based MGs are being encouraged. However, the problem with the commercially available inverter system for solar power applications is that it works only as backup power during the load shedding or blackout hours. Moreover, when both grid and solar powers are available, these inverters prefer grid power to be supplied for the loads, and the stored energy of the battery including its charging from SPV is kept as reserve energy. Consequently, solar energy gets underutilized.

To resolve these issues, a new control method has been proposed and tested which works in conjunction with conventional grid-connected or grid-tied inverters. The proposed controller helps to maximize the utilization of SPV based MG system and provides an uninterrupted power supply to loads. The proposed controller takes the battery voltage as the main reference voltage in addition to the absence and presence of solar and grid powers. The loads are categorized into two categories: normal loads and emergency loads. As per various conditions of input variables, different static switches operate to maximize the utilization of solar power and to provide uninterrupted power to loads during load shedding. An attractive feature of the proposed controller is its flexibility to connect and disconnect (islanding) the MG from the mains/grid.

The proposed control strategy is validated on the MATLAB/Simulink software to conceptualize the designed methodology. Different types of loads and power sources are taken to verify the performance of the proposed controller, and the system is simulated for 24 hours for real-time analysis. The experimental setup of the proposed controller is also realized and tested successfully. The proposed setup comprises a commercial inverter system, solar PV module, solar dc-dc charge controller, battery, and different types of loads and sources. The control strategy is implemented using Arduino Mega (2560) microcontroller. The results show that solar power can be fully utilized, particularly when the grid and solar power are both presents. Moreover, the proposed controller provides continuous power to loads such that even in the worst scenario when both grid and solar powers are not available and the battery is not sufficiently charged, the emergency load receives power.

The future extension of this work could be along the lines of extending it to a hybrid microgrid system with different renewable energy sources. Moreover, an electric vehicle system can also be integrated to provide power during peak load hours and load shedding.

Acknowledgments

The authors extend their appreciation to the Deanship of Scientific Research at King Khalid University for supporting this work through Research Groups Program under grant number (RGP.2/81/43).

References

[1] S. M. Shariff et al., System design and realization of a solar-powered electric vehicle charging station. 14 (2) (2020) 2748–2758.

[2] Y. Liu et al., Application of small-sized SMES in an EV charging station with DC bus and PV system. 25 (3) (2015).

[3] F. Marra, et al., EV charging facilities and their application in LV feeders with photovoltaics, IEEE Trans. Smart Grid 4 (3) (2013) 1533−1540. Available from: https://doi.org/10.1109/TSG.2013.2271489.

[4] N. Chen, M. Wang, N. Zhang, X. Shen, Energy and information management of electric vehicular network: a survey, IEEE Commun. Surv. Tutor. 22 (2) (2020) 967−997. Available from: https://doi.org/10.1109/COMST.2020.2982118. Institute of Electrical and Electronics Engineers Inc.

[5] D. Ton, E. Bryan, C. Marnay, Office of electricity delivery & energy reliability. Microgrids and best practices, Power Systems Engineering Research and Development (2015).

[6] Understanding India's Power Capacity: Surplus or not, and for how long? (2019) <https://www.brookings.edu/research/understanding-indias-power-capacity-surplus-or-not-and-for-how-long/> (accessed 13.01.21).

[7] The Office of Electricity Delivery and Energy Reliability (OE) drives electric grid modernization and resiliency in the energy infrastructure. OE Mission Grid. Modernization. (2016).

[8] A. Hirsch, Y. Parag, J. Guerrero, Microgrids: a review of technologies, key drivers, and outstanding issues, Renew. Sustain. Energy Rev. 90 (2018) 402−411. Available from: https://doi.org/10.1016/j.rser.2018.03.040.

[9] Countries with highest primary energy consumption | Statista. (2021) <https://www.statista.com/statistics/263455/primary-energy-consumption-of-selected-countries/> (accessed 15.02.21).

[10] NS Energy Staff Writer, Profiling the top five largest electricity-generating countries in 2021, 2021 May 20.

[11] Power Sector at a Glance ALL INDIA | Government of India | Ministry of Power." (2022) <https://powermin.gov.in/en/content/power-sector-glance-all-india> (accessed 28.01.22).

[12] India's renewable energy generation capacity addition estimated at 16 GW in FY23, says Icra. (2022) <https://www.livemint.com/industry/energy/indias-renewable-energy-generation-capacity-addition-in-fy23-estimated-at-16-gw-11641807620654.html> (accessed 28.01.22).

[13] India plans to produce 175 GW of renewable energy by 2022 - United Nations Partnerships for SDGs platform. (2022) <https://sustainabledevelopment.un.org/partnership/?p = 34566> (accessed 28.01.22).

[14] PM Narendra Modi to launch 'power for all' scheme on September 25: power Minister R K Singh. (2017) <https://economictimes.indiatimes.com/industry/energy/power/pm-narendra-modi-to-launch-power-for-all-scheme-on-september-25-power-minister-r-k-singh/articleshow/60805044.cms> (accessed 29.01.22).

[15] Rooftop Solar Subsidy in India 2022: How much, how & where to get. <https://www.loomsolar.com/blogs/collections/solar-panel-subsidy-in-india> (accessed 29.01.22).

[16] Riding on Sunshine - Nation News - Issue Date: Feb 8, 2021. <https://www.indiatoday.in/magazine/nation/story/20210208-riding-on-sunshine-1763814-2021-01-30> (accessed 29.01.22).

[17] I. Tank & S. Mali, Renewable based DC microgrid with energy management system, in: Proceedings of the IEEE International Conference on Signal Processing, Informatics, Communication and Energy Systems (SPICES), 2015 February, pp. 1−5. Available from: https://doi.org/10.1109/SPICES.2015.7091542

[18] F. Martin-Martínez, A. Sánchez-Miralles, M. Rivier, A literature review of microgrids: a functional layer based classification, Renew. Sustain. Energy Rev. 62 (2016) 1133−1153. Available from: https://doi.org/10.1016/j.rser.2016.05.025.

[19] R. Zamora, A.K. Srivastava, Controls for microgrids with storage: review, challenges, and research needs, Renew. Sustain. Energy Rev. 14 (7) (2010) 2009–2018. Available from: https://doi.org/10.1016/J.RSER.2010.03.019.

[20] A. Molderink, V. Bakker, M.G.C. Bosman, J.L. Hurink, G.J.M. Smit, Management and control of domestic smart grid technology, IEEE Trans. Smart Grid 1 (2) (2010) 109–119. Available from: https://doi.org/10.1109/TSG.2010.2055904.

[21] A. Mcrabet, Control of simulated solar PV microgrid operating in grid-tied and islanded modes, IECON 2018 - 44th Annu. Conf. IEEE Ind. Electron. Soc. 1 (2018) 1729–1734.

[22] B. Belvedere et al., A microcontroller-based power management system for standalone microgrids with hybrid power supply. 3 (3) (2012) 422–431.

[23] H.S. Das, M.M. Rahman, S. Li, C.W. Tan, Electric vehicles standards, charging infrastructure, and impact on grid integration: a technological review, Renew. Sustain. Energy Rev. 120 (2020). Available from: https://doi.org/10.1016/j.rser.2019.109618. no. February 2019.

[24] J. G. De Matos, F. S. F. Silva, S. Member, and L. A. D. S. Ribeiro, Power control in AC isolated microgrids with renewable energy sources and energy storage systems. 62 (6) (2015) 3490–3498. Available from: https://doi.org/10.1109/TIE.2014.2367463.

[25] S. A. Pourmousavi, R. K. Sharma, B. Asghari, A framework for real-time power management of a grid-tied microgrid to extend battery lifetime and reduce cost of energy, in: Proceedings of the IEEE PES Innovative Smart Grid Technologies, 2011, pp. 1–8. Available from: https://doi.org/10.1109/ISGT.2012.6175707.

[26] Z. Chen, Energy management strategy for grid-tied microgrids considering the energy storage efficiency. 65 (12) (2018) 9539–9549, Available from: https://doi.org/10.1109/TIE.2018.2818660.

[27] M. Babakmehr, F. Harirchi, A. Alsaleem, A. Bubshait, M.G. Simoes, Designing an intelligent low power residential PV-based Microgrid, 2016 IEEE Ind. Appl. Soc. Annu. Meet. (2016) 1–8. Available from: https://doi.org/10.1109/IAS.2016.7731878.

[28] T.K. Roy, et al., Nonlinear adaptive backstepping controller design for islanded DC microgrids, IEEE Trans. Ind. Appl. 54 (3) (2018) 2857–2873. Available from: https://doi.org/10.1109/TIA.2018.2800680.

[29] K. Ananda-Rao, R. Ali, S. Taniselass, N. H. Baharudin, Microcontroller based battery controller for peak shaving integrated with solar photovoltaic. 5 (6.) (2017). Available from: https://doi.org/10.1049/cp.2016.1262.

[30] D. B. Licea, M. Ghogho, Low complexity closed-loop energy manager for a grid-tied PV system with battery, in: Proceedings of the IEEE International Conference on Communications, Control, and Computing Technologies for Smart Grids, 2018, pp. 1–6.

[31] S. Muhammad, M. Saad, M.S.J. Asghar, F. Ahmad, Low cost residential microgrid system based home to grid (H2G) back up power management Low cost residential microgrid system based home to grid (H2G) back up power management, Sustain. Cities Soc. 36 (2017) 204–214. Available from: https://doi.org/10.1016/j.scs.2017.10.016. no. November.

[32] S.M. Amrr, M.S. Shemami, H.K.M. Irfan, M.S.J. Asghar, Design and operation of a low-cost microgrid-integrated EV for developing countries: a case study, Electr. Veh. Integr. a Smart Microgrid Env. (2021) 335–357. Available from: https://doi.org/10.1201/9780367423926-15.

[33] Battery university, Charging of lead acid battery, 2021.

[34] G. Tie, On grid system (grid export/net metering): principles of operation and the schematic diagram 2012.

Electrochemical energy storage part II: hybrid and future systems

21

Udita Bhattacharjee, Shuvajit Ghosh*, Madhushri Bhar* and Surendra Kumar Martha*
Department of Chemistry, Indian Institute of Technology Hyderabad, Sangareddy, Telangana, India

Chapter Outline

Highlights

- An outline of the hybrid electrochemical energy storage (EES) devices developed for optimal energy-power output
- A brief overview of their active materials, electrochemistry, and feasibility
- Future directions and applicability and comparability of the newly developed EES systems

Nomenclature

3D	Three dimensional
CIB	Chloride ion battery
CNT	Carbon nanotube
COF	Covalent organic framework
DCB	Dual-carbon battery
DEC	Diethyl carbonate
DIB	Dual-ion battery

* Equal contributing authors.

Emerging Trends in Energy Storage Systems and Industrial Applications. DOI: https://doi.org/10.1016/B978-0-323-90521-3.00023-5

DMC	Dimethyl carbonate
DME	Dimethoxy ethane
DOL	1,3-dioxolane
EC	Ethylene carbonate
EES	Electrochemical energy storage
EV	Electric vehicle
FEC	Fluoroethylene carbonate
FIBs	Fluoride-ion batteries
HTSSFIB	High-temperature solid-state FIB
HUCs	Hybrid ultracapacitors
KIBs	Potassium-ion batteries
LCO	Lithium cobalt oxide
LHCE	Localized high concentrated electrolyte
LIBs	Lithium-ion BatteriesPON
LIPON	Lithium phosphorous oxynitride
LISICON	Lithium superionic conductors
LTO	Lithium titanate
MF	Metal fluoride
MIB	Mg-ion batteries
MOF	Metal-organic framework
Mr	Membrane reforming
NASICON	Sodium superionic conductors
OER	Oxygen evolution reaction
ORR	Oxygen reduction reaction
PAN	Polyacrylonitrile
PC	propylene carbonate
PEO	Polyethylene oxide
PMMA	Poly methyl methacrylate
POC	Poly octamethylene citrate
PVDF	Polyvinylidene fluoride
PVDF-HFP	Polyvinylidene fluoride-co-hexafluoropropylene
RTFIB	Room-temperature FIB
SEI	Solid electrolyte interface
SHE	Standard hydrogen electrode
SIBs	Sodium-ion batteries
SSE	Solid state electrolyte
SSWB	Sodium seawater battery
TEGDME	Tetraethylene glycol dimethyl ether
WiBS	Water in bisalt
WiS	Water in salt
XRD	X-ray diffraction

21.1 General introduction

Electrochemical energy storage (EES) as discussed in the Chapter 6 has evolved a lot since its inception. Conventional EES systems are either energy- or power-dense devices and are thus limited by the performance of specific applications. Therefore, research is driven toward manifesting high energy-power output and longer cycle life in a single device. Also, pushing the energy and power limit and enhancing the feasibility through newer electrochemical redox combinations and engineering has been of keen interest.

Various EES systems have been hybridized together to utilize their advantages and simultaneously mitigate their disadvantages. The hybrid electrochemical systems were conceptualized to minimize the hustle of using different devices for different applications and combining all the utilities in a compact space. The majorly explored hybrid EES devices include metal-air batteries (a combination of batteries and fuel cell) and hybrid ultracapacitors (a combination of batteries and supercapacitors). The metal-air batteries have theoretical energy densities of 3−30 times higher than conventional batteries. But the instability and low reversibility of the products formed limit their practical realization. The research on metal-air batteries started long back with the development of primary Zn-air batteries in 1878 by Maiche. Aqueous Fe, Al, and Mg metal-air batteries were subsequently developed, leading the path to the development of nonaqueous alkali (Li/Na) metal-air batteries. The other type of hybrid EES system which is newly being explored is the hybrid electrochemical capacitor which combines two mechanistically different electrodes (one battery type and one capacitor type). The battery type electrodes may be lead-acid, Ni-MH, or LIB/SIB types. The main issue of hybrid EES is kinetically and thermodynamically balancing different mechanisms so that the optimum performance is maintained.

Further, advanced and sustainable EES devices with higher energy output are being studied but the longevity and feasibility of these devices still limit their applications. This chapter will emphasize briefly the following aspects:

- The mechanistic and material perspective of the hybrid EES systems as well as the future of EES.
- The applicability and feasibility of the new generation EES devices
- Also, the hurdles in the development of the hybrid and newer EES systems will be discussed.

21.2 Hybrid electrochemical systems

21.2.1 *Metal-air batteries*

The major obstacle behind electric propulsion is the inadequate energy storing capability of current batteries. The energy density of gasoline is 13 kWh kg^{-1}. Considering ~12.6% of the US fleet's average gasoline tank-to-wheel efficiency, practical energy density becomes ~1700 Wh kg^{-1}. Current LIB technology relies on 200−250 Wh kg^{-1}. The cell-level energy density needs to be improved by approximately 10 times to match the gasoline level since the battery-to-wheel efficiency ratio is ~90% for electric vehicles [1].

Aqueous metal (Zn, Al, Mg, and Fe)-air batteries are a promising alternative to lithium-ion batteries (LIBs). Zn-air, in particular, has drawn more attention by virtue of its low cost (<\$100 kWh^{-1}), high theoretical energy density (1350 Wh kg^{-1}

without oxygen), and inherent safety. It is made of a porous cathode through which air can permeate, a metal anode, a membrane separator, a concentrated alkaline electrolyte, and a bifunctional electrocatalyst that can simultaneously perform oxygen reduction reaction (ORR) and oxygen evolution reaction (OER). During discharge, oxygen is reduced at the electrocatalyst surface, complemented by oxidation of metallic zinc to soluble $[Zn(OH)_4]^{2-}$. During charge, zinc is plated back on the anode, and the OER occurs at the cathode. The first concept of the Zinc-air battery was introduced way back in the 19th century. Primary Zinc-air batteries were commercialized in the 1930s, and the rechargeable version was in 2012 by NantEnergy, USA. Unfortunately, despite several promises, the product could not capture the market due to its limited practical capacity. Problems still exist for this system are bifunctionality of the electrocatalyst, sluggishness of the multiple proton-coupled electron transfer reactions, large electrode polarization due to gas release, and dendrite formation on the metal anode.

The Lithium-air battery having the highest theoretical energy density (11,640 Wh kg^{-1} for a nominal voltage of 3 V) can satisfy the demand and may be projected as a future for electric aviation systems.

The electrochemical mechanism of the Li-air battery is different from the Li-ion battery. During discharge, O_2, Li^+, and electrons combine at the cathode to form Li_2O_2 as a discharge product through ORR. Alternately, Li_2O_2 decomposes to evolve oxygen gas through OER during charge [2]. The electrochemical reaction is as follows

$$2Li^+ + O_2 + 2e^- \leftrightarrow Li_2O_2(s);\ E^0 = 2.96\ \text{V versus}\ Li/Li^+ \qquad (21.1)$$

Lithium metal is the best choice as an anode. In nonaqueous aprotic solvents, decomposition products formed due to Li metal and electrolyte reaction accumulates on the anode surface Solid electrolyte interface (SEI). It protects the surface from further electrolyte decomposition and can inhibit the reaction kinetics when it becomes thick. In aqueous electrolytes, vigorous reactions between water and lithium create a problem. Various ways are attempted to overcome the issues such as the formation of protective polymer layers (Polystyrene, PEO-Li salt mixture, glass-ceramic materials, etc.), passivation of the surface by thin Li-alloy, uniform lithium deposition in subsequent cycles by controlling temperature and pressure, etc. [3].

The cathode is composed of a catalyst layer, gas diffusion layer, and current collector. The cathode is the limiting factor as most of the unexpected side reaction occurs here. Current collectors can be metallic (Porous Ni or Cu foam etc.) or nonmetallic (Carbon paper, cloth, etc.). The gas diffusion layer serves as a bridge between the air and catalyst layers. It provides oxygen diffusion channels between air and catalyst, inhibits water to seep into the battery, and protect electrolyte from leakage. Catalyst is used to lower the overpotential of ORR and OER and to control the kinetics. The microstructure of porous carbon cathode is a major factor. Uniformly ordered and interconnected pores can act as catalyst support by simultaneously inner pores for gas diffusion and outer pores for electrolyte access. Heteroatom doping and morphology are other ways to enhance efficiency. A large voltage gap of ~1 V between charge and discharge leads to a round-trip efficiency inferior to LIBs [4]. Various types of catalysts such as transition metal oxides (Co_3O_4, $CoFe_2O_4$, CuO, NiO, Fe_2O_3, etc.), carbon matrix (Graphene, CNT, heteroatom doped analogs, etc.), precious metals (Pd > Pt > Ru ~Au for Li^+-ORR activity, Pt/C, Ru/C, etc.), etc. are extensively studied [5]. Different types of electrolytes—aqueous, nonaqueous, hybrid, solid-state, etc., have been reported in the literature. Aqueous electrolyte avoids cathode clogging because reaction products are water-soluble. Therefore, the aqueous system has a higher discharge potential than the aprotic. However, the vigorous reaction between water and lithium is a problem. A nonaqueous electrolyte is a combination of organic solvents like ether, ester, and carbonates that can solvate lithium salts like $LiPF_6$, LiTFSI, LiFSI, etc. The main discharge product Li_2O_2 is insoluble in the nonaqueous electrolytes. Its continuous buildup on the cathode surface leads to pore clogging and volume expansion which in turn increases impedance and degrades battery performance. O_2 transport and solubility in the electrolyte are critical issues, unlike LIBs, as higher oxygen solubility increases discharge capacity. Oxygen solubility increases either by lowering electrolyte viscosity or increasing the partial pressure of oxygen. However, low-dense carbonate electrolytes tend to decompose. Low polarity ether and glymes can easily wet the less polar porous carbon surface leading to better electrolyte infiltration, oxygen diffusion, and ORR activity. Solid-state electrolytes, ionic liquids, and hybrid design are the choices that have been investigated [6].

Most of the investigations are performed utilizing ultrapure oxygen (> 99.999%). Replacing oxygen with ambient air causes many critical issues. Li anode reacts with nitrogen (~78%), water vapor (~1%–2%), and CO_2 (~0.04%) from moist air to form Li_3N, LiOH, and Li_2CO_3, respectively, which are neither easily dissolved nor decompose and accumulates on cathode surface and reduces the efficiency [4]. Development of redox mediators (LiI, TEMPO, Vitamin K2, etc.) that actively take part in charge-discharge redox reactions and decrease the overpotential are crucial for long-term cyclability. Long-distance green EVs, grid support, electric aviation, etc., are potential future applications for Li-air batteries.

The biggest problem that Lithium-air battery is the rechargeability in the practical cell. The reversibility is severely hampered due to the reaction between lithium metal and oxygen gas. Oxygen reacts with lithium to form lithium superoxide (LiO_2). The superoxide radical is highly unstable and reacts with electrolytes and other cell components to finally form lithium peroxide (Li_2O_2). The irreversibility stems from the following reduction or disproportion reactions of lithium superoxide:

Reduction: $LiO_2 + e^- + Li^+ \rightarrow Li_2O_2$

$$\text{Disproportion: } 2LiO_2 \rightarrow Li_2O_2 + O_2$$

However, one of the main causes of irreversibility can be mitigated by using sodium analog. Sodium and lithium belong to the same group in the periodic table, and they share similar chemical reactivity toward oxygen. But the sodium superoxide (NaO_2) formed during discharge at the cathode side is thermodynamically more stable than LiO_2 and does not get converted to peroxide species via further reactions. Therefore, the reversible formation and decomposition of NaO_2 during discharge and charge, respectively, is a more energy-efficient and kinetically favored process at room temperature. Besides this mechanistic boost, the basic mechanism of lithium-air and sodium-air are similar. Sodium, being earth-abundant metal, is expected to reduce production costs to some extent. However, in the sodium analog, the theoretical energy density (1605 Wh kg^{-1}) is decreased to half of the lithium energy density. But it is still higher than the conventional Li-ion battery [7].

21.2.2 Hybrid ultracapacitors

This category of hybrid devices includes the combination of rechargeable batteries and electrochemical capacitors. HUCs are composed of two different types of electrodes immersed in an electrolyte (usually the electrolyte used in the battery counterpart). One electrode works on the principle of a rechargeable battery (a fast kinetics faradaic reaction), and the other electrodes work through the capacitive phenomenon. The HUCs bridge the energy density and power density disparity between the rechargeable batteries and supercapacitors. Thus, they combine the higher cyclability advantage of supercapacitors with the energy density of rechargeable batteries. Depending on the type of battery counterpart, the common types of HUCs explored are lead-acid hybrid capacitors (4.5 M H_2SO_4 electrolyte), and alkaline Ni-MH hybrid capacitors (KOH electrolyte), alkali metal-ion hybrid capacitors (organic carbonate-based electrolytes, ionic liquids, polymer electrolytes). The designation of anode and cathode is specific to a specific system. For instance, in lead-acid hybrid capacitors, the activated carbon or pseudocapacitive electrode serves as the anode and the conventional PbO_2 electrode as the cathode. In Li/Na-ion hybrid capacitors, the activated carbon capacitive electrode usually acts as the cathode and metal-ion battery-type anodes. HUCs can deliver capacities till ~30 Wh kg^{-1} and cycle life > 100,000 cycles [8–10]. The energy output of a HUC is limited by the capacitive activated carbon electrode whereas the power capabilities and cyclability is limited by the battery-type electrode.

Lead carbon (Pb-C) HUCs utilize activated carbon as an anode, PbO_2 as a cathode, and aqueous sulfuric acid as electrolytes. This kind of HUC eliminates the sulfation issue in lead-acid batteries as the lead anode is replaced by an EDLC-type activated carbon electrode. The main issue with these devices is the lower power capabilities and slower reversibility of the PbO_2 cathodes. Besides, cell balancing between the cathode and anode is an important parameter as the commercial PbO_2 electrode cannot be used in Pb-C HUCs as both the electrodes have huge capacitance differences. The current collectors at both the electrodes need to be improvised to obtain long cycle life and energy. The highest deliverable voltage for lead-carbon HUCs is 2 V and can be operated between 1 and 2.3 V.

Another type of aqueous hybrid supercapacitor that has been explored in the literature is AC//NiOOH, where activated carbon is used as an anode, NiOOH as cathode, and KOH solution as electrolyte. Similar to Pb-C HUCs, modification of the battery type electrode for better charge acceptability can improve the device's overall power and cyclability output [9].

Due to the limitations of upper cut-off voltages for the aqueous electrolytes, similar nonaqueous systems are being sought after. The state-of-art LIBs have promising energy and voltage output and, thus, have dominated the market share of batteries since their commercialization. Thus, by coupling a LIB-type electrode with a capacitive electrode, researchers have aimed at pushing the energy-power output of HUC. There are a total of four LIC types on the basis of their electrode configurations. In the first type, LIB anodes like graphite, $Li_4Ti_5O_{12}$, and metal oxides are combined with capacitive activated carbon cathode to form lithium-ion hybrid ultracapacitors (LiHUC). The second type of LiHUC involves capacitive carbon anode and LIB-type cathode like $LiCoO_2$, and $LiFePO_4$. The third type of LIHUC involves a LIB-type anode and hybrid cathode of EDLC-type activated carbon and LIB-type cathode. The fourth type of LIHUC is a symmetrical one with both electrodes as graphene type material which can provide both battery type and EDLC type phenomenon. A representation of the different types of LiHUC is provided in Fig. 21.1. The LICs are being commercialized. A device-level deliverable energy density of $\sim 25\ Wh\ kg^{-1}$ with 100,000 cycles is achieved. Further, material level modification and engineering are being done to push the performance matrix. The more sustainable alternatives to LIBs like sodium-ion batteries (SIBs) or potassium-ion batteries have also been hybridized with double-layer capacitors to generate feasible HUCs with similar performance matrices [11,12].

HUCs aim to provide the dual benefits of two different electrochemical systems in a single unit cell. Thus, they can be considered to suffice the advanced energy storage and conversion requirements in the future.

21.3 Future electrochemical energy storage

21.3.1 Multi-ion batteries

Convention LIBs function on the "Rocking-chair model," where Li^+ rocks back and forth between two electrodes during charge and discharge. In a multi-ion battery, movements of more than one ion are utilized for storing energy. The first concept of dual-ion battery (DIB) was introduced in dual-carbon battery (DCB) way before LIBs was conceptualized. In 1938, Rüdorff and Hofmann demonstrated reversible intercalation of HSO_4^- into graphite. However, the demonstration of the staging mechanism of PF_6^- intercalation into graphite via in situ XRD by the Dahn group in 2000 laid the foundation stone of modern Li-based DCB systems. In 2012, Placke et al. first coined the term "dual ion batteries" (DIBs) for these systems, and several variants of DIBs have been reported in the literature to date [13,14].

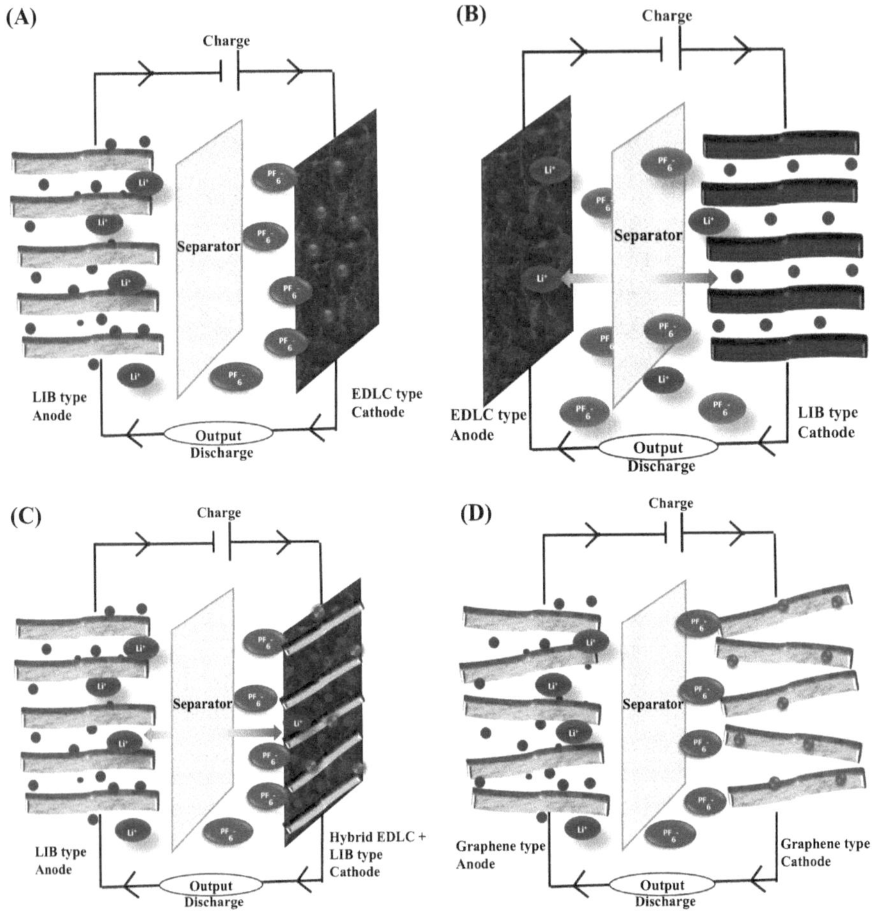

Figure 21.1 Schematic representation of (A) Type I (LIB anode/EDLC cathode), (B) Type II (EDLC anode/LIB cathode), (C) Type III (LIB anode//Hybrid EDLC + LIB cathode), and (D) Type IV (Graphene anode//Graphene cathode) LiHUC.

21.3.1.1 Mono-cation/mono-anion batteries

Rechargeable batteries that function on mono-cation/mono-anion are mostly dual-carbon types. Both cathodes and anodes are made up of similar or dissimilar types of carbon materials (e.g., Graphite, graphene, soft carbons, hard carbons, activated carbons, various microstructures, composites, etc.). Electrolyte-salt A^+B^- (e.g., $A^+ = Li^+$, Na^+, K^+, Zn^{+2}, Mg^{2+}, Ca^{2+}, etc., and $B^- = PF_6^-$, ClO_4^-, BF_4^-, FSI^-, $TFSI^-$, $BETI^-$, $DFOB^-$, ionic liquids, etc.) is the main reservoir of active ions. During charge, A^+ and B^- simultaneously move from electrolyte to cathode and anode, respectively, and intercalates into the carbon host. During discharge, A^+, and B^- return to the electrolyte. Unlike rocking-chair configuration, electrolyte here acts as a medium for ion migration and supplies active ions. Therefore, the concentration of the electrolyte is directly related to the experimental capacity. Specific capacity is limited to 60–100 mAh g^{-1} (150–200 mAh g^{-1} for typical LIBs) at 50–200 mA g^{-1} current density at 25°C. The most attractive feature of DCBs is the high operating potential ($>$4.5 V for Li-based salts) as anion intercalation within the carbon host occurs at a high voltage, which pushes the energy density beyond 200 Wh kg^{-1} for a full cell. The coulombic efficiency (hardly 95% in maximum cases) and cycling stability are relatively low with respect to LIBs as bulky anion intercalation makes carbon structure more susceptible to rupture. DCBs are expected to have high power density as active ions cover less distance during one charge cycle (Distance between anode and cathode for solvated Li^+ in LIBs *vs.* distance between electrode and electrolyte for solvated ions in DCBs). DCBs mainly address the issue of cost, sustainability, recyclability, etc., as carbon materials are cheaper and the system is devoid of any transition metal ions, unlike LIBs. Therefore, grid storage, regenerative braking, and low-power tools are the projected domains of DCBs' applications. However, technical challenges at the full pouch/prismatic level are plenty. Firstly, the high voltage range (3.0–5.0/5.2 V) of operation tends to corrode most of the cell components, including the separator, metal foils, Al or steel case, etc. Special care like ceramic coating on polyolefin separators and steel casings and carbon-coating on metal foils are employed as protective measures. Secondly, highly concentrated electrolyte (3.0–5.0 M of electrolyte salt, unlike 1.0–1.2 M in standard LIBs) increases capacity, forms inorganic-compound dominated stable SEI and aids cycling stability. However, low ionic conductivity decreases power density, and high-cost limits use for commercial purposes. Thirdly, the thickness of electrolyte-soaked separator ranges between 20–25 μm for state-of-the-art LIBs to support high active mass loading (70–80 μm thickly coated foil with 20–35 mg cm^{-2} active mass loading) for energy-dense batteries. In contrast, the soaked separator in the electrolyte-flooded dual-carbon cell reaches a thickness $>$ 100 μm, which demands a different optimized cell design for DIBs to accommodate more electrolytes. Fourthly, the counter-anion of electrolyte salt dictates cell characteristics. PF_6^- is considered to be the best choice for carbonate-based solvents to trade-off among several properties like compatibility, oxidative and reductive stability, gallery height, solvation-desolvation energy, etc. Instead of carbon in both electrodes, other intercalative materials such as metal-organic frameworks (MOF), organic materials, Mxenes, polymers, etc., alloying materials (Sn, Si, etc.), and insertion-type materials (TiO_2, MoO_3, MoS_2, T-Nb_2O_5, etc.) are used. Similar advantages and disadvantages of moving from standard LIBs to Li-analogs also occur in DIBs [15,16].

21.3.1.1.1 Dual cation

Different types of dual-action configurations have also been reported. In Type I, A-cation moves from cathode to electrolyte, and B-cation advances toward anode from electrolyte during charge. The process is reversed during discharge. Mg-ion batteries (MIB) are of particular interest due to the dendrite-free electrodeposition of Mg-anode. However, higher charge density results in sluggish kinetics of Mg^{+2} into cathode materials. To circumvent this issue, a dual Mg-Li/Na-ion battery is proposed. The battery consists of Mg-metal as an anode, high-performance Li or Na-ion host as cathode, and Mg/Na, Mg/Li dual ion electrolyte. Although this strategy solves MIB cathode shortcomings, the concentration of Mg and Na/Li ions in the electrolyte and low oxidative stability (2 V vs Mg^{2+}/Mg) are the limiting factors. In Type II, both the cations undergo synergistic movement from or to one electrode. This type of mechanism is mainly predominant in aqueous Zn and K-ion batteries, where cointercalation of H^+ contributes to additional capacity, and enhances energy density and rate capability. However, large viscosity, high price, and environmental concerns of "Water-in-salt" (WiS) or "Water-in-bisalt" (WiBS) electrolyte hinder large-scale energy storage [17].

21.3.1.1.2 Triple-ion battery

Triple-ion battery incorporates two cations and one anion or vice versa. Sn as an anode expanded graphite as a cathode and $LiPF_6$-$NaPF_6$ hybrid electrolyte are proposed for a novel triple-ion battery. During charge, Li^+ and Na^+ moved to the anode and formed an alloy with Sn (e.g., Li_2Sn_5 and NaSn, respectively) and PF_6^- moved to the anode. A high working voltage of 4 V along with 93 mAh g^{-1} capacity is obtained at a 5 C rate. 95% capacity is retained over 500 cycles at a 5 C rate [18]. The configuration and electrochemical performances of a few multi-ion batteries are presented in Table 21.1.

21.3.1.2 Anion shuttle batteries

Rechargeable alkali-metal-ion batteries' current research focuses mainly to utilize lightweight cations (Li^+, Na^+, K^+, Ca^{+2}, Mg^{+2}, etc.) for charge transfer. However, secondary batteries that store energy via shuttling of anions (mainly halides—F-, Cl-, Br-, I-) have also been investigated in the last decade. Fluorine is the smallest (71 pm of F vs 99 pm for Cl vs 114 pm for Br vs 133 pm for I), the lightest element among Group 17 halides in the periodic table highest electronegative (3.98 on Pauling scale) draws the most attention. Multifaceted benefits of F-doping and coating are leveraged in LIBs or post LIBs. The chemical reaction of fluorine (F) with metals (M) leads to the formation of a strongly ionic M-F bond with a large change in free energy resulting in high theoretical voltage in electrochemical cells. Fluoride-ion batteries (FIBs) to date can be classified into two different categories—High-temperature solid-state FIB (HTSSFIB) and room-temperature FIB (RTFIB). First FIB was reported in 2011 by the Fichtner group using metal fluoride (MF) cathode, a metal anode (M′), and solid fluoride conductors as an electrolyte [23]. The reactions that occur during discharge are as follows:

$$\text{At cathode: } MF_x + xe^- \rightarrow M + xF^- \tag{21.2}$$

$$\text{At anode: } xF^- + M' \rightarrow M'F_x + xe^- \tag{21.3}$$

HTSSFIBs need solid-state F-ion conductors. Composite of fluorite (CaF_2, SrF_2, BaF_2, etc.) and tynosite (LaF_3, CeF_3, etc.) are efficient fluoride-ion conductors at

Table 21.1 Configuration and electrochemical performances of multi-ion batteries.

Type	Cathode/anode	Electrolyte	Capacity ($mAh\ g^{-1}$)/ current density ($mA\ g^{-1}$)	Nominal voltage in V (voltage range)	Capacity retention (%)/no. of cycles	Coulombic efficiency (%)
Mono-cation/mono-anion	Graphite/Graphite [19]	5 M KFSI/EC/DMC	90 /100	4.7 (3.2–5.25)	89%/300	93–99
Mono-cation/mono-anion	MCMB/EG [20]	0.7 M $Ca(PF_6)_2$ in EC/DMC/EMC	66/2 C	4.6 (3.0–5.2)	94%/300	~90
Dual cation (Type I)	$Na_{1.5}VPO_{4.8}F_{0.7}$/Mg [21]	2 M $NaBH_4$ + 0.2 M $Mg(BH_4)_2$ in tetraglyme at anode and 1 M $NaClO_4$ in PC at cathode	98/1 C	3.2 (2.4–3.8)	100%/20	99
Dual cation (Type II)	Prussian blue/ $Ti_3C_2C_x$ Mxene [22]	1 M KNO_3 (aq.)	120/0.8	–	74%/3000	~100
Triple ion	Graphite/Sn [18]	1 M $NaPF_6$/$LiPF_6$ (7:3) in EC/EMC/DMC/PC	93.6/5 C	4.0 (2.0–4.8)	95%/500	95%

~150°C. Nanocrystalline $La_{1-x}Ba_xF_{3-x}$ ($0 \leq x \leq 0.15$) having conductivity in the range of $2-3 \times 10^{-4}$ S cm^{-1} at 160°C (highest for $x = 0.1$) are mostly used as electrolyte. One electrode should be raw metal (Ce, Bi, Mg, Cu, etc.), and another should be MF (BiF_3, CuF_2, SnF_2, CeF_3, etc.) [24]. The cell using Ce metal as an anode, BiF_3 solid solution cathode, $La_{0.9}Ba_{0.1}F_{2.9}$ as an electrolyte shows a first cycle discharge capacity of 197 mAh g^{-1} (theoretical specific capacity: 302 mAh g^{-1}) with an average voltage of 2.15 V and retains only 25% of the capacity at the end of 38th cycle via conversion mechanism [23]. These poor capacity retentions result from high volume changes of materials and subsequent loss of interparticle contact. Thin-film electrodes can mitigate the problem to a certain extent. Recently, an intercalation-based La_2NiO_{4+d} cathode along with Pb/PbF_2 or Zn/ZnF_2 anode and $La_{0.9}Ba_{0.1}F_{2.9}$ as an electrolyte HTSSFIB can deliver a reversible capacity of 30 mAh g^{-1} with 95% coulombic efficiency and stable capacity over 200 cycles [25]. On the other hand, liquid electrolyte-based RTFIBs need to contain F^- or F_2H^-, electrolyte salts should be soluble in a nonaqueous solvent, and must not decompose to HF. A system composed of fluorinated polyaniline cathode, PbF_2 anode, and fluorinated ionic liquid electrolyte showed a first discharge capacity of around 103 mAh g^{-1} [26]. Although some promising initial results, FIBs suffer from safety issues. Most of the MFs or F-precursors are skin irritants, and the chances of toxic HF release are far higher than LIBs. Comparatively, a chloride-ion battery (CIB) working on a similar electrochemical mechanism is much more sustainable. In a CIB, MCl_x (M = Co, Ni, V, Bi, etc.) or oxychloride (VOCl, FeOCl, BiOCl, etc.) based cathode, Li or Mg metal anode, and chloride-based ionic liquids are used as the electrolyte [27]. A reported CIB system utilizing VOCl as the cathode, and Li metal as anode demonstrates a maximum of 50 cycles with a capacity of 150 mAh g^{-1} and 1.6 V nominal voltage [28]. Rechargeable metal-bromine and the metal-iodine battery are also reported in the literature [29,30].

21.3.1.3 Sodium-seawater batteries

Seawater is a never-ending source of sodium. 69.2% of the earth's surface is covered by saline water, and each liter of it contains 10.7 g of sodium ions. Although this amount varies with geography, climate, temperature, rainfall, etc., of a particular region, Na^+ is the second-most abundant ion dissolved in seawater after Cl^- and causes salinity of seawater. The presence of dissolved ions provides a high ionic conductivity (~0.05 μS cm^{-1} for deionized water *vs.* ~50 mS cm^{-1} for saline water at 20°C), pushes the pH to the basic region (~8.1 for saline water vs ~7 for normal water) and contains dissolved gases like O_2, CO_2, etc. Employing seawater as a battery electrolyte began around the 1940s in primary batteries like Mg-AgCl, Mg-$PbCl_2$, Mg-Hg_2Cl_2, Mg-CuCl, etc. During the 1990s, primary metal-seawater batteries were developed for marine applications. These systems using metal (e.g., Mg, Mg alloy, Zn, Al, etc.) anode and seawater as cathode never reached a voltage beyond 1.8 V and suffered from a high discharge rate. Later, a rechargeable sodium-seawater battery was first demonstrated in 2014. Progress in

SSWB research accelerated of late due to the simultaneous development of SIBs and metal-oxygen batteries [31].

Prototype rechargeable SSWB comprises an anode compartment, an open-structured cathode, and a solid electrolyte membrane separating these two compartments. The Anode side is made up of efficient sodium storing material along with nonaqueous electrolytes. Open-structured cathode containing seawater is exposed to ambient air, and the current collector-electrocatalyst combination facilitates the cell reactions. The solid electrolyte membrane represents the most critical component of SSWB, which physically separates nonaqueous electrolyte and seawater catholyte. The solid electrolyte membrane should be highly selective toward Na^+, blocking the rest of the dissolved ions of seawater. During the charge, seawater supplies Na^+ to the anode side. The current collector-electrocatalyst combination provides necessary sites for oxygen evolution reaction (OER), and the energy is stored at the anode side as sodium metal. During discharge, dissolved oxygen gets reduced via ORR, Na^+ returns to seawater, and the stored chemical energy is released as electricity. The standard half-cell reactions are given below [32].

$$\text{At cathode side (in Seawater): } 4OH^- \leftrightarrow O_2 + 2H_2O + 4e^-; \quad E^0 = 0.77\ \text{V versus SHE}$$

$$\text{At anode side (in the nonaqueous electrolyte): } 4Na^+ + 4e^- \leftrightarrow 4Na; \quad E^0 = -2.71\ \text{V versus SHE} \tag{21.5}$$

$$\text{Overall cell reaction: } 4Na^+ + O_2 + 2H_2O \leftrightarrow 4NaOH; E_{cell} = 3.48\ \text{V} \tag{21.6}$$

The slightly alkaline pH of seawater gives rise to additional issues in the cathode compartment. During charge, a hypochlorite (ClO^-) formation reaction can also occur as seawater contains 19.3 g of Cl^- per litre. However, OER is slightly thermodynamically favorable, according to the Pourbaix diagram. During discharge, four-electron ORR reaction competes with a two-electron ORR ($E = 0.21$ V *vs.* SHE) and hydrogen evolution reaction ($E = -0.47$ V *vs.* SHE), both of which drops the total voltage of the cell below 3 V. The former reaction can be avoided by using selective electrocatalysts and the latter by maintaining a threshold level of dissolved oxygen in seawater [31].

SSWBs put forward several challenges from cell construction, material selection, and electrochemical testing perspectives as traditional battery testing protocols cannot be followed. To avoid seawater leakage and large charge-discharge polarization (~ 1 V) particular cell setup needs to be arranged. The most vital part of the construction, the solid electrolyte membrane, must-have features like high sodium ion selectivity, high ionic conductivity, high chemical stability against seawater and nonaqueous electrolyte, and high interfacial stability in wide voltage range, etc. NASICON pellet having a 3D Na^+ diffusion pathway and high ionic conductivity (1×10^{-3} S cm^{-1}) is found to be the most appropriate electrolyte, stable in seawater, and 1 M $NaCF_3SO_3$ in tetraethylene glycol dimethyl ether (TEGDME). It also avoids the hydration reaction like inorganic sulfide electrolytes (Na_3PS_4, Na_3SbS_4, etc.) and other side reactions. The sluggishness of OER and ORR reactions limits the electrochemical output of SSWBs [33]. The current collector-electrocatalyst combination must provide uniformly distributed reaction sites, electronic pathway, lightness in weight, and cost-efficiency. Metal current collector losses protective oxide layer by an attack of Cl^- ions and gets corroded easily. Carbon-based materials with surface functional groups providing hydrophilicity are the best option. An electrocatalyst that bridges the large charge-discharge polarization gap should be bifunctional to facilitate OER/ORR simultaneously. Highly porous metal oxide nanoparticles $Co_xMn_{3-x}O_4$, $Co_3V_2O_8$, etc., and metal-free heteroatom (S, N, P, O, etc.) doped carbon materials lower the voltage drop by ~0.5 V and provides round-trip voltage efficiency of >80% [34,35]. Moreover, binder-free 3D microporous carbon sponges are reported to perform the reaction as a current collector-electrocatalyst assembly efficiently. The anode compartment can be designed following SIBs. Hard carbon is the best performing anode, as similar problems have been encountered by using sodium metals and alloying materials as anode like in SIBs. Hybrid materials like Sn/C, P/C, etc., have shown to be a way out of obtaining high capacity. Besides, standard SIB electrolyte 1 M $NaClO_4$ or 1 M $NaPF_6$ in EC/PC/DEC solvents, 1 M $NaCF_3SO_3$ in ether solvents, and ionic liquids also perform well. Some of the obtained results are summarized in Table 21.2.

Researchers have successfully upscaled coin-type cells to pouch-type cells and combined cells in modules or racks for large-scale applications over the past few years. IBM Corporation, already working to develop this system's commercial prototypes [38]. SSWBs are mainly suitable for marine sector applications. Basic designs need to be manipulated a little bit depending upon the flexibility of the battery holding matrix. Life jackets, light buoys, underwater machinery, automatic surface vehicles, offshore and seaside energy storage, etc., are the possible potential applications for rechargeable SSWBs. Furthermore, the concept is extended to develop desalination batteries and desalination carbon capture systems. Currently, SSWBs perform well only under low current densities and incorporate costly items in cell construction. Future challenges lie in improving power density, utilizing sodium metal directly as the anode, developing an electrocatalyst that can support long-term energy storage, and lowering the cost to make it competitive in the commercial market [32].

21.3.1.4 Solid-state and metal batteries

Lithium metal was one of the earliest developed anodes for lithium-based batteries, which became obsolete due to dendrite growth and subsequent safety concerns. However, the need for an energy-dense battery pack is at its peak due to the critical demands of EV mileage and cost. Substituting graphite with lithium metal can push the energy density from 250 to 440 Wh kg^{-1}. The main challenge remains in ensuring long-term cyclability and safety. Highly reactive lithium surfaces form dendritic depositions due to the repeated stripping-plating process during cycling. These dendrites can pierce the separator and cause a short circuit. Numerous ways are investigated to suppress metallic surface overgrowth [39]. Some of these include—(a)

Table 21.2 Comparison of various sodium seawater systems reported.

Anode	Nonaqueous electrolyte	Current collector/ electro catalyst	Separator	Capacity (mAh g^{-1}) @ current density (mA cm^{-2})	Voltage polarization in V
Hard carbon [33]	1 M $NaClO_4$ in EC-PC	Carbon paper	$Na_2Zr_2Si_2PO_{12}$	115 @ 0.05	1.2
Hard carbon [33]	1 M $Na_3CF_3SO_3$ in TEGDME	Carbon paper	$Na_2Zr_2Si_2PO_{12}$	126 @ 0.05	1.08
Na [36]	1 M $Na_3CF_3SO_3$ in TEGDME	Carbon paper/porous carbon	$Na_2Zr_2Si_2PO_{12}$	191 @ 0.05	0.47
Red P/ carbon [37]	1 M $NaClO_4$ in EC-DEC + 5% FEC	Carbon felt	$Na_2Zr_2Si_2PO_{12}$	920 @ 100 mA g^{-1}	–

alloying lithium with other metals or using a noncrystalline glassy form of lithium to buffer volume change, (b) confining lithium metal into a 3D carbon matrix to hinder infinite volume change, (c) Functionalisation of the host with lithiophilic substance (e.g., Cu_2O, Au, polar-group containing polymer, etc.) for preferential lithium nucleation in the inner surface rather outer surface, (d) sandwiching lithium metal with mechanically robust materials for uniform deposition and nullifying the possibility of forming spiny projections, (e) stabilizing metal-separator interface by protective coating (e.g., Li_3PO_4, Li_3N, Al_2O_3, polymeric films like PVDF-HFP, etc.) on the metal surface, (f) decorating conventional separators with functional groups for maintaining smooth lithium-ion flux through it or with conductive coating for avoiding vertically directed deposition, and (g) incorporating additives in small amounts or using novel electrolyte mixture for thin and robust SEI formation. Another way to circumvent these safety issues is by employing solid-state electrolytes (SSEs) in the system. Being mechanically strong and having a high Li^+ transference number SSEs are believed to thwart dendrite growth. It overcomes chemical stability, spillage, and flammability issues of liquid electrolytes. SSEs can be broadly classified into two categories—inorganic ceramic electrolytes and organic polymeric electrolytes. Inorganic sulfides (e.g., Li_2S-P_2S_5, $Li_{10}GeP_2S_{12}$, etc.) provide high room-temperature ionic conductivity (10^{-2}–10^{-4} S cm^{-1}) due to strong Lithium-sulfur (Li-S) bonds, good mechanical strength (Especially, glassy β-Li_3PS_4), and flexibility. However, chemical instability, including sensitivity toward O_2 and H_2O, obstacles their commercial production. Inorganic oxides solve these environmental issues and, therefore, most studied and realistic for upscaling [40]. They can be subcategorized into following types—perovskite ($Li_{3x}La_{2/3-x}TiO_3$), NASICON-type ($Na_{1+x}Zr_2P_{3-x}Si_xO_{12}$, $Li_{1+x}M_xTi_{2-x}(PO_4)_3$, etc.), LISICON-type (e.g., $Li_{3.25}Si_{0.25}P_{0.75}O_4$, $Li_{3.6}Ge_{0.8}S_{0.2}O_4$, etc.), garnet ($Li_7La_3Zr_2O_{12}$, $Li_{6.55}La_3Zr_2Ga_{0.45}O_{12}$, etc.), LiPON ($Li_{2.9}PO_{3.3}N_{0.46}$), etc. However, their mechanical property and interfacial property are inferior. Inorganic halides (e.g., argyrodite-type Li_6PS_5X, where X = Cl, Br, or I) are another class of electrolytes [41]. The other class, that is, organic polymers (PEO, PMMA, PVDF, PAN, etc.) are more flexible in nature and more compatible with electrodes than inorganic ones. But they suffer from low room-temperature ionic conductivity (10^{-5}–10^{-6} S cm^{-1}), poor thermal stability and mechanical weakness, etc. [42]. To avoid these shortcomings, composite electrolytes such as inorganic fillers in the polymer matrix, heterogeneously sandwiched architectures, inorganic frameworks with filled polymers, and open framework compositing (MOFs, COFs, POCs, etc.) are investigated [43]. Although various exciting experimental reports and advanced computational studies are needed in Li-based systems and beyond lithium batteries. All-solid-state metal batteries are still away from gaining widespread commercial acceptance.

21.3.1.5 Lithium-sulfur batteries

A lithium-sulfur battery is a promising rechargeable system due to the high elemental abundance of sulfur, the high theoretical capacity of ~1600 mAh g^{-1}, and high energy density of 2600 Wh kg^{-1}. The cell voltage varies nonlinearly in the range between 2.5–1.7 V during discharge. During discharge, the sulfur cathode is reduced to lithium sulfide, and the cathode half-cell reaction can be expressed as,

$$S + 2Li^+ + 2e^- \leftrightarrow Li_2S \tag{21.7}$$

It involves intermediate polysulfide formation across cathode during charging, such as,

$$Li_2S \rightarrow Li_2S_2 \rightarrow Li_2S_4 \rightarrow Li_2S_6 \rightarrow Li_2S_8 \rightarrow S_8 \tag{21.8}$$

The main challenges of Li-S battery are: (i) dissolution of polysulfide intermediates (Li_2S_x, $6 \le x \le 8$) into the electrolyte causing loss of active sulfur material, (ii) high volume expansion (~80%) during lithiation, (iii) irreversible side reactions due to the nucleophilic attack of polysulfides toward the carbonate-based electrolyte solvents, (iv) low electronic conductivity (10^{-30} S cm^{-1}) of sulfur and (v) dendrite formation on Li-metal anode. These problems can be mitigated by electrode and electrolyte modifications. Li_2S or Li_2S-carbon composite is used as cathode instead of bare sulfur suppressing the polysulfide shuttle effect. The electrolytes are ether-based, such as DOL (1,3-dioxolane), DME (Dimethoxy ethane), and glycol. 1 M LiTFSI in DOL-DME solvent with 1% $LiNO_3$ additive is commonly used as an electrolyte for Li-S battery [44–46]. Research is still in progress on these future generation LIBs to make them commercially viable in real systems.

21.3.1.6 Metal-ion-based aqueous energy storage systems

Battery-powered devices are ready to run in the future. However, charging the batteries, which still rely on power-plant-supplied electricity, remains the bottleneck of completely transitioning from nonrenewable to renewable energy sources. Therefore, decarbonization of grids is the ultimate solution to entirely shrug off the dependence on fossil fuels. The principal criteria of large format energy storage systems are—low cost of construction, installation, and maintenance, long cycle life, high round-trip efficiency, environmental benignity, safety, etc. Batteries are a suitable fit as it meets all the demands. Although nonaqueous LIBs are currently dominating the portable electronics market, the electrolyte's flammable nature poses a severe threat to sustainability. Several fire accidents have already been reported in LIB plants. Substituting nonaqueous organic electrolytes with aqueous materials can provide manifold benefits. Ubiquitous water can cut down the material cost, eliminate all stringent cell manufacturing requirements, solve safety concerns, and the high ionic conductivity of water can boost the power density by a factor of two. The first report on an aqueous rechargeable battery utilizing $LiMn_2O_4$ cathode, VO_2 (B) anode ("B" refers to crystal geometry of VO_2), an aqueous solution containing 5.0 M $LiNO_3$ and 0.001 M LiOH as an electrolyte, and microporous polypropylene separator (Celgard 3500) was registered in 1994, which

Table 21.3 Comparison of metal-ion-based aqueous energy storage systems reported.

Cathode	Anode	Aqueous electrolyte	Initial capacity (mAh g^{-1}) @ Current density	Capacity retention (%) @ No. of cycles	Working Potential (V)
Porous $LiMn_2O_4$ [48]	Activated carbon	0.5 M Li_2SO_4	112 @ 90 C	93% @ 10000	0.0–1.8
$Na_3V_2(PO_4)_3$ [49]	$NaTi_2(PO_4)_3$	1 M Na_2SO_4	58 @ 10 A g^{-1}	50% @ 50	0.5–1.6
Pyrolytic graphite foil [50]	Al foil	$AlCl_3$/[EMIm]Cl	60–66 @ 4 A g^{-1}	100% @ 7500	0.5–2.4
δ-MnO_2/carbon molecular sieves [51]	VO_2 Nanowire	1.0 M $MgSO_4$	112.2 @50 mA g^{-1}	46.9% @100	0.0–1.8
$ZnMn_2O_4$@C [52]	Zn	3.0 M $Zn(CF_3SO_3)_2$	85 @ 500 mA g^{-1}	94% @ 500	0.8–1.9

showed energy density of 75 Wh kg^{-1} with excellent cycling stability [47]. The cell reactions are as follows:

$$\text{At cathode: } LiMn_2O_4 \leftrightarrow Li_{1-x}Mn_2O_4 + xLi^+ + xe^- \quad (21.9)$$

$$\text{At Anode: } VO_2(B) + xLi^+ + xe^- \leftrightarrow Li_xVO_2 \quad (21.10)$$

$$\text{Overall cell reaction: } LiMn_2O_4 + VO_2(B) \leftrightarrow Li_{1-x}Mn_2O_4 + Li_xVO_2(B);\ E_{cell} = 1.5V \quad (21.11)$$

Since then, several analogs are reported. Some of the best results are summarized in Table 21.3.

However, these systems could not draw the commercial market's attention due to several factors. Firstly, the electrochemical stability window of water is restricted by H_2 and O_2 evolution reaction, that is, 1.23 V. Kinetic limitation and pH dependence of electrolysis can expand the stability up to 2 V at maximum. This fact narrows down the choice of electrode materials. The cathode and anode materials that are active between 3–4 V and 2–3 V, respectively, can be used as no protective SEI layer formation occurs on the electrode surface, unlike nonaqueous systems via decomposition of organic solvents. Moreover, the water-splitting ability of the materials like delithiated spinel $LiMn_2O_4$ and $Li_2MnP_2O_7$ exacerbates the situation. Secondly, the violent reaction between water and lithium again restricts material selection. Protective surface coatings on materials can reduce the side reactions to some extent. Thirdly, the dissolution of active materials, especially nanomaterials due to their exposed surface, such as decomposition of $LiFePO_4$ into Fe_3O_4 around pH 10 and the instability of Prussian blue type materials in alkaline pH remains an issue for long-term cycling. Surface stabilizers or protective coatings can be useful in this respect. Fourthly, coinsertion of protons can block the guest atom sites in the host material, which has been

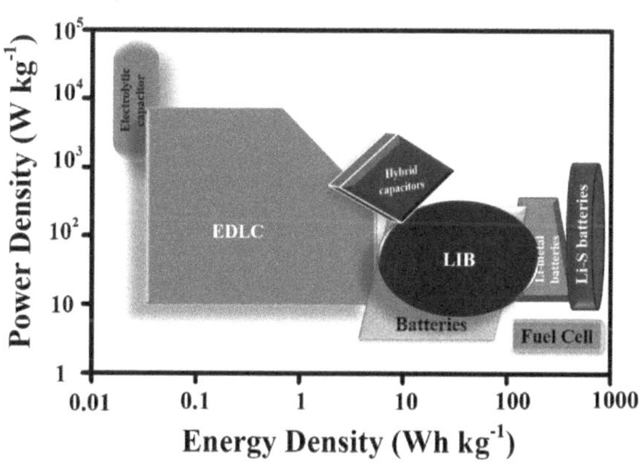

Figure 21.2 Ragone plot showing the different electrochemical energy storage systems [58].

observed in $LiCoO_2$ and MnO_2. Cycling at higher pH can avoid the problem but triggers several side reactions. Therefore, the pH dependence of electrolysis reaction, concentration of intercalated lithium in the host, and interfacial instability curbs the usage of some of the best performing nonaqueous LIB materials in the aqueous system [53]. Superconcentrated aqueous electrolytes have emerged recently to address several issues simultaneously. WiS and WiBS electrolytes are reported to extend the voltage limit to ~ 3 V and beyond, respectively, for aqueous LIBs without compromising the inherent benefits of water [54,55]. Yamada et al. demonstrated that $Li(TFSI)_{0.7}(BETI)_{0.3} \cdot 2H_2O$ widens the operational voltage window of LCO/LTO and LNMO/LTO with high C-rate performance [56]. Various salt-solvent mixtures, along with the development of localized high concentrated electrolyte where the addition of diluents drives ionic conductivity and lowers viscosity, are developed in recent years. In another report, Tarascon and coworkers have revealed that formed SEI in the superconcentrated electrolyte is not protective enough to avoid electrolysis reactions. An elevated temperature operation leads to parasitic reactions and unexpected cell drying-out due to continuous electrolyte consumption during cycling, and self-discharge may hinder commercial aspects [57].

21.4 Conclusions

The electrochemical energy storage systems are reservoirs of energy that can be utilized when in need and thereby decreasing human civilization's dependence on nonrenewable sources. Some of the hybrids of conventional EES systems and future EES devices have been discussed briefly in this chapter, and their comparative energy density-power density relationship can be realized from the Ragone plot in Fig. 21.2.

- Apart from the applications of the conventional EES devices, the hybrid configuration with batteries and fuel cells is being considered for EVs, load-leveling systems, and the hybrid devices with a combination of batteries and supercapacitors are being considered for backups for a longer period and also, EVs.
- Various new generation electrochemical energy storage devices (e.g., multi-ion batteries, Sodium-seawater, Li-S, metal-ion aqueous batteries, etc.) and hybrid systems (e.g., Metal-air batteries and metal-ion hybrid supercapacitors) are being explored to optimize the combined effects of higher energy density, power density, and long cyclability. However, surpassing the present status of LIBs might require decades of research and development in the concerned field.

Acknowledgments

Udita and S. K. Martha acknowledge DST-SERB (Sanction Order: CRG/2018/003543), Shuvajit acknowledges CSIR (letter no. 09/1001(0067)/2019-EMR-I), and Madhushri acknowledges DST-INSPIRE (code: IF180708) for fellowship and financial support to this work.

References

[1] G. Girishkumar, B. McCloskey, A.C. Luntz, S. Swanson, W. Wilcke, Lithium-air battery: promise and challenges, J. Phys. Chem. Lett. 1 (14) (2010) 2193–2203.
[2] C. Wang, Y. Yu, J. Niu, Y. Liu, D. Bridges, X. Liu, et al., Recent progress of metal–air batteries—a mini review, Appl. Sci. 9 (14) (2019) 2787.
[3] M.A. Rahman, X. Wang, C. Wen, A review of high energy density lithium–air battery technology, J. Appl. Electrochem. 44 (1) (2014) 5–22.
[4] J.H. Kang, J. Lee, J.W. Jung, J. Park, T. Jang, H.S. Kim, et al., Lithium–air batteries: air-breathing challenges and perspective, ACS Nano 14 (11) (2020) 14549–14578.
[5] Y.C. Lu, H.A. Gasteiger, Y. Shao-Horn, Catalytic activity trends of oxygen reduction reaction for nonaqueous Li-air batteries, J. Am. Chem. Soc. 133 (47) (2011) 19048–19051.
[6] M. Balaish, A. Kraytsberg, Y. Ein-Eli, A critical review on lithium–air battery electrolytes, Phys. Chem. Chem. Phys. 16 (7) (2014) 2801–2822.
[7] P. Hartmann, C.L. Bender, M. Vračar, A.K. Dürr, A. Garsuch, J. Janek, et al., A rechargeable room-temperature sodium superoxide (NaO2) battery, Nat. Mater. 12 (3) (2013) 228–232.
[8] M. Mounika, G.D. Kumar, D. Danalakshmi, Evolution of hybrid super capacitors and its future pathway. IJEAST 4 (6) (2019) 10.
[9] Y. Wang, Y. Song, Y. Xia, Electrochemical capacitors: mechanism, materials, systems, characterization and applications, Chem. Soc. Rev. 45 (21) (2016) 5925–5950.
[10] Y. Shao, M.F. El-Kady, J. Sun, Y. Li, Q. Zhang, M. Zhu, et al., Design and mechanisms of asymmetric supercapacitors, Chem. Rev. 118 (18) (2018) 9233–9280.
[11] L. Jin, C. Shen, A. Shellikeri, Q. Wu, J. Zheng, P. Andrei, et al., Progress and perspectives on pre-lithiation technologies for lithium ion capacitors, Energy Environ. Sci. 13 (8) (2020) 2341–2362.
[12] J. Ding, W. Hu, E. Paek, D. Mitlin, Review of hybrid ion capacitors: from aqueous to lithium to sodium, Chem. Rev. 118 (14) (2018) 6457–6498.
[13] S. Chen, Q. Kuang, H.J. Fan, Dual-carbon batteries: materials and mechanism, Small. 16 (40) (2020) 2002803.
[14] J.A. Seel, J.R. Dahn, Electrochemical intercalation of PF 6 into graphite, J. Electrochem. Soc. 147 (3) (2000) 892.
[15] M. Wang, Y. Tang, A review on the features and progress of dual-ion batteries, Adv. Energy Mater. 8 (19) (2018) 1703320.
[16] T. Placke, A. Heckmann, R. Schmuch, P. Meister, K. Beltrop, M. Winter, Perspective on performance, cost, and technical challenges for practical dual-ion batteries, Joule 2 (12) (2018) 2528–2550.
[17] Q. Liu, H. Wang, C. Jiang, Y. Tang, Multi-ion strategies towards emerging rechargeable batteries with high performance, Energy Storage Mater. 23 (2019) 566–586.
[18] C. Jiang, Y. Fang, W. Zhang, X. Song, J. Lang, L. Shi, et al., A multi-ion strategy towards rechargeable sodium-ion full batteries with high working voltage and rate capability, Angew. Chem. Int. (Ed.) 57 (50) (2018) 16370–16374.
[19] K.V. Kravchyk, P. Bhauriyal, L. Piveteau, C.P. Guntlin, B. Pathak, M.V. Kovalenko, High-energy-density dual-ion battery for stationary storage of electricity using concentrated potassium fluorosulfonylimide, Nat. Commun. 9 (1) (2018) 1–9.
[20] S. Wu, F. Zhang, Y. Tang, A novel calcium-ion battery based on dual-carbon configuration with high working voltage and long cycling life, Adv. Science. 5 (8) (2018) 1701082.

[21] K.V. Kravchyk, M. Walter, M.V. Kovalenko, A high-voltage concept with sodium-ion conducting β-alumina for magnesium-sodium dual-ion batteries, Commun. Chem. 2 (1) (2019) 1–6.
[22] Y. Zhu, Y. Lei, Z. Liu, Y. Yuan, H.N. Alshareef, An unconventional full dual-cation battery, Nano Energy 81 (2020) 105539.
[23] M.A. Reddy, M. Fichtner, Batteries based on fluoride shuttle, J. Mater. Chem. 21 (43) (2011) 17059–17062.
[24] F. Gschwind, G. Rodriguez-Garcia, D.J.S. Sandbeck, A. Gross, M. Weil, M. Fichtner, et al., Fluoride ion batteries: theoretical performance, safety, toxicity, and a combinatorial screening of new electrodes, J. Fluor. Chem. 182 (2016) 76–90.
[25] M.A. Nowroozi, K. Wissel, M. Donzelli, N. Hosseinpourkahvaz, S. Plana-Ruiz, U. Kolb, et al., High cycle life all-solid-state fluoride ion battery with La 2 NiO 4 + d high voltage cathode, Commun. Materials. 1 (1) (2020) 1–16.
[26] I. Darolles, C.M. Weiss, M.M. Alam, A. Tiruvannamalai, S.C. Jones, Fluoride ion battery compositions. US20120164541A1 (2012) 13. https://patents.google.com/patent/US20120164541A1/en.
[27] F. Gschwind, H. Euchner, G. Rodriguez-Garcia, Chloride ion battery review: theoretical calculations, state of the art, safety, toxicity, and an outlook towards future developments, Eur. J. Inorg. Chem. 2017 (21) (2017) 2784–2799.
[28] P. Gao, M.A. Reddy, X. Mu, T. Diemant, L. Zhang, Z. Zhao-Karger, et al., VOCl as a cathode for rechargeable chloride ion batteries, Angew. Chem. Int. (Ed.) 55 (13) (2016) 4285–4290.
[29] M. Xing, Z.Z. Zhao, Y.J. Zhang, J.W. Zhao, G.L. Cui, J.H. Dai, Advances and issues in developing metal-iodine batteries, Mater. Today Energy. 18 (2020) 100534.
[30] H. Li, M. Li, X. Zhou, T. Li, H. Zhao, A novel rechargeable bromine-ion battery and the induction of bromine ions on metal electrodes, Sustain. Energy Fuels 4 (8) (2020) 3871–3878.
[31] S.T. Senthilkumar, W. Go, J. Han, L.P.T. Thuy, K. Kishor, Y. Kim, et al., Emergence of rechargeable seawater batteries, J. Mater. Chem. A. 7 (40) (2019) 22803–22825.
[32] S.M. Hwang, J.S. Park, Y. Kim, W. Go, J. Han, Y. Kim, et al., Rechargeable seawater batteries—from concept to applications, Adv. Mater. 31 (20) (2019) 1804936.
[33] H. Kim, J.S. Park, S.H. Sahgong, S. Park, J.K. Kim, Y. Kim, Metal-free hybrid seawater fuel cell with an ether-based electrolyte, J. Mater. Chem. A. 2 (46) (2014) 19584–19588.
[34] M. Abirami, S.M. Hwang, J. Yang, S.T. Senthilkumar, J. Kim, W.S. Go, et al., A metal-organic framework derived porous cobalt manganese oxide bifunctional electrocatalyst for hybrid na-air/seawater batteries, ACS Appl. Mater. Interfaces 8 (48) (2016) 32778–32787.
[35] K.H. Shin, J. Park, S.K. Park, P. Nakhanivej, S.M. Hwang, Y. Kim, et al., Cobalt vanadate nanoparticles as bifunctional oxygen electrocatalysts for rechargeable seawater batteries, J. Ind. Eng. Chem. 72 (2019) 250–254. Apr.
[36] S.T. Senthilkumar, S.O. Park, J. Kim, S.M. Hwang, S.K. Kwak, Y. Kim, Seawater battery performance enhancement enabled by a defect/edge-rich, oxygen self-doped porous carbon electrocatalyst, J. Mater. Chem. A. 5 (27) (2017) 14174–14181.
[37] Y. Kim, S.M. Hwang, H. Yu, Y. Kim, High energy density rechargeable metal-free seawater batteries: a phosphorus/carbon composite as a promising anode material, J. Mater. Chem. A. 6 (7) (2018) 3046–3054.
[38] Free of Heavy Metals, New Battery Design Could Alleviate Environmental Concerns [Internet]. IBM Research Blog. <https://www.ibm.com/blogs/research/2019/12/heavy-metal-free-battery/>, 2019 (accessed 13.03.21).

[39] X. Zhang, Y. Yang, Z. Zhou, Towards practical lithium-metal anodes, Chem. Soc. Rev. 49 (10) (2020) 3040–3071.

[40] Y. Zheng, Y. Yao, J. Ou, M. Li, D. Luo, H. Dou, et al., A review of composite solid-state electrolytes for lithium batteries: fundamentals, key materials and advanced structures, Chem. Soc. Rev. 49 (23) (2020) 8790–8839.

[41] J.C. Bachman, S. Muy, A. Grimaud, H.H. Chang, N. Pour, S.F. Lux, et al., Inorganic solid-state electrolytes for lithium batteries: mechanisms and properties governing ion conduction, Chem. Rev. 116 (1) (2016) 140–162.

[42] J.W. Fergus, Ceramic and polymeric solid electrolytes for lithium-ion batteries, J. Power Sources 195 (15) (2010) 4554–4569.

[43] X. Fu, D. Yu, J. Zhou, S. Li, X. Gao, Y. Han, et al., Inorganic and organic hybrid solid electrolytes for lithium-ion batteries, CrystEngComm. 18 (23) (2016) 4236–4258.

[44] J. Scheers, S. Fantini, P. Johansson, A review of electrolytes for lithium–sulphur batteries, J. Power Sources 255 (2014) 204–218. Jun.

[45] W. Kang, N. Deng, J. Ju, Q. Li, D. Wu, X. Ma, et al., A review of recent developments in rechargeable lithium–sulfur batteries, Nanoscale. 8 (37) (2016) 16541–16588.

[46] H.J. Peng, J.Q. Huang, X.B. Cheng, Q. Zhang, Review on high-loading and high-energy lithium–sulfur batteries, Adv. Energy Mater. 7 (24) (2017) 1700260.

[47] W. Li, J.R. Dahn, D.S. Wainwright, Rechargeable lithium batteries with aqueous electrolytes, Science. 264 (5162) (1994) 1115–1118.

[48] Q. Qu, L. Fu, X. Zhan, D. Samuelis, J. Maier, L. Li, et al., Porous LiMn2O4 as cathode material with high power and excellent cycling for aqueous rechargeable lithium batteries, Energy Environ. Sci. 4 (10) (2011) 3985–3990.

[49] Q. Zhang, C. Liao, T. Zhai, H. Li, A high rate 1.2V aqueous sodium-ion battery based on all NASICON structured NaTi2(PO4)3 and Na3V2(PO4)3, Electrochim. Acta. 196 (2016) 470–478. Apr.

[50] M.C. Lin, M. Gong, B. Lu, Y. Wu, D.Y. Wang, M. Guan, et al., An ultrafast rechargeable aluminium-ion battery, Nature 520 (7547) (2015) 324–328.

[51] H. Zhang, D. Cao, X. Bai, High rate performance of aqueous magnesium-ion batteries based on the δ-MnO2@carbon molecular sieves composite as the cathode and nanowire VO2 as the anode, J. Power Sources 444 (2019) 227299.

[52] N. Zhang, F. Cheng, Y. Liu, Q. Zhao, K. Lei, C. Chen, et al., Cation-deficient spinel ZnMn2O4 cathode in Zn(CF3SO3)2 electrolyte for rechargeable aqueous Zn-ion battery, J. Am. Chem. Soc. 138 (39) (2016) 12894–12901.

[53] J. Liu, C. Xu, Z. Chen, S. Ni, Z.X. Shen, Progress in aqueous rechargeable batteries, Green. Energy Environ. 3 (1) (2018) 20–41.

[54] Y. Yamada, J. Wang, S. Ko, E. Watanabe, A. Yamada, Advances and issues in developing salt-concentrated battery electrolytes, Nat. Energy 4 (4) (2019) 269–280.

[55] O. Borodin, J. Self, K.A. Persson, C. Wang, K. Xu, Uncharted waters: super-concentrated electrolytes, Joule 4 (1) (2020) 69–100.

[56] Y. Yamada, K. Usui, K. Sodeyama, S. Ko, Y. Tateyama, A. Yamada, Hydrate-melt electrolytes for high-energy-density aqueous batteries, Nat. Energy 1 (10) (2016) 1–9.

[57] L. Droguet, A. Grimaud, O. Fontaine, J.M. Tarascon, Water-in-Salt Electrolyte (WiSE) for aqueous batteries: a long way to practicality, Adv. Energy Mater. 10 (43) (2020) 2002440.

[58] B. Kale, S. Chatterjee, Electrochemical energy storage systems: India perspective, Bull. Mater. Sci. 43 (1) (2020) 96.

Electric vehicles: a step toward sustainability

22

Manoj Goswami[1,2], Satendra Kumar[1,2], Hafsa Siddiqui[1], Viplov Chauhan[1], Netrapal Singh[1,2], N. Sathish[1,2], Mohammad Ashiq[1,2] and Surender Kumar[1,2]
[1]CSIR—Advanced Materials and Processes Research Institute (AMPRI), Bhopal, Madhya Pradesh, India, [2]Academy of Scientific and Innovative Research (AcSIR), Ghaziabad, Uttar Pradesh, India

Chapter Outline

Novelty

Electric vehicles (EVs), which can reduce CO_2 emissions, air pollutants, and noise from automobiles are a potential technology for sustainability. Electric power trains are significantly more energy-efficient than traditional power trains, and the electricity used might come from renewable energy sources. Hence, road transport electrification is expected to play a crucial role in achieving these goals. In this chapter, we looked at a variety of technologies that are used in the development of EVs.

Emerging Trends in Energy Storage Systems and Industrial Applications. DOI: https://doi.org/10.1016/B978-0-323-90521-3.00024-7

The function of a hybrid electrochemical storage system was also covered, as well as how energy storage devices work in EVs.

Highlights

This chapter gives an overview of the current international renewable and clean energy challenge. The impact of EV technologies, starting from the birth of the EV to a hybrid EV (HEV) is discussed in Section 22.1, whereas different configurations and current state of the art of these EV technologies are presented in Sections 22.2. Then, it explores the research field of HEVs and the science behind them. In addition to this, novel concepts of nanotechnology for an HEV are briefly explained. Furthermore, Section 22.3 discusses several energy generation technologies that are widely utilized to expand the range of EVs. Varying operating energies, charger designs, and power conversion technologies are covered in Section 22.4. The chapter ends with a section on summary and outlook, which presents the authors' viewpoint on the emerging scenario toward sustainability.

22.1 Introduction

Transportation-related carbon-based emissions are a key contributor to climate change and have a severe impact on human health. The vast volume of traffic formed by traditional combustion systems has serious environmental repercussions [1]. In 2020, road transport accounts for 21% of worldwide energy consumption and 17% of global carbon dioxide (CO2) emissions. Carbon emissions from road transportation have been steadily rising and will continue until it is gradually decoupled from fossil fuels. Increased fuel efficiency and the positioning of alternative fuels in road transportation, such as biofuels, electricity, and hydrogen, will be required to keep global warming below 2°C compared to preindustrial levels [2]. rising concerns about climate change persuaded EU members to agree to cut emissions by 80% by 2050 to limit CO2 levels in the atmosphere at 450 parts per million and keep global temperatures below 2°C. The road transportation industry is likely to lower its emissions by 95% as a result of these agreements. Furthermore, because urban pollution caused by vehicle use causes major health concerns, developing a low-carbon transportation system is critical for the long-term future. EVs are a promising technology in this approach, as they can reduce CO_2 emissions from passenger automobiles and light commercial vehicles, as well as noise [3]. Electric power trains are significantly more energy-efficient than traditional power trains, and the electricity used might come from renewable energy sources. Hence, road transport electrification is expected to play a crucial role in achieving these goals. Governments all over the world are implementing various initiatives, regulations, and programmers to encourage more people to buy EVs. The expansion of charging infrastructure, as well as community awareness of the benefits of EVs, is one of the steps taken to promote them [3]. The European Union (EU) plans to follow through on its strategic choice with a regulatory framework that combines the agreed-upon CO_2 levels for 2025 and 2030 with a technology-neutral effective framework for low-emission vehicles. The incentive of 50 g CO_2 per km applies to low-emission automobiles. Under this scheme, manufacturers who achieve a higher share of low-emission vehicles at are agreed-upon standard level for 2025 and 2030 (recommended reference values are 15% and 30% for 2025 and 2030, respectively) will be recognized with a less restrictive CO_2 goal. EV technologies have seen a series of technological advancements since the 19th century, intending to become competitive with the dominating internal combustion engine (ICE) cars [4]. Electric cars have benefited from the invention of the electric motor. Between 1832 and 1839, nonrechargeable primary cells were used. The first workable electric automobile was introduced in the United States by William Morrison in 1890 [5]. This six-passenger vehicle with a top speed of 14 mph was essentially an electrified wagon, but it was noteworthy in terms of attracting interest in EVs. Dr. Ferdinand Porsche created the world's first serial HEV in 1900, and the Pieper business of Liege, Belgium, reported the first hybrid car in 1899. Due to high initial prices and limited autonomy, EVs were taken out of the market in the 1930s, and it wasn't until the 1970s that interest in the technology reappeared. EVs have been revived in the 21st century via technological advancement and a greater focus on renewable energy, as well as the potential to reduce transportation's impact on climate change, air pollution, and other environmental issues. According to European and US efforts for the development of clean technologies to reduce CO_2 emissions over the last 10 years, EVs are one of the best existing solutions for tackling climate change. Benefits to promoting EV consumption were first offered in the late 2000s, particularly in the United States and the EU, which culminated in a thriving market for EVs in the 2010s. The electric car market is likely to be considerably boosted by greater awareness and structural advantages, such as those to include in the COVID-19 pandemic's sustainable recovery. Lockdowns have lowered the number of greenhouse gases produced by gasoline or diesel cars during the COVID-19 outbreak. Governments should do more to accomplish climate targets, according to the International Energy Agency, which started in 2021 that regulations for large electric cars should be implemented [6]. The fundamental types of EVs, and their properties, charging, and design are covered in this chapter.

22.2 Technologies for electric vehicles

EVs could either run on electricity or feature an ICE. The motor and additional power are provided by the battery with grid electricity charging and braking energy recovery. Electricity can be stored in batteries, fuel cells, and uninterruptible capacitors and can originate from both producing plants and renewable energy sources. Unlike traditional automobiles, EVs use a more efficient source of power and electrical motor than ICEs [7,8]. The usage of regenerative braking and thermoelectric generators reduces energy waste. The thermoelectric generator transfers heat from the engine and machine systems to electricity automatically, whereas regenerative braking absorbs energy, changes it to electrical energy, and then returns it to the batteries. Because electric motors provide a lot of torque across a wide speed range, electric cars usually don't need a gearbox. Propulsion, energy source, and auxiliary are the three main subsystems that make up an EVs drive train. The electric propulsion subsystem includes the vehicle controller, power electronic converter, electric motor, mechanical transmission, and driving wheels. The energy source subsystem includes the energy source, the energy management unit, and the energy refilling unit. The emergency backup subsystem consists of the power steering unit, the hotel climate control unit, and the auxiliary supply unit. Based on the control inputs from the accelerator and brake pedals, the vehicle controller provides suitable control signals to the electronic power converter, which regulates the power flow between the electric motor and the energy source. The reverse power flow is caused by the EV's regenerative braking, and this regenerated energy can be returned to the energy source if it is receptive. Regenerative energy can be accepted by most EV batteries, as well as UCs and flywheels [9]. It also links with the energy refilling unit to operate the recharging unit and keep track of how much energy is being used. The auxiliary power supply, which operates at multiple voltage levels, provides power to all of the EV auxiliaries, particularly the climate control and power steering components. Depending on the power supplement and propulsion characteristics, EVs have a range of drivetrain topologies, allowing for differing mobility and energy management features. These systems differ in terms of how onboard energy is generated and/or replenished, as well as how the internal electric motor and the combustion engine are coupled. Because of the mix of battery capacities, charging capabilities, and technological sophistication, consumers have a choice of options in terms of automobile ranges, refueling options, and pricing. These are categorized are shown in Fig. 22.1.

The battery-EVs is powered entirely by electricity from the power storage unit, with an electric motor providing propulsion. The electric motor and engine are

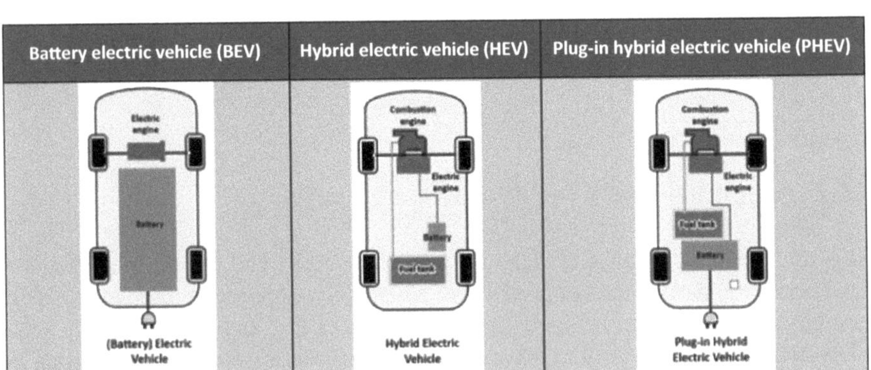

Figure 22.1 Different electric vehicle categories.

combined in the HEVs drive system, and the power sources include both electricity and gasoline or diesel.

22.3 Emerging electric motor technologies

Electric motors are the heart and soul of EVs [10]. It is a critical part of EV technologies because they allow the automobile to move by converting electrical energy from the battery into mechanical energy. During regenerative action, it also serves as a generator, sending energy back to the source. Because of their inefficiency, large structure, and lack of reliability, direct current (DC), motor drives are not a popular choice. Different types of electric motors (AC and DC) have been developed over time to meet the needs of the automotive industry; the most prominent being induction and permanent magnet (PM) motors.

Due to their simple control electronics, DC motors were once thought to be the best technology for EVs; but, because they require frequent maintenance, they are not ideal for mass-scale EV adoption. Because the power source—batteries—provides DC, AC motors necessitate intricate and expensive power electronics, including an inverter. AC motors have a greater overall cost. Their advantages include increased power density, which is important in an automotive application since it allows for smaller and lighter motors, and higher efficiency, which increases the range for a given battery capacity. As illustrated in Fig. 22.2, electric motors suitable for the aforementioned EV technologies can be split into numerous categories:

Many ongoing projects aim to improve hybrid and plug-in electric car motors, with a particular focus on reducing the use of rare earth minerals, which are currently used in PM motors. Electrical energy stored in a battery is converted into mechanical energy by an electric motor in an electric drive system [11]. An electric motor is made up of a rotor (the moving component of the motor) and a stator (the stationary part of the motor) (the stationary part of the motor). In a PM motor, an air gap separates a rotor with a set of magnets and a current-carrying stator (typically in the shape of an iron ring). There are three types of electric motors to choose from in hybrid or plug-in electric automobile traction drive systems [12–14].

Some of the most typical motor and drive technology needs are as follows: (1) strong power and torque density; wide speed range; (2) consistent torque and constant power (CP) operations; (3) high efficiency across a wide speed range; (4) exceptional durability and robustness; (5) everything at affordable price Induction machines is used in EVs, HEVs, and FCVs because of their simplicity, durability, and wide speed range. The efficiency of an IM is lower than that of a PM because of the inherent rotor loss. As a result, an IM's size is bigger than that of a PM's (of the same power and speed rating). PMs are notable for their high efficiency, high speed, and enhanced power density, making them an appealing option for many EVs. PM motors, however, have a limited CP range due to the presence of the PM field, which can only be reduced by providing a stator field component that opposes the rotor magnetic field. In the event of a stator winding short circuit malfunction,

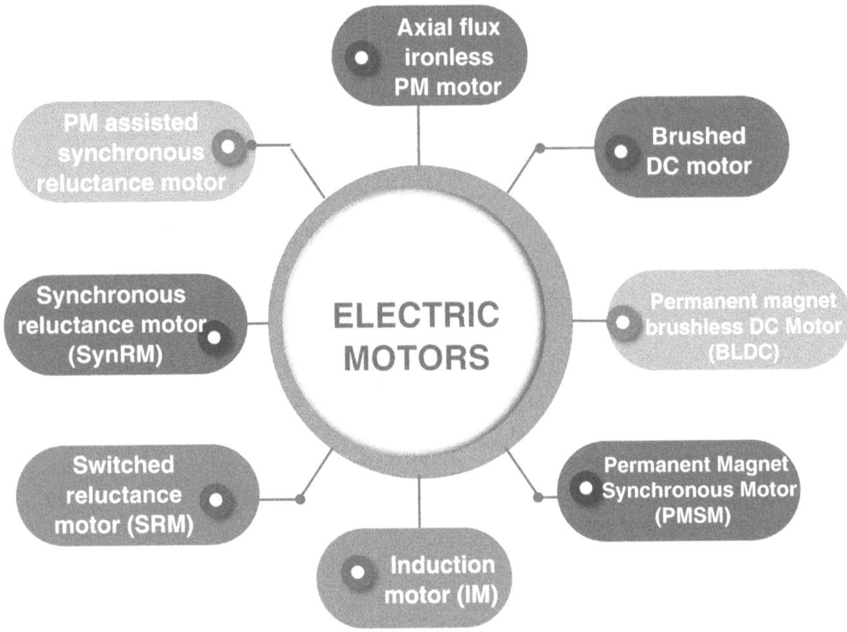

Figure 22.2 Various types of electric motors suited for EV technology.

the system may face problems due to the presence of a rotor PM field. The SRM can be used for electric propulsion in EVs, HEVs, and FCVs due to its simple and durable design, simple control, ability to run at extremely high speeds, and hazard-free operation.

22.4 Battery electric vehicles

The battery electric car is depicted in Fig. 22.3. A battery electric vehicle (BEV) relies solely on chemical energy stored in rechargeable battery packs (e.g., hydrogen fuel cell, ICE, etc.) for propulsion [15]. In BEVs, an electric motor powered by a battery replaces the ICEV and the tank [16].

BEVs also known as full electric vehicles or all-electric vehicles, are automobiles with an electric motor and an electric battery for energy storage. A DC/DC converter manages the power supplied to an electric motor by adjusting the voltage of the electric current supplied to a 2-quadrant inverter or 4-quadrant reversible inverter with a devoted electronic power controller. The battery is usually recharged via a plug and battery charging equipment, which can be carried on board or installed at the charging location. Once the vehicle is moving, the massive battery packs may be recharged by regenerative braking and then laterally charged when the vehicle comes to a stop. The electronic power controller manages the power provided to the motor, and hence controls the vehicle's forward and reverse speeds,

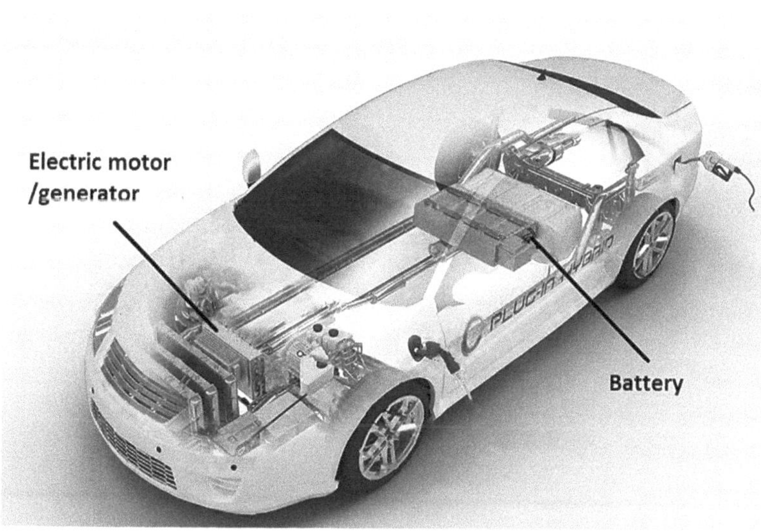

Figure 22.3 Battery electric car.

as well as the regenerative braking (zero friction braking) procedure mentioned above. EVs are powered entirely by electricity and operate in a charge-depleting mode at all times [17]. They don't have an ICE, a fuel cell, or a gasoline tank since they rely only on battery packs for power. BEVs include motorcycles, bicycles, scooters, skateboards, railcars, boats, forklifts, buses, trucks, and vehicles (Fig. 22.4). Depending on how the electric power generated by the electric motor is sent to the driving wheel, there are two types of BEVs. (1) The ICE is replaced by an electric motor. The transmission is responsible for getting the power from the electric motor to the wheels. (2) Each driving wheel has an electric motor fitted. Overall efficiency suffers as a result of the gearbox's use (due to inherent friction in the mechanisms that compose it). To reduce the aforementioned mechanical losses, regenerative braking technology can be used. Both DC and AC electric motors can be used. A DC motor can provide 20 to 30 kW of power and requires a battery voltage of up to 200 V. A DC motor is normally connected to a DC controller with a power rating of 40 to 60 kW. An AC motor with a higher rated power uses a three-phase AC motor with a 300 V battery voltage. Because BEVs rely on the energy stored in their battery packs to power them, their range is related to the battery's capacity. BEVs can go 100–250 km on a single charge, with top-tier versions capable of 300–500 km. Driving range is affected by a variety of factors including driving style, vehicle configurations, road conditions, climate, battery type, and age. As a result, choosing the correct drive and putting in place a good control system are crucial for maximizing battery energy management.

BEVs offer many benefits: they are effective because the EVs strong power is delivered to the wheels, leading to faster acceleration; they don't create noise while running; they are ecologically beneficial since they emit no exhaust; and they can be charged

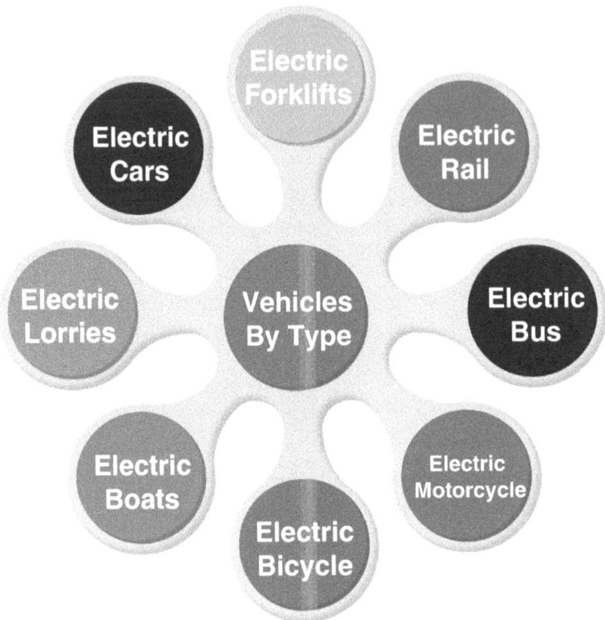

Figure 22.4 Electric vehicles by type.

24 hours a day on low-cost electricity from any sort of power plant, including renewable power. However, there are also disadvantages to using BEVs, including the high cost of production, limited mobility as well as driving speed, recharging times, relatively smaller size (particularly in comparison to ICE engines), and the absence of electric engine sound, which can end up causing traffic fatalities.

22.5 Hybrid electric vehicles

HEVs are operated by a combination of two power stations (a traditional ICE and a battery/electric motor). The electric battery gives the capacity to the engine, permitting the ICE to work all the more real. The battery of HEV can be charged by the engine or by kinetic brake energy recovered during regenerative braking. Fig. 22.5 displays the prototype of a Hybrid electric car.

Several variations of the HEV technology have been developed since 1995 when rival manufacturers renewed their interest in HEVs. Micro HEVs, mild HEVs, full HEVs, and PHEVs are among them.

22.5.1 Full hybrid electric vehicles

The electric engine and batteries in full HEVs are significantly bigger than those in mild and micro HEVs. Thus, the electric engine can be utilized as the main power

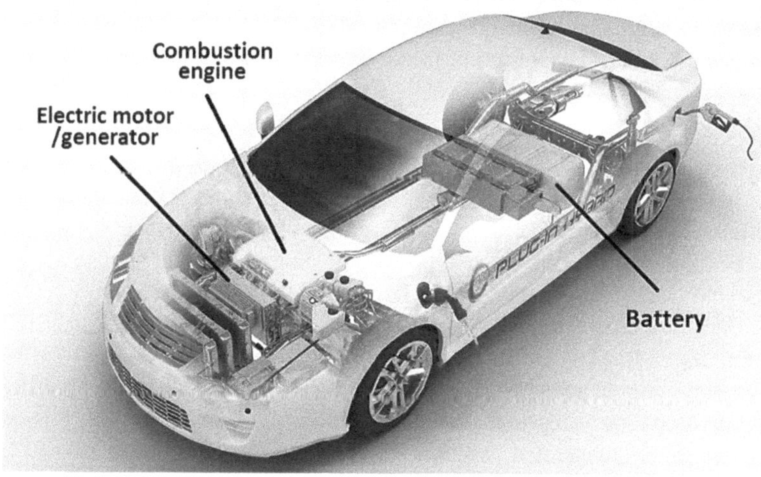

Figure 22.5 Hybrid electric vehicle.

source, contingent upon the vehicle's power interest. Full HEVs have a lot of more modest motors than mild and micro HEVs and in this way require more modern energy in the executive's framework. In real-world driving, complete HEVs provide better fuel economy by 40%–45% as compared to other nonhybrid vehicles. There are some examples of full HEVs such as Toyota Prius, Chevrolet Tahoe, Ford C-Max, Honda CR-Z, and Kia Optima which are performing on the road today. Synergy Hybrid and Power Hybrid automobiles are two types of full hybrid vehicles. Synergy Hybrid makes sacrifices in terms of performance in driving, efficiency, and reduction in the emission of CO_2 [18]. The engine in this class is smaller than in typical automobiles, such as the Toyota Prius. The goal of a HEVs is to provide a better driving experience, so the engine is not dissipated, and the vehicle, when combined with the motor, will have higher driving performance than a normal car, such as the Toyota Highlander.

22.5.2 Mild hybrid electric vehicles

A mild HEV is the same as a micro HEV, but with a larger integrated alternator/starter motor and a battery that allows for power assistance during vehicle acceleration. During driving, moderate HEVs often improve fuel economy by 20%–25% when compared to other nonhybrid vehicles. The Mercedes-Benz S400, BMW 7 Series, Chevrolet Malibu, and Honda Civic are good examples of mild HEVs on the market [18].

22.5.3 Micro hybrid electric vehicles

In the micro HEV an inbuilt electric motor act as an inverter that employs a start/stop function for the engine. At the point when the vehicle reaches a standstill and restarts when the driver delivers the brake pedal. The vehicle is just moved by the ICE while cruising. The Mercedes-Benz A, BMW 1 and 3 series, Peugeot Citroen C3, and Fiat 500 are all examples of micro hybrids on the road today [18].

22.5.4 *Plug-in hybrid electric vehicles*

PHEVs have the same basic layout as complete HEVs, but it has an outer electric framework charging outlet, substantially larger electrical components, and a smaller engine. PHEVs can run on electric power for lengthy periods due to their high-capacity electrical components. The Ford C-Max Energi, Chevy Volt, Fusion Energi, Porsche Panamera, and Toyota Prius Plug-in are all examples of PHEVs on the road today [18].

HEVs have some advantages, including fuel savings due to the capacity to deliver kinetic power using both the electric motor and the ICE. With this, kinetic braking energy is regenerated, which is often wasted as heat to mechanical brakes in traditional vehicles. Furthermore, HEVs have the option of using the electric motor to turn the engine, resulting in a better cranking technique (inertia cranking). Moreover, the usage of the engine shutoff/start-up function in full HEVs helps eliminate fuel use when idling. On the other side, some people are concerned about this.

22.6 Nanotechnology for a hybrid electric vehicle

The biggest issue that EVs face is their batteries. They are extremely expensive and have a limited charging time. Nanotechnology could improve the performance of an EVs battery. This technology has the potential to fill in the gaps in e-mobility while also providing significant benefits. Furthermore, by employing nanoparticles and nanocomposite materials, specific properties of lithium batteries could be improved [19]. Polymer nanocomposites, nano-enhanced sensors, and nanostructured metals could improve battery performance aspects such as lightweight, energy efficiency, and cyclic stability with improved control. Nanotechnology can be used to attain certain battery performance criteria. Negative electrode graphite powder has traditionally been utilized as an intercalation material in lithium-ion batteries. The rate of intercalation/de-intercalation of lithium, as well as the battery capacity, can be enhanced if the micrometer-sized powder is substituted with carbon nanomaterials (e.g., carbon nanotubes). Because the diffusion path for lithium ions is reduced, their mobility is increased, and electrical conductivity is increased, the current density of electrodes will increase. As a result, the electrochemical reaction occurs more quickly, and the battery's efficiency improves as a result of the usage of nanomaterials. Because of their large surface area, carbon nanotubes may bind to significantly larger amounts of lithium [20]. 1-D nanowires constructed of vanadium oxide (V_2O_5), titanium dioxide (TiO_2), or tin oxide (SnO) will also be used as electrode materials. Though commercial fabrication of these electrode materials is still in its initial stages, $Li(NiCoAl)O_2$, $Li(AlMn)_2O_4$, $Li(NiMnCo)O_2$, $LiMn_2O_4$, and $LiCo_2O_4$ are the most commercially available oxide materials to date. However, none of these materials are cost-effective and safe [21–25]. These materials' intercalation capability has been shown to be greatly improved by nanostructuring. Research is still on to determine the best compositions for achieving higher performance in batteries that are cost-effective, light, and compact. As a result, future EV research must focus on improving battery charging times, nanomaterials to enhance battery efficiency by lowering the cost of nanomaterials for batteries as well as the size and weight of battery packs, and commercial commercialization of these nanoparticles. This will also improve the battery management and thermal management systems of an EVs battery pack.

22.7 Hybrid energy storage system for e-vehicles

There are a few giant companies that are working on the development of HEVs. They also look at the supercapacitors as a hope for a faster and more cost-effective solution to the problems faced by using batteries. The energy requirement of the e-vehicles differs from those of the other battery-operated appliances, like cell phones, laptops, watches, etc. its high consumption, and the key problem is that this high energy is not necessarily consumed in a monotonous fashion. The frequent changes during the discharging of a battery affect its shelf life and the batteries are not able to handle a huge sudden surge in the demand by the vehicles as it speeds up or kicks starts heavy duty work like forklifts or accelerating on uphill drives or continuously applying breaks in a busy [26–29]. Though we still make do by employing huge battery packs in these scenarios but driving with half a ton unbalanced load may not give the best driving experience on turns and even with the low center of gravity of most electric cars. To tackle this problem, active roll control systems are used which reduce the body roll at turns and on rough surfaces [30]. This is done by using roll bars, hydraulic actuators were used before the supercapacitors, on both axles which can stabilize the car body. The advantage of using the supercapacitors is the faster response and a high-power supply where the actuators used to fail. The heavy weight of the body tends to unbalance the vehicle at turns but electrically controlling the stiffness roll bars using super-capacitive active roll control systems helps the vehicle to take smooth turns and that too on the fly.

The use of many compact supercapacitors as the energy source for such applications has been demonstrated by simulations as a proof of concept but the practical existence of high energy density supercapacitors is still a much sought-after dream. Supercapacitors may not replace the batteries completely but they can still be used to overcome their drawbacks. The vehicles in which the supercapacitors are used along with the batteries for the supply of energy are called HEVs and the employed system is termed a hybrid energy storage system. These systems mainly consist of multiple batteries and supercapacitors connected in different series and parallel connections which can exchange charges using bidirectional DC-DC converters have classified these systems into three categories, viz. passive configuration, semipassive configuration, and fully active configuration [31]. The use of these systems can improve the overall battery performance that suffers during peak current usage. The supercapacitors are not only efficient in aiding the batteries during the supply time but also when the kinetic energy of the moving vehicle is itself charging the battery.

Though both the batteries and supercapacitors are electrochemical systems used for energy storage, the energy storing principle differs in both, batteries store the energy in the chemical form while the supercapacitors make use of an electric field to store the energy. The chemical energy can be gradually drawn at a constant rate for a long time and maintains a high energy density while supercapacitors have high rate capabilities even for fast charging and discharging resulting in a higher power supply. The hybrid energy storage system makes use of the advantages of both these devices to achieve higher efficiency and usability. Another advantage of using a supercapacitor in the hybrid energy storage system is that it can effectively operate a lot longer than any other energy storage system where the supercapacitors themselves can be in working conditions to last enough for the vehicle's lifetime.

The use of supercapacitors in the hybrid energy storage system is the same, but not limited to, that of classic capacitors in electronic devices. When a vehicle is accelerating or decelerating it demands a high current. The supercapacitors can be used to meet this high demand and supply electricity instantaneously; saving the energy of the battery which can then be used for other purposes. The supercapacitors can compensate when a high peak of usage for a short time is required, something which the modern batteries are not capable of doing. The supercapacitor not only shields the battery during current spikes but also acts as an efficient energy storage device.

The hybrid energy storage system is used in regenerative braking systems in which they are used to store energy. This is possible for the electric traction systems as they can convert the change in the kinetic energy, during deceleration, to electrical energy. The use of supercapacitors in this conversion is twofold. As, the conversion can be huge the amount of current received can damage the batteries the charge from this process can rather be stored in the supercapacitors first, avoiding the overcharging of the batteries and allowing regenerative braking even when the batteries are fully charged. The supercapacitors then store energy in the field generated by the accumulation of this charge which is then reused in accelerating the vehicle. As the hybrid energy storage system used here has two components with very different characteristics the regenerative braking system usually uses DC-DC buck/boost converters as shown in Fig. 22.6.

In this circuit, the buck/boost converter acts as the buck converter during the deceleration while during acceleration as the boost converter. The systems of the supercapacitors and the batteries are the respective devices connected in series connections. The number of components varies as per the specifications of the components. As in practical applications not all the supercapacitors are alike and thus tend to charge unevenly due to slight differences in the capacitance values. Thus, to avoid any accidents a voltage cell balancing unit is also connected to the system of supercapacitors. The control system plays an important role in detecting the changes in the rotational speed of the motor with a reference potential rate change.

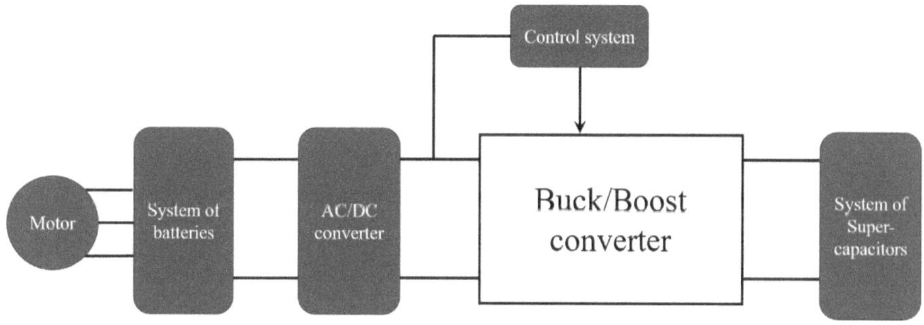

Figure 22.6 Block diagram of the hybrid energy storage system.

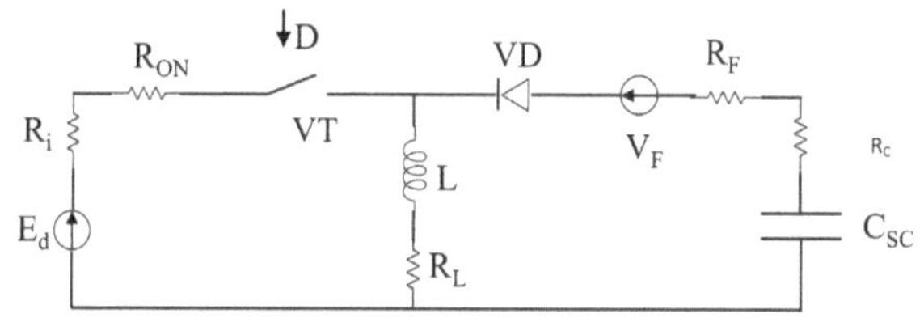

Figure 22.7 DC/DC converter.

The transistor VT, shown in Fig. 22.7, is controlled with the help of the control system consisting of a microprocessor that decides by comparing the voltages of the battery and the supercapacitor. During the boost mode, the required amount of energy is transferred from the supercapacitors to the inductor which is then transferred to the battery pack while during the deceleration (buck mode) the energy is transferred from the motors to the supercapacitors.

Applying Kirchhoff's voltage law to the DC/DC converter one can better understand its working [32,33] Here The voltage source E_d has an internal resistance R_i. This source is connected to a transistor *VT* with an on-state internal resistance represented by R_{ON}. The inductor L is the filter inductance whose active resistance is R_L. The supercapacitors with capacitance C_{SC} and internal resistance R_C are connected to the dynamic resistance R_F and a forward potential V_F which is connected to the diode VD.

For different stages of the converter, we can write the mathematical expressions as follows

a. When the transistor is in the conducting state, then

$$L(i_L,\theta)\frac{di_L}{dt} = E_d - (R_L + \quad R_{ON} + \quad R_i)i_L \tag{22.1}$$

$$C_{SC}\frac{dU_{SC}}{dt} = 0$$

Where i_L is the inductance and the U_{SC} is the voltage across the supercapacitors.

b. When the transistor switch is turned off

$$L(i_L, \theta)\frac{di_L}{dt} + (R_L + R_F + R_{SC})i_L + U_{SC} = 0 \tag{22.2}$$

$$C_{SC}\frac{dU_{SC}}{dt} = i_L$$

When the motor acts as the generator, that is, during the deceleration, then the speed V of the EV provides the voltage at the output of the rectifier and it can be written as $E_d = kV$. If the breaks are applied with a constant force then we have, dV/dt = constant, and E_d will be a linear function of time which can be written as

$$E_d = E_{d\ max} - kt$$

Now, if it is assumed that the inductance is invariant to the changes in the current and temperature, that is, L = const., and $L << C_{SC}$ so that U_{SC} is ~ const. for $t \in [nt_p, n(t_p + D)]$, then, for the system of the Eq. (22.1), the switch is on condition, the solutions will be

$$i_L\left((n + D)t_p\right) = i_L\left(nt_p\right)e^{-a/LDt_p} + \left(1 - e^{-a/LDt_p}\right)b$$

$$U_{SC}(n + D)t_p = U_{SC}(nt_p)$$

$$a = R_L + R_{ON} + R_i$$

$$b = \frac{E_d}{a}$$

Where D is the Duty cycle.

And when the transistor is in the off condition, then we can write the solutions for the system of Eq. (22.2) as

$$i_L\left((n + 1)t_p\right) = i_L\left((n + D)t_p\right)e^{-a/LDt_p} + \left(1 - e^{-a/LDt_p}\right)$$

$$U_{SC}((n + D))t_p = U_{SC}((n + 1)t_p)$$

$$a = R_L + R_{SC} + R_F$$

$$b = \frac{E_d}{a}$$

The values of U_{SC} give us the values of the voltage of the supercapacitor for starting or accelerating the vehicle and collecting the energy from the braking, respectively. Also, the minimum necessary power required to accelerate a vehicle, of mass m, to the final velocity V_{max} can be given as

$$E = \frac{mV_{max}^2}{2\eta}$$

where η the efficiency of the with which the energy released from the source is converted to the received by the wheels. If this total energy, the energy stored in the supercapacitor is $E_{SC} = kE$, $k \in [0,1]$. The value of k gives the partition of the energy and $U_{SC\ max} = E_{d\ max}$ and thus we get the minimum value of capacitance required for the worst-case scenario to be

$$C_{SC} > \frac{2kE}{U_{max}^2}$$

For the energy obtained by the supercapacitors, we can write $E_{SC} = E_{CS} + E_{RB}$ where E_{CS} is the energy charged in the supercapacitor at a constant velocity and E_{RB} is the energy that it gains regenerative braking. E_{CS} is also the initial charging voltage after which the regenerative braking starts and determines the initial voltage across the supercapacitor as

$$U_{SC\quad initial} = \sqrt{\frac{2E_{CS}}{C_{SC}}}.$$

This is only one type of the possible five hybrid systems. There are a total of five possible configurations as listed by [139]

1. Systems without the converters
2. With one bidirectional DC-DC converter placed in series after either the battery of the supercapacitor
3. With two bidirectional DC-DC converters, again in series and one controlling the battery while the other controls the supercapacitor
4. With the parallel connection of the two bidirectional DC-DC converter
5. With a zero-voltage switching capability and multiinput bidirectional DC-DC converter

In any given configuration the supercapacitors must have a lower voltage than the batteries in the charging or deceleration mode to make the batteries safe.

Other than these the supercapacitors are also the better choice for working at wide temperature ranges, as the batteries work based on chemical reactions which get immensely affected by temperatures and tend to become inactive at lower temperatures while behaving unpredictably at high temperatures. But it should also be kept in mind that the temperature variation can affect the capacitance value of the supercapacitors also, which in terms affects the performance and thus highly sensitive particularly faster temperature detection systems need to be used [34].

22.8 Simulation and experimental results of Hybrid energy storage system for light electric vehicle

S.Devi Vedhya and M Balaji from the SSN College of Engineering, Kalavakkam, India present the modeling, power management, and design of HESS for light EVs [35]. In this study, the authors interfaced a bi-directional converter (BDC) with a battery/ultracapacitor to operate an EV. The used block diagram electric drive system by using dual-energy sources is shown in Fig. 22.8.

Based on the proposed modeling, strategies, and components, a simulation study was carried out using MATLAB®/Simulink. The overall layout and the control of the proposed system are shown in Fig. 22.9.

In this simulation, a battery-alone system is considered, which provides the required power to the system. During the first Indian drive cycle, the battery draws a current of 140 A with a state of charge (SOC) dropping from 80% to 77.8% (Fig. 22.10A–C). When an ultracapacitor with a battery (hybridized) is applied to the proposed system, the battery provides a low-frequency power while the ultracapacitor provides a high-frequency power component. The addition of UC with a battery supplied the required power and provide a current of 50 A. Fig. 22.10D-I shows the obtained results of hybridized UC/battery system. The battery current is lower for hybridized systems and the voltage of UC is maintained between 16 V (minimum) to 32 V (maximum) with 88% of SOC. During the acceleration and de-acceleration in drive cycles, the SOC falls and rises and reaches 78%.

The authors validated the simulation results by performing a small-scale experiment. The experimental setup is shown in Fig. 22.11A. To perform the experiment a battery bank (main source), composed of four lead-acid batteries in series connection (48 V) was used with a lead-acid battery (auxiliary source), instead of UC. A throttle is used to control the speed and the rotation of the brushless direct current (BLDC) motor. The speed profile is shown in Fig. 22.11B. At the time of acceleration, the hybridized system provides sufficient current, and thus BDC works as a

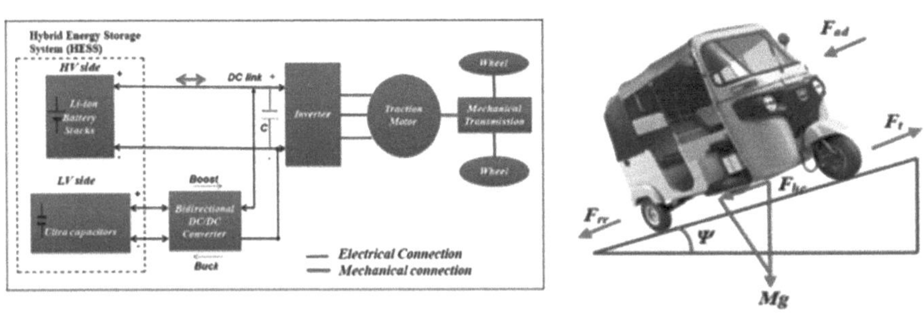

Figure 22.8 Proposed block diagram of an electric drive system (A) and vehicle load model (B).
Source: Reproduced with permission S.D. Vidhya, M. Balaji, Modelling, design and control of a light electric vehicle with hybrid energy storage system for Indian driving cycle. Meas. Control. 52 (9–10) (2019) 1420–1433.

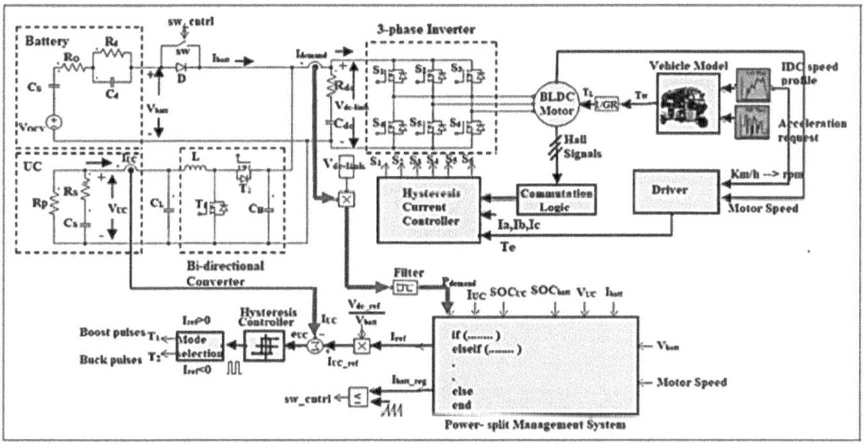

Figure 22.9 Overall schematic representation of the system.
Source: Reproduced with permission S.D. Vidhya, M. Balaji, Modelling, design and control of a light electric vehicle with hybrid energy storage system for Indian driving cycle. Meas. Control. 52 (9–10) (2019) 1420–1433.

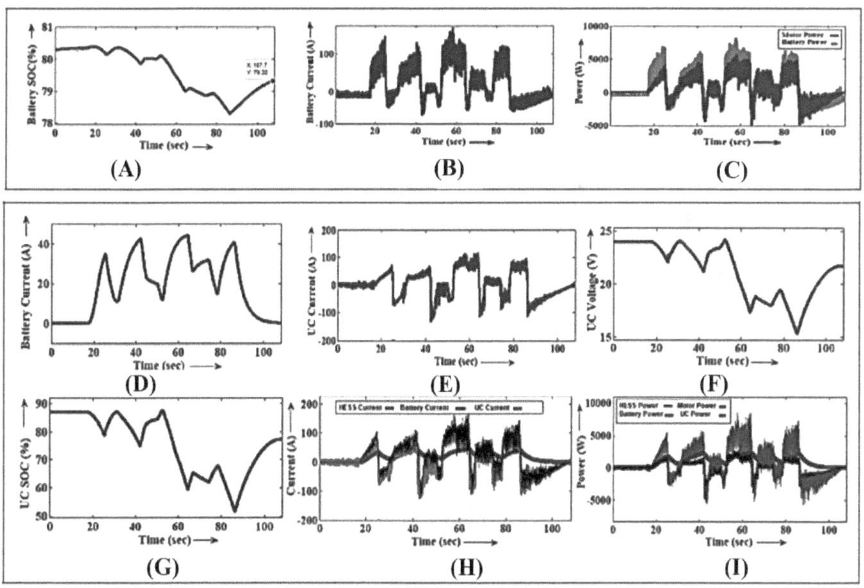

Figure 22.10 Battery alone system: (A) battery SOC, (B) battery current and (C) distribution of motor and battery power. Hybridized system: (D) battery current, (E) UC current, (F) UC voltage, (G) UC SOC, (H) distribution of current, and (I) distribution of power.
Source: Reproduced with permission S.D. Vidhya, M. Balaji, Modelling, design and control of a light electric vehicle with hybrid energy storage system for Indian driving cycle. Meas. Control. 52 (9–10) (2019) 1420–1433.

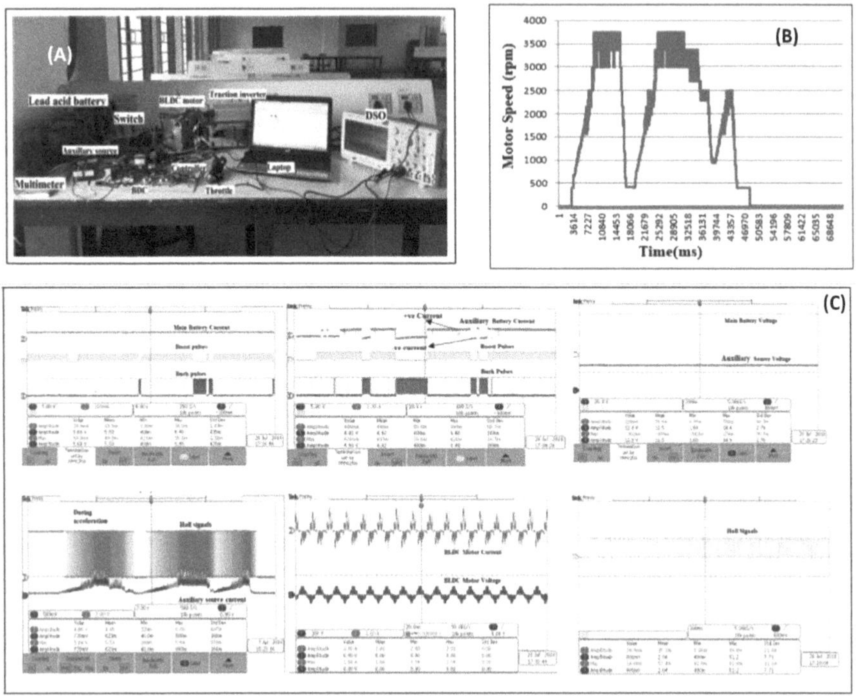

Figure 22.11 The experimental setup (A), the speed profile (B), and the current and voltage of the main and auxiliary battery during buck and boost mode of BDC operation with corresponding hall signal, phase voltage, and BLDC motor current (C).
Source: Reproduced with permission S.D. Vidhya, M. Balaji, Modelling, design and control of a light electric vehicle with hybrid energy storage system for Indian driving cycle. Meas. Control. 52 (9–10) (2019) 1420–1433.

boost converter. While, for the constant speed mode, the only main source provides the demanded current and some charging current for the auxiliary source. In this condition, BDC works as a buck converter. Fig. 22.11A shows the current and voltage of the main and auxiliary battery during the buck and boost mode of BDC operation. The corresponding hall signal, phase voltage, and BLDC motor current are also shown in Fig. 22.11C. The obtained results from the experiments significantly verified the simulation results.

22.9 Case studies on hybrid energy storage system

A lot of studies and projects are carried out on the HESS in different regions of the world in the last few years. Most of the real-life applications are frequency control, backup power, and renewable energy sources. Here we mentioned some examples of real-time scenarios of HESS:

- DUK energy, North Carolina, United States installed a 1.2 MW Supercapacitor/battery HESS in 2016 to control peak demand, load shifting, and support.
- Varel Niedersachsen, Germany installed a 11.5 MW Li-ion, NaS battery HESS in 2018. This was a joint project of Hitachi chemicals, NGK insulators, Hitachi power solutions, and its Japanese partners. The main aim of the project is to balance out frequency flections from renewable energy sources.
- redT energy storage solutions, Monash University, Australia developed a 1 MWh VF/Li-ion battery HESS in 2018. This project demonstrates storing and dispatching energy from a 1 MW solar panel.
- AEG power solution, Berman Germany developed a 20 MW power to heat/battery. The main function of this HESS is to control the frequency and power operations and renewable energy sources integration.

22.10 Opportunities

HESS can be predominantly deployed in the following sectors:

- **Power sectors**
  In the power sector, HESS can be used for frequency regulation and providing backup power for the grid. Battery/supercapacitor and battery/fly-wheel is the most commonly used hybrid combination in the power sector. By using this combination in microgrids, the life cycles of batteries increased due to less stress during the charging and discharging. Overall, the HESS can be used in power sector for grid stability, frequency regulation, backup, and power quality services.
- Transport sectors
  In any EV sudden acceleration take a lot of energy from batteries, which reduces the power supply duration and life cycle of the batteries. The hybrid combination of battery/supercapacitor will manage the large fluctuations from acceleration and regenerative braking. In electronic trains, the HESS is used to flatten the fluctuations from the accelerating and deaccelerating cycles. The HESS can be deployed in the trains or railway substations.
- Renewable energy sectors
  In Renewable energy sectors, HESS is used to improve the power quality and efficiency of the systems. It can be a better solution in terms of cost-effectiveness, durability, and practicability for complete system implementation. It is used to improve the storage and lifetime of the system by reducing dynamic stress on the high-energy storage unit.

22.11 Summary and outlook

EVs have a lot of potential to substitute ICEVs in a variety of on-road applications. EVs can help reduce reliance on petroleum, improve local air quality, reduce GHG emissions, and improve the driving experience, among other things. Vehicle electrification is in line with broader electrification and decarbonization trends, and it works in tandem with other mobility innovations like urban micro-mobility, automation, and mobility-as-a-service. With EVs capable of assisting power-system planning and operations in a variety of ways, efficient integration of EVs into power systems gives several prospects for synergistic enhancement of electromobility and electric power systems' efficiency and economics. Because both technologies are expected to be widely deployed in the future, fully utilizing the synergies between EVs and VRE sources offers a path toward affordable and sustainable energy and mobility for all. Investments in charging framework, consumer awareness, effective and safe VGI initiatives, and proper business models that support all aspects of vehicle electrification are all critical components in enabling such future technological advancement. In most nations, the coronavirus pandemic is having a negative influence on LDV sales, and EV sales in 2020 are likely to be lesser than in 2019, This is the first drop in a decade. However, ICEV sales are expected to drop faster but in 2020, EV sales might reach a record share. EV's long-term potential is unaffected by these short-term difficulties. Several studies predict that EVs will play a significant role in the future, as seen by enormous investments in car development, charging facilities, and further technological advancements, particularly in battery and supply chain technologies. Consumer acknowledgment and progress, alongside mechanical headways, create an upright self-supporting circle of innovation part enhancements and cost decreases that could empower far-reaching reception. It's still difficult to predict the future, especially when it comes to technology adoption. Nonetheless, this in-depth analysis provides a bright picture of the future of EVs for on-road transportation. The authors are optimistic that technological, legislative, sociological, behavioral, and business-model constraints will be overcome over time, allowing everyone to benefit from cleaner, more efficient, and economical transportation options.

References

[1] M.A. Baloch, I. Ozturk, F.V. Bekun, D. Khan, Modeling the dynamic linkage between financial development, energy innovation, and environmental quality: does globalization matter? Bus. Strat. Environ. 30 (1) (2021) 176–184.

[2] IEA, Global EV Outlook 2021—Analysis. <https://www.iea.org/reports/global-ev-outlook-2020>, 2021.

[3] F. Knobloch, S.V. Hanssen, A. Lam, H. Pollitt, P. Salas, U. Chewpreecha, et al., Net emission reductions from electric cars and heat pumps in 59 world regions over time, Nat. Sustain. 3 (2020) 437–447.

[4] L. Rotaris, M. Giansoldati, M. Scorrano, The slow uptake of electric cars in Italy and Slovenia. Evidence from a stated-preference survey and the role of knowledge and environmental awareness, Transport. Res. Pol. Pract. 144 (2021) 1–18.

[5] H. Climent, B. Pla, P. Bares, V. Pandey, Exploiting driving history for optimising the energy management in plug-in hybrid electric vehicles, Energy Convers. Manag. 234 (2021) 113919.

[6] IEA, Global EV Outlook 2021—Policies to promote electric vehicle deployment. <https://www.iea.org/reports/global-ev-outlook-2021/policies-to-promote-electric-vehicle-deployment>, 2021.
[7] T. Markel, Plug-in hybrid electric vehiclescurrent status, long-term prospects and key challenges. Tech. rep., National Renewable Energy Laboratory (2006).
[8] AccessScience, Electric Vehicles and Range Anxiety, AccessScience (McGraw-HillEducation), 2014. Available from: http://doi.org/10.1036/1097-8542.BR1031141.
[9] R.T. Doucette, M.D. Mcculloch, A comparison of high-speed flywheels, batteries, and ultracapacitors on the bases of cost and fuel economy as the energy storage system in a fuel cell-based hybrid electric vehicle, J. Power Sources 196 (2011) 1163–1170.
[10] A. Sciarretta, L. Guzzella, Control of hybrid electric vehicles, IEEE Control. Syst. 27 (2007) 60–70.
[11] G. Paganelli, T.M. Guerra, S. Delprat, J.-J. Santin, M. Delhom, E. Combes, Combes Simulation and assessment of power control strategies for a parallel hybrid car, Proc. Inst. Mech. Eng. Part. D. J. Autom. Eng 214 (2000) 705–717.
[12] M.S. Rafaq, J.-W. Jung, A comprehensive review of state-of-the-art parameter estimation techniques for permanent magnet synchronous motors in wide speed range, IEEE Trans. Ind. Inf. 16 (2020) 4747–4758.
[13] L.S. FeiXu, Y. Li, The weighted vector control of speed irrelevant dual induction motors fed by the single inverter, IEEE T. Power Electron. 28 (2013) 5665–5672.
[14] G.F. Lukman, J.W. Ahn, Torque ripple reduction of switched reluctance motor with non-uniform air-gap and a rotor hole, Machines 9 (2021) 348.
[15] E. Abotalebi, M.D. Scott, M.R. Ferguson, Can Canadian households benefit economically from purchasing battery electric vehicles? Transp. Res. Part. D. 77 (2019) 292–302. Available from: https://doi.org/10.1016/j.trd.2019.10.014 (2019).
[16] R. Hou, P. Magne, B. Bilgin, Low-voltage electrical systems for nonpropulsion loads, in: A. Emadi (Ed.), Advanced Electric Drive Vehicles, CRC Press, Boca Raton, FL, 2014, pp. 317–328. 978-1-4665–9769-3 (Chapter 9).
[17] R.M. Dell, P.T. Moseley, D.A.J. Rand, Progressive Electrification of Road Vehicles, Academic Press Towards Sustainable Road Transport, 2014, pp. 157–192 (Chapter 5). Available from: https://doi.org/10.1016/B978-0-12–404616-0.00005-0.
[18] Technology CfAA, HEV levels. <http://autocaatorg/Technologies/Hybrid_and_Battery_Electric_Vehicles/HEV_Levels/>, 2018 (accessed January 2019).
[19] B. Dunn, H. Kamath, J.M. Tarascon, Electrical energy storage for the grid: a battery of choices, Science 334 (2011) 928–935.
[20] V. Etacheri, R. Marom, R. Elazari, G. Salitra, D. Aurbach, Challenges in the development of advanced Li-ion batteries: a review, Energy. Environ. Sci. 4 (2011) 3243–3262.
[21] P.G. Bruce, S.A. Freunberger, L.J. Hardwick, J.-M. Tarascon, Li–O2 and Li–S batteries with high energy storage, Nat. Mater. 11 (2012) 19–29.
[22] Z. Chen, Z. Lu, J.R. Dahn, Staging phase transitions in Li*x*CoO2, J. Electrochem. Soc. 149 (2002) A1604–A1609.
[23] S.Y. Chung, J.T. Bloking, Y.M. Chiang, Electronically conductive phospho-olivines as lithium storage electrodes, Nat. Mater. 1 (2002) 123–128.
[24] R. Malik, F. Zhou, G. Ceder, Kinetics of non-equilibrium lithium incorporation in $LiFePO_4$, Nat. Mater. 10 (2011) 587–590.
[25] A. Yuan, L. Tian, W. Xu, Y. Wang, Al-doped spinel LiAl0.1Mn1.9O_4 with improved high-rate cyclability in aqueous electrolyte, J. Power Sources 195 (2010) 5032–5038.

[26] T. Paul, T. Mesbahi, S. Durand, D. Flieller, W. Uhring, Sizing of lithium-ion battery/ supercapacitor hybrid energy storage system for forklift vehicle, Energies. 13 (2020) 4518.
[27] J.R. Miller, Engineering electrochemical capacitor applications, J. Power Sources 326 (2016) 726–735.
[28] K.T. Chau, Y.S. Wong, Hybridization of energy sources in electric vehicles, Energy Convers. Manag. 42 (2001) 1059–1069.
[29] Z.M. Anderson, S. Avadhany, M.D. Cole, R. Driscoll, J. Giarratana, M. Giovanardi, et al., Active vehicle suspension system, 2017.
[30] F. Ju, Q. Zhang, W. Deng, J. Li, Review of structures and control of battery-supercapacitor hybrid energy storage system for electric vehicles, in: 2014 IEEE Int. Conf. Autom. Sci. Eng., IEEE, 2014, pp. 143–148.
[31] J.W. Dixon, M.E. Ortuzar, Ultracapacitors + DC-DC converters in regenerative braking system, IEEE Aerosp. Electron. Syst. Mag. 17 (2002) 16–21.
[32] H. Douglas, P. Pillay, Sizing ultracapacitors for hybrid electric vehicles, in: 31st Annu. Conf. IEEE Ind. Electron. Soc. 2005. IECON 2005, IEEE, 2005, 6 pp.
[33] S.F. Tie, C.W. Tan, A review of energy sources and energy management system in electric vehicles, Renew. Sustain. Energy Rev. 20 (2013) 82–102.
[34] G. Yifeng, Analysis and design of the super capacitor monitoring system of hybrid electric vehicles, Procedia Eng. 15 (2011) 90–94.
[35] S.D. Vidhya, M. Balaji, Modelling, design and control of a light electric vehicle with hybrid energy storage system for Indian driving cycle, Meas. Control. 52 (9–10) (2019) 1420–1433.

Appendix 1

Table A1 Specifications of power electronic devices.				
Semiconductor device	**Voltage**	**Current**	**Switching frequency**	**Power capability**
Power diode	Up to 10 kV	Up to 10 kA	Variable	Medium
SCR	8 kV	2.5 kA	50 or 60 Hz	High
Power BJT	1.4 kV	0.4 kA	10–10 kHz	Medium
Power MOSFET	4.5 kV	1.6 kA	25–500 kHz	Low
MCT	1.5 kV	0.1 kA	Up to 500 Hz	Medium
GTO	10 kV	6.5 kA	Up to 500 Hz	High
IGBT	6.5 kV	2.4 kA	10–20 kHz	Medium

Table A2 The values of c_p for different materials.			
No.	**Material**	**Phase**[a]	c_p **(kJ kg^{-1} K^{-1})**
1	Air (typical room conditions)	(G)	1.012
2	Air [sea level, dry, 0°C (273.15K)]	(G)	1.005
3	Aluminum	(S)	0.897
4	Ammonia	(L)	4.700
5	Animal tissue (incl. human)	Mixture	3.500
6	Antimony	(S)	0.207
7	Argon	(G)	0.520
8	Arsenic	(S)	0.328
9	Asphalt	(S)	0.920
10	Beryllium	(S)	1.820
11	Bismuth	(S)	0.123
12	Brick	(S)	0.840
13	Cadmium	(S)	0.231
14	Carbon dioxide (CO_2)	(G)	0.839
15	Chromium	(S)	0.449
16	Concrete	(S)	0.880
17	Copper	(S)	0.385
18	Diamond	(S)	0.509
19	Ethanol	(L)	2.440
20	Gasoline (octane)	(L)	2.220
21	Glass	(S)	0.840
22	Glass, crown	(S)	0.670

(Continued)

Table A2 (Continued)

No.	Material	Phase[a]	c_p (kJ kg^{-1} K^{-1})
23	Glass, flint	(S)	0.503
24	Glass, pyrex	(S)	0.753
25	Glass, silica	(S)	0.840
26	Gold	(S)	0.129
27	Granite	(S)	0.790
28	Granite	(S)	0.790
29	Graphite	(S)	0.710
30	Gypsum	(S)	1.090
31	H_2O @ −10°C (ice)	(S)	2.050
32	H_2O @ 100°C (saturated liquid)	(L)	4.181
33	H_2O @ 100°C (saturated vapor or steam)	(G)	2.080
34	H_2O @ 25°C	(L)	4.181
35	Helium	(G)	5.193
36	Hydrogen	(G)	14.300
37	Hydrogen sulfide (H_2S)	(G)	1.015
38	Iron	(S)	0.412
39	Lead	(S)	0.129
40	Lithium	(S)	3.580
41	Lithium @ 181°C	(L)	4.379
42	Magnesium	(S)	1.020
43	Marble, mica	(S)	0.880
44	Mercury	(L)	0.140
45	Methane at 2°C	(G)	2.191
46	Methanol	(L)	2.140
47	Molten salt (142°C–540°C)	(L)	1.560
48	Neon	(G)	1.030
49	Nitrogen	(G)	1.040
50	Oxygen	(G)	0.918
51	Paraffin wax ($C_{25}H_{52}$)	(S)	2.500
52	Polyethylene (rotomolding grade)	(S)	2.303
53	Sand	(S)	0.835
54	Silica (fused)	(S)	0.703
55	Silver	(S)	0.233
56	Sodium	(S)	1.230
57	Soil	(S)	0.800
58	Steel	(S)	0.466
59	Tin	(S)	0.227
60	Titanium	(S)	0.523
61	Tungsten	(S)	0.134
62	Uranium	(S)	0.116
63	Water	(L)	4.181
64	Wood	(S)	1.700 (from 1.200 to 2.900)
65	Zinc	(S)	0.387

Although the values were extracted from Ref. [24], they are available in some other textbooks like Refs. [25–27].
[a](S): solid; (L): liquid; (G): gas.

Table A3 Table of h_{fg} for different materials.

No.	Substance	Organic	Melting point, T_m (°C)	h_{fg} (kJ kg^{-1})
1	1-Cyclohexylooctadecane	✓	41	218
2	2-Heptadecanone	✓	48	218
3	3-Heptadecanone	✓	48	218
4	4-Heptadacanone	✓	41	197
5	9-Heptadecanone	✓	51	213
6	Acetamide	✓	81	241
7	Acetic acid	✓	17	184
8	α-Chloroacetic acid	✓	61	130
9	Acrylic acid	✓	68	115
10	Actanilide	✓	119	222
11	Alpha napthol	✓	96	163
12	Aluminum	×	660	397
13	α-Nepthylamine	✓	59	93
14	Azobenzene	✓	67	121
15	Bees wax	✓	62	177
16	Benzamide	✓	127	169
17	Benzoic acid	✓	122	143
18	Bromcamphor	✓	77	174
19	Camphene	✓	50	238
20	Camphenilone	✓	39	205
21	Capric acid	✓	36	152
22	Caprilic acid	✓	16	149
23	Caprylone	✓	40	259
24	Catechol	✓	104	207
25	Cetyl acid	✓	49	141
26	Chloroacetic acid	✓	56	130
27	Copper	×	1085	209
28	Cyanamide	✓	44	209
29	Dinto toluent (2,4)	✓	70	111
30	Diphenyl amine	✓	53	107
31	Docasyl bromide	✓	40	201
32	Durene	✓	79	156
33	Elaidic acid	✓	47	218
34	Formic acid	✓	8	247
35	Glautaric acid	✓	98	156
36	Glycerin	✓	18	199
37	Glycolic acid	✓	63	109
38	Glyolic acid	✓	63	109
39	Gold	×	1064	64
40	Heptadecanone	✓	41	201
41	Heptaudecanoic acid	✓	61	189
42	Hydrocinnamic acid	✓	48	118
43	Hypophosphoric acid	✓	55	213

(Continued)

Table A3 (Continued)

No.	Substance	Organic	Melting point, T_m (°C)	h_{fg} (kJ kg^{-1})
44	Iron	✗	1538	247
45	KNO_3	✗	337	116
46	KNO_3 (10%)/$NaNO_3$	✗	290	170
47	KNO_3/KBr (4.7%)/KCl (7.3%)	✗	342	140
48	KNO_3/KCl (4.5%)	✗	320	150
49	KOH	✗	360	167
50	Lauric acid	✓	44	212
51	Lead	✗	327	23
52	Lithium	✗	181	432
53	Methyl behenate	✓	52	234
54	Methyl bromobenzoate	✓	81	126
55	Methyl eicosanate	✓	45	230
56	Methyl fumarate	✓	102	242
57	Methyl palmitate	✓	29	205
58	$Mn(NO_3)_2 \cdot 6H_2O$/$MnCl_2 \cdot 4H_2O$ (4%)	✗	15–25	126
59	Myristic acid	✓	58	199
60	$Na_2SiO_3 \cdot 5H_2O$	✗	72	267
61	NaCl (26.8%)/NaOH	✗	370	370
62	NaCl (42.5%)/KCl (20.5%)/$MgCl_2$	✗	385–393	410
63	NaCl (5.0%)/$NaNO_3$	✗	282	212
64	NaCl (5.7%)/$NaNO_3$ (85.5%)/Na_2SO_4	✗	287	176
65	NaCl/KCl (32.4%)/LiCl (32.8%)	✗	346	281
66	NaCl/$NaNO_3$ (5.0%)	✗	284	171
67	$NaCl \cdot Na_2SO_4 \cdot 10H_2O$	✗	18	286
68	$NaNO_2$	✗	310	174
69	$NaNO_3$	✗	282	212
70	NaOH	✗	318	158
71	NaOH/Na_2CO_3 (7.2%)	✗	283	340
72	Nitro naphthalene	✓	57	103
73	O-Nitroaniline	✓	50	93
74	Oxolate	✓	54	178
75	O-Xylene dichloride	✓	55	121
76	Palmitic acid	✓	55	163
77	Paraffin 14-carbons	✓	6	228
78	Paraffin 15-carbons	✓	10	205
79	Paraffin 16-carbons	✓	17	237
80	Paraffin 17-carbons	✓	22	213
81	Paraffin 18-carbons	✓	28	244
82	Paraffin 19-carbons	✓	32	222
83	Paraffin 20-carbons	✓	37	246
84	Paraffin 21-carbons	✓	40	200

(Continued)

Table A3 (Continued)

No.	Substance	Organic	Melting point, T_m (°C)	h_{fg} (kJ kg^{-1})
85	Paraffin 22-carbons	✓	44	249
86	Paraffin 23-carbons	✓	48	232
87	Paraffin 24-carbons	✓	51	255
88	Paraffin 25-carbons	✓	49	238
89	Paraffin 26-carbons	✓	56	256
90	Paraffin 27-carbons	✓	59	236
91	Paraffin 28-carbons	✓	62	253
92	Paraffin 29-carbons	✓	63	240
93	Paraffin 30-carbons	✓	65	251
94	Paraffin 31-carbons	✓	68	242
95	Paraffin 32-carbons	✓	70	170
96	Paraffin 33-carbons	✓	74	268
97	Paraffin 34-carbons	✓	76	269
98	p-Bromophenol	✓	64	86
99	p-Dichlorobenzene	✓	53	121
100	Pentadecanoic acid	✓	53	178
101	Phenol	✓	41	120
102	Phenylacetic acid	✓	77	102
103	p-Joluidine	✓	43	167
104	p-Lattic acid	✓	26	184
105	Polyethylene glycol 600	✓	20	146
106	p-Xylene dichloride	✓	100	139
107	Quinone	✓	115	171
108	Silver	✗	962	105
109	Sodium sulfate ($Na_2SO_4 \cdot 10H_2O$)	✗	32	252
110	β-Chloroacetic acid	✓	56	147
111	Stearic acid	✓	69	199
112	Stibene	✓	124	167
113	Succinic anhydride	✓	119	204
114	Thiosinamine	✓	77	140
115	Thymol	✓	52	115
116	Titanium	✗	1668	296
117	TME (63%)/H_2O (37%)	✓	30	218
118	Trimyristin	✓	33	201
119	Tristearin	✓	56	191
120	Water	✗	0	334
121	Zinc	✗	420	112

Although the values were extracted from Ref. [34], they are available in some other textbooks like Ref. [35].

Table A4 Commissioned and under construction VS-PSPP in all of the world.

Unit	Country	HPT (MW)	MG (MVA)	Power electronic		Speed range (rpm)	Rated speed (rpm)	Man.[b]	Year
				Type[a]	Size (MVA)				
Narude	Japan	22	22	1	—	190–210		H	1987
Yagisawa	Japan	2 × 82	2 × 85	1	25.8	130–156		T	1990
Takami	Japan	105	140	2	24.3	209–254		M	1992
Ohkawachi	Japan	392	395	1	72	330–390 (± 8.3%)		H	1993
Shiobara	Japan	330	360	1	51.1	356–394 (± 8%)		T	1995
Okukiyotsu	Japan	340	345	2	31.5	408–450 (± 5%)		T	1996
Yanbaru	Japan	30	31.5	2	3.96	423–477 (± 5%)		T	1999
Goldisthal	Germany	2 × 265	2 × 340	1	2 × 57.6	300–347 (−10/ + 4%)	333	V	2003
Avce	Slovenia	185	195	3	10	576–636 (± 4%)	600	ABB,M	2008
Omarugawa III	Japan	2 × 330	2 × 350	3	2 × 27.4	576–624 (± 4%)	600	M	2008
		2 × 330	2 × 370	1				H	2010
Grimsel II[c]	Switzerland	80	100	4	2 × 43.3	680–765 (−10/ + 2%)	750	ABB	2010
Okutataragi	Kansai Electric Power Co.	2 × 320	—	—	—	—	300	H	2014
Kyogoku	Japan	228	230	2		—		T	2015
Linthal	Switzerland	4 × 250	4 × 280	3	—	470–530 (± 6%)	500	A	2015

Frades II	Portugal	2 × 390	2 × 440	3	—	350–381 (+2%, −7%)	364	Vo, S	2015
Vende Nova III	Portugal	2 × 380	2 × 420	—	45	—	—	Vo, S	2015
Le cheylas	France	1 × 250	—	—	—	(± 7%)	300		
Tehri	India	4 × 250	4 × 278	3		214–256 (± 6%)	230.8	GE, T	2016
Kazunogawa	Japan	2 × 475	2 × 475	2	25	480–520	—	T	2017
Iowa Hill	USA	3 × 133	3 × 133	2	—	—	—	—	2018
Nant de Drance	Switzerland	6 × 157	6 × 175	3	—	399–459 (± 7%)	428.6	A	2019
Kozjak	Slovenia	2 × 220	2 × 230	—	—	—	600	V	—
Grimsel III	Switzerland	—	—	—	100	—	—	ABB	2020
Fengning II	China	2 × 300	2 × 330	—	—	—	—	GE	2021

[a]Power Electronic Type: 1: Cyclo-converter, 2: GTO- back to back (VSI), 3: IGCT- back to back (VSI), 4: full converter (VSI).
[b]Manufacturer: H: HITACHI, T: TOSHIBA, M: MITSUBISHI, V: VATECH, Vo: VOITH, A: ALSTOM, GE: General Electric, S: Siemens.
[c]Only pumping mode operation.

Index

Note: Page numbers followed by "*f*" and "*t*" refer to figures and tables, respectively.

FSC
www.fsc.org
MIX
Papier | Fördert gute Waldnutzung
FSC® C083411

Kolibri 360 GmbH
Ferdinand-Jühlke-Straße 7
99095 Erfurt, Deutschland
produktsicherheit@kolibri360.de